The Origin of Stars and Planetary Systems

NATO Science Series

A Series presenting the results of activities sponsored by the NATO Science Committee. The Series is published by IOS Press and Kluwer Academic Publishers, in conjunction with the NATO Scientific Affairs Division.

A. **Life Sciences**	IOS Press
B. **Physics**	Kluwer Academic Publishers
C. **Mathematical and Physical Sciences**	Kluwer Academic Publishers
D. **Behavioural and Social Sciences**	Kluwer Academic Publishers
E. **Applied Sciences**	Kluwer Academic Publishers
F. **Computer and Systems Sciences**	IOS Press
1. **Disarmament Technologies**	Kluwer Academic Publishers
2. **Environmental Security**	Kluwer Academic Publishers
3. **High Technology**	Kluwer Academic Publishers
4. **Science and Technology Policy**	IOS Press
5. **Computer Networking**	IOS Press

NATO-PCO-DATA BASE

The NATO Science Series continues the series of books published formerly in the NATO ASI Series. An electronic index to the NATO ASI Series provides full bibliographical references (with keywords and/or abstracts) to more than 50000 contributions from internatonal scientists published in all sections of the NATO ASI Series.
Access to the NATO-PCO-DATA BASE is possible via CD-ROM "NATO-PCO-DATA BASE" with user-friendly retrieval software in English, French and German (WTV GmbH and DATAWARE Technologies Inc. 1989).

The CD-ROM of the NATO ASI Series can be ordered from: PCO, Overijse, Belgium

Series C: Mathematical and Physical Sciences – Vol. 540

The Origin of Stars and Planetary Systems

edited by

Charles J. Lada

Harvard-Smithsonian Center for Astrophysics,
Cambridge, MA, U.S.A.

and

Nikolaos D. Kylafis

Department of Physics,
University of Crete,
Heraklion, Crete, Greece

Kluwer Academic Publishers

Dordrecht / Boston / London

Published in cooperation with NATO Scientific Affairs Division

Proceedings of the NATO Advanced Study Institute on
The Physics of Star Formation and Early Evolution – II,
Crete, Greece
24 May–5 June 1998

A C.I.P. Catalogue record for this book is available from the Library of Congress.

ISBN 0-7923-5908-9 (HB)
ISBN 0-7923-5909-7 (PB)

Published by Kluwer Academic Publishers,
P.O. Box 17, 3300 AA Dordrecht, The Netherlands.

Sold and distributed in North, Central and South America
by Kluwer Academic Publishers,
101 Philip Drive, Norwell, MA 02061, U.S.A.

In all other countries, sold and distributed
by Kluwer Academic Publishers,
P.O. Box 322, 3300 AH Dordrecht, The Netherlands.

Printed on acid-free paper

Table of Contents

Molecular Clouds
Leo Blitz and Jonathan P. Williams

The Dynamical Structure and Evolution of Giant Molecular Clouds
Christopher F. McKee

Physical Conditions in Nearby Molecular Clouds
Philip C. Myers

Models and Observations of the Chemistry Near Young Stellar Objects

Ewine F. Van Dishoeck and Michiel R. Hogerheijde

Bipolar Molecular Outflows
R. Bachiller & M. Tafalla

Herbig-Haro Flows
Bo Reipurth & A.C. Raga

The Nature of Young Solar-Type Stars
François Ménard & Claude Bertout

The Evolution of Pre-Main-Sequence Stars
Francesco Palla

OB Associations

A.G.A. Brown, A. Blaauw and R. Hoogerwerf, J.H.J. De Bruijne and P.T. De Zeeuw

Circumstellar Disks
Steven V.W. Beckwith

Accretion Disks and Eruptive Phenomena
Scott J. Kenyon

The Formation of Planets
Steven P. Ruden

Extrasolar Planets: *Techniques, Results, and the Future*
G.W. Marcy and R.P. Butler

PREFACE

A few years after the publication of *The Physics of Star Formation and Early Stellar Evolution*, we received a request from the publisher for an updated second edition of this popular reference book. As originally intended, the volume had proved to be a useful "text" book for graduate astronomy courses and seminars which dealt with topics related to stellar origins. The book was based on a series of lectures delivered by a distinguished group of leading researchers at a NATO Advanced Study Institute (ASI) held in May 1990 on the island of Crete, Greece. The primary goal of the ASI was in fact to produce a book which "would simultaneously provide a broad and systematic overview of, as well as a rigorous introduction to, the fundamental physics and astronomy at the heart of modern research in star formation and early stellar evolution."

However, by 1995 concern had arisen among those who used the text as a reference for graduate seminars and courses that the book would need to be updated to stay abreast of the discoveries and progress in this rapidly evolving field. After some discussion we concluded that a new edition of the book was warranted and that the goal of producing a new edition would be best accomplished by organizing a second ASI in Crete to review the progress in star formation research. This was not only because there had been substantial new advances in this area of research but also because a new generation of young scientists had entered the field since the original ASI was held. For such young researchers participating in a summer school which provided an intensive and comprehensive introduction to this important field of astrophysics would be of great benefit.

The second NATO Advanced Study Institute on Star Formation and Early Stellar Evolution was held between 24 May and 5 June 1998 at the beautiful Knossos Royal Village Hotel in Crete Greece. Modeled after our first ASI, the institute was conducted as an advanced school for graduate students and young researchers. The ASI brought together an international group of 20 distinguished researchers to critically review and update in a systematic fashion the current state of knowledge concerning the entire scope of our present understanding of star formation from the origin of molecular clouds to the formation of planets around young stars. These lecturers were given the task of preparing a series of graduate level tutorial review lectures for the school. These lectures formed the basis of the individual chapters of the presesnt book: *The Origins of Stars and Planetary Systems*. Approximately half the lecturers contributed to the first book and were invited to revise and update their original chapters for this second book. The other half of the lecturers provided new chapters and material reflecting some of the most important advances in the field since the original ASI was held nine years ago. It is particularly gratifying to us that several of the new authors of the present volume participated in our first ASI as students or young researchers. This is a small measure of the success of the original ASI and its overall concept.

The present book consists of twenty chapters which are divided into four sections: I - The Nature of Molecular Clouds and Their Relation to Star Formation; II - The Origin and Early Evolution of Low Mass Stars; III - The Role of Clustering in Star Formation and the Origin of Massive Stars; and IV - The Physics of Circumstellar Disks and Planet Formation. The topics of the individual chapters were selected so that taken together they would provide a systematic overview of the star formation process from the origin of molecular clouds to the formation of planets. At the same time, each chapter represents an independent and comprehensive review of a given subject area and as a result the reader can expect to find occasional overlap in subject matter between the various chapters.

The format of the second ASI was closely modeled after that of the first ASI. The second ASI consisted of ten working days spread out over a two week period during the last week of May and the first week of June 1998. Each working day was divided into three classroom sessions of two hours duration. Two sessions were held in the morning and one in the evening. All participants shared breakfast before the first session and dinner after the last session. There was a 4.5 hour break each day for lunch, recreation and special discussions. This time period was extremely beneficial as it allowed for much interaction between all the participants including the senior lecturers and students. This was particularly effective when combined with the circumstance that all participants, senior and junior, were present for the entire two week duration of the ASI.

On most days each of the three classroom sessions consisted of a ninety minute invited review lecture with an additional 30 minutes for questions and discussion. Discussion, both during and outside the lectures was an important component of this ASI. Once during each week a mid-morning session was devoted to formal poster presentations. As in the first ASI, poster sessions were an integral part of the school. Although there was only one formal poster session each week, posters were displayed in a room immediately adjacent to the main lecture hall for an entire week, giving ample time for participants to view and discuss the posters before and after lectures and during coffee breaks. As with the previous ASI, substantial prizes were awarded to the most outstanding poster presentations (based on both the quality of science and the quality of presentation). The posters were reviewed by two panels of distinguished judges. Professors Leo Blitz, Steve Beckwith and Scott Kenyon reviewed the posters of the first week and Professors George Field, Frank Shu and Bo Reipurth reviewed the posters during the second week. Three prizes and a number of honorable mentions were awarded to participants from each week's poster session. The first place poster prizes of the Crete II ASI were awarded to Claudia Lavalley and Joao Alves; second place prize winners were Jon Williams and Jos de Bruijne, and the third place prize winners were Mike Meyer and Paul Kalas. Honorable mentions were presented to David Wilner, Marc Pound, Konstantinos Pavlakis and Michiel Hogerheijde. The prizes were presented in a special ceremony at the closing banquet. In addition, the closing banquet also featured a special award presentation by the student participants in recognition of those main lecturers whose lectures were judged to be the most outstanding in terms of clarity and pedagogical presentation. The three student prizes for outstanding lectures were awarded to Ewine van Dishoeck, Geoff Marcy and Steve Beckwith.

During the middle of the ASI a special one day session was devoted to a series of shorter invited talks describing recent research in certain areas which for one reason or another were not covered in adequate depth in the primary lectures, but nonetheless were deemed important for inclusion in any comprehensive review of the field. These special lectures were delivered by Philippe Andre, Richard Crutcher, Neal Evans, David Hollenbach, Antonella Natta, Anneila Sargent , Malcolm Walmsley, Erick Young and Hans Zinnecker, and they covered a range of topics including the nature of protostars, photoevaporating circumstellar disks, disks around Herbig AeBe stars, infall in star forming clouds, the physical conditions in dense cores, ground-based interferometric and HST observations of star formation, and pre-main sequence binary stars. Due to limitations of space the contents of these lectures could not be included in this book.

In addition to the scientific sessions and activities, two magnificent Greek banquets were held at the hotel with outstanding cuisine and lively music and dancing. Special field trips were also organized for the free weekend in the middle of the conference. These included excursions to the Palace of Knossos, the Archeology Museum in Heraklion, the famous Sammaria Gorge on the southern shore of Crete and a cruise to the beautiful island of Santorini.

We are grateful to the NATO Scientific Affairs Division for its generous funding of this Advanced Study Institute. We are also grateful to the University of Crete for valuable support with the overall organization of the conference. The company MITOS secured for us low cost accommodation and wonderful excusions and provided critical assistance in the organization of the meeting, including establishing and maintaining the WEB site for meeting information, applications and registration. The National Science Foundation provided travel grants that enabled 4 students from the US to attend the ASI. Nancy Poole at the Smithsonian Astrophysical Observatory provided valuable assistance in editing and preparing the manuscript for this book. We are also grateful to Scott Kenyon for contributions to the editorial effort. We also thank Joao Alves, Aurore Bacmann, Harold Butner, James DiFrancesco, Spyridon Kitsionas, Liz Lada, Erick Young, and Hans Zinnecker for providing the set of pictures of participants and events from which those appearing throughout this book were selected. The scientific organizing committee for this ASI consisted of Nick Kylafis, Charlie Lada, Frank Shu and Steve Beckwith.

Finally the editors owe particular thanks to the authors of this book who successfully discharged their task of preparing in-depth tutorial reviews of the fundamental problems in star formation research. We hope that researchers and students interested in undertaking serious study of star formation will find this book as useful and stimulating a guide to the field as the earlier volume appears to have been.

Charles J. Lada   Needham, Massachusetts

Nikolaos D. Kylafis   Heraklion, Crete

March 31, 1999

Participants

Joao Alves
Harvard Smithsonian CfA
60 Garden St., MS 42
Cambridge MA 02138
USA
jalves@cfa.harvard.edu

Philippe Andre
CEA-Saclay-Service d' Astro
709 F 91191
Gif sur Yvette Cedex
France
pandre@cea.fr

Hector G. Arce
Harvard University
60 Garden St, MS 42
Cambridge MA 02138
USA
harce@cfa.harvard.edu

Rafael Bachiller
National Astro. Obs.
Apartado 1143
E-28800 Alcala de Henares
Spain
bachiller@oan.es

Aurore Bacmann
CEA-Saclay
709 F 91191
Gif Sur Yvette Cedex
France
abacmann@cea.fr

Shantanu Basu
Canadian Inst. for Theoret. Astroph.
60 St. George Street
Toronto Ontario M5S 3H8
Canada
basu@cita.utoronto.ca

Steven Beckwith
Space Telescope Sci. Inst.
3700 San Martin Dr.
Baltimore MD 21218
USA
svwb@stsci.edu

Raoul E. Behrend
Observatoire de Geneve
CH-1290 Sauverny
Switzerland
Raoul.Behrend@obs.unige.ch

Robbins K. Bell
NASA Ames Res. Center
NASA/ARC 245-3
Moffett Field CA 94035
USA
bell@cosmic.arc.nasa.gov

Maria Teresa Beltran
University of Barcelona
Dept. of Astronomy
Av. Diagonal 647
E-08028 Barcelona
Spain
maite@fareb1.am.ub.es

Nina Beskrovnaya
Pulkovo Observatory
Pulkovo 65/1
196140 St. Petersburg
Russia
beskr@pulkovo.spb.su

Leo Blitz
Univ. of California
Astronomy Dept.
Berkeley CA 94720
USA
blitz@gmc.berkeley.edu

xxiv

Christian Boily
Astonomisches Rechen - Institut
Moenchhofstrasse 12 -14
D-69120 Heidelberg
Germany
cmb@ari.uni-heidelberg.de

Ian A. Bonnell
Institute of Astronomy
Madingley Road
Cambridge CB3 0HA
U.K.
bonnell@ast.cam.ac.uk

Sylvain Bontemps
Stocholm Observatory
SE-13336 Saltsjobaden
Sweden
bontemps@astro.su.se

Annemieke M.S. Boonman
Leiden Observatory
P.O. Box 9513
2300 RA Leiden
The Netherlands
boonman@strw.leidenuniv.nl

Jean C. Bouret
Observatoire Midi-Pyrenees
14 Ave. Edouard Belin
F-31400 Toulouse
France
bouret@obs-mip.fr

Crystal Brogan
University of Kentucky
177 Chem-Phys Bldg.
Lexington KY 40506-0055
USA
brogan@pa.uky.edu

Kate Brooks
University of NSW
Sydney 2052 NSW
Australia
kbrooks@wodin.phys.unsw.edu.au

Anthony Brown
UNAM
Apdo. Postal 877
28800 Ensenada B.C.
Mexico
brown@bufadora.astrosen.unam.mx

Harold M. Butner
SMTO Steward Obs.
University of Arizona
Tucson AZ 85721
USA
hbutner@as.arizona.edu

Ed Churchwell
University of Wisconsin
Astronomy Dept.
Madison WI 53706
USA
churchwell@astro.wisc.edu

Glenn E. Ciolek
Univ. of Chicago
Dept. of Astron. & Astroph.
5640 S. Ellis Avenue
Chicago IL 60637
USA
ciolek@jets.uchicago.edu

Eric Copet
DESPA - Obs. De Paris
5 Place J. Janssen
F-92195 Meudon Cedex
France
copet@deniseyg.obspm.fr

Vitor Costa
Universidade do Porto
Rua do Campo Alegre 823
4150 Porto
Portugal
vcosta@astro.up.pt

Richard M. Crutcher
University of Illinois
Astronomy Dept.
1002 W. Green Street
Urbana IL 61801
USA
crutcher@uiuc.edu

Jos H. J. de Bruijne
Leiden Observatory
P.O. Box 9513
2300 RA Leiden
The Netherlands
debruyne@strw.leidenuniv.nl

James Di Francesco
Center for Astrophysics
60 Garden St, MS 42
Cambridge MA 02138
USA
jdifran@cfa.harvard.edu

Neal Evans
University of Texas
Astronomy Dept.
Austin TX 78712-1083
USA
nje@astro.as.utexas.edu

Markus Feldt
Max-Planck-Institut fuer Astronomie
Koenigstuhl 17
D-69117 Heidelberg
Germany
mfeldt@mpia-hd.mpg.de

George Field
Center for Astrophysics
60 Garden St, MS 51
Cambridge MA 02138
USA
gfield@cfa.harvard.edu

Martin Fricke
Jena University
Schillergaesschen 2-3
07743 Jena
Germany
fricke@astro.uni-jena.de

Sabine Frink
Astron. Rechen Inst.
Moenchhofstrasse 12-14
D-69120 Heidelberg
Germany
sabine.frink@urz.uni-heidelberg.de

Philip Gladwin
University of Wales
P.O. Box 913
Cardiff Wales CF2 3YB
U.K.
P.Gladwin@astro.cf.ac.uk

Dimitrios Gouliermis
National Observatory of Athens
P.O. Box 20048
118 10 Athens
Greece
dgoulier@astro.noa.gr

Nicolas Grosso
CEA Servive d' Astro.
Ormes des Merisiers
91191 Gif-Sur-Yvette Cedex
France
grosso@discovery.saclay.cea.fr

xxvi

Karl Jr. Haisch
University of Florida
211 SSRB, P.O. Box 112055
Gainesville FL 32611-2055
USA
haisch@astro.ufl.edu

Catrina Hamilton
Wesleyan University
Connecticut College
New London CT 06320
USA
cmham@oak.cc.conncoll.edu

Michiel Hogerheijde
Leiden Observatoty
P.O. Box 9513
2300 RA Leiden
The Netherlands
michiel@strw.leidenuniv.nl

David Hollenbach
NASA Ames Research Center
MS 245-3
Moffett Field CA 94035-1000
USA
hollenbach@warped.arc.nasa.gov

Ronnie Hoogerwerf
Leiden Observatory
Postbus 9513
2300 RA Leiden
The Netherlands
hoogerw@strw.leidenuniv.nl

Nuria Huelamo Bautista
University of Madrid
Avda. Complutense s/n
280 40 Madrid
Spain
nuria@orion.mat.ucm.es

Nazar Ikhsanov
Universitaets Sternwarte Muenchen
Scheinerstr. 1
D-81679 Muenchen
Germany
ikhsanov@usm.uni-muenchen.de

Doug Johnstone
Canadian Inst. for Theoret. Astroph.
60 St. George str.
Toronto Ontario M5S 3H8
Canada
johnstone@cita.utoronto.ca

Anlaug Amanda Kaas
Stockholm Observatory
SE-133 36 Saltsjoebaden
Sweden
amanda@astro.su.se

Paul Kalas
Space Telescope Sci. Inst.
3700 San Martin Dr.
Baltimore MD 21218
USA
kalas@stsci.edu

Scott Kenyon
Smithsonian Astro. Obs.
60 Garden str.
Cambridge MA 02138
USA
skenyon@cfa.harvard.edu

Spyridon Kitsionas
University of Wales
P.O. Box 913
Cardiff CF2 3YB Wales
U.K.
Spyridon.Kitsionas@astro.cf.ac.uk

Hubertus H. Klahr
Astrophysikalisches Institut
Schillergaesschen 3
D-07745 Jena
Germany
klahr@astro.uni-jena.de

Randolf Klein
University of Jena
Schillergaesschen 2
D-07745 Jena
Germany
rklein@astro.uni-jena.de

Ralf Klessen
Max Planck Institute for Astronomy
Koenigstuhl 17
D-69117 Heidelberg
Germany
klessen@mpia-hd.mpg.de

Pavel Kroupa
Univ. Heidelberg
Inst. fuer Theo. Astro.
Tiergartenstr. 15
D-69121 Heidelberg
Germany
pavel@ita.uni-heidelberg.de

Nick Kylafis
University of Crete
Physics Dept.
P.O. Box 2208
710 03 Heraklion Crete
Greece
kylafis@physics.uch.gr

Charles Lada
Smithsonian Astro. Obs.
MS 72
60 Garden str.
Cambridge MA 02138
USA
clada@bok.harvard.edu

Elizabeth Lada
University of Florida
211 Bryant Space Sci. Bld.
Gainesville FL 32611
USA
lada@astro.ufl.edu

Ralf Launhardt
MPIfR Bonn
Auf dem Huegel 69
D-53121 Bonn
Germany
launh@astro.uni-jena.de

Claudia Lavalley
Observatoire de Grenoble
414 rue de la Piscine
Domaine Universitaire
BP 53, F-38041 Grenoble
France
lavalley@obs.ujf-grenoble.fr

Mayra E. Lebron Santos
University of Mexico
Jose Peon del Valle 19 Esq J.J. Tablada
Morelia Mich.
Mexico
mlebron@astrosmo.unam.mx

Jung Kyu Lee
Univ. of N. S. Wales
Dept. of Astrophysics
Sydney 2052
Australia
jklee@roen.phys.unsw.edu.au

Joanna Levine
University of Florida
211 SSRB
P.O. Box 112055
Gainesville FL 32601
USA
levine@astro.ufl.edu

Fotini Maragoudaki
University of Athens
Section of Astronomy
Panepistimiopolis Zografos
Greece
fmarag@atlas.uoa.gr

Geoffrey Marcy
Univ. of California
Astronomy Dept.
Berkeley CA 94720
USA
gmarcy@etoile.berkeley.edu

Clare E. Martin
University of St. Andrews
Dept. of Mathematics
North Haugh
St. Andrews Fife KY16 9SS
U.K.
clarem@dcs.st-and.ac.uk

Christopher D. Matzner
Physics & Astronomy Depts.
Univ. of California
601 Campbell Hall
Berkeley CA 94720
USA
matzner@arkham.berkeley.edu

Christopher McKee
Univ. of California
Dept. of Astronomy
Berkeley CA 94720
USA
cmkee@astro.berkeley.edu

Thomas Megeath
Harvard Smithsonian CfA
60 Garden St., MS-72
Cambridge MA 02138
USA
tmegeath@cfa.harvard.edu

Francois Menard
Obser. de Grenoble
414 Rue de la Piscine
F-38041 Grenoble
France
menard@obs.ujf-grenoble.fr

Michael R. Meyer
Steward Observatory
933 N. Cherry Ave.
Tucson AZ 85721
USA
mmeyer@as.arizona.edu

Vincent Minier
Onsala Space Obs.
439 92 Onsala
Sweden
vincent@oso.chalmers.ce

Sergej Moiseenko
Space Research Institute
Profsouyznaya Str. 84/32
Moscow 117810
Russia
moiseenko@mx.iki.rssi.ru

Sergio Molinari
IPAC/Caltech
MS 100-22
Pasadena CA 91125
USA
molinari@ipac.caltech.edu

Miguel C. Moreira
Univ. of Lisbon
Dept. de Fisica
Campo Grande
Edif C1 Piso 4
Portugal
miguelm@delphi.cc.fc.ul.pt

Telemachos Ch. Mouschovias
Univ. of Illinois
Dept. of Astronomy
Urbana IL 61801
USA
tchm@astro.uiuc.edu

August A. Muench
Harvard-Smithsonian CfA
60 Garden Street, MS 42
Cambridge MA 02138
USA
gmuench@cfa.harvard.edu

Philip Myers
Harvard-Smithsonian CfA
60 Garden Street
Cambridge MA 02138
USA
pmyers@cfa.harvard.edu

Fumitaka Nakamura
Niigata University
8050 Ikarashi 2
Niigata 950 2181
Japan
fnakamur@ed.niigata-u.ac.jp

M. S. Nanda Kumar
Physical Research Lab.
Navrangapura
Ahmedabad 380009
India
nanda@prl.ernet.in

Antonella Natta
Osservatorio di Arcetri
Largo E. Fermi 5
50125 Firenze
Italy
natta@arcetri.astro.it

Alberto Noriega-Crespo
IPAC
Caltech 100-22
Pasadena CA 91125
USA
alberto@ipac.caltech.edu

Dieter Nuernberger
IRAM
300 Rue de la Piscine
F-38406 St. Martin - d' Heres
France
nurnberg@iram.fr

Elena Ortiz Garcia
University of Madrid
C-XI Facultad de Ciencias
28049 Madrid
Spain
elena@astro1.ft.uam.es

Mayra C. Osorio Gutierrez
University of Mexico
CP 58090 Col. Santa Maria De Guido
Morelia Michoacan
Mexico
mayra@astrosmo.unam.mx

Javier Palacios
University of Madrid
Campus de Cantoblanco
Madrid 28049
Spain
javier@xiada.ft.uam.es

Francesco Palla
Osserv. Astrofisico di Arcetri
Largo E. Fermi 5
50125 Firenze
Italy
palla@arcetri.astro.it

xxx

Konstantinos Pavlakis
University of Leeds
Woodhouse Lane
Leeds LS2 9JT
UK
pavlakis@ast.leeds.ac.uk

Marc W. Pound
University of Maryland
College Park MD 20742
USA
mpound@astro.umd.edu

Bo Reipurth
University of Colorado
Center for Astrophysics
Boulder CO 80309
USA
reipurth@casa.colorado.edu

Naomi A. Ridge
Liverpool John Moores University
Byrom Street
Liverpool KL3 3AF
U.K.
nar@astro.livjm.ac.uk

Steven Ruden
University of California
Dept. of Physics
Irvine CA 92672
USA
spruden@uci.edu

Carlos Antunes Santos
Observatorio Astronomico de Lisboa
Tapada da Ajuda
1300 Lisboa
Portugal
csantos@delphi.cc.fc.ul.pt

Paolo Saraceno
CNR
Via Fosso del Cavaliere
00133 Roma
Italy
saraceno@ifsi.rm.cnr.it

Anneila I. Sargent
Caltech 105-24
Pasadena CA 91125
USA
afs@astro.caltech.edu

Nicola Schneider
University of Cologne
Zuelpicher Str. 77
50937 Koeln
Germany
schneider@ph1.uni-koeln.de

Richard D. Schwartz
University of Missouri-St. Louis
Dept. of Physics and Astronomy
8001 Natural Bridge Road
St. Louis MO 63121
USA
schwartz@newton.umsl.edu

Frank Shu
University of California
Dept of Astronomy
Berkeley CA 94720-3411
USA
fshu@astro.berkeley.edu

Stephen L. Skinner
JILA
Univ. of Colorado
Boulder CO 80309-0440
USA
skinner@jila.colorado.edu

Luigi Spinoglio
Ist. di Fisica dello Spazio Interplanetario
Via Fosso del Cavaliere 100
00133 Roma
Italy
luigi@ifsi.rm.cnr.it

Leonardo Testi
Caltech 105-24
Passadena CA 91125
USA
lt@astro.caltech.edu

Christoph Trojan
University of Cologne
Zuelpicher Str. 77
50937 Koeln
Germany
trojan@zeus.ph1.uni-koeln.de

Floris van der Tak
Sterrewacht Leiden
Postbus 9513
2300 RA Leiden
The Netherlands
vdtak@strw.leidenuniv.nl

Johan van der Walt
Potchefstroom University
Private Bag x6001
Potchefstroom 2520
South Africa
fskdjvdw@pukrs1.puk.ac.za

Ewine van Dishoeck
Leiden Observatory
P.O. Box 9513
2300 RA Leiden
The Netherlands
ewine@strw.leidenuniv.nl

Gerd Jan Van Zadelhoff
Leiden Observatory
Postbus 9513
2300 RA Leiden
The Netherlands
zadelhof@strw.leidenuniv.nl

Anja E. Visser
University of Cambridge
Institute of Astronomy
Madingley Road
Cambridge CB3 0HA
UK
anja@mrao.cam.ac.uk

Malcolm Walmsley
Osservatorio di Arcetri
Largo Fermi 5
50125 Firenze
Italy
walmsley@arcetri.astro.it

Francis P. Wilkin
NASA Ames Research Centre
MS 245-3
Moffett Field CA 94035
USA
wilkin@warped.arc.nasa.gov

Jonathan Williams
Harvard-Smithsonian CfA
60 Garden Street
MS 42
Cambridge MA 02138
USA
jpw@cfa.harvard.edu

David J. Wilner
Harvard Smithsonian CfA
60 Garden Street
Cambridge MA 02138
USA
dwilner@cfa.harvard.edu

Grace Wolf-Chase
University of California
Physics Dept.
Riverside CA 92521
USA
wolf@fun.ucr.edu

Emmanuel Xilouris
University of Crete
Physics Dept.
P.O. Box 2208
710 03 Heraklion Crete
Greece
xilouris@physics.uch.gr

Erick T. Young
University of Arizona
Astronomy Dept.
Tucson AZ 85721
USA
eyoung@as.arizona.edu

Ka Chun Yu
CASA
Campus Box 389
Boulder CO 80309
USA
kachun@casa.colorado.edu

Joao Yun
Observatorio Astronomico de Lisboa
Tapada da Ajuda
1300 Lisboa
Portugal
yun@delphi.cc.fc.ul.pt

Hans Zinnecker
Astrophysikalisches Inst. Potsdam
An deer Sternwarte 16
D-14482 Potsdam
Germany
hzinnecker@aip.de

NATO Advanced Study Institute

The Physics of Star Formation
and
Early Stellar Evolution

May 24 - June 5, 1998
Knossos Royal Village Beach Hotel, Crete, Greece

Topics:
Origin, Structure & Evolution of GMCs, OB Associations, Embedded Clusters, The IMF, Protostars, Outflows, Jets, Disks, Masers, Collapse Theory, Planet Formation, & Detection of Extra-Solar Planets.

For Information & Applications:
http://www.mitos.com.gr/conf/starASI98/index.html
email: mitos@stepc.gr Deadline: 15 March 1998

Directors:
N.D. Kylafis, University of Crete
C.J. Lada, Smithsonian Astrophysical Obs.

I – THE NATURE OF MOLECULAR CLOUDS AND THEIR RELATION TO STAR FORMATION

George Field masters the water slide.

MOLECULAR CLOUDS

LEO BLITZ
Astronomy Department, University of California
Berkeley, CA USA
blitz@gmc.berkeley.edu

AND

JONATHAN P. WILLIAMS
National Radio Astronomy Observatory
Tucson, AZ USA
jpwilliams@nrao.edu

1. Introduction

All known star formation is thought to occur in molecular clouds. The association of molecular clouds with star formation is so strong that it is generally assumed that wherever there are young stars, one will always be able to find molecular gas, even when there is evidence to the contrary (e.g. the TW Hya association; Rucinski & Krautter 1983). Indeed, when sufficiently sensitive observations are made, the association of star formation with molecular clouds is observed in all environments, galactic or extragalactic. Evidence for star formation in molecular clouds has even been observed as far back as z = 4.7 (Omont *et al.* 1996), and sensitive new instruments such as the Millimeter Array and the Square Kilometer Array should make it possible to detect star forming molecular clouds at the earliest times.

The general goal of molecular cloud studies is to determine how the interstellar medium produces molecular clouds, especially the Giant Molecular Clouds (GMCs; $M > 10^4$ $M_\odot$) which are responsible for the vast majority of all star formation, and how the molecular clouds in turn produce stars and clusters. We take the view that once a core within a molecular cloud can no longer support itself against gravity, beginning the inexorable process of collapse into a single star or binary, the study of the molecular gas falls into a different regime. The inexorability of the star formation pro-

3

C.J. Lada and N.D. Kylafis (eds.), The Origin of Stars and Planetary Systems, 3–28.
© *1999 Kluwer Academic Publishers. Printed in the Netherlands.*

cess also occurs at the GMC level; there is only one GMC of dozens known within about 3 kpc of the Sun with scant evidence of star formation (see §3.3).

We wish to obtain answers to the four fundamental questions below:

1) How do GMCs form?
2) How do single stars (and binaries) form?
3) How do clusters form?
4) What determines the Initial Mass Function (IMF)?

We will begin below by discussing some of the observational progress that has been made in answering the first question. We will deal with the second question only in passing; it will be dealt with extensively by articles in this volume by Shu et al., Lada, Myers, and others. We will address steps leading to the answers to the third and fourth questions by looking for clues from the analysis of the structure of molecular clouds.

This review is meant primarily to address the progress made in the study of molecular clouds since the first Crete meeting on star formation, *The Physics of Star Formation and Early Evolution* (Blitz 1991), which discusses work on the subject through 1990. Other useful reviews include the article in this volume by McKee, which gives a good theoretical picture of the physics of molecular clouds, and the article in *Protostars and Planets IV* by Williams, Blitz & McKee (1999) which discusses progress in molecular cloud studies since *Protostars and Planets III* (Blitz 1993).

2. Formation of Molecular Clouds

2.1. GALAXY SCALE ISSUES

The physics of the formation of GMCs is one of the major unsolved problems of the interstellar medium. Although many papers have been written on the subject, especially in the late 1970s and early 1980s, it is not yet known what the dominant formation mechanism is, or even what the relative importance of gravity, radiation and magnetic fields are in the cloud formation process. For example, GMCs are known to be self-gravitating because their mean internal pressures exceed that of the general ISM by about an order of magnitude (e.g. Blitz 1991). Since they are known to be at least as old as the oldest stars identified to have formed from them (e.g. ~20 My in the case of Orion; Blaauw 1964), they must be stable for at least that long. It is then reasonable to conclude that gravity is one of the key elements in the formation of self-gravitating, relatively stable clouds. However, molecular clouds found at high galactic latitude (*high latitude clouds*; HLCs) also generally have turbulent pressures greater than the mean mid-plane ISM pressure, but have masses a few orders of magnitude smaller than the GMCs. Typically they are far from being self-gravitating

(Magnani, Blitz & Mundy 1985; Reach, Wall & Odegard 1998) and gravity cannot have been a factor in their formation. Is gravity important only in forming clouds that exceed a certain minimum mass?

Mechanisms proposed for molecular cloud formation can be divided into three general categories: collisional agglomeration of smaller clouds (e.g. Kwan 1979; Scoville & Hersh 1979; Stark 1979; Cowie 1980; Kwan & Valdez 1983), gravi-thermal instability (e.g. Parker 1966; Mouschovias, Shu & Woodward 1974; Shu 1974; Elmegreen 1982a,b), and the pressurized accumulation in shocks, either in supernovae (Öpik 1954; Herbst & Assousa 1977) or in Galactic shocks (e.g. Woodward 1976). These mechanisms were reviewed by Elmegreen (1990; and references therein) who concluded that the most likely formation mechanism is the gravi-thermal instability applied to the *cloudy* ISM [his italics]. However, direct evidence has been very hard to come by and it has not been possible to apply the ideas to other galaxies or even to other parts of the Milky Way such as the Galactic Center. There may be clues in the angular momentum distribution of GMCs (Blitz 1993) but this has received little attention; it is not known, for example, whether GMCs in the disk rotate faster than they do in the solar vicinity, as expected from gravitationally formed clouds in a differentially rotating disk, or whether the counter-rotating clouds in the local neighborhood are pathological, or common throughout the disk. In one of the few papers on the subject, Phillips (1999) has analyzed the role of angular momentum in GMC support and formation: isolated clouds tend have angular velocity vectors perpendicular to the Galactic plane suggesting that their spin arises from Galactic shear, but internal structures have more randomly oriented spin axes due, possibly to turbulence, or dynamical interactions.

Nevertheless, two questions regarding molecular clouds which give us important clues for how GMCs form, have been settled in the last decade: 1) GMCs have lifetimes of $2 \times 10^7 < \tau < 1 \times 10^8$y, considerably shorter than a Galactic rotation period; 2) In galaxies with strong, well-defined spiral arms, molecular clouds are generally confined to the arms.

The question of cloud lifetimes, while hotly debated in the early 1980s, seems to have been settled by two sets of observations. The first is a good calibration of the H_2/HI surface density ratio, $\Sigma(H_2)/\Sigma(HI)$, everywhere in the Milky Way (Dame 1993). This result coupled with a downward revision of the $I(CO)/N(H_2)$ ratio from the EGRET experiment on GRO (Hunter et al. 1997), implies that there is no radius in the Milky Way where $\Sigma(H_2)/\Sigma(HI)$ is significantly greater than 1. Thus the mass flow arguments suggesting long cloud lifetimes ($\gtrsim 10^9$ y; e.g. Solomon and Sanders 1980) are no longer applicable in the disk (see however §2.3 below for a discussion of the Galactic Center). The second is that the depletion time for the molecular gas due to star formation over the entire Galaxy is about

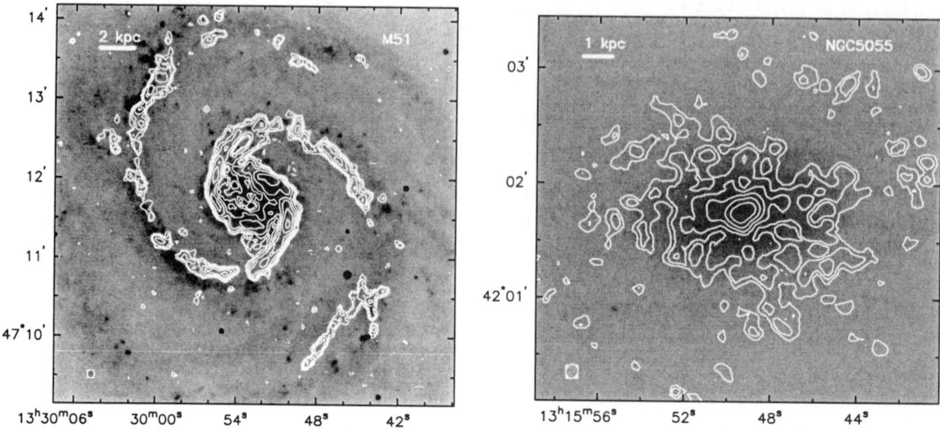

Figure 1. *Left*: CO in M51 from the BIMA Survey of Nearby Galaxies (SONG), overlaid on an optical image of the galaxy. The image contains zero spacing data and thus samples all of the spatial frequencies to the resolution limit (shown by the small box in the lower left). The CO is very strongly concentrated to the spiral arms and lies upstream of most of the ionized gas. *Right*: BIMA SONG image of the CO in NGC 5055 overlaid on an optical image showing that the galaxy is devoid of large-scale spiral structure at visible wavelengths. Some of the off-nuclear CO peaks are associated with weak spiral arms seen in the near infrared. The difference in CO morphology between these two galaxies is rather striking.

$2 - 5 \times 10^8$ yr independent of radius (Lacey & Fall 1985; Blitz 1995), setting a strict upper limit to the lifetime of the molecular gas (rather than just the molecular clouds) in the Galaxy.

The degree of confinement of the molecular gas to the spiral arms has long been understood to be a direct test of GMC lifetimes, and the work of Cohen et al. (1980) has recently been improved with more sensitive data from the same telescope by Digel et al. (1996) and the new FCRAO outer Galaxy CO survey (Heyer & Terebey 1998). The CO integrated intensity contrast between the arm and interarm regions in the outer Galaxy is at least a factor of 28, indicating that the gas that enters a molecular arm is overwhelmingly atomic. Some of the results are discussed in greater detail below. With the improvement in resolution and sensitivity of the millimeter-wave interferometers it has now become possible to determine the degree of confinement of the molecular gas to the spiral arms in a wide variety of other galaxies. Figure 1a shows an image of the CO emission from M51 with the BIMA Array. This image contains data from all spatial scales down to the resolution limit and therefore contains all of the CO flux within the region surveyed. In M51, as in most of the other galaxies with strong spiral arms in the BIMA galaxy survey, the CO is closely confined to the spiral arms. However, in NGC 5055 (Figure 1b), there is a much smaller tendency for the CO to lie in spiral arms. The optical image shows no large scale spiral structure at all, though near infrared images show weak underlying spiral arms (Thornley 1996). Nevertheless NGC 5055 does show evidence for spirality, short incoherent spiral arm segments, that may play a role in GMC formation. Based on Hα images of the galaxy, It appears that the molecular clouds in NGC 5055 are not significantly different in their star forming properties from those in the Milky Way (Thornley 1996), because the star formation rate does not appear to be suppressed with respect to that measured in our own Galaxy.

We do not know, however, whether the physical properties of the individual molecular clouds in NGC 5055 differ from those in the Milky Way; the combination of sensitivity and resolution needed to find out will probably require the MMA. So, while the role of spiral arms in GMC formation is amply demonstrated in M51, M100, the Milky Way (see below) and other galaxies, it is unclear to what degree the spiral arms are *necessary* for GMC formation. Furthermore, there are some grand design spirals such as M81 that are weak in CO, so much so that it is not known to what degree the molecular clouds are confined to the spiral arms. More sensitive observations can settle the issue in M81 with present instruments.

2.2. THE CHAFF

We now wish to address the question of GMC formation in the disk of the Milky Way. In particular, recent observational progress has made it possible to address directly the role of molecular cloud agglomeration in the formation of GMCs.

New sensitive surveys of the molecular gas in the second Galactic quadrant with the FCRAO telescope (Heyer et al. 1998) and around one GMC (Mon OB1; Oliver, Masheder & Thaddeus 1996) have shown that there is a great deal of low surface brightness (i.e. low surface density) molecular gas in the spiral arms in the Galaxy and in the vicinity of GMCs. This gas bears a striking resemblance to the HLCs discovered in the mid-1980s (Blitz, Magnani & Mundy 1984). We may ask, is this molecular "chaff" seen in the high sensitivity surveys the same as the HLCs and can this molecular gas agglomerate to form the GMCs in the Milky Way?

We first note that like the HLCs, most of the low column density gas in the Oliver et al. study is not self-gravitating; the dynamical masses typically exceed the luminous masses by more than an order of magnitude. The derived luminous masses of the clouds that compose the Mon OB 1 chaff are higher than the typical HLC masses, but this may be because Oliver et al. artificially placed all the clouds at a distance of about 1 kpc even though the clouds might be at any distance along the line of sight from 0 to about

1 – 1.5 kpc. The Oliver et al. map bears a striking resemblance to the FCRAO maps of Heyer et al. (1998), suggesting that the chaff permeates the entire outer Galaxy but with a decreased surface density between the arms. An example of the FCRAO survey in the range $102° < l < 142°$ is shown in Figure 2a. A longitude-velocity plot over the same range is shown in Figure 2b.

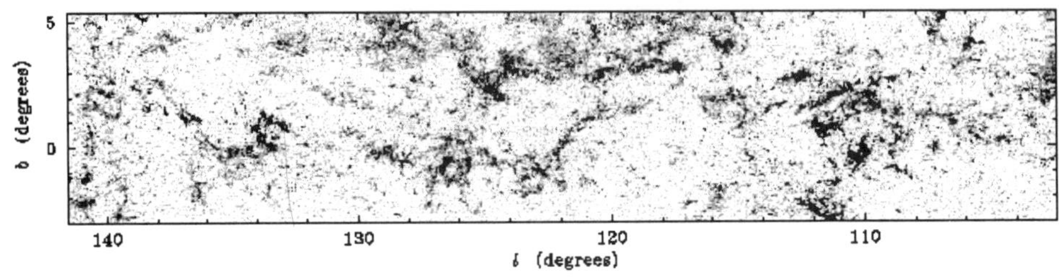

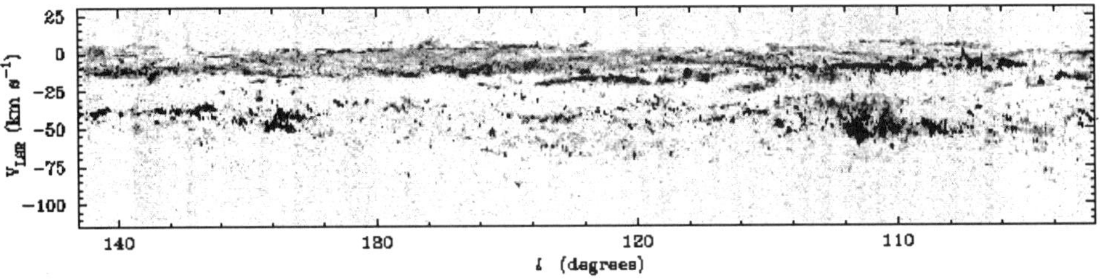

Figure 2.   CO emission in the outer Galaxy (from Heyer et al. 1998). Top panel is a velocity integrated $l - b$ plot, lower panel is a $l - v$ plot (integrated over the latitude range of the survey) showing the local and Perseus spiral arms. Note the almost complete absence of molecular gas between the two spiral arms at about 0 – -10 km s^{-1} and -40 km s^{-1}. The high spatial dynamic range of this survey shows the large scale distribution of molecular gas in the ISM in exceptional detail.

Heyer & Terebey (1998) have determined the molecular cloud spectrum $(dN/d \ln M)$ for more than 1500 clouds in the Perseus Arm and found a power law exponent of -0.75, similar to the value -0.6 found for inner Galaxy clouds (e.g. Sanders, Scoville, & Solomon 1985). Furthermore, their measurements included clouds with masses as low as $\sim 100\ M_\odot$, well into the mass range of HLCs. This measurement has several implications. First, it suggests that the chaff (M $< 10^3\ M_\odot$) is part of the same population as the self-gravitating GMCs, and may very well share the same formation process: there is no obvious change in slope of the mass spectrum at the point where the clouds are no longer self-gravitating. Second, Heyer (1999) argues that, like the low column density clouds near MonOB1, the clouds that make up the chaff are not self-gravitating. Third, the slope implies that as for GMCs, most of the mass in chaff is contributed by the highest mass clouds. Thus even though Oliver et al. have probably erred on the masses of many individual clouds as the result of incorrect distance assumptions, the total estimated mass is probably not too much in error.

Oliver et al. (1996), find that the total mass of chaff near Mon OB1 is 3×10^4 $M_\odot$. This is contained within a Galactic surface area of about $1 - 2 \times 10^5$ pc^2, thus the surface density of the chaff is $0.15 - 0.3$ $M_\odot$ pc^{-2}. The surface density of the HLCs is found to be $0.1 - 0.2$ $M_\odot$ pc^{-2} (Magnani, Blitz & Mundy 1985; Magnani, Lada, & Blitz 1986; Reach et al. 1998; Magnani, Hartmann & Thaddeus 1999), close to the Oliver et al. determination. The HLCs thus have the same surface density as the chaff, and they are each non-self-gravitating clouds of comparable mass. The chaff therefore appears to be the Galaxy-wide identification of what were first identified as high-latitude molecular clouds.

Figure 2 suggests that we may extrapolate the Oliver et al. values to the Galaxy as a whole. Assuming a constant surface density, we find that the total mass of the chaff is $7.5 - 15 \times 10^7$ $M_\odot$, and that the total mass in GMCs is $\sim 9 \times 10^8$ $M_\odot$ (Dame 1993 – scaled to the Hunter 1997 recalibration of $I(CO)/N(H_2)$). This difference implies that there is not enough mass in small molecular clouds to form the GMCs through collisional agglomeration. That is, if the number of GMCs is roughly in steady state and the formation time equals the dissolution time, then, averaged over the Galaxy, there would have to be approximately equal gas masses in GMCs and in the chaff. As Cowie (1980) showed, this need only to be true in the spiral arms, since orbit crowding can enhance the surface density of small clouds in the arms. However, the mass of chaff is well below that of the GMCs by about an order of magnitude, even in the spiral arms. It therefore appears that GMCs in the disk of the Milky Way *must* form by condensation from the HI, rather than from pre-existing molecular gas.

What does this imply for GMC formation and evolution? First, the short lifetime of molecular clouds implies that molecular gas recycles through the ISM fairly rapidly, and that the instantaneous recycling approximation often used in Galactic chemical evolution calculations is a good one. Second, it suggests that the collisional agglomeration of pre-existing molecular clouds as a formation mechanism for GMCs seems to be untenable in any form. Third, although spiral shock induced formation of GMCs seems to be indicated in many galaxies, it does not explain why there is extensive molecular gas and star formation in galaxies with weak or absent spiral arms. The formation of molecular clouds therefore seems to occur by condensation from the HI in conjunction with some other mechanism. This conclusion has already been discussed previously (e.g. Elmegreen 1990, Blitz 1991,

1993); the radius from which the H_2 forms in the solar vicinity is about 150 pc for a typical GMC with radius ~ 30 pc. Although these conclusions seem secure within the Galactic disk, and are probably also true even at the peak of the molecular ring, the situation is dramatically different for GMCS near the Galactic Center where the ratio $\Sigma(H_2)/\Sigma(HI) \gtrsim 100$ (e.g. Liszt & Burton 1996).

2.3. THE ASSOCIATION OF ATOMIC AND MOLECULAR GAS

HI envelopes around molecular clouds are quite common (e.g., Moriarty-Schieven, Andersson & Wannier 1997; Williams & Maddalena 1996). Figure 3 shows an example of such a cloud around the Rosette Molecular Cloud (Williams, Blitz, & Stark 1995). In the solar vicinity the mass of the molecular envelopes seem to be about the same as the mass of molecular gas in a GMC (Blitz 1990), but the atomic gas is considerably more extended. These envelopes may be remnants of the atomic clouds that condensed to form the GMCs or may be the photodissociated gas from the GMCs, since the extinction is $\sim 0.25 - 0.5$ A_V, just what is needed to shield the CO from the interstellar UV field. Most probably, they are a combination of both. In the solar vicinity, the separation between the HI envelopes is considerably larger than their diameters: the mean distance between GMCs is about 500 pc, and the molecular cloud/HI envelope complexes have diameters of about $150 - 200$ pc (Blitz 1990). The complexes are thus distinct from the background HI, indeed that is how they are identified.

In the inner Galaxy the situation is somewhat different. Near the peak of the molecular ring, the surface density of molecular gas is about equal to that of the atomic gas (Dame – 1993 modified by the new calibration of Hunter et al. 1997). If the GMCs in the ring have atomic envelopes with the HI mass equal to the H_2 mass, as is true locally, then all of the atomic gas in the molecular ring would be associated with GMC envelopes, leaving little atomic gas for a true intercloud medium. This may not be a problem, however: the molecular gas surface density is a factor of $\sim 5 - 6$ greater in the molecular ring than locally so the mean distance between GMCs $\sim 500/\sqrt{5} \simeq 200$ pc, about the same as the diameter of the CO/HI GMC complexes. It may be that the envelopes merge and form a general background within which the GMCs are located. This might explain why it has been more difficult to associate atomic hydrogen clouds with GMCs in the molecular ring than it is locally.

Within several hundred parsecs of the Galactic center, the situation is drastically different. First, the gas is almost entirely molecular; only about 1% of the surface density of the gas is atomic (e.g. Liszt & Burton 1996). In this region of the Galaxy, gas that becomes neutral must quickly become

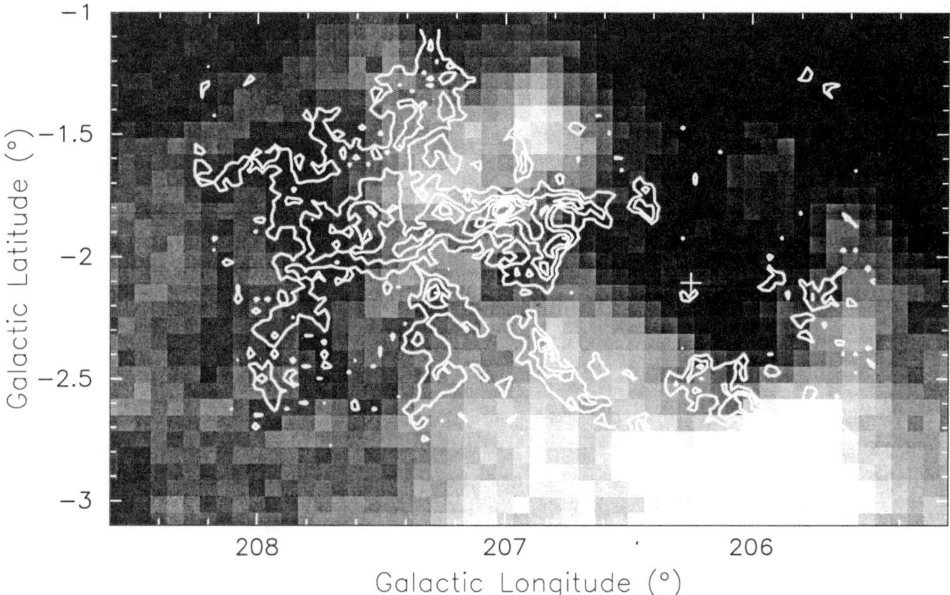

Figure 3. An HI envelope around the Rosette molecular cloud. The grayscale, range 450 to 620 K km s^{-1}, shows HI data from Arecibo observations by Kuchar & Bania (1993). Contours, beginning from and with increment 18 K km s^{-1}, are CO emission from Bell Labs observations by Blitz & Stark (1986). Emission has been summed over a velocity range $v = 4 - 25$ km s^{-1}. The cross marks the OB association that lies at the center of the Rosette nebula and has cleared out the neutral gas. The regions of strong HI emission (lighter colors) lie on the CO cloud boundaries, forming an envelope around the cloud.

molecular, and if the formation and dissolution of molecular clouds goes through an atomic phase, that phase must be very brief.

The relation of the atomic to molecular gas must also be very different in the central 300 pc of the Galaxy because of the large interstellar gas pressure due to the deep stellar potential of the bulge/bar. In the bulge, the interstellar gas pressure is both predicted and measured to be almost three orders of magnitude greater than it is locally (Spergel & Blitz 1992), and both the atomic and molecular gas are highly overpressured compared to the solar vicinity. The HI is particularly problematic because at a pressure $P/k \simeq 10^7$ Kcm^{-3}, and a maximum temperature of 10^4 K, the mean density of the atomic gas near the center must be $\geq 10^3$ cm^{-3}, a density that is as high as the typical molecular gas density in local GMCs. How can this be? Apparently, the HI can only exist as either very small dense knots, in which case it would be difficult to avoid turning molecular, or as thin, high density sheets on the surfaces of the molecular clouds at the center. Because the molecular gas already has a low volume filling fraction, the ionized gas must make up almost all of the volume in the bulge/bar. This picture of the relationship between the atomic and molecular gas has not been directly verified by observation, but a combination of VLA observations with archival molecular cloud data should be able to do so.

The effect on star formation of this high pressure is hard to predict without better knowledge of the power law γ relating the density and pressure of the gas. For example, for a perfect gas, the Jeans mass of clumps is

$$M_J = 10 \frac{v_{rms}^4}{(G^3 P_0)^{1/2}} = 10^6 \left(\frac{v_{rms}}{5 \mathrm{kms}^{-1}} \right)^4 \left(\frac{P_0}{5 \times 10^6 K \mathrm{cm}^{-3}} \right)^{-1/2} M_\odot. \quad (1)$$

The large increase in pressure at the Galactic center suggests that the typical self-gravitating mass would significantly decrease at the center. However, since the molecular clouds are known to be both hotter and more turbulent at the center (Güsten 1989), the strong dependence of the Jeans mass on the velocity dispersion of the gas can be more than offset by the increase in pressure. The largest uncertainly in applying Equation (1), is knowing what to take for v_{rms} for the individual star forming cores.

Astonishingly, even though the gas pressure is apparently quite high, the rate of star formation in the nuclear region (which can only be estimated for the massive stars) does not seem to be significantly different from that of the Galactic disk (Güsten 1989). This is remarkable given that the surface pressure on the molecular clouds is nearly three orders of magnitude greater than locally. If true, it suggests that the star forming properties of a GMC may be determined only locally within a cloud and that the ambient conditions may have to be even more extreme than those at the Galactic Center to have a significant effect on the IMF. If so, it will be necessary to look at the internal structure of a GMC to see how it organizes itself to form stars. We turn to that question in the following section.

3. CLOUD STRUCTURE

3.1. CATEGORIZATION

Although molecular clouds are, by definition, regions in which the gas is primarily molecular by *mass*, much of the *volume* of such a cloud is not molecular. That is, the filling fraction of molecular gas is low, $\lesssim 20\%$ (Blitz 1993), and the cloud is highly structured with large density variations from one location to another. The structure (in volume density, column density, and velocity) of molecular clouds reflects the conditions from which they form, acts as a signpost to their evolution, and may, at sufficiently high densities, be related to the mass scale of stars and the slope of the IMF.

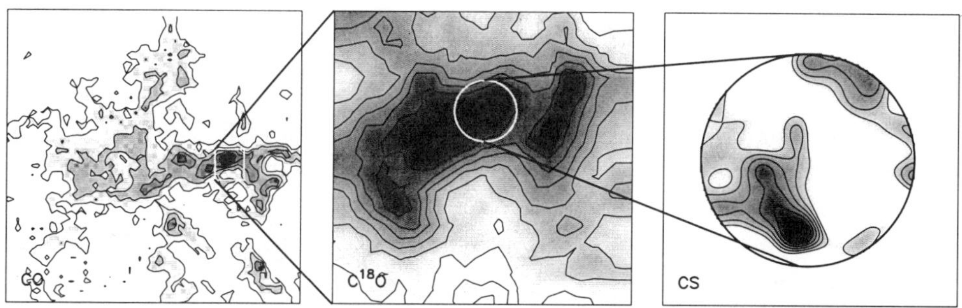

Figure 4.   Hierarchical cloud structure. The three panels show a representative view from cloud to clump to core. The bulk of the molecular gas (cloud; left panel) is best seen in CO which, although optically thick, faithfully outlines the location of the H$_2$. Internal structure (clumps; middle panel) is observed at higher resolution in an optically thin line such as C^{18}O. With a higher density tracer such as CS, cores (right panel) stand out. The observations here are of the Rosette molecular cloud and are respectively, Bell Labs (90''), FCRAO data (50''), and BIMA data (10'').

In this section, we discuss techniques to analyze cloud structure and the results and implications of such analyses. First, we define an operational categorization into clouds, clumps, and cores. The scales of interest are illustrated in Figure 4. This categorization is not inconsistent with the fractal models for cloud structure that are discussed in §3.4, although we argue that gravity introduces scales that limit the range of validity of the fractal description.

Clouds are regions in which the gas is primarily molecular as stated above. Almost all known molecular clouds in the Galaxy are detectable in CO. Giant molecular clouds, with masses $\gtrsim 10^4\ M_\odot$, are generally gravitationally bound, and may contain several sites of star formation. However, there are also many small molecular clouds with masses $\lesssim 10^2\ M_\odot$, such as the unbound high latitude clouds discovered by Blitz, Magnani, & Mundy (1984), the chaff discussed in §2.2, and the small gravitationally bound molecular clouds in the Galactic plane cataloged by Clemens & Barvainis (1988). A small number of low mass stars are observed to form in some of these clouds but their contribution to the total star formation rate in the Galaxy is negligible (Magnani et al. 1995). Heyer et al. (1999) suggest that most of the clouds with M $< 10^3$ M$_\odot$ are not self-gravitating; the Clemens & Barvainis clouds are presumably a small but unknown fraction of mass of the low mass chaff.

Clumps are coherent regions in $l - b - v$ space, generally identified from spectral line maps of molecular emission. Star-forming clumps are the massive clumps out of which stellar clusters form. Although most clusters are unbound, the gas out of which they form is bound (Williams et al. 1995). Clumps may be blended together at low intensities, particularly in low density molecular tracers such as CO and its isotopes. In this case, several techniques exist to decompose the emission into its constituent clumps (§3.2).

Cores are regions out of which single stars (or multiple systems such as binaries) form and are necessarily gravitationally bound. Not all material that goes into forming a star must come from the core; some may be accreted from the surrounding clump or cloud as the protostar moves through it (Bonnell et al. 1997).

3.2. STRUCTURE ANALYSIS TECHNIQUES

Molecular cloud structure can be mapped via radio spectroscopy of molecular lines (e.g., Bally et al. 1987), continuum emission from dust (e.g., Wood, Myers, & Daugherty 1994), or stellar absorption by dust (Lada et al. 1994). The first gives kinematical as well as spatial information and results in a three dimensional cube of data, whereas the latter two result in two dimensional datasets. Many different techniques have been developed to analyze these data which we discuss briefly here.

Stutzki & Güsten (1990) and Williams, de Geus, & Blitz (1994) use the most direct approach and decompose the data into a set of discrete clumps, the first based on recursive tri-axial gaussian fits, and the latter by identifying peaks of emission and then tracing contours to lower levels. The resulting clumps can be considered to be the "building blocks" of the cloud and may be analyzed in any number of ways to determine a size-linewidth relation, mass spectrum, and variations in cloud conditions as a function of position (Williams et al. 1995). There are caveats associated with each method of clump deconvolution, however. Since the structures in a spectral line map of a molecular cloud are not, in general, gaussian, the recursive fitting method of Stutzki & Güsten (1990) will tend to find and subsequently fit residuals around each clump, which results in a mass spectrum that is steeper than the true distribution. On the other hand, the contour tracing method of Williams et al. (1994) has a tendency to blend small features with larger structures and results in a mass spectrum that is flatter than the true distribution.

Heyer & Schloerb (1997) use principal component analysis to identify differences in line profiles over a map. A series of eigenvectors and eigenimages are created which identify ever smaller velocity fluctuations and their spatial distribution, resulting in the determination of a size-linewidth relation. Langer, Wilson, & Anderson (1993) use Laplacian pyramid transforms (a generalization of the Fourier transform) to measure the power on different size scales in a map; as an application, they determine the mass spectrum in the B5 molecular cloud. Recently, Stutzki et al. (1998) have described a closely related Allan-variance technique to characterize the fractal structure of 2-dimensional maps. Houlahan & Scalo (1992) define an algorithm that constructs a structure tree for a map; this retains the spatial relation of the individual components within the map but loses information regarding their shapes and sizes. It is most useful for displaying and ordering the hierarchical nature of the structures in a cloud.

Adams (1992) discusses a topological approach to quantify the difference between maps. Various "output functions" (e.g., distribution of density, volume, and number of components as a function of column density; see Wiseman & Adams 1994) are calculated for each cloud dataset and a suitably defined metric is used to determine the distance between these functions and therefore to quantify how similar clouds are, or to rank a set of clouds.

A completely different technique was pioneered by Lada et al. (1994). They determine a dust column density in the dark cloud IC 5146 by star counts in the near-infrared and mapped cloud structure over a much greater dynamic range ($A_V = 0 - 32$ mag) than a single spectral line map. The effective resolution, $\gtrsim 30''$, is determined by the sensitivity of the observations.

The most striking result of applying these various analysis tools to molecular cloud datasets is the identification of self-similar structures characterized by power law relationships between, most famously, the size and linewidth of features (Larson 1981), and the number of objects of a given mass (e.g., Loren 1989). Indeed, mass spectra are observed to follow a power law with nearly the same exponent, $x = 0.6 - 0.8$, where $dN/d\ln M \propto M^{-x}$ from clouds with masses up to 10^5 $M_\odot$ in the outer Galaxy to features in nearby high-latitude clouds with masses as small as 10^{-4} $M_\odot$ (Heyer & Terebey 1998; Kramer et al. 1998a; Heithausen et al. 1998). Since a power law does not have a characteristic scale, the implication is that clouds and their internal structure are scale-free. This is a powerful motivation for a fractal description of the molecular ISM (Falgarone et al. 1991, Elmegreen 1997a). On the other hand, molecular cloud maps do have clearly identifiable features, especially in spectral line maps when a velocity axis can be used to separate kinematically distinct features along a line of sight (Blitz 1993). These features are commonly called clumps, but there are also filaments (e.g., Nagahama et al. 1998), and rings, cavities, and shells (e.g., Carpenter et al. 1995). Both the discrete (clump) and fractal description of clouds can be used as tools of analysis and both reveal much about cloud physics and star formation. We discuss each in turn in the following subsections.

3.3. CLUMPS

Clump decomposition methods such as those described above by Stutzki & Güsten (1990) and Williams et al. (1994) can be readily visualized and have an appealing simplicity. In addition, as for all automated techniques, these algorithms offer an unbiased way to analyze datasets, and are still a valid and useful tool for cloud comparisons even if one does not subscribe to the notion of clumps within clouds as a physical reality (Scalo 1990).

In a comparative study of two clouds, Williams et al. (1994) searched for differences in cloud structure between star forming and non-star forming GMCs. The datasets they analyzed were maps of ^{13}CO(1–0) emission with similar spatial (0.7 pc) and velocity resolution (0.68 km s^{-1}) but the two clouds, although of similar mass $\sim 10^5$ $M_\odot$, have very different levels of star formation activity. The first, the Rosette molecular cloud, is associated with an HII region powered by a cluster of 17 O and B stars and also contains a number of bright infrared sources from ongoing star formation deeper within the cloud (Cox, Deharveng, & Leene 1990). The second cloud, G216-2.5, originally discovered by Maddalena & Thaddeus (1985), contains no IRAS sources from sites of embedded star formation and has an exceptionally low far-infrared luminosity to mass ratio (Blitz 1990), $L_{\rm IR}/M_{\rm cloud} < 0.07$ $L_\odot/M_\odot$, compared to more typical values of order unity (see Williams & Blitz 1998).

Almost 100 clumps were cataloged in each cloud, and sizes, linewidths, and masses were calculated for each. These basic quantities were found to be related by power laws with the same index for the two clouds, but with different offsets (Figure 5) in the sense that for a given mass, clumps in the non-star forming cloud are larger, and have greater linewidths than in the star forming cloud. The similarity of the power law indices suggests that, on these scales, $\sim$ few pc, and at the low average densities, $\langle n_{\rm H_2} \rangle \sim 300$ cm^{-3}, of the observed clumps, the principal difference between the star forming and non-star forming cloud is the change of scale rather than the collective nature of the structures in each cloud.

Figure 5 shows that the kinetic energy of each clump in G216-2.5 exceeds its gravitational potential energy, and therefore no clump in the cloud is self-gravitating (although the cloud as a whole is bound). On the other hand, Williams et al. (1995) show that, for the Rosette molecular cloud, star formation occurs only in the gravitationally bound clumps in the cloud. Therefore, the lack of bound clumps in G216-2.5 may explain why there is little or no star formation currently taking place within it.

Even in the Rosette cloud, most clumps are not gravitationally bound (and do not form stars). These unbound clumps have similar density profiles, $n(r) \propto 1/r^2$, as the bound clumps (Williams et al. 1995), but contain

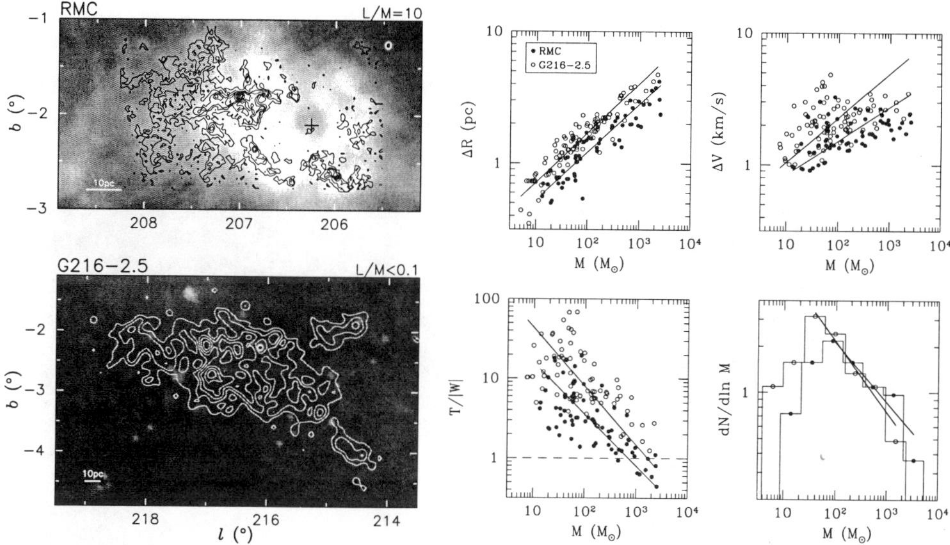

Figure 5. Structure in the Rosette and G216-2.5 molecular clouds. The two left panels show contours of velocity integrated CO emission (levels at 15 K km s^{-1} for the RMC, 1.8 K km s^{-1} for G216-2.5) overlaid on a grayscale image of the IRAS 100 μm intensity (1.1 to 2.5 MJy sr^{-1}, same for both clouds). The Rosette cloud is infrared bright, indicative of its high star formation rate, but G216-2.5 has a very low infrared luminosity due to a lack of star formation within it. The four rightmost panels show power law relations between clump mass and size, linewidth, energy balance (i.e. the ratio of kinetic energy, $\mathcal{T} = 3M(\Delta v/2.355)^2/2$, to gravitational potential energy, approximated as $\mathcal{W} = -3GM^2/5R$), and number (i.e. clump mass spectrum) for the two clouds. The solid circles represent clumps in the RMC, and open circles represent clumps in G216-2.5. Each relationship has been fit by a power law: note that the power law exponent is approximately the same for the clumps in each cloud despite the large difference in star formation activity.

relatively little dense gas as traced by CO(3–2) or CS(2–1) (Williams & Blitz 1998). The unbound clumps are "pressure confined" in that their internal kinetic pressure, which is primarily turbulent, is comparable to the mean pressure of the ambient GMC (Blitz 1991; Bertoldi & McKee 1992). Simulations suggest that these clumps are transient structures (Ostriker, Gammie, & Stone 1999).

The nature of the interclump medium remains unclear: Blitz & Stark (1986) found low intensity broad line wings in their CO and ^{13}CO(1–0) observations of the Rosette molecular cloud which, they postulated, resulted from a pervasive low density molecular interclump medium, but Schneider et al. (1996) found that the line wings are also apparent in the $J = 3 - 2$ transitions with a line ratio similar to the core emission possibly indicating a density not much different from the bulk of the cloud, and therefore not interclump. On the other hand, substantial HI envelopes exist around molecular clouds (Blitz 1990, 1993) and may also pervade the volume between the clumps within a cloud (Figure 6). The turbulent pressures of the atomic and molecular components, $\rho\sigma^2$, where ρ is the density and σ is the velocity dispersion, are comparable (Williams et al. 1995) and show that the HI can indeed confine the CO clumps.

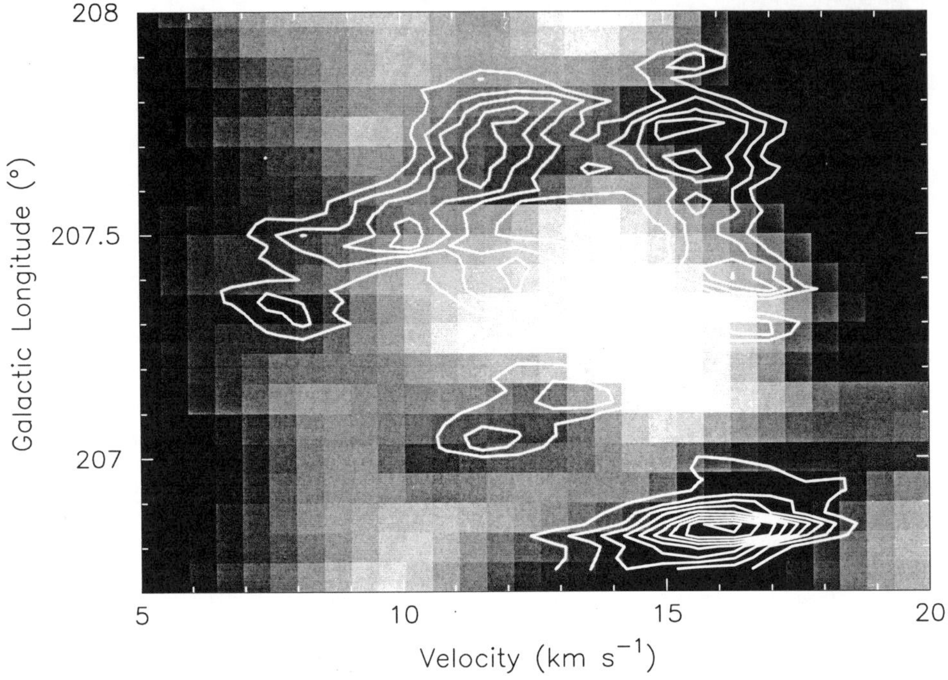

Figure 6. Detailed anti-correlation between HI and ^{13}CO in the Rosette molecular cloud. The grayscale (range 5 to 7 K °) is HI emission (Kuchar & Bania 1993), and the contours (starting level and increment is 0.06 K °) are ^{13}CO data from Blitz & Stark (1986). Each are a slice in latitude integrated over $b = -2°1$ to $-1°9$. The ^{13}CO map shows a number of clumps, most of which are gravitationally unbound, but they tend to lie in valleys in the HI which may act as a pressure-confining medium.

Cloud, clump, and core density profiles are reflections of the physics that shape their evolution, but the density profiles of clouds and clumps have received scant attention. For clouds, which often are quite amorphous without a clear central peak, the density profile is often difficult to define observationally. For clumps, Williams et al. (1995) showed that surface density profiles of pressure bound, gravitationally bound, and star-forming clumps all have similar power law indices close to 1. Formally, the fits range from -0.8 to -1.2, but these differences appear to be only marginally significant. For a spherical cloud of infinite extent, $\Sigma(r) \propto r^{-1}$ implies $\rho(r) \propto r^{-2}$, suggesting that the (turbulent) pressure support is spatially constant. However,

McLaughlin & Pudritz (1996) argued that for finite spheres, the volume density distribution can be considerably flatter than that inferred for infinite clumps. Density distributions inferred from observations also require consideration of beam-convolution effects. It is nevertheless astonishing that both strongly self-gravitating clumps and those bound by external pressure have such similar, perhaps identical density distributions. Why this should be so is unclear.

3.4. FRACTAL STRUCTURES

An alternate description of the ISM is based on fractals. High spatial dynamic range observations of molecular clouds, whether by millimeter spectroscopy (e.g., Falgarone et al. 1998), IRAS (Bazell & Désert 1988), or using the Hubble Space Telescope (e.g., O'Dell & Wong 1996; Hester et al. 1996) show exceedingly complex patterns that appear to defy a simple description in terms of clouds, clumps and cores; Scalo (1990) has argued that such loaded names arose from lower dynamic range observations and a general human tendency to categorize continuous forms into discrete units.

As we have discussed above, it seems that however one analyzes a molecular cloud dataset, one finds self-similar structures. Moreover, the highly supersonic linewidths that are observed in molecular clouds probably imply turbulent motions (see discussion in Falgarone & Phillips 1990), for which one would naturally expect a fractal structure (Mandelbrot 1982).

The fractal dimension of a cloud boundary, D, can be determined from the perimeter-area relation of a map, $P \propto A^{D/2}$. Many studies of the molecular ISM find a similar dimension, $D \simeq 1.4$ (Falgarone et al. 1991 and references therein). In the absence of noise, $D > 1$ demonstrates that cloud boundaries are fractal. That D is invariant from cloud to cloud (star-forming or quiescent, gravitationally bound or not) is perhaps related to the similarity in the mass spectrum index in many different molecular clouds (Kramer et al. 1998a; Heithausen et al. 1998). Fractal models have been used to explain both the observed mass spectrum of structures (Elmegreen & Falgarone 1996) and the stellar IMF (Elmegreen 1997b).

Conclusions about physical processes in molecular clouds that are drawn from the perimeter-size relation should be treated with caution, however, since in column density maps such as from IRAS observations or integrated intensity spectral line maps, the observed structures are projections of an inherently three-dimensional distribution which, for sufficiently high filling factor, results in multiple overlapping of unrelated objects. Such overlapping is found to mimic the observed fractal perimeter-size relation in Monte-Carlo simulations of clumpy media by Witt & Gordon (1996). That is, in an inhomogeneous cloud the overlap of discrete objects from projection can be a fractal even if the individual objects are not. Furthermore, noise can produce a non-integral exponent in the perimeter-area relation giving the appearance of a fractal, but the degree to which noise affects the published results has not yet been investigated.

Probability density functions (PDFs) may be used to describe the distribution of physical quantities (such as density and velocity) in a region of space without resorting to concepts of discrete objects such as clouds, clumps and cores. Falgarone & Phillips (1990), for example, have analyzed the PDF of the velocity field of several clouds at different scales. The low-level, broad line wings that are observed in non-star forming regions show that the probability of rare, high-velocity motions in the gas are greater than predicted by a normal (gaussian) probability distribution. This *intermittent* behavior is expected in a turbulent medium, and the detailed analysis of Falgarone & Phillips shows that the deviations from the predictions for Kolmogorov turbulence are small. Miesch & Scalo (1995) calculate velocity centroid PDFs from ^{13}CO observations of star-forming regions and also report non-gaussian behavior. Lis et al. (1996) compare their results with a similar analysis applied to simulations of compressible turbulence; such work may be a promising avenue for exploring the role of turbulence in molecular clouds.

Detailed simulations of 3-D hydromagnetic turbulence including gravity have now become possible and may ultimately help to determine how the structure in molecular clouds forms. The best simulations to date are those by Ostriker et al. (1999) and the next few years should show a great deal of progress in the application of codes to molecular clouds. The utility of the clump finding algorithms such as Clumpfind (Williams et al. 1994) applied to both the simulations and the observations will be a good test of how closely the simulations match reality.

3.5. DEPARTURES FROM SELF-SIMILARITY

The universal self-similarity that is observed in all types of cloud, over a wide range in mass and star forming activity is remarkable, but a consequence of this universality is that it does not differentiate between clouds with different rates of star formation (or those that are not forming stars at all) and therefore it cannot be expected to explain the detailed processes by which a star forms. Star formation must be preceded by a departure from structural self-similarity.

There have long been suggestions that the thermal Jeans mass gives a scale that determines the characteristic mass of stars (Larson 1985). In order to determine whether such a scale is important in molecular clouds, Blitz & Williams (1997) examined how the structural properties of a large

scale, high resolution ^{13}CO map of the Taurus molecular cloud obtained by
Mizuno et al. (1995) varied as the resolution was degraded by an order of
magnitude. In their work, they use the temperature PDF of the dataset to
compare the cloud properties as a function of resolution. This is the most
basic statistic and requires minimal interpretation of the data. In Figure 7
we show the temperature PDF for the Taurus dataset at two resolutions
and four other ^{13}CO maps of molecular clouds.

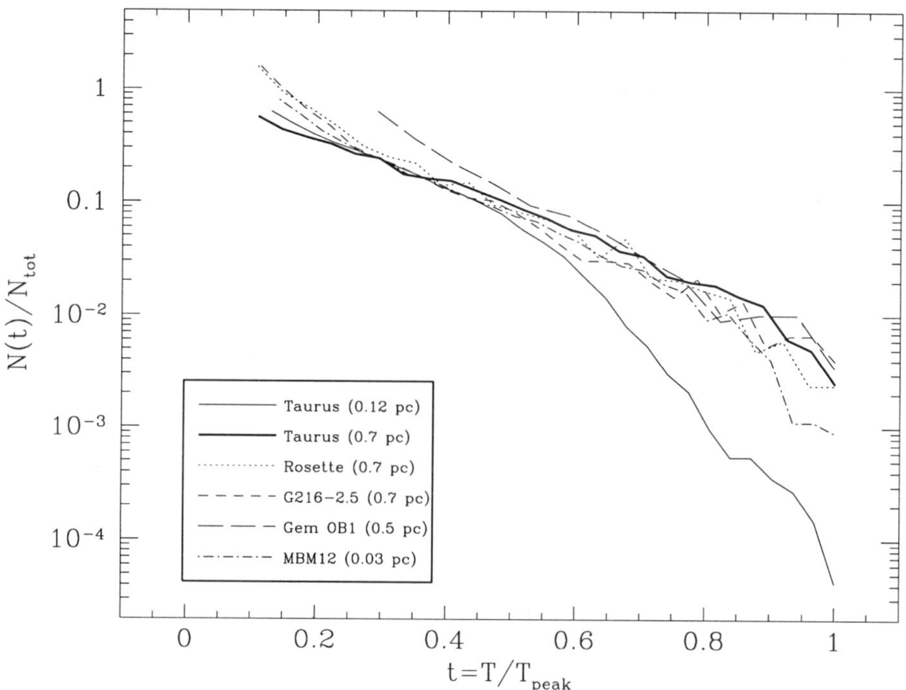

Figure 7.   Intensity PDF in five molecular clouds, all from ^{13}CO(1–0) maps, except Gem
OB1 which is CO(1–0). Temperatures have been normalized by the peak value in each
dataset and then binned. The fraction of data points with $T/T_{\mathrm{peak}} > 0.3$ is plotted as a
function of normalized temperature. Note the similarity between the different datasets,
except for the high resolution Taurus map which has a lower fraction of data points with
temperatures within 60% of the peak.

To compare the different cloud PDFs, Figure 7 shows temperatures
that have been normalized by the peak, T_{peak}, of each map. Each PDF
has also been truncated at $T/T_{\mathrm{peak}} \simeq 0.15 - 0.25$ to show only those points
with high signal-to-noise. Within the Poisson errors (not shown for clarity),
the PDFs of the different clouds are all the same, except for the higher
resolution Taurus PDF for which there is a lower relative probability of

having lines of sight with $T/T_{\mathrm{peak}} \gtrsim 0.7$. The low intensities of the ^{13}CO emission imply that the optical depth is small along all lines of sight in the map, and is not responsible for this effect. Rather, from examination of the integrated intensity maps, Blitz & Williams (1997) show that this is due to a steepening of the column density profiles at small size scales (see also Abergel et al. 1994).

There are two immediate implications from Figure 7. First, the common exponential shape for the temperature PDF is another manifestation of the self similar nature of cloud structure. It is a simple quantity to calculate and may provide a quick and useful test of the fidelity of cloud simulations. Second, since the behavior of the Taurus dataset changes as it is smoothed, it cannot be described by a single fractal dimension over all scales represented in the map since the intensity PDF for a fractal is invariant under smoothing (Figure 8). There is other evidence for departures from self-similarity at similar size scales. Goodman et al. (1998) examine in detail the nature of the size-linewidth relation in dense cores as linewidths approach a constant, slightly greater than thermal, value in a central "coherent" region ~ 0.1 pc diameter (Myers 1983). Also, Larson (1995) finds that the two-point angular correlation function of T Tauri stars in Taurus departs from a single power law at a size scale of 0.04 pc (see also Simon 1997).

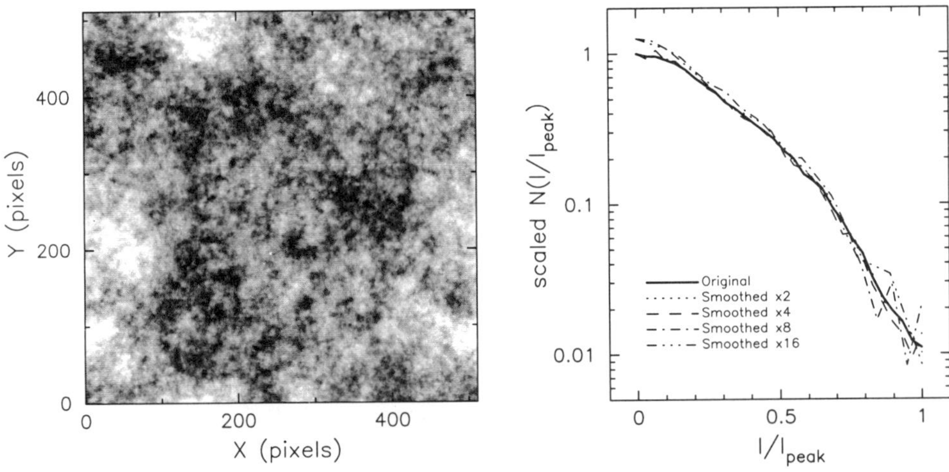

Figure 8.   A fractal cloud and the effect of smoothing on its intensity PDF. The left panel shows a grayscale image of a model fractal cloud from $T = 0$ to 1 K (courtesy of Chris Brunt). The right panel shows the normalized temperature histogram for the original dataset (heavy solid line) and then for the same dataset after binning by 2, 4, 8, and 16 pixels. The intensity PDF is independent of resolution as expected for this self-similar structure.

For gas of density $n_{\mathrm{H_2}} \sim 10^3$ cm^{-3}, these size scales correspond to masses of order $\sim 1\, M_\odot$, close to the thermal Jeans mass at a temperature

$T = 10$ K. It is important to note that the above evidence for characteristic scales comes from studies of gravitationally bound, star forming regions: self-similarity in unbound clouds continues to much smaller scales. Figure 7 shows that the temperature PDF of the unbound, high latitude cloud MBM12 is identical, at a resolution of 0.03 pc, to the other lower resolution PDFs of star forming GMCs. Similarly, the mass spectra of other high latitude clouds follow power laws, $dN/d\ln M \propto M^{-x}$ with $x \simeq 0.6 - 0.8$, down to extremely low masses, $M \simeq 10^{-4}$ $M_\odot$ (Kramer et al. 1998a; Heithausen et al. 1998). It appears to be the action of gravity that creates the observed departures from self-similarity. Note, however, that if the Jeans mass is physically relevant to the structure of molecular clouds, then clouds must be magnetically supercritical (i.e., magnetic fields cannot prevent dynamical collapse) since, as explained in the chapter by Shu et al., the Jeans mass has no meaning in subcritical clouds (such clouds are supported by magnetic fields against gravitational collapse no matter how highly compressed).

4. THE RELATION BETWEEN CLOUD STRUCTURE AND THE IMF

The spectrum, lifetime, and end state of a star are primarily determined by its mass. Consequently, the problem of understanding how the mass of a star is determined during its formation, and the origin of the IMF, has a very wide application in many fields from galaxy evolution to the habitability of extrasolar planets. The form of the IMF is typically assumed to be invariant, but since it is directly measureable only locally, knowing how it comes about can help us predict how it might vary under different astrophysical environments.

Many explanations for the form of the IMF use as their starting point the mass spectrum of clouds and clumps as revealed by molecular line emission, $dN/d\ln M \propto M^{-x}$ with $x \simeq 0.5$. Most such structures, however, are not forming stars: the majority of stars form in clusters in a few of the most massive clumps in a cloud. An understanding of the origin of the IMF can only come about with a more complete picture of the formation of star-forming clumps and the fragmentation of these clumps down to individual star forming cores.

The cores that form an individual star (or multiple stellar system) have typical average densities $n_{H_2} \sim 10^5$ cm^{-3} and can be observed in high excitation lines or transitions of molecules with large dipole moments (Benson & Myers 1989), or via dust continuum emission at millimeter and sub-millimeter wavelengths (Kramer et al. 1998b). Because of their high densities, the surface filling fraction of cores is low, even in cluster forming environments. Therefore searches for cores have generally followed signs of star formation activity, e.g. IRAS emission, outflows, etc. and there have been few unbiased searches (e.g. Myers & Benson 1983). However, increases in instrument speed have now made it possible to survey millimeter continuum emission over relatively large areas of the sky. There have been two very recent results in this regard, the first by Motte, André, & Neri (1998) at 1.3 mm using the array bolometer on the IRAM 30 m telescope, and the second by Testi & Sargent (1998) at 3 mm using the OVRO interferometer.

Motte et al. (1998) mapped the ρ Ophiuchus cloud, the closest rich cluster forming region and Testi & Sargent (1998) mapped the Serpens molecular cloud, a more distant, but richer star forming region. In each case, the large scale, high resolution observations reveal a large number of embedded young protostars and also starless, dense condensations. Both studies find that the mass spectrum of the cores is significantly steeper, $x > 1.1$ (where $dN/d\ln M \propto M^{-x}$) than clump mass spectra, $x \simeq 0.6 - 0.8$. The core mass spectra approach the slope, $x = 1.35$, of the stellar (Salpeter) IMF. These intriguing results suggest that the masses of stars that form in a cloud are directly linked to its structure. However, these investigations could not determine whether the clumps are self-gravitating, a necessary condition to show that these continuum cores actually form stars. Recall, for example, that the lowest mass molecular clouds, as well as the low mass clumps within a GMC are not, in general, self-gravitating. Surveys of other cluster forming regions should be made and these studies should be followed up with spectral line observations to determine whether the starless cores are self-gravitating, or even collapsing (Williams & Myers 1999), and therefore likely progenitors of future stars.

It is worth noting also that, since the filling fraction of the cores is much lower than that of the lower density CO clumps, the projection effects that we cautioned about above and simulated by Witt & Gordon (1996) are much less severe. Thus, it is not entirely clear that the observed departure from self-similarity is due to real physical processes such as gravity and/or the dissipation of turbulence, or is simply a result of the lower filling fraction.

As high resolution studies of individual cores in cluster environments become more commonplace, the relationship between stellar mass and core mass will be determined more precisely. If the core mass spectrum is indeed similar to the stellar IMF, then the fraction of a cores mass that goes into a star (the star formation efficiency of the core) must be approximately independent of mass and the stellar IMF is determined principally by the cloud fragmentation processes. By measuring the core mass spectrum in different clusters in a variety of different molecular clouds, the influence of the large scale structure and environment on the IMF can be quantified.

5. Summary

Considerable progress has been made in addressing many of the issues concerning molecular clouds since the first Crete conference and new results tentatively suggest a direct link between cloud structure and the IMF.

It seems clear that molecular clouds form through the condensation of HI since very little molecular gas exists between the spiral arms. Collisional agglomeration in any form appears to be ruled out because of the small total mass of low mass clouds (the chaff). However, many details of the formation process, such as the role of Galactic shocks and magnetic fields for instance, remain to be understood.

A close association between atomic and molecular gas is evident in the solar neighborhood and detailed maps suggest that the interclump medium within clouds is predominantly atomic. In the Galactic Center, however, the formation of molecular clouds and their association with the atomic gas must be very different because the pressure is more than 2 orders of magnitude higher, and the atomic mass fraction more than 2 orders of magnitude lower, than at the solar circle. Understanding cloud and star formation in this environment is a step toward understanding how stars form in even more extreme environments such as starbursts.

The structure within clouds reflects their formation from an inhomogeneous atomic ISM and, at moderate densities, $n_{H_2} \lesssim 10^3$ cm^{-3}, is self-similar up to a scale set by self-gravity. In this regime, scaling laws such as clump mass spectra have similar power law indices independent of the star-forming nature of the cloud.

At higher densities, and smaller sizes, as linewidths approach their thermal value, structures depart from the same self-similar description. This departure may mark the boundary between cloud evolution and star formation. Clusters of individual star forming cores, with a mass spectrum that approaches the Salpeter IMF, are observed in the ρ Ophiuchus and Serpens clouds. The study of the structure, dynamics, and distribution of these cores will lead to a better understanding of the relationship between the structure and evolution of molecular clouds and the initial mass function of stars.

Funding has been provided by a grant from the NSF. JPW is supported by a Jansky fellowship. We thank Tamara Helfer for providing Figure 1, Mark Heyer for Figure 2, and Loris Magnani for results prior to publication. We would also like to acknowledge discussions with Mark Heyer, Tam Helfer, Chris McKee, Tom Sodroski and Michele Thornley, about various topics covered in this review.

References

1. Abergel, A., Boulanger, F., Mizuno, A., & Fukui, Y. 1994, ApJ Lett., 423, L59.
2. Adams, F.C. 1992, ApJ, 387, 572
3. Bally, J., Stark, A.A., Wilson, R.W., & Langer, W.D. 1987, ApJ Lett., 312, L45
4. Bazell, D., & Désert, F.X. 1988, ApJ, 333, 353
5. Benson, P.J., & Myers, P.C. 1989, ApJ Suppl., 71, 89
6. Bertoldi, F., & McKee, C.F. 1992, ApJ, 395, 140
7. Blaauw, A. 1964, Ann. Rev. Astron. Astrophys., 2, 213
8. Blitz, L. 1990, *The Evolution of the Interstellar Medium*, ed. L. Blitz, (ASP Press: San Francisco), 273
9. Blitz, L. 1991, *The Physics of Star Formation and Early Stellar Evolution*, ed. C. J. Lada & N. D. Kylafis (Dordrecht: Kluwer), 3
10. Blitz, L. 1993, *Protostars and Planets III*, eds. E.H. Levy and J.I. Lunine, (Tucson: Univ. of Arizona Press), 125
11. Blitz, L. 1995 in *CO: 25 Years of Millimeter-wave Spectroscopy* eds. W.B Latter et al., (Kluwer:Dordrecht), 11
12. Blitz, L. Magnani, L. and Mundy, L. 1984, ApJ Lett., 282, L9
13. Blitz, L., & Stark, A.A. 1986, ApJ Lett., 300, L89
14. Blitz, L., & Williams, J.P. 1997, ApJ Lett., 488, L145
15. Bonnell, I.A., Bate, M.R., Clarke, C.J., & Pringle, J.E. 1997, MNRAS, 285, 201
16. Carpenter, J.M., Snell, R.L., & Schloerb, F.P. 1995, ApJ, 445, 246
17. Clemens, D.P., & Barvainis, R. 1988, ApJ Suppl., 68, 257
18. Cohen, R.S., Cong, H.-I., Dame, T.M. & Thaddeus, P. 1980, ApJ Lett., 239, L53
19. Cowie, L.L. 1980, ApJ, 236, 868
20. Cox, P., Deharveng, L. & Leene, A. 1990, A&A, 230, 181
21. Dame, T.M. 1993, in *Back to the Galaxy*, S.S. Holt & F. Verter, eds. (AIP Press:New York), 267
22. Digel, S.W., Lyder, D.A., Philbrick, A.J., Puche, D., & Thaddeus, P. 1996, ApJ, 458, 561
23. Elmegreen, B.G. 1982a, ApJ, 253, 634
24. Elmegreen, B.G. 1982b, ApJ, 253, 655
25. Elmegreen, B.G. 1990, in *The Evolution of the Interstellar Medium*, ed. L. Blitz, (ASP Press: San Francisco), 247
26. Elmegreen, B.G. 1997a, ApJ, 477, 196
27. Elmegreen, B.G. 1997b, ApJ, 486, 944
28. Elmegreen, B.G., & Falgarone, E. 1996, ApJ, 471, 816
29. Falgarone, E., Panis, J.-F., Heithausen, A., Pérault, M., Stutzki, J., Puget, J.-L., & Bensch, F. 1998, A&A, 331, 669
30. Falgarone, E., & Phillips, T.G. 1990, ApJ, 359, 344
31. Falgarone, E., Phillips, T.G., & Walker C.K. 1991, ApJ, 378, 186
32. Goodman, A.A., Barranco, J.A., Wilner, D.J., & Heyer, M.H. 1998, ApJ, 504, 223
33. Güsten, R. 1989, in *The Center of the Galaxy*, ed. M. Morris, (Kluwer:Dordrecht), 89
34. Herbst, W. & Assousa, G.E. 1977, ApJ, 217, 473
35. Heyer, M.H., & Schloerb, F.P. 1997, ApJ, 475, 173
36. Heyer, M.H., & Terebey, S. 1998, ApJ, 502, 265
37. Heyer, M.H., Brunt, C., Snell, R.L., Howe, J.E., and Schloerb, P. 1998, ApJ Suppl., 115, 241
38. Heyer, M.H. 1999, in *New Perspectives of the Interstellar Medium*, eds: A.R.Taylor & T.L. Landecker, ASP Conference Series, in press.
39. Heithausen, A., Bensch, F., Stutzki, J., Falgarone, E., & Panis, J.-F. 1998, ApJ Lett., 331, L65
40. Hester, J.J., et al. 1996, AJ, 111, 2349
41. Houlahan, P., & Scalo, J. 1992, ApJ, 393, 172

42. Hunter, S.D. et al. 1997, ApJ, 481, 205
43. Kramer, C., Stutzki, J., Röhrig, R., Corneliussen, U. 1998a, A&A 329, 249
44. Kramer, C., Alves, J., Lada, C., Lada, E., Sievers, A., Ungerechts, H., & Walmsley, M. 1998b, A&A Lett., 329, L33
45. Kuchar, T.A., & Bania, T.M. 1993, ApJ, 414, 664
46. Kwan, J. 1979, Apj, 229, 567
47. Kwan, J. & Valdez, F. 1983, 271, 604
48. Lacey, C.G. & Fall, S.M. 1985, ApJ 290, 154
49. Lada, C.J., Lada, E.A., Clemens, D.P., & Bally, J. 1994, ApJ, 429, 694
50. Langer, W.D., Wilson, R.W., Anderson, C.H. 1993, ApJ Lett., 408, L45
51. Larson, R.B. 1995, MNRAS, 272, 213
52. Larson, R.B. 1985, MNRAS, 214, 379
53. Larson, R.B. 1981, MNRAS, 194, 809
54. Lis, D.C., Pety, J., Phillips, T.G., & Falgarone, E. 1996, ApJ, 463, 623
55. Liszt, H.S. & Burton, W.B. 1996, in *Unsolved Problems of the Milky Way*, eds. L. Blitz, & P. Teuben, (Kluwer:Dordrecht), 297
56. Loren, R.B. 1989, ApJ, 338, 902
57. Maddalena, R., & Thaddeus, P., 1985, ApJ, 294, 231
58. Magnani, L., Blitz, L. & Mundy 1985, ApJ, 295, 402
59. Magnani, L., Lada, E.A. & Blitz, L. 1986, ApJ, 301, 395
60. Magnani, L., Caillault, J.-P., Buchalter, A., & Beichman, C.A. 1995, ApJ Suppl., 96, 159
61. Magnani, L. Hartmann, D. & Thaddeus, P. 1999, in preparation
62. Mandelbrot, B.B. 1982, *The Fractal Geometry of Nature* (San Francisco: Freeman)
63. McLaughlin, D.E., and Pudritz, R.E. 1996, ApJ, 469, 194
64. Miesch, M.S., & Scalo, J.M. 1995, ApJ, 429, 645
65. Mizuno, A., Onishi, T., Yonekura, Y., Nagahama, T., Ogawa, H., & Fukui, Y. 1995, ApJ Lett., 445, L161
66. Moriarty-Schieven, G.H., Andersson, B.-G., & Wannier, P.G. 1997, ApJ, 475, 642
67. Motte, F., André, Ph., & Neri, R. 1998, A&A, 336, 150
68. Mouschovias, T., Shu, F. & Woodward, P. 1974, A&A, 33, 73
69. Myers, P.C. 1983, ApJ, 270, 105
70. Myers, P.C., & Benson P.J. 1983, ApJ, 266, 309
71. Nagahama, T., Mizuno, A., Ogawa, H., & Fukui, Y. 1998, AJ, 116, 336
72. O'Dell, C.R., & Wong, S.K. 1996, AJ, 111, 846
73. Oliver, R.J., Masheder, M.R.W. & Thaddeus, P. 1996, A&A, 315, 578
74. Omont, A., Pettijean, P., Guilloteau, S., McMahon, R.G., Solomon, P.M. & Pecontal, E., 1996, Nature, 382, 428
75. Öpik, E. 1953, Irish AJ, 2, 219
76. Ostriker, E.C., Gammie, C.F., & Stone, J.M. 1999, ApJ, in press
77. Parker, E.N. 1966, ApJ, 145, 811
78. Phillips, J.P.. 1999, A&A Suppl., 134, 241
79. Reach, W.T., Wall, W.F. & Odegard, N, 1998, ApJ, 507, 507
80. Rucinski, S.M., & Krautter, J. 1983, A&A, 121, 217
81. Sanders, D.B., Scoville, N.Z. & Solomon, P.M. 1985, ApJ, 289, 323
82. Scalo, J. 1990, *Physical Processes in Fragmentation and Star Formation*, eds. R. Capuzzo-Dolcetta et al. (Dordrecht: Kluwer) 151
83. Schneider, N., Stutzki, J., Winnewisser, G., & Blitz, L. 1996, ApJ Lett., 468, 119
84. Scoville, N.Z. & Hersh, K. 1979, ApJ, 229, 578
85. Shu, F. 1974, A&A, 33, 55
86. Simon, M. 1997, ApJ Lett., 482, 181
87. Solomon, P.M. & Sanders, D.B. 1980 in *Giant Molecular clouds*, eds. P.M. Solomon & M.G. Edmunds, (Pergammon: Oxford) 41
88. Spergel, D.N. & Blitz, L. 1992, Nature, 357, 665

89. Stark, A.A. 1979, Ph.D. Dissertation, Princeton University
90. Stutzki, J., Bensch, F., Heithausen, A., Ossenkopf, V., & Zielinsky, M. 1998, A&A, 336, 697
91. Stutzki, J., & Güsten, R., 1990, ApJ, 356, 513
92. Testi, L., & Sargent, A.I. 1998, ApJ Lett., 508, 91
93. Thornley, M.D. 1996, ApJ Lett., 469, L45
94. Williams, J.P., & Blitz, L. 1998, ApJ, 494, 657
95. Williams, J.P., Blitz, L., & Stark, A.A. 1995, ApJ, 451, 252
96. Williams, J.P., de Geus, E.J., & Blitz, L. 1994, ApJ, 428, 693
97. Williams, J.P., Blitz, L. & McKee, C. F. 1999, in *Protostars and Planets IV* eds. V. Mannings & A. Boss, (University of Arizona Press), in press
98. Williams, J.P., & Maddalena, R.J. 1996, ApJ, 464, 247
99. Williams, J.P., & Myers, P.C. 1999, ApJ Lett., submitted
100. Wiseman, J.J., & Adams, F.C. 1994, ApJ, 435, 708
101. Witt, N.A., & Gordon, K.D. 1996, ApJ, 463, 681
102. Wood, D.O.S., Myers, P.C., & Daugherty, D.A. 1994, ApJ Suppl., 95, 457
103. Woodward, P.R. 1976, ApJ, 207, 466

James Di Francesco and Phil Myers intently look on as Leo Blitz raises, Frank Shu folds and Jon Williams suddenly feels thirsty.

THE DYNAMICAL STRUCTURE AND EVOLUTION OF GIANT MOLECULAR CLOUDS

CHRISTOPHER F. MCKEE

Institute for Advanced Study
Princeton NJ 08540
and
Departments of Physics and Astronomy
University of California, Berkeley CA 94720[†]

1. Introduction: The Observed Characteristics of GMCs

The interstellar medium (ISM) of galaxies contains gas that spans a wide range of physical conditions, from hot X-ray emitting plasma to cold molecular gas. The molecular gas is of particular importance because it is believed to be the site of all the star formation that occurs in galaxies. In the Milky Way, molecular gas constitutes about half the total mass of gas within the solar circle. Much of this gas is concentrated in large aggregations called giant molecular clouds (GMCs), which have masses $M \gtrsim 10^4 \, M_\odot$. Smaller molecular clouds are also observed, such as the high latitude clouds discovered by Blitz et al. [9] and the small molecular clouds in the Galactic plane cataloged by Clemens & Barvainis [19]. GMCs have internal structure, and I shall follow the terminology of Williams et al. [114] in describing this: *Clumps* are coherent regions in longitude–latitude–velocity space that are generally identified from spectral line maps of molecular emission. *Star-forming clumps* are the massive clumps out of which stellar clusters form. Finally, *cores* are the regions out of which single stars (or multiple stellar systems like binaries) are formed. These characteristics, together with the important observational properties of GMCs, are reviewed elsewhere in this volume by Blitz. In this review I shall first briefly summarize some of the key properties of GMCs and then attempt to account for the dynamical properties theoretically.

[†]Permanent address

C.J. Lada and N.D. Kylafis (eds.), The Origin of Stars and Planetary Systems, 29–66.

1.1. CHEMICAL AND THERMAL PROPERTIES

Molecular clouds are composed primarily of H_2, but this is relatively difficult to observe. The next most abundant molecule is generally CO, which can be readily observed in both emission and absorption. It is surprising that such molecules can exist in the harsh environment of interstellar space, particularly because of the destructive effects of ultraviolet radiation. Atomic hydrogen (H^0) shields most interstellar gas from EUV photons (those with energies 100 eV $\gtrsim h\nu \geq$ 13.6 eV), but FUV photons (those with 13.6 eV$> h\nu \gtrsim 5$ eV) are far more penetrating. These photons ionize atoms such as C, Mg, S, and Fe, and photodissociate molecules. The photodissociation of H_2 and CO occurs in a two step process: first, the molecule undergoes an electronic excitation by absorption of a resonance line photon; then some fraction of the molecules radiate into a state in the vibrational continuum and fly apart [23].

A significant column density of molecules can build up only when the absorption lines become optically thick, or, if this is inadequate, when the FUV radiation is sufficiently attenuated by dust. Under typical conditions in the local ISM, observations show that the extinction to the cloud surface must exceed about 0.1 mag in order for a significant column density of H_2 to be observed [11] (the extinction through the entire cloud is then twice this, or about 0.2 mag). Since CO is less abundant than H_2, a significantly larger column density is required in order for the carbon to become incorporated into molecules. The calculations of van Dishoeck & Black [109] show that a GMC in the local ISM has a layer of C^+ and C^0 corresponding to a column density $N_H = 1.4 \times 10^{21}$ cm^{-2}. (Note that we shall measure all densities in terms of the total hydrogen density, $n_H = 2n_{H_2}$ for molecular gas, and similarly for column densities.) For the dust to gas ratio observed in the local ISM, the relation between extinction and hydrogen column is [102]

$$A_V = \frac{N_H}{2.0 \times 10^{21} \text{ cm}^{-2}},\tag{1}$$

so this column is equivalent to an extinction of 0.7 mag.

An interstellar cloud with a mean extinction significantly greater than 2×0.7 mag is thus expected to have a thin outer layer of H^0 ($\Delta A_V \simeq$ 0.1 mag), a thicker layer of H_2, C^+, and C^0 ($\Delta A_V \simeq 0.6$ mag), and an interior that is nearly fully molecular. The temperature of the outer atomic layer is of order 50-100 K. In the deep interior, where the gas is fully molecular, the temperature of about 10 K is set by the balance between heating due to cosmic ray ionization and cooling due to CO emission. Because the chemical and thermal structure of the edge of the cloud is dominated by photodissociation, it is termed a photodissociation region [42].

The ionization in most of the volume of a molecular cloud is due to FUV photons; under typical conditions cosmic rays dominate the ionization only for regions in a cloud such that the extinction to the surface exceeds about 4 mag [60]. The ionization in the cosmic ray ionized region of a molecular cloud can be expressed as

$$x_e \equiv \frac{n_e}{n_{\rm H}} = \frac{C_i}{n_{\rm H}^{1/2}}. \qquad (2)$$

Williams et al. [112] found $C_i = (0.5 - 1.0) \times 10^{-5}$ cm$^{-3/2}$ from chemical modeling of observations of a number of low mass cores, where the factor of two uncertainty arises from uncertainties in the chemical reaction rates. In arriving at this result, they estimated that the typical density in the cores they observed is $n_{\rm H} \simeq 2 - 6 \times 10^4$ cm^{-3} and that the cosmic ray ionization rate is $\zeta_{\rm H} = 2.5 \times 10^{-17}$ s^{-1}. This value for the ionization is in good agreement with theoretical expectations [65].

1.2. DYNAMICAL PROPERTIES

Some of the most salient characteristics of GMCs were summarized in 1981 by Larson [48], and these results are sometimes referred to as "Larson's laws". The first relation is the *line width–size relation*: molecular clouds are supersonically turbulent with line widths Δv that increase as a power of the size, $\Delta v \propto R^p$. Larson himself estimated that $p \simeq 0.38$, close to the value 1/3 appropriate for turbulence in incompressible fluids. Subsequent work has distinguished between the relation valid for a collection of GMCs and that valid within individual GMCs or parts of GMCs. For different GMCs inside the solar circle, Solomon et al [100] found

$$\sigma = (0.72 \pm 0.07) R_{\rm pc}^{0.5 \pm 0.05} \quad \text{km s}^{-1}, \qquad (3)$$

where σ is the one–dimensional velocity dispersion; it is related to the full-width half-maximum of the line by $\sigma = \Delta v / 2.355$. By comparison, the thermal velocity dispersion of molecular gas is $\sigma_{\rm th} = 0.188(T/10 \text{ K})^{1/2}$ km s^{-1}. Within low–mass cores, Caselli & Myers [15] found that the non–thermal velocity dispersion (i.e, what remains after eliminating the thermal contribution) is

$$\sigma_{\rm nt} \simeq 0.55 R_{\rm pc}^{0.51} \quad \text{km s}^{-1}, \qquad (4)$$

which is quite similar to that found for GMCs by Solomon et al. For "high–mass cores", however, Caselli & Myers found

$$\sigma_{\rm nt} \simeq 0.77 R_{\rm pc}^{0.21} \quad \text{km s}^{-1}, \qquad (5)$$

a substantially flatter relation. Although these cores are forming more massive stars than the low mass cores, they are not forming OB stars. Plume et al [87] have surveyed clumps in which OB star formation is believed to be occurring. They found that such clumps do not obey the line width–size relation, and furthermore, that σ_{nt} is greater than indicated by equation (5).

Larson's second result was that GMCs are gravitationally bound. We shall discuss this in §2.2 below. He concluded that clumps within GMCs are also gravitationally bound, but for ^{13}CO clumps this appears to be true only for the most massive ones [5].

His third conclusion was that all GMCs have about the same column density. As he pointed out, only two of these three conclusions are independent; any one of them can be derived from the other two. For example, if the clouds are gravitationally bound, then $\sigma^2 \propto GM/R \propto \bar{N}_H R$, where $\bar{N}_H = M/(\pi \mu_H R^2)$ is the mean column density of the cloud, and $\mu_H = 2.34 \times 10^{-24}$ g is the mean mass per hydrogen. The line width–size relation for GMCs, $\sigma^2 \propto R$ [100] then implies $\bar{N}_H = \text{const}$. For GMCs inside the solar circle, Solomon et al [100] found

$$\bar{N}_{H22} = (1.5 \pm 0.3) R_{\mathrm{pc}}^{0.0 \pm 0.1} \quad \mathrm{cm}^{-2}, \tag{6}$$

where $\bar{N}_{H22} \equiv \bar{N}_H/(10^{22} \text{ Hydrogen nuclei cm}^{-2})$; this corresponds to an extinction $\bar{A}_V = 7.5$ mag with the local gas to dust ratio. This result does not apply to unbound clumps in GMCs, which can have lower column densities [5], nor to the OB star–forming clumps studied by Plume et al [87], which have $\bar{N}_{H22} \sim 60$.

There are several other characteristics of GMCs that we must take note of. First, GMCs appear to have magnetic fields that are dynamically significant [41]; this will be discussed in §2.4 below. Second, GMCs are highly clumped, in that the typical density n_H—i.e., the local density around a typical molecule—is significantly greater than the volume–averaged density $\bar{n}_H$ in the cloud. Liszt [51] has summarized studies of the typical density of clouds in the Galactic plane as inferred from their excitation, and cites values of n_H ranging from 10^3 to 1.2×10^4 cm^{-3}; in other words, $n_H \simeq 3000$ cm^{-3} ± 0.5 dex. However, the mean density of gas in the GMCs, $\bar{n}_H$, is considerably less, at least for the massive ones: since $M \propto \bar{n}_H R^3$ and $\bar{N}_H \propto \bar{n}_H R$, we have

$$\bar{n}_H = \frac{84}{M_6^{1/2}} \left(\frac{\bar{N}_{H22}}{1.5} \right)^{3/2} \quad \mathrm{cm}^{-3}, \tag{7}$$

where $M_6 \equiv M/(10^6 \ M_\odot)$. The filling factor of the gas in GMCs is then typically

$$f \equiv \frac{\bar{n}_{\rm H}}{n_{\rm H}} = \frac{0.084}{n_{\rm H3} M_6^{1/2}} \left(\frac{\bar{N}_{\rm H22}}{1.5} \right)^{3/2}, \tag{8}$$

where $n_{\rm H3} = n_{\rm H}/(10^3 \ {\rm cm}^{-3})$. Clouds of mass $M \lesssim 10^4 \ M_\odot$ must have $n_{\rm H} > 10^3 \ {\rm cm}^{-3}$ if they are to have the typical column density found by Solomon et al [100]. The nature of the low density interclump medium is uncertain; for example, it is not even known whether it is atomic or molecular [114]. A possible explanation for the small filling factor of the gas in GMCs will be given below, where we discuss models of turbulence in these clouds.

Finally, GMCs have a power-law mass distribution with a relatively sharp cutoff. Let $d\mathcal{N}_c(M)$ be the number of GMCs with mass between M and $M + dM$. Then the observations of GMCs inside the solar circle (but excluding the Galactic Center) are consistent with the mass distribution [115]

$$\frac{d\mathcal{N}_c}{d \ln M} = 63 \left(\frac{6 \times 10^6 \ M_\odot}{M} \right)^{0.6} \qquad (M \leq 6 \times 10^6 \ M_\odot), \tag{9}$$

$$= 0 \qquad (M > 6 \times 10^6 \ M_\odot). \tag{10}$$

The mass distribution is often represented in terms of $d\mathcal{N}_c/dM$ instead of $d\mathcal{N}_c/d \ln M$, which leads to an exponent of 1.6 instead of 0.6. The form adopted here has the advantage that $d\mathcal{N}_c/d \ln M$ represents an actual number of clouds. The fact that the coefficient 63 is so much larger than unity means that it must have a physical significance [62]: If there were no cutoff to the distribution, one would expect $63/0.6 \simeq 100$ GMCs more massive than $6 \times 10^6 \ M_\odot$ in the Galaxy; in fact, there are none.

1.3. STAR FORMATION IN GMCS

GMCs are the sites of most of the star formation in the Galaxy. A crucial fact about star formation is that it is usually very inefficient: In the absence of support, GMCs would collapse to very high densities and presumably form stars in a free fall time

$$t_{\rm ff} = \left(\frac{3\pi}{32 G \bar{\rho}} \right)^{1/2} = \frac{1.37 \times 10^6}{\bar{n}_{\rm H3}^{1/2}} \quad {\rm yr.} \tag{11}$$

(As we shall see in §2.7, simulations indicate that it is very difficult to maintain the turbulence that supports GMCs.) Now, the total mass of

GMCs inside the solar circle is 10^9 $M_\odot$ [115], and the mean density in GMCs is about 10^2 cm^{-3} (eq. 7). If GMCs collapsed in a free fall time and formed stars, the Galactic star formation rate would be

$$\dot{M}_* \simeq \frac{10^9 \; M_\odot}{4 \times 10^6 \; \text{yr}} = 250 \; M_\odot \; \text{yr}^{-1}, \tag{12}$$

far greater than the observed rate in this part of the Galaxy of 3 $M_\odot$ yr^{-1} [62]. This gross disparity between the potential star formation rate and the actual one was pointed out many years ago by Zuckerman and Evans [116], who concluded that the supersonic motions observed in GMCs do not reflect gravitational infall. Any successful theory of star formation must account for this inefficiency.

2. Dynamical Structure of GMCs

2.1. THE VIRIAL THEOREM

Some insight into the structure of GMCs can be gained from the virial theorem, as shown in the classic studies of self–gravitating gas clouds by McCrea [59] and by Mestel & Spitzer [68]. Define the quantity $I = \int r^2 dm$, which is proportional to the trace of the inertia tensor. Next, evaluate $\ddot{I} \equiv d^2 I/dt^2$ from the equation of motion,

$$\rho \frac{d\mathbf{v}}{dt} = -\nabla P_{\text{th}} + \frac{1}{4\pi}(\nabla \times \mathbf{B}) \times \mathbf{B} + \rho \mathbf{g}, \tag{13}$$

where $\mathbf{g}$ is the acceleration due to gravity. The result is the virial theorem,

$$\frac{1}{2}\ddot{I} = 2(\mathcal{T} - \mathcal{T}_s) + \mathcal{M} + \mathcal{W}. \tag{14}$$

Alternatively, this relation can be derived by taking the dot product of $\mathbf{r}$ with the equation of motion and integrating over a volume corresponding to a fixed mass. We shall now consider each of the terms in this equation.

The term on the LHS reflects variations in the rate of change of the size and shape of the cloud. This term is usually neglected, but it may be significant for a turbulent cloud. In contrast to the terms on the RHS of the equation, it can be of either sign, and as a result its effects can be averaged out either by applying the virial theorem to an ensemble of clouds or by averaging over a time long compared with the dynamical time.

The first two terms on the RHS contain the effects of kinetic energy, including thermal energy; they do *not* include the energy associated with internal degrees of freedom. The thermal energy density inside the cloud is $(3/2)\rho c^2 = (3/2)P_{\text{th}}$, where c is the isothermal sound speed, and the bulk kinetic energy density is $(1/2)\rho v^2$. The total kinetic energy inside the cloud is then

$$\mathcal{T} = \int_{V_{cl}} \left(\frac{3}{2}P_{th} + \frac{1}{2}\rho v^2\right) dV \equiv \frac{3}{2}\bar{P}V_{cl}, \qquad (15)$$

where V_{cl} is the volume of the cloud and $\bar{P}$ is the mean pressure of the gas (i.e., it does not include the pressure associated with the magnetic field). Note that the kinetic energy associated with rotation is included in $\mathcal{T}$ and therefore $\bar{P}$; rotation is generally not a dominant effect in molecular clouds, however [39] [65]. The mean pressure can be related to the 1D velocity dispersion in the cloud σ, which is observable:

$$\sigma^2 \equiv \frac{1}{M} \int \left(c^2 + \frac{1}{3}v^2\right) dM = \frac{\bar{P}}{\bar{\rho}}. \qquad (16)$$

The second kinetic energy term is a surface term,

$$\mathcal{T}_s \equiv \frac{1}{2} \int_S P_{th}\mathbf{r} \cdot \mathbf{dS}, \qquad (17)$$

where the integral is over the surface of the cloud. If the thermal pressure in the ambient medium is constant at P_s, then $\mathcal{T}_s = (3/2)P_s V_{cl}$. As a result, the two kinetic energy terms combine to give

$$2(\mathcal{T} - \mathcal{T}_s) = 3(\bar{P} - P_s)V_{cl}. \qquad (18)$$

Thus, it is the difference between the energy in the cloud and that in the background medium that enters the virial theorem. If the ambient medium is turbulent and has a density much smaller than that of the cloud, one can show that P_s is somewhat less than the total pressure far from the cloud [63]: The stress exerted by the ambient medium normal to the cloud is entirely thermal (and is significantly larger than the thermal pressure far from the cloud if the turbulence in the ambient medium is supersonic), but it is reduced below the value it otherwise would have by flow of intercloud gas along the cloud surface.

The magnetic term in the virial theorem is fairly complicated,

$$\mathcal{M} = \frac{1}{8\pi} \int_V B^2 dV + \frac{1}{4\pi} \int_S \mathbf{r} \cdot \left(\mathbf{BB} - \frac{1}{2}B^2\mathbf{I}\right) \cdot \mathbf{dS}, \qquad (19)$$

where $\mathbf{I}$ is the unit tensor. If the cloud is immersed in a low density ambient medium and if the stresses due to MHD waves in the ambient medium are negligible, then the field outside the cloud will be approximately force free. In this case, one can show that the magnetic term becomes [63]

$$\mathcal{M} = \frac{1}{8\pi} \int (B^2 - B_0^2)dV, \qquad (20)$$

where B_0 is the field strength in the ambient medium far from the cloud. Thus, $\mathcal{M}$ is the difference between the total magnetic energy with the cloud and that in the absence of the cloud.

Finally, we consider the gravitational term $\mathcal{W}$. In the absence of an external gravitational field, this is the gravitational energy of the cloud [102],

$$\mathcal{W} = \int \rho \mathbf{r} \cdot \mathbf{g} dV = -\frac{3}{5} a \left(\frac{GM^2}{R} \right), \tag{21}$$

where a is a numerical factor of order unity that has been evaluated by Bertoldi and McKee [5]. In order to relate this term to the other terms in the virial theorem, we define the "gravitational pressure" P_G by

$$\mathcal{W} \equiv -3P_G V_{\mathrm{cl}}; \tag{22}$$

intuitively, P_G is just the mean weight of the material in the cloud. The gravitational pressure can be evaluated as

$$P_G = \left(\frac{3\pi a}{20} \right) G\Sigma^2 \rightarrow 1.39 \times 10^5 \bar{N}_{\mathrm{H}22}^2 \quad \mathrm{K\ cm}^{-3}, \tag{23}$$

where $\Sigma \equiv M/\pi R^2 \equiv \mu_{\mathrm{H}} \bar{N}_{\mathrm{H}}$ is the mean surface density of the cloud. The numerical evaluation is for a spherical cloud with a $1/r$ density profile.

These results enable us to express the steady–state, or time–averaged, virial theorem (eq. 14 with $\ddot{I} = 0$) as

$$\bar{P} = P_s + P_G \left(1 - \frac{\mathcal{M}}{|\mathcal{W}|} \right). \tag{24}$$

In this form, the virial theorem has an immediate intuitive meaning: the mean pressure inside the cloud is the surface pressure plus the weight of the material inside the cloud, reduced by the magnetic stresses.

2.2. ARE GMCS GRAVITATIONALLY BOUND?

With these results in hand, we can now address the issue of whether molecular clouds and their constituents are gravitationally bound. We assume that the cloud is large enough that the motions are highly supersonic (§1.2), and as a result the energy in internal degrees of freedom is negligible. The total energy is then $E = \mathcal{T} + \mathcal{M} + \mathcal{W}$, which can be expressed as

$$E = \frac{3}{2} \left[P_s - P_G \left(1 - \frac{\mathcal{M}}{|\mathcal{W}|} \right) \right] V_{\mathrm{cl}}. \tag{25}$$

with the virial theorem (eq. 24). In the absence of a magnetic field, the condition that the cloud be bound (i.e., $E < 0$) is simply $P_G > P_s$. We shall use this criterion even for magnetized clouds, bearing in mind that using the total ambient gas pressure (thermal plus turbulent) for P_s is an overestimate and that our analysis is approximate because we have used the time-averaged virial theorem. In the opposite case, in which $P_s \gg P_G$, the cloud is said to be *pressure–confined*. In the typical case in which the pressure is largely turbulent, the "pressure–confined cloud" is likely to be transient.

For GMCs, the surface pressure is that of the ambient ISM. In the solar vicinity, the total interstellar pressure is about 2.8×10^4 K cm^{-3}, which balances the weight of the ISM [14]. Of this, about 0.7×10^4 K cm^{-3} is due to cosmic rays; since they pervade both the ISM and a molecular cloud, they do not contribute to the support of a cloud and may be neglected. The magnetic pressure is about 0.3×10^4 K cm^{-3} [40], leaving $P_s \simeq 1.8 \times 10^4$ K cm^{-3} as the total ambient gas pressure.

What is the minimum value of P_G for a molecular cloud? According to van Dishoeck and Black [109], molecular clouds exposed to the local interstellar radiation field have a layer of C$^+$ and C^0 corresponding to a visual extinction of 0.7 mag (§1.1). If we require at least $1/3$ of the carbon along a line of sight through a cloud to be in the form of CO in order to term the cloud "molecular", then the total visual extinction must be $\bar{A}_V > 2$ mag (allowing for a shielding layer on both sides). According to equation (23), this gives $P_G \gtrsim 2 \times 10^4$ K cm$^{-3} \sim P_s$, verifying that molecular clouds as observed in CO are at least marginally bound. Larson's "second law" is thus seen to be a consequence of the relationship between the column density required for CO to be significant and the pressure in the ISM. Note that if we defined molecular clouds as having a significant fraction of H$_2$ rather than CO, the minimum column density required would be substantially less and the clouds might not be bound. Furthermore, the conclusion that CO clouds are bound depends on the metallicity, the interstellar pressure and the strength of the FUV radiation field, so that CO clouds may not be bound everywhere in the Galaxy or in other galaxies [28].

GMCs in the solar neighborhood typically have mean extinctions significantly greater than 2 mag, and as a result P_G is generally significantly greater than P_s. Indeed, observations show $P_G \sim 2 \times 10^5$ K cm^{-3}, an order of magnitude greater than P_s [5] [8] [113]. Thus, if GMCs are dynamically stable entities (the crossing time for a GMC is about 10^7 yr, smaller than the expected lifetime [10] [115]), then GMCs must be self-gravitating. In the inner galaxy, where P_s is expected to be greater, the typical GMC linewidths also appear to be somewhat greater than those found locally [90], and thus P_G is still comfortably greater than P_s.

In order to determine whether clumps within GMCs are gravitationally bound, it is convenient to work in terms of the *virial parameter* [5] [76]

$$\alpha \equiv \frac{5\sigma^2 R}{GM}, \tag{26}$$

which is readily determined from observation. With the aid of equations (21) and (22), this parameter can be expressed as

$$\alpha = a\left(\frac{2\mathcal{T}}{|\mathcal{W}|}\right) = a\left(\frac{\bar{P}}{P_G}\right). \tag{27}$$

Bertoldi & McKee [5] show how this result can be applied to ellipsoidal clouds; if α is interpreted as an average over the orientation of the cloud, then the factor a is generally within a factor of about 1.3 of unity. In terms of the virial parameter, the net energy of the cloud is

$$E = |\mathcal{W}|\left[\frac{\alpha}{2a} - \left(1 - \frac{\mathcal{M}}{|\mathcal{W}|}\right)\right]. \tag{28}$$

A clump is therefore bound for $\alpha \lesssim 2$, although the exact value depends on the strength of the field. For $\alpha \gg 1$, clumps are pressure–confined.

Whether a clump is pressure–confined or gravitationally bound is directly related to its surface density [7]. The surface pressure on a clump is just the mean pressure inside the GMC,

$$P_s(\text{clump}) \simeq P_G(\text{GMC}) \propto \Sigma^2(\text{GMC}), \tag{29}$$

where we have assumed that the GMC is strongly bound, so that the pressure acting on it can be ignored. As a result, the virial theorem for a clump becomes

$$\bar{P}(\text{clump}) \propto \Sigma^2(\text{GMC}) + \Sigma^2(\text{clump}), \tag{30}$$

where we have assumed that the mass of an individual clump is small, so that it does not significantly affect the surface density of the cloud. Pressure–confined clumps have $\Sigma(\text{clump}) \ll \Sigma(\text{GMC})$ since $P_G \ll P_s$. On the other hand, gravitationally bound clumps have $\Sigma(\text{clump}) \gtrsim \Sigma(\text{GMC})$. The studies of the stability of gas clouds discussed below show that P_G cannot be too much greater than P_s if the cloud is to be gravitationally stable; correspondingly, $\Sigma(\text{clump})$ cannot be much greater than $\Sigma(\text{GMC})$ if the clump is stable. We conclude that gravitationally bound clumps have column densities similar to that of the GMC in which they are embedded, $\Sigma(\text{clump}) \sim \Sigma(\text{GMC})$, as is often observed [7].

From an analysis of ^{13}CO clumps in Ophiuchus, Orion B, the Rosette, and Cepheus OB3, Bertoldi & McKee [5] concluded that most clumps are pressure–confined; however, most of the mass is in clumps that appear to be gravitationally bound, or nearly so. Pressure–confined clumps do not satisfy the line width–size relation; instead, the velocity dispersion and mean density in the clumps are about constant. As a result, the virial parameter scales with mass as $M^{-2/3}$, with the most massive clumps having $3 \gtrsim \alpha \gtrsim 1$ (with the exception of Cepheus, for which the data are of low resolution). Furthermore, the observable star formation appears to be confined to these massive clumps. Thus, not only are GMCs as a whole gravitationally bound, but there is a mass scale $M(\mathrm{SFC})$, the mass of the star–forming clumps, such that structures with $M \sim M(\mathrm{SFC})$ are bound as well.

On still a smaller scale, cloud cores (out of which individual stars or stellar systems like binaries form) are also observed to be gravitationally bound. These cores exist in both star–forming clumps and, in star–forming clouds like Taurus, in relative isolation. Thus, there appears to be a hierarchy of bound structures: GMCs, the star–forming clumps within them, and cloud cores (cf [33]). As we shall see in §2.4.1, this hierarchy may be mirrored in the magnetic properties of GMCs.

2.3. ISOTHERMAL CLOUDS

Molecular gas is often observed to be at a temperature of about 10 K. The simplest model for a molecular cloud is thus an isothermal cloud. In fact, the cores of molecular clouds often have thermal pressures that are greater than the nonthermal pressures, so the model of an isothermal cloud may apply approximately to such cores. (For non–isothermal models, see [12] and [32].) For now, we shall neglect the effect of a static magnetic field.

Using the virial theorem, we can infer the basic properties of an isothermal, self–gravitating cloud [59] [101]. Setting $\mathcal{M} = 0$ in equation (24) and evaluating P_G with the aid of equation (23), we find

$$P_s = \frac{3M\sigma^2}{4\pi R^3} - \frac{3aGM^2}{20\pi R^4}, \tag{31}$$

where we have written the mean surface density as $\Sigma = M/\pi R^2$. If the cloud radius R is large, the second term on the RHS is negligible, and the mean pressure in the cloud (represented by the first term) is approximately equal to the ambient pressure P_s. Now consider a sequence of equilibria of smaller and smaller R. Initially, reducing R requires a higher ambient pressure P_s. Eventually, the second term on the RHS, which is P_G, becomes comparable to the first and the increase in P_s is halted. This occurs when the escape velocity is comparable to the velocity dispersion, $GM/R \sim \sigma^2$, as can be seen by comparing the two terms on the RHS of equation (31). Further reductions in R lead to a decrease in surface pressure, which is unstable. The point at which P_s is a maximum therefore represents a *critical point* that separates stable from unstable solutions. Using this result for the radius at the critical point, we find that the maximum pressure is of order

$$P_{\mathrm{cr}} \propto \frac{M\sigma^2}{(GM/\sigma^2)^3} \propto \frac{\sigma^8}{G^3 M^2}. \tag{32}$$

Equivalently, this relation can be interpreted as giving the maximum mass that can be supported against gravity in a medium of pressure P_{cr}. This critical mass is termed the Bonnor-Ebert mass after the two individuals who first worked out the structure of isothermal spheres [13] [25]:

$$M_{\mathrm{BE}} = 1.18 \frac{\sigma^4}{(G^3 P_s)^{1/2}}, \tag{33}$$

$$= 1.15 \frac{(T/10 \text{ K})^2}{(P_s/10^5 \text{ K cm}^{-3})^{1/2}} \ M_\odot. \tag{34}$$

The numerical value is for the conditions typical of a low mass core, with $n \sim 10^4$ cm^{-3} and $T \sim 10$ K, and it is significant that the result is of order the typical stellar mass.

The maximum central density of a stable isothermal sphere is only $\rho_c = 14.0\rho_s$, where ρ_s is the density at the surface. The mean density is $2.5\rho_s$. Equilibria at lower mass have significantly lower central concentrations: for example, the central density is only $2.2\rho_s$ for $M = 0.5M_{\mathrm{BE}}$.

2.4. MAGNETIC FIELDS VS GRAVITY

Magnetic fields are believed to play a crucial role in the structure and evolution of molecular clouds. Theorists concluded that magnetic fields are important even before molecular clouds were discovered because simple estimates showed that interstellar gas is far more highly magnetized than stars are (e.g., [68]), and they therefore turned their attention to how this excess flux could be lost. The ongoing efforts to observe magnetic fields in molecular clouds will be discussed in §2.4.5 below.

2.4.1. *Magnetic Critical Mass*

The simplest case of a magnetically supported cloud is one in which the field is purely poloidal and the kinetic energy vanishes ($\mathcal{T} = 0$); since the gas is cold and there are no bulk motions, it settles into a thin disk. We assume the field is fully connected with that in the ambient medium. (The effects of closed field lines are discussed in reference [65].) For a thin disk, the volume of the cloud vanishes, and P_G goes to infinity; the virial theorem (24) yields

$$\mathcal{M} = |\mathcal{W}|. \tag{35}$$

Let the magnetic flux threading the cloud be

$$\Phi = \int 2\pi r B \, dr \equiv \pi R^2 \bar{B}. \tag{36}$$

Now consider a sequence of equilibria in which the flux is constant and the functional form of the mass–to–flux ratio is constant, but the value of the mass–to–flux ratio increases. At some point the mass–to–flux ratio will become large enough that gravity will overwhelm the magnetic stresses, and the cloud will collapse. The point at which the mass–to–flux ratio reaches the maximum is the critical point. We write the net magnetic energy there as

$$\mathcal{M}_{\mathrm{cr}} = \frac{1}{8\pi} \int (B^2 - B_0^2) dV \Big|_{\mathrm{cr}} \equiv \left(\frac{b}{3}\right) \bar{B}^2 R^3 = \left(\frac{b}{3\pi^2}\right) \frac{\Phi^2}{R}, \tag{37}$$

where b is a numerical factor of order unity. We can evaluate the *magnetic critical mass* M_Φ by inserting this relation into equation (35) and using equation (21):

$$M_\Phi = \left(\frac{5b}{9\pi^2 a}\right)^{1/2} \frac{\Phi}{G^{1/2}} \equiv c_\Phi \left(\frac{\Phi}{G^{1/2}}\right). \tag{38}$$

So long as the magnetic flux is frozen to the matter, M_Φ is a constant. For $M < M_\Phi$, the cloud is said to be *magnetically subcritical*: the mass–to–flux ratio is small enough that magnetic stresses always exceed gravity, so such a cloud can never undergo gravitational collapse. Conversely, if $M > M_\Phi$, the cloud is *magnetically supercritical*, and magnetic fields cannot prevent gravitational collapse.

If the cloud has a constant mass–to–flux ratio (which Shu and Li [96] term "isopedic"), then the numerical factor $c_\Phi = 1/2\pi$ [81]. Isopedic disks are highly idealized: they can be finite in extent only if the field vanishes outside the disk, and they can be in equilibrium only if they are critical ($M = M_\Phi$). Precisely because they are so idealized, it is possible to obtain useful results even if they are non-axisymmetric and time-dependent [96]. The ratio M_Φ/M is constant in such a disk. The distinction between magnetically subcritical and supercritical disks is particularly clear in this case since the ratio of the magnetic force to the gravitational force on a mass M in the disk is (in our notation) $(M_\Phi/M)^2$.

The field strength in a cloud can be expressed in terms of its column density by noting that $M/M_\Phi \propto \Sigma/\bar{B} \propto \bar{N}_{\mathrm{H}}/\bar{B}$, so that

$$\bar{B} = 50.5 \left(\frac{\bar{N}_{\mathrm{H22}}}{M/M_\Phi}\right) \mu G = 10.1 \left(\frac{\bar{A}_V}{M/M_\Phi}\right) \mu G. \tag{39}$$

It is sometimes convenient to have an alternative expression for the magnetic critical mass that is independent of the mass of the cloud. Mouschovias

& Spitzer [71] showed that such an alternative critical mass, denoted M_B, is related to M_Φ by

$$\frac{M_B}{M} \equiv \left(\frac{M_\Phi}{M}\right)^3.$$

(40)

For an ellipsoidal cloud of size $2Z$ along the axis of symmetry and radius R normal to the axis, we have [5]

$$M_B = 512 \left(\frac{R}{Z}\right)^2 \frac{\bar{B}_{1.5}^3}{\bar{n}_{H3}^2} \quad M_\odot,$$

(41)

where $\bar{B}_{1.5} \equiv \bar{B}/(10^{1.5}\ \mu G)$. This form for the magnetic critical mass is similar in form to the Bonnor–Ebert mass, since

$$M_B \propto \frac{\bar{B}^3}{\bar{n}^2} \propto \frac{1}{\bar{B}} \left(\frac{\bar{B}^4}{\bar{\rho}^2}\right) \propto \frac{v_A^4}{P_B^{1/2}},$$

(42)

where $v_A \equiv \bar{B}/(4\pi\bar{\rho})^{1/2}$ is the Alfvén velocity; here v_A plays the role of the isothermal sound speed and the magnetic pressure $P_B \equiv \bar{B}^2/8\pi$ that of the external pressure.

Equation (41) for the magnetic critical mass naturally suggests three very different mass scales for molecular clouds [5]: (1) On the largest scales, clouds are formed from compression of the diffuse ISM [26] [67], which has a mean density $\bar{n}_H \sim 1$ cm^{-3} and a field $\bar{B} \sim 3\ \mu G$; this gives $M_B \sim 5 \times 10^5$ $M_\odot$, a typical mass for a GMC (cf. eq. 10). Since M_B is constant so long as the mass is constant and flux freezing holds, this value will be preserved as the gas is compressed and the GMC forms. (2) On intermediate scales, the clumps within GMCs might well have originated as the diffuse clouds in the gas that formed the GMC [26]. In this case, the density is about 30–100 times greater than the average interstellar density, but the field is about the same, since diffuse clouds are not gravitationally bound. As a result, we expect $M_B \sim 50 - 500\ M_\odot$ in clumps, which is $\lesssim$ the mass of star–forming clumps [M(SFC)] in nearby GMCs [5]. On the other hand, M_B is substantially greater than the mass of a typical star: Stellar mass clumps at the density of star–forming clumps are magnetically subcritical. (3) Finally, in regions in which the density of the gas can grow so that the thermal pressure is comparable to the magnetic pressure, either by ambipolar diffusion or by flow along field lines, then $M_B \sim M_{BE} \sim 1 M_\odot$ from equation (34).

2.4.2. *Toroidal Fields*

Toroidal fields can provide a confining force, thereby reducing the magnetic critical mass [36] [107]. To see how this occurs, consider a current $I(r)$ in the z direction, which generates a toroidal field $\mathbf{B}_\phi = 2I/cr$. The resulting force per unit volume is $\mathbf{J} \times \mathbf{B}/c = -(J_z B_\phi/c)\hat{\mathbf{r}}$, which indeed provides a pinching force. To determine the contribution to the magnetic term in the virial theorem, it is best to recall that the virial theorem can be derived by taking the dot product of the equation of motion with $\mathbf{r}$ and integrating over the volume; the contribution of the toroidal field to the magnetic term is then

$$\mathcal{M}_{\text{toroidal}} = \frac{1}{c} \int (\mathbf{J}_z \times \mathbf{B}_\phi) \cdot \mathbf{r} dV \simeq -\frac{LI^2}{c}, \tag{43}$$

where L is the length of the cylinder in which $\mathcal{M}$ has been evaluated. Thus, whereas poloidal fields give a positive definite contribution to $\mathcal{M}$ (provided the field decreases outward), toroidal fields give a negative definite contribution.

Several caveats about toroidal fields should be kept in mind: First, the ratio of the toroidal field to the poloidal field cannot become too large without engendering instabilities (e.g., [44]). Second, the current that generates the toroidal field must return to where it started, since currents do not have sources or sinks in MHD [44]. Once the current density reverses direction, so does the force, and the toroidal field ceases to be confining. If the virial theorem is applied to a volume large enough to encompass the entire return current, then $I = 0$ and the toroidal field has no net effect. In astrophysical MHD, it is convenient to think of the current as being generated by the field rather than vice versa; it follows that if the toroidal field is restricted to a finite volume, then it has no net effect on the virial theorem applied to any larger volume. Finally, a purely toroidal field is subject to dissipation by magnetic reconnection, which can occur with reasonable efficiency according to Lazarian & Vishniac [49]. However, such reconnection can be avoided if there is a sufficiently strong poloidal field along the axis, as in the protostellar wind model of Shu et al [98].

2.4.3. Clouds Supported by both Magnetic and Gas Pressure

Let us return to the case of a poloidal field, and ask, what is the critical mass for a cloud supported by gas pressure as well as magnetic stresses? We can use the same approach that we adopted to estimate the Bonnor–Ebert mass [71]. If we retain the magnetic term in the virial theorem (24), the same steps that led to equation (31) give

$$P_s = \frac{3M\sigma^2}{4\pi R^3} - \frac{3aGM^2}{20\pi R^4} \left(1 - \frac{\mathcal{M}}{|\mathcal{W}|} \right). \tag{44}$$

For a cloud at the critical point, $\mathcal{M} = \frac{3}{5} aGM_\Phi^2/R$ and $|\mathcal{W}| = \frac{3}{5} aGM^2/R$, so that $\mathcal{M}/|\mathcal{W}| = (M_\Phi/M_{\text{cr}})^2$ there. As discussed in reference [65], this remains a good approximation even for subcritical clouds. We then have

$$1 - \frac{\mathcal{M}}{|\mathcal{W}|} \simeq 1 - \left(\frac{M_\Phi}{M_{\mathrm{cr}}}\right)^2 \equiv k_m. \tag{45}$$

Generalizing the argument that led to the Bonnor–Ebert mass, we consider a sequence of equilibria in which the ambient pressure P_s is increased while the mass and flux remain constant. The maximum value of P_s is reached when the two terms become comparable, which occurs at a radius such that $\sigma^2 \sim GMk_m/R$. The maximum pressure is then of order

$$P_{\mathrm{cr}} \propto \frac{M\sigma^2}{(GMk_m/\sigma^2)^3} \propto \frac{\sigma^8}{G^3 k_m^3 M^2}. \tag{46}$$

For a given ambient pressure, this gives a cubic equation for the critical mass. If the gas is isothermal, one obtains [71]

$$M_{\mathrm{cr}} = c_1 M_{\mathrm{BE}} \left[1 - \left(\frac{M_\Phi}{M_{\mathrm{cr}}}\right)^2\right]^{-3/2}, \tag{47}$$

where c_1 is a numerical constant and where we have substituted back for k_m from its definition in equation (45). Tomisaka et al [106] find that their numerical results are best fit by $c_1 = 1.18$, quite close to the value estimated by Mouschovias & Spitzer [71] many years earlier.

An approximate solution to equation (47) for M_{cr} is $M_{\mathrm{cr}} \simeq M_{\mathrm{BE}} + M_\Phi$, which is accurate to within about 5% for $M \lesssim 8M_\Phi$ [60]; for weaker fields, this approximation predicts that $M_{\mathrm{cr}} \to M_{\mathrm{BE}}$, which is presumably more accurate than the solution of equation (47). We anticipate that this result will remain approximately valid in the more general case in which the gas is not isothermal. Defining the Jeans mass M_{J} as the critical mass associated with thermal and nonthermal motions of the gas, we then have

$$M_{\mathrm{cr}} \simeq M_{\mathrm{J}} + M_\Phi. \tag{48}$$

Axisymmetric numerical models for magnetized clouds were first constructed by Mouschovias [69] [70]. He focused on the case of a mass–to–flux distribution corresponding to a uniform field threading a uniform, spherical cloud. For this case, Mouschovias & Spitzer [71] found $c_\Phi \simeq 0.126$; subsequent calculations by Tomisaka et al. [106] found $c_\Phi \simeq 0.12$, which we adopt for numerical estimates. Tomisaka et al. showed that for a range of mass–to–flux distributions it is the mass–to–flux ratio on the central flux tube that controls the stability, with the critical value being $(dM_\Phi/d\Phi)_c \simeq 0.17/G^{1/2}$. (For Mouschovias' case, they found a coefficient $0.18 = 1.5 \times 0.12$; the factor 1.5 is just the ratio of the central mass–to–flux ratio to the mean.)

Tomisaka et al's result is quite close to the value for an isopedic disk, $1/(2\pi G^{1/2}) \simeq 0.16/G^{1/2}$.

2.4.4. *Are Clouds Magnetically Supercritical? Theory*

With the framework we have developed, we can ask the question, are GMCs magnetically supercritical or subcritical? McKee [60] argued that GMCs are magnetically supercritical based on the following line of argument: GMCs must be approximately critical ($M \simeq M_{cr}$) since they are highly pressured relative to their environment, and calculations show that this is possible only for nearly critical clouds [70] [106]. The large nonthermal motions observed in GMCs and the fact that the clouds do not appear to be highly flattened imply that M_J is not small compared to M_Φ. Hence, from equation (48), we have $M \simeq M_{cr} \simeq M_J + M_\Phi > M_\Phi$. This argument does not set a lower bound on M_Φ, but if initially a cloud had $M_J \gg M_\Phi$ and the gas pressure was dominated by nonthermal motions, then the field would be amplified into approximate equipartition, giving $M_\Phi \sim M_J$. Altogether, then, one expects GMCs to have $M \sim 2M_\Phi$ on theoretical grounds. This argument was extended to the gravitationally bound clumps within GMCs by Bertoldi & McKee [5]. These conclusions are not universally accepted, however [73].

Nakano [80] has used similar reasoning to conclude that the observed cores in molecular clouds should also be magnetically supercritical. The general argument that applies to all three cases—GMCs, star–forming clumps, and cores—is that if a cloud is clearly gravitationally bound, then its mass must be nearly equal to the critical mass; if furthermore, the cloud has more than one significant source of pressure support, then it is supercritical with respect to each of the sources of support individually. Nakano then used this argument to question the standard paradigm of low–mass star formation, which is based on ambipolar diffusion in magnetically subcritical clumps [72] [95]. However, this criticism may be unwarranted: about half the observed cores already contain embedded stars [4], so in the conventional interpretation such cores would have already experienced substantial ambipolar diffusion. The issue that remains to be resolved by observation is whether the *protocores*—i.e., the precursors of the observed cores—are magnetically subcritical.

2.4.5. *Observation of Magnetically Supercritical Clouds*

What do observations say about the strength of magnetic fields in molecular clouds? The data on magnetic field strengths come from observations of Zeeman splitting of molecular lines, which determine $B_\parallel$, the component of the magnetic field along the line of sight (see Heiles et al. [41] for a discussion of techniques for measuring magnetic fields). Myers & Goodman

[76] summarized the data available on magnetic field strengths in 1988, and concluded that the data were consistent with approximate equipartition among magnetic, kinetic, and gravitational energies. Their results are consistent with $M \simeq 2M_\Phi$ [60]. Since in their sample the line widths vary substantially less than the density in the clouds, the approximate equality between kinetic and magnetic energies ($\rho v_{\mathrm{rms}}^2/2 \sim B^2/8\pi$) implies the approximate relation $B \propto \rho^{1/2}$, as advocated by Mouschovias [73]. This in turn means that the Alfvén velocity is independent of the density of the cloud; Heiles et al. [41] found that $\langle v_A \rangle \sim 2$ km s^{-1}. Subsequently, Crutcher et al. [21] studied the cloud B1 in some detail. They found that the inner envelope was marginally magnetically subcritical, whereas the densest region was somewhat supercritical. The observational results were shown to be in good agreement with a numerical model that, however, did not include the observed nonthermal motions.

The most comprehensive study of magnetic fields in molecular clouds available to date is that by Crutcher [20]. He concludes that molecular clouds are generally magnetically supercritical: His sample, which tends to focus on the central regions of clouds, has no clear case in which a cloud is magnetically subcritical. In order to reach this conclusion, he had to allow for projection effects: If the magnetic field makes an angle θ with respect to the line of sight, then the observed field $B_\parallel$ is related to the true field B by $B_\parallel = B \cos\theta$, so that on average $\langle B_\parallel \rangle = B/2$. After allowing for this, he finds that $\langle M/M_\Phi \rangle \simeq 2.4$ for clouds with measured fields. Twelve of the 27 clouds in his sample have only upper limits on the field strength; these clouds, which typically have both lower densities and lower column densities than the clouds with measured field strengths, are also magnetically supercritical. If the clouds are flattened along the field lines, then the observed area is smaller than the true area by factor $\cos\theta$ as well, so that $M/M_\Phi \propto \cos^2\theta$; on average, this is a factor $1/3$. However, since clouds are observed to have substantial motions, they are unlikely to be highly flattened along field lines, so Crutcher concludes that $\langle M/M_\Phi \rangle \simeq 2$, consistent with the theoretical arguments advanced above. Many of the upper limits come from a study of the Zeeman effect in OH; if these data apply to the "protocores" discussed at the end of §2.4.4, then they indicate that the fields there are supercritical as well, contrary to the assumption underlying many theories of low mass star formation. The clouds with detected fields are gravitationally bound, with a mean value for the virial parameter $\alpha \simeq 1.4$. He also finds that the Alfvén Mach number of the turbulent motions, $m_A = \sigma_{\mathrm{nt}}\sqrt{3}/v_A$, is about unity in the clouds with measured fields, as inferred previously by Myers & Goodman [76] on the basis of less complete data. Two points should be kept in mind, however: First, these data do not address the issue of the strength of the field on large scales (i.e., the field threading an entire GMC). Second, since the data for the detected regions deal with dense regions in molecular clouds, it is possible that the observed mass–to–flux ratio has been altered by ambipolar diffusion.

2.5. MHD WAVES IN MOLECULAR CLOUDS

The velocity dispersions in molecular clouds are typically $\sigma \sim 1 - 2 \text{ km s}^{-1}$, whereas the thermal velocity dispersion is only about 0.2 km s^{-1} at a temperature of 10 K. The fact that the motions in molecular clouds are highly supersonic led Arons & Max [1] to suggest that the motions in molecular clouds are MHD waves. This was a prescient suggestion, made before the magnetic field measurements discussed above. Subsequent discussions of MHD waves in molecular clouds have been given by Zweibel & Josefatsson [117], Falgarone & Puget [33], Pudritz [88], and McKee & Zweibel [64].

2.5.1. *Wave Pressure*

There are three types of MHD waves: fast, slow, and Alfvén. Alfvén waves are particularly simple because they have an *isotropic* pressure [24]. At first sight, this result is surprising, since the stress exerted by a static magnetic field is anisotropic. The isotropy of the pressure due to the Alfvén waves can be understood as follows [64]: In an Alfvén wave the perturbation in the field $\delta\mathbf{B}$ is orthogonal to the background field; the stress associated with δB gives a pressure $\delta B^2/8\pi$ in the direction of the background field $\mathbf{B}_0$ and in the direction orthogonal to both $\mathbf{B}_0$ and $\delta\mathbf{B}$. In the direction of $\delta\mathbf{B}$, the net stress is $-\delta B^2/8\pi$ due to the tension in the field. However, it is just in this direction that the motion of the gas contributes a dynamic pressure $\rho\delta v^2$. Since the kinetic and magnetic energies are in equipartition for MHD waves [118], we have $\rho\delta v^2/2 = \delta B^2/8\pi$; the net stress along the direction of the perturbed field is then $-\delta B^2/8\pi + \rho\delta v^2 = \delta B^2/8\pi$, which is identical to that in the other two directions. The wave pressure is then

$$P_{\text{w}} = \frac{\delta B^2}{8\pi} = \frac{3}{2}\rho\sigma_{\text{nt}}^2, \tag{49}$$

where in the last step we have assumed that δB has a random orientation with respect to the observer so that $\delta v^2 = 3\sigma_{\text{nt}}^2$.

Understanding the dynamics of Alfvén waves in molecular clouds is a complex problem in radiation magnetohydrodynamics. However, we can gain some insight into the general problem by considering simple limiting cases. First, consider the question of how the wave pressure varies during an adiabatic compression [64]. For an adiabatic process, we have $P_{\text{w}} \propto \rho^{\gamma_{\text{w}}}$ for some γ_{w}. The pressure is related to the energy density u_{w} by $P_{\text{w}} =$

$(\gamma_{\rm w} - 1)u_{\rm w}$. Now, equipartition implies

$$u_{\rm w} = \frac{1}{2}\rho\delta v^2 + \frac{\delta B^2}{8\pi} = \frac{\delta B^2}{4\pi} = 2P_{\rm w}, \tag{50}$$

which in turn implies

$$\gamma_{\rm w} = \frac{3}{2}. \tag{51}$$

Thus, in a medium of uniform density $\rho(t)$, the wave pressure varies as $P_{\rm w}(t) \propto \rho(t)^{3/2}$. Since $\gamma_{\rm w}$ is greater than unity, the waves heat up during a compression (i.e., the wave frequency increases).

Next consider how the Alfvén wave pressure varies with position in a static cloud with a density $\rho(\mathbf{r})$. We assume a steady state, with no sources, sinks, or losses by transmission through the surface of the cloud. (The latter approximation is reasonably good if there is a large density drop at the cloud surface, since then the transmission coefficient is small.) We anticipate that $P_{\rm w} \propto \rho^{\gamma_{\rm pw}}$, where the polytropic index $\gamma_{\rm pw}$ may differ from the adiabatic index $\gamma_{\rm w}$. The answer to this problem for electromagnetic radiation is simple: the radiation pressure would be constant ($\gamma_{\rm pw} = 0$). To determine the answer for Alfvén waves, we use the energy equation for Alfvén waves [24] [64] (the $\pm$ determines the direction of propagation),

$$\frac{\partial u_w}{\partial t} + \nabla \cdot u_{\rm w}(\mathbf{v} \pm \mathbf{v}_A) + \frac{1}{2}u_{\rm w}\nabla \cdot \mathbf{v} = 0. \tag{52}$$

In a steady state in a static cloud, this reduces to

$$\nabla \cdot u_{\rm w}\mathbf{v}_A = \nabla \cdot \frac{u_{\rm w}\mathbf{B}}{(4\pi\rho)^{1/2}} = 0, \tag{53}$$

which implies

$$\mathbf{B} \cdot \nabla \frac{P_{\rm w}}{\rho^{1/2}} = 0 \tag{54}$$

since $\nabla \cdot \mathbf{B} = 0$ and $P_{\rm w} \propto u_{\rm w}$. Based on this argument, McKee & Zweibel [64] concluded that $P_{\rm w}(\mathbf{r}) \propto \rho(\mathbf{r})^{1/2}$ along any field line. If the constant of proportionality is the same for all the field lines, then the Alfvén wave pressure satisfies a polytropic relation $P_{\rm w}(\mathbf{r}) \propto \rho(\mathbf{r})^{\gamma_{\rm pw}}$ with

$$\gamma_{\rm pw} = 1/2. \tag{55}$$

This result is consistent with that of Fatuzzo & Adams [34], who studied the particular case of Alfvén waves in a self-gravitating slab threaded by a uniform vertical magnetic field. The fact that $\gamma_{\rm pw}$ is less than unity means that the velocity dispersion increases as the density decreases: the surface of the cloud is "hotter" than the center, which is consistent with the observed line width–size relation. The fact that the polytropic index $\gamma_{\rm p}$ differs from the adiabatic index γ introduces a complication into modeling molecular clouds, as we shall see below.

2.5.2. *Wave Damping*

MHD waves are subject to both linear and nonlinear damping (see §2.7 for the latter). The linear damping is due to ion–neutral friction, the same process that governs ambipolar diffusion. At sufficiently low frequencies, the ions and neutrals are well coupled so that the Alfvén velocity is determined by the density of the entire medium, $\rho = \rho_n + \rho_i$. For transverse waves the equation of motion for the neutrals is

$$\rho_n \frac{d\mathbf{v}_n}{dt} = \rho_n \nu_{ni} \mathbf{v}_D, \tag{56}$$

where ν_{ni} is the neutral–ion collision frequency and $\mathbf{v}_D$ is the velocity of the ions with respect to the neutrals. For a linear wave of frequency ω and velocity amplitude δv, this yields $v_D = \omega \delta v / \nu_{ni}$. The specific energy of the waves (including both kinetic and magnetic energy) is $\epsilon = \delta v^2$; the rate at which this energy is damped out is $\dot\epsilon = \nu_{ni} v_D^2$. In terms of the damping rate for the wave amplitude Γ, we have $\dot\epsilon = -2\Gamma\epsilon$, so that

$$\Gamma = \frac{\omega^2}{2\nu_{ni}}. \tag{57}$$

This heuristic argument, which implicitly assumes $\Gamma \ll \omega$, suggests that the waves are critically damped (i.e., $\Gamma = \omega$) at a frequency $\omega_{\rm cut} = 2\nu_{ni}$, corresponding to a wavenumber $k_{\rm cut} = \omega_{\rm cut}/v_A = 2\nu_{ni}/v_A$. A more precise calculation [54] shows that the real part of the frequency vanishes at $k = k_{\rm cut}$. MHD waves in which the motion of the ions and neutrals is coupled cannot propagate if $k > k_{\rm cut}$, and they therefore cannot provide pressure support on scales smaller than about $R_{\rm cut} \equiv \pi/k_{\rm cut}$.

The numerical value of $R_{\rm cut}$ depends on the neutral–ion collision frequency, $\nu_{ni} = n_i \langle \sigma v \rangle = 1.5 \times 10^{-9} n_i$ s^{-1} [79]. Using equation (2) for the ionization, we find that for a cloud of radius R and mass–to–flux ratio governed by M/M_Φ,

$$\frac{R}{R_{\rm cut}} = 7.7 \left(\frac{C_i}{10^{-5} \ {\rm cm}^{-3/2}} \right) \frac{M}{M_\Phi}. \tag{58}$$

Thus, for typical levels of ionization produced by cosmic rays (§1.1), clouds large enough that they cannot be supported by static magnetic fields alone ($M > M_\Phi$) are large enough to support a modest spectrum of MHD waves ($R \gg R_{\rm cut}$) [65] [112].

50 CHRISTOPHER F. MCKEE

2.6. POLYTROPIC MODELS FOR MOLECULAR CLOUDS

Polytropic models, in which the pressure varies as a power of the density,

$$P = K_{\mathrm{p}}\rho^{\gamma_{\mathrm{p}}}, \tag{59}$$

have long been used to model stars. The power γ_{p} is often expressed in terms of an index n,

$$\gamma_{\mathrm{p}} \equiv 1 + \frac{1}{n}. \tag{60}$$

Prior to the advent of computers, polytropic models were the best available for studying stellar structure, and even now they are useful for gaining insight. Models of molecular clouds are decades behind those for stars, with computational models only now beginning to be developed (§2.7). Thus, polytropic models can be expected to be of use here too, in order to shed light on the density structure of molecular clouds, the line width—size relation, the precollapse conditions for star formation, and the relation between the properties of GMCs and the medium in which they are embedded.

The isothermal Bonnor–Ebert models discussed above are the simplest examples of polytropes. Non-isothermal polytropes ($\gamma_{\mathrm{p}} \neq 1$) have been discussed by Shu et al. [97], Viala & Horedt [111], and Chiéze [17]. Maloney [57] pointed out that the line width—size relation demanded a "negative-index" polytrope, in which $\gamma_{\mathrm{p}} < 1$ so that n is negative. In order to treat the nonthermal motions in molecular clouds, Lizano & Shu [52] assumed that the pressure associated with these motions is proportional to the logarithm of the density—the limiting case of a negative index polytrope in which $\gamma_{\mathrm{p}} \to 0$. This "logatropic" form for the turbulent pressure gives a sound speed $(dP/d\rho)^{1/2} \propto 1/\rho^{1/2}$, which is consistent with Larson's laws (§1.2); on the other hand, it as yet has no physical basis. The logatropic equation of state has been studied further by Gehman et al. [38] and, in a different form, by McLaughlin & Pudritz [66].

2.6.1. *Structure of Polytropes*
The structure of a polytrope is controlled by the value of γ_{p}. Some insight into the behavior of spherical polytropes can be gained by considering the limiting case of singular polytropic spheres, which have power law solutions. From the equation of hydrostatic equilibrium

$$\frac{dP}{dr} = -\frac{GM\rho}{r^2}, \tag{61}$$

one readily finds [16]

$$\rho \propto r^{-2/(2-\gamma_{\mathrm{p}})}, \quad P \propto r^{-2\gamma_{\mathrm{p}}/(2-\gamma_{\mathrm{p}})}, \quad c \propto r^{(1-\gamma_{\mathrm{p}})/(2-\gamma_{\mathrm{p}})}, \tag{62}$$

where $c^2 = P/\rho$ is the generalized isothermal sound speed. A singular isothermal sphere [94], for example, has $\gamma_p = 1$ so that $\rho \propto P \propto 1/r^2$ and $c =$const. For a cloud supported by Alfvén waves, we have $\gamma_p = 1/2$ so that $\rho \propto r^{-4/3}$ and $c \propto r^{1/3}$. In the limit as $\gamma_p \to 0$, which is an approximation for a logatrope, we have $\rho \propto 1/r$ and $c \propto r^{1/2}$. Note that in the last two cases the velocity dispersion increases outward, as observed. However, these simple power law models cannot be used to determine the nature of the pressure support in molecular clouds, both because actual polytropes are not power laws and because actual clouds have more than one source of support.

Since the Bonnor–Ebert mass scales as $c^3/(G^3\rho^{1/2})$, it is convenient to introduce a dimensionless mass [61] [103]

$$\mu \equiv \frac{M}{c^3/(G^3\rho)^{1/2}} = \frac{M}{c^4/(G^3P)^{1/2}}. \tag{63}$$

For stars, which have $\gamma_p > 4/3$, this quantity can go to infinity at the surface: stars are supported by the hot gas in their interiors. However, the sources of support for molecular clouds have $\gamma_p \leq 4/3$, and for such clouds there is an upper limit on μ of 4.555, so that stable clouds satisfy [61]

$$M < 4.555 \left(\frac{c_s^4}{G^{3/2}P_s^{1/2}} \right), \tag{64}$$

where the subscript "s" emphasizes that the sound speed and pressure are evaluated at the surface of the cloud. Thus, the mass of a molecular cloud is limited by conditions at its *surface*. By contrast, the maximum mass of a star is set by conditions at or near its center. This upper limit on μ is a monotonically increasing function of γ_p; for negative index polytropes ($\gamma_p < 1$), it must be less than the Bonnor–Ebert value, $\mu = 1.18$.

2.6.2. *Stability of Polytropes: Locally Adiabatic Components*

In order to determine the stability of a cloud, we need to know how it will respond to a perturbation. We shall assume that the perturbation can be modeled as being adiabatic, in the sense that there is no heat exchange between pressure components (e.g., we ignore wave damping) and there is no heat exchange with the environment (e.g., we ignore the loss of wave energy by transmission into the ambient medium). McKee & Holliman [61] distinguish two cases: if the heat flow associated with the given pressure component is very inefficient, then the component is *locally adiabatic*. In the opposite limit of very efficient heat flow, the component is *globally adiabatic*. An adiabatic gas in conventional parlance is locally adiabatic. For a locally adiabatic component, the entropy parameter

$$K_i \equiv \frac{P_i}{\rho^{\gamma_i}} \tag{65}$$

remains constant during the perturbation, where the adiabatic index γ_i is distinct from the polytropic index $\gamma_{\mathrm{p}i}$. (The actual entropy is proportional to the logarithm of the entropy parameter.) The magnetic field can be modeled approximately as a locally adiabatic component with $\gamma_B = 4/3$; the corresponding entropy parameter is $K_B \propto B^2/\rho^{4/3} \propto M_B^{2/3}$, which indeed is constant so long as flux freezing holds.

If the adiabatic and polytropic indexes are the same ($\gamma_i = \gamma_{\mathrm{p}i}$), then the gas is *isentropic* since the entropy parameter $K_i \propto \rho^{\gamma_{\mathrm{p}i} - \gamma_i}$ is spatially and temporally constant. Even if the gas is subject to heating and cooling, it may be possible to model it as an isentropic gas: If the heating rate scales as nT^a and the cooling rate as $n^2 T^b$, where a and b are constant, then the gas can be modeled as isentropic with $\gamma_i = \gamma_{\mathrm{p}i} = 1 + 1/(a - b)$. A discussion of the value of γ_i based on molecular cooling curves and allowing for variable b is given by Scalo et al. [92]. Note that if the gas is not isentropic, then perturbation of a polytrope leads to a configuration that is not polytropic.

A polytrope supported by a locally adiabatic pressure component is stable for $\gamma_i > 4/3$. For $\gamma_i < \gamma_{\mathrm{p}i}$, the polytrope is convectively unstable. Along the line $\gamma_i = \gamma_{\mathrm{p}i}$ (isentropic polytropes), the critical point that divides unstable clouds from stable ones lies at the maximum value of μ; equation (64) thus gives an upper limit on M_{cr}. As γ_i increases above $\gamma_{\mathrm{p}i}$, the value of μ at the critical point (μ_{cr}) changes, but it remains close to, and somewhat less than, the maximum value of μ. The magnitude of the density contrast between the center of the cloud and the surface, ρ_c/ρ_s, increases dramatically as γ_i increases, and can become infinite even if γ_i is less than $4/3$ [61].

For isothermal polytropes, M_{cr} is reduced somewhat by an increase in the external pressure, $M_{\mathrm{cr}} \propto P_s^{-1/2}$ (eq. 64). For negative–index polytropes, however, M_{cr} can decrease much more sharply with P_s due to the decrease in c_s [97]. Since the decrease in the temperature is bounded (it is difficult to cool below 10 K in a typical molecular cloud, for example), it is more convenient to express the critical mass in terms of quantities after the compression,

$$M_{\mathrm{cr}} = \mu_{\mathrm{cr}} \left(\frac{c_{s,f}^4}{G^{3/2} P_{s,f}^{1/2}} \right), \tag{66}$$

where $c_{s,f}$ is the final value of c_s, etc. This result shows that the reduction in the critical mass due to cooling, which reduces c_s, can be large, but the reduction due to the compression is limited by the weak $P_{s,f}^{-1/2}$ dependence. For example, in a radiative shock the final pressure is related to the initial value $P_{s,i}$ by $P_{s,f} = (v_{\text{shock}}/c_{s,i})^2 P_{s,i}$. In spherical implosions even higher compressions, and correspondingly greater reductions in M_{cr}, are possible [105], although in practice it may be difficult to maintain the high degree of spherical symmetry required to achieve very large compressions.

2.6.3. *Stability of Polytropes: Globally Adiabatic Components*

MHD waves are not locally adiabatic since they can move in response to a perturbation. In the absence of damping and losses, or in the case in which sources balance damping and losses, it is the wave action integrated over the cloud that is conserved [24], and the waves can be said to be "globally adiabatic." The wave action is related to the entropy parameter for the waves [64], and McKee & Holliman [61] have determined how to treat the stability of a globally adiabatic pressure component. This is a generalization of the problem of the stabiity of globular clusters considered by Lynden–Bell & Wood [53], with the complication that the gas is not isothermal. Lynden–Bell & Wood modeled globular clusters as polytropes with $\gamma_{\text{p}} = 1$ and $\gamma = 5/3$. Whereas locally adiabatic polytropes are stable for $\gamma > 4/3$, this is not the case for globally adiabatic polytropes. If the density contrast between the center and edge becomes too great, the cluster is subject to core collapse, in which the stars carry heat from the core to the envelope and allow the core to collapse while the envelope expands. This is a generic property of globally adiabatic polytropes: for $\gamma_{\text{p}} < 6/5$, such polytropes are unstable for arbitrary values of γ, with global collapse occurring for $\gamma < 4/3$ and core collapse for $\gamma > 4/3$. Since $\gamma_{\text{w}} = 3/2$, a polytrope supported by Alfvén waves is subject to core collapse. Note that for $\gamma > 4/3$, the cloud heats up and becomes more stable if it is compressed; it is *decompression* that leads to instability.

Because γ_{pw} is only 0.5, the critical mass for a cloud supported by Alfvén waves is smaller than that for an isothermal cloud. Calculations show that the critical mass for such a cloud is [61]

$$M_{\text{w}} = 0.39 \left(\frac{\sigma_{\text{nt},s}^3}{G^{3/2} \rho_s^{1/2}} \right) = 0.65 \left(\frac{\langle \sigma_{\text{nt}}^2 \rangle^{3/2}}{G^{3/2} \rho_s^{1/2}} \right). \tag{67}$$

A cloud supported by the pressure of both an isothermal gas and Alfvén waves has a critical mass [61]

$$M_{\text{J}} = 1.18 \left(\frac{\sigma_{\text{eff}}^3}{G^{3/2} \rho_s^{1/2}} \right), \tag{68}$$

where the effective velocity dispersion is

$$\sigma_{\text{eff}}^2 \equiv c_{\text{th}}^2 + 0.67 \langle \sigma_{\text{nt}}^2 \rangle. \tag{69}$$

Insofar as it is accurate to represent the nonthermal motions in clouds as Alfvén waves, they are less effective at supporting clouds than thermal motions because they tend to concentrate in the low density envelope, as indicated by their polytropic index $\gamma_{\mathrm{pw}} = 1/2$.

2.6.4. *Multi–Pressure and Composite Polytropes*

Real clouds are supported by thermal pressure, magnetic stresses, and wave pressure. Polytropic models with multiple components can be classified into three types [61]: *Composite polytropes*, in which the pressure components are spatially separated (an example is the core–envelope model for red giant stars of Schönberg & Chandrasekhar [93]); *multi–fluid polytropes*, in which the different components interact only gravitationally, as in the case of a molecular cloud with an embedded star cluster; and *multi–pressure polytropes*, in which there is a single self–gravitating fluid with several pressure components,

$$P(r) = \dot{\Sigma} P_i(r) = \Sigma K_{\mathrm{pi}} \rho^{\gamma_{\mathrm{pi}}}. \tag{70}$$

Lizano & Shu [52] developed the first multi–pressure polytropic model for molecular clouds. They treated the axisymmetric magnetic field exactly, and modeled the gas pressure as consisting of an isothermal component for the thermal pressure and a logatropic component for the turbulent pressure. Gehman et al. [38] studied the stability of logatropes in both planar and cylindrical geometries; they effectively assumed an isentropic equation of state, which does not allow for either the stiffness or the mobility of the Alfvén waves. McLaughlin & Pudritz [66] studied a variant of the logatrope in spherical geometry. To assess the stability of the clouds, they adopted the boundary condition proposed by Maloney [57] in which the central temperature is held constant. While this is plausible for the thermal pressure, no justification has been advanced for using it for the wave pressure.

An approximate alternative to the multi–pressure polytrope has been developed by Myers & Fuller [75] and Caselli & Myers [15]. In the "TNT" model, the density is assumed to obey

$$n \propto \left(\frac{r_0}{r}\right)^2 + \left(\frac{r_0}{r}\right)^p, \tag{71}$$

where the two terms represent the effect of thermal and nonthermal motions, respectively. This form for the density is inserted into the equation of hydrostatic equilibrium to determine the density at r_0 and the velocity dispersion as a function of r. In most cases it is possible to obtain a good fit to data on the line width as a function of r by fitting the characteristic length scale r_0 and the exponent p. Caselli & Myers find that "massive cores" have significantly smaller values of r_0 (0.01 pc vs. 0.3 pc) and larger mean extinctions ($A_V = 15$ mag vs. 3.6 mag) than "low mass cores" (although the values of the masses are not discussed).

A preliminary study of multi–pressure polytropes including thermal pressure, magnetic pressure, and Alfvén waves has been given in John Holliman's thesis [43]. An important result from this work is that when allowance is made for the layer of gas in which the C is atomic, which has a thickness of 0.7 mag [109], stable clouds can have a mean gas pressure in the CO up to about 8 times the gas pressure acting on the surface of the cloud. (Elmegreen [27] had previously considered the pressure due to the smaller layer of HI, and did not address the stability of his model.) Applying the virial theorem (eq. 24) to the CO, we have $\bar{P}(\mathrm{CO}) \simeq P_G(\mathrm{CO})$. Since $P_G(\mathrm{CO}) \simeq 1.4 \times 10^5 \bar{N}_{\mathrm{H}22}^2(\mathrm{CO})$ K cm^{-3} from equation (23), the mean column density associated with the CO is

$$\bar{N}_{\mathrm{H}22}(\mathrm{CO}) \simeq 1.0 \left(\frac{P_s}{2 \times 10^4 \text{ K cm}^{-3}}\right)^{1/2}. \tag{72}$$

Allowing for the somewhat higher ambient pressure in the inner Galaxy and in regions of active star formation, and for the atomic gas associated with the GMCs, we infer

$$0.4 \lesssim \bar{N}_{\mathrm{H}22} \lesssim 2, \tag{73}$$

where the lower limit is set by the condition that there be a significant column density of CO (§1.1). This argument, which builds on previous work by Chiéze [17] and Elmegreen [27], provides an explanation for Larson's third law: All molecular clouds have about the same column density since if the column is too low they are not molecular and if it is too high they are not stable. The argument near the end of §2.2 shows why Larson's law then applies to typical star–forming clumps in GMCs as well. OB star–forming clumps have far larger extinctions ($\bar{N}_{\mathrm{H}22} \sim 60$) [87], and so do not satisfy Larson's third law; their stability needs investigation.

As discussed in §1.2, any one of Larson's laws can be derived from the other two. To obtain the line width—size relation, we write the velocity dispersion in the cloud in terms of the virial parameter α (eq. 26),

$$\sigma = 0.55 \left(\alpha \bar{N}_{\mathrm{H}22} R_{\mathrm{pc}}\right)^{1/2} \quad \text{km s}^{-1}. \tag{74}$$

For clouds that are gravitationally bound ($\alpha \sim 1$), the fact that the mean column density is restricted to a fairly narrow range of values then leads to the observed line width—size relation, $\sigma \propto R^{1/2}$. The line width also depends on the magnetic field strength: Since $\bar{A}_V \propto \bar{B}/(M/M_\Phi)$ from equation (39), we see that $\sigma \propto \bar{B}^{1/2}$ [74] [76] provided M/M_Φ is about the same for all the clouds as indicated in §§2.4.4 and 2.4.5.

A limitation of the polytropic models described above is that they do not allow for the damping of MHD waves on small scales. Curry & McKee [22] have developed composite polytropic models to deal with this problem. A simple example of a composite polytrope is one in which a cold isothermal core is embedded in a hot isothermal envelope; in this case, it is possible to have a total pressure drop between the center of the cold core and the surface of the hot envelope of up to $14^2 = 196$. To model molecular cloud cores, they assume that the inner region is dominated by thermal motions whereas the outer region is dominated by nonthermal motions and a static field. By carefully allowing for projection effects, they are able to get good agreement with the observed line width profiles.

2.7. MODELING TURBULENCE IN MOLECULAR CLOUDS

The most important recent development in the theoretical study of molecular clouds is the advent of sophisticated simulations of turbulent, magnetic, self–gravitating clouds [37], [55], [56], [83], [84], [85], [104]. Many of the issues associated with turbulence in molecular clouds and the results from these simulations have been reviewed by Vazquez–Semadeni et al. [110]. Here I shall mention only two results that are of particular relevance to our discussion.

First, the simulations show that when the turbulent velocities are supersonic, the gas becomes clumped. As yet, there is no agreement on exactly how the clumping factor depends on the physical conditions. Vazquez–Semadeni et al [110] summarize the existing multi–dimensional, isothermal calculations as finding that the mean mass–averaged value of $\log(\rho/\bar{\rho})$ scales with the logarithm of the sonic Mach number. Since the Mach number increases with scale, this result suggests that the clumping factor in clumps is smaller than in GMCs as a whole. To date, there is insufficient dynamical range in space to study hierarchical structure in GMCs (cores in star–forming clumps in GMCs), or in time to study how the initial conditions in the interstellar medium might affect the clump structure of GMCs.

The second and more important result is that the waves damp remarkably quickly [55], [56], [104]. For example, Stone et al. [104] find that the energy dissipation time for forced MHD turbulence is about $0.75L/v_{\rm rms}$, where L is the scale on which the turbulence is supplied and $v_{\rm rms} = \sqrt{3}\sigma_{\rm nt}$ is the rms velocity of the turbulence. None of the groups find a significant difference between magnetized and unmagnetized turbulence. This is completely contrary to the expectation that magnetic fields would reduce dissipation by "cushioning" the flow. Furthermore, since circularly polarized Alfvén waves of arbitrary amplitude can propagate without dissipation in a uniform medium, it had been thought that they could survive more than an eddy turnover time in a non–uniform medium. Stone et al estimate that the energy dissipation rate per unit mass is $\dot{\epsilon}_{\mathrm{diss}} \simeq v_{\mathrm{rms}}^3/L$. The motions observed in molecular clouds must therefore be continually rejuvenated, presumably by energy injection from newly formed stars [82].

Although several groups have independently found that the waves damp extremely rapidly, this result must be treated with caution. From a technical standpoint, the numerical resolution is as yet inadequate to resolve the waves that occur in the clumps that form within the simulation volume; this could be important since such waves could resist compression of the clumps. To date, simulations have been done for gas that fills a rectangular volume; it has not been possible to simulate an isolated GMC. More importantly, there are several problems from an observational standpoint. It is very difficult to understand how a GMC such as G216-2.5, which has no visible star formation, can have a level of turbulence that exceeds that in the Rosette molecular cloud, which has an embedded OB association [114]. On a smaller scale, cores are observed to have comparable levels of non-thermal motions whether there is an embedded star or not (e.g., [15]). One of the striking results of the simulations is that the rate of dissipation appears to be insensitive to whether the turbulence is super–Alfvénic or not, yet observations show that the Alfvén Mach number is of order unity [20]. Although there is considerable scatter in the properties of molecular clouds, the existence of regularities such as those found by Larson [48] is difficult to understand if the turbulence decays on the dynamical time scale whereas the stars that are supposed to support the clouds form on a considerably slower time scale, as discussed in §1.3.

3. Evolution of Molecular Clouds and Star Formation

3.1. FORMATION OF GMCS

The formation of GMCs is a rich and complex topic that has been reviewed by Elmegreen [29]. As in the case of galaxy formation, there are two countervailing views: in the "bottom–up" picture, small objects coagulate to form large ones, whereas in the "top–down" picture, large objects form first and fragment into small objects. In the case of GMCs, the bottom–up scenario corresponds to the growth of clouds by collisions. In order to generate the observed power–law mass distribution, a number of generations of collisions are necessary. However, as Elmegreen [29] points out, the short lifetime of molecular clouds ([10] [115]) makes this very difficult.

In the top–down scenario of GMC formation, clouds form in spiral arms, where the low shear and high densities allow gas to accumulate along the arm [30]. The volume from which the gas in a GMC is accumulated is quite large [67]. If we assume that the accumulation volume prior to compression by the spiral arm is an ellipsoid with a axial diameter L_0 and a radius R_0, that the mean density in this volume is initially n_0, and that the magnetic field (measured in μG) is initially $B_{0\mu}$, then one finds [65]

$$L_0 = 96 \left(\frac{B_{0\mu}}{n_0}\right) \frac{M}{M_\Phi} \quad \text{pc}, \tag{75}$$

$$R_0 = 380 \left(\frac{M_6}{B_{0\mu}}\right)^{1/2} \left(\frac{M}{M_\Phi}\right)^{-1/2} \quad \text{pc}. \tag{76}$$

For typical conditions ($n_0 \sim 1$ cm^{-3}, $B_{0\mu} \sim 3$ and $M \simeq 2M_\Phi$, as discussed in §2.4.4), this gives $L_0 \sim 600$ pc and $R_0 \sim 150 M_6^{1/2}$ pc. This is large enough to contain many diffuse clouds, which could subsequently become clumps in the GMC [26].

3.2. DYNAMICAL EVOLUTION OF GMCS

Once a GMC has formed, it will be supported against gravity by a combination of static magnetic fields and turbulent pressure; since the observed motions are highly supersonic, thermal pressure is relatively unimportant. Because the waves are damped (§2.7), the cloud will contract. The contraction of the cloud will adiabatically compress the waves, tending to counteract the damping. In order to stop the contraction, energy injection is required, and this is provided by protostellar outflows [82].

The resulting evolution of the cloud can be described with the "cloud energy equation" [60]. Let $\epsilon \equiv E/M$ be the total cloud energy per unit mass. Since the motions observed in GMCs are highly supersonic, we shall assume that the energy in internal degrees of freedom is negligible, as in §2.2. The rate of change of ϵ can be written symbolically as

$$\frac{d\epsilon}{dt} = \mathcal{G} - \mathcal{L}, \tag{77}$$

where $\mathcal{G}$ is the rate of energy gains per unit mass and $\mathcal{L}$ is the rate of energy losses per unit mass.

Energy is lost by wave damping, $\mathcal{L} = -\dot{E}_w/M$, where the wave energy E_w includes the energy in both motions and in fluctuating fields. For motions coupled to the field, the fluctuating field energy is in equipartition with the kinetic energy [118]. If we assume that the motions are isotropic, then equipartition applies to 2 of the 3 directions and $E_w = (5/3)\mathcal{T}$. In fact, Stone et al. [104] find $E_w \simeq 1.6\mathcal{T}$ when the gas is strongly magnetized ($\beta \equiv P_{\text{th}}/P_B$ in the range 0.01-0.1). Let η be the ratio of the damping time to the free fall time t_{ff}; we then have

$$\mathcal{L} \simeq 1.6 \frac{\mathcal{T}/M}{\eta t_{\text{ff}}} = 1.6 \frac{v_{\text{rms}}^2/2}{\eta t_{\text{ff}}}. \tag{78}$$

The simulations discussed in §2.7 suggest that $\eta \sim 1$, but for the reasons discussed there $\eta \sim$ a few seems more reasonable.

As pointed out by Norman & Silk [82], energy injection by newly formed stars is the dominant source of kinetic energy for molecular clouds. The rate at which protostellar outflows inject energy can be written as

$$\mathcal{G} = \epsilon_{\mathrm{in}} \left(\frac{\dot{M}_*}{M} \right) \equiv \frac{\epsilon_{\mathrm{in}}}{t_{g*}}, \tag{79}$$

where ϵ_{in} is the energy injected per unit stellar mass and t_{g*} is the *star formation time scale*, defined as the time to convert the gas entirely into stars. An outflow drives a shock into the surrounding medium and a reverse shock is driven back into the outflow. Under the assumption that both shocks are radiative, momentum conservation implies $m_w v_w = M_{sw} v_{sw}$, where m_w and v_w are the mass and velocity of the outflowing wind, and M_{sw} and v_{sw} are the mass and velocity of the swept up cloud. The swept up material merges with the ambient cloud when v_{sw} drops to about the effective sound speed $(P_{\mathrm{tot}}/\rho)^{1/2}$, which is of order v_{rms} in a highly turbulent cloud. Introducing a factor $\phi_w \sim 1$ that allows for the uncertainty in our estimate of the energy injection per outflow, we then have

$$m_* \epsilon_{\mathrm{in}} = 1.6\phi_w \times \frac{1}{2} M_{sw} v_{\mathrm{rms}}{}^2 = 1.6 \left(\frac{\phi_w m_w v_w}{v_{\mathrm{rms}}} \right) \frac{v_{\mathrm{rms}}{}^2}{2}, \tag{80}$$

where we have included a factor 1.6 for the energy stored in fluctuating magnetic fields. Magnetic energy stored in the wind could do work on the ambient medium, which would make ϕ_w greater than unity; on the other hand, the protostellar jets could escape from the cloud, which would reduce ϕ_w. Inserting this result into equation (79) and using equations (77) and (78), we find that the cloud energy equation becomes

$$\frac{d\epsilon}{dt} = 1.6 \left[\left(\frac{\phi_w m_w v_w}{m_* v_{\mathrm{rms}}} \right) \frac{1}{t_{g*}} - \frac{1}{\eta t_{\mathrm{ff}}} \right] \frac{v_{\mathrm{rms}}{}^2}{2}. \tag{81}$$

This equation shows that the cloud will contract if the star formation rate is too small (for a bound cloud, ϵ will become more negative) and expand if the rate is too large. If the cloud contracts quasistatically (so that $\ddot{I}$ is negligible), and if the cloud is strongly bound ($P_s \ll \bar{P}$, which may be difficult to achieve in practice), then the virial theorem implies $E = \mathcal{T} + \mathcal{M} + W = -\mathcal{T}$, or $\epsilon = -v_{\mathrm{rms}}{}^2/2$. Thus, contraction of a bound cloud leads to an *increase* in the velocity dispersion of the cloud. This "virialization" might account for the motions observed in clouds that are not actively forming stars [60].

As the density of the contracting cloud rises, the star formation rate is expected to rise as well. If the rate of energy injection by newly formed stars balances the damping rate, the star formation is said to be *dynamically regulated* [7]:

$$\frac{t_{g*\mathrm{DR}}}{t_{\mathrm{ff}}} = \left(\frac{m_w v_w}{m_* v_{\mathrm{rms}}}\right) \phi_w \eta. \tag{82}$$

Theoretically, we expect the momentum per unit stellar mass to be

$$\left(\frac{m_w}{m_*}\right) v_w \simeq \frac{1}{3} \times 200\,\mathrm{km\ s^{-1}} \simeq 70\,\mathrm{km\ s^{-1}} \tag{83}$$

[78]. Observations suggest a somewhat lower value of about 40 km s^{-1} [7]. With the latter value, and taking $v_{\mathrm{rms}} \sim 2$ km s^{-1}, we have $t_{g*\mathrm{DR}}/t_{\mathrm{ff}} \simeq 20\phi_w\eta \gg 1$. Thus, dynamically regulated star formation is inefficient, consistent with observation.

To obtain a more precise comparison between the dynamically regulated star formation rate and observation, we write the dynamically regulated star formation rate as

$$\dot{M}_{*\mathrm{DR}} = \frac{M}{t_{g*\mathrm{DR}}} = \left(\frac{m_* v_{\mathrm{rms}}}{m_w v_w}\right) \frac{M}{\phi_w \eta t_{\mathrm{ff}}}. \tag{84}$$

Noting that $v_{\mathrm{rms}}/t_{\mathrm{ff}} \propto \alpha^{1/2}\bar{N}_{\mathrm{H}}$ and using 40 km s^{-1} for the momentum injection per unit stellar mass, we find

$$\dot{M}_{*\mathrm{DR}} \simeq 2.7 \times 10^{-8} \left(\frac{\alpha^{1/2}\bar{N}_{\mathrm{H22}}M}{\phi_w\eta}\right) \quad M_\odot\ \mathrm{yr^{-1}}. \tag{85}$$

First, apply this equation to the entire Galaxy. The mass of H$_2$ inside the solar circle is $1.0 \times 10^9\ M_\odot$ [115], of which about half is actively forming stars [99]. With an average column density $\bar{N}_{\mathrm{H22}} \simeq 1.5$ [100] and with $\alpha \simeq 1$ [5], we have $\dot{M}_* \simeq 20/(\phi_w\eta)\ M_\odot$ yr^{-1}. This is consistent with the observed value of 3 $M_\odot$ yr^{-1} [62] for $\phi_w\eta \simeq 7$. McKee [60] adopted a momentum per unit stellar mass of 70 km s^{-1} instead of 40 km s^{-1} (which is equivalent to $\phi_w = 1.7$) and took $\eta = 5$, and so concluded that energy injection by low mass stars could supply the turbulent energy needed to support GMCs. On the other hand, if $\eta \sim 1$ as suggested by the simulations discussed in §2.7, then either winds are much more efficient at energizing clouds ($\phi_w \gg 1$) or most GMCs are energized by some other source, such as massive stars.

The concept of dynamically regulated star formation can be applied to individual star–forming clumps as well. Consider the star–forming clumps associated with the young star clusters NGC 2023, 2024, 2068, 2071 in Orion B [46] [47] [45]. The clumps all have masses of order 400 $M_\odot$, although the number of young stars associated with the clumps ranges from 21 in NGC 2023 to 309 in NGC 2024. With a mean stellar mass of 0.5 $M_\odot$ [91], $\alpha \sim 1$, $\bar{N}_{H22} \sim 1.5$, the dynamically regulated star formation rate in one of these clouds is $\dot{\mathcal{N}}_{*DR} \simeq 32/(\phi_w \eta)$ stars Myr^{-1}. Because of the small masses of the clumps involved, large fluctuations in the star formation rate may be expected. Furthermore, ϕ_w might be significantly less than unity if the outflows can escape from the clumps [58], as has been observed in some cases [89]. Thus, the dynamically regulated rate appears to be within the range observed in these clumps assuming that the star formation has been occurring for the past 1-3 Myr. The OB star–forming clumps observed by Plume et al [87] are an order of magnitude more massive than these, and it is not known if their properties are also consistent with dynamically regulated star formation.

3.3. PHOTOIONIZATION–REGULATED STAR FORMATION

To this point, we have discussed the star formation rate in terms of how many stars must form in order to provide adequate energy input to prevent molecular clouds from collapsing, but we have not discussed the physical mechanism that actually determines the rate at which stars form. For low mass stars, this time scale is believed to be controlled by ambipolar diffusion [68] [72] [95]. As we have seen in §2.4.1, the magnetic critical mass for clumps in molecular clouds is larger than the typical mass of a star. In order for gravity to overcome the force due to the magnetic field, gas must either accumulate along the field lines or across the field; the latter can occur only via ambipolar diffusion. If the gas is to accumulate along the field, then for $M/M_\Phi \simeq 2$ (§2.4.3) about half the mass along a given flux tube would have to be concentrated into a single star. A model in which the accumulation of gas along flux tubes is regulated by wave damping has been proposed recently by Myers & Lazarian [77]. Here we shall focus on the role of ambipolar diffusion in low mass star formation.

Most of the gas in a molecular cloud is neutral, and does not interact directly with the magnetic field. On the other hand, at typical densities in molecular clouds, the charged particles are well coupled to the field. As a result, if the neutral gas attempts to collapse under its own self gravity, it will be restrained by collisions with the charged particles. Balancing the force of gravity against that of friction gives

$$\frac{GM\rho}{R^2} \simeq n_i \langle \sigma v \rangle \rho v_{AD}, \tag{86}$$

where v_{AD}, the ambipolar diffusion velocity, is the relative velocity between the ions and the neutrals. The time scale for ambipolar diffusion then depends only on the ionization of the gas [102],

$$t_{\mathrm{AD}} \simeq \frac{R}{v_{\mathrm{AD}}} = \left(\frac{3\langle \sigma v \rangle}{4\pi G \mu_{\mathrm{H}}} \right) x_e. \tag{87}$$

The numerical coefficient in the relation between t_{AD} and x_e depends on the geometry of the cloud and the nature of the non–magnetic forces. To be specific, we identify t_{AD} as the time for ambipolar diffusion to initiate the formation of a very dense core. Fiedler & Mouschovias [35] simulated an axisymmetric cloud in which the thermal pressure slowed the ambipolar diffusion significantly; their results give $t_{\mathrm{AD}}/x_e \simeq 0.8 \times 10^{14}$ yr. Ciolek & Mouschovias [18] found a similar result for the case of a thin disk, $t_{\mathrm{AD}}/x_e \simeq 1.0 \times 10^{14}$ yr. We adopt

$$t_{\mathrm{AD}} = 1.0 \times 10^{14} \phi_{\mathrm{AD}} x_e \quad \text{yr}, \tag{88}$$

where ϕ_{AD} is a constant of order unity that allows for deviations from the typical value. In gas that is shielded from FUV radiation, the ionization is due to cosmic rays (eq. 2), which implies

$$\frac{t_{\mathrm{AD}}}{t_{\mathrm{ff}}} = 23\phi_{\mathrm{AD}} \left(\frac{C_i}{10^{-5} \ \mathrm{cm}^{-3/2}} \right). \tag{89}$$

For $C_i \sim 0.7 \times 10^{-5}$ (§1.1), $t_{\mathrm{AD}} \simeq 15 t_{\mathrm{ff}}$, so that ambipolar diffusion is quite slow compared to gravitational collapse.

When the magnetic critical mass M_B significantly exceeds the typical stellar mass, as appears to be the case in Galactic molecular clouds (§2.4.1), then the rate at which stars form is set by the average rate of ambipolar diffusion in the cloud [60]:

$$\dot{M}_* = \frac{M}{t_{g*}} \simeq \int \frac{dM}{t_{\mathrm{AD}}(x_e)}. \tag{90}$$

The ionization in the outer parts of molecular clouds is relatively high due to photoionization of the metals. Regions in the clouds that are shielded by an extinction $A_V > A_V(\mathrm{CR}) \simeq 4$ are ionized primarily by cosmic rays; as a result, they have a lower ionization (eq. 2) and correspondingly lower ambipolar diffusion time. As a result, the star formation rate in a molecular cloud is approximately that in the cosmic ray ionized region,

$$\dot{M}_* \simeq \frac{M(A_V > A_{\mathrm{CR}})}{t_{\mathrm{AD}}(x_{e\mathrm{CR}})}. \tag{91}$$

Since the rate at which stars form is governed by the mass that is shielded from photoionizing FUV radiation, McKee termed this *photoionization–regulated star formation* [60]. This process is naturally inefficient, both because only a fraction of the cloud is in the cosmic ray ionized region (models give a typical value of about 10% [60]) and because the ambipolar diffusion time is much greater than the free fall time (eq. 89).

How does this rate compare with that observed in the Galaxy? If we adopt $n_{\rm H} \simeq 3000$ cm^{-3} from §1.2, we find $t_{\rm AD} \simeq 2.4 \times 10^7$ yr. For 10% of the mass in cosmic ray ionized regions, this gives $t_{g*} \simeq 2.4 \times 10^8$ yr. Since the total mass of molecular gas inside the solar circle is about 1.0×10^9 $M_\odot$ [115], the predicted rate of photoionization regulated star formation there is $\dot{M}_* = 10^9 M_\odot / 2.4 \times 10^8$ yr $\simeq 4$ $M_\odot$ yr^{-1}, quite close to the observed value of 3 $M_\odot$ yr^{-1}.

One of the key predictions of this theory is that star formation is restricted to regions of relatively high extinction, $A_V \gtrsim A_{\rm CR} \simeq 4$ mag. This has been tested in a study of the L1630 region of the Orion molecular cloud, and indeed all the star formation was found to be concentrated in regions in which the extinction was greater than this. McKee [60] also predicted that molecular clouds in the Magellanic Clouds would have comparable extinctions, and therefore higher column densities, than Galactic molecular clouds in order that star formation be able to provide the energy needed to prevent the molecular clouds from collapsing. Pak et al [86] have found that the column densities of molecular clouds in the LMC and SMC do in fact scale approximately inversely with the metallicity, consistent with an approximately constant extinction.

4. Conclusion

We have seen that it is possible to understand a number of the observed properties of molecular clouds in terms of a model in which GMCs and the clumps within them are modeled as clouds in approximate hydrostatic equilibrium, with the turbulence treated as a separate pressure component. In particular, we have seen why the clouds are generally gravitationally bound; why they have approximately constant column densities; why they have line widths that increase with size; why, along with the star–forming clumps and cores within them, they are somewhat magnetically supercritical; why they are the sites of star formation in the Galaxy; and why star formation is inefficient. Of course, this model is a drastically oversimplified picture of the real situation. Other researchers, looking at the same data, have developed completely different models: For example, Elmegreen & Falgarone [31] have developed a fractal model for structure in the interstellar medium in which the concept of pressure plays no role, and Ballesteros–Paredes et al [3] have argued that pressure balance is irrelevant in a turbulent medium such as the ISM. By the time of the next Crete meeting, there may be some synthesis between these differing viewpoints or perhaps an entirely new idea. The challenge facing us is formidible, for we must attempt to extend our understanding to regions of OB star formation as well.

5. Acknowledgments

I am deeply grateful to both Nick Kylafis and Charlie Lada for providing me the opportunity to learn more about star formation in such a stimulating environment. My thanks to Dick Crutcher for sharing his results prior to publication, and to Frank Bertoldi, Charles Curry, Chris Matzner, and Dean McLaughlin for comments on the manuscript. My work is supported in part by NSF grant AST95-30480, a grant from the Guggenheim Foundation, a NASA grant that supports the Center for Star Formation Studies, and a grant from the Sloan Foundation to the Institute for Advanced Study.

References

1. Arons, J., & Max, C. E. 1975, *ApJ*, **196**, L77
2. Arquilla, R. and Goldsmith, P. F. 1985, *ApJ*, **297**, 436
3. Ballesteros–Paredes, J., Vazquez–Semadeni, E., & Scalo, J. 1999, *ApJ*, **000**, 000
4. Beichman, C.A., Myers, P.C., Emerson, J.P., Harris, S., Mathieu, R.D., Benson, P.J., & Jennings, R.E. 1986, *ApJ*, **307**, 337
5. Bertoldi, F., & McKee, C. F. 1992, *ApJ*, **395**, 140
6. Bertoldi, F., & McKee, C.F. 1996, in *Amazing Light, A Volume Dedicated to Charles Hard Townes on His 80th Birthday*, ed. R.Y. Chiao. Springer, New York. p. 41
7. Bertoldi, F., & McKee, C.F. 1999, in preparation
8. Blitz, L. 1991, in *The Physics of Star Formation and Early Stellar Evolution*, ed. C. J. Lada & N. D. Kylafis, Kluwer, Dordrecht p. 3
9. Blitz, L. Magnani, L. and Mundy, L. 1984, *ApJ*, **282**, L9
10. Blitz, L., & Shu, F. H. 1980, *ApJ*, **238**, 148
11. Bohlin, R.C., Savage, B.D., & Drake, J.F. 1978, *ApJ*, **224**, 132
12. Boland, W., & deJong, T. 1984, *A&A*, **134**, 87
13. Bonnor, W. B. 1956, *MNRAS*, 116, 350
14. Boulares, A. & Cox, D. P. 1990, *ApJ*, **365**, 544
15. Caselli, P., & Myers, P. C. 1995, *ApJ*, **446**, 665
16. Chandrasekhar, S. 1939, An Introduction to the Study of Stellar Structure (New York: Dover)
17. Chièze, J. P. 1987, *A&A*, **171**, 225
18. Ciolek, G. E., & Mouschovias, T. Ch. 1994, *ApJ*, 425, 142
19. Clemens, D.P., & Barvainis, R. 1988, *ApJ Supp.*, **68**, 257
20. Crutcher, R.M. 1999, *ApJ*, **000**, 000
21. Crutcher, R.M., Mouschovias, T.M., Troland, T.H., & Ciolek, G.E. 1994, *ApJ*, **427**, 839
22. Curry, C., & McKee, C.F. 1999, in preparation
23. Dalgarno, A. 1987, in *Physical Processes in the Interstellar Medium*, ed. G. Morfill and M. Scholer. Kluwer, Dordrecht, p. 219.
24. Dewar, R. L. 1970, Phys. Fluids, **13**, 2710
25. Ebert, R. 1955, *Zeitschrift für astrophysik*, **37**, 216
26. Elmegreen, B.G. 1985, in *Protostars & Planets II*, ed. D. Black & M. Matthews. University of Arizona Press, Tucson, p. 33
27. Elmegreen, B. G. 1989, *ApJ*, **338**, 178
28. Elmegreen, B.G. 1993a, *ApJ*, **411**, 170

29. Elmegreen, B.G. 1993b, in *Protostars and Planets III*, ed. E. H. Levy & J. I. Lunine. University of Arizona Press, Tucson, 97
30. Elmegreen, B.G. 1994, *ApJ*, **433**, 39
31. Elmegreen, B.G., & Falgarone, E. 1996, *ApJ*, **471**, 816
32. Falgarone, E., & Puget, J. L. 1985, *A&A*, 142, 157
33. Falgarone, E., & Puget, J. L. 1986, *A&A*, 162, 235
34. Fatuzzo, M., & Adams, F. C. 1993, *ApJ*, **412**, 146
35. Fiedler, R. A., & Mouschovias, T. Ch. 1993, *ApJ*, 415, 680
36. Fiege, J., & Pudritz, R.E. 1999, *ApJ*, **000**, 000
37. Gammie, C. F., & Ostriker, E. C. 1996, *ApJ*, **466**, 814
38. Gehman, C. S., Adams, F. C., Fatuzzo, M., & Watkins, R. 1996, *ApJ*, **457**, 718
39. Goodman, A. A., Benson, P. J., Fuller, G. A., & Myers, P. C. 1993, *ApJ*, 406, 528
40. Heiles, C. 1996, in *Polarimetry of the Interstellar Medium*, ed. W. Roberge and D.C.B. Whittet, ASP, San Francisco, p. 457
41. Heiles, C., Goodman, A. A., McKee, C. F., & Zweibel, E. G. 1993 in *Protostars and Planets III*, ed. E. H. Levy & J. I. Lunine. University of Arizona Press, Tucson, 279
42. Hollenbach, D.J., & Tielens, A.G.G.M. 1999, *Rev Mod Phys*, **000**, 000
43. Holliman, John H. 1995, *The Structure and Evolution of Self-Gravitating Molecular Clouds*, PhD Thesis, UC Berkeley
44. Jackson, J.D. 1975. *Classical Electrodynamics*, 2nd ed. Wiley, New York.
45. Lada, E. 1999, this volume.
46. Lada, E.A., Evans, N.J., DePoy, D.L., & Gatley, I. 1991, *ApJ*, **371**, 171
47. Lada, E.A., Bally, J., & Stark, A.A. 1991, *ApJ*, **368**, 432
48. Larson, R. B. 1981, *MNRAS*, **194**, 809
49. Lazarian, A., & Vishniac, E. 1999, *ApJ*, **000**, 000
50. Li, W., Evans, N.J., & Lada, E.A. 1997, *ApJ*, **488**, 277
51. Liszt, H.S. 1995, *ApJ*, **442**, 163
52. Lizano, S., & Shu, F. H. 1989, *ApJ*, **342**, 834
53. Lynden-Bell, D., & Wood, R. 1968, *MNRAS*, **138**, 495
54. Kulsrud, R.M., & Pearce, W.P. 1969, *ApJ*, **156**, 445
55. Mac Low, M.-M. 1998, *ApJ*, submitted.
56. Mac Low, M.-M., Klessen, R.S., Berkert, A., & Smith, M.D. 1998, *Phys Rev. Lett.*, **80**, 2754
57. Maloney, P. 1988, *ApJ*, **334**, 761
58. Matzner, C.D., McKee, C.F., & Bertoldi, F. 1999, in preparation
59. McCrea, W.H. 1957, *MNRAS*, **117**, 562
60. McKee, C. F. 1989, *ApJ*, **345**, 782
61. McKee, C.F., & Holliman, J.H. 1999, *ApJ*, **000**, 000
62. McKee, C.F., & Williams, J.P. 1997, *ApJ*, **476**, 144
63. McKee, C. F., & Zweibel, E. 1992, *ApJ*, **399**, 551
64. McKee, C. F., & Zweibel, E. 1995, *ApJ*, **440**, 686
65. McKee, C. F., Zweibel, E. G., Goodman, A. A., and Heiles, C. 1993, in *Protostars and Planets III*, ed. E. H. Levy & J. I. Lunine University of Arizona Press, Tucson, 327
66. McLaughlin, D.E., & Pudritz, R.E. 1996, *ApJ*, **469**, 194
67. Mestel, L. 1985, in *Protostars & Planets II*, ed. D. Black & M. Matthews. University of Arizona Press, Tucson, p. 320
68. Mestel, L., & Spitzer, L. 1956, *MNRAS*, **116**, 503
69. Mouschovias, T. Ch. 1976a, *ApJ*, 206, 753
70. Mouschovias, T. Ch. 1976b, *ApJ*, 207, 141
71. Mouschovias, T. Ch., & Spitzer, L 1976, *ApJ*, 210, 326
72. Mouschovias, T. Ch. 1987, in Physical Processes in Interstellar Clouds, ed. G. E. Morfill & M. Scholer (Dordrecht: Reidel), 453
73. Mouschovias, T. Ch. 1991, in *The Physics of Star Formation and Early Stellar*

Evolution, ed. C. J. Lada & N. D. Kylafis. Kluwer, Dordrecht, p 449
74. Mouschovias, T. Ch., & Psaltis, D. 1995, *ApJ*, **444**, L105
75. Myers, P. C., & Fuller, G. A. 1992, *ApJ*, **396**, 631
76. Myers, P. C., & Goodman, A. A. 1988, *ApJ*, **326**, L27
77. Myers, P.C. & Lazarian, A. 1998, *ApJ Letters*, **507**, L157
78. Najita, J.R., & Shu, F.H. 1994, *ApJ*, **429**, 808
79. Nakano, T. 1984, *Fund. Cos. Phys.*, **9**, 139
80. Nakano, T. 1998, *ApJ*, **494**, 587
81. Nakano, T., & Nakamura, T. 1978. *Publ. Astron. Soc. Japan*, **30**, 681
82. Norman, C.A., & Silk, J.I. 1980, *ApJ*, **238**, 158
83. Ostriker, E. C., Gammie, C. F., & Stone, J. M. 1998, *ApJ*, 000, 000
84. Padoan, P. Jones, B.J.T., & Nordlund, A.P. 1997 *ApJ*, **474**, 730
85. Padoan, P., & Nordlund, A.P. 1998 *ApJ*, **000**, 000
86. Pak, S., Jaffe, D.T., van Dishoeck, E.F., Johansson, L.E.B., & Booth, R.S. 1998, *ApJ*, **498**, 735
87. Plume, R., Jaffe, D.T., Evans, N.J., Martin-Pintado, J., Gomez-Gonzalez, J. 1997, *ApJ*, **476**, 730
88. Pudritz, R. E. 1990, *ApJ*, 350, 195
89. Reipurth, B., Bally, J., & Devine, D. 1997, *AJ*, **114**, 2708
90. Sanders, D.B., Scoville, N.Z., & Solomon, P.M. 1985, *ApJ*, **289**, 373
91. Scalo, J. 1986, *Fund. Cos. Phys.*, **11**, 1
92. Scalo, J., Vasquez-Semadeni, E., Chappell, D., & Passot, T. 1998, *ApJ*, **504**, 835
93. Schönberg, M, & Chandrasekhar, S. 1942, *ApJ*, 96, 161
94. Shu, F. H. 1977, *ApJ*, **214**, 488
95. Shu, F. H., Adams, F. C., & Lizano, S. 1987, *Ann. Rev. Astron. Astrophys.*, **25**, 23
96. Shu, F.H., & Li, Z.-Y. 1997, *ApJ*, **475**, 251
97. Shu, F.H., Milione, V., Gebel, W., Yuan, C., Goldsmith, D.W., & Roberts, W.W. 1972, *ApJ*, 173, 557
98. Shu, F.H., Najita, J., Ostriker, E., Wilkin, F., Ruden, S.P., & Lizano, S.1994 *ApJ*, **429**, 781
99. Solomon, P. M., & Rivolo, A. R. 1989, *ApJ*, **339**, 919
100. Solomon, P. M., Rivolo, A. R., Barrett, J. W., & Yahil, A. 1987, *ApJ*, **319**, 730
101. Spitzer, L. 1968 *Diffuse Matter in Space* (Interscience: New York)
102. Spitzer, L. 1978 *Physical Processes in the Interstellar Medium* (Wiley: New York)
103. Stahler, S.W. 1983, *ApJ*, **268**, 165
104. Stone, J.M. Ostriker, E.C. & Gammie, C.F. 1998, *ApJ*, **000**, 000
105. Tohline, J.E., Bodenheimer, P.H., & Christodoulou, D.M. 1987, *ApJ*, **322**, 787
106. Tomisaka, K., Ikeuchi, S., & Nakamura, T. 1988, *ApJ*, **335**, 239
107. Tomisaka, K. 1991, *ApJ*, **376**, 190
108. Turner, B. E. 1993, *ApJ*, 405, 229
109. van Dishoeck, E.F., & Black, J.H. 1988. *ApJ*, 334, 771
110. Vazquez–Semadeni, E., Ostriker, E., Passot, T., Gammie, C., and Stone, J. 1999, in *Protostars and Planets IV*, ed. V. Mannings, A. Boss, & S. Russell, University of Arizona Press, Tucson, in press.
111. Viala, Y., & Horedt, G. P. 1974, *A&A Supp*, **16**, 173
112. Williams, J.P., Bergin, E.A., Caselli, P., Myers, P.C., & Plume, R. 1998, *ApJ*, **503**, 689
113. Williams, J.P., Blitz, L., & Stark, A.A. 1995, *ApJ*, **451**, 252
114. Williams, J.P., Blitz, L., & McKee, C.F. 1999, in *Protostars and Planets IV*, ed. V. Mannings, A. Boss, & S. Russell, University of Arizona Press, Tucson, in press.
115. Williams, J.P., & McKee, C.F. 1997, *ApJ*, **476**, 166
116. Zuckerman, B., & Evans, N. J., II 1974, *ApJ*, **192**, L149
117. Zweibel, E.G., & Josafatsson, K. 1983, *ApJ*, **270**, 511
118. Zweibel, E. G., & McKee, C. F. 1995, *ApJ*, 439, 779

PHYSICAL CONDITIONS IN NEARBY MOLECULAR CLOUDS

PHILIP C. MYERS
Harvard-Smithsonian Center for Astrophysics
60 Garden St.
Cambridge, MA 02138
USA

1. Introduction

In this chapter we describe properties of the nearest star-forming complexes, and physical processes which lead to, and are associated with, star formation. We present basic ideas and recent results, primarily from an observational viewpoint, with emphasis on spectroscopy. We present in §2 general properties of molecular clouds, in §3 properties of dense cores, in §4 the role of turbulent cores in forming star clusters, in §5 evidence for inward motion in dense cores, and in §6 some models of inward motion.

2. Nearby Molecular Clouds

Astronomers call a "molecular cloud" an object which can be detected and mapped in its brightest line, the $J = 1 - 0$ rotational transition of the CO molecule at 2.6 mm wavelength. Molecular clouds are the most massive objects in galaxies, with masses up to more than 10^6 $M_\odot$, and are the birthplaces of stars. They are called "molecular" because their main constituent, hydrogen, is primarily in molecular form, H_2, while their atomic hydrogen component is less abundant by a factor $\sim 10^3$. Because the H_2 molecule has no permanent dipole moment, its emission is much weaker than that of CO, despite the relatively low abundance of CO, $\sim 10^{-4}$.

The nearest molecular clouds, within a few hundred pc of the Sun, are well-studied and well-resolved by molecular line observations. These complexes in Taurus, Ophiuchus, Lupus, Chamaeleon, Perseus, and Orion are evident by their extinction of background starlight, their far infrared emission, and their emission in numerous molecular lines. Many of their features are illustrated in Figure 1, a map of CO emission from the Orion complex

C.J. Lada and N.D. Kylafis (eds.), The Origin of Stars and Planetary Systems, 67–96.
© 1999 *Kluwer Academic Publishers. Printed in the Netherlands.*

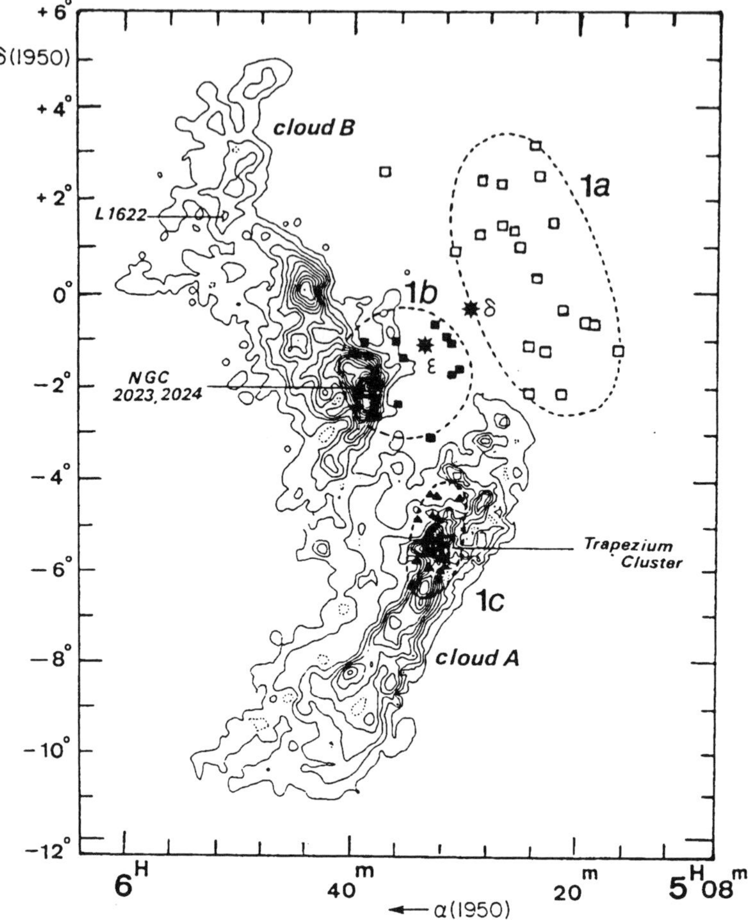

Figure 1. CO line emission from the Orion A and B giant molecular cloud complexes (Maddalena et al 1986), in relation to their associated young star clusters (Blaauw 1991). The overall extent of each complex is about 60 pc.

in relation to its associated star clusters (Blaauw 1991). As revealed by CO line maps, a typical molecular cloud complex has an irregular shape, with "blobs" of low aspect ratio (1-2) and "filaments" of high aspect ratio (5-10). The map as a whole is considered a "complex" because it contains more than one component with closed contours. These components are considered to be physically associated with each other because of their close proximity in projected position and in radial velocity. The individual map structures are called "clouds" or "clumps," and their star forming condensations are known as "cores" or "dense cores."

The overall extent of CO maps of nearby molecular cloud complexes is typically 10-30 pc, and their enclosed mass is 10^{4-6} $M_\odot$. The largest regions, called "giant molecular clouds" or "GMCs," have mass exceeding 10^6 $M_\odot$, and are the birthplaces of massive stars and rich star clusters. More detailed descriptions of GMCs and of massive star formation are given in the chapters by Blitz and Churchwell, respectively. The Orion complex is the nearest GMC, while the smaller nearby complexes, such as in Taurus and Ophiuchus, are often called "dark clouds" because of their obscuration of background starlight.

The density structure of molecular clouds is highly nonuniform, and its inference from observation requires caution. The mean density depends strongly on the size scale over which the average is taken, and each probe used to map a cloud is sensitive only to a range of density. Thus the mean density over a complex is an average over $\sim$10 pc, or roughly 10^2 cm^{-3}. This density is well sampled by observations of the J=1-0 line of CO. The mean density over a "clump" is an average over $\sim$1 pc, or 10^3 cm^{-3}; this density is traced by the J=1-0 line of ^{13}CO, an isotopomer of ^{12}CO. The mean density over a dense core is an average over $\sim$0.1 pc, or 10^4 cm^{-3}, traced by the (J, K) = (1,1) line of NH_3. This diversity of density implies that no single observable tracer can completely represent the structure of a molecular cloud, regardless of the spatial resolution used: observations of the 1-0 line of CO do not reveal dense cores, and observations of the (1,1) line of NH_3 do not reveal cloud complexes.

2.1. MOLECULAR CLOUD CONSTITUENTS

The main constitutents of molecular clouds are molecules, dust grains, ions, and stars. Here we briefly describe the role each plays in molecular cloud physics.

More than 100 species of interstellar molecules are known, and more than 1000 spectral lines are known at centimeter, millimeter, and submillimeter wavelengths. The simplest molecules are diatomic species such as OH, CH, CO, CS, and their isotopomers. The more complex molecules are carbon chains and rings, and indeed most interstellar molecules are organic species, some of which were first discovered in space and then confirmed by laboratory spectroscopy. The largest interstellar molecule known, as of October 1998, is the carbon chain $HC_{11}N$ (Bell et al 1997), and still longer chains, such as $HC_{17}N$, have been found in the laboratory with sensitive spectroscopic techniques (McCarthy et al 1998). A review of interstellar molecules, and their rich chemistry, is in the chapter by van Dishoeck.

Figure 2 illustrates the variety of molecular lines, observed at a wavelength 0.8 mm in the richest source of bright lines, the Kleinmann-Low nebula in the Orion molecular cloud (Schilke et al 1997). Note that many

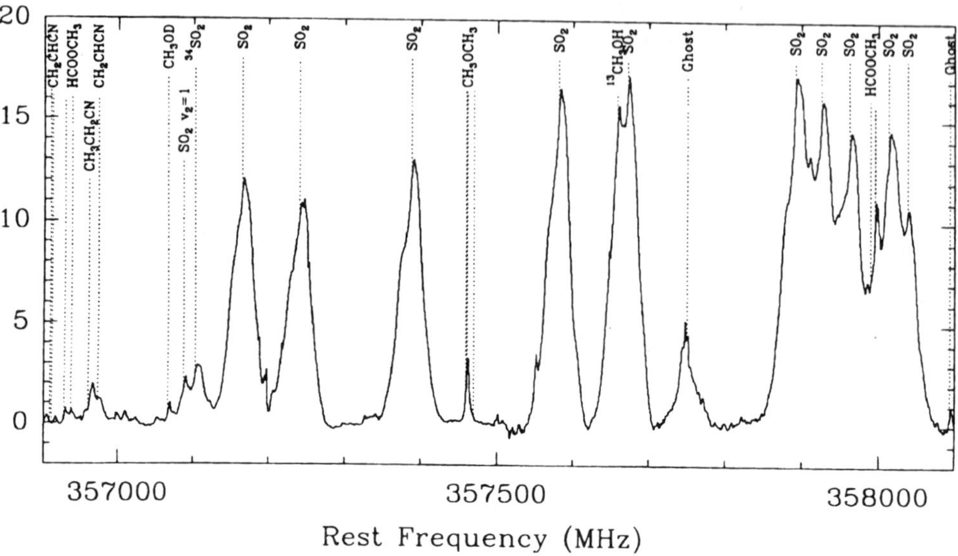

Figure 2. Molecular line emission from the Kleinmann-Low Nebula in Orion, in a spectral window about 1 GHz wide, centered at freqency 357.5 GHz or wavelength 0.84 μm (Schilke et al 1997). Note the high density of lines, and their origin in many complex molecular species.

lines are due to complex organic species, and in some portions of the spectrum there is more flux in lines than between the lines.

The molecular species CO, H_2O, and their isotopomers are especially important because their rotational lines are the main coolants for nearly all of the gas in molecular clouds. This cooling radiation balances the heating, due mainly to cosmic rays and external UV photons, giving an equilibrium temperature of about 10 K. Without these cooling species, molecular clouds would be substantially warmer, approaching the temperatures of diffuse clouds, $\sim$100 K.

Dust grains are solid particles condensed in the atmospheres of giant stars and blown into the interstellar medium by their stellar winds. Dust grains are believed to consist mostly of graphitic carbon and silicates, probably with amorphous, as opposed to crystalline, internal structure. Grains have a typical size of 0.1 μm, and a broad distribution of sizes extending down toward the regime of large molecules. In molecular clouds grains are believed to become the main coolant as the gas density exceeds $\sim$10^5 cm^{-3}. Grain surfaces are thought to catalyze the formation of H_2 from H. In dense cores grains are expected to have "mantles," layers of molecular material which are deposited as molecules which are sufficiently polar collide with the grain, and bond to the surface. When such grains are heated by radiation from a recently formed star, many of their complex mantle molecules evaporate into the "hot core" gas. Recent reviews of grain properties are in Roberge & Whittet (1996).

Molecular clouds are not entirely neutral, but instead have a low level of ionization, with electron fraction $x_e < 10^{-5}$, due to a combination of cosmic ray ionization and photoionization by ultraviolet photons from distant OB stars. If an OB star forms in the molecular cloud, its UV photons fully ionize the gas in a small portion of the cloud (an "H II region"), while most of the surrounding molecular cloud remains largely neutral. Molecular cloud ionization is important both chemically and physically. Most of the gas-phase chemical reactions in a molecular cloud are driven by ionized species, whose high reaction cross sections make them influential despite their low abundance. Also, the trace ionization in molecular clouds is sufficient to couple, by ion neutral collisions, the motions of the magnetic field to those of the neutral gas. This mechanism allows propagation of magnetohydrodynamic (MHD) waves and turbulence, whereby the magnetic field and the neutral gas move together as a single fluid. These MHD motions are thought to help support molecular clouds against their self-gravity.

Stars are the most interesting and important constituent of molecular clouds. Molecular clouds are believed to be the formation site of all known young stars, and our understanding of how stars form is based on the principle that molecular clouds provide the "intitial conditions" for the process of star birth. Molecular clouds host young stellar objects (YSOs) in a wide range of evolutionary states, from "Class 0 protostars" some 0.01 Myr old, deriving most of their luminosity from gravitational infall, to "T Tauri" stars a few Myr old, deriving their luminosity from quasi-static contraction. Molecular clouds also host stars in a wide range of spatial groupings - from isolated single stars as in Taurus, having no known neighbors within a few pc, to rich star clusters as in Orion, having more than 1000 stars within a few pc. The masses of stars in molecular clouds range from about 0.1 to 30 $M_\odot$, nearly the whole range of known stellar masses. Indeed the initial mass function (IMF), or the distribution of stellar masses at birth, is indistinguishable between stars in molecular clouds and field stars. Reviews of low-mass and massive star formation, from observational and theoretical viewpoints, are given in the chapters by Lada, Churchwell, Shu, and Bonnell.

Embedded stars contribute a major part of the turbulent kinetic energy in molecular clouds, through their winds and outflows. The winds, outflows and radiation from the most massive stars probably limit the lifetime of molecular clouds by unbinding and dispersing their gas.

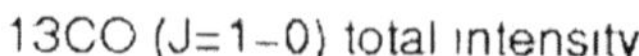

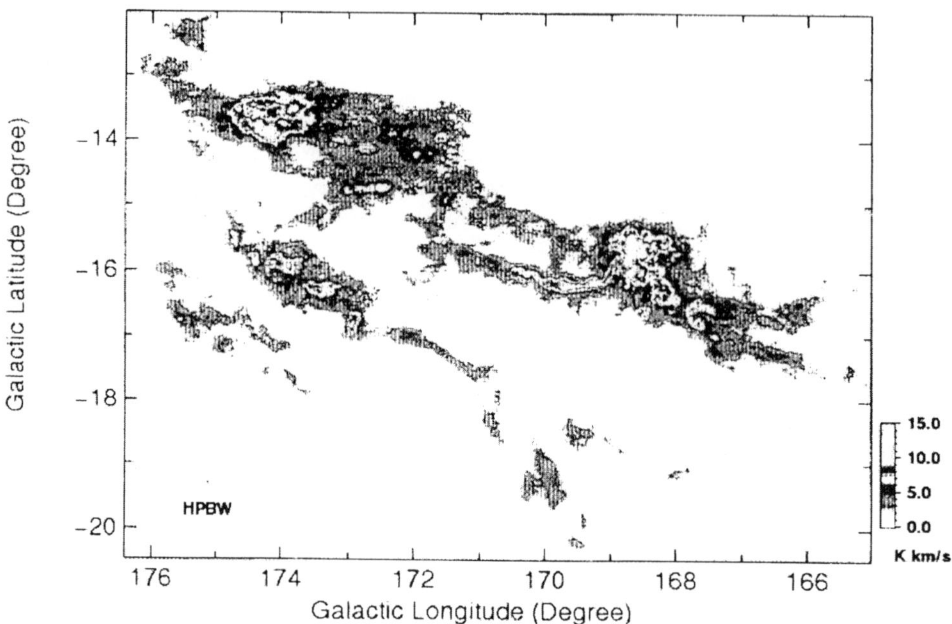

Figure 3. The Taurus molecular cloud complex, in the 2.7 mm wavelength J=1-0 line of ^{13}CO (Mizuno et al 1995). The longest dimension is about 25 pc, substantially less than in Orion (Figure 1). Note the highly elongated, clumpy structure.

2.2. CLOUD DIVERSITY

Molecular clouds differ in numerous ways, and many of their properties vary together with increasing size from region to region (Myers 1991, Harju, Walmsley & Wouterloot 1993, Jijina, Myers & Adams 1998). The smallest molecular cloud complexes, as in Chamaeleon III, have length $\sim$10 pc, maximum visual extinction of a few magnitudes, turbulent line widths of 1 km s^{-1}, temperature 10 K, relatively few stars, no massive stars, and no clusters. The largest complexes, as in Orion, have extent $\sim$50 pc, visual extinction of $\sim$100 magnitudes, turbulent line widths of several km s^{-1}, gas temperature typically 20 K, and thousands of stars in dense clusters, including numerous OB stars. The more massive molecular clouds are responsible for forming most stars, even though the lower-mass clouds are more numerous. Figure 3 shows a map of the Taurus complex in the J=1-0 line of ^{13}CO (Mizuno et al 1995). In relation to other nearby complexes, Taurus is intermediate in its star-forming activity (Chen et al 1995, 1997).

2.3. DENSITY PROBES

Among the hundreds of molecular lines detected in molecular clouds, fewer than ten have been used extensively, because they are relatively bright, because their frequencies are suitable for observation with many telescopes, and because they probe characteristic regimes of density. A spectral line "traces" a particular range of density of hydrogen molecules, which collide with the emitting species and help excite the observed transition. For a simple two-level system in statistical equilibrium, the characteristic density needed for excitation is the "critical density" n_{cr}, at which the rate of spontaneous emission equals the rate of collisional excitation:

$$n_{cr} = \frac{64\pi^4 \mid \mu \mid^2 \nu^3}{3hc^3 < \sigma v >} \tag{1}$$

Here $\mid \mu \mid$ is the transition dipole moment, ν is the transition frequency, h is Planck's constant, c is the speed of light, and $< \sigma v >$ is the collision rate coefficient (Townes & Schawlow 1975; Rohlfs & Wilson 1996). Some representative values of n_{cr} are 10^3 cm^{-3} for the J=1-0 line of CO at 2.6 mm wavelength, 10^4 cm^{-3} for the (J,K) = (1,1) line of NH$_3$ at 13 mm wavelength, and 10^5 cm^{-3} for the J=2-1 line of CS at 3.0 mm (Ungerechts et al 1997).

If the transition optical depth τ is substantially greater than 1, the critical density in eq. (1) must be modified to account for radiative trapping. In this optically thick case, a molecule has substantial probability of absorbing a transition photon emitted by another molecule in the cloud. Then excitation can occur by both absorption and collisions, and the density needed to achieve a given excitation temperature is reduced. For a plane-parallel cloud and for the "large velocity gradient" approximation to the radiative transfer, n_{cr} in eq. (1) is multiplied by the "trapping factor"

$$\beta = \frac{1 - e^{-3\tau}}{3\tau} \tag{2}$$

(e.g. Rohlfs & Wilson 1996). For the J=1-0 transitions of the isotopomers CO, ^{13}CO, and C^{18}O, the typical critical densities are smaller than in eq. (1) by a factor β^{-1} or 300, 9, and 1.5, respectively. This behavior allows lines of isotopomers to probe different density regimes of the same cloud with essentially the same angular and spectral resolution.

The critical density in eq. (1) is the density at which an optically thin transition has substantial excitation. The degree of excitation is described by the excitation temperature T_{ex}, defined in analogy with the Boltzmann distribution by

$$\frac{n_u}{n_l} = \frac{g_u}{g_l}\exp\left(-\frac{h\nu}{kT_{ex}}\right) \qquad (3)$$

Here n_u and n_l are the densities of emitting molecules in the transition upper and lower states, respectively, and g_u and g_l are the corresponding statistical weights. To illustrate the role of the critical density, consider the case where the radiation temperature is lower than the kinetic temperature, as shown in Figure 4. This situation arises, for example, when the radiation temperature, $T_R = 2.7$ K, is that of the cosmic microwave background, and the kinetic temperature, $T_k = 10$ K, is due to cooling by CO lines and heating by cosmic rays and UV photons (Goldsmith & Langer 1978). As the gas density decreases below the critical value in eq. (1), the excitation is "subthermal," and the excitation temperature T_{ex} approaches T_R. When the density equals the critical value, T_{ex} is about halfway between T_R and T_k. When the density increases above n_{cr}, the excitation becomes "thermal" and T_{ex} approaches T_k.

The excitation of a transition above its background level is substantial for all densities $n \geq n_{cr}$, but the effective range of density traced by a transition is typically a factor of a few, because the distribution of gas density in molecular clouds decreases rapidly with increasing density. For example, a spherically symmetric cloud having density decreasing with increasing radius as $n \sim r^{-1}$ has differential mass distribution $dm/dn \sim n^{-3}$. Thus observation of a cloud in a particular tracer images the cloud in a limited range of density around n_{cr} – limited at lower densities by the sharp increase of excitation with increasing density, and limited at higher densities by the sharp decrease of cloud mass with increasing density.

3. Dense Cores

We define a "dense core" as a molecular cloud condensation with mean gas density 10^4 cm^{-3}. It is useful to specify 10^4 cm^{-3} as a fiducial density, for several reasons. First, several easily observable molecular lines have effective critical density of order 10^4 cm^{-3}, including the $(J, K) = (1,1)$ line of NH$_3$ at 1.3 cm and the J=2-1 line of CS at 3.0 mm. These probes have provided a wealth of information about the gas in molecular clouds with density within a factor of a few of 10^4 cm^{-3} (e.g., Linke & Goldsmith 1980; Ho & Townes 1983; Jijina, Myers & Adams 1998). Second, the mass of gas with density 10^4 cm^{-3} in Taurus and other nearby complexes is generally of order 1 M$_\odot$ (Myers & Benson 1983, Benson & Myers 1989). These dense cores have mass comparable to that of the typical low-mass star, and such cores and stars are closely associated, according to observations at far-infrared (Beichman et al 1986) and near-infrared (Myers et al 1987) wavelengths. This association

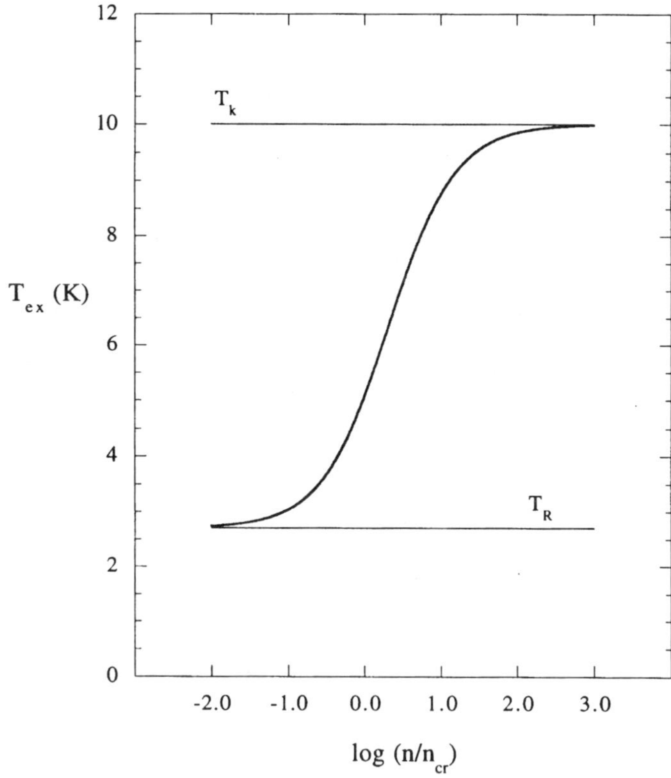

Figure 4.   Increase of excitation temperature T_{ex} of a spectral line transition with increasing density n of collision partners. As n increases, T_{ex} increases from its minimum at the radiation temperature T_R to its maximum at the kinetic temperature T_k, reaching appreciable excitation at the critical density n_{cr}.

indicates a close link between the physical properties of the core gas and its evolution to form a star, and provides observational support for models of isolated, low-mass star formation (Shu, Adams & Lizano 1987). Third, maps in lines of CS and NH_3 in regions of more massive star formation tend to be associated with groups and clusters of young stars, suggesting that properties of such cores are related to the formation of groups and clusters (Lada, Bally & Stark 1991; Harju, Walmsley & Wouterlout 1993).

In this section we describe dense core properties as indicated by observations of lines of NH_3, CS, N_2H^+, and HCO^+, which trace densities $\geq 10^4$ cm^{-3}, but not by lines of CO, ^{13}CO, or $C^{18}O$, which trace lower density. First we describe general properties which are independent of the degree of turbulence evident in the line width, and then those which vary with core turbulence.

3.1. STAR FORMATION

In nearby complexes, cores selected by their optical obscuration, without knowledge of their stellar association, have been shown by infrared surveys to have a significant number of associated YSOs (Benson, Myers & Wright 1984, Beichman et al 1986, Myers et al 1987). In Taurus and similar complexes, about half of the 20 cores which have been mapped in NH_3 lines have associated YSOs (Jijina, Myers & Adams 1998, hereafter JMA). Conversely, YSOs with IRAS emission which is sufficiently bright and red, indicating a high column density of cold circumstellar dust, have a high incidence of NH_3 line emission (Harju, Walmsley, & Wouterloot 1991). The association of cores and YSOs is more frequent for redder YSO spectra, supporting the idea that cores are the birth environment of YSOs, and that more evolved YSOs lose their association with their parent cores, probably by a combination of dispersal through outflows (Tafalla & Myers 1997) and relative motion of the core and YSO.

3.2. ELONGATION

Core maps show a significant elongation, with typical aspect ratio of 2, and a clear departure from the circular shape expected for a map of a spherically symmetric system (Benson & Myers 1989, Lada, Bally & Stark 1991, Wood, Myers, & Daugherty 1994). Elongated cores are seen in quiescent, low-mass complexes such as Taurus (Benson & Myers 1989), in intermediate complexes such as Perseus (Ladd, Myers & Goodman 1994), and in massive, turbulent complexes such as Orion (Harju, Walmsley & Wouterlout 1993, Cesaroni & Wilson 1994). Statistics of core map elongation have been interpreted as indicating that most cores have 3-dimensional shape which is prolate, rather than oblate (Myers et al 1991, Ryden 1996). The location of elongated cores in more extended filamentary structures, indicated by their visual extinction and by maps of lower-density tracers, strongly favors the prolate over the oblate interpretation (Myers et al 1991). A detailed discussion of the structure of filamentary clouds is given in Lada (1999). On the other hand, models of core evolution in the context of a relatively strong magnetic field predict oblate structure, with flattening along the direction of the field, and with projected aspect ratio similar to that observed (Lizano & Shu 1989, Basu & Mouschovias 1994, Li & Shu 1996).

Resolution of this apparent discrepancy may require a better understanding of how cores form, and more detailed observations of magnetic field directions in filamentary clouds. For example cores which form from filamentary clouds may have elongated maps because they are prolate, if the magnetic field lies along the filament, or because they are triaxial, if the magnetic field lies perpendicular to the filament. Alternatively, elongated cores may arise if the magnetic field has a helical structure wound around the filamentary parent cloud (Feige & Pudritz 1998).

3.3. VIRIAL BALANCE

Cores have line widths, column densities, and map sizes which appear close to the condition of virial balance, in which the gravitational potential energy and kinetic energy are approximately equal (Myers et al 1991, Harju, Walmsley & Wouterloot 1993). For an isolated uniform sphere of mass M, radius R, and velocity dispersion σ, this condition is

$$\frac{GM}{5R} = \sigma^2 \tag{4}$$

(Spitzer 1978). Figure 5 shows a comparison of dense core data with several virial equilibrium models, indicating that the typical core agrees with most models within the scatter of data, provided that both thermal and nonthermal motions are included. If as usual for molecular clouds virial balance is understood as a statement of approximate hydrostatic equilibrium, then the supporting pressure is $mn\sigma^2$, where σ includes both thermal and nonthermal motions.

The nonthermal motions evident in core line widths span a wide range, from subsonic by a factor ~ 4 in small quiescent cores such as L1512 (Fuller & Myers 1993) to supersonic by a factor > 10 in large turbulent cores associated with compact H II regions (Ho & Townes 1983). The ratio of nonthermal to thermal motions varies continuously from core to core, but it is convenient to divide cores into "thermal" and "turbulent" regimes, separated by the critical case where the velocity dispersion of the nonthermal motions equals the velocity dispersion of the thermal motions of the molecule of mean mass (Myers, Ladd & Fuller 1991). Here we denote nonthermal motions as "turbulent," without ascribing any particular physical model of turbulence to them. The critical case is easily derived from observations of the (J, K) = (1,1) and (2,2) lines of NH_3 because these observations yield accurate estimates of both the kinetic temperature and the observed line width (Ho & Townes 1983). Denoting the FWHM of the observed line profile, corrected for optical depth and resolution effects, as Δv_{obs}, we may write

$$\Delta v_{obs}^2 = \Delta v_{turb}^2 + 8\ln 2 kT/m_{obs} \tag{5}$$

where Δv_{turb}, the FWHM of the turbulent motions, is independent of mass, and m_{obs} indicates the mass of the observed species. In the critical case $\Delta v_{turb}^2 = 8(ln2)kT/m_{obs}$, where m is the mean molecular mass, or $2.3m_H$,

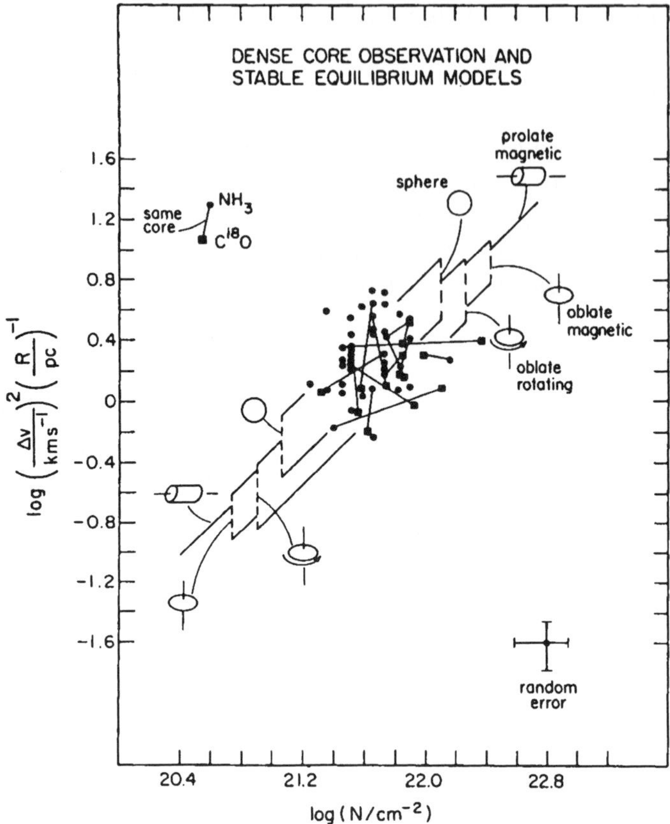

Figure 5. Comparison of dense core line widths, map sizes, and column densities with models of virial equilibrium, based on observations of lines of NH₃ and C¹⁸O (Myers et al 1991). The data indicate consistency with most models of core virial equilibrium, although they do not clearly favor any particular model.

assuming one He atom for every five H_2 molecules. Combining this relation with eq. (5) gives the critical line width

$$\Delta v_{obs,cr} = [\frac{8\ln 2kT}{m}(1 + \frac{m}{m_{obs}})]^{1/2} \tag{6}$$

For observations of 10 K gas in a line of NH_3, for which $m_{obs} = 17m_H$, the critical line width is 0.47 km s^{-1}.

In the less massive cores, observed NH_3 line widths are less than $\Delta v_{obs,cr}$, so they are thermal, and models of their support against self-gravity can safely ignore their turbulent motions, as in the standard model of isolated star formation (Shu, Adams & Lizano 1987). In the more massive molecular

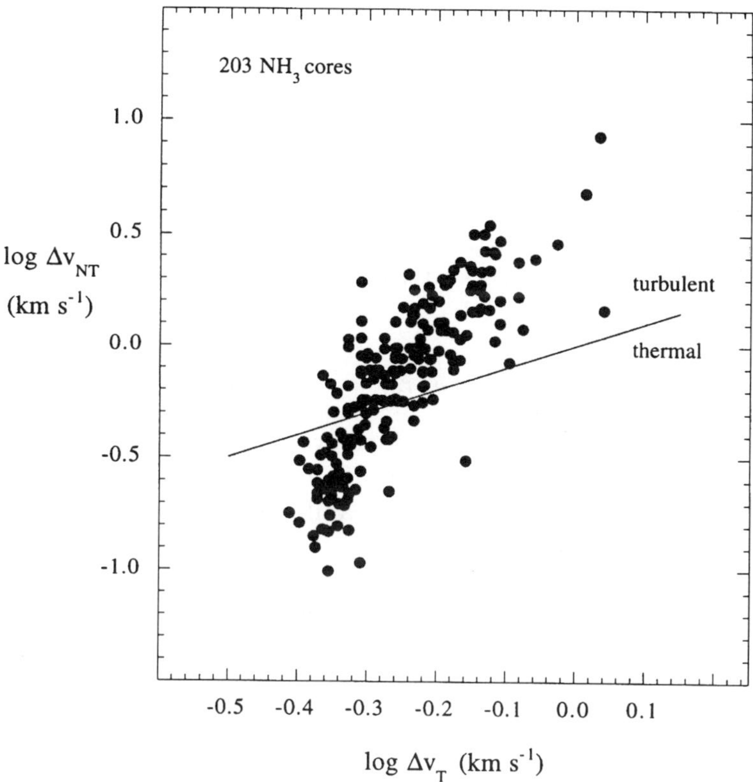

Figure 6.   Nonthermal and thermal line widths in NH₃ dense cores. The FWHM of the distribution of nonthermal motions, Δv_{NT}, increases rapidly with the corresponding FWHM of the distribution of thermal motions of the molecule of mean mass, Δv_T. Thus turbulent cores tend to be warm, and quiescent cores tend to be cool (JMA).

cloud cores, observed NH₃ line widths generally exceed $\Delta v_{obs,cr}$, so their turbulent motions exceed their thermal motions. Since these cores approximately satisfy virial balance, eq. (4), their turbulent motions play a key role in supporting them against their self-gravity. Understanding the nature of the turbulent motions remains a major problem in molecular cloud physics.

Figure 6, adapted from JMA, shows that the nonthermal and thermal components of the line width observed in NH₃ cores tend to correlate, so that cores which exceed their critical line width tend to be warmer than cores whose line widths are subcritical (JMA).

3.4. IONIZATION AND FIELD-NEUTRAL COUPLING

The low level of ionization in cores is important because it determines how well the core magnetic field can couple to the neutral gas through ion neutral collisions. The ions are essentially tied to magnetic field lines because the period of ion gyration around the field is much shorter than neutral-ion collision times. If the field fluctuates with a period shorter than the time for a neutral to be hit by an ion, this motion cannot be transmitted to the neutrals, so the field and neutral gas cannot move together as a single-fluid system. This decoupling occurs at the size scale of the MHD cutoff wavelength

$$\lambda_0 = \pi v_A \tau_{ni} \tag{7}$$

where v_A is the Alfven speed in the combined system of ions and neutrals, and τ_{ni} is the time for a neutral to be hit by an ion (Kulsrud & Pearce 1969). Such decoupling can limit the cloud's ability to adjust its structure on small scales.

The ionization structure of a molecular cloud is expected to be dominated by ultraviolet (UV) photons from external OB stars in its outer parts, where the visual extinction to the cloud surface is about 4 mag (McKee 1989, Bertoldi & McKee 1992). At higher extinctions, the ionizing flux of UV photons is attenuated below the level of cosmic-ray particles, so cosmic ray ionization dominates. The dominant ion is expected to be HCO^+, arising from ionization of CO.

Attempts to measure the ionization in dense cores have relied on line observations of HCO^+, $H^{13}CO^+$, DCO^+, and $C^{18}O$, and on models of cloud ion-neutral chemical reactions, to estimate the electron fraction x_e (Guelin, Langer & Wilson 1982, Wootten, Loren & Snell 1982). These early studies in a few low-mass cores, led to the estimate that $x_e \approx 10^{-7}$ in cores with density $n \approx 10^4$ cm^{-3}, in close agreement with estimates based on cosmic-ray ionization. More recent studies have cosmic-ray observed several tens of cores, both isolated low-mass cores (Williams et al 1998, Caselli et al 1998) and more massive cores (Bergin et al 1999), and have employed more detailed chemical models. These more recent studies have confirmed the earlier conclusion that core ionization is consistent with simple cosmic-ray ionization, with ionization rates of order 5×10^{-17} s^{-1}.

Figure 7 shows results from the ionization study of 20 low-mass cores by Williams et al (1998), indicating no significant difference in ionization between cores with and without embedded low-luminosity stars.

With these results it follows that the field-neutral coupling in dense cores of radius r is sufficient to allow propagation of MHD waves above their cutoff wavelength. The strength of the coupling is quantified by a

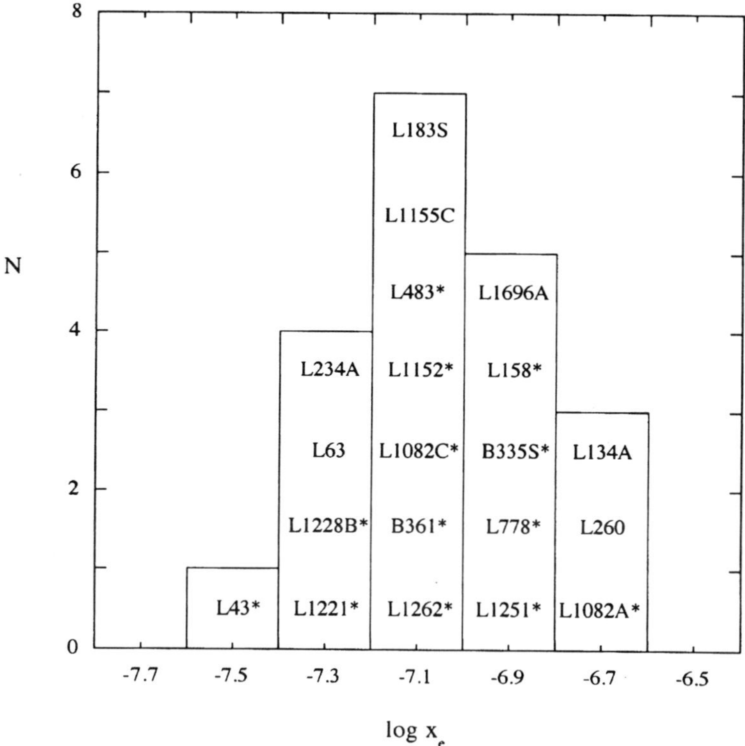

Figure 7.   Distribution of electron fraction x_e in 20 low-mass dense cores, based on line observations and chemical models (Williams et al 1998a). Asterisks denote cores with embedded low-mass stars. The distribution peaks at log $x_e \approx$ -7, as expected for ionization dominated by cosmic rays, and shows no significant difference between cores with and without embedded stars.

factor $W \equiv r/\lambda_0$ (Myers & Khersonsky 1995; see also Shu 1983). The foregoing ionization estimates indicate that in the typical dense cores $W \approx 10$. Thus dense cores are not likely to have substantial field-neutral decoupling. However their coupling is significantly weaker than in their surrounding UV-ionized gas, and so they can attenuate externally incident MHD waves (Nakano 1998, Myers & Lazarian 1998).

3.5. LINE WIDTH - SIZE RELATIONS

When cores and their surrounding gas are mapped in tracers such as the lines of NH_3, $C^{18}O$, and ^{13}CO, which are sensitive to a wide range in density, they show a pronounced increase of line width Δv with map size R (Ho et al 1977, Snell 1981, Larson 1981, Fuller & Myers 1992, Caselli & Myers 1995, Barranco & Goodman 1998). This relation is represented as a power law, $\Delta v \sim R^p$, usually with $0.3 < p < 0.7$. This trend is a property of cloud turbulence, since most of the variation in Δv is in its nonthermal, rather than its thermal part. Line width-size relations for many clouds are consistent with hydrostatic equilibrium and with a "logatropic" density structure, in which the density decreases with radius r as r^{-1} (McLaughlin & Pudritz 1996). Many versions of line width-size relation have been described, depending on the number of tracers used and the number of clouds in the sample (Goodman et al 1998a), but in each case the dense core is a local minimum of nonthermal motions, compared to its surrounding gas. A closely related property is the tendency for a low-mass core to be more "coherent" than its surroundings, i. e. for the line widths to show greater uniformity within the core than outside it (Goodman et al 1998b).

In addition to the general properties described above, many aspects of dense cores vary strongly with their level of turbulent motions. For convenience we contast the "thermal" and "turbulent" cores described above as two distinct groups, each represented by cores in Taurus and Orion, respectively.

3.6. ASSOCIATED STARS

Thermal cores tend to have fewer associated stars than do turbulent cores. In Taurus, all of the 20 known NH_3 cores are thermal; about half of these have no associated YSO, and those which have associated YSOs generally have only one. In Orion A and B, nearly all of the 80 known cores are turbulent, only about 10 % are starless, and many are associated with groups and clusters of at least 5 YSOs. This difference in turbulence between Taurus and Orion cores is not simply a consequence of the associated stars and their winds and outflows. Even starless cores in Orion have more turbulence than their Taurus counterparts, and their turbulent line widths are not significantly different from their neighboring cores in Orion having associated stars (JMA). These differences suggest that turbulence is part of the core initial conditions for star formation.

3.7. CORE SIZE

Thermal cores tend to be smaller than turbulent cores by a typical factor of 2. The FWHM diameter of thermal NH_3 cores in Taurus is typically 0.05 pc while the size of the turbulent cores is typically 0.1 pc in Orion (JMA). This result is significant because it means that he volume of gas denser than 10^4 cm^{-3} is about 8 times greater for Orion cores than for Taurus cores. Thus Orion cores have significantly greater mass available to make groups and clusters than do Taurus cores.

3.8. TEMPERATURE

Thermal cores are cooler than turbulent cores: in Taurus the typical temperature is 10 K (Benson & Myers 1989) while in Orion it is more nearly 20 K (Harju, Walmsley & Wouterloot 1993, Cesaroni & Wilson 1996, JMA). This difference in temperature cannot be attributed solely to a greater incidence of stars embedded in the Orion cores, since even starless cores in Orion tend to be substantially warmer than in Taurus. Instead, Orion cores are probably warmer than Taurus cores because they are heated by the much greater number of stars which lie in the Orion complex but outside the cores themselves (JMA).

3.9. EXTERNAL EXCITATION OF CORE TURBULENCE

The similarity of line widths in cores with and without embedded YSOs, and the sharp increase in line width from starless cores not near clusters to those near clusters, together suggest that turbulent cores have substantial external excitation of their turbulence. The environments of Taurus and Orion cores differ in the much greater number of cluster stars outside Orion cores than around Taurus cores, suggesting that the high incidence of winds and outflows from stars around Orion cores helps to maintain the turbulence of Orion cores at a much greater level than is possible for Taurus cores (JMA). For a detailed review of molecular outflows and their environmental effects, see Bachiller (1996) and the chapter in this volumne by Bachiller. A schematic view of external excitation of core turbulence by winds and outflows is shown in Figure 8.

4. Cluster-Forming Cores

The discussion of turbulent cores in Section 3 indicates that star formation in groups and clusters is a major mode of star formation, which may make significantly more stars than the isolated mode. More detailed discussion of cluster observations are in the chapter by E. Lada, and in Lada & Lada (1995). Despite its importance, the clustered mode has been the subject of much less theoretical analysis than the isolated mode, which is well-described by the "standard model" (Shu, Adams & Lizano 1987). In this section we describe recent discussions of clustered star formation.

4.1. MODELS OF CLUSTERED STAR FORMATION

A major goal of clustered star formation models is to account for the characteristic mass ($\sim$1 $M_\odot$) and spacing ($\sim$0.05 - 0.1 pc) of stars observed in young clusters, consistent with the dense core conditions where young

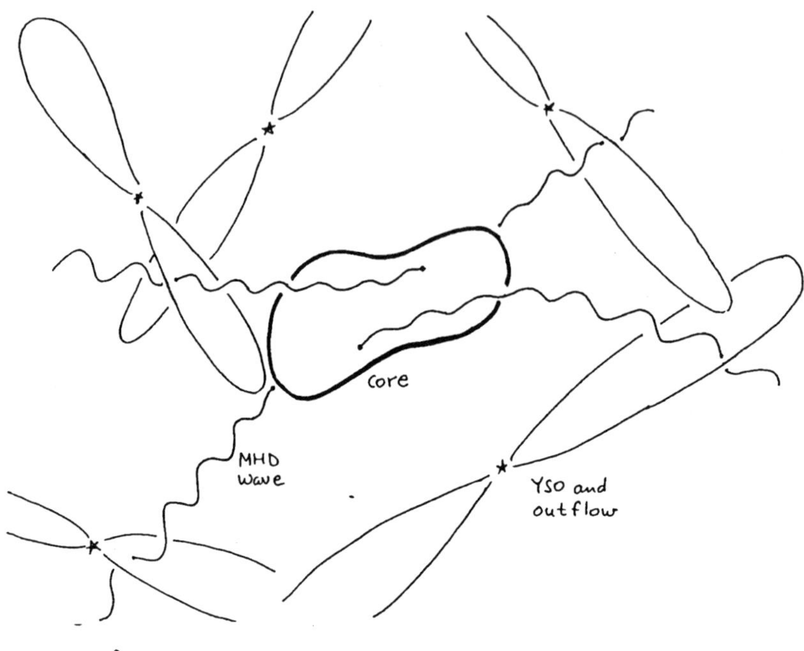

Figure 8.   A schematic illustration of how stars outside a starless core may contribute to core turbulence, through excitation of MHD waves by stellar winds and outflows.

clusters are found. A second goal is to account for the number distribution of massive stars in clusters, a power-law decrease of the number of stars N with increasing stellar mass M, $dN/dM \sim M^{-p}$, with $p \approx 2$. For a detailed review of this subject, see Larson (1995). Here we describe two easily distinguished pictures of clustered star formation.

The "orbiting seed fragments" model of Bonnell et al (1997) envisions small bound condensations which travel through the gas of the protocluster dense core, accreting mass as they execute many orbits. These fragments compete for mass, somewhat as planetesimals do in models of the solar nebula, resulting in a wide range of masses. The more massive members sink to the densest part of the gravitational potential well, where they accrete mass more rapidly and can collide with each other. This picture accounts for the distribution of stellar masses but not for the particular conditions of dense cores where clusters are found.

The "Jeans mass" models of Larson (1985) and Mouschovias (1991) predict the existence of protostellar condensations with the mass and size needed to match the typical mass and spacing of stars in clusters, but not having the number distribution with mass needed to match the IMF. In these models, self-gravitating condensations form in a medium of fixed

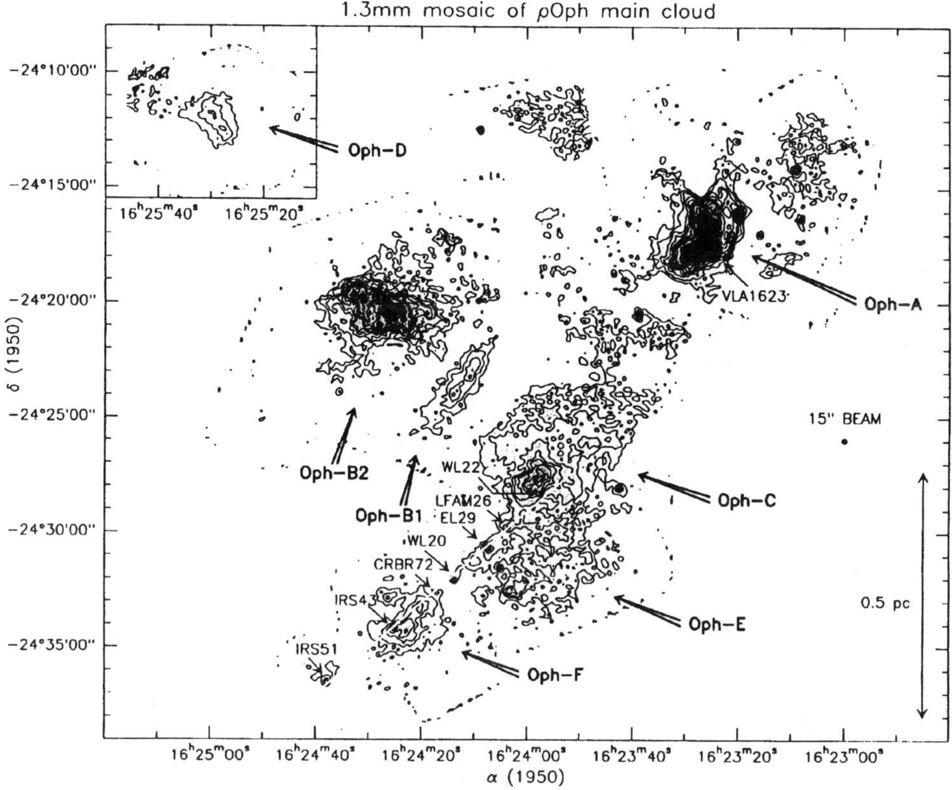

Figure 9.   Emission at 1.3 mm from the L1688 cloud in the Ophiuchus dark cloud complex, showing numerous small condensations associated with the embedded stellar cluster (Motte, Andre & Neri 1998).

external pressure, as in the Bonnor-Ebert problem (Bonnor 1957, Ebert 1956). In the nonmagnetic case the external pressure is specified by the observed temperature and density (Larson 1985). In the magnetic case the formation of the condensation is motivated by the requirement that the condensation be cut off from external MHD waves (Mouschovias 1991).

4.2. OBSERVATIONAL STUDIES OF YOUNG CLUSTERS

Recent high-resolution observations of young cluster-forming regions show numerous small structures having the spacing and mass distribution expected for cluster members. Motte, Andre & Neri (1998) imaged the Ophiuchus cluster in the 1.3 mm continuum with a bolometer array on the IRAM 30-m telescope, as shown in Figure 9.

Motte et al (1998) found ~50 sources, most of them pointlike, distributed among the young stars known in the cluster region, with about the same projected spacing as the stars. The mass distribution of these objects, although limited by small number statistics, has a slope more similar to that of stars in clusters than to that of interstellar clumps derived from observations of ^{13}CO line emission. A similar result was found in the Serpens cluster by Testi & Sargent (1998), using the OVRO interferometer at 3 mm wavelength. These continuum sources also have spacing consistent with young stars in dense clusters, and have a mass distribution similar to that of the IMF.

4.3. KERNEL MODEL

An extension of the cut-off Jeans mass model, in the context of a turbulent cluster-forming core, was proposed by Myers (1998). In this picture the external pressure is provided by the combination of thermal and MHD turbulent motions thought to account for the observed thermal and non-thermal line widths in cores. In that case a simple estimate indicates that a small condensation within the core - a "kernel" - can exist only when the ratio of Alfven and sound speeds in the surrounding dense core gas is greater than about 3, as illustrated in Figure 10, and when the column density of the core gas is greater than about 10 magnitudes. These conditions add specificity to the Jeans models because they allow precursors of cluster members to form only in the turbulent, high-extinction cores where they are observed, and not in the quiescent, low extinction cores where they are not observed.

A key property of quiescent kernels in turbulent cores is their line width contrast, as shown in Figure 11. A turbulent core with no kernel should have a substantially broader spectral line than would an otherwise identical core with an embedded kernel.

It is desirable to observe the condensations in cluster-forming regions with spectral lines which reveal the condensations' motions in the cloud via their line velocities, and their internal motions via their line widths. Observations of the core "Orion B9" (Caselli & Myers 1995) in the 1-0 line of N_2H^+ with the BIMA interferometer indicates a condensation with HM extent about 0.03 pc, and a line profile with FWHM 0.25 km s-1 − both consistent with expected kernel properties. Observations of the core Oph E mm2d of Motte, Andre & Neri (1998) in the same line with the 30-m telescope indicate a FWHM line width 0.3 km s^{-1} (Motte & Andre 1998). These encouraging observations indicate that it will soon be possible to thoroughly investigate the motions of small condensations in cluster regions, and to test various models of cluster formation.

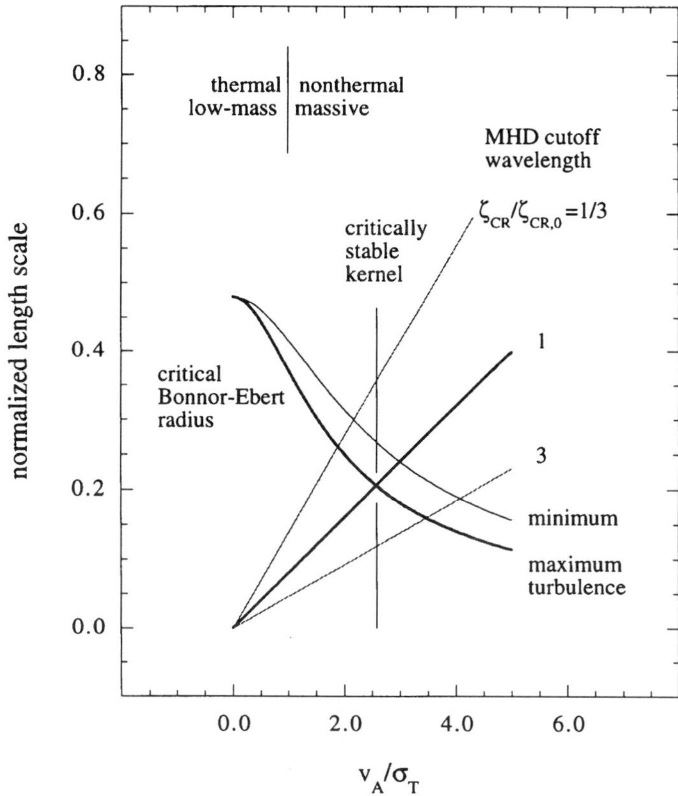

Figure 10. Kernel size scales as a function of Alfven Mach number in the surrounding dense core. The kernel is critically stable when the Bonnor-Ebert radius and MHD cutoff wavelength in the dense core are equal. The size scales, each normalized by the core Jeans length, cross when the Alfven Mach number is $\sim$3, in the regime of massive turbulent cores, as opposed to low-mass, thermal cores (Myers 1998).

5. Inward Motions in Dense Cores

The study of star formation has highly detailed observational knowledge of "initial conditions" and of "young stars" but much less understanding of the physical processes of condensation and collapse which link these two evolutionary states. In the last several years it has become possible to use spectroscopic observations to discern the degree of inward motion in dozens of dense cores with and without associated stars. In this section we briefly review the history, techniques, and recent observations in this area, with emphasis on the earliest phase - the starless cores. More detailed discussion of infall in cores with associated stars is given in the reviews by Myers,

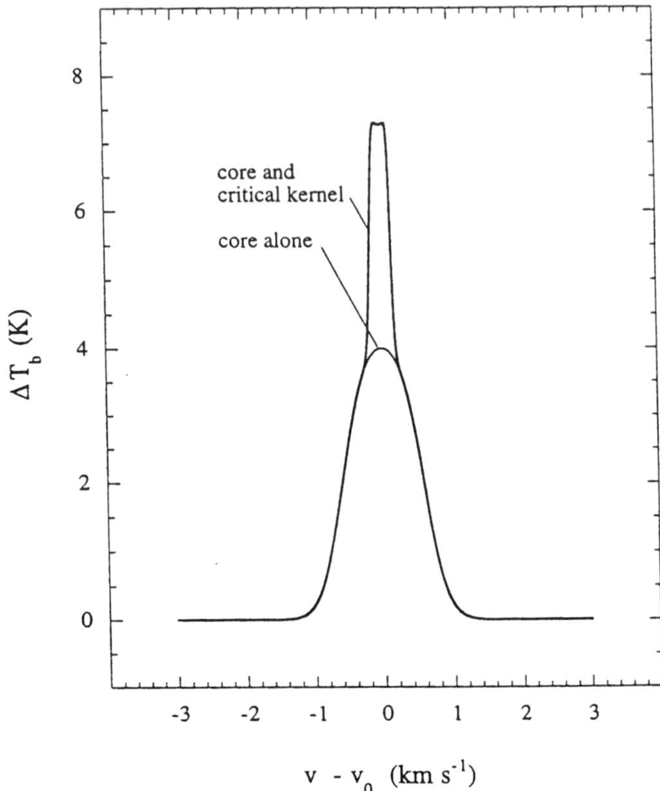

Figure 11.   Predicted profiles of emission in the N_2H^+ 1-0 line from a turbulent core with and without a critical kernel (Myers 1998).

Evans & Ohashi (1999) and by Evans (1999).

5.1. INFALL ASYMMETRY

The basic technique used to infer inward motions in millimeter-wavelength spectral lines is the observation of "infall asymmetry," the skewing of an optically thick spectral line to the blue of an otherwise similar optically thin line. This effect, first explained by Hummer & Rybicki (1968) in the context of lines from expanding H II regions, was used to analyze molecular cloud lines in terms of infall by Lucas (1976), Leung & Liszt (1976), Leung & Brown (1977), and more recently by Walker et al (1986), Adelson & Leung (1988), Zhou (1992), Zhou et al (1993), Walker, Narayanan & Boss (1994), Choi et al (1995), and Myers et al (1996), among others.

Infall asymmetry arises when a foreground region of lower excitation

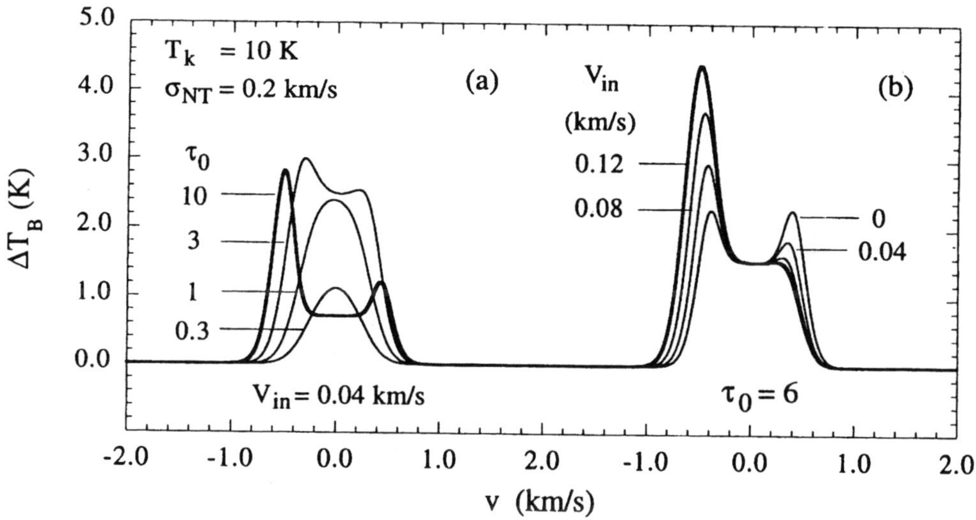

Figure 12. Predicted spectral line profiles from two approaching layers as (a) the peak optical depth increases at fixed approach speed, and (b) the approach speed increases at fixed optical depth. In each case the "infall asymmetry," or skewing of the line profile to the blue, becomes more pronounced (Myers et al 1996).

temperature absorbs photons from a background region of higher excitation temperature, and when the two regions have inward relative motion. The absorption occurs preferentially on the red side of the line profile, yielding a line shape with a peak skewed to the blue, or with a blue peak and a red shoulder, or with two peaks, and with the blue peak brighter than the red peak. The infall asymmetry is evident provided the excitation temperature gradient is sufficiently steep, the foreground optical depth is sufficiently great, and the characteristic infall speed is comparable to the velocity dispersion (Leung & Brown 1979).

The infall asymmetry becomes more pronounced as the optical depth increases, and as the infall speed increases, as illustrated in Figure 12 (Myers et al 1996).

Observation of infall asymmetry has a controversial history, in part because such observation can be taken as evidence of star-forming gravitational infall - a long-sought process of great physical importance. As described in the chapter by Evans, and by Myers, Evans & Ohashi (1998), claims of infall have been challenged or doubted because of confusion by outflows, by inconsistency of infall asymmetry from tracer to tracer, and by a mixture of infall and outflow asymmetry within a map, among other reasons. The climate of opinion began to change with the pioneering work of Zhou et al (1993), who reported infall asymmetry in the dense core B335 containing an extremely red "class 0 protostar" (Andre, Ward-Thompson & Barsony 1993), and who analyzed the observations of CS and H_2CO lines with radiative transfer models based on the "inside-out" collapse model of Shu (1977). Since then the case for infall in cores with embedded candidate protostars has strengthened, through the survey work of Gregersen et al (1997) and Mardones et al (1997).

5.2. INFALL ASYMMETRY IN STARLESS CORES

"Starless" cores are dense cores with no evidence of associated YSOs, based primarily on the IRAS survey (Neugebauer et al 1984) and secondarily on observations, if available, at optical and near- infrared wavelengths. For IRAS observations of the nearest star-forming complexes, lack of detection implies an upper limit on the luminosity of an associated embedded source of about 0.1 $L_{\odot}$ (Myers et al 1987). Starless cores are also called "pre-protostellar" (Ward-Thompson et al 1994) although it is not clear that every starless core will make a protostar.

The best-studied case of infall asymmetry in a starless core is in L1544, an opaque dark cloud on the eastern edge of the Taurus complex. This core has been mapped at high spectral resolution in lines of CS, H_2CO, C_3H_2, N_2H^+, and $C^{34}S$ (Tafalla et al 1998). It shows infall asymmetry in all of these lines except the optically thin $C^{34}S$ 2-1 line, and the spatial extent of the infall asymmetry appears to increase with the transition optical depth. For example the infall asymmetry extends over 0.02 pc in the N_2H^+ 1-0 line and over 0.15 pc in the CS 2-1 line. The remarkably large extent in CS 2-1 is much greater than the 0.04 pc extent of the HM intensity contour of the N_2H^+ 1-0 emission, often taken as a measure of the extent of the dense core. This pattern of infall asymmetry is shown in Figure 13.

On the smallest scale, the asymmetry in the N_2H^+ 1-0 line has been observed with 10 arcsec resolution using the BIMA interferometer, in combination with data from the IRAM 30-m telescope (Williams et al 1998b). The extent of the asymmetry agrees with the single-dish map, but the combined map shows more spatial detail - an elongated structure with a velocity gradient along the long axis, and with the strongest infall asymmetry at the position where the line is brightest. No continuum point source is evident.

Remarkably, analysis of the line profiles of CS on the large scale and N_2H^+ on the small scale shows about the same characteristic infall speed in each case - about 0.1 km s^{-1}. There is no obvious increase in infall speed with decreasing spatial scale, as would be expected for a strongly centrally condensed source.

The large extent of infall asymmetry in this starless core appears to rule out consistency with the inside-out collapse model of Shu (1977), since an

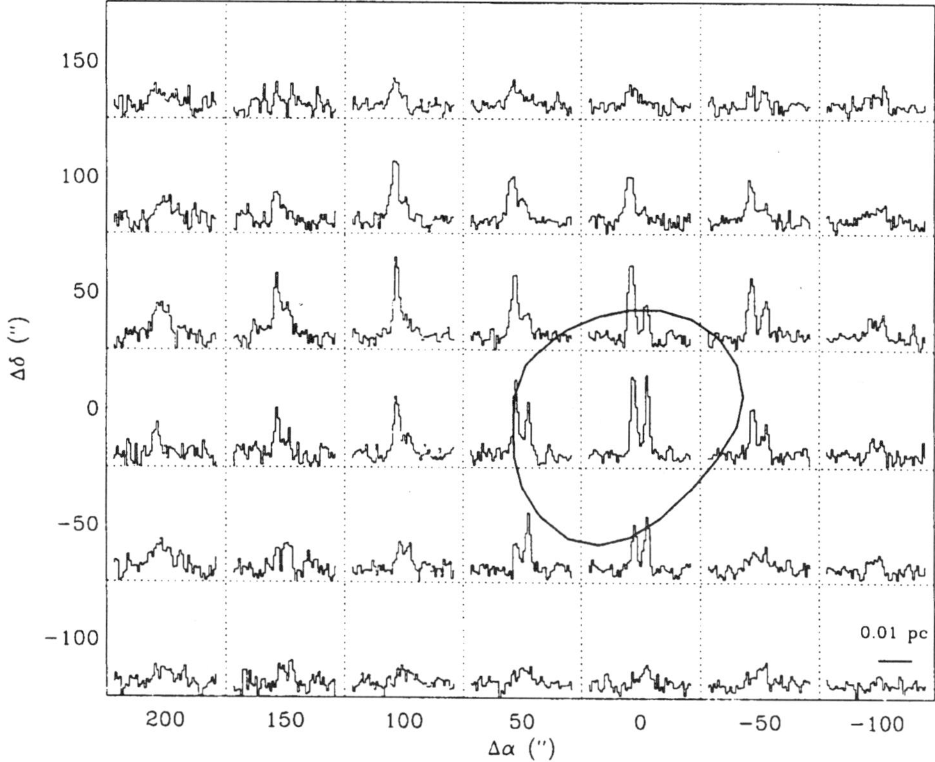

Figure 13.   Map of profiles of the CS 2-1 line from the starless core L1544 in relation to the HM contour of the 1-0 line of N_2H^+, a tracer of dense core emission. The CS profiles show infall asymmetry extended over $\sim$0.15 pc, much greater than the extent of the dense core (Tafalla et al 1998).

inside-out collapse of such large radius would have already formed a point source of several $L_\odot$, whereas no protostellar source is known, down to a luminosity limit of about 0.1 $L_\odot$. Similarly, the infall speed appears too fast to match most models of core formation by ambipolar diffusion, which predict infall speeds less than about 0.03 km s^{-1} at the radius and density of observation (e.g., Ciolek & Mouschovias 1995).

The number of starless cores with such spatially extended infall asymmetry is not yet known, but surveys indicate that tens of starless cores show infall asymmetry in at least one tracer (Lee, Myers & Tafalla 1998), and maps indicate asymmetry in CS lines, extended over at least 0.1 pc in L1498 and L429-1 (Lee, Plume & Myers 1998). Evidently spatially extended infall asymmetry in starless cores is not limited to L1544.

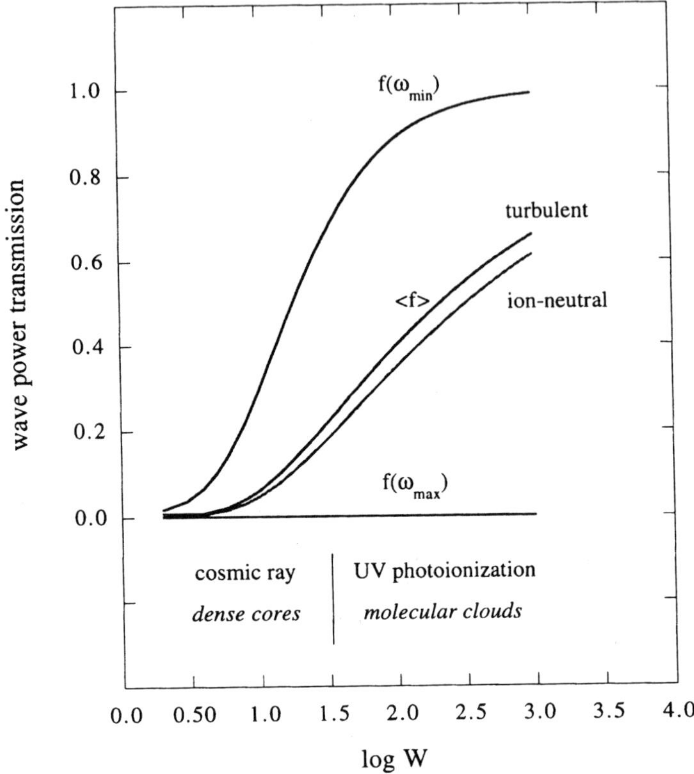

Figure 14. MHD wave power transmission in a region of radius r and MHD cutoff wavelength λ_0, whose ion-neutral friction is described by the coupling parameter $W \equiv r/\lambda_0$. As W decreases from the regime of photoionized molecular cloud to the cosmic-ray-ionized dense core, ion-neutral friction attenuates the MHD waves. This attenuation increases the pressure contrast between the outer and inner regions, driving an inward flow (Myers & Lazarian 1998)

5.3. TURBULENT COOLING FLOWS

In addition to the starless cores described above, some cores with stars also show similarly extended infall asymmetry, including L1251B, Serpens, and NGC 1333 (Mardones et al 1998). These examples suggest that inward motions may in some cases be driven by forces acting on larger scales than expected for gravitational infall. A possible pressure-driven inward flow has been suggested in the context of localized dissipation of turbulent motions (Myers & Khersonsky 1995, Myers & Lazarian 1998).

In this picture, a cooling instability arises as localized dissipation of turbulence decreases the local turbulent pressure, leading to an inward pressure gradient, followed by inward flow along field lines, increasing density, and further dissipation of turbulence. The resulting inward flow is called a "turbulent cooling flow" in rough analogy with cooling flows following thermal cooling in clusters of galaxies (Cowie & Binney 1977).

Figure 14, from Myers & Lazarian (1998), shows that one mechanism of turbulent dissipation, ion-neutral friction, should lead to substantial damping of MHD waves for the level of ionization expected in and near dense cores. This result supports the suggestion of Nakano (1998) that MHD waves and turbulence should soon dissipate following core formation, if no new turbulence is excited, for example by winds from young stars.

The speed of a turbulent cooling flow should be less than the effective speed of the turbulent motions, and estimates for the environments of dense cores indicate that the flow speed has a maximum close to the sound speed, or 0.2 km s^{-1} for a cloud at temperature 10 K (Myers 1998). This suggests that turbulent cooling flows can match the flow speeds indicated by observations of extended infall asymmetry. If so, such flows can condense matter onto dense cores substantially faster than ambipolar diffusion, and can modify and enhance the process of gravitational infall. However more detailed observations and calculations are needed to explore this and other types of pressure-driven flow in molecular clouds.

6. Summary

This chapter reviews physical conditions in nearby molecular clouds, describing basic properties of molecular clouds, dense cores, and their inward motions, in the context of star formation. We focus on the role of turbulence in the support of molecular clouds, in the formation of the precursors of stars in clusters, and in the generation of pressure-driven flows through the dissipation of turbulent motions.

I thank the organizers of the NATO Advanced Study Institute on which this book is based, Charlie Lada and Nick Kylafis, for their invitation and their hospitality, and the other participants for a lively and interesting meeting. Lincoln Greenhill, Judy Lees, and Lee Simmons provided valuable assistance with the figures and with the preparation of the manuscript.

References

Adelson, L. M., & Leung, C. M. 1988, MNRAS, 235, 349

Andre, P., Ward-Thompson, D., & Barsony, M. 1993, ApJ, 406, 122

Bachiller, R. 1996, ARAA, 34, 111

Basu, S., & Mouschovias, T. Ch. 1994, ApJ, 432, 720

Barranco, J. A., & Goodman, A. A. 1998, ApJ, 504, 207

Bell, M. B., Feldman, P. A., Travers, M. J., McCarthy, M. C., Gottlieb, C. A., & Thaddeus, P. 1997, ApJ, 483, L61

Beichman, C. A., Myers, P. C., Emerson, J. P., Harris, S., Matheiu, R., Benson, P. J., & Jennings, R. E. 1986, ApJ, 307, 337

Benson, P. J., & Myers, P. C. 1989, ApJS, 71, 89 Benson, P. J., Myers, P. C., & Wright,
    E. L. 1984, ApJ, 279, L27
Bergin, E. A., Plume, R., Williams, J. P., & Myers, P. C. 1999, ApJ, in press
Bertoldi, F., & McKee, C. F. 1992, ApJ, 395, 140
Blaauw, A. 1991, in *The Physics of Star Formation and Early Evolution*, eds. C. J. Lada
    and N. D. Kylafis, Kluwer Acad. Pub., Dordrecht, Netherlands, p. 125
Bonnell, I. A., Bate, M. R., Clarke, C. J., & Pringle, J. E. 1997, MNRAS, 285, 201
Bonnor, W. B. 1956, MNRAS, 116, 351
Caselli, P., & Myers, P. C. 1995, ApJ, 446, 665
Caselli, P., Walmsley, C. M., Terzieva, R., & Herbst, E. 1998, ApJ, 499, 234
Cesaroni, R., & Wilson, T. L. 1994, A&A, 281, 209
Chen, H., Myers, P. C., Ladd, E. F., & Wood, D. O. S. 1995, ApJ, 445, 377
Chen, H., Grenfell, T. G., Myers, P. C., & Hughes, J. D. 1997, ApJ, 478, 295
Choi, M., Evans, N. J., Gregresen, E. M., & Wang, Y. 1995, ApJ, 448, 742
Ciolek, G. E., & Mouschovias, T. Ch. 1995, ApJ, 454, 194
Cowie, L. L., & Binney, J. 1977, ApJ, 215, 723
Ebert, R. 1955, Z. Astrophys., 37, 217
Evans, N. J. 1999, ARAA, 37, in press.
Feige, J., & Pudritz, R. 1998, ApJ, submitted
Fuller, G. A., & Myers, P. C. 1992, ApJ, 384, 523
Fuller, G. A., & Myers, P. C. 1993, ApJ, 418, 273
Goldsmith, P. F., & Langer, W. D. 1978, ApJ, 222, 881
Goodman, A. A., Barranco, J. A., Wilner, D. J., & Heyer, M. H. 1998, ApJ, 504, 223
Gregersen, E., Evans, N. J., Zhou, S., & Choi, M. 1997, ApJ, 484, 256
Guelin, M., Langer, W. D., & Wilson, R. W. 1982, A&A, 107, 107
Harju, J., Walmsley, C. M., & Wouterloot, J. G. 1991, A&A, 245, 643
Harju, J., Walmsley, C. M., & Wouterloot, J. G. 1993, A&AS, 98, 51
Ho, P. T. P., & Townes, C. H. 1983, ARAA, 21, 239
Ho, P. T. P, Martin, R. N., Myers, P C., & Barrett, A. H. 1977, ApJ, 215, L29
Hummer, D. G., & Rybicki, G. B. 1968, ApJ, 153, L107
Jijina, J., Myers, P. C., & Adams, F. C. 1998, ApJ, submitted (JMA)
Kulsrud, R. E., & Pearce, W. P. 1969, ApJ, 156, 445
Lada, C. J. 1999, in *The Physics and Chemistry of the Interstellar Medium*, eds. V.
    Ossenkopf & J. Stutzki (Aachen: Shaker-Verlag), in press
Lada, E. A., & Lada, C. J. 1995, AJ, 109, 1682
Lada, E. A., Bally, J., & Stark, A. A. 1991, ApJ, 368, 432
Ladd, E. F., Myers, P. C., & Goodman, A. A. 1994, ApJ, 433, 117
Larson, R. B. 1985, MNRAS, 214, 379
Larson, R. B. 1981, MNRAS, 194, 809
Larson, R. B. 1995, MNRAS, 272, 213
Lee, C. W., Myers, P. C., & Tafalla, M. 1998, in preparation
Lee, C. W., Plume, R., & Myers, P. C. 1998, in preparation
Leung, C. M., & Brown, R. B. 1977, ApJ, 214, L73
Leung, C. M., & Liszt, H. S. 1976, ApJ, 208, 732
Li, Z., & Shu, F. H. 1996, ApJ, 472, 211
Linke, R. A., & Goldsmith, P. F. 1980, ApJ, 235, 437
Lizano, S., & Shu, F. H. 1989, ApJ, 342, 370
Lucas, R. 1976, A&A, 46, 473
Maddalena, R. J., Morris, M., Moscowitz, J., & Thaddeus, P. 1986, ApJ, 303, 37
Mardones, D., Myers, P. C., Tafalla, M., Wilner, D. J., Bachiller, R., & Garay, G. 1997,
    ApJ, 489, 721
Mardones, D., Bachiller, R., Garay, G., Myers, P., Tafalla, M., Wilner, D. J., & Williams,
    J. P. 1998, in preparation
McLaughlin, D. E., & Pudritz, R. E. 1996, ApJ, 469, 194
McCarthy, M. C., Travers, M. J., Chen, W., Gottlieb, C. A., & Thaddeus, P. 1998, ApJ,

498, L89

McKee, C. F. 1989, ApJ, 345, 782

Mizuno, A., Onishi, T., Yonekura, Y., Nagahama, T., Ogawa, H., & Fukui, Y. 1995, ApJ, 445, L161

Motte, F., Andre, P., & Neri, R. 1998, A&A, 336, 150

Mouschovias, T. Ch. 1991, ApJ, 373, 169

Myers, P. C. 1991, in *Fragmentation of Molecular Clouds and Star Formation*, eds. E. Falgarone, F. Boulanger, & G. Duvert (Dordrecht: Kluwer), p. 221.

Myers, P. C. 1998, ApJ, 496, L109

Myers, P. C., & Benson, P. J. 1983, ApJ, 266, 309

Myers, P. C., & Khersonsky, V. K. 1995, ApJ, 442, 186

Myers, P. C., & Lazarian, A. 1998, ApJ, 507, L157

Myers, P. C., Evans, N. J., & Ohashi, N. 1998, to appear in *Protostars and Planets IV*, eds. V. Mannings, A. Boss, & S. Russell (Tucson: U. of Arizona Press)

Myers, P. C., Fuller, G. A., Mathieu, R. D., Beichman, C. A., Benson, P. J., Schild, R. E., & Emerson, J. P. 1987, ApJ, 319, 340

Myers, P. C., Fuller, G. A., Goodman, A. A., & Benson, P. J. 1991, ApJ, 376, 561

Myers, P. C., Ladd, E. F., & Fuller, G. A. 1991, ApJ, 372, L95

Myers, P. C., Mardones, D., Tafalla, M., Williams, J. P., & Wilner, D. J. 1996, ApJ, 465, L133

Nakano, T. 1998, ApJ, 494, 587

Neugebauer, G. et al 1984, ApJ, 278, L1

Roberge, W. G., & Whittet, D. C. B. 1996, eds., *Polarimetry of the Interstellar Medium*, (San Francisco: ASP Conference Series, v. 97)

Rohlfs, K., & Wilson, T. L. 1996, *Tools of Radio Astronomy* (Berlin: Springer)

Ryden, B. 1996, ApJ, 471, 822

Schilke, P., Groesbeck, T. D., Blake, G. A., & Phillips, T. G. 1997, ApJS, 108, 301

Shu, F. H. 1977, ApJ, 214, 488

Shu, F. H. 1983, ApJ, 273, 202

Shu, F. H., Adams, F. C., & Lizano, S. 1987, ARAA, 25, 23

Snell, R. L. 1981, ApJS, 45, 121

Spitzer, L. 1978, *Physical Processes in the Interstellar Medium* (New York: Wiley), p. 241

Tafalla, M., & Myers, P. C. 1997, ApJ, 491, 653

Tafalla, M., Mardones, D., Myers, P. C., Caselli, P., Bachiller, R., & Benson, P. 1998, ApJ, 504, 900

Testi, L., & Sargent, A. I. 1998, ApJ, 508, L91

Townes, C. H., & Schawlow, A. H. 1975, *Microwave Spectroscopy* (New York: Dover)

Walker, C. K., Lada, C. J., Young, E. T., Maloney, P. R., & Wilking, B. A. 1986, ApJ, 309, L47

Walker, C. K., Narayanan, G., & Boss, A. P. 1994, ApJ, 431, 767

Ward-Thompson, D., Scott, P. F., Hills, R. E., & Andre, P. 1994, MNRAS, 268, 276

Williams, J. P., Bergin, E. A., Caselli, P., Myers, P. C., & Plume, R. 1998a, ApJ, 503, 689

Williams, J. P., Di Francesco, J., Myers, P. C., & Wilner, D. J. 1998b, ApJ, in press

Wood, D. O., Myers, P. C., & Daugherty, D. 1994, ApJS, 95, 45

Wootten, H. A., Loren, R. B., & Snell, R. L. 1982, ApJ, 255, 160

Zhou, S. 1992, ApJ, 394, 204

Zhou, S., Evans, N. J., Kompe, C., & Walmsley, C. M. 1993, ApJ, 404, 232

Ewine van Dishoeck and Floris van der Tak enjoy the first week's banquet.

MODELS AND OBSERVATIONS OF THE CHEMISTRY NEAR YOUNG STELLAR OBJECTS

EWINE F. VAN DISHOECK
Leiden Observatory P.O. Box 9513
2300 RA Leiden, The Netherlands
e-mail: ewine@strw.leidenuniv.nl

AND

MICHIEL R. HOGERHEIJDE
Astronomy Department, Univ. of California
Berkeley, CA 94720, USA
e-mail: michiel@astro.berkeley.edu

1. Introduction

The study of the chemical evolution of gas and dust from pre-stellar dense cores to circumstellar disks around young stars forms an essential part of understanding star- and planet formation. Throughout the collapse- and protostellar phases, simple and complex molecules are formed, many of which deplete onto cold grains and are eventually incorporated into the icy planetesimals of new solar systems (see Figure 1). Tracing this chemical evolution provides a wealth of information, not only about the chemical processing in primitive solar nebulae, but also about physical processes which occur in the immediate surroundings of young stellar objects (YSOs).

Interstellar chemistry has been an active field of research for more than 60 years, ever since the detection of the first molecules in 1937–1941. Nearly 120 different gas-phase species have been identified (see Table 1), not including isotopic varieties, with abundances down to 10^{-11} with respect to H_2. The majority of the species have been detected through their rotational transitions at millimeter wavelengths. Most of the early observations were performed with typical beam sizes of $1'$, corresponding to linear scales of nearly 10,000 AU (0.04 pc) in the nearest star-forming regions in Taurus and Ophiuchus. With the advent of large single-dish submillimeter telescopes and millimeter interferometers, it has become possible to study the dense

C.J. Lada and N.D. Kylafis (eds.), The Origin of Stars and Planetary Systems, 97–140.
© 1999 *Kluwer Academic Publishers. Printed in the Netherlands.*

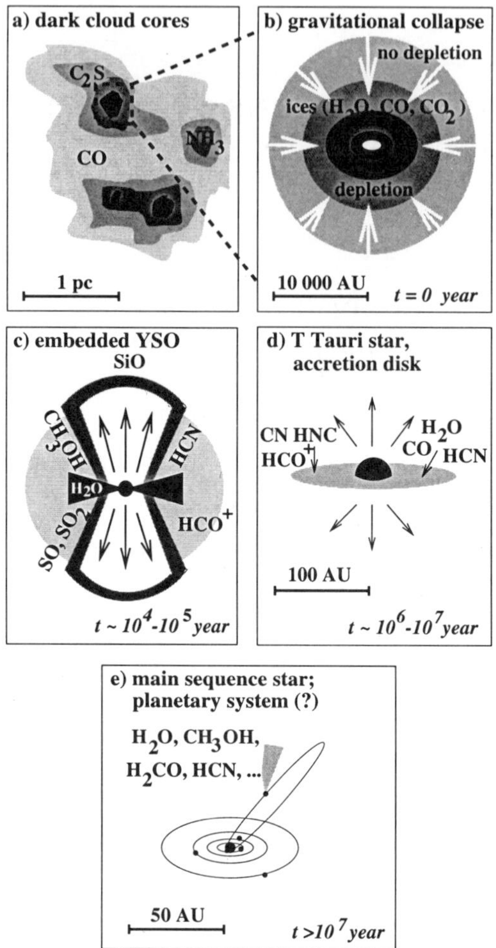

Figure 1. A schematic view of the characteristic molecules at different stages of low-mass star formation. (a) Dark cloud cores, where radicals and unsaturated carbon chains are prominent; (b) Collapse stage, where high levels of depletion of molecules have been inferred; (c) Deeply embedded YSO phase, where heating and supersonic outflows result in evaporation of ices and create a variety of chemical environments; (d) Young T Tauri star with a residual protoplanetary accretion disk containing gas-phase molecules and ices; (e) Mature planetary system with icy bodies such as comets (Figure by M.R. Hogerheijde, after Shu et al. 1987; from: van Dishoeck & Blake 1998).

envelopes around YSOs more directly on scales of $\sim 2''$–$20''$ (300–3000 AU at 150 pc) (Blake 1997) (see Figure 2).

Although much effort has focused on gas-phase chemistry, water ice was identified already in 1973 through its vibrational absorption toward bright infrared sources. The enormous advances in infrared instrumentation in the last decade, both from the ground and from space with the *Infrared Space Observatory* (ISO), have resulted in the detection of several other solid-state species and allow a complete inventory to be made (Tielens & Whittet 1997). The development of realistic models which include both gas-phase and grain-surface chemistry provides a challenge to theorists.

The combination of new observations and models has lead to the following scenario for the chemistry during star formation (see Figure 1). In the cold molecular cores prior to star formation, the chemistry is dominated by low-temperature gas-phase reactions leading to the enhanced formation of small radicals and unsaturated molecules. Long carbon-chains are produced if the gas is initially atomic-carbon rich. During the cold pre-stellar and collapse phases, many molecules accrete onto the grains and form an icy mantle. Here surface chemistry and processing by ultraviolet photons, X-rays and cosmic rays can modify the composition. Once the new star starts to warm the surrounding envelope, the ices are heated and molecules evaporate back into the gas phase, probably in a sequence according to their sublimation temperatures. In addition, the outflows from the young star penetrate the envelope, creating high temperature shocks and lower temperature turbulent regions in which both volatile and refractory material containing silicon can be returned. These freshly evaporated molecules then drive a rich chemistry in the 'hot cores' for a period of 10^5 yr. Finally, the envelope is dispersed by winds and/or ultraviolet photons, leading to the appearance of photon-dominated regions.

Part of the gas-phase and icy material is incorporated into the circumstellar disks, where they survive for a few $\times 10^6$ yr before being dispersed or assembled into new planetary bodies. Indeed, the connection between interstellar chemistry and solar system material has been greatly strengthened by observations of comets Hyakutake and Hale–Bopp, which reveal a remarkable similarity between the composition of interstellar and cometary ices (e.g., Ehrenfreund et al. 1997, Irvine et al. 1999).

Molecular spectroscopy is also a very powerful tool to probe the enormous range in temperatures from 10 to 10,000 K and densities from 10^4 to 10^{13} cm^{-3} that accompany the star-formation process. In addition, molecular lines provide the only means to trace the velocity fields in the highly-extincted circumstellar environment. In fact, the two analyses go hand in hand. On the one hand, an accurate physical model is a prerequisite for deriving reliable molecular abundances. On the other hand, it is important to know which molecule traces which physical component if its emission is to be used as a probe of the physical parameters.

This chapter first discusses the basic physical and chemical processes that play a role during star formation. Subsequently, the observational methods and problems in deriving of abundances are discussed. Finally,

Table 1 Identified interstellar and circumstellar molecules

Simple Hydrides, Oxides, Sulfides, Halogens and related molecules

H_2 (IR)	CO	NH_3	CS	$NaCl^*$
HCl	SiO	SiH_4^* (IR)	SiS	$AlCl^*$
H_2O	SO_2	C_2 (IR)	H_2S	KCl^*
N_2O	OCS	CH_4 (IR)	PN	AlF^*
HF				

Nitriles and Acetylene derivatives

C_3 (IR,UV)	HCN	CH_3CN	HNC	$C_2H_4^*$ (IR)
C_5^* (IR)	HC_3N	CH_3C_3N	HNCO	C_2H_2 (IR)
C_3O	HC_5N	CH_3C_5N ?	HNCS	
C_3S	HC_7N	CH_3C_2H	HNCCC	
C_4Si^*	HC_9N	CH_3C_4H	CH_3NC	
	$HC_{11}N$	CH_3CH_2CN	HCCNC	
	HC_2CHO	CH_2CHCN		

Aldehydes, Alcohols, Ethers, Ketones, Amides and related molecules

H_2CO	CH_3OH	HCOOH	CH_2NH	CH_2CC
H_2CS	CH_3CH_2OH	$HCOOCH_3$	CH_3NH_2	CH_2CCC
CH_3CHO	CH_3SH	$(CH_3)_2O$	NH_2CN	
NH_2CHO	$(CH_3)_2CO$	H_2CCO	CH_3COOH	

Cyclic Molecules

C_3H_2	SiC_2	c-C_3H	CH_2OCH_2

Molecular Ions

CH^+ (VIS)	HCO^+	$HCNH^+$	H_3O^+	HN_2^+
HCS^+	$HOCO^+$	HC_3NH^+	HOC^+	H_3^+ (IR)
CO^+	H_2COH^+	SO^+		

Radicals

OH	C_2H	CN	C_2O	C_2S
CH	C_3H	C_3N	NO	NS
CH_2	C_4H	$HCCN^*$	SO	SiC^*
NH (UV)	C_5H	CH_2CN	HCO	SiN^*
NH_2	C_6H	CH_2N	MgNC	CP^*
HNO	C_7H	NaCN	MgCN	
C_6H_2	C_8H	C_5N^*		

From Ohishi (1997) with detections as of October 1998 added; Species denoted with * have only been detected in the circumstellar envelope of carbon–rich stars. Most molecules have been detected at radio and millimeter wavelengths, unless otherwise indicated (IR, VIS or UV).

recent results on observations and models of the chemistry in the various star-formation phases are presented. This discussion focuses on those species whose abundances are particularly enhanced or decreased at a certain phase. These species often have only minor abundances in terms of overall composition —less than 10^{-7} with respect of hydrogen—, but provide important 'signposts' or 'clocks' of the evolutionary state of an object. This chapter is based on the recent reviews by van Dishoeck (1998a) on

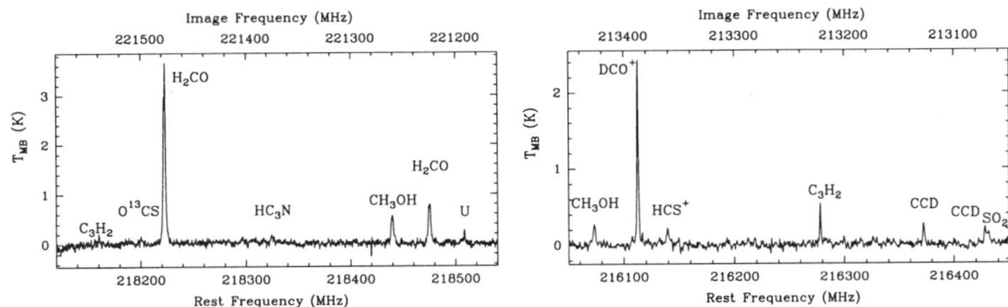

Figure 2. JCMT submilllimeter spectra of the deeply embedded low-mass YSO IRAS 16293−2422, showing lines of characteristic molecules such as H_2CO, CH_3OH and deuterated species (from: van Dishoeck et al. 1995).

the chemistry in diffuse and dense clouds, and by van Dishoeck & Blake (1998), van Dishoeck (1998b) and Langer et al. (1999) on the chemistry in star-forming regions. Other reviews on this subject include Hartquist et al. (1998), Irvine (1998) and van Dishoeck et al. (1993). For an excellent overview of the physics of molecular clouds, see Genzel (1992).

2. Gas-phase chemistry

2.1. BASIC MOLECULAR PROCESSES

The networks developed to describe the gas-phase chemistry in star-forming regions contain thousands of reactions between several hundreds of species (Millar et al. 1997a, Lee et al. 1996a). These thousands of reactions involve only a handful of different basic molecular processes, however, which have been described in detail in reviews by Dalgarno (1987) and van Dishoeck (1988). At the low densities in the interstellar gas, three-body processes are unimportant, so that only two-body reactions need to be considered. The rate of a reaction between species X and Y is given by $kn(X)n(Y)$ in cm^{-3} s^{-1}, where k is the reaction rate coefficient in cm^3 s^{-1} and n is the concentration in cm^{-3}.

There are two basic mechanisms by which molecular bonds can be formed (see Table 2). The first process is radiative association of gas-phase atoms or molecules, in which the new molecule is stabilized by the emission of a photon. The second mechanism is the formation of molecules on the surfaces of grains, in which the grain carries off the released energy corresponding to the molecular bond. The rate coefficients for the formation processes range from less than 10^{-17} up to 10^{-13} cm^3 s^{-1} and have considerable uncertainties, often up to an order of magnitude.

Molecules are readily destroyed by the absorption of ultraviolet pho-

TABLE 2. Some basic microscopic processes

Process	Name
Chemical Processes	
$X + Y \rightarrow XY + h\nu$	Radiative association
$X + Y{:}grain \rightarrow XY + grain$	Grain-surface formation
$XY + h\nu \rightarrow X + Y$	Photodissociation
$XY^+ + e \rightarrow X + Y$	Dissociative recombination
$X^+ + YZ \rightarrow XY^+ + Z$	Ion–molecule reaction
$X^+ + YZ \rightarrow X + YZ^+$	Charge–transfer reaction
$X + YZ \rightarrow XY + Z$	Neutral–neutral reaction
Heating and Cooling Processes	
$grain \text{ or } PAH + h\nu \rightarrow grain^+ \text{ or } PAH^+ + e^*$	Photoelectric heating
$H_2 + cosmic\ ray \rightarrow H_2^+ + e^*$	Cosmic ray heating
$CO(J) + coll. \rightarrow CO(J^*) \rightarrow CO(J') + h\nu$	CO line cooling
$O(^3P_2) + coll. \rightarrow O(^3P_1) \rightarrow O(^3P_2) + h\nu$	[O I] line cooling
$C^+(^2P_{1/2}) + coll. \rightarrow C^+(^2P_{3/2}) \rightarrow C^+(^2P_{1/2}) + h\nu$	[C II] line cooling
$gas + grain \rightarrow gas' + grain'$	Gas–grain heating or cooling

tons. This process is very effective at the edges of dark clouds and in clouds located close to young stars (so-called photon-dominated regions (PDRs), Hollenbach & Tielens 1997, 1999). For molecules such as H_2 and CO, the photodissociation can take place only at very short wavelengths between 912 and 1100 Å, whereas most other species are dissociated by radiation out to 3000 Å. The ultraviolet photons can also ionize atoms with ionization potentials less than 13.6 eV (e.g., C, S, Si, Fe, Mg, ...), thereby increasing the electron abundance inside the cloud. In the unshielded interstellar radiation field such as given by Draine (1978), typical rates are 10^{-9}–10^{-10} s^{-1}, corresponding to lifetimes against photodissociation or photoionization of only 10^2–10^3 yr.

Inside a cloud, the ultraviolet radiation is reduced because of absorption and scattering by grains (e.g., Roberge et al. 1991). Deep inside dark clouds (> 5 mag), little of the ambient radiation penetrates, but a weak ultraviolet field is maintained by cosmic-ray induced photons, resulting from the excitation of H_2 by secondary electrons produced by cosmic ray ionization of H_2 (Gredel et al. 1989). For a typical cosmic ray ionization rate ζ of a few $\times 10^{-17}$ s^{-1}, the resulting photorates are four to five orders of magnitude lower than those in the unshielded interstellar radiation field at the edge.

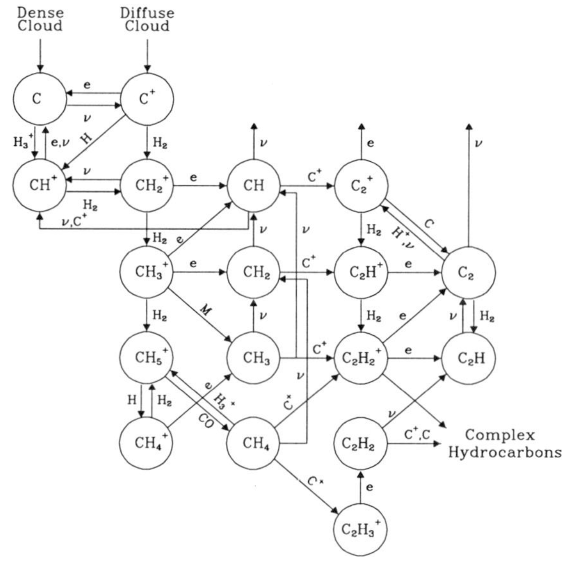

Figure 3.   Initial steps in the gas-phase carbon chemistry in diffuse and dark clouds.

Molecular ions are efficiently destroyed by the process of dissociative recombination. Rate coefficients in 10–30 K gas are typically 10^{-7} to 10^{-6} cm^3 s^{-1}, with the lower values representative for H_3^+ —a key species in the chemistry (Dalgarno 1994, Larsson 1997). A major uncertainty in the models is the branching ratio to the various products. Complete experimental data are only just becoming available, and recent results indicate that three-body product channels (e.g., $H_3O^+ + e \rightarrow OH + H + H$ or $O + H_2 + H$) have a much larger probability than thought previously (e.g., Vejby-Christensen et al. 1997).

In dense clouds, destruction of neutral molecules can also occur through chemical reactions, in particular with He^+ (e.g., $He^+ + CO \rightarrow He + C^+ + O$). Collisional dissociation is only important in regions of very high temperature (> 3000 K) and density, such as found in shocks.

Once molecular bonds have been formed, they can be rearranged by chemical reactions leading to more complex species. The first generation of models (e.g., Herbst & Klemperer 1973, Prasad & Huntress 1980) included primarily ion-molecule reactions. The majority of these reactions are known to be fast at low temperatures owing to the long-range attraction between the ion and the molecule. It has recently been realized that neutral–neutral reactions involving radicals and/or atoms can also have rate coefficients as large as $\sim 10^{-10}$–10^{-9} cm^3 s^{-1} at low temperatures (e.g., Smith 1997, Kaiser et al. 1997).

2.2. GAS-PHASE NETWORKS

The gas-phase chemical networks describing the most important formation routes of various molecules based on the above processes have been
discussed extensively in the literature, see reviews by, e.g., Millar (1990),
Herbst (1995) and van Dishoeck (1998a). The abundances of the elements
play a key role in the chemistry, especially the absolute and relative amounts
of gas-phase carbon and oxygen in atomic or molecular form. It has recently
become clear that the interstellar abundances of most elements are $\sim 30\%$
lower than the solar abundances (see Meyer 1997 for an overview). The
fraction of heavy elements in the gas phase is large (at least 30%) in diffuse
clouds, except for the most refractory elements (Si, Fe, ..), which are 99%
in solid form. For dense clouds, the depletions of species like C, N, O and S
are expected to be more significant, but are ill constrained observationally.

Because hydrogen is so much more abundant than any other element,
reactions with H and H_2 dominate the networks if they are exothermic.
This is only the case for small ions at low temperatures; most reactions of
neutrals and large ions with H or H_2 have substantial barriers. H_2 itself is
produced predominantly on the surfaces of grains (Hollenbach & Salpeter
1971), but once formed it is essential for the gas-phase chemistry of other
species. Ionization of H_2 by cosmic rays at a rate ζ leads to H_2^+, which
reacts with H_2 to form the stable H_3^+ ion. This ion plays a pivotal role in
the subsequent ion-molecule chemistry through proton transfer. Its recent
detection in interstellar space provides strong observational support for the
gas-phase chemical networks (Geballe & Oka 1996, McCall et al. 1998).

The production of complex hydrocarbons in dense clouds occurs via
three types of pathways (Herbst 1995): (i) carbon insertion reactions (e.g.
$C^+ + CH_4 \rightarrow C_2H_2^+ + H_2$ or $C + C_2H_2 \rightarrow C_3H + H$); (ii) condensation
reactions (e.g. $CH_3^+ + CH_4 \rightarrow C_2H_5^+ + H_2$ or $C_2H + C_2H_2 \rightarrow C_4H_2 + H$);
and (iii) radiative association reactions (e.g. $C^+ + C_n \rightarrow C_{n+1}^+ + h\nu$) (see
also Figure 3). In general, carbon insertion with C^+ is thought to be the
dominant route. Since this leads to loss of one hydrogen and because the
larger ions $C_nH_m^+$ do not react rapidly with H_2, low-temperature gas-phase
chemistry produces strongly unsaturated hydrocarbons, in agreement with
observations of dark clouds. The necessary C^+ ions are produced in small
amounts through destruction of CO by He^+. Reactions with C may be competitive if they are as rapid as suggested by recent laboratory experiments.
Once carbon is locked up in the very stable CO molecule, the formation of
more complex hydrocarbons ceases.

At high temperatures of $\sim 200\text{--}2000$ K such as encountered in the 'hot
core' regions surrounding massive young stars and in shocks, gas-phase
reactions with H and H_2 become significant. Specifically, the reactions O +

$H_2 \rightarrow OH + H - 4480$ K and $OH + H_2 \rightarrow H_2O + H - 2100$ K start to proceed at temperatures of a few hundred K (Ceccarelli et al. 1996, Charnley 1997), and drive most of the available gas-phase oxygen not locked up in CO into H_2O. Similar reactions of S with H_2 convert the gas-phase sulfur into H_2S. However, the back reactions of OH and H_2O with H have comparable energy barriers, so that the balance between O, OH and H_2O depends also on the H/H_2 ratio in the warm gas.

When most of the oxygen has been driven into H_2O at high temperatures, little O, OH and O_2 are available in the gas, preventing the formation of, for example, SO_2. Since O and O_2 also destroy reactive species such as CN, the CN abundance is enhanced at higher temperatures leading to enhanced formation of HC_3N through reaction with C_2H_2.

2.3. GAS-PHASE MODELS

2.3.1. *Depth-dependent vs. time-dependent models*

The calculation of the abundances in star-forming regions requires a physical model in which the temperature, density, radiation field etc. are specified as functions of position and/or time. In general, two different classes of models are considered: (i) Steady-state, depth-dependent models, in which the physical parameters and molecular abundances do not change with time, but are functions of depth into the cloud. Models of the translucent outer envelopes of clouds (e.g., van Dishoeck 1998a) and of dense ultraviolet photon- or X-ray dominated regions (PDRs or XDRs, see e.g., Hollenbach & Tielens 1997, Sternberg et al. 1997, Maloney et al. 1996) fall in this category. (ii) Time-dependent, depth-independent models, in which the concentrations are computed as functions of time at a single position deep into the cloud. Models of dark pre-stellar cores (e.g., Lee et al. 1996a, Millar et al. 1997b), collapsing envelopes (e.g., Bergin & Langer 1997) and hot cores near massive young stars (e.g., Charnley et al. 1992) fall into this category. The time scale for reaching chemical equilibrium depends on the density, temperature and ionization fraction and is typically 10^5–10^7 yr.

In both cases, the parameters that enter the models are (1) the elemental abundances of C, O, N, S, metals...; and (2) the cosmic ray ionization rate ζ in s^{-1}. In the steady-state, depth-dependent models, additional parameters are (3) the geometry (e.g. plane-parallel, spherical, ...); (4) the density $n_H = n(H) + 2n(H_2)$ as a function of position; (5) the incident radiation field, specified by a factor I_{UV} times the standard interstellar radiation field as given by, e.g., Draine (1978); and (6) the grain parameters, i.e., the extinction curve, albedo and scattering function. The temperatures of the gas and dust as functions of position in the cloud can be obtained self-consistently from the balance of the heating and cooling processes listed in Table 2. Alternatively, they can be constrained from observations and provided as additional input parameters (see §4.2).

In the time-dependent models, the additional parameters besides (1) and (2) are: (3) the density, usually taken to be constant with time (so-called pseudo time-dependent models); (4) the visual extinction A_V at the position in the cloud, usually taken to be so large that external photons can be neglected; and (5) the initial abundances of the various species at $t=0$, often taken to be atomic except for H_2. The temperature can be obtained from the thermal balance, but is almost always set at 10 K for both the gas and dust, typical of a dark cloud shielded from ultraviolet radiation and heated by cosmic rays only. In comparison with observations, the ratios of the local concentrations (in cm^{-3}) are taken to be equal to the ratios of the column densities integrated over depth (in cm^{-2}). This assumption is accurate for molecules whose abundances peak in the center of the cloud, but not necessarily for species such as radicals whose abundances peak in the outer part of the envelope (see Figure 5 of van Dishoeck 1998a).

2.3.2. *The $C^+ \rightarrow C \rightarrow CO$ transition*

The principal chemical characteristics of the depth- and time-dependent models are governed by the transition of carbon from atomic to molecular form. In the depth-dependent case, C^+ recombines to C around $A_V \approx 1$ mag, followed by the conversion to CO around $A_V \approx 2$ mag. The CO photodissociation rate as a function of depth is crucial in this transition (e.g., van Dishoeck & Black 1988, Lee et al. 1996b). The chemistry of other species follows the $C^+ \rightarrow CO$ transition. At the edge, only the simplest diatomic molecules are found. Around $A_V = 1$–2 mag, the presence of both C and C^+ and simple hydrides results in an increase in the abundance of hydrocarbon molecules such as CN and C_2H. Deeper inside the cloud, atomic carbon is no longer available, and destruction by O becomes more important than photodissociation. Stable species such as CH_4, C_2H_2 and HCN become dominant.

Many of the same features are observed in the pseudo time-dependent models, if depth is replaced by time. These models usually start with the assumption that dark clouds originate from diffuse gas, so that all species except H_2 are initially in atomic form, with carbon present as C^+. On a time scale of $\sim 10^3$ yr, C^+ recombines to C, which subsequently transforms to CO after $\sim 10^5$ yr. Since the presence of atomic carbon is essential to build up more complex organic species such as HC_3N, they are abundant only at early times, but not at steady-state (see Figure 4).

The main time-dependent aspects of the other elemental pools —O, N and S—, depend on how their chemistry is linked to that of carbon. For example, the abundance of CS has a different time behavior than that of

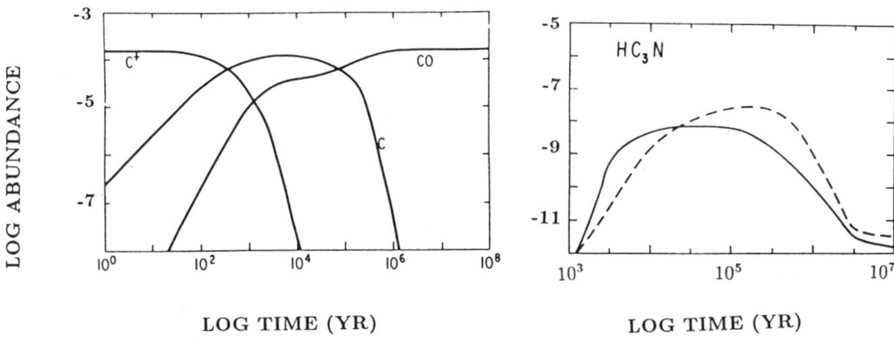

Figure 4. Time dependence of the abundances of C^+, C and CO (left) and HC_3N (right) for a cloud with $n_H = 2 \times 10^4$ cm^{-3}. The latter behavior is typical of all complex carbon–bearing molecules (from: Herbst & Leung 1986).

SO, because CS is formed by reactions with carbon-hydrides, whereas SO is destroyed by atomic carbon. This leads to two classes of molecules: (i) species like CN, HCN, CS and complex carbon chains, whose production is linked to the carbon network and which have larger abundances at early times; and (ii) molecules such as N_2, NH_3, N_2H^+ and SO that are independent or destructively linked to the carbon chemistry and which exhibit higher abundances at equilibrium.

2.4. FRACTIONATION

At the low temperatures of pre-stellar cores and the outer envelopes of YSOs, significant enhancement of minor isotopes can occur as a result of so-called 'fractionation'. The most relevant case for star-forming regions is the deuterium fractionation, initiated by the isotope exchange reactions $H_3^+ +$ HD $\rightarrow$ $H_2D^+ + H_2$ and $CH_3^+ + HD \rightarrow CH_2D^+ + H_2$ (Wootten 1987, Millar et al. 1989). These reactions occur preferentially in the forward direction at low temperatures, because the heavier H_2D^+ and CH_2D^+ are slightly more stable than H_3^+ and CH_3^+ due to their lower zero-point vibrational energy. H_2D^+ and CH_2D^+ subsequently transfer the deuterium to, for example, CO or N to form DCO^+ or $DCNH^+$. The latter ion leads to DCN through dissociative recombination. The observed DCO^+/HCO^+ and DCN/HCN abundance ratios are factors of 100 to >1000 larger than the overall [D]/[H] ratio of $\sim 1.6 \times 10^{-5}$.

2.5. SUCCESSES AND PROBLEMS

Pure gas-phase chemistry models have been remarkably successful in explaining many observed features of interstellar chemistry (Herbst 1995, Turner et al. 1998). These include the abundances of simple species in diffuse and translucent clouds; the presence of H_3^+ and the related high abundances of protonated species such as HCO^+ and N_2H^+ in dark clouds; the high abundances of unsaturated and metastable molecules in dark clouds such as $l-$ and $c-C_3H$ or HNC; and the large isotopic fractionation.

Most of the problems with gas-phase models stem from uncertainties in the basic reaction rates. For example, if reactions of bare carbon chains C_n with O or N are rapid, the observed abundances of carbon-chain molecules like HC_7N cannot be reproduced, even at early times. Another complication is that time-dependent models have more than one solution in certain regions of parameter space (e.g., le Bourlot et al. 1995, Lee et al. 1998). It is not yet clear, however, what the astrophysical consequences of this 'bistability' phenomenon are. Finally, gas-phase models of dark clouds predict that a substantial fraction of the oxygen is driven into O_2 at low temperatures. Recent limits on the O_2/CO ratio of less than 0.03 are difficult to accomodate in such models (Olofsson et al. 1998, Marechal et al. 1997, Melnick et al. 1998).

3. Grain-surface chemistry

3.1. BASIC SURFACE PROCESSES

The chemistry on the surfaces of interstellar grains has received ample discussion in the literature (e.g., Tielens & Allamandola 1987, Herbst 1993, Tielens & Charnley 1997, Langer et al. 1999). Four different steps can be distinguished: (i) accretion; (ii) diffusion; (iii) reaction; and (iv) ejection or evaporation. The time scale for a molecule to collide with a grain and accrete is given by $t_{\rm acc} \approx 2 \times 10^9/n_H y_S$ yr where y_S is the sticking probability which is thought to lie between 0.1 and 1.0 for most species (see Williams 1993 for a review). Thus, in envelopes around YSOs with densities of order 10^5 cm^{-3} or larger, the time scale for depletion is shorter than the collapse time (Walmsley 1991). The timescales for diffusion and evaporation increase exponentially with the binding energies of the species to the grain, which in turn scale with its mass in the case of physical adsorption. Under most interstellar circumstances, the rate-limiting step is the accretion of new species rather than the diffusion over the surface.

Because atomic hydrogen is so abundant and mobile, grain-surface chemistry leads primarily to hydrogenated species such as H_2O, NH_3 and CH_4. Although many of the surface reactions have energy barriers, they can still

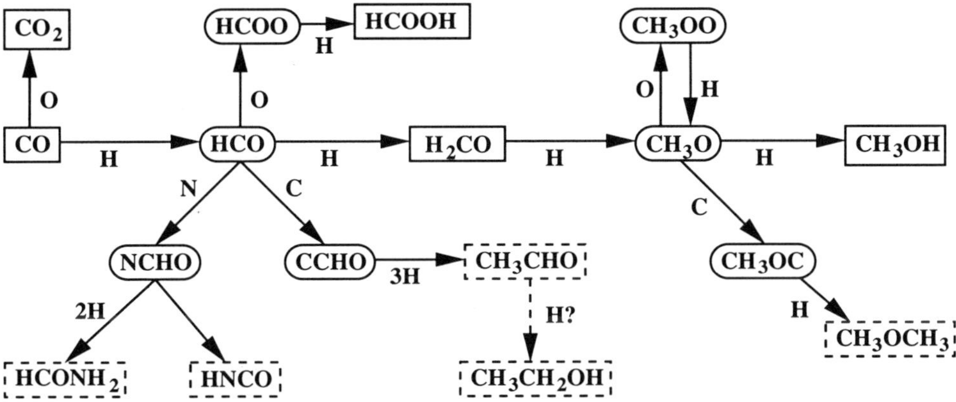

Figure 5. Grain-surface chemistry routes involving CO. Solid rectangular boxes contain molecules which have been identified in interstellar ices, whereas dashed boxes indicate molecules detected in the gas phase in hot cores (figure by Tielens, based on Tielens & Hagen 1982).

proceed because of the relatively long time scale (~ 1 day) available for reaction before another species is delivered from the gas to the grain.

The dominant molecule accreted from the gas is CO, and it therefore forms a key in the surface chemistry. Hydrogenation of CO leads to HCO, H_2CO and eventually CH_3OH (see Figure 5). The intermediate HCO (formyl) and CH_3OH (methoxy) radicals can also react with various other accreted species, leading to more complex organic molecules such as CH_3CHO, HCOOH and $HCONH_2$. The identification of many of these species in molecular clouds, directly in ices or indirectly in hot cores where the ices have evaporated off the grains, provides support for this scheme.

Hydrogenation of molecules is particularly effective in the outer regions of YSO envelopes where the atomic hydrogen abundance is high. At densities greater than $\sim 10^5$ cm^{-3} shielded from ultraviolet radiation, atomic oxygen becomes more abundant than atomic hydrogen, leading to oxidation of various accreted molecules. For example, CO can react with O to form CO_2, and O_2 with O leads to O_3.

Deuterium fractionation of molecules can also occur through grain-surface chemistry, because the gas-phase atomic D concentration inside dense clouds is enhanced compared with atomic H through dissociative recombination of DCO$^+$ (Tielens 1983). The atomic D can then react with molecules on the surface to form deuterated species. Doubly-deuterated molecules such as D_2CO and NHD_2, which have been detected in the envelopes around low- and high-mass stars, are thought to be formed by this route (e.g., Ceccarelli et al. 1998).

TABLE 3. Interstellar ice composition and evaporation temperatures

Species	Abundance[a]	T_{ev}[b] (K)
H_2O	100	90
CO	16	16
CO_2	24	50
CH_4	2	18
CH_3OH	4	80
XCN	2	?
OCS	...	85
NH_3	10[c]	55
N_2	...	13
O_2	< 30[d]	13

[a] Abundances are relative to H_2O=100, and refer to NGC 7538 IRS9 (Whittet et al. 1996). Abundances are variable from source to source due to differential outgassing and thermal processing. The H_2O abundance is typically 10^{-4} with respect to H_2.

[b] Evaporation temperatures for pure ices under interstellar conditions; see text for mixed ices.

[c] Tentative identification by Lacy et al. (1998).

[d] Vandenbussche et al. (1999).

3.2. THERMAL PROCESSING AND EVAPORATION

In star-forming regions, the ice mantles formed in the cold pre-stellar and collapse phases can be heated by the young star. This leads to restructuring of the ice matrix and outgassing when the temperature is near its sublimation point. Table 3 summarizes the sublimation temperatures of various species under interstellar conditions (Sandford & Allamandola 1990, 1993); these are typically lower than found in laboratory experiments because of the lower pressures in space. Interstellar ices are not pure, but in mixed ices, the evaporation of each component is largely determined by its own sublimation behavior, unless the abundance of one of the species is less than $\sim 5\%$. Thus, an ice mixture of $H_2O/CO/CO_2$=2/1/1 shows release of CO around 20 K, CO_2 around 50 K and H_2O around 90 K. However, a mixture of H_2O/CH_4=100/1 shows evaporation of both H_2O and CH_4 around 90 K. Only a small fraction of the CO or CO_2 stays behind, and is released when the whole H_2O ice rearranges from amorphous to crystalline form.

Ices containing CH_3OH show particularly interesting behavior upon heating, because CH_3OH has a sublimation temperature above that of the phase transformation of H_2O. Thus, H_2O/CH_3OH ices heated to $\sim$ 80–

90 K will segregate into rather pure H_2O- and CH_3OH domains (Blake et al. 1991). Recent experiments indicate that CO_2/CH_3OH mixtures show similar segregation behavior (Ehrenfreund et al. 1998), and evidence for this process is seen in the observed profiles of interstellar solid CO_2 toward massive YSOs (Boogert et al. 1999, Gerakines et al. 1999).

3.3. NON-THERMAL DESORPTION

Thermal evaporation is only efficient at dust temperatures T_{dust} higher than 20 K. Some desorption can also occur by non-thermal processes in the cold pre-stellar cores and outer envelopes with $T_{dust} \approx 10$ K. These processes have been reviewed most recently by Schutte (1996), and include cosmic ray spot heating, heating due to the energy liberated by the formation of molecules on grains, heating due to grain–grain collisions, and explosive heating due to exothermic reactions between stored radicals on grains. The desorption time scales depend strongly on the binding energies of the molecules at the surfaces and the surfaces involved (silicates, carbonaceous material, ices). All of them are particularly effective for non-polar ices containing CO, O_2 and N_2, and less for H_2O-rich ices which contain strong hydrogen bonds.

3.4. ENERGETIC PROCESSING

The cosmic-ray induced ultraviolet photons and the radiation produced by the young star can further process the ices. Ultraviolet photolysis of H_2O, CO, NH_3 and CH_4 produces radicals which can react with each other and with the parent molecules to form more complex molecules like H_2CO, HCOOH and $HCOCH_3$ (e.g., Bernstein et al. 1995, Gerakines et al. 1996). Much of the outcome of this chemistry is very similar to that of the hydrogenation of CO, and further quantitative studies are needed to assess their relative importance.

Photolysis of ice mixtures containing H_2O, CO, NH_3 and CH_3OH often leaves a non-volatile organic residue containing a variety of complex molecules (e.g., Schutte et al. 1993, Bernstein et al. 1995). Some of these chemical effects are also found in experiments in which ices are bombarded with highly-energetic charged particles, analogous to cosmic rays or X-rays (e.g., Moore et al. 1983, Kaiser & Roessler 1997, Teixeira et al. 1998). The photochemical and high-energy bombardment processes are often referred to as 'energetic' processing in the literature, without discrimination.

3.5. POLAR AND APOLAR ICES

The ices in interstellar clouds are generally not homogeneous, but consist of different layers or domains (e.g., Tielens et al. 1991, Ehrenfreund et al. 1998). Two broad classes of ices can be distinguished. Polar ices are dominated by H_2O ice and contain minor amounts of CO_2, CO, CH_4 and CH_3OH. Apolar ices are dominated by species such as CO and perhaps some O_2 and N_2, but contain very little H_2O. The two phases can be distinguished by their line profiles, since the force constants of CO molecules embedded in an H_2O-matrix will be slightly different from those of CO surrounded by other CO molecules.

The different ice phases can arise from a combination of at least two processes, caused by the density and temperature gradient in YSO envelopes (see Schutte 1999 for review). First, the density gradient results in a steep gradient in the gas-phase H/CO ratio. At low densities or at the edge of the cloud, the high concentration of H results in polar ices. Deep inside the envelope at high densities, mostly apolar ices of accreted CO, O_2 and N_2 form. Layered ices can be produced in a collapsing envelope with the polar ices condensing first and the non-polar species forming a volatile 'crust'. Second, the evaporation processes can shape the composition of the ice mantles, because the desorption mechanisms are much more efficient for volatile non-polar species like CO than for non-volatile material like H_2O and CH_3OH. This 'distillation' effect decreases the apolar ices compared with the polar ices at higher temperatures.

3.6. GAS–GRAIN MODELS

Based on the above processes, two different chemical regimes can be distinguished in models which take both gas-phase and grain-surface processes into account. In the 'accretion-limited' regime, the diffusion time is much shorter than the accretion time so that a species can diffuse on the surface until it finds a co-reactant. The chemistry is limited by rate at which new species are delivered to the surface. In the 'reaction-limited' regime, the opposite holds so that many reactive species are present on a grain surface and the reaction is controlled by surface concentrations as well as kinetic parameters. Most of the gas–grain chemical models have been formulated in the 'reaction-limited' regime using rate equations which mirror those used for gas-phase chemistry (e.g., Hasegawa & Herbst 1993). This approach is only accurate when a large number of reactive species exist on a single grain surface, since only average abundances are calculated. This condition is usually not met in interstellar clouds, since the accretion times are long, grains are small, and reactions are fast, so that at most one reactive species is present on a grain at any time. The surface chemistry is therefore in the accretion-limited regime and can only be properly treated by a Monte-Carlo method which determines the likelihood of two such species arriving from the gas in succession onto a particular grain in a steady-state model (e.g., Tielens & Hagen 1982). Recently, ad-hoc modifications of the rate equations have been proposed to correct the shortcomings of the reaction-limited approach (Caselli et al. 1998, Shalabiea et al. 1998).

3.7. SUCCESSES AND PROBLEMS

The observations of H_2 in diffuse clouds and of large amounts of ices in dense clouds can only be explained in models in which these molecules are formed on the surfaces of interstellar grains and/or accreted from the gas. Thus, gas–grain interactions play an essential role in interstellar chemistry. The main problem is an accurate, quantitative description. Many basic grain-surface reactions with H or O at low temperatures are still poorly understood theoretically or ill-constrained by laboratory experiments, partly because the details of the surfaces of interstellar grains are still poorly characterized.

4. Determination of molecular abundances

The comparison of observations and chemical model calculations ultimately depends on the translation of spectral-line data into molecular abundances. In the following, the observational techniques will be discussed, and the methods for constraining the physical structure and subsequently the chemical abundances will be outlined.

4.1. OBSERVATIONAL TECHNIQUES

The energy level structure of a molecule can be decomposed into an electronic part $E_{e\ell}$, a component due to the vibrational motion of the nuclei E_{vib}, and a component describing the overall rotation of the molecule in space E_{rot}. In general, $E_{e\ell} >> E_{vib} >> E_{rot}$: transitions between two electronic levels lie typically in the visible/ultraviolet part of the spectrum, those between two vibrational levels in the infrared part, and those between two rotational levels at (sub-)millimeter wavelengths. Because of the high extinction, only infrared and millimeter observations are feasible in star-forming regions.

4.1.1. *Rotational line emission at (sub-)millimeter wavelengths*

Most molecules listed in Table 1 are observed by their rotational transitions within the ground electronic and vibrational state (Figure 6). The sensitivity of large single-dish (sub-)millimeter telescopes allows the detection of

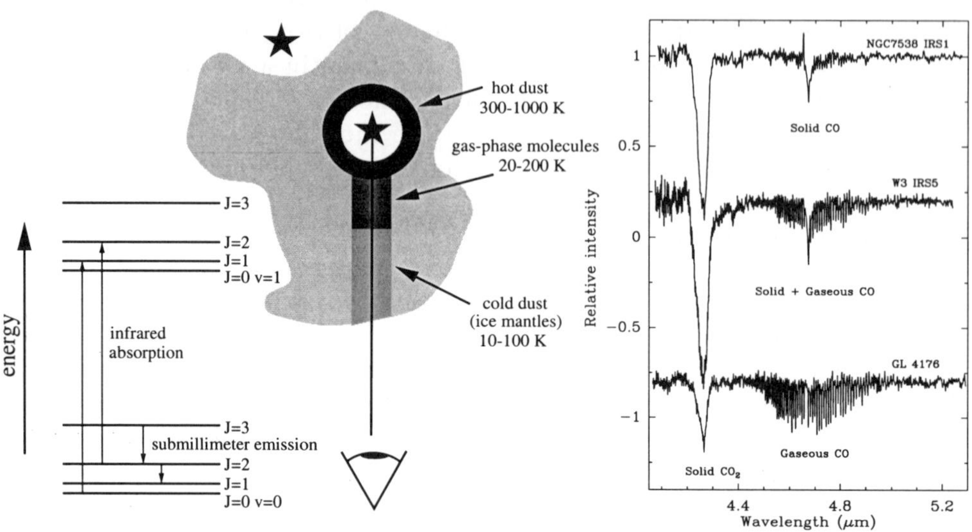

Figure 6.   Left: Schematic illustration of infrared absorption line observations of gas and dust toward embedded or background sources. The infrared continuum is provided by the hot dust at 300–1000 K. Right: Normalized ISO–SWS spectra toward three massive young stellar objects embedded in dense molecular clouds. The strong, broad absorption at 4.27 μm is due to solid CO_2, whereas the characteristic ro-vibrational P- and R-branch structure at 4.4–4.9 μm indicates the presence of warm, gaseous CO along the line of sight (van Dishoeck et al. 1998).

weak lines and rare chemical species and isotopes down to abundances of 10^{-11} with respect to hydrogen. However, not all molecules can be observed by this method. Symmetric molecules lack the permanent dipole moment required for rotational emission lines, so that important species like H_2, N_2, CH_4 and C_2H_2 cannot be observed in this way. In addition, the Earth's atmosphere prevents ground-based observations of low-lying transitions of O_2, CO_2 and H_2O. Because of its chemical stability, CO often serves as a tracer of H_2, assuming either a fiducial CO/H_2 abundance ratio or a conversion factor between integrated CO intensity and H_2 column density (see van Dishoeck & Black 1987, van Dishoeck 1998a for reviews). The only direct measurement of $CO/H_2 = 2.7 \times 10^{-4}$ in the warm dense cloud toward NGC 2024 IRS2 (Lacy et al. 1994) is a factor of 3 higher than the value of 8×10^{-5} commonly used in dark clouds (Frerking et al. 1982). Submillimeter continuum emission from dust can also act as a standard when values are adopted for the dust emissivity, its temperature, and the dust-to-gas mass ratio.

Clouds can be mapped at angular resolutions between 10″ (1500 AU at 150 pc) and a few arcminutes with single-dish instruments. Even in the nearest star-forming clouds, regions of different physical or chemical characteristics (envelope, outflow, disk, ...) are likely to fall within one beam, complicating the derivation of molecular abundances. Much higher angular resolution of the order of one arcsecond is offered by interferometers. However, the sensitivity of these instruments is lower than that of single-dish telescopes, and their interpretation is complicated because they filter out all emission on extended scales defined by the shortest baselines (cf. §8.1).

A major advantage of heterodyne (sub-)millimeter spectroscopy is the very high spectral resolution, typically better than a few tenths of a km s^{-1} and covering the full extent of the lines up to a few hundred km s^{-1}. This allows separation of regions with different physical conditions within one beam if they have different velocities or line profiles. A drawback of rotational emission spectroscopy is that the frequency separation between different transitions of the same molecule is usually too large to be obtained in a single setting, and multiple observations are needed to constrain the excitation, often involving different telescopes or beam sizes.

4.1.2. *Vibrational absorption at infrared wavelengths*

Absorption of infrared emission into (ro-)vibrationally excited levels of molecules offers a complementary view of the chemical content (see Figure 6). Background stars or embedded objects of sufficient infrared brightness can usually be found in star-forming regions, often the YSO under study itself. Absorption lines probe molecular gas along a pencil beam to the infrared source with very high angular resolution, but mapping is not possible and different conditions still exist along the line of sight. The velocity resolution is generally lower, so that the line profiles are often not resolved.

Because of the low spectral resolution and the intrinsic strengths of the transitions, the abundance limit of infrared observations is two to four orders of magnitude higher than that at millimeter wavelengths. On the other hand, symmetric species like C_2H_2 and CH_4 can be observed, because the vibrational modes induce a temporary dipole moment. Different transitions of a molecule often fall in a single infrared spectrum, providing direct constraints on the molecular excitation up to high energy levels. By their very nature, absorption lines also directly measure the line opacity, if the profiles are resolved. Another major advantage is that infrared studies are not limited to the gas phase. Molecules frozen onto dust grains show a markedly different line profile, since the rotational degrees of freedom are collapsed into a single, broad feature which is slightly shifted from the position of that of the gas-phase molecule (Figure 6). This offers the possibility to study the relative amounts of a species present in the gas-phase and in the condensed phase, and to probe the ice mantles as formation sites of various molecules like CH_3OH and H_2CO. The *Short Wavelength Spectrometer* (SWS) on ISO is particularly well suited for such studies (see §6.2, Figure 6 and 10).

4.2. CONSTRAINING THE PHYSICAL STRUCTURE

In order derive molecular column densities or abundances from the observations, a good physical model of the region is a prerequisite. Observations of different rotational lines provide very useful probes, because they are collisionally excited by H_2 for densities of 10^2–10^7 cm^{-3} and temperatures of 5–200 K (Evans 1980, Walmsley 1987, Genzel 1992).

The statistical equilibrium equations for the level populations n_i due to radiative and collisional processes are given by

$$n_i \sum_{j<i} A_{ij} + n_i \sum_{j}(B_{ij}J + C_{ij}n_{\mathrm{col}}) = \sum_{j>i} n_j A_{ji} + \sum_{j} n_j(B_{ji}J + C_{ji}n_{\mathrm{col}}),$$

where the left-hand side represents the processes leading to the loss of population of level i and the right-hand side the gain of population. Here A_{ij} and B_{ij} are the Einstein A and B coefficients for spontaneous and stimulated emission and absorption, J is the intensity of the radiation field averaged over all directions and integrated over the line profile, C_{ij} is the collisional rate coefficient between levels i and j, and n_{col} is the density of collision partners, primarily H_2. The upward and downward rate coefficients are related through detailed balance by $C_{ji} = (g_i/g_j)C_{ij}\exp(-h\nu/kT)$. The values of C_{ij} for collisions of different molecules with H_2 $J=0$ have largely been derived from detailed quantum mechanical calculations (Green 1975, Flower 1990), and form one of the largest uncertainty in analyzing molecular excitation. Information on collisional rate coefficients at high temperatures >200 K is still scarce, and collisions with H_2 $J=1$ or higher are generally not taken into account explicitly.

The critical density for a given transition is the density at which the rates for collisional processes become comparable to those for radiative processes, resulting in substantial population of level i. In formula, $n^i_{\mathrm{crit}} = A_{ij}/\sum_j C_{ij}$. If the transition becomes optically thick, the critical density is lowered by $1/\tau$, since part of the emitted photons are re-absorbed. Here τ is the optical depth of the line. Note that higher frequency transitions have higher critical densities, since $A_{ij} \propto \nu^3$. Also, molecules with large dipole moments μ have high critical densities, since $A_{ij} \propto \mu^2$. Thus, by choosing the appropriate molecule and transition, a large range of densities can be probed. This is schematically indicated in Figure 7, which shows the range of physical parameters for which different line ratios are most sensitive.

The contribution of the line photons to the excitation is taken into account by a variety of formalisms, including the large velocity gradient,

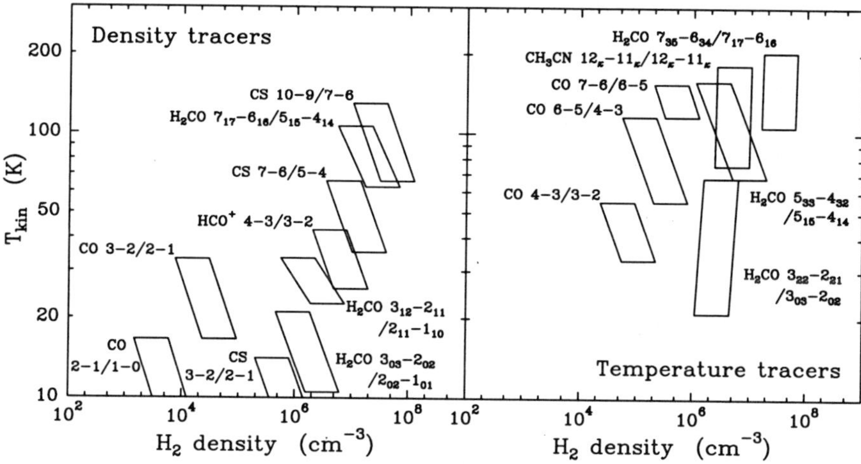

Figure 7. Examples of molecular line ratios which are used to constrain the density and temperature structure of YSO envelopes.

Sobolev, microturbulent, or escape probability methods (e.g., Sobolev 1960, Habing 1988, Osterbrock 1989 Appendix 2).

4.3. FROM OBSERVATIONS TO ABUNDANCES

Figure 8 gives a schematic overview of the steps involved in the determination of abundances from observations. Correcting for atmospheric effects and telescope losses yields the antenna temperature. This is the actual sky brightness averaged over the main forward beam of the telescope, or, for interferometric observations, the sky brightness averaged over the synthesized beam size after spatial filtering by the array baselines.

Relating the antenna temperature to the underlying brightness distribution on the sky requires assumptions about the source structure. Figure 8 illustrates two extreme approaches to this problem. The 'homogeneous' method, shown on the left, uses a single set of physical conditions and a given size, or beam-filling factor, for the source, which transforms the beam-averaged antenna temperature into brightness temperature on the sky. The molecular excitation and opacity relate this brightness temperature to the column density of the molecule. The line opacity follows from the intensity ratio of lines of different isotopic varieties. The excitation can either be assumed to be in local thermodynamic equilibrium, valid if the density exceeds the critical density for the observed transition, or can be derived from statistical equilibrium and radiative transfer calculations as described above once the physical parameters have been constrained. Repeating the procedure for an appropriate abundance standard like CO yields the molecular abundance relative to H_2.

More complex source structures can be described with multiple 'homogeneous' components, each with its own beam-filling factor. This approach works well if the number of different components is limited, for example, if the cloud is described as a collection of dense clumps embedded in a more diffuse interclump medium (e.g., Hogerheijde et al. 1995). When the beam is filled with a large number of different physical regimes, as is the case, e.g., for the power-law density and temperature distributions often used to describe YSO envelopes, the multi-component 'homogeneous' approach no longer suffices. The right-hand side of Figure 8 shows a more 'detailed' method. Starting with a fiducial source model, the molecular excitation and the radiative transfer throughout the cloud is calculated. Monte-Carlo techniques are often used in this step (e.g., Bernes 1979). The emergent sky brightness distribution is convolved with the telescope beam, or sampled at the interferometer's spatial frequencies. Matching the observed and model antenna temperature constrains the molecular abundance.

The interpretation of interferometric observations, and the derivation of molecular abundances from them, requires that spatial filtering by the interferometer is taken into account explicitly. Even if the data sets of the molecule in question and that of a standard like $C^{18}O$ contain the same spatial frequencies, both lines can have very different distributions of intensity over those frequencies if they trace different physical regimes. Different levels of flux are resolved out by the interferometer, resulting in abundance estimates deviating by factors of a few or more if not accounted for properly.

5. Chemistry in pre-stellar cores

In the following sections, an overview will be given of recent observations and models of star-forming regions at different evolutionary stages, using the material presented in §1-4 as general background information. The discussion focuses on those species whose abundances are most affected at a particular evolutionary state. These molecules are often only minor components in terms of the overall composition of the gas or dust. Recent overviews of the major reservoirs of the principal elements C, N and O are given, e.g., in van Dishoeck et al. (1993) and van Dishoeck & Blake (1998). In brief, most of the oxygen is contained in solid silicates, oxides and ices, as well as gas-phase O, CO and, in warm regions, H_2O. Most of the carbon is in some solid carbonaceous form, with the remainder in gas-phase CO. Nitrogen is mostly locked up in gas-phase N_2 and N, but the amount in

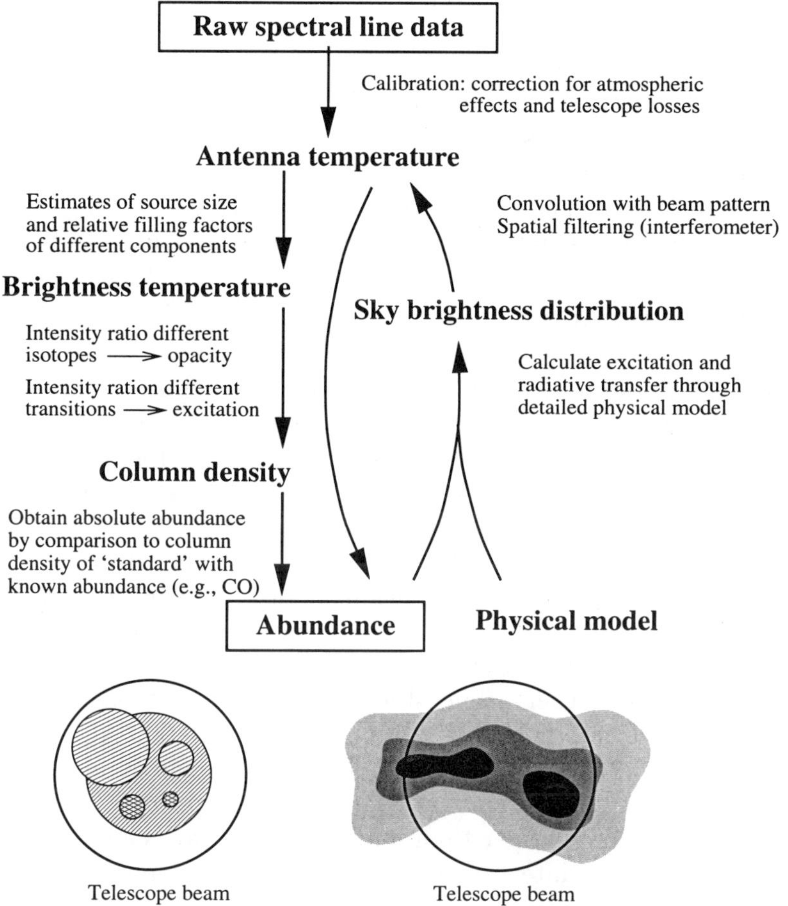

Figure 8. The steps involved in deriving molecular abundances from spectral line data. The left-hand side shows the 'homogeneous' approach, the right-hand side the 'detailed' approach.

solids is poorly constrained.

5.1. TRANSLUCENT CLOUDS

In order to study the effect of YSOs on the chemistry and test the basic chemical networks, the abundances in quiescent clouds prior to star formation need to be constrained. The best clouds for this purpose are the translucent and high-latitude clouds with visual extinctions of a few mag and densities of a few thousand cm^{-3}. These clouds have been studied by optical absorption lines against reddened stars (e.g., Gredel et al. 1993), by millimeter absorption lines against distant radio sources (e.g., Lucas & Liszt 1997) and by millimeter emission observations (e.g., Turner 1996, 1998 and references cited). No signs of star formation have been found in these regions. The chemistry is characterized by simple diatomic and triatomic molecules, radicals and ions. Detailed models by Turner et al. (1998) show that the observed abundances of most species are well reproduced by the ion-molecule networks. The main exceptions are formed by NH_3, H_2CO and H_2S, whose formation is probably dominated by grain-surface chemistry.

5.2. DARK CLOUD CORES

More than 100 dark cores have been identified through optical extinction and molecular line studies, at least half of which have not yet formed any stars (Myers & Benson 1983, see chapter by Myers). Most chemical studies have been directed toward TMC-1, a small elongated, dense ridge of $\sim 0.6 \times 0.06$ pc in Taurus with a mass of less than 40 $M_\odot$. This clump shows a particularly rich chemistry, with a large chemical gradient across the core. NH_3 and other 'late' molecules peak in the northern part, whereas long carbon-chain molecules such as HC_7N and C_4H are most abundant in the southern part (Olano et al. 1988, Pratap et al. 1997). High spectral and spatial resolution observations show that there are at least three different velocity components in this region on less than 10,000 AU scales, with different intensity ratios among the molecules (Langer et al. 1997). The abundances of the carbon-chains range from 10^{-11} up to 10^{-8} and are several orders of magnitude larger than found in other dark clouds. Such large abundances are usually interpreted with pseudo time-dependent models at 'early times' $t \approx 10^5$ yr, assuming that the cloud evolved from a diffuse cloud phase in which carbon was initially in atomic form (e.g., Lee et al. 1996a).

In order to investigate whether TMC-1 is chemically peculiar, Suzuki et al. (1992) and Benson et al. (1998) performed a systematic study of a few characteristic molecules (C_2S, HC_3N, HC_5N, N_2H^+ and NH_3) in a large set of dark cores. The abundances of the carbon chains are found to correlate well with each other, but not with NH_3 or N_2H^+, consistent with the case of TMC-1. The observed C_2S/NH_3 abundance has been reproduced quantitatively in models which start from diffuse gas and form dense cores over a period of 10^5 to 2×10^6 yr (see Figure 9). In this scenario, TMC-1S is one of the youngest clouds. There are numerous other non-equilibrium processes, however, such as penetration of ultraviolet radiation, outflows or turbulence, which can lead to the breakdown of CO into C or C^+ and therefore give the cloud core a chemically 'young' appearance. Thus, the 'chemical age' does not necessarily measure the true age of the core, but

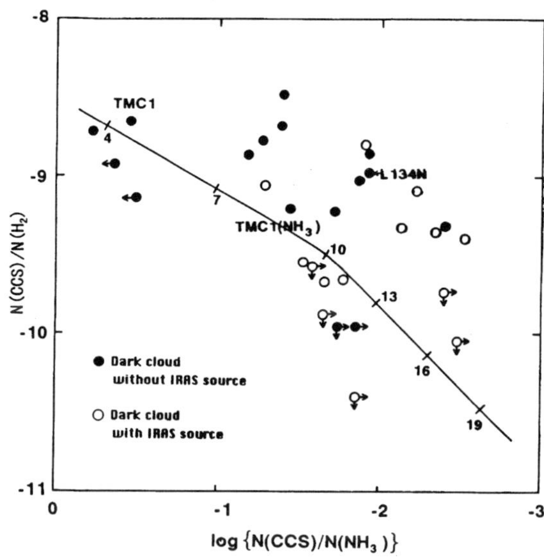

Figure 9. Observed CCS abundance versus the CCS/NH_3 abundance ratio in dark clouds with and without young stars. The solid line indicates the model abundance ratios for $T=10$ K and $n(H_2)=10^4$ cm^{-3}. The numbers along the line indicate the 'age' of the cloud in 10^5 yr, assuming that the gas is initially atomic-carbon rich (from: Suzuki et al. 1992).

indicates the time since some dynamical event produced an atomic-carbon rich phase.

The model results are also sensitive to the $[C]_g/[O]_g$ elemental abundance ratio in the gas. If this value is increased from the canonical ratio of 0.4 to higher values, improvement with observations for the complex molecules is obtained (Bergin et al. 1997, Terzieva & Herbst 1998). Pratap et al. (1997) show that the observed chemical gradient in TMC-1 can be reproduced by changes in $[C]_g/[O]_g$ along the ridge caused by a density gradient.

This point is further illustrated by high spatial resolution observations of the L1498 pre-stellar core by Kuiper et al. (1996), which show a chemically differentiated structure with NH_3 peaking in the inner region and C_2S in the atomic-carbon rich (clumpy) outer part. Further chemical studies of such cores, especially those which have a more centrally concentrated density structure and appear to be on the verge of collapse (e.g., Ward-Thompson et al. 1994, Motte et al. 1998), are warranted.

The ice mantles in quiescent dark clouds can be probed through the observation of field stars behind the clouds (e.g., Chiar et al. 1995, Whittet et al. 1998). H_2O, CO and CO_2 ices have been detected, which indicate that up to 40% of the heavy elements may be frozen out at densities of a few $\times 10^4$ cm^{-3} (Schutte 1999). Evidence for significant CO depletion deep inside dark clouds ($A_V > 20$ mag) also comes from comparison of C^{18}O observations with extinction maps derived from infrared star counts (Lada et al. 1994, Kramer et al. 1999).

5.3. IONIZATION FRACTION

The ionization fraction $x(e)=n(e)/n(H_2)$ is an important parameter in the dynamical evolution of star-forming regions (see chapters by McKee, Shu, and Mouschovias). In model calculations, $x(e)$ drops from its high value of $\sim 10^{-4}$ at the edge of the core to $\sim 10^{-8}$ in the center, and scales roughly with $(\zeta/n_H)^{1/2}$ (Millar 1990, Bergin & Langer 1997). At the edge, C$^+$ is the main supplier of electrons, whereas deep inside metal ions such as Mg$^+$ and Fe$^+$, as well as molecular ions such as H$_3^+$, HCO$^+$ and H$_3$O$^+$ dominate.

Recent observations of DCO$^+$/HCO$^+$ in a sample of dark cores give ionization fractions of a few $\times 10^{-7}$, for an adopted cosmic ray ionization rate of 5×10^{-17} s^{-1} (Plume et al. 1998, Williams et al. 1998). Surprisingly, no systematic trends between pre-stellar cores and cores with stars have been found, but this may be due to the large scale of the observations ($\sim$10,000 AU). A somewhat lower ionization fraction of 10^{-8}–10^{-7} is found for more massive cores (Bergin et al. 1999, de Boisanger et al. 1996).

6. Chemistry in cold envelopes around YSOs

6.1. MODELS

In the initial stages of collapse, the density increases strongly whereas the temperature stays low, $T \approx 10$ K. The principal prediction of the models is enhanced freeze-out of molecules onto the cold grains under these conditions. The only exceptions are H$_2$, He, H$_3^+$ and perhaps N$_2$. Models appropriate for the cold outer envelopes have been made by Rawlings et al. (1992), Willacy et al. (1994), Bergin & Langer (1997) and Shalabiea & Greenberg (1995), using parametrized fits to the density profiles in simple collapse models such as that of Shu (1977) or Basu & Mouschovias (1994). They differ strongly in the adopted mechanisms that return molecules to the gas, ranging from none in Rawlings et al.'s to efficient desorption in Shalabiea & Greenberg's. The results are also sensitive to the adopted binding energies on the icy mantles, especially whether the outer layer is H$_2$O-rich or CO-rich. The ions HCO$^+$ and N$_2$H$^+$ are predicted to be good tracers of cold envelopes, because their abundances remain high owing to the increase in the H$_3^+$ abundance when its main removal partners (CO, O, ...) are depleted. The use of HCO$^+$ to trace the envelope structure has been demonstrated by Hogerheijde et al. (1997, 1998).

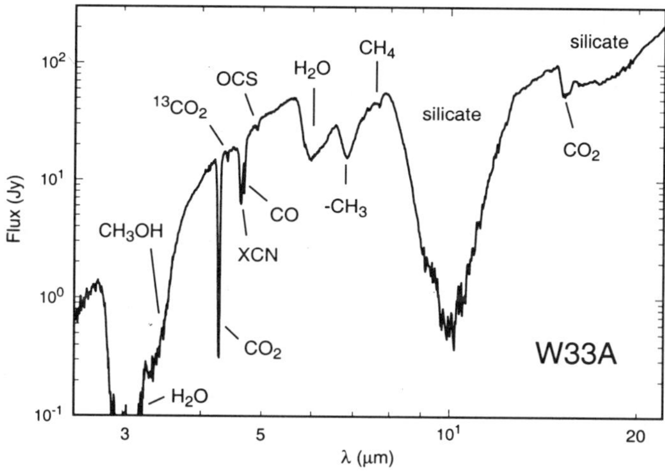

Figure 10. ISO–SWS spectrum of the deeply embedded massive YSO W 33A. Various absorption features due to silicate grain cores and icy mantles are indicated (from: Gibb et al. 1999).

6.2. OBSERVATIONS

Observationally, depletion in collapsing cores is very difficult to prove, because every cloud has a 'skin' in which the abundances are close to normal (Mundy & McMullin 1997). Even if the skin amounts to only a few % of the total column density, its factor of 10–100 higher abundances can effectively mask any depletions deep inside. The dust obscuration in the deeply embedded class 0 stage is still so high that the young stars are too weak at near- and mid-infrared wavelengths for direct observations of ices. Careful modeling of the line and dust emission appears the only way to probe the abundances. One of the best-studied cases is that of NGC 1333 IRAS4, where depletions of more than a factor of 10 have been inferred (Blake et al. 1995). This phase of high depletion appears short-lived, however, since for other class 0 objects like Serpens SMM 1 much less freeze-out has been inferred (see §8.1).

At the more evolved, but still deeply embedded class I stage, infrared absorption line studies of the ice mantles in the cold envelopes become possible, both for low- and high-mass objects (Whittet 1993, Tielens & Whittet 1997, Chiar et al. 1998, Teixeira et al. 1998). Ground-based and ISO studies allow a nearly complete inventory of these ices to be obtained for deeply embedded massive young stars such as NGC 7538 IRS9, W33 A and RAFGL 7009S (Whittet et al. 1996, d'Hendecourt et al. 1996, Ehrenfreund et al. 1997) (see Figure 10). The molecules that have been identified and their typical abundances are summarized in Table 3. Although only abundances down to $\sim 0.5\%$ of H_2O-ice ($\sim 10^{-7}$ with respect to H_2) can be probed, the inferred mantle composition is remarkably simple, consisting of species resulting from the hydrogenation and oxidation of O, C, N and CO. The total column density of gas-phase CO is larger than that of solid CO integrated along the line of sight for all sources studied so far.

7. Chemistry in warm envelopes around newly-formed stars

When a young star begins to heat the inner surrounding envelope by radiation and/or shocks above ~ 100 K, various distinct chemical regions occur, including the hot core region, the region of interaction of the outflow with the surrounding envelope, and the more extended radiatively-heated warm envelope (see Figure 11).

7.1. WARM ENVELOPES

Self-consistent models of the thermal balance, chemistry and radiative transfer in spherical envelopes heated by an internal source have been constructed for low-mass YSO's by Ceccarelli et al. (1996) and for high-mass YSOs by Doty & Neufeld (1997). Calculations of the radial dust temperature distribution include those of Campbell et al. (1995), Kaufman et al. (1998), and van der Tak et al. (1999). For a $\sim 10^5$ $L_\odot$ source, the region where $T_{\rm dust} \approx T_{\rm gas} > 90$ K extends to $\sim 10^{16}$ cm, resulting in evaporation of H_2O and CH_3OH ices and an enhancement of the gas-phase abundances of these species by a factor of 100–1000 (see Figure 11). The chemistry in this inner hot core is discussed in §7.2. The region where $T > 20$ K extends to $\sim 10^{17}$ cm. An increase in the temperature from 10 to 100 K does not have a significant effect on the gas phase chemistry, but does result in evaporation of volatile species like CO.

Observational data on molecules in star-forming regions are scattered throughout the literature, using a variety of (sub-)millimeter telescopes with different beam sizes. Comprehensive single-dish studies of a number of molecules have been performed for only a few sources, including W 3 IRS5 (Helmich & van Dishoeck 1997, see §8.2) and NGC 2264 IRS1 (Schreyer et al. 1997). More detailed physical models are needed to disentangle the contributions of changing temperatures and densities in the envelope from possible radial abundance gradients.

Important complementary information is obtained from high-resolution infrared observations. Mitchell et al. (1990) showed the presence of both cold ($T < 60$ K) and hot ($T = 120$–1000 K) gas along the lines of sight toward massive YSO's from ^{12}CO and ^{13}CO observations. The hot gas contains high abundances of C_2H_2, HCN and CH_4 (Lacy et al. 1989, 1991;

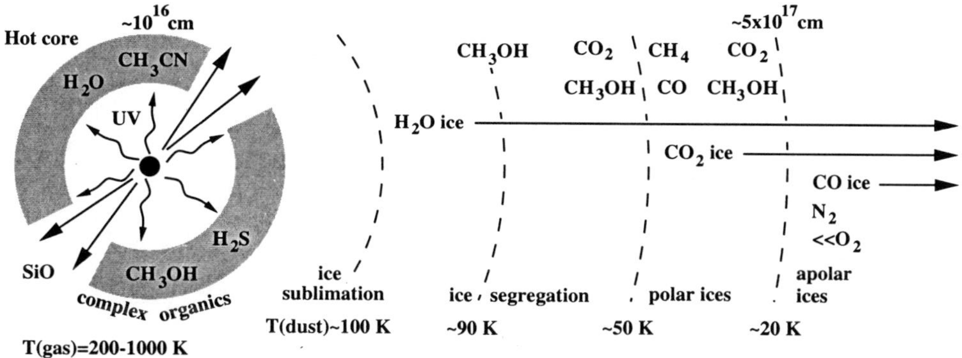

Figure 11. Schematic illustration of the physical and chemical environment of massive YSOs. The variation in the structure of the ice mantles due to heating and thermal desorption is shown (based on Tielens et al. 1991, Williams 1993, van Dishoeck & Blake 1998).

Evans et al. 1991; Carr et al. 1995; Boogert et al. 1998; Lahuis & van Dishoeck 1999), which are enhanced by at least 1–2 orders of magnitude compared with the colder envelope. Gas-phase CO_2 has a surprisingly low abundance (van Dishoeck et al. 1996, Dartois et al. 1998, van Dishoeck 1998b). Clear variations in the gas/solid ratios are seen for various objects, indicating the development of a hot core in the inner envelopes.

7.2. HOT CORE REGIONS

Hot cores are small (<0.1 pc), dense ($n(H_2)> 10^6$ cm^{-3}) and warm ($T \approx$ 200 K or higher) regions located close to massive young stars (see Walmsley & Schilke 1993, Ohishi 1997 for reviews). Observationally, hot cores are characterized by high abundances of fully hydrogenated molecules such as NH_3, H_2S, and H_2O (e.g., Gensheimer et al. 1996), as well as saturated complex organic molecules like CH_3OH, CH_3CN, CH_3OCH_3 and $HCOOCH_3$ (e.g., Hatchell et al. 1998). Examples of hot cores include the Orion hot core and compact ridge (e.g., Blake et al. 1987, Sutton et al. 1995), SgrB2(N) (e.g., Kuan & Snyder 1996), W 3(H_2O) (e.g., Helmich & van Dishoeck 1997) and objects near ultracompact H II regions such as G34.3+0.15 (Macdonald et al. 1996; see chapter by Churchwell). The observed abundances of the species can vary significantly from region to region.

The rich chemistry in hot cores is thought to be driven by the evaporation of icy mantles due to heating near the young star. Charnley et al. (1992) showed that the observed abundances in the Orion hot cores can be reproduced if a mixture of simple ices containing H_2O, CO, CH_3OH, NH_3 and/or HCN is evaporated into the warm gas. These molecules sub-

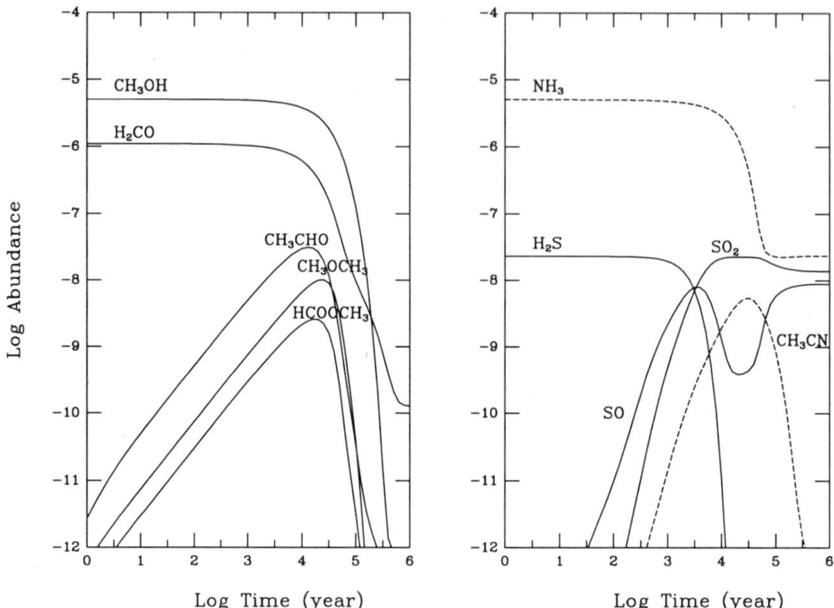

Figure 12. Models of the gas-phase chemical evolution during the 'hot core' phase for $T=200$ K. At $t=0$, molecules such as CH_3OH, NH_3 and H_2S are evaporated from the grain mantles into the gas phase, where they drive a rapid chemistry leading to more complex organic molecules such as CH_3OCH_3, $HCOOCH_3$ and CH_3CN. The abundances peak after $10^4 - 10^5$ years (based on Charnley et al. 1992, Charnley 1997).

sequently drive a rapid gas-phase chemistry leading to complex organic molecules (e.g., Charnley et al. 1992, Caselli et al. 1993, Charnley 1997). The abundances of the complex species peak after $\sim 10^4$ yr, thus offering a chemical 'clock' since the time of formation of the hot core (see Figure 12). At later times, the normal ion-molecule chemistry takes over, resulting in destruction of the complex organic molecules after $\sim 10^5$ yr.

Since methanol is one of the most abundant ice mantle consitutents, it plays a crucial role in the hot core chemistry. Specifically, complex oxygen-containing organics can be formed by protonation of CH_3OH by H_3^+, H_3O^+ or HCO^+, followed by transfer of a $-CH_3$ group to a neutral molecule, e.g., $CH_3OH_2^+ + CH_3OH \rightarrow (CH_3)_2OH^+ + H_2O$. Dissociative recombination of the ion then produces di-methyl ether, CH_3OCH_3. The detailed chemistry of interstellar alcohols leading to large esters and ethers has been discussed by Charnley et al. (1995). Nitrogen-containing organics can be formed through reactions of carbon ions with evaporated NH_3, or through reactions of CN with evaporated C_2H_2. In general, the time scale for formation of nitrogen-rich molecules is somewhat longer than that of oxygen-bearing species.

The H_2S/SO_2 ratio may be a particularly sensitive clock (Charnley 1997, Hatchell et al. 1998). Most of the sulfur is thought to be in the form of H_2S in the ices, although this has not yet been confirmed observationally. Upon evaporation, H_2S is destroyed by reactions with atomic H to atomic S at temperatures of a few hundred K. S can subsequently react with OH and O_2 to form SO and SO_2 on a timescale of $\sim 10^4$ yr. The observed high D/H ratios in hot core molecules (e.g., Millar & Hatchell 1998) largely reflect the high D/H ratios in the ices, which result from the cold dark cloud phase prior to star formation.

7.3. OUTFLOW REGIONS

The earliest stages of star formation are accompanied by powerful outflows, which originate within the young star/accretion disk boundary region. As the high velocity gas flows outward, it strikes the envelope and creates shocks, which alter the chemistry. Recent reviews of the physical and chemical structure of shocks can be found in Draine & McKee (1993), Hollenbach (1997) and Bachiller (1997, this volume).

If the shocks have high velocities (>50 km s^{-1}), the temperature increases to such high values ($> 10^4$ K) that most molecules are collisionally dissociated. In addition, the ultraviolet radiation from such J-type shocks can dissociate molecules both within and ahead of the shock. The molecules reform slowly in a lengthy, warm zone in the wake of the shock (Neufeld & Dalgarno 1989a,b).

For lower shock speeds in clouds with low fractional ionization ($x_e < 10^{-6}$), the shock is of C-type and the temperatures reach peak values of only 2000–3000 K, insufficient to dissociate the molecules. Reactions with energy barriers such as $O + H_2$ and $S + H_2$ rapidly drive the available gas-phase oxygen and sulfur into H_2O and H_2S.

In addition to these gas-phase processes, grain cores and mantles are affected by shocks. The high-velocity J-shocks can cause destruction or thermal sputtering of the cores, resulting locally in much enhanced gas-phase abundances of Si, Fe, ... The lower velocity C-shocks can inject a variety of refractory and volatile species into the gas phase through non-thermal sputtering (e.g., Pineau des Forêts & Flower 1997). The released Si can then react with OH and O_2 to form SiO, one of the principal diagnostics of shocks (e.g., Schilke et al. 1997).

Outflows have been studied observationally in a variety of molecules, including H_2 near-infrared emission, CO and SiO. One of the best studied regions chemically is L1157 (Bachiller & Peréz Gutiérrez 1997). Both refractory (SiO, SO, ...) and volatile mantle species (HCN, H_2CO, CH_3OH,

TABLE 4. Physical regions of class 0 YSOs traced by various observations

Region	Observation
Bulk of the (cold) envelope	Interferometer: Continuum on short spacings
	Single-dish: Lines tracing $T_{kin} < 40$ K
	(e.g., $C^{18}O$ 2–1; $C^{17}O$ 3–2; $H^{(13)}CO^+$ 3–2, 4–3)
Warm, inner regions of the envelope	Interferometer: Continuum on long spacings
	Single-dish: Lines tracing ~ 100 K
	(e.g., ^{13}CO 6–5; H_2CO 3_{22}–2_{21})
Outflow: gradual entrainment (?)	Interferometer: V or U shaped structures
	(e.g., HCO^+, HCN)
	Single-dish: line wings (e.g., CO, HCN)
Outflow: direct impact	Single-dish and interferometer: Products of
	ice mantle evaporation and grain destruction
	(e.g., SiO, SO)

...) are found to be enhanced in the gas by factors ranging from a few to 10^6. Strong H_2O lines have been observed in outflow regions from the ground (e.g., Cernicharo et al. 1994) and with the ISO satellite, indicating H_2O abundances of a few times 10^{-5} (e.g., Liseau et al. 1996, Ceccarelli et al. 1998, van Dishoeck et al. 1998).

8. Examples

8.1. AN INTERMEDIATE-MASS CLASS 0 YSO IN SERPENS

Recent work by Hogerheijde et al. (1999) of the class 0 YSO Serpens SMM 1 (FIRS 1) illustrates the use of molecular-line and continuum observations with single-dish and aperture-synthesis instruments to constrain the physical and chemical conditions in the envelope of a low- to intermediate-mass protostar. The Serpens molecular cloud ($d \approx 400$ pc) is in the process of forming a loosely bound cluster of more than 50 stars (Eiroa & Casali 1992). SMM 1 is one of the submillimeter-continuum class 0 YSOs identified by Casali et al. (1993) with $L_{bol} = 77$ $L_\odot$ and $M_\star = 0.7$–3.9 $M_\odot$. Table 4 gives an overview of the different physical components traced by the various observations.

The continuum observations are useful to constrain both the mass and the density structure of the envelope. The total mass is derived from the single dish data, whereas the density variation of the envelope is best obtained from the spatially resolved emission in the interferometer. The method is illustrated on the left-hand side of Figure 13. The dependence of the flux on

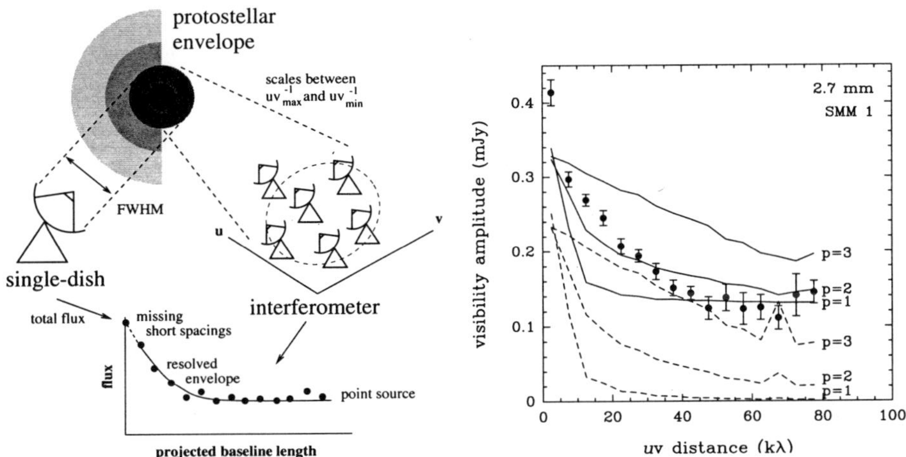

Figure 13.   Left: Schematic illustration of single-dish versus interferometer observations. Right: Comparison of model visibility amplitudes at 2.7 mm to observations of SMM 1. The observed vector-averaged amplitudes are indicated by the filled symbols and their 1σ error bars. The dashed lines are models without a central point source, and density power-law slopes of -1.0 (*lower curve in each panel*), -2.0 (*middle curve*), and -3.0 (*upper curve*). Solid lines are models with a point source flux of 0.13 Jy at 2.7 mm and a spectral slope of 2.0.

projected baseline length for SMM 1 (right hand side of Figure 13) indicates a power-law density distribution; different prescriptions like, e.g., a Gaussian distribution, are excluded by the data. A power-law index of -2.0 ± 0.5 gives a good fit to the observations, which agrees well with models for protostellar collapse (e.g., Shu 1977). The fluxes on long baselines suggest the presence of extra, unresolved emission, probably associated with the inner few hundred AU of the envelope where the temperature is likely to exceed the adopted $T_{\mathrm{dust}} \propto r^{-0.4}$ distribution. The absolute temperature scale is $\sim$27 K at a characteristic radius of 1000 AU, constrained by the spectral energy distribution over millimeter to far-infrared wavelengths. The total envelope mass is 8.7 $M_\odot$, with an uncertainty of a factor of 2–3 due to the adopted dust emissivity (Ossenkopf & Henning 1994). The fact that the inferred envelope mass is larger than the stellar mass confirms SMM 1's nature as a class 0 YSO.

Using this description for the physical structure of the envelope, the molecular abundances can be determined using models of the line emission. Hogerheijde et al. (1999) adopt a Monte-Carlo technique to solve the radiative transfer and the molecular excitation throughout the envelope. In addition to the physical conditions, the molecular line calculations re-

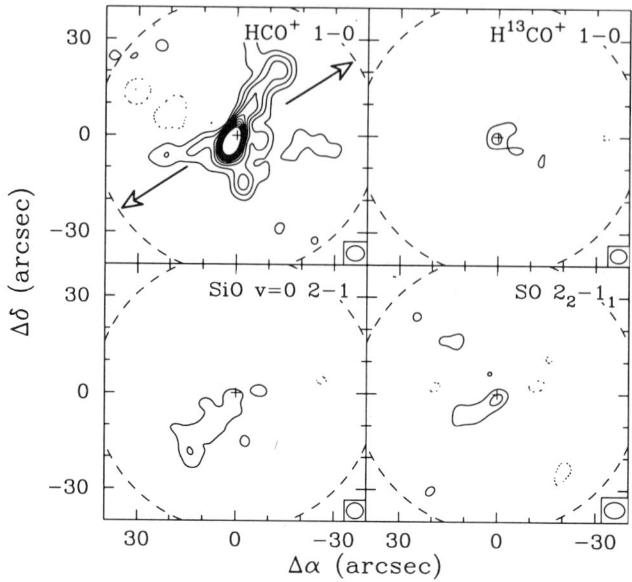

Figure 14. Cleaned, naturally weighted images of molecular line emission observed with OVRO toward SMM 1. Contours are drawn at 3σ intervals, of 1 K km s^{-1} for HCO$^+$ and SO, 8 K km s^{-1} for H^{13}CO$^+$, and 4 K km s^{-1} for SiO. The synthesized beam size is indicated in each panel. The dashed circle shows the primary beam size. The arrows in the HCO$^+$ panel indicate the position angle of the radio jet from Rodríguez et al. (1989) (from: Hogerheijde et al. 1999).

quire knowledge of the velocity structure. Systematic velocity fields like infall and outflow can be included, but in the simplest analysis a turbulent line width of $\sim$1.4 km s^{-1} independent of radius is adopted, based on optically thin C^{17}O 3–2 and H^{13}CO$^+$ 3–2 lines. The neglect of systematic velocity fields influences the results for optically thick lines like ^{12}CO and HCO$^+$. Note that the lines themselves can in principle also constrain the density structure (e.g., Hogerheijde et al. 1997), but are less accurate for low- to intermediate-mass YSOs because of lack of spatial resolution in the single-dish data and of signal-to-noise in the interferometer observations.

The envelope mass can also be constrained from the optically thin CO emission. If the 'standard' dark cloud CO/H$_2$ abundance of 10^{-4} is adopted, the model overestimates the emission in the C^{18}O 2–1 and C^{17}O 3–2 lines by a factor of 2. Apart from adjusting the CO abundance, a gas kinetic temperature of 0.6 times the dust temperature reconciles this discrepancy; $T_{\rm kin} < T_{\rm dust}$ is expected because of imperfect thermal coupling between the dust and the gas, and line cooling of the gas. Alternatively, CO can freeze out onto dust grains when the temperature drops below its sublimation temperature of $\sim$20 K (see Table 3), lowering the emission. A CO depletion of a factor of 6 in the cold outer layers of the envelope provides a good fit to the $C^{17}O$ 3–2 and $C^{18}O$ 2–1 lines. These lines are particularly good tracers of depletion below 20 K, since the 2–1 lines, tracing $T_{kin} \sim 16$ K gas, are stronger affected than the 3–2 lines, tracing $T_{kin} \sim 30$ K gas. The inferred depletion of a factor of ~ 6 for SMM 1 is smaller than found toward other class 0 envelopes (e.g., 10–20 for NGC 1333 IRAS 4, Blake et al. 1995), suggesting that the evolutionary phase characterized by large depletion factors may be short-lived, or that local environment (e.g., heating of the envelope from the outside) influences the depletion.

Most of the observed submillimeter transitions of other molecules trace material in excess of ~ 30 K, and the derived abundances do not depend critically on the exact value of T_{kin} or possible depletion below 20 K. Table 5 lists the inferred abundances for the parameters of the envelope derived from the dust emission. Compared to other well-studied class 0 YSOs such as IRAS 16293$-$2422 (van Dishoeck et al. 1995), the abundances of HCO^+, HCN, and H_2CO are very similar, while C_3H_2, CN, and HC_3N are enhanced by an order of magnitude in SMM 1. The values for SiO and SO are lower by a factor of 10, most likely because the $14''$–$19''$ single-dish beam of the SMM 1 observations does not include the tip of the outflow where these species peak (see Figure 14).

Many but not all lines are well fit by this model. High-excitation lines like ^{13}CO 6–5 and H_2CO 3_{22}–2_{21} indicate the presence of an additional amount of warm, ~ 100 K, material. The estimated column density of this gas amounts to less than 1% of the envelope mass, but dominates in emission of these highly excited lines. It is likely that this material corresponds to the inner few hundred AU of the envelope, where the temperature may exceed the adopted distribution. Some of the emission in the other lines may originate in this component as well. Table 5 includes the abundances derived under the assumption that all emission originates in this warm gas. These values are larger by typically an order of magnitude, illustrating that care needs to be taken to separate these two contributions.

Figure 14 shows the OVRO images of the low-excitation lines of various molecules. The emission in the line wings of optically thick lines of CO, HCO^+ and HCN is clearly associated with the bipolar outflow, as is that of SiO 2–1. HCO^+ outlines the outflow cavity, possibly tracing the slow entrainment of material into the flow. SiO coincides with the axis of the outflow, likely revealing destruction of dust particles by the direct impact of the outflow on ambient material. To derive abundances from these interferometer observations, the modeling has to include spatial filtering. For example, only 0.25 $M_\odot$ is traced by the optically thin $C^{18}O$ 1–0 interferometer observations, but if the resolving-out of extended emission is taken into account, this value is consistent with the total envelope mass

TABLE 5. Derived molecular abundances in the envelope of Serpens SMM 1[a]

Species	Envelope	Warm gas[b]	IRAS 16293−2442[c]
^{12}CO	$\equiv 1(-4)$	$\equiv 1(-4)$	...
HCO^+	$1(-9)$	$2(-8)$	$2(-9)$
HCN	$2(-9)$	$5(-8)$	$2(-9)$
H_2CO	$8(-10)$	$9(-9)$	$7(-10)$
C_3H_2	$2(-10)$	$3(-9)$	$4(-11)$
CN	$5(-9)$	$3(-8)$	$1(-10)$
HC_3N	$2(-10)$	$9(-10)$	$3(-11)$
SiO	$1(-11)$	$1(-10)$	$1(-10)$
SO	$2(-10)$	$2(-9)$	$4(-9)$

[a] From Hogerheijde et al. (1999).

[b] Abundances derived under the assumption that *all* emission originates in warm 100 K gas.

[c] From van Dishoeck et al. (1995).

of 8.7 $M_\odot$. Firmer conclusions about abundance variations on small scales requires more realistic, >1D models of the physical structure, including a correct treatment of the velocity field.

8.2. THE W 3 MASSIVE STAR-FORMING REGION

The W 3 massive star-forming region at $\sim$ 2.3 kpc provides an excellent opportunity to study the chemistry of massive YSOs ($L \approx 10^5 L_\odot$) at different evolutionary stages originating from the same parent cloud. Helmich et al. (1994) and Helmich & van Dishoeck (1997) performed an unbiased single-dish submillimeter spectral survey of three YSOs at 15″ (0.15 pc) resolution: IRS5, IRS4 and W 3(H_2O). Although such observations lack the spatial resolution of the data on nearby low-mass objects discussed above, they provide a useful global picture of the chemistry. In addition, the envelope mass associated with these objects is significantly larger than for low-mass YSOs, so that even minor species can be detected.

Figure 15 summarizes the spectra toward the three objects in the 345 GHz atmospheric window. Clear physical and chemical differences are found between the three sources, in spite of their similar luminosities. The beam-averaged densities are at least 10^6 cm^{-3} and the temperatures range from $\sim$ 55 K for IRS4 to at least 220 K for W 3(H_2O). Toward W 3(IRS5), silicon- and sulfur-bearing molecules such as SiO and SO_2 are prominent. This source has a powerful, massive outflow and is presumably the youngest of

CHEMICAL EVOLUTION IN THE W 3 MASSIVE STAR-FORMING REGION

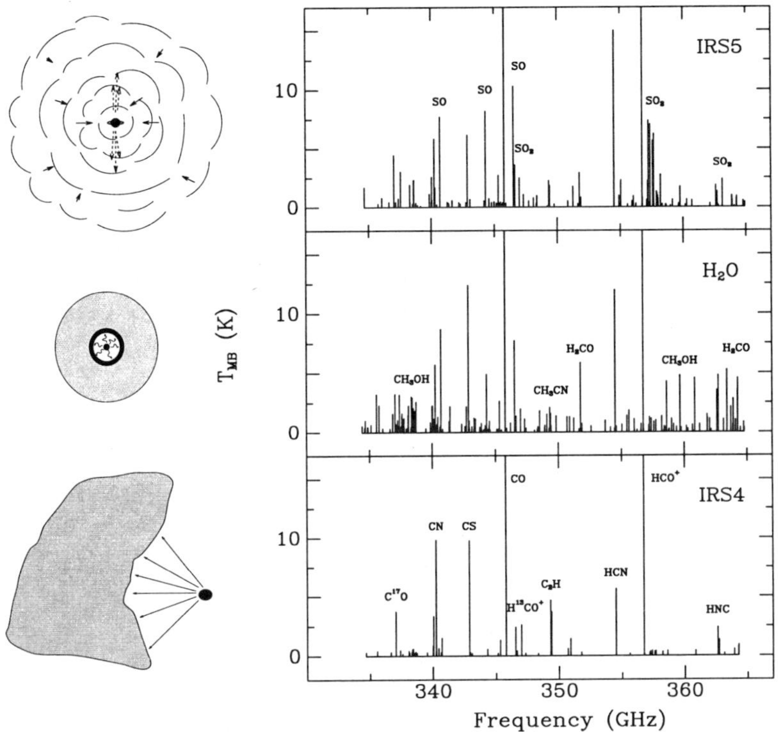

Figure 15.   Summary of the JCMT 335–365 GHz line survey of three massive star–forming regions in the W 3 molecular cloud. The spectra were constructed from the observed line intensities obtained in double–side-band mode. Strong lines in common in the three spectra are labelled in the W 3 IRS4 spectrum only. Large physical and chemical differences are found between the three regions, which are attributed to different evolutionary stages (Based on Helmich & van Dishoeck 1997).

the three sources, showing also strong ice absorption features in its infrared spectra. Toward W 3(H₂O), discovered by Turner & Welch (1984), organic molecules like CH₃OH, CH₃OCH₃, and CH₃OCHO are at least one order of magnitude more abundant, indicating that it is well into the hot core phase. Finally, only simple molecules and radicals are found toward IRS4. One possible interpretation is that this object is more evolved and has already broken free from the parent cloud, setting up an H II region and PDR on the back side of the cloud.

More detailed physical models, which constrained by multi-line observations of high-dipole molecules like CS, H₂CO, HCN and their isotopes, are needed to infer reliable abundances. Recent analyses of the massive YSO GL 2591 suggest a power-law density distribution with an index of -1.0 to -1.5, somewhat shallower than found for low-mass YSOs such as SMM 1 (Carr et al. 1995, van der Tak et al. 1999).

9. Chemistry in circumstellar disks

Most of this chapter has focused on the chemistry in the envelopes of deeply-embedded YSOs. Young stars with ages $1 - 5$ Myr have dispersed their envelopes but are still surrounded by circumstellar disks with masses of $\sim 10^{-3} - 10^{-1}$ $M_{\odot}$ and sizes of ~ 100 AU, comparable to those inferred for our primitive solar nebula (see Beckwith & Sargent 1996 for a review). Imaging of the gas in disks around T Tauri and Herbig Ae stars has so far been limited to CO and its isotopes, for reasons of sensitivity. Observations of other molecules are only just beginning. Single-dish surveys of various molecules have been carried out for a few objects, in particular DM Tau and GG Tau (Dutrey et al. 1997), TY Hya (Kastner et al. 1997) and LkCa 15, MWC 480 and HD 163296 (Thi et al., in preparation). Millimeter interferometers are now also capable of imaging the more abundant molecules in disks, showing interesting morphological differences (Qi et al., in preparation).

The detection of species like HCO^+, CN and HNC indicates that both ion-molecule, photon-dominated and gas-grain chemistry play a role. The amount of depletion is high because of the high densities (typically $10^6 - 10^9$ cm^{-3} in the outer disk) and appears to be species-dependent, with the most volatile molecules remaining in the gas furthest from the central star. Chemical models which couple the gas-dust chemistry with the dynamical evolution as the material is transported inwards have been developed recently by Aikawa et al. (1997, 1998) and Willacy et al. (1998). At small disk radii, irradiation from the central star and stellar X-rays may also affect the chemistry. These studies are important first steps to determine the connection between interstellar and solar nebula processes in the formation of icy planetesimals such as comets and Kuiper-Belt objects.

10. Concluding remarks

Significant progress has been made in the last decade on models and observations of the chemistry in star-forming regions. New (sub-)millimeter data at higher angular resolution and frequencies allow the dense and warm components of the YSO envelopes to be probed directly. Ground- and spaced-based infrared observations provide the first complete inventory of solid-state species along the line of sight and put constraints on gas/solid ratios of abundant species. The large variations in observed abundances illustrate that the chemistry clearly responds to the enormous changes in temperature and density during star-formation. Molecules freeze out onto grains in the cold pre-stellar cores and outer envelopes surrounding YSOs, where grain-surface chemistry produces new species. They are returned to the gas by heating due to radiation from the young stars and by shocks due to outflows impacting on the envelope. A rich chemistry can ensue in the hot gas, leading to abundant complex organic molecules and driving much of the available oxygen into water. Thus, these molecules provide important chemical and temporal diagnostics of the YSO environment.

The derivation of abundances from spectral line data is far from simple, however, and invariably includes assumptions about the underlying source structure. This physical structure can be constrained by analyzing spatially resolved continuum data, as illustrated for the case of Serpens SMM 1, and by using line ratios of appropriate molecules to infer temperatures and densities. Different approaches, ranging from the "homogeneous" to the "detailed" method, offer optimal strategies for different cases. Interferometric data require special care, because the instrument is a spatial filter which acts differently on lines tracing different physical conditions.

The development of comprehensive gas–grain models has made it possible to put the observational results into a coherent framework of the evolution of gas and solid species throughout the formation of protostars to their incorporation into circumstellar disks and eventually comets and planetesimals. Such models are rapidly progressing beyond the (pseudo) time-dependent method, and attempts are being made to couple the chemistry with (multi-)dimensional hydrodynamics. Accurate information on the basic molecular processes entering such models remains a prerequisite.

Future instrumentation, such as the next generation submillimeter arrays and air- or space-borne infrared telescopes such as SOFIA, SIRTF, FIRST and NGST, will allow these studies to be extended to much higher sensitivity and/or smaller scales of only a few AU. Together they should provide a much clearer view on the connection between matter found in interstellar clouds and in new planetary systems, and on the origin of organic matter found in primitive objects in our own solar system.

Acknowledgements

The authors are grateful to many colleagues for discussions and preprints of their work. They are indebted to the Leids Kerkhoven-Bosscha Fonds and the organizers for financial support. Research on Astrochemistry in Leiden is supported by the Netherlands Organization for Scientific Research (NWO). MRH is supported by the Miller Institute for Basic Research in Science.

References

Aikawa, Y., Umebayashi, T., Nakano, T., Miyama, S.M. (1997), *ApJ*, **486**, p. L51
Aikawa, Y., Umebayashi, T., Nakano, T., Miyama, S.M. (1998), *Faraday Disc.*, **109**, p. 281
Bachiller, R. (1997), in *Molecules in Astrophysics: Probes and Processes*, IAU Symposium 178, ed. E.F. van Dishoeck. Kluwer Academic Publishers, Dordrecht, p. 103
Bachiller, R. and Peréz Gutiérrez, M. (1997), *ApJ*, **487**, p. L93
Basu, S. and Mouschovias, T.C. (1994), *ApJ*, **432**, p. 720
Beckwith, S.V.W. and Sargent, A.I. (1996), *Nature*, **383**, p. 139
Benson, P.J., Caselli, P. and Myers, P.C. (1998), *ApJ*, **506**, p. 743
Bergin, E.A. and Langer, W.D. (1997), *ApJ*, **486**, p. 316
Bergin, E.A., Goldsmith, P.F., Snell, R.L. and Langer, W.D. (1997), *ApJ*, **482**, p. 285
Bergin, E.A., Plume, R., Williams, J.P. and Myers, P.C. (1999), *ApJ*, in press
Bernes, C. (1979), *A&A*, **73**, p. 67
Bernstein, M.P., Sandford, S.A., Allamandola, L.J., Chang, S. and Scharberg, M.A. (1995), *ApJ*, **454**, p. 327
Blake, D., Allamandola, L., Sandford, S., Hudgins, D., Freund, F. (1991), *Science*, **254**, pp. 548
Blake, G.A. (1997), in *Molecules in Astrophysics: Probes and Processes*, IAU Symposium 178, ed. E.F. van Dishoeck, Kluwer Academic Publishers, Dordrecht, p. 31
Blake, G.A., Sandell, G., van Dishoeck, E.F., Groesbeck, T.D., Mundy, L.G. and Aspin, C. (1995), *ApJ*, **441**, p. 689
Blake, G.A., Sutton, E.C., Masson, C.R. and Phillips, T.G. (1987), *ApJ*, **315**, p. 621
Boogert, A.C.A., Helmich, F.P., van Dishoeck, E.F., Schutte, W.A., Tielens, A.G.G.M. and Whittet, D.C.B. (1998), *A&A*, **336**, p. 352
Boogert, A.C.A., Ehrenfreund, P., Gerakines, P., et al. (1999), *A&A*, in press
Campbell, M.F., Butner, H.M., Harvey, P.M., Evans, N.J., Campbell, M.B. and Sabbey C.N. (1995), *ApJ*, **454**, p. 831
Carr, J.S., Evans, N.J., Lacy, J.L. and Zhou, S. (1995), *ApJ*, **450**, p. 667
Casali, M.M., Eiroa, C. and Duncan, W.D. (1993), *A&A*, **275**, p. 195
Caselli, P., Hasegawa, T.I. and Herbst, E. (1993), *ApJ*, **408**, p. 548
Caselli, P., Hasegawa T.I. and Herbst, E. (1998), *ApJ*, **495**, p. 309
Ceccarelli, C., Hollenbach, D.J. and Tielens, A.G.G.M. (1996), *ApJ*, **471**, p. 400
Ceccarelli, C., Castets, A., Loinard, L., Caux, E. and Tielens, A.G.G.M. (1998), *A&A*, **338**, p. L43
Cernicharo, J., González-Alfonso, E., Alcolea, J., Bachiller, R. and John, D. (1994), *ApJ*, **432**, p. L59
Charnley, S.B. (1997), *ApJ*, **481**, p. 396
Charnley, S.B, Kress, M.E., Tielens, A.G.G.M. and Millar, T.J. (1995), *ApJ*, **448**, p. 232
Charnley, S.B., Tielens, A.G.G.M. and Millar, T.J. (1992), *ApJ*, **399**, p. L71
Chiar, J.E., Adamson, A.J., Kerr, T.H. and Whittet, D.C.B. (1995), *ApJ*, **455**, p. 234
Chiar, J.E., Gerakines, P.A., Whittet, D.C.B., Pendleton, Y.J., Tielens, A.G.G.M., Adamson, A.J. and Boogert, A.C.A. (1998), *ApJ*, **498**, p. 716
Dalgarno, A. (1987), in *Physical Processes in Interstellar Clouds*, eds. G. Morfill and M.S. Scholer, D. Reidel, Dordrecht, p. 219
Dalgarno, A. (1994), *Adv. At. Mol. & Opt. Phys.*, **32**, p. 57
Dartois, E., d'Hendecourt, L., Boulanger, F., Jourdain de Muizon, M., Breitfellner, M., Puget, J.-L., and Habing, H.J. (1998), *A&A*, **331**, p. 651
de Boisanger, C., Helmich, F.P. and van Dishoeck, E.F. (1996), *A&A*, **310**, p. 315
d'Hendecourt, L., Jourdain de Muizon, M., Dartois, E., Breitfellner, M., Ehrenfreund, P., Benit, J., Boulanger, F., Puget, J.L. and Habing H.J. (1996), *A&A*, **315**, p. L365
Doty, S.D. and Neufeld, D.A. (1997), *ApJ*, **489**, p. 122
Draine, B.T. (1978), *ApJS*, **36**, p. 595
Draine, B.T. and McKee, C.F. (1993), *ARA&A*, **31**, p. 373

Dutrey, A., Guilloteau, S. and Guélin, M. (1997), *A&A*, **317**, p. L55

Eiroa, C. and Casali, M.M. (1992), *A&A*, **262**, p. 468

Ehrenfreund, P., d'Hendecourt, L., Dartois, E., Jourdain de Muizon, M., Breitfellner, M., Puget, J.L. and Habing, H.J. (1997), *Icarus*, **130**, p. 1

Ehrenfreund, P., Dartois, E., Demyk, K. and d'Hendecourt, L. (1998), *A&A*, **339**, p. L17

Evans, N.J. (1980), in *Interstellar Molecules*, Proc. IAU 87, ed. B.H. Andrew, D. Reidel, Dordrecht, p. 1

Evans, N.J., Lacy, J.H. and Carr, J.S. (1991), *ApJ*, **383**, p. 674

Flower, D. (1990). Molecular Collisions in the Interstellar Medium, Cambridge University Press

Frerking, M.A., Langer, W.D. and Wilson, R.W. (1982), *ApJ*, **262**, p. 590

Geballe, T.R. and Oka, T. (1996), *Nature*, **384**, p. 334

Gensheimer, P.D., Mauersberger, R. and Wilson, T.L. (1996), *A&A*, **314**, p. 281

Genzel, R. (1992), in *The Galactic Interstellar Medium*, eds. D. Pfenniger and P. Bartholdi, Saas Fee Advanced Course 21, Springer, Berlin, p. 275

Gerakines, P.A., Schutte, W.A. and Ehrenfreund, P. (1996), *A&A*, **312**, p. 289

Gerakines, P.A., Whittet, D.C.B., Ehrenfreund, P. et al. (1999), *ApJ*, in press

Gibb, E., Whittet, D.C.B., et al. (1999), in preparation

Gredel, R., van Dishoeck, E.F. and Black, J.H. (1993), *A&A*, **269**, p. 477

Gredel, R., Lepp, S., Dalgarno, A. and Herbst, E. (1989), *ApJ*, **347**, p. 289

Green, S. (1975), in Atomic and molecular physics and the interstellar matter, ed. R. Balian et al., Elsevier North Holland, Amsterdam, p. 83

Habing, H.J. (1988), in *Millimetre and Submillimetre Astronomy*, ed. R.D. Wolstencroft and W.B. Burton , Kluwer Academic Publishers, Dordrecht, p. 207

Hartquist T.W., Caselli, P., Rawlings, J.M.C., Ruffle, D.P., Williams, D.A. (1998), in *The Molecular Astrophysics of Stars and Galaxies*, ed. T.W. Hartquist and D.A. Williams, Oxford, Oxford University, p. 101

Hasegawa, T.I. and Herbst, E. (1993), *MNRAS*, **261**, p. 83

Hatchell, J., Thompson, M.A., Millar, T.J. and Macdonald, G.H. (1998), *A&AS*, **133**, 29

Helmich, F.P. and van Dishoeck, E.F. (1997), *A&AS*, **124**, p. 205

Helmich, F.P., Jansen, D.J., de Graauw, Th., Groesbeck, T.D. and van Dishoeck, E.F. (1994), *A&A*, **283**, p. 626

Herbst, E. (1993), in *Dust and Chemistry in Astronomy*, eds. T.J. Millar and D.A. Williams, IOP, Bristol, p. 183

Herbst, E. (1995), *Ann. Rev. Phys. Chem.*, **46**, p. 27

Herbst, E. and Leung, C.M. (1986), ApJ, **310**, 378.

Herbst, E. and Klemperer, W. (1973), *ApJ*, **185**, p. 505

Hogerheijde, M.R., Jansen, D.J. and van Dishoeck, E.F. (1995), *A&A*, **294**, p. 792

Hogerheijde, M.R., van Dishoeck, E.F., Blake, G.A. and van Langevelde, H.J. (1997), *ApJ*, **489**, p. 293

Hogerheijde, M.R., van Dishoeck, E.F., Blake, G.A. and van Langevelde, H.J. (1998), *ApJ*, **502**, p. 315

Hogerheijde, M.R., van Dishoeck, E.F., Salverda, J.M. and Blake, G.A. (1999), *ApJ*, in press

Hollenbach, D.J. (1997), in *Herbig-Haro Objects and the Birth of Low-mass Stars, IAU Symposium 182*, ed. B. Reipurth & C. Bertout, Kluwer Academic Publishers, Dordrecht, p. 181

Hollenbach, D.J. and Salpeter, E.E. (1971), *ApJ*, **163**, p. 155

Hollenbach, D.J. and Tielens, A.G.G.M. (1997), *ARA&A*, **35**, p.179

Hollenbach, D.J. and Tielens, A.G.G.M. (1999), *Rev. Mod. Phys.*, in press

Irvine, W.M. (1998), in *Origins of Life and Evolution of the Biosphere*, **28**, p. 365

Irvine W.M., Schloerb, F.P., Crovisier, J., Fegley, B., and Mumma, M.J. (1999), *Protostars & Planets IV*, eds. V.G. Mannings and S. Russell, University of Arizona, in press

Kaiser, R.I. and Roessler, K. (1997), *ApJ*, **475**, p. 487

Kaiser, R.I., Stranges, D., Lee, Y.T. and Suits, A.G. (1997), *ApJ*, **477**, p. 982

Kastner, J.H., Zuckerman, B., Weintraub, D.A. and Forveille, T. (1997), *Science*, **277**, p. 67

Kaufman, M.J., Hollenbach, D.J. and Tielens, A.G.G.M. (1998), *ApJ*, **497**, p. 276

Kramer, C., Alves, J., Lada, C.J., Lada, E.A., Sievers, A., Ungerechts, H. and Walmsley, C.M. (1999), *A&A*, **329**, p. L33

Kuan, Y.J. and Snyder, L.E. (1996), *ApJ*, **470**, p. 981

Kuiper, T.B.H., Langer, W.D. and Velusamy, T. (1996), *ApJ*, **468**, p. 761

Lacy, J.H., Carr, J.S., Evans, N.J., Baas, F., Achtermann, J.M. and Arens, J.F. (1991), *ApJ*, **376**, p. 556

Lacy, J.H., Evans, N.J., Achtermann, J.M., Bruce, D.E., Arens, J.F. and Carr, J.S. (1989), *ApJ*, **342**, p. L43

Lacy, J.H., Knacke, R., Geballe, T.R. and Tokunaga, A.T. (1994), *ApJ*, **428**, p. L69

Lacy, J.H., Faraji, H., Sandford, S.A. and Allamandola, L.J. (1998), *ApJ*, **501**, p. L105

Lada, C.J., Lada, E.A., Clemens, D. and Bally, J. (1994). *ApJ*, **429**, p. 694

Lahuis, F. and van Dishoeck, E.F. (1999), *A&A*, to be submitted

Langer, W.D., Velusamy, T., Kuiper, T.B.H., Peng, R., McCarthy, M.C., Travers, M.J., Kovacs, A., Gottlieb, C.A. and Thaddeus, P. (1997), *ApJ*, **480**, p. L63

Langer, W.D., van Dishoeck, E.F., Blake, G.A. et al. (1999), *Protostars & Planets IV*, eds. V.G. Mannings and S. Russell, University of Arizona, in press

Larsson, M. (1997), *Ann. Rev. Phys. Chem.*, **48**, 151

le Bourlot, J., Pineau des Forêts, G., Roueff, E. and Flower, D.R. (1995), *A&A*, **302**, p. 870

Lee, H.-H., Bettens, R.P.A. and Herbst, E. (1996a), *A&AS*, **119**, p. 111

Lee, H.-H., Herbst, E., Pineau des Forêts, G., Roueff, E. and le Bourlot, J. (1996b), *A&A*, **311**, p. 690

Lee, H.-H., Roueff, E., Pineau des Forêts, G., Shalabiea, O., Terzieva, R. and Herbst, E. (1998), *A&A*, **334**, p. 1047

Liseau, R., Ceccarelli, C., Larsson, B. et al. (1996), *A&A*, **315**, p. L181

Lucas, R. and Liszt, H. (1997), in *Molecules in Astrophysics: Probes and Processes*, IAU Symposium 178, ed. E.F. van Dishoeck, Kluwer Academic Publishers, Dordrecht, p. 421

Macdonald, G.H., Gibb, A.G., Habing, R.J. and Millar, T.J. (1996), *A&AS*, **119**, p. 333

Maloney, P.R., Hollenbach, D.J. and Tielens, A.G.G.M. (1996), *ApJ*, **466**, p. 561

Marechal, P., Pagani, L., Langer, W.D. and Castets, A. (1997), *A&A*, **318**, p. 252

McCall, B.J., Hinkle, K.H., Geballe, T.R. and Oka, T. (1998), *J. Chem Soc. Far. Disc.*, **109**, p. 267

Melnick, G., Stauffer, J.R., Ashby, M. et al. (1998), BAAS, 193, 72.01

Meyer, D.M. (1997), in *Molecules in Astrophysics: Probes and Processes*, IAU Symposium 178, ed. E.F. van Dishoeck, Kluwer Academic Publishers, Dordrecht, p. 407

Millar, T.J. (1990), in *Molecular Astrophysics —A volume honoring Alexander Dalgarno*, ed. T.W. Hartquist, Cambridge University Press, p. 115

Millar, T.J., Bennett, A. and Herbst, E. (1989), *ApJ*, **340**, p. 906

Millar, T.J., Farquhar, P.R.A. and Willacy, K. (1997a), *A&AS*, **121**, p. 139

Millar, T.J., Macdonald, G.H. and Gibb, A.G. (1997b), *A&A*, **325**, p. 1163

Millar, T.J. and Hatchell, J. (1998), *J. Chem Soc. Far. Disc.*, **109**, p. 15

Mitchell, G.F., Maillard, J.-P., Allen, M., Beer, R. and Belcourt, K. (1990), *ApJ*, **363**, p. 554

Moore, M.H., Donn, B., Khanna, R. and A'Hearn, M.F. (1983), *Icarus*, **54**, p. 388

Motte, F., André, P. and Neri, R. (1998), *A&A*, **336**, p. 150

Mundy, L. and McMullin, J.P. (1997), in *Molecules in Astrophysics: Probes and Processes*, IAU Symposium 178, ed. E.F. van Dishoeck, Kluwer Academic Publishers, Dordrecht, p. 183

Myers, P.C. and Benson, P.J. (1983), *ApJ*, **266**, p. 309

Neufeld, D.A. and Dalgarno, A. (1989a), *ApJ*, **340**, p. 869

Neufeld, D.A. and Dalgarno, A. (1989b), *ApJ*, **344**, p. 251

Ohishi, M. (1997), in *Molecules in Astrophysics: Probes and Processes*, IAU Symposium 178, ed. E.F. van Dishoeck, Kluwer Academic Publishers, Dordrecht, p. 61

Olano, C.A., Walmsley, C.M. and Wilson, T.L. (1988), *A&A*, **196**, p. 194

Olofsson, G., Pagani, L., Tauber, J., et al. (1998), *A&A*, **339**, p. 81

Ossenkopf, V. and Henning, T. (1994), *A&A*, **291**, p. 943

Osterbrock (1989), *Astrophysics of Gaseous Nebulae and Active Galactic Nuclei*, University Science Books, Mill Valley

Pineau des Forêts, G. and Flower, D.R. (1997), in *Molecules in Astrophysics: Probes and Processes*, IAU Symposium 178, ed. E.F. van Dishoeck, Kluwer Academic Publishers, Dordrecht, p. 113

Plume, R., Bergin, E.A., Williams, J.P. and Myers, P.C. (1998), *J. Chem Soc. Far. Disc.*, **109**, p. 47

Prasad, S.S. and Huntress, W.T. (1980), *ApJS*, **43**, p. 1

Pratap, P., Dickens, J.E., Snell, R.L. et al. (1997), *ApJ*, **486**, p. 862

Rawlings, J.M.C., Hartquist, T.W., Menten, K.M. and Williams, D.A. (1992), *MNRAS*, **255**, p. 471

Roberge, W.G., Jones, D., Lepp, S. and Dalgarno, A. (1991), *ApJS*, **77**, p. 287

Rodríguez, L.F., Curiel, S., Moran, J., Mirabel, I.F., Roth, M. and Garay, G. (1989), *ApJ*, **346**, p. L85

Sandford, S.A. and Allamandola, L.J. (1990), *Icarus*, **87**, p. 188

Sandford, S.A. and Allamandola, L.J. (1993), *ApJ*, **417**, p. 815

Schilke, P., Walmsley, C.M., Pineau des Forêts, G. and Flower, D.R. (1997), *A&A*, **321**, p. 293

Schreyer, K., Helmich, F.P., van Dishoeck, E.F. and Henning, T. (1997), *A&A*, **326**, p. 347

Schutte, W.A. (1996), in *The Cosmic Dust Connection*, ed. J.M. Greenberg, Kluwer Academic Publishers, Dordrecht, p. 1

Schutte, W.A. (1999), in *Laboratory astrophysics and space missions*, eds. P. Ehrenfreund et al., Kluwer Academic Publishers, Dordrecht, p. 69

Schutte, W.A., Tielens, A.G.G.M. and Allamandola, L.J. (1993), *ApJ*, **415**, p. 397

Shalabiea, O. and Greenberg, J.M. (1995), *A&A*, **296**, p. 779

Shalabiea, O.M., Caselli, P. and Herbst, E. (1998), *ApJ*, **502**, p. 652

Shu, F.H. (1977), *ApJ*, **214**, p. 488

Shu, F.H., Adams, F.C., and Lizano, S. (1987), *ARA&A*, **25**, p. 23

Smith, I.W.M. (1997), in *Molecules in Astrophysics: Probes and Processes*, IAU Symposium 178, ed. E.F. van Dishoeck, Kluwer Academic Publishers, Dordrecht, p. 253

Sobolev, V.V. (1960), *Moving Envelopes of Stars*, Harvard University Press, Cambridge

Sternberg, A., Yan, M. and Dalgarno, A. (1997), in *Molecules in Astrophysics: Probes and Processes*, IAU Symposium 178, ed. E.F. van Dishoeck, Kluwer Academic Publishers, Dordrecht, p. 141

Sutton, E.C., Peng, R., Danchi, W.C., Jaminet, P.A., Sandell, G., and Russell, A.P.G. (1995), *ApJS*, **97**, p. 455

Suzuki, H., Yamamoto, S., Ohishi, M. et al., (1992), *ApJ*, **392**, p. 551

Teixeira, T.C., Emerson, J.P. and Palumbo, M.E. (1998), *A&A*, **330**, p. 711

Terzieva, R. and Herbst, E. (1998), *ApJ*, **501**, p. 207

Tielens, A.G.G.M. (1983), *A&A*, **119**, p. 177

Tielens, A.G.G.M. and Allamandola, L.J. (1987), in *Interstellar Processes*, eds. D. Hollenbach and H.A. Thronson, D. Reidel, Dordrecht, p. 379

Tielens, A.G.G.M. and Charnley, S.B. (1997), *Origin of Life*, **27**, p. 23

Tielens, A.G.G.M. and Hagen, W. (1982), *A&A*, **114**, p. 245

Tielens, A.G.G.M., Tokunaga, A.T., Geballe, T.R. and Baas, F. (1991), *ApJ*, **381**, p. 181

Tielens, A.G.G.M. and Whittet, D.C.B. (1997), in *Molecules in Astrophysics: Probes and Processes*, IAU Symposium 178, ed. E.F. van Dishoeck, Kluwer Academic Publishers,

Dordrecht, p. 45

Turner, B.E. (1996), *ApJ*, **468**, p. 694.

Turner, B.E. (1998), *ApJ*, **495**, p. 804

Turner, B.E., Lee, H.H. and Herbst, E. (1998), *ApJS*, **115**, p. 91

Turner, J.L. and Welch, W.J. (1984), *ApJ*, **287**, p. L81

Vandenbusssche, B., Ehrenfreund, P., Boogert, A.C.A., et al. (1999), *A&A*, submitted

van der Tak, F., van Dishoeck, E.F., Evans, N.J., Bakker, E. and Blake, G.A. (1999), *ApJ*, submitted

van Dishoeck, E.F. (1988), in *Millimetre and Submillimetre Astronomy*, ed. R.D. Wolstencroft and W.B. Burton, Kluwer Academic Publishers, Dordrecht, p. 117

van Dishoeck, E.F. (1998a), in *The Molecular Astrophysics of Stars and Galaxies*, ed. T.W. Hartquist and D.A. Williams, Oxford, Oxford University, p. 53

van Dishoeck, E.F. (1998b), *Faraday Disc.*, **109**, p. 31

van Dishoeck, E.F. and Black, J.H. (1987), in *Physical Processes in Interstellar Clouds*, eds. G. Morfill and M.S. Scholer, D. Reidel, Dordrecht, p. 241

van Dishoeck, E.F. and Black, J.H. (1988), *ApJ*, **334**, p. 771

van Dishoeck, E.F. and Blake, G.A. (1998), *ARA&A*, **36**, p. 317

van Dishoeck, E.F., Blake, G.A., Draine, B.T., Lunine, J.I. (1993), in *Protostars and Planets III*, eds. E.H. Levy and J.I. Lunine, Univ. of Arizona, p. 163

van Dishoeck, E.F., Blake, G.A., Jansen, D.J. and Groesbeck, T.D. (1995), *ApJ*, **447**, p. 760

van Dishoeck, E.F., Helmich, F.P., de Graauw, Th. et al. (1996), *A&A*, **315**, p. L349

van Dishoeck, E.F., Wright, C.M., Cernicharo, J., et al. (1998), *ApJ*, **502**, p. L173

van Dishoeck, E.F., Helmich, F.P., Schutte, W.A., et al. 1998, in *Star Formation with the Infrared Space Observatory*, eds. J. Yun and R. Liseau, ASP vol. 132, p. 54

Vejby-Christensen, L., Andersen, L.H., Heber, O., Kella, D., Pedersen, H.B., Schmidt, H. and Zajfman, D. (1997), *ApJ*, **483**, p. 531

Walmsley, C.M. (1987), in *Physical Processes in Interstellar Clouds*, eds. G. Morfill and M.S. Scholer, D. Reidel, Dordrecht, p. 161

Walmsley, C.M. (1991), in *Fragmentation of molecular clouds and star formation*, IAU Symposium 147, eds. E. Falgarone et al., Kluwer Academic Publishers, Dordrecht, p. 161

Walmsley, C.M. and Schilke, P. (1993), in *Dust and Chemistry in Astronomy*, eds. T.J. Millar and D.A. Williams, IOP Publishing, Bristol, p. 37

Ward-Thompson, D., Scott, P.F., Hills, R.E. and André, P. (1994), *MNRAS*, **268**, p. 276

Whittet, D.C.B. (1993), in *Dust and Chemistry in Astronomy*, eds. T.J. Millar and D.A. Williams, IOP Publishing, Bristol, p. 9.

Whittet, D.C.B., Gerakines, P.A., Tielens, A.G.G.M. et al. (1998), *ApJ*, **498**, p. L159

Whittet, D.C.B., Schutte, W.A., Tielens, A.G.G.M. et al. (1996), *A&A*, **315**, p. L357

Willacy, K.R., Rawlings, J.M.C. and Williams, D.A. (1994), *MNRAS*, **269**, p. 921

Willacy, K.R., Klahr, H.H., Millar, T.J. and Henning, Th. (1998), *A&A*, **338**, p. 995

Williams, D.A. (1993), in *Dust and Chemistry in Astronomy*, eds. T.J. Millar and D.A. Williams, IOP Publishing, Bristol, p. 143

Williams, J.P., Bergin, E.A., Caselli, P., Myers, P.C. and Plume, R. (1998), *ApJ*, **503**, p. 689

Wootten, A. (1987), in *Astrochemistry*, eds. M.S. Vardya and S.P. Tarafdar, Kluwer Academic Publishers, Dordrecht, p. 311

II – THE ORIGIN AND EARLY EVOLUTION
OF LOW MASS STARS

Claudia Lavalley receives her first place poster prize from Professors Field and Lada.

THE FORMATION OF LOW MASS STARS

An Observational Overview

CHARLES J. LADA
Smithsonian Astrophysical Observatory
60 Garden Street
Cambridge, MA 02138 USA

1. Introduction: The Continuing Process of Star Formation

The question of the origin of stars is one of the most fundamental of astronomy. Yet, despite thousands of years of stellar observation, and early speculations by Newton and Laplace, it has only been in the latter part of the present century that the investigation of star formation has become an active discipline of astrophysical research. That the origin of stars has remained so mysterious for so long is largely due to the fact that the process of star formation has never been directly observed either with the naked eye or the with most powerful telescopes. Moreover, prior to the twentieth century neither the energy source nor the bulk composition of stars were known. Indeed, without knowledge of the physical nature of stars it was very difficult to develop an understanding of their origin. At the end of the nineteenth century, the nature of the mystery surrounding stellar origins was nicely summed up by the fictional character Huck Finn (in the book The Adventures of Huck Finn by Mark Twain) when he observed the stars in the night sky and wondered *"...did they just happen or was they made?"*

Have the stars been around since creation or were they being made in the heavens? As far as anybody could tell at dawn of this century, the entire universe consisted of a single stellar system, stars lived forever and the question of the origin of stars was a question of cosmology. As such the problem of star formation was not susceptible to detailed scientific investigation. That is, even the most basic hypotheses concerning stellar origins could not be directly tested by observation. Fortunately, over the last half century astronomical research has lead to the fundamental realization that star formation has been a continuous, ongoing process throughout the history of the Galaxy and the universe. This critical fact has been demonstrated by both theory and observation.

143

C.J. Lada and N.D. Kylafis (eds.), The Origin of Stars and Planetary Systems, 143–192.

Perhaps the most important step contributing to this realization was the discovery of the nature of stars as natural thermonuclear reactors which fuse hydrogen, the primary product of the Big Bang, into helium and then ultimately into the heavier elements which make up the periodic table. This process is elegantly explained by the theory of stellar structure and evolution, perhaps the greatest theoretical achievement of twentieth century astronomy. Although this great theory does not in any way account for or predict the formation of stars (and is in this sense incomplete), it has successfully explained the basic physical properties of stars as well as the processes of stellar evolution and death once stars exhaust their nuclear fuel reserves. In particular, stellar evolution theory demonstrated that certain luminous stars, OB stars, burn their nuclear fuel at such prodigious rates that they can live for only a small fraction of the lifetime of our galaxy. The existence of such stars clearly indicates that star formation has occured in the present epoch of Galactic history.

The relative youth of OB stars was also clearly demonstrated by observations of their spatial distribution on the sky. In 1947 the Armenian astronomer V.A. Ambartsumian showed that OB stars were almost always members of stellar groupings he termed OB Associations. Furthermore he found that the space densities of stars in OB associations were well below the threshold necessary to prevent their disruption by Galactic tidal forces. Ambartsumian calculated dynamical ages for the associations that were much less than the age of the galaxy. These dynamical ages turned out to be in good agreement with the nuclear ages of the stars and independently provided evidence that star formation is still an active process in the Galaxy.

The discovery of the interstellar medium of gas and dust during the early part of the twentieth century provided a crucial piece of corroborating evidence in support of the concept of present epoch Galactic star formation. Subsequent observations of interstellar material established that clouds of interstellar gas and dust had roughly stellar composition and were considerably more massive than a single star or group of stars. This revealed that the raw material to make new stars was relatively abundant in the Galaxy. These three pieces of evidence, 1) stellar evolution theory, 2) expanding OB associations and 3) the interstellar medium, constitute three basic "proofs" of ongoing star formation in the Milky Way.

Because star formation is occurring in the present epoch, the question of the origin of stars can be investigated by direct observation. If this were not the case, the prospects for completing the theory of stellar structure and evolution with a description of how stars form would be bleak. Yet, for most of the century, direct observation of the star formation process and the development of a theory to explain it, were severely hampered by the fact that most stars form in dark clouds and during their formative stages are invisible optically. Fortunately, advances in observational technology over the last quarter century opened the infrared and millimeter-wave windows to astronomical investigation and enabled direct observations of star forming regions and this has significantly expanded our knowledge of the star formation process. As a result the foundations for a coherent theory of star formation and early evolution are being laid.

In this chapter I will attempt to present a general overview of the critical observations which form the basis of our current understanding of the origins of low mass, sunlike, stars. The present review significantly updates earlier treatments of this topic which I wrote for the first Crete volume [71] and later the Proceedings of the Seventh Guo Shoujing Summer School on Astrophysics [135].

2. Stellar Observations: The Fossil Record

2.1. ASSOCIATIONS AND CLUSTERS

Stars in the disk of our galaxy can be categorized by their spatial distribution as being either members of the general Galactic field population or members of associations or clusters. An association or cluster is defined as a group of stars of the same physical type whose surface density significantly exceeds that of the field for stars of the same physical type [72]. Associations differ from clusters in their space densities: associations have space densities well below that (i.e., 0.1 $M_\odot$ pc^{-1}) [21] required for stability against galactic tides while clusters have space densities (e.g., $\geq$ 1.0 $M_\odot$ pc^{-1}) considerably in excess of the tidal stability limit. The spatial extents of associations are typically very large, 50 - 150 parsecs, while clusters are typically small in size ($\approx$ 1 parsec in diameter). Only a very small fraction of the stars in the disk of the Galaxy are presently members of associations or clusters. However, virtually all known O stars are members of OB associations. This indicates that all O stars formed in OB associations and results from the fact that the dynamical ages of OB associations (i.e., approximately 10^7 years) are greater than the lifetimes of the individual O stars ($\sim 10^6$ yrs.). About 10% of all B stars are members of associations. But when one takes into account that the lifetime of a B star is an order of magnitude larger than the lifetime of an association, it is clear that all present day B stars probably formed in OB associations. [18, ?]

OB associations contain considerably more low mass stars than OB stars, but the spatial density of these fainter stars is not that much greater than that of faint field stars. Consequently, the low mass component of OB associations is very difficult to identify from stellar density counts alone. If a stellar association is sufficiently young ($\leq 2 - 3 \times 10^6$ yrs.) then many of its low mass stars are likely to be emission-line (T Tauri type) stars. Emission-line stars are relatively rare in the field and groups of young, low mass T-Tauri stars can be easily identified as T associations, even if their surface densities are relatively low. In most instances T associations spatially coexist with OB associations [35] and trace the low mass component of such stellar groups. As will be discussed later, observations of very young associations (those embedded in molecular clouds) suggest the ratio of OB stars to low mass stars is roughly given by the initial mass function for local field stars. Consequently, it follows that OB associations can account for the formation of nearly all stars being born in the present epoch of galactic history. However, it is interesting to note that if OB associations were producing stars over the age of the Galaxy at the rate currently observed they would account for only about 10% of the stellar mass of the Milky Way. Open clusters are stellar systems which are stable against galactic tides and have considerably longer lifetimes (10^8 yrs.) than associations. Roughly 10% of all stars being formed end up as members of bound open clusters [109]. However most open clusters themselves probably formed inside OB associations.

The internal radial velocity dispersions of OB association members are typically on the order of 2-3 $\mathrm{km\,s^{-1}}$. It is important to bear in mind that even in the absence of galactic tidal forces, these motions are too large to allow the associations to be bound by self-gravity. It has long been recognized that OB associations could provide important information about the star formation process since they are so young that the individual members have not had enough time to move very far from their places of birth. For example, the present sizes of OB associations must closely reflect the sizes of the clouds which spawned them [18, 34, 35]. In addition, the structure of OB associations can provide important fossil clues about the structure and even the temporal evolution of a star forming complex. One common property of OB associations is that they are sub-structured, often consisting of sub-groupings of sequentially differing age [17]. Such sub-structure has suggested that star formation proceeds through a star forming region in an ordered temporal sequence. The chain reaction like nature of the spatial-temporal structure of OB subgroups has lead to the suggestion that OB subgroups are formed by a process of sequential triggering [35].

2.2. MULTIPLICITY

The tendency for stars to form in groups is not limited to associations and clusters. On much smaller, more intimate size scales (i.e., 1-100 AU) we find that most stars are members of binary or multiple star systems. For example, radial velocity measurements of field G dwarf stars indicate that

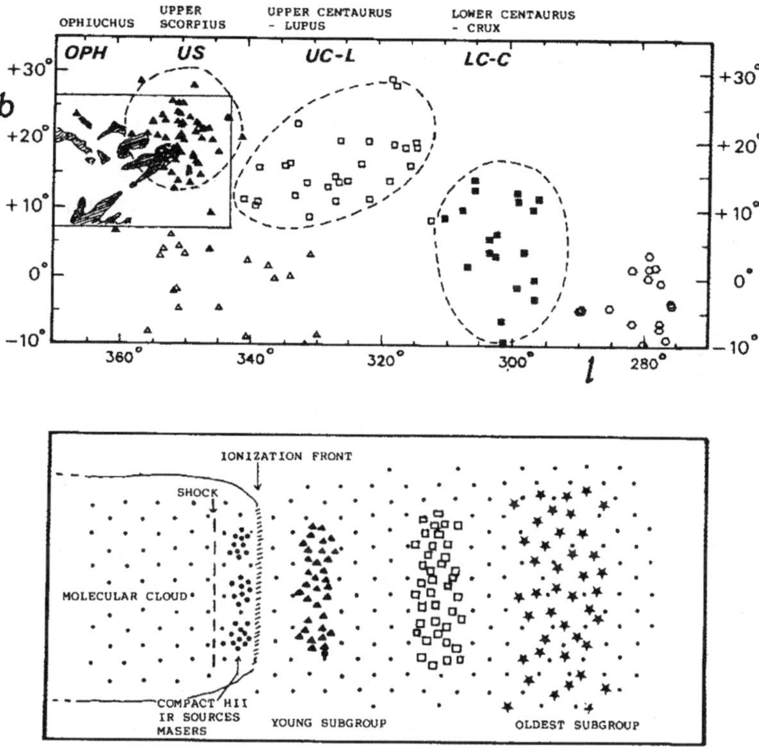

Figure 1. *(Top) Positions of OB stars in the Oph-Sco-Cen Association. The subgroups are outlined and a sketch of the dark clouds in the Ophiuchus region is also shown. (Bottom) A schematic diagram illustrating the Elmegreen-Lada sequentially triggered star formation model for OB subgroups. Figure adapted from Blaauw (1991).*

roughly 57% of all such stars have stellar companions [1, 32]. The binarity fraction does not appear to depend too strongly on spectral type as studies of both M dwarfs [38] and B stars [126] have indicated. Since some binaries may have been disrupted since formation, these fractions could be lower limits to the actual fraction of binary stars produced in the star formation process. Binary stars provide a strong constraint for star formation theories. In this regard, it is interesting to note that the mass of binary companions seems to be independent of the mass of their primaries [32].

2.3. THE INITIAL MASS FUNCTION

A fundamental consequence of the theory of stellar evolution is that the life history of a star is almost entirely pre-determined by its initial mass.

Consequently, to understand star formation and the consequent luminosity evolution of the galaxy requires a detailed knowledge of both the initial distribution of stellar masses at birth and how this quantity varies through space and time within the galaxy. Unfortunately, stellar evolution theory is not able to predict the Initial Mass Function (IMF) of stars. This quantity must be derived from observations. However, to do so is not straightforward because stellar mass is not itself an observable quantity. Stellar radiant flux or luminosity is the most readily observed property of a star. Therefore, the starting point of any attempt to determine a stellar mass spectrum is the determination of a stellar luminosity function. Then, an appropriate mass-luminosity relation is required to transform the luminosity function into a stellar mass function.

The first attempt to derive an empirical initial mass spectrum was carried out by Salpeter [111] who used as a starting point the general (or van Rhijn) luminosity function determined for field stars within the neighborhood of the sun. The field star luminosity function, $\Phi(M_V)$, is the relative number of stars per unit magnitude interval per unit volume of space as a function of absolute magnitude, M_V, in the solar neighborhood. (The determination of the field star luminosity function is itself a difficult task and discussion of this is beyond the scope of this chapter, and the reader who desires more information on this topic is referred to the text by Mihalas and Binney [94]). In principle, the field star luminosity function counts all presently visible field stars in the neighborhood of the sun. Because stars more massive than the sun have main sequence lifetimes shorter than that of the galaxy they are under represented in the field star luminosity function with respect to low mass stars, whose lifetimes are in excess of the age of the galaxy. Therefore, the present day field star luminosity function is biased toward low luminosity stars and must be corrected for the effects of post–main–sequence stellar evolution to obtain a luminosity function more representative of that produced at stellar birth.

Salpeter introduced the concept of an initial luminosity function, $\Psi(M_V)$, which is the relative number of stars per unit volume in each absolute magnitude interval originally produced by the star formation process. He derived $\Psi(M_V)$ from the field star luminosity function as follows:

$$\Psi(M_V) = \Phi(M_V) \quad for \quad \tau_{MS} \geq \tau_{MW}$$

and

$$\Psi(M_V) = \frac{\tau_{MW}}{\tau_{MS}}\Phi(M_V) \quad for \quad \tau_{MS} < \tau_{MW}$$

where τ_{MW} is the age of the galaxy and τ_{MS} is the main sequence lifetime of a star under consideration. Here, it has been assumed for simplicity that the stellar birthrate and the form of $\Psi(M_V)$ has been constant with time over the age of the galaxy. To derive an initial distribution of stellar mass we need to apply a mass luminosity relation to $\Psi(M_V)$. With the additional assumption that $\Psi(M_V)$ counts only main sequence stars, we can use the empirical mass-luminosity relation for main sequence stars (i.e., $L_* \sim m_*^p$, where $p \approx 3.45$) [116, 5]. Because $\Psi(M_V)$ is a distribution of stellar (absolute, visual) magnitudes and not luminosities, it is convenient to use a form of the mass-luminosity relation that relates M_V to stellar mass. Since $L_* \sim m_*^p$, and $M_V \sim \log L_*$, then $M_V \sim \log m_*$. Consequently it is useful to introduce the concept of the initial mass function $\xi(logm)$ which is the relative number of stars formed per unit volume per unit *logarithmic* mass interval and is straightforwardly related to $\Psi(M_V)$ as follows:

$$\Psi(M_V)dM_V = \xi(logm_*)dlogm_*.$$

The shape of the initial mass function is usually characterized by a spectral index, $\beta \equiv \partial log\xi(logm_*)/\partial logm_*$, which is the slope of the function in a log-log plot. We note that the initial mass *spectrum*, $f(m_*)$, (i.e., the differential frequency distribution of stellar masses at birth) is related to the mass function by:

$$\xi(logm_*) = \frac{m_* f(m_*)}{0.434}.$$

The spectral index of the mass spectrum, γ, is equal to $\beta - 1$.

Salpeter found that the initial mass function (IMF) could be reasonably well represented by a simple power-law form viz:

$$\xi(logm_*) \sim m_*^{-1.35}$$

In other words, a constant spectral index over the range of stellar mass that was considered, between approximately 0.4 and 10 $M_\odot$. In addition, the value of the spectral index (being less than -1.0) indicated that more stellar mass was contained in low mass stars than in high mass stars.

However, more recent and detailed studies [113, 65, 66, 115] suggest that the field star IMF is not characterized by a single spectral index. In particular, at masses around 1 $M_\odot$, there is a break in the power-law slope and the IMF becomes flatter at lower masses (i.e., $\xi(logm_*) \sim m_*^{-0.2}$). The presence of such a break in the IMF is significant and suggests that there may be a characteristic mass of star formation of around a solar mass or even less. However, it is not clear whether the IMF remains flat or decreases at the lowest masses. In fact, for stellar masses near and below the hydrogen burning limit (i.e. ≤ 0.08 $M_\odot$, the brown dwarf regieme) the IMF is very poorly constrained by existing observations. In any event, it is still clear that most stars that form in the galaxy are low mass objects.

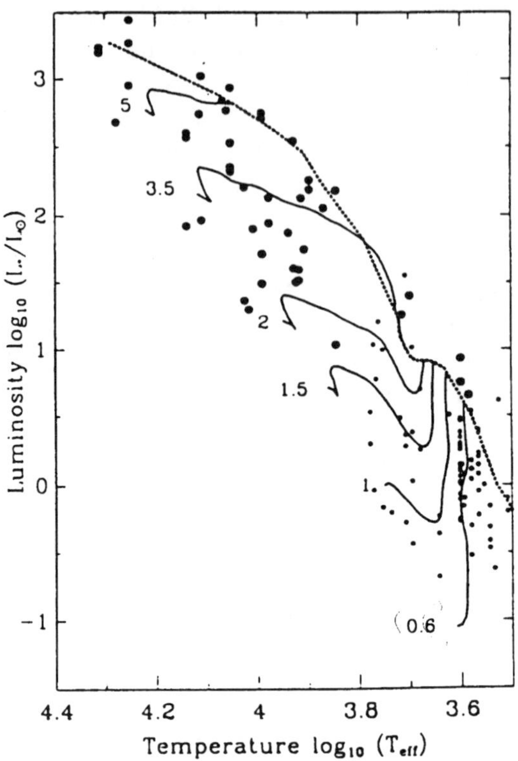

Figure 2.   *The HR diagram for a selection of pre-main sequence stars taken from a paper by Palla and Stahler (1993). Theoretical tracks (solid curves) for selected stellar masses are shown along with the birthline (dotted curve).*

2.4. THE HR DIAGRAM AND EARLY STELLAR EVOLUTION

The HR diagram is a very powerful tool for studying stellar evolution. Throughout most of its life a star will appear on the main sequence (the locus of positions in the HR diagram for stable hydrogen burning stars). The location of a star on the main sequence is determined almost solely by its mass. The main sequence lifetime of a star is also a very strong function of its mass. The more massive a star the shorter its main sequence lifetime. The utility of the HR diagram for investigating stellar evolution is best illustrated by its application to stellar clusters. In particular, the locations of cluster stars in the HR diagram with respect to the main sequence are determined primarily by the age of the cluster. For example, nearly 40 years ago a photometric study of the young cluster NGC 2264 by Walker [128] showed that most of the stars in this group, specifically all stars with spectral types later than about A0, were characterized by luminosities and effective temperatures which placed them above the main sequence in the subgiant region of the HR diagram (just below the locations one would expect to find red giants and long period variables). That the more massive stars with earlier spectral types were found to be on the main sequence strongly suggested that the population of subgiant stars in this cluster were pre-main sequence stars rather than post-main sequence stars. That is, these stars were not yet hydrogen burning objects. Many of these pre-main sequence stars turned out to be emission-line, T-Tauri type stars. This observation thus provided strong confirming evidence in support of the notion that T associations are very young.

Stellar evolution theory predicts that pre-hydrogen burning stars will lie above the main sequence and describes in detail the early evolution of such stars in the HR diagram. Figure 3 shows the HR diagram for a selection of pre-main sequence stars with masses between roughly 0.5-5 $M_\odot$ [103]. These pre-main sequence stars form a band stretching across the HR diagram parallel the main sequence which forms it lower boundary. The fact that this band has a distinct upper envelope is very significant. This upper boundary, termed the "birthline" [120], marks the positions of young stars when they first appear on the HR diagram. For a star of a given mass this position corresponds to a specific size or stellar radius. In other words, a star first becomes observable and appears on the HR diagram only after it obtains a specific maximum size, evidently pre-determined by its prior protostellar evolution. The birthline represents the initial condition for pre-main sequence (quasi-static) contraction. It is the dividing line between the protostellar and pre-main sequence stages of stellar evolution. Stellar evolution theory does not predict the existence of the birthline. Indeed, when evolving off the main sequence a star will expand to a size well beyond its birthline position. The physics which determine the birthline are the mysterious physics of protostellar evolution. Understanding the nature of protostars is, at present, the major frontier of star formation research.

Once a star reaches the birthline its subsequent evolution in the HR diagram depends on its mass. The theoretical evolutionary trajectories or tracks of 6 stars ranging in mass from 0.6 $M_\odot$ to 5 $M_\odot$ are plotted in Figure 3. Low mass stars start out as fully convective stars and descend to the main sequence along nearly vertical trajectories or convective tracks [51]. The lowest mass stars remain fully convective until they reach the main sequence. Stars with masses of about 0.8 $M_\odot$ or greater develop radiative interiors before reaching the main sequence. That is, the conditions in the interiors of these stars make it easier for them to transport energy outward by radiation than convection. According to the theory, once a star is radiative its luminosity is primarily dependent on its mass. This holds for any radiative star whether or not it is on the main sequence or is a post-

or pre-main sequence object. Thus, such radiative stars evolve to (or away from) the main sequence with roughly constant luminosity on more or less horizontal (radiative) tracks on the HR diagram [55]. Stars with masses greater than about 3 $M_\odot$ miss the (pre-main sequence) convective phase altogether. Pre-main sequence evolutionary tracks are useful not only for determining the masses of pre-main sequence stars but also for determining their ages *subsequent to their appearance on the birthline.* Prior to their appearing on the birthline, stars are in the protostellar stage of evolution and for reasons to be discussed later cannot be easily placed in the HR diagram. Finally, we point out that only low mass stars are expected to have a pre-main sequence phase of evolution. This is illustrated by the fact that the birthline intersects the main sequence for stars of roughly 8 $M_\odot$. Stars more massive than this begin to burn hydrogen and reach the main sequence before their protostellar stage of evolution ends. The physical reason for this becomes apparent if one compares the timescale for pre-main sequence evolution with that of protostellar collapse. The timescale for the gravitational collapse of a cloud core, the free-fall time, is determined largely by ρ, the density of the cloud:

$$\tau_{ff} = \sqrt{\frac{3\pi}{32G\rho}}$$

For the typical mean density (n $\approx 10^4$ cm^{-3}) of a cloud core (of either low or high mass) the free–fall time is about 4 x 10^5 years. The time scale for pre-hydrogen burning evolution is the Kelvin-Helmotz time:

$$\tau_{KH} \approx \frac{GM_*^2}{R_* L_*}$$

which is very rapid for a high mass star (i.e., $\approx 10^4$ years for $M_* = 50\ M_\odot$) and relatively slow for a low mass star (i.e., $\approx 3\text{x}10^7$ years for $M_* = 1\ M_\odot$). More importantly for high mass stars $\tau_{KH} < \tau_{ff}$ and these stars begin burning hydrogen and reach the main sequence before the termination of the infall or collapse phase of protostellar evolution. On the other hand, for low mass stars $\tau_{KH} > \tau_{ff}$ and low mass stars have an *observable* pre-main sequence stage of stellar evolution.

3. Giant Molecular Clouds: Sites Star Formation

3.1. OB ASSOCIATIONS AND GMCS

As mentioned earlier, stars in OB associations are still very near the locations where they were born. It is therefore not surprising that OB associations are invariably associated with visible manifestations of interstellar gas

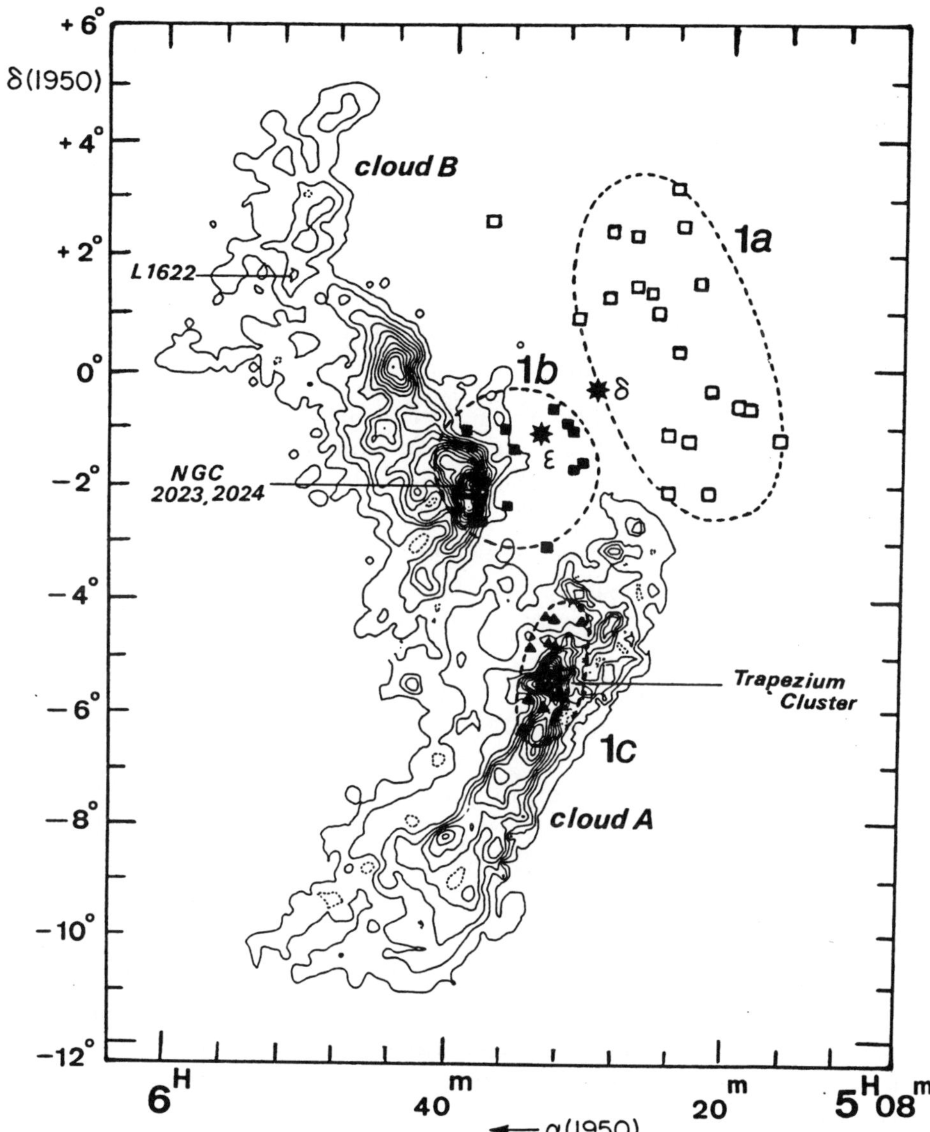

Figure 3. The ORI OB1 association and its Giant Molecular Cloud Complex. The 4 subgroups of sequentially differing age are enclosed by dashed lines. The oldest (1a) is significantly more expanded than the youngest (Trapezium Cluster). Figure taken from Blaauw (1991).

and dust [17]. In particular, virtually all known OB associations are spatially associated with Giant Molecular Clouds [34, 35, 19]. A good example of this is shown in Figure 3 which plots the positions of the OB subgroups in the Ori I association along with a map of the Orion GMC. The Orion region illustrates three important properties of OB associations and their associated GMCs: 1) The dimensions of the GMC are comparable to if not larger than those of the OB association. 2) The OB subgroups differ in size and age with the youngest (in this case the Trapezium cluster) usually the smallest and often exciting an HII region (in this case the famous Orion Nebula) or reflection nebula and therefore demonstrating a direct *physical* relation to the GMC. 3) The mass of the GMC ($\approx 2 - 3 \times 10^5 M_\odot$) is considerably larger than that of the OB association ($\approx 10^3 \ M_\odot$). From these three considerations it follows that that OB associations, and consequently most all stars currently forming in the galaxy, are born in GMCs.

The large difference between the stellar mass of the OB association and the molecular cloud (3) illustrates a fundamental property of star formation in GMCs. Namely, that the star formation efficiency (SFE=stellar mass/(stellar + gaseous mass)) in GMCs is low. For example, in the λ Ori OB association a direct comparison of the stellar and molecular mass of the star forming complex yielded an observed efficiency of only 0.3 % [31]. Although this is probably a lower limit to the true efficiency, it likely falls short by no more than a factor of 5. Thus we expect typical SFEs of at most a few % in GMCs which when combined with the estimates of 10^9 $M_\odot$ for the total mass of GMCs in the Galaxy, and 10^7 yrs. for the typical lifetimes of GMCs, results in a star formation rate that is consistent with the overall rate for the galaxy of roughly 2-3 $M_\odot \ \text{yr}^{-1}$, derived from other observations [138].

The low efficiency of star formation in GMCs is of central importance for understanding the dynamical nature of OB associations. The unbound state of stellar associations is a natural consequence of star formation with a low conversion efficiency of gas to stars followed by a rapid removal of the unprocessed gas from the system [31, 68]. Figure 4 is a schematic sketch which depicts the evolution of star formation in a GMC and the creation of an expanding association [68]. First, low mass stars form in dense cores located throughout the cloud converting roughly 1-3% of the total gaseous mass into stars. At some point massive O stars form in the cloud and heat, ionize and disrupt the molecular gas. (We note here that it is not clear at the present time whether most of the low mass stars are formed prior to the O stars or whether the episode of O star formation is also accompanied by an increase in the production of low mass stars.) In a relatively short time ($\approx 10^6$ years), the O stars disrupt the entire complex and remove the vast majority of the original binding mass of the system. (Calculations show that O stars can disrupt an entire GMC if only 4% of the total cloud mass is converted into stars with a normal IMF [131].) The stars in the cloud, which were originally orbiting in virial equilibrium with the deep potential

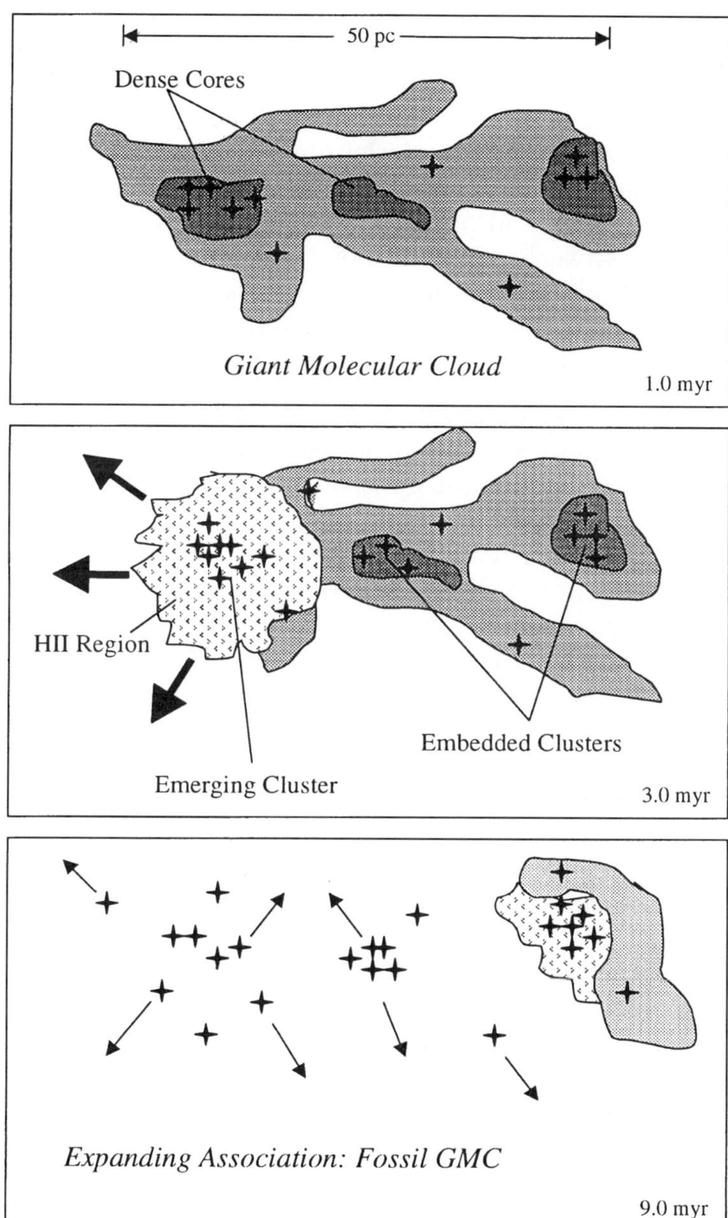

Figure 4.   A schematic diagram illustrating the evolution of an OB association.

well of the massive GMC, respond to the rapid removal of the majority of the binding mass by freely expanding into space with their initial virial velocities. This hypothesis predicts that the velocity dispersion of stars in an OB association should be on the same order of that of the molecular gas in a GMC (i.e., 2-4 $km\,s^{-1}$) which agrees with observations as discussed earlier [18]. In addition it is interesting to note that the angular sizes of OB subgroups within an association increase with age of the subgroup in a manner which is consistent with typical expansion velocities (of about 2 $km\,s^{-1}$)[17, 18] (see also Figure 3). The very few associations that are not near a GMC, such as Cas-Tau or Lac OB1 are associated with smaller molecular clouds and moreover have ages of $\approx$ 20 million years, making them among the oldest OB associations known [18]. They have probably existed long enough to have completely disrupted their parental GMCs.

With sizes on the order of 100 pc GMCs are the largest known objects in the Galaxy. Moreover, with masses in excess of 10^5 $M_\odot$, GMCs rival globular clusters as the most massive objects in the Milky Way. Molecular clouds are characterized by gas kinetic temperatures in the range between 10-50 K making them the coldest objects in the universe. There are a few thousand GMCs in the Galaxy and they account for a substantial fraction of all interstellar matter. Giant molecular clouds are composed primarily of hydrogen which is mostly molecular in form due to the low temperatures which characterize the gas. However, two factors make molecular hydrogen generally unobservable in molecular clouds: first, because it is a homonuclear molecule molecular hydrogen lacks a permanent dipole moment and its rotational transitions are extraordinarily weak; second, it is so light in weight that its lowest rotational transitions are at mid-infrared wavelengths and are too energetic to be collisionally excited at the temperatures that characterize GMCs. Consequently GMCs are best traced by emission lines of molecules such as CO, CS or NH_3 and much of what we know about the physical conditions in GMCs is derived from observations of these molecules.

3.2. PHYSICAL PROPERTIES OF GMCS

Molecular emission lines from GMCs are typically characterized by linewidths in the range 2-4 $km\,s^{-1}$. Given the masses and sizes of these clouds, their linewidths are close to what one would predict if the clouds are in virial equilibrium, (i.e., $\Delta V \approx (GM/R)^{-0.5}$) which is consistent with the idea that the clouds are bound. At the same time however, the linewidths are considerably greater than that expected for thermal motion

Physical Properties of A Standard GMC	
Mass	$2 - 3 \times 10^5$ M$_\odot$
Maximum Linear Extent	50-100 pc
Mean Density(n(H$_2$))	200-300 cm^{-3}
Mean N(H$_2$)	$3 - 6 \times 10^{21}$ cm^{-2}
Typical Linewidth	2-4 km s^{-1}
Kinetic Temperature	10 K
Sound Speed	0.2 km s^{-1}
Typical Magnetic Field Strength	30 μG
Alfven Speed	3 km s^{-1}

of the emitting gas molecules (of mass m):

$$\Delta V_{FWHP} = 2.354(kT/m)^{0.5} = 0.05 \ kms^{-1} \text{ for CO.}$$

This indicates that the overall dynamical state of a GMC is characterized by supersonic, non-thermal, bulk motions. If these motions are turbulent in nature, they should be highly dissipative, [42] yet GMCs maintain their supersonic dynamical states throughout their lifetimes, estimated to be on the order of 10^7 to 10^8 years. Although they are gravitationally bound, GMCs cannot be in a state of overall dynamical collapse because the resulting star formation rate ($\sim$ 250 M$_\odot$ yr^{-1}) would be considerably in excess of observed values [138]. Moreover, the free-fall time for the collapse of a GMC is short compared to the its lifetime. Therefore it is clear that GMCs are being supported against global dynamical collapse by some mechanism. However, GMCs are too cold to be supported by thermal pressure. Exactly what does support GMCs against collapse has been a long standing problem in star formation research.

Molecular clouds are permeated by magnetic fields with strengths in the range of 10 to 1000 μG [53]. The largest values arise in extremely dense, very compact regions which exhibit OH maser emission. More typical field strengths are on the order of about 30 μG. The Alfven speed, $v_A = B/(4\pi\rho)^{\frac{1}{2}}$ is therefore about 3 km s^{-1}. Thus the fluid motions within a GMC are sub-Alfvenic and perhaps less dissipative than is suggested by their otherwise supersonic dynamical state. Moreover these field strengths are sufficiently strong that they may retard or prevent the global collapse of a GMC [96, 97]. The typical physical properties of GMCs in the solar neighborhood are summarized in Table 1

3.3. THE STRUCTURE OF GMCS: DENSE CORES

The molecular mass of a GMC is spatially distributed in a non-uniform manner. These clouds are highly structured, consisting of numerous filaments, clumps and dense cores [20]. In fact GMCs are so complex that it is very difficult to provide a meaningful quantitative description of their structure [112]. Molecular clouds have fractal geometries with a typical fractal dimension of 1.5, but it is not at all obvious what the significance of this fact is [114, 30]. A crude but useful way to characterize molecular cloud structure is to determine the fraction of cloud mass at various densities. The densities in GMCs range from a few hundred to a few million hydrogen molecules per cubic centimeter. Most (80-90%) of the material is at densities between 100-1000 cm^{-3}. About 10 % of the gas is contained in clumps and cores with mean densities of 10^4 cm^{-3} or slightly more. The distribution of dense gas through a GMC is generally not a well determined quantity. However, studies of the Orion clouds provide some indication of how such gas might be distributed [77, 122]. Dense gas is traced by molecular transitions whose excited states require high densities to be collisionally populated and which are optically thin. The most extensively used tracers for this purpose have been the the J =2-1 line of CS and the (1-1) metastable transition of NH_3, although other transitions and other lines can be equally useful.

In Figure 5 is displayed the CS(J=2-1) map of the Orion B molecular cloud [77]. Here the regions of brightest CO emission (see Figure 4) have been completely surveyed by Lada, Bally and Stark [77](hereafter LBS) for CS emission. The distribution of the CS emission and the high density gas is very clumpy. Moreover, the map appears to be dominated by a few very large clumps or cores. More than 40 dense cores were identified by LBS within the mapped region. These cores were found to have radii which ranged from 0.1 pc (the limit of telescopic resolution) to 0.5 pc. The cores were typically elongated and not circular in shape with a mean aspect ratio of 2.0 which is similar to dense cores studied in other regions [28, 15, 84]. These cores were found to range in mass between 8 - 500 $M_\odot$. The mass spectrum of these cores is displayed in Figure 6. For cores with masses in excess of 20 $M_\odot$ LBS found that the spectrum is well represented by a power-law with an index of -1.6. Since the spectral index of the cloud core mass spectrum is greater than -2, most of the dense gas in this GMC is contained in the most massive cores. Indeed, more than 50% of the dense gas detected was found to be contained within the 5 most massive cores [77]. In this regard the mass spectra of molecular cloud cores is qualitatively different that that (i.e., $\gamma_* = -2.35$) of stars which presumably form from them [69]. Since the definition of a core is arbitrary one might question the

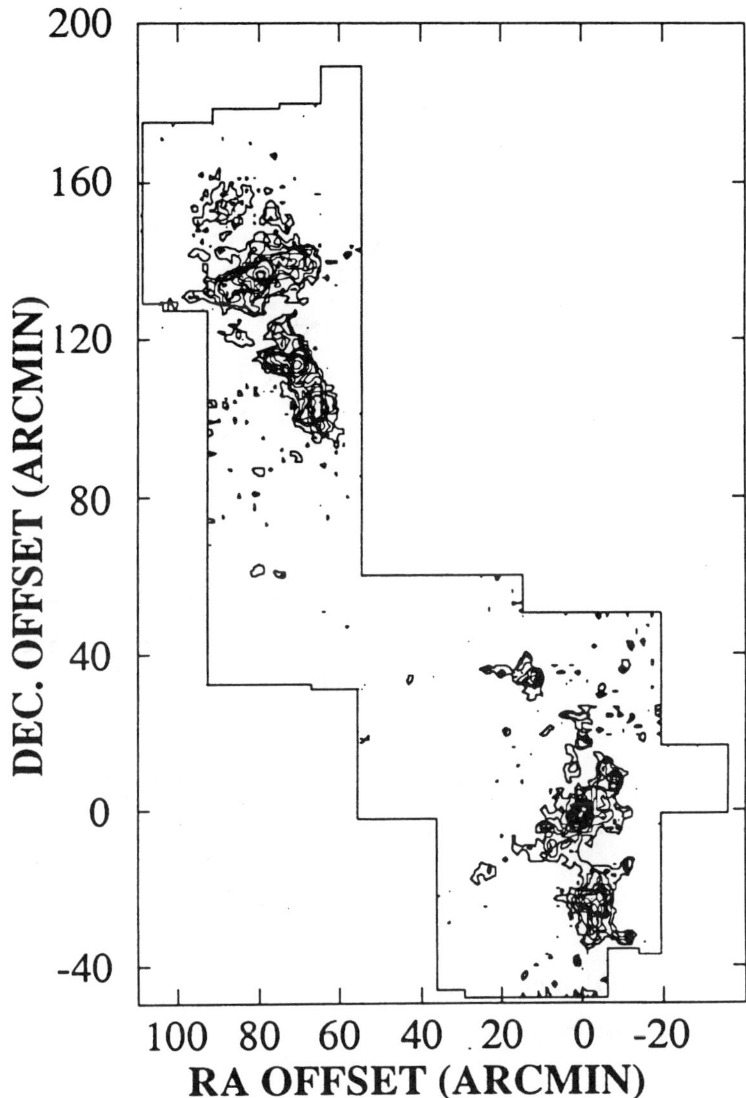

Figure 5.   Distribution of dense gas in the L1630 (Ori cloud B, see Figure 3) molecular cloud as traced by CS emission. Taken from LBS.

significance of the derived mass spectrum. However, LBS derived the mass spectrum for the Orion B dense gas using two different clump definitions and got the same result. Moreover, it is interesting that observations of lower density clumps in 5 other GMCs as well as the Orion A cloud yield clump mass spectra with very nearly identical spectral indices despite the

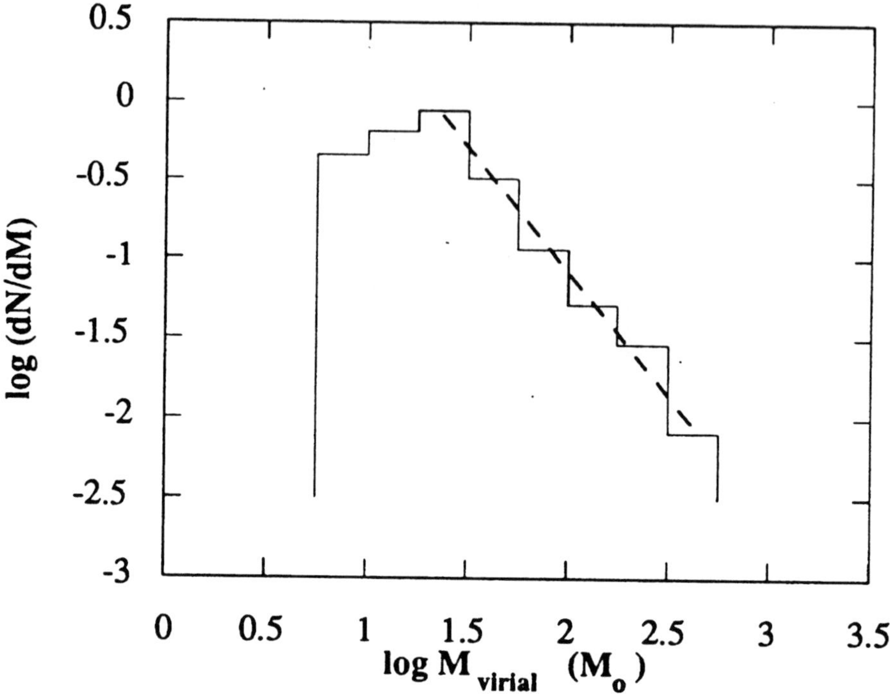

Figure 6. The mass spectrum of dense cores in the L 1630 cloud from the study of LBS. The spectrum can be characterized by a power law with an index of -1.6.

fact that different definitions of clumps were employed in each case [20, 122].

4. The Young Stellar Objects

4.1. CLUSTERING

Stars form in GMCs and so in this sense we expect the youngest stars to be found in groups, that is, to be spatially confined to the well defined boundaries of a GMC [68]. On the other hand, what one should expect about the spatial distribution of stars within a GMC is not at all obvious. The positions of these stars are essentially the positions of the sites of star formation activity. Knowledge of the spatial distribution of these sites and its relation to other physical properties of a star forming region (e.g., cloud structure, chemistry, temperature, and dynamical state, presence of HII regions, supernova remnants, shocks, winds, etc.) is clearly important for developing a theory of star formation. Unfortunately, stars located within a GMC are heavily extincted and difficult to observe. As a result, until recently, little was known about the distribution of young stellar objects (YSOs) within molecular clouds. The best (and only) way to get a good census of the stellar content of a GMC is through infrared observations and such observations have been only practical during the last few years. The IRAS satellite provides a data base of completely spatially sampled far-infrared observations of the sky and can be used to obtain flux limited counts of far-infrared sources in nearby clouds. However, the majority of young stars in a cloud are weak far-infrared emitters and cannot generally be detected by IRAS. Moreover, the angular resolution of IRAS was relatively coarse so that the results are often compromised by source confusion even in the nearest clouds. Near-infrared imaging surveys are much better suited to detect such stars, although such observations are themselves often limited by background source confusion.

The first cloud to be systematically surveyed at near-infrared wavelengths was the L1630 (Orion B) cloud [78, 76] which makes up the northern half of the Orion GMC complex (see Figures 3,5). A large region of the cloud surrounding the all the regions of significant CS emission was observed at 2.2 μm. Roughly half the area surveyed contained CS emission and dense gas. The observations were sensitive enough to observe a large span of the IMF and most of the stars observed had masses comparable to or less than that of the sun. The spatial distribution of these stars showed a strong tendency for the stars to be clustered. More than half ($\sim$ 58%) of all the infrared sources detected were found to be contained within three rich clusters which together subtended an angular size less than 18% of the entire area surveyed. After correcting for the presence of background/foreground star contamination, it was found that 96% of the sources *physically associated* with the cloud were contained in these 3 clusters.

Figure 7 shows the relation of the clusters to the dense gas in the cloud. The three rich clusters are spatially associated with 3 of the 5 most massive dense cores in the cloud [76]. These observations vividly demonstrate that in this cloud star formation occurs almost exclusively in: 1) rich clusters and 2) dense molecular gas. Indeed, the first phenomenon likely follows from the second. That is, if dense gas is a necessary condition for star formation then we expect most stars to form in a few rich clusters. This follows because the mass spectrum of dense gas in the cloud indicates that most of the dense gas is contained in the few most massive cores. The fact that two of the massive cores in L1630 did not produce rich clusters indicates that although dense gas may be a necessary condition for star formation it is not a sufficient condition for star formation.

Since the core mass spectra of nearby GMCs are so similar, we also ex-

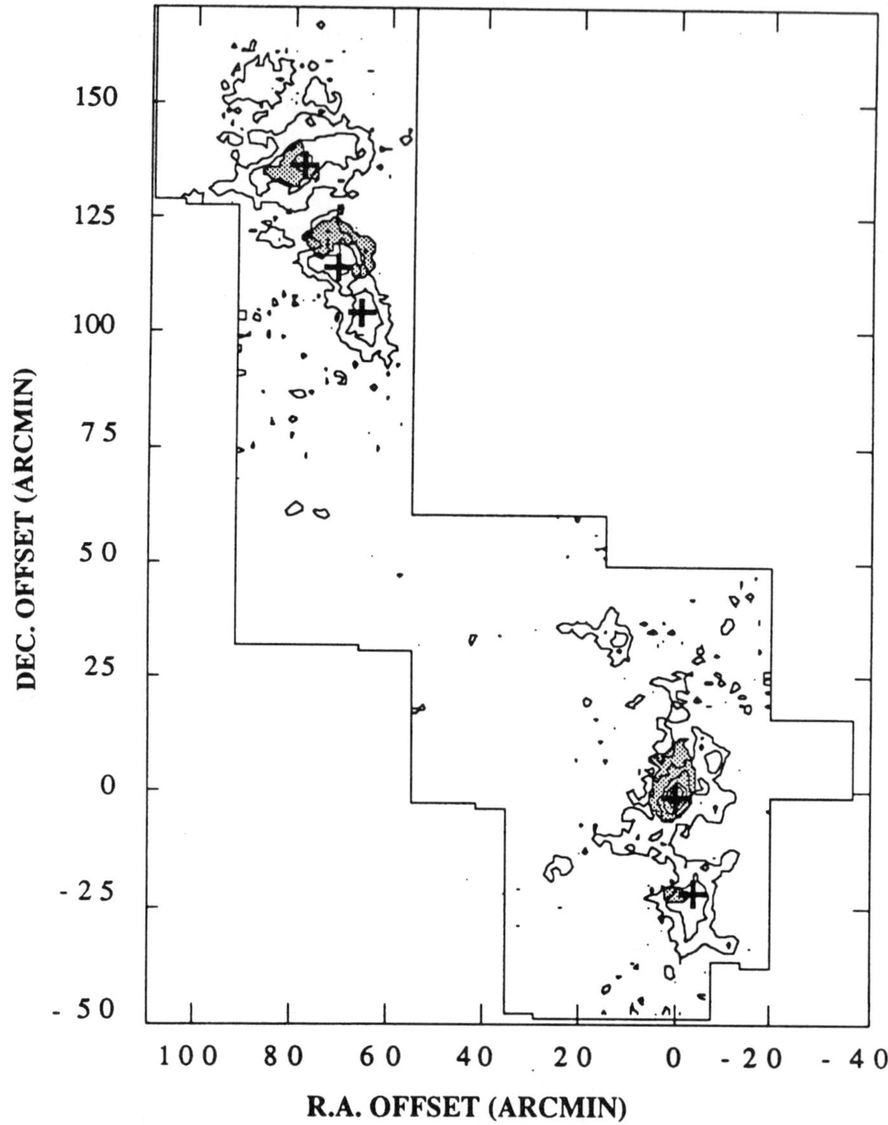

Figure 7. The locations and extents of four embedded infrared clusters (shaded contours)
compared to the distribution of dense molecular gas (solid contours) in the L 1630 cloud.
Figure from Lada (1992).

pect that these clouds should also produce most of their stars in embedded
clusters as well. Indeed, infrared imaging observations have uncovered rich
embedded clusters in almost every GMC studied (e.g., [105, 137]). Figure 8
shows a K-band infrared image of one such embedded cluster. However, the

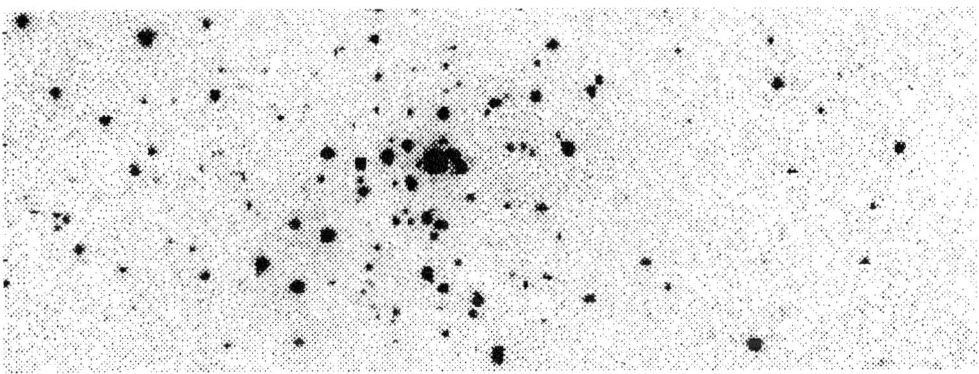

Figure 8.   Example of a partially embedded infrared cluster. Deep K-band image of the young cluster NGC 2282 (Horner et al. 1996)

existing observations for most clouds are not extensive enough to determine whether clusters account for the bulk of all stars formed within them. An exceptional case is the λ Ori OB association, a fossil association which has recently emerged from and cleared away its parental GMC [31]. A systematic survey of emission line stars over the entire association shows that the majority of stars detected are also clustered [31]. The clusters are about a factor of 4 more extended than those in L1630. However, these sizes are what would be expected if the clusters started out with dimensions similar to those of L1630 (i.e., 1 pc) but became unbound after gas removal from the association (i.e., the velocity dispersion of stars in the association is measured to be $\sim$ 2 km s^{-1}and the age of the OB cluster in its center is about 3 x 10^6 years [89, 101, 43]).

On the other hand, observations of the L1641 cloud (the Orion A cloud) present a somewhat different picture for that cloud. A spatially complete emission line survey [133] and an extensive (but not spatially complete) infrared survey [121] of the cloud suggest that in addition to stars forming in rich embedded clusters and small groups [27] there is a more distributed population of stars which could account for a significant fraction (30-40%) of all stars formed in the cloud [93, 43, 6]. The origin of this relatively large distributed population is at present unclear. However, such observations have given rise to the idea that there are two modes of star formation: a clustered mode and an isolated or distributed mode [69, 81]. In the latter circumstance, individual stars (primarily low mass stars) are formed in individual low mass cores distributed throughout a molecular cloud [68]. The prototypical example of the isolated mode of star formation is the nearby Taurus complex. With a mass of $\sim$ 10^4 M$_\odot$, the Taurus dark cloud is a relatively small molecular cloud complex. The complex consists of a number of filamentary structures which contain small dense cores. These dense cores span a small range of the clump mass spectrum from about 1–40 $M_\odot$. About half of these cores contain newly formed stars [15] but typically there is only one star within an individual core. The Taurus clouds have held a very important place in star formation research. It is here that the formation of individual sun-like stars can be closely scrutinized with minimal confusion from neighboring star forming activity. Understanding the formation of individual stars is a fundamental step for the development of a comprhensive theory of star formation. However, the Taurus clouds likely do not represent the typical environment (i.e., a GMC) in which most stars form in our galaxy.

In summary it appears that most stars form in groups with a spectrum of richness ranging from relatively poor clusters with only a few stars to very rich clusters with hundreds of stars. Again, to the extent that the (stellar) mass spectra of clusters follow the mass spectra of cores from which they form, we would expect most stars in a GMC to originate in relatively rich clusters rather than in poor clusters or in isolation. We note here that most of the rich clusters that form in GMCs must be disrupted soon after they emerge from the cloud, otherwise there would be considerably more (bound) open clusters in the field than are observed [72]. This suggests that the star formation efficiency in cluster forming cores rarely reaches values as high as 50% [72, 73].

4.2. MULTIPLICITY

As is the case for field stars, there is a high frequency of multiplicity, namely binarity, among pre-main sequence stars [90]. Indeed, for stars in the Taurus T association the binary frequency is found to be higher (perhaps by a factor of 3) in the mass range surveyed than it is in the field [41, 83]. This is illustrated in Figure 9 which shows the binary frequency of pre-main sequence stars in three nearby star forming regions (Taurus, Ophiuchus and Corona Australis) compiled by Mathieu (1994). The excess of binaries at intermediate periods is due to stars in the Taurus-Auriga clouds which account for all the intermediate period binaries in Mathieu's sample. On the other hand, this excess in the pre-main sequence binary frequency is not universal. Observations of binary stars in clusters such as the rich Trapezium cluster [106] and the Pleaides [23] indicate a binary frequency distribution very similar to that of the field. Overall, studies of young stars confirm the field star results that most stars form in binary systems. Moreover, they may also suggest that most field stars were formed in environments typical of rich clusters like the Trapezium rather than in poor groups such as in Taurus (e.g., [64]). It is possible that binaries form with the same frequency

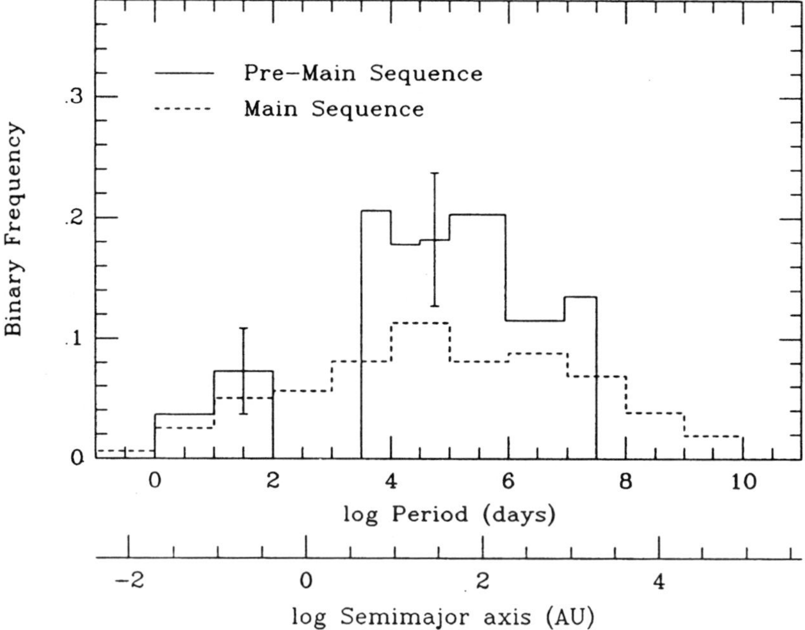

Figure 9. *The binary frequency distribution for pre-main sequence stars in nearby clouds from Mathieu (1994).*

both in clusters and in isolation, but in the environment of a rich cluster many binaries become disrupted very early in their evolution, while those formed in relative isolation survive for longer times.

Finally, it is important to note that binaries have been found among pre-main sequence stars of all ages, including stars near the birthline on the HR diagram. This indicates that binary systems are formed in the protostellar phase, prior to the pre-main sequence phase of evolution [90]. Indeed, this has been confirmed by the observation that a number of embedded, protostellar objects turn out to be binary systems (e.g., [134, 123]).

4.3. THE INITIAL LUMINOSITY AND MASS FUNCTIONS

The field star IMF is a globally averaged quantity, averaged over both the lifetime of the galaxy and a over a specific volume of space (i.e, the solar neighborhood). However, derivation of the IMF from observations of field stars has been hampered by two serious issues. First, observations have not been sensitive enough to adequately sample the mass function at low stellar mass, particularly near and below the hydrogen burning limit. Second, the detailed form of the IMF, particularly at masses near a solar mass and greater, is very sensitive to the uncertain history of galactic star formation [115]. Although it is often assumed that the IMF is universal in both time and space, this has never been demonstrated in any completely satisfactory way. However, in order to develop comprehensive theories for star formation and galactic evolution, it is crucial to know whether the detailed form of the IMF is universal. Are there spatial and/or temporal variations in the initial conditions and other important astrophysical parameters that can alter the process of star formation and the form of the IMF? Or is star formation such a robust process that the outcome is always an IMF of the same form?

In principle, observation of large enough groups of young or newly forming stars in different parts of the galaxy should be able to provide fundamental constraints on these questions. The smallest spatial size scale over which a meaningful determination of a luminosity function can be made is that characteristic of an open cluster. Clusters are important laboratories for studying the initial luminosity function because they consist of statistically significant groups of stars who share the common heritage of forming from the same parental cloud at the same epoch in time. Young embedded clusters are particularly useful since they are not old enough to have lost significant numbers of members due to stellar evolution or dynamical effects such as evaporation or violent relaxation [73, 72]. Thus, unlike the case for field stars, the present day stellar mass function of a young cluster is its initial mass spectrum. Moreover, low mass stars in a young cluster are brighter than at any other time prior to their evolution off the main sequence. Indeed, for very young and nearby clusters ($\leq$ a few million years) even stars with masses significantly below that of the hydrogen burning limit (i.e., brown dwarves) are readily detected at infrared wavelengths. However, study of very young clusters suffers from two disadvantages: 1) the clusters are often heavily obscured and cannot be easily observed at optical wavelengths; 2) the stars in such a cluster are mostly pre-main sequence stars and uncertain corrections for pre-main sequence evolution and non-coevality must be applied to the members to derive mass spectra from luminosity functions. Only in the last few years have advances in infrared detectors enabled the direct observation of such embedded and often obscured clusters and helped to minimize the first disadvantage. The second disadvantage requires modeling and is more difficult to overcome [137, 37, 79, 80, 100].

Two methods have been generally employed to derive the initial distribution of stellar masses from observations of young clusters. The first method involves modeling the luminosity function of an embedded cluster to derive the underlying mass function (e.g., [137, 79, 100]). The second method involves determining individual masses for the entire stellar population of a cluster by comparison of the locations of its individual member stars on the HR diagram with the predictions of PMS evolutionary models (e.g., [58, 85]). Both methods have advantages and disadvantages, but in general are complimentary.

The first method requires observation of the luminosity function of a stellar population and knowledge of its star forming history. The monochromatic luminosity function is directly obtained from photometric measurements of the members of an embedded stellar population. The measurement of a monchromatic stellar brightness is the most basic observable property of a star. With digital imaging detectors, such as CCDs and infrared array cameras, such observations generally provide the most complete set of data which can be used to investigate the form of a cluster IMF. This method for deriving the stellar mass function relies on the fact that the observed luminosity function is a product of the underlying stellar mass function and the mass-luminosity relation for the cluster stars. For a monochromatic luminosity function constructed from infrared (e.g., K band) observations:

$$dN/dm_K = dN/dlogm_* \times dlogm_*/dm_K$$

where m_k is the stellar (K) magnitude, and m_* is the stellar mass. The first term on the right hand side of the equation is the stellar mass function and the second term the mass-luminosity relation. This method is essentially that originally employed by Salpeter to derive the field star IMF. However, PMS stars, which account for most of the stars in the a young cluster, can not be characterized by a unique mass-luminosity relation. Indeed, the mass-luminosity relation for any such star is a function of time. Consequently, one must model the time-varying mass-luminosity relation for all the stars using PMS evolutionary models and knowledge of the star formation history of the cluster. The latter quantity typically can be derived by placing the stars on an HR diagram, which in turn requires multiwavelength photometry or spectroscopy of a representative sample of the cluster members. Given PMS evolutionary models and knowledge of the star formation history of the cluster, one can then model the luminosity function to derive the mass function.

The second method requires simultaneous knowledge of both the luminosities and effective temperatures of all the stars in a cluster so that they can be individually placed on the HR diagram. This in turn requires both photometry and either colors or a spectrum of each star. For extincted clusters, spectroscopy is the preferred method of obtaining a stellar effective temperature. The advantage of this method is that the final product is the set of individual masses for all stars for which both both spectra and photometry were obtained. In other words this procedure provides a more direct and precise determination of the mass function than the first method.

Figure 10a shows the K-band luminosity function derived for the Trapezium or inner Orion Nebula cluster [137, 91]. For reference note that at the distance of Orion a one million year old star at the hydrogen burning limit would have a K magnitude of only about 12th magnitude. Also plotted are two model luminosity functions derived for the cluster using recent PMS models [29] and the mean age and age history for the cluster derived from the observations of Hillenbrand [58]. The first model is the luminosity function corresponding to Salpeter IMF, a single power-law extrapolated to below the hydrogen burning limit. Clearly, the Salpeter IMF is not consistent with the observations at low luminosities as is the case with the field stars. The underlying mass function must depart from the Salpeter form at low stellar mass. The second model luminosity function is that obtained by Meunch Lada and Lada (1999) [100] using an underlying IMF which produces the simplest best-fit to the observed luminosity function of the Trapezium cluster. This best-fit model IMF is a function consisting of three power-law segments. Below 5 $M_\odot$ the IMF rises with decreasing stellar mass, between 0.5 and 0.1 $M_\odot$, the slope of the IMF is relatively flat and below 0.1 $M_\odot$ the IMF decreases with decreasing stellar mass. (Note that at the very faint end of the luminosity function there are more stars observed than predicted by the model. This is a result of the fact that the model did not take into account contamination from reddened background stars which likely populate the lowest luminosity bins in the observed luminosity function).

The IMF derived from the modeling of luminosity function is shown in Figure 10b along with the mass function derived for the entire Orion Nebula Cluster by Hillenbrand [58]. This latter mass function was derived using optical spectra to place stars on the HR diagram and individually extract their masses. Because of sensitivity limitations, spectra were obtained for a nearly complete sample of stars with masses down to but not below the hydrogen burning limit. The two methods of deriving the IMF returned two mass functions which are in general agreement for stellar masses above the hydrogen burning limit. Moreover, the derived mass functions are, within the uncertainties, consistent with the most recent estimations of the field star IMF for masses above the hydrogen burning limit [115, 66]. Although neither the spectroscopic mass function nor the field star IMF provide much information about the form of the IMF at substellar masses, the IMF below the hydrogen burning limit is clearly constrained by the luminosity function. Current indications suggest that this brown dwarf mass function is falling with decreasing mass. The overall significance of this result is that there appears to be a peak in the initial stellar mass function at about 0.25 $M_\odot$.

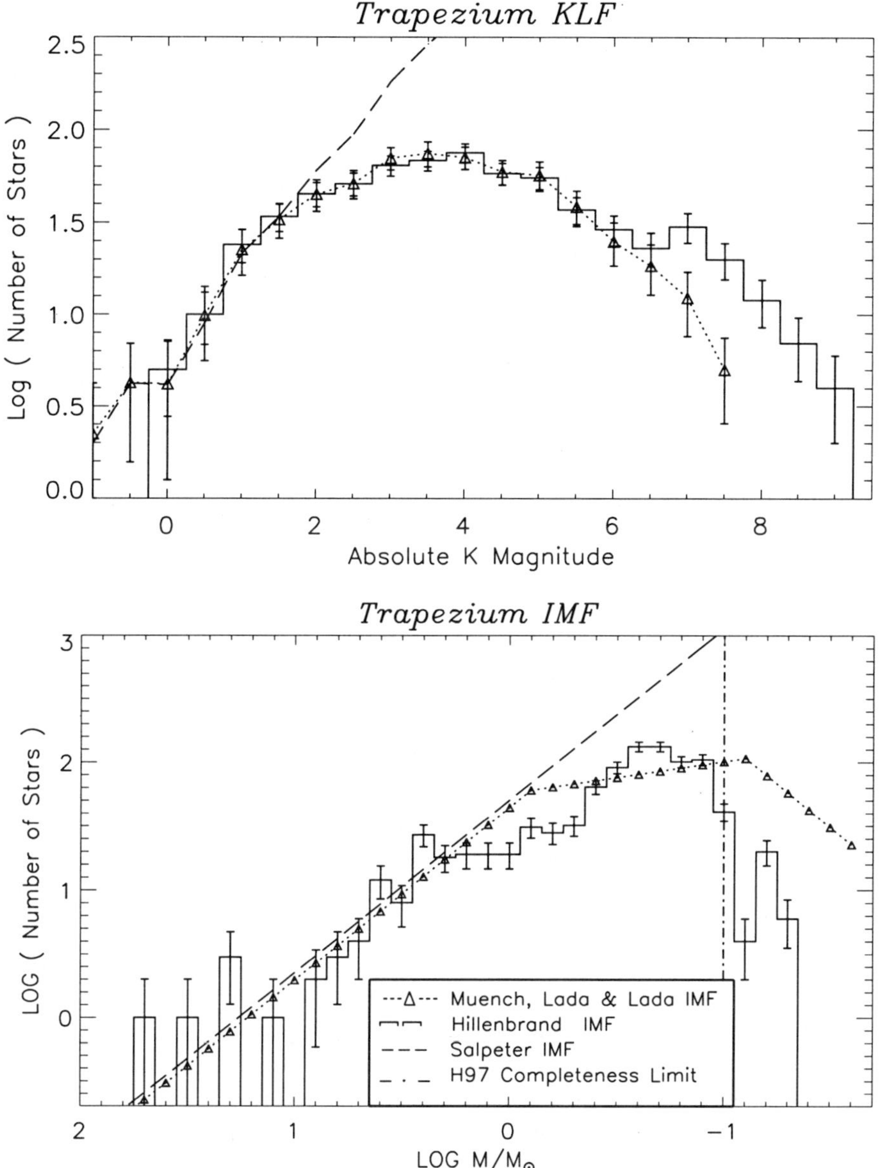

Figure 10.   a) *The K band luminosity function of the Trapezium cluster in Orion. (histogram: from Zinnecker, McCaughrean and Wilking 1993 and McCaughrean et al 1996). Also plotted are two model luminosity functions. The first (dashed line) was derived assuming an underlying IMF given by the Salpeter function and is clearly a poor match to the data. The second corresponds to the three power-law mass function of Muench, Lada and Lada (1999) which was selected from a large suite of model IMFs to provide a best-fit to the observed data. b) The IMF for the larger Orion Nebula cluster derived by Hillenbrand (1997) from optical spectroscopic observations (histogram) and its completeness limit. Also plotted are the Muench et al. three power-law IMF which best fit the observations of the Trapezium luminosity function (a) and the Salpeter IMF.*

This in turn suggests that there is a characteristic mass produced by the star formation process. That that mass is close to the mass of a star at the hydrogen burning limit is very curious. Finally, it appears that the IMF for nearby young clusters is in some general sense universal and even similar to the field star IMF. This in itself suggests a robust quality to the star formation process in the galaxy, since the field star IMF is a globally averaged IMF, averaged over the age of the galaxy and a large volume of space around the sun, while the embedded cluster IMFs are instantaneous IMFs produced in very localized volumes and over relatively small spatial scales (i.e., r $\sim$ 1 pc). However, observations of older more distant open clusters suggest that important variations in the form of the IMF may be found, particularly for the more massive stars [115] and the issue of the universality of the IMF is still not satisfactorily resolved. The existing observations do suggest that significant progress on this problem can be made once complete luminosity functions are obtained for a larger sample of young embedded clusters.

4.4. SPECTRAL CLASSIFICATION: THE EVOLUTIONARY STATUS AND NATURE OF YOUNG STELLAR OBJECTS

The evolutionary status of a normal, low-mass star is essentially determined by its placement on the HR diagram which requires measurements of two quantities: luminosity and effective temperature. It is therefore no surprise that stars are most usefully classified by these two parameters (i.e., spectral type and luminosity class). The luminosity of a star is straightforwardly determined from its observed flux and its distance. The effective temperature of a star is determined from its colors or observations of temperature sensitive spectral features. This method works very well as long as a star is well behaved spectrally, that is, as long as it produces a black-body like spectrum and can be characterized by a single effective temperature. However, in this sense most young stellar objects are not normal stars. A good example is the famous young star FU Ori. The spectral type of this star is a function of the wavelength at which it is observed. Consequently, FU Ori cannot be characterized by a single effective temperature. Its placement on the HR diagram is ambiguous and the classical theory of stellar evolution cannot explain its nature.

This "abnormal" characteristic of young stellar objects is a result of their intimate physical association with the natal gas and dust from which they are formed. New stars are born within dense molecular cloud cores and throughout their formation and early evolution they are associated with varying amounts of molecular gas and dust. This circumstellar gas and dust can absorb and reprocess substantial amounts of the radiation emitted by the embedded stars significantly altering their spectral appearance. Indeed, at optical wavelengths the youngest objects are rendered completely invisible by the obscuration of opaque circumstellar dust and a significant fraction, if not all, of their luminous energy is radiated in the infrared portion of the spectrum. In general the circumstellar gas and dust associated with young stellar objects has a spatial extent considerably greater than that of a stellar photosphere. Consequently, emitting circumstellar dust, which is in radiative equilibrium with the stellar radiation field of the buried star, will exhibit a wide range of (effective) temperatures and the emission that emerges will have a spectral distribution much wider than that of a single temperature blackbody. For this reason it can be very difficult, if not impossible, to meaningfully place a YSO on an HR diagram.

Clearly, the evolutionary status of a young stellar object cannot be derived solely from knowledge of its luminosity or a single effective temperature. More complete knowledge of it spectrum, particularly at infrared wavelengths, is required. The shape of the broad–band infrared spectrum of a young stellar object depends both on the *nature* and *distribution* of the surrounding material. Clearly then, we expect that the shape of the spectrum will be a function of the state of evolution of a YSO. The earliest (protostellar) stages, during which an embryonic star is surrounded by large amounts of infalling circumstellar matter, should have a very different infrared signature than the more advanced (pre–main sequence and main sequence) stages, where most of the original star forming material has already been incorporated into the young star itself. Observations of low mass young stellar objects (mostly in isolated regions) have confirmed this expectation and such young stellar objects can be meaningfully classified by the *shapes* of their optical-infrared spectral energy distributions [68] in an empirical sequence of protostellar and early stellar evolution [3].

The spectral energy distributions of the vast majority of known (low mass) young stellar objects fall typically into one of four broad classes designated 0, I, II and III. Figure 11 illustrates spectral energy distributions (SEDs) representing each of these classes. Class 0 and I sources are characterized by SEDs which peak in the sub-millimeter and far-infrared portions of the spectrum indicating that the SEDs are dominated by emission from cold dust. These objects are the most deeply embedded of the YSOs. They are thought to be protostellar in nature and therefore the least evolved of the YSOs. Class II and III sources are characterized by energy distributions which peak in the optical and near-infrared regions of the spectrum where emission from stellar photospheres is expected to dominate. These objects are much less reddened and correspond to the more evolved pre-main sequence and ZAMS stars.

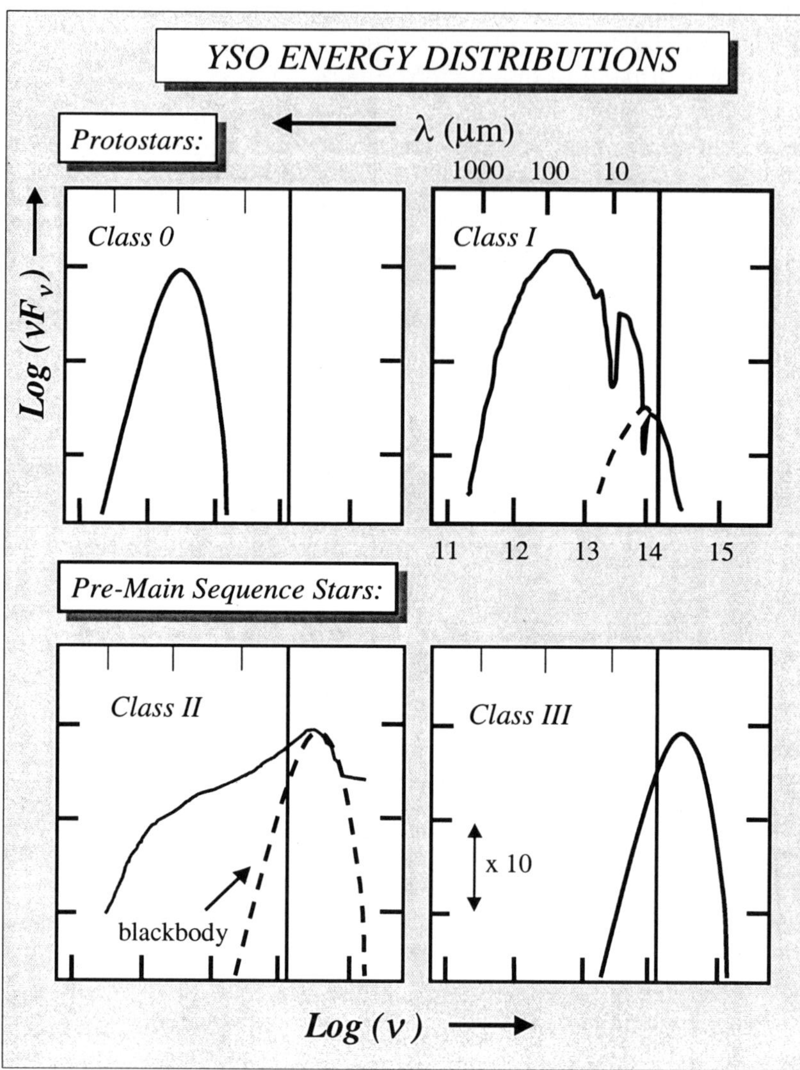

Figure 11. *The empirical classification scheme for YSO spectral energy distributions. A vertical line appears at a wavelength of 2.2μm for fiducial reference in each panel. Class 0 and Class III sources have distributions whose widths are similar to single temperature blackbody functions. Class II and Class I sources display infrared excess which produces energy distributions which are broader than a single blackbody function.*

However, knowledge of the source energy distribution of a YSO is, by itself, not sufficient to fully describe its the physical nature. Such objects are associated to varying degrees with a range of interesting astrophysical phenomena, such as bipolar outflows, jets, masers, HH-objects, variability, and optical/infrared emission lines. A full description or understanding of the physical natures of YSOs also requires knowledge of their relation to these various phenomena.

All things considered, YSO evolution can be divided into two primary phases. The earliest stage of evolution is the embedded phase during which protostellar evolution takes place. In this phase a star acquires the bulk of its mass through accretion of surrounding circumstellar material onto an embryonic stellar core. Throughout most of this phase the embryonic star is heavily obscured and invisible at optical wavelengths. The second phase of YSO evolution is the revealed phase, which for low mass stars corresponds to the pre-main sequence phase of early stellar evolution. By the beginning of this phase the YSO has acquired most of its final mass and has evolved to become visible at optical and near-infrared wavelengths.

4.4.1. *The Embedded Phase: Protostars*

<u>Class I Sources</u>. Class I SEDs are broader than a single blackbody function and peak at far-infrared or sub-millimeter wavelengths. Longward of two microns these SEDs usually rise with increasing wavelength producing a huge "excess" of infrared emission compared to that expected from a normal stellar photosphere. They typically exhibit the silicate absorption feature at 10 microns wavelength. Class I sources derive their large infrared excesses from the presence of large amounts of circumstellar dust. These sources are usually deeply embedded in dense molecular cloud cores and rarely exhibit detectable emission in the optical portion of the spectrum. However, they are detected at near-infrared (i.e., 2.2 μm) wavelengths and frequently associated with small (infrared) reflection nebulae. Indeed, a significant fraction, in some cases all, of the near-infrared emission from a Class I source is scattered light. Roughly half of the Class I sources exhibit atomic emission lines in the infrared, similar to Class II sources (e.g., [46]). Otherwise, the infrared spectra of these objects are found to be featureless and heavily veiled [46, 25, 26]. Class I sources are often associated with bipolar molecular outflows, indeed the driving sources of the vast majority of known bipolar molecular outflows are Class I objects. Class I sources are often, but not always, more luminous than Class II sources. For example, in Ophiuchus, Class I sources have luminosities in the range $\sim$ 0.2-50 $L_\odot$, with a median value of $\sim$ 3 $L_\odot$, while Class II sources range between 0.1 - 4 $L_\odot$ with a median value of $\sim$ 0.6 $L_\odot$ [132, 45]. On the other hand, in Taurus, Class I sources have a median luminosity of only 0.7 $L_\odot$, the same as that of Class II sources in that cloud [60]. Class I sources are relatively rare among young stellar objects in molecular clouds and statistical arguments suggest ages of these sources of order 1-5 x 10^5 years [99, 132, 63].

Flat-Spectrum Sources. A subset of the Class I sources, these objects merit additional comment. Flat-spectrum sources have SEDs intermediate between Class I and II sources. Unlike the standard Class I source, these objects are often visible stars which display extreme T-Tauri characteristics. Indeed, the most well known member of this group is T Tauri itself. At optical wavelengths, the spectra of these objects display little in the way of photospheric absorption features and are heavily veiled most likely by excess continuum radiation from accretion shocks at the stellar surface. Like Class I sources, these objects are also veiled at infrared wavelengths, but not as strongly. Roughly half the flat-spectrum sources display atomic and molecular absorption lines at 2.2 μm. This is significant because these are the only embedded objects for which the inner stellar component can be spectrally classified and placed on the HR diagram. These objects are found to be characterized by late-type (M) photospheres whose luminosities place them near or above the birthline ([47, 62]) similar to the youngest known Class II sources. Flat-spectrum sources appear to be objects in transition between the Class I and II phases and can be considered "optical protostars".

Class 0 Sources. These objects appear to be extreme examples of Class I objects. They are considerably more extincted and embedded than Class I sources. Their energy distributions peak at submillimeter wavelengths and most are not detected at wavelengths shortward of 20 μm. Their energy distributions have widths similar to single temperature black body functions, and unlike Class I sources they can be placed on the HR diagram. However, they are characterized by extremely low temperatures, 20-30 K! All are associated with energetic, very highly collimated bipolar molecular outflows which are typically more energetic than those associated with Class I objects [22]. On the other hand as a whole Class 0 sources are not significantly more luminous that Class I sources [22]. In addition, observations of the Ophiuchi dark cloud indicate that Class 0 sources emit significantly more submillimeter radiation than do Class I objects [7]. In particular, the circumstellar mass traced by submillimeter measurements within 1000 AU of a Class I source is usually found to amount to a fraction of a stellar mass, while for a Class 0 source this mass can be comparable to that of the central star. Based on these considerations Andre, Ward-Thompson and Barsony [9] have suggested that these sources constitute a distinct class of YSO. Whether Class 0 sources are precursors of Class I sources or represent a separate branch of early stellar evolution is not yet clear. Class 0 sources are relatively rare making up roughly 10% of embedded sources. If they are precursors of Class I sources, then their lifetimes are on the order of $\sim 10^4$

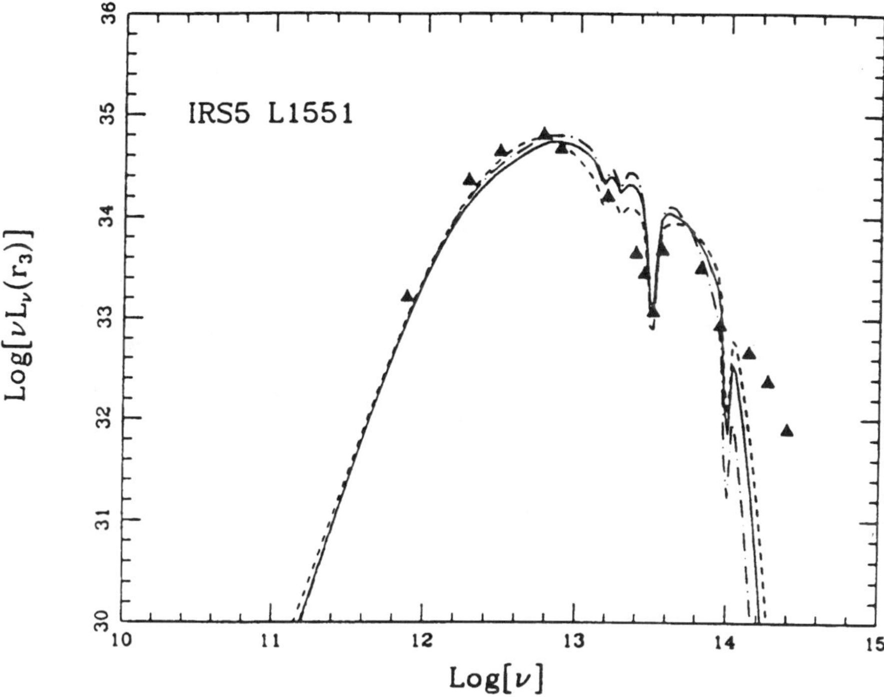

Figure 12.   *SED for the Class I source L 1551 IRS 5, a protostellar source in the Taurus molecular cloud. Solid and dashed lines show various rotating-collapsing isothermal protostellar models fit to the data. (From Adams, Lada and Shu 1987).*

years.

On The Protostellar Nature of the Embedded Sources.

Protostars can be defined as *objects in the process of accumulating into a stellar-like configuration the bulk of the material they will ultimately contain as main sequence stars.* The identification of protostars is fundamental to the eventual development of a theory of star formation. There are four important pieces of evidence which suggest that embedded sources are indeed protostellar in nature.

First, Class I and Flat-Spectrum SEDs have been successfully modeled as systems consisting of embryonic stellar cores surrounded by circumstellar disks and massive envelopes of gas and dust whose density structure is the same as that theoretically predicted for rotating, infalling protostellar cloud cores [3, 24]. As an example, Figure 12 shows the SED of the Class I source L1551 along with protostellar models for the object. Such modelling

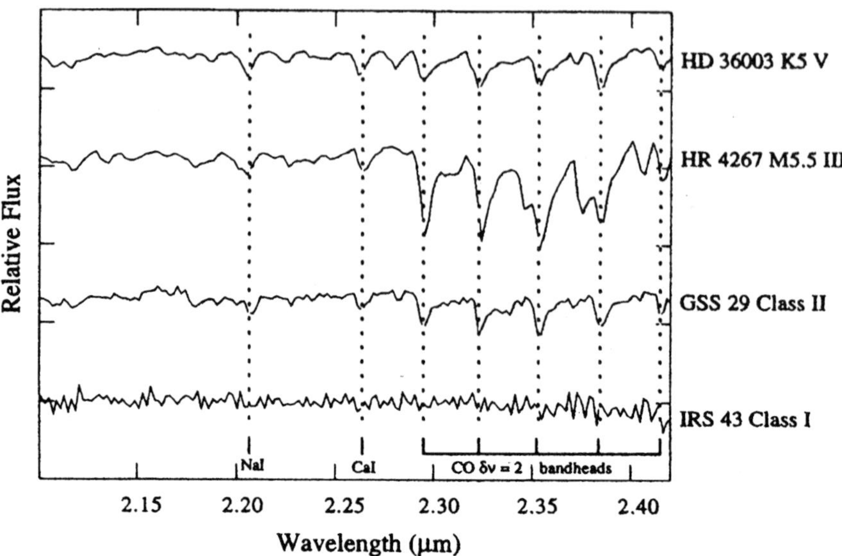

Figure 13. *Infrared spectra of a low-mass Class I source, a Class II source and late type dwarf and giant spectral standard stars. (from Greene & Lada 1996).*

suggests mass infall rates of $\sim 5 \times 10^{-6}$ M$_\odot$ yr^{-1} for typical Class I sources. Consequently, a star can acquire a significant fraction of a solar mass over the duration ($\sim 10^5$ yrs.) of the Class I phase.

<u>Second</u>, as mentioned above, low resolution infrared spectroscopy has revealed that the majority of Class I sources are largely featureless as a result of the presence of strong continuum veiling [46, 25, 26]. As an example figure 13 shows the K-band, 2.2 μm spectra of a Class I source, a Class II source, a late-type field dwarf and giant, respectively. The spectrum of the Class II source exhibits absorption features which are very similar to those of the dwarf, as expected, while that of the Class I source is featureless. The lack of absorption lines in the Class I spectrum indicates the presence of an additional source of continuum flux besides that provided by the protostellar photosphere. This excess continuum flux acts to dilute or veil the absoption lines fluxes. In order to completely wash out the features at the resolution observed, the excess flux must be significantly stronger (about a factor of 5) than the continuum flux emitted by the protostellar photosphere. Moreover, this excess continuum flux must originate very close (i.e., < 0.2 AU) to the stellar surface. The most likely source of this excess continuum is either hot dust in the inner portion of an infalling protostellar envelope or emission from the inner regions of a luminous accretion disk. However, because the inner regions of a luminous accretion disk would be

expected to produce its own system of detectable absorption lines (at 2μm), it is most likely that the veiling observed in Class I sources is produced by a luminous and dusty inner envelope. Dust in such an inner envelope is unlikely to be in stable orbits around the central star due to the outward pressure of radiation close to the stellar surface. However such close in material could be contained in a dynamically infalling envelope. Indeed, the same infall models which fit the SEDs of Class I sources, can produce sufficient K-band excess to completely veil protostellar infrared spectra via reprocessed light from their inner envelopes [46].

Third, essentially all Class 0 and Class I sources drive bipolar molecular outflows. The only viable source for the enormous energies which characterize the gas in molecular outflows is the release of gravitational potential energy by material falling deep into the protostellar potential well. That is, material falling directly onto the surface of an embryonic star from either an infalling circumstellar envelope or from a circumstellar accretion disk [67]. Exactly how the energy from infall and accretion is tapped to drive an energetic protostellar wind is unclear. However, the energetics of outflows are such that average protostellar mass accretion rates required to generate them must fall in the range $\dot{M}_{acc} \sim 10^{-6} - 10^{-5}$ M$_\odot$ yr^{-1} (for Class I and 0 sources, respectively [22]).

Fourth and most significant, direct kinematic evidence for infall motions appears to have been observed in a number of Class 0 sources. Under favorable circumstances, millimeter-wave molecular spectral-lines from collapsing clouds will display a kinematic signature of infall motion. This signature takes the form of an infall asymmetry in the line profile in which the red-shifted portion of an optically thick emission line is depressed relative to the corresponding blue-shifted portion. This is shown in Figure 14 In order to produce such an asymmetry, there must be an inwardly increasing excitation temperature gradient and an infall velocity whose magnitude is sufficiently large to be well resolved by the spectrometer. The first detections of this infall signature were observed toward the Class 0 sources IRAS 1629 [127] and B335 [136]. These studies further showed that the shape of the asymmetry in the line profile could be well matched by predictions using the standard infall models (e.g., [124]) to describe the velocity and density fields. Subsequent surveys of embedded sources have revealed the infall symmetry towards a substantial number ($\sim 1/3$) of Class 0 sources [48, 87]). Although few Class I sources are known to display a kinematic signature of infall, the most compelling case for infall and a protostellar status for an embedded object may be that of the Class I source HL Tau. This flat-spectrum source not only exhibits kinematic evidence for infall in millimeter-wave observations of molecular lines [52] but also in infrared observations of molecular carbon absorption-lines which originate much closer

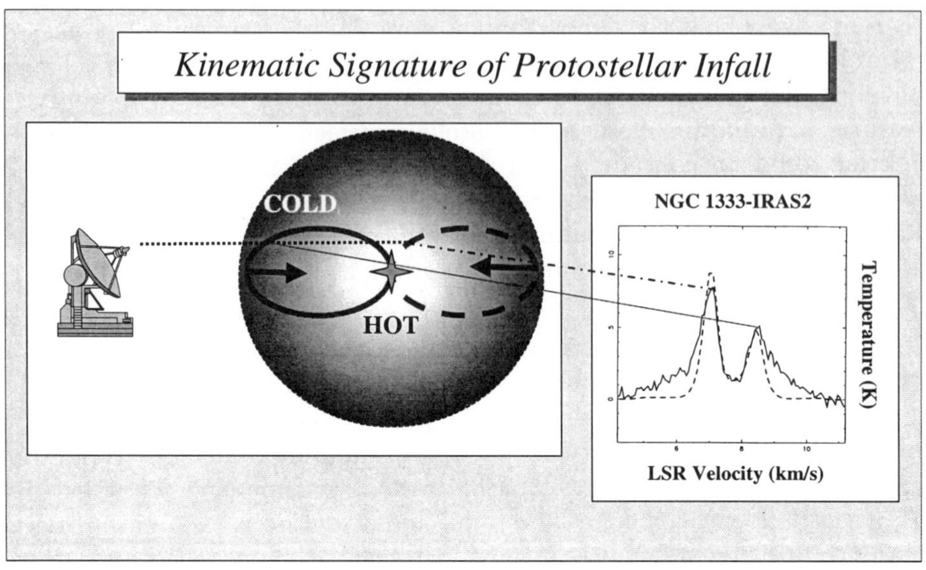

Figure 14.   The kinematic signature of infall expected around a protostellar source. Oval curves represent lines of constant infall radial velocity. The dashed line represents blue-shifted gas, the solid line, red-shifted gas. Along a single line-of-sight the telescope beam intercepts the $\tau = 1$ surface in the outer, colder regions of the cloud at redshifted velocities and the inner hotter regions at blue-shifted velocities. The spectrum of NGC 1333-IRAS2 is from Ward-Thompson et al 1996.

to the stellar surface are found to be highly redshifted [44].

4.4.2.   The Revealed Phase: Pre-Main Sequence Stars

Class II Sources. Class II SEDs peak at visible or near-infrared wavelengths. Like Class I sources the Class II SEDs are broader than a single blackbody function. Longward of two microns Class II SEDs fall with increasing wavelength usually in a power-law like fashion [75, 110, 4]. This results in an infrared excess which, though significant, is much smaller than that exhibited by Class I sources. The infrared excess indicates the presence of circumstellar material associated with the star. Figure 15 shows the composite SED of seven Class II sources. The dereddened and averaged data points are indicated by triangles and a theoretical fit to the spectrum (in which the infrared portion is modeled by a power–law function) is indicated by a solid curve.

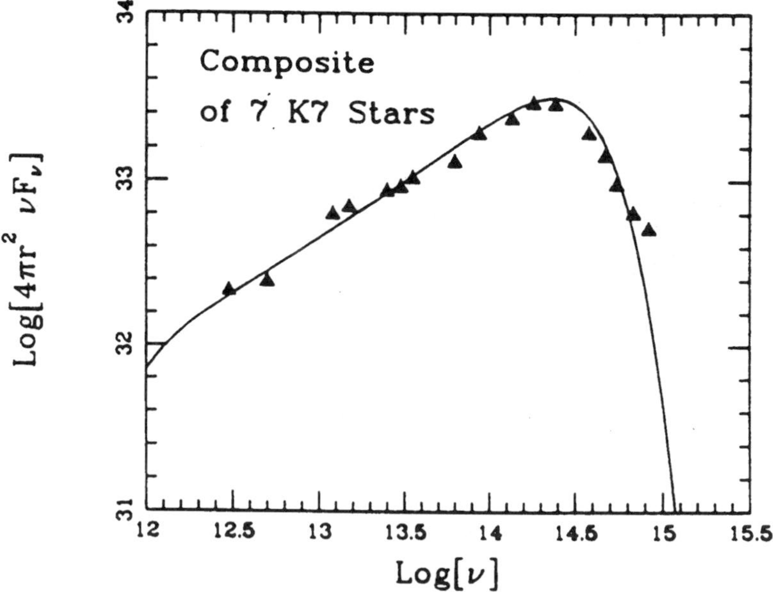

Figure 15. *The composite SED of seven Class II stars along with that (solid line) of a model circumstellar disk. (From Adams, Lada and Shu 1988).*

More than two decades ago Lynden-Bell and Pringle [86] predicted that T Tauri stars would display such energy distributions if they were surrounded by luminous accretion disks. Consider an optically thick and spatially thin disk that surrounds a young star and radiates everywhere like a blackbody. Imagine the disk to be composed of concentric annuli as illustrated in Figure 16 with radial dimension ΔR and area $2\pi R\Delta R$. Each annulus radiates as a blackbody of temperature $T(R)$. The emergent spectrum of the disk will then be the superposition of a series of blackbody curves of varying T(R) as shown in Figure 16. Now if $T(R) \sim R^{-n}$, the Wien laws tells us that the frequency of maximum emission scales as $\nu \sim T(R) \sim R^{-n}$. The luminosity radiated in each annulus is given by:

$$L_\nu d\nu = 2\pi R dR \sigma T(R)^4 \sim R^{2-3n}d\nu \sim \nu^{3-\frac{2}{n}}$$

Therefore, if the temperature gradient in the disk is characterized by a radial power–law, the emergent spectrum will also be characterized by a power–law slope in frequency or wavelength. For an SED, $\nu L_\nu \sim \nu^{4-\frac{2}{n}}$ or $\alpha = \frac{2}{n} - 4$ where α is the slope of the SED when plotted as a function of $\log\lambda$.

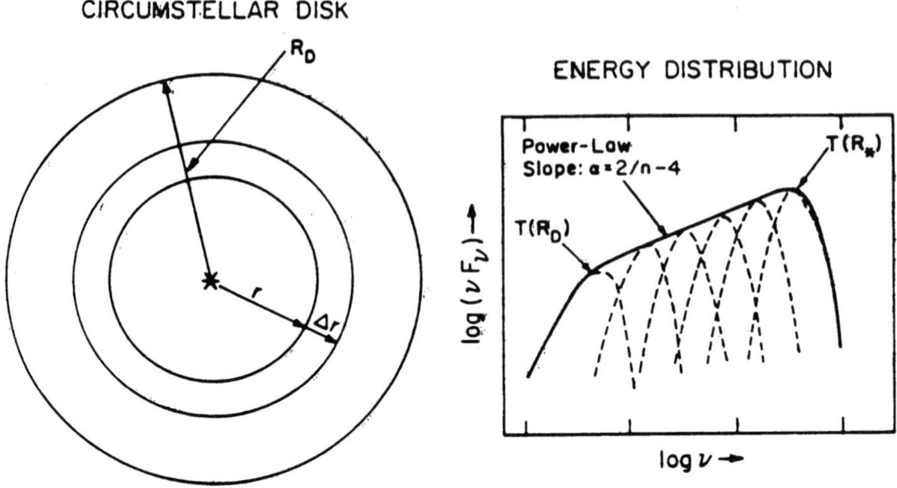

Figure 16.   Schematic diagram of a spatially thin, optically thick disk and its emergent spectral energy distribution. The disk spectrum is composed of a superposition of blackbodies of varying temperature.

Thus the power-law shape of the infrared portion of Class II SEDs strongly suggests that the infrared excess arises in an optically thick circumstellar disk. The slope of the SED is directly related to the temperature gradient in the disk. The slopes of Class II SEDs longward of 2 microns wavelength are observed to have values in the range between -0.7 and -1.3 corresponding to a range in n, the index of the disk temperature gradient, of 0.6 to 0.75. A viscous accretion disk is predicted to produce a temperature gradient characterized by n = 0.75 [86] which corresponds to α = -1.33. This also turns out to be the same temperature gradient and spectral slope predicted for a flat purely passive disk which derives its luminosity from the reprocessing and re-radiation of light it has absorbed from the central star [2, 39]. The majority of Class II sources have shallower slopes (typically $\alpha \approx -0.7$) which suggests that they are surrounded by flared (passive) disks [61]. The excellent agreement between the predictions of disk models and observations as indicated in Figure 15 suggests that the most likely interpretation of the nature of Class II sources is that they represent young stars surrounded by circumstellar disks. They differ from Class I objects in that they lack large, massive (infalling) envelopes of gas and dust. However, it is interesting to note that the infrared to millimeter excess emission from Class II sources is sufficiently large that if the emitting material were spherically distributed and not confined to a highly flattened structure, such as a disk, the star would suffer significantly more extinction than is

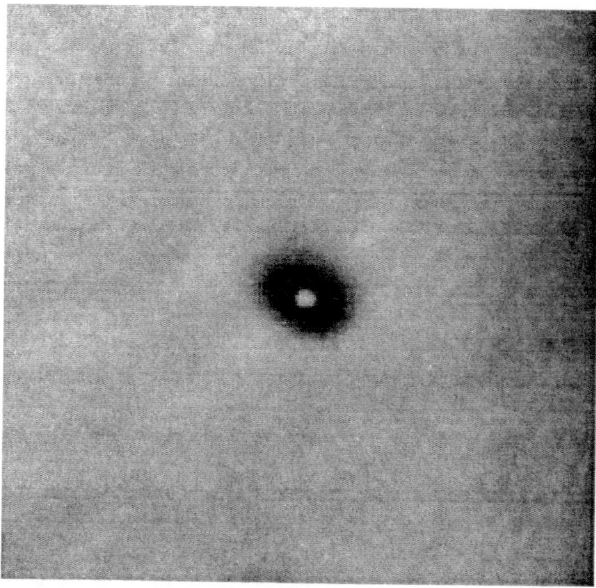

Figure 17. *This HST image of a resolved circumstellar disk seen in silhouette against the Orion nebula provides dramatic confirmation of the disk hypothesis for the SEDs of Class II sources. From McCaughrean and O'Dell 1996.*

observed [3, 14]. Indeed, disk masses derived from detection of optically thin continuum millimeter-wave emission range between 0.01–0.1 solar masses [14].

Class II sources can be observed at optical as well as infrared wavelengths. Therefore, considerably more is known about the nature of these objects than is known about Class I or Class 0 sources. When observed optically Class II sources typically exhibit the characteristics of Classical T-Tauri stars (CTTS). Conversely, most all CTTS stars possess Class II SEDs. Classical T-Tauri stars are low mass, pre-main sequence, emission-line variable stars. In addition to excess infrared continuum emission, these stars also exhibit excess emission at ultraviolet wavelengths. The optical spectra of CTTS contain Hydrogen emission lines and frequently various forbidden emission lines as well. The forbidden lines are believed to arise in stellar winds originating near the surface of the star. Typically these lines are observed to be blue-shifted and the absence of red-shifted emission is interpreted as strong supporting evidence for the presence of an occulting disk close to the stellar surface [33, 10]. The origin of the Balmer emission lines is more mysterious. Analysis of these lines provides evidence for both mass loss and mass accretion in these objects [16]. However, both the mass accretion and ultraviolet excess are believed to be consequences of disk accretion onto the stars [16]. The accretion rates are typically $\sim 10^{-8}$ $M_\odot$

yr^{-1} (e.g., [49]) and though significant, are relatively low compared to the typical infall rates encountered in protostellar evolution. Because they are visible, CTTS can be placed on the HR diagram, and when their positions are compared to theoretical PMS tracks one derives ages for CTTS usually between 10^6 and 4 x 10^6 years. However, because the theoretical tracks are calculated assuming diskless PMS stars, it is not clear how they actually relate to the positions of accreting CTTS on the HR diagram. Class II sources are relatively common in star formation regions where they typically outnumber Class I sources by more than 10-to-1.

In summary, Class II sources have been observed over three decades in frequency and over this entire range their SEDs are well matched by the predictions of model stars with circumstellar disks. UV observations and optical spectral-line observations provide additional supporting evidence for the presence of such circumstellar disks and further indicate that these stars are actively accreting material from their disks.

Class III Sources. Class III SEDs typically peak at visible and infrared wavelengths for low mass stars and decrease longward of two microns more steeply than Class II sources. Since their shapes are more or less similar to single temperature blackbodies, the energy distributions of Class III sources are easily interpreted as arising from extincted or unextincted photospheres of young stars. By definition these stars display no infrared excess. However, their light still could be substantially extinguished by foreground dust.

Extinction attenuates the spectrum of a YSO by a factor of $\exp(-\tau_\nu)$ and since $\tau_\nu \sim \nu$, for interstellar dust, then $F_\nu \sim \exp(-\nu)$ and it is difficult to distinguish such a spectrum from the high frequency side of a blackbody function where $F_\nu \sim \exp(-\frac{h}{kT}\nu)$. Thus an extincted star has an energy distribution similar to a blackbody but with the high frequency end exponentially extinguished and the peak displaced toward lower frequencies thus giving the appearance of being "reddened".

Class III sources are basically free of any significant amounts of circumstellar gas and dust and can be readily placed on the HR diagram. Class III sources are found to lie above the main sequence and can be thought of as "classical" pre-main sequence stars in the sense that their positions on the HR diagram can be unambiguously compared to predictions of theoretical PMS tracks [129]. Comparison with such theoretical tracks shows that the ages of Class III sources range from roughly 10^6 years to more than 10^7 years. Although most Class III sources have ages > 5 x 10^6 years, and are likely candidates for post-T Tauri stars (PTTS), a significant number have ages which overlap with those of Classical T Tauri Stars (CTTS) [129]. Since they lack infrared and ultraviolet excesses and are not strong emission-line emitters, these stars are difficult to distinguish from foreground and background stars in star formation regions. Thus it is difficult to obtain a census of their relative numbers. However, compared to field stars, Class III sources are relatively strong (but variable) X-Ray sources and can be identified in X-Ray surveys. In the Taurus region such observations suggest that the population of Class III sources is at least comparable in size to that of Class II sources. Older Class III objects are also found to extend well beyond the boundaries of star formation regions in X-ray surveys [102]. Class III sources typically produce little or no Hα line emission. All Class III sources are therefore Weak-lined T Tauri stars (WTTS). (Optical astronomers classify PMS stars with Hα equivalent widths less than 10 Angstroms as WTTS and PMS stars with Hα equivalent widths greater than 10 Angstroms as CTTS.)

5. The YSO Evolutionary Sequence

Although the SED classes (0, I, II, III) discussed above correspond to distinct physical classes of YSOs, the variation in the shapes of the energy distributions from Class 0 to III is quasi continuous and for some purposes can be usefully parameterized by an SED spectral index [68] or a "bolometric" temperature [98]. This variation in SED shape represents an evolutionary sequence corresponding to the gradual dissipation of gas and dust envelopes around newly formed stars [68]. Adams, Lada and Shu [3] were able to theoretically model this empirical sequence as a more or less continuous sequence of early stellar evolution from protostar to young main sequence star using the self-consistent physical theory of rotating collapsing isothermal spheres as the initial condition (see also [117, 54]). In the current version of this theoretical picture Class 0 and Class I sources are the youngest and least evolved objects, indeed, true protostars, objects assembling the bulk of the mass they will ultimately contain when they arrive on the main sequence. They consist of a central embryonic (hydrostatic) stellar core surrounded by a circumstellar disk and are steadily gaining mass from infall and accretion of surrounding matter. Class III sources are the most evolved objects. For these objects the vast majority of the original star forming material has been either all incorporated into the star or removed from its vicinity. The vast majority of low mass stars probably pass through these stages during their formation. However, the fact that some Class III sources have similar ages as Class II sources, suggests that the duration of the various phases may be different for different stars, even those of similar mass.

5.1. THE PIVOTAL ROLE OF BIPOLAR OUTFLOWS

To evolve from Class 0 to I to II requires the removal or dissipation of the circumstellar material in the protostellar infalling envelope. To evolve

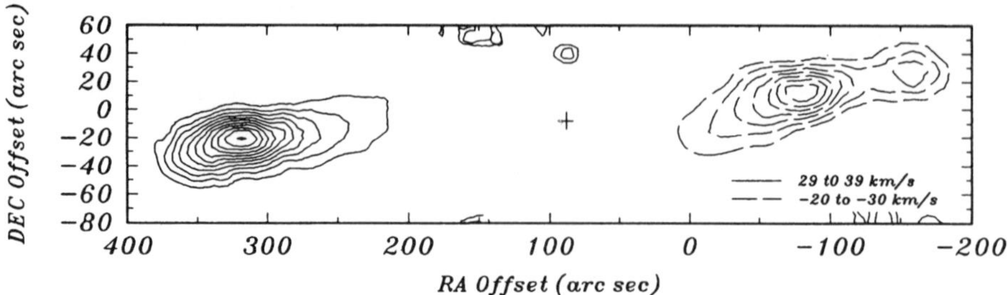

Figure 18.   An example of a bipolar molecular outflow. Shown here are the high velocity lobes of the NGC 2264G outflow traced by emission in the J=2→1 transition of CO. A cross marks the location of a deeply embedded Class 0 protostellar object. From Lada and Fich (1996).

from Class II to Class III requires clearing of the circumstellar disk. In principle, the clearing of circumstellar gas and dust could be accomplished by accreting all the surrounding material onto the star itself. This might occur in massive cloud cores which form clusters via fragmentation. There is some recent evidence to suggest that such fragmentation may produce clumps with a mass spectrum similar to the stellar IMF [95, 125]). In which case a star could evolve from the Class I to III phase once it simply accreted the entire mass of the fragment form which it was born. On the other hand, this possibility conflicts with the observation that star formation occurs with considerable inefficiency: the cores from which stars form contain much more mass than the stars themselves, and most of the surrounding material cannot end up inside the star. This suggests that at some point very early on in the evolution of a young stellar object the cloudy material from which it grows must be physically removed by some active agent.

This agent is most likely the energetic bipolar outflow which is ignited early in the protostellar stage of evolution [119, 12, 67, 13, 40, 11]. Bipolar outflows and associated jets, masers and Herbig-Haro objects, are generated almost exclusively by embedded objects (Class 0 and I). They are very rarely associated with Class II or III stars. Bipolar molecular outflows are individually energetic enough to disrupt cloud cores and collectively even powerful enough to have a significant impact on the dynamics and structure of an entire GMC [88]. In fact, the molecular outflows generated by a population of embedded YSOs may be able to generate the turbulent pressure that keeps GMCs from global collapse, thereby solving one of the outstanding problems of cloud dynamics.

Essentially all Class I and Class 0 sources drive molecular outflows, yet these objects are supposed to be protostellar in nature, their dynamics dominated by infall and accretion. This presents a paradox: how is a star

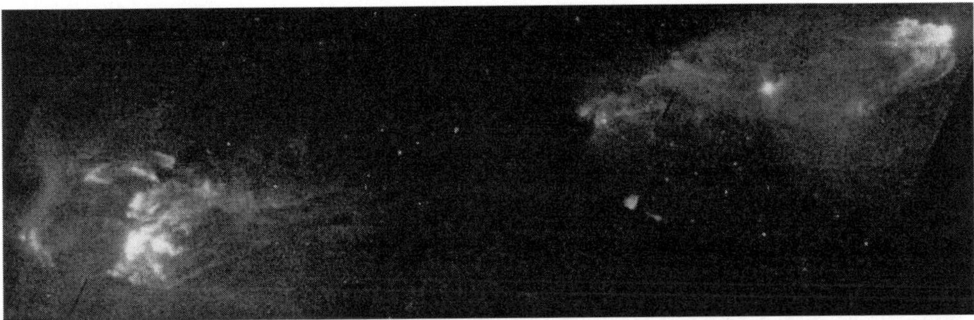

Figure 19.   An example of a bipolar outflow visible at optical wavelengths. This is an HST image of the Herbig-Haro objects HH 1 and HH 2. The visible emission originates in shocks resulting from the impact of high velocity jets on ambient cloud material. A deeply buried Class I source, located in the dark cloud core between the two lobes, drives this well known outflow.

simultaneously a source of both outflow and infall, how does a star grow by losing mass? The resolution of this paradox, as mentioned earlier, must be that outflows are driven by infall and accretion.

Magnetic fields and angular momentum may play critical roles in enabling the transfer of energy from infalling to outflowing gas. To visualize how this might work, consider a rotating infalling protostellar envelope. Because of the conservation of angular momentum the trajectories of infalling material become curved as they near the growing stellar object at the center of the potential well. As a result, most of the infalling gas misses the star and falls into the equatorial plane forming an extended disk. Consequently for a rotating protostar most of the mass that ends up on the star must be accreted from the surrounding disk. In order for material to flow through the disk and onto the protostar, the material must lose both energy and angular momentum. If the mass of the disk is not much larger than that of the central object, the material in the disk should rotate differentially in Keplerian fashion. Gas falling through such a disk will reach the surface of the central star with an orbital velocity and specific angular momentum which is relatively high compared to that in the star. If this material is added to the star it will spin up the star. The star will quickly reach break up equatorial velocities at which point material can no longer be added to it. A centrifugal barrier prevents the further growth of the protostar. Thus the process of star formation can only proceed if the incoming gas somehow can lose additional angular momentum in the process of accreting onto the star or if the star can somehow spin down while accretion is taking place.

Angular momentum can be carried away from a star by a stellar wind. *Consequently, a protostar may be able to gain mass only if it simultaneously*

loses mass. To allow star formation to continue the rate of mass loss from
the wind should be a fraction of the mass accretion rate i.e.,

$$\dot{M}_{wind} = f\dot{M}_{accretion}$$

where the fraction f is determined by the physics of the wind generating mechanism. The ideal protostellar wind is one that carries away little
mass but lots of angular momentum. A number of recent investigations
have shown that centrifugally–driven hydromagnetic winds are potentially
capable of doing the job [107]. Such winds could be driven from either magnetized circumstellar disks [107] or from the surfaces of central protostars
[118]. Thus the natural consequence of star formation in rotating, magnetic
cloud cores may be the formation of protostar-disk systems which generate
powerful outflows and in doing so resolve the paradox of the protostar which
gains mass by losing mass. An additional consequence of such "accretion
driven" outflow is that as a protostar evolves, its outflow disrupts the surrounding core ultimately limiting the mass which can fall onto the central
star. As the protostellar envelope disperses, the reservoir of infalling material available to drive disk accretion and mass outflow also diminishes and
the YSO evolves to progressively less active states (i.e., Class I-II-III), consistent with observations and the evolutionary spectral sequence described
earlier.

5.2. EPISODIC ACCRETION?

Finally, it is presently uncertain whether the infall/accretion/outflow process is steady over time. The observation that outflows from Class 0 sources
are more energetic than those from Class I sources suggests that the in-
fall/accretion declines from the Class 0 to Class I phases [22]. Clearly there
is a dramatic decrease in outflow activity by the Class II and III stages
suggesting an equally drastic drop in the rate of infall/accretion. Moreover,
there is a growing body of evidence indicating that episodic mass ejection
may be an important aspect of the protostellar evolution of low mass stars
[57, 108, 50]. There is a subset of YSOs (*FU Ori objects*), originally identified by Herbig, [57] which are known to undergo intense outbursts of light.
These outbursts, which can increase the luminosity of the star by a factor
as much as 100 in a few years, are most likely due to dramatic, apparently
episodic, increases in disk accretion onto the star [50]. Although most often
identified by specific optical spectral characteristics [50], these objects typically turn out to be Class I sources. It is possible that FU Ori outbursts
occur throughout protostellar evolution. Although infall rates for protostars
are high, the relatively low luminosities (~ 1 $L_\odot$) of some Class I sources
suggest they are simultaneously characterized by significantly lower disk

accretion rates [60]. The infall/accretion luminosity for a protostar is given by:

$$L_{accretion} = \frac{1}{2} \times \frac{GM_* \dot{M}}{R_*}$$

For nominal Class I parameters this can be expressed as:

$$L_{accretion} = 5.5 \times \left[\frac{M_*(M_\odot)}{0.25}\right]\left[\frac{\dot{M}(M_\odot \mathrm{yr}^{-1})}{2 \times 10^{-6}}\right]\left[\frac{10^{11}}{R_*(\mathrm{cm})}\right]\ L_\odot$$

This is the luminosity due to infall and accretion assuming all the infalling material reaches the surface of the star. If most of the infalling material lands on the disk and does not accrete through it at the same rate as the infall onto it (i.e., $\dot{M}_{acc} < \dot{M}_{infall}$), then $L_{accretion}$ would be smaller than the steady state value. However, in this situation the disk would grow in mass. This could not continue indefinitely and eventually the disk would have to dump significant amounts of mass onto the surface of the central protostar. Such a discharge would presumably result in an FU Ori outburst. The resulting accretion rate would be expected to be somewhere in the range 10^{-4} $M_\odot$ yr^{-1} and 10^{-5} $M_\odot$ yr^{-1} (corresponding respectively to the inferred accretion rates of FU Ori and L1551, two representative FU Ori objects). Depending on which rate is typical, we would expect anywhere from 5% to 50% of Class I sources to be in outbust at any one epoch of history (since the lifetime of a Class I source is roughly 10^5 yrs and the final mass of its central star is roughly 0.5 $M_\odot$). Spectroscopic surveys of the Ophiuchus and Taurus clouds [46] suggest that approximately 3% of the Class I sources in those clouds display the spectroscopic signature of an FU Ori star. Perhaps this favors luminous, short lived outbursts throughout the late protostellar phase. On the other hand, many protostellar sources appear to be reasonably luminous and do not display the spectroscopic signature of an FU Ori object. These sources may be actively accreting at the steady-state rate. Clearly more observations of embedded objects are required to determine how significant FU Ori type outbursts are for protostellar evolution.

6. Concluding Remarks

Over the last few years a coherent picture of protostellar and early stellar evolution (for at least isolated, low mass objects) has been derived from observations and buttressed by theory. Still, much remains to be done in order to achieve the important goal of a complete and predictive theory of star formation. In particular, our understanding of the origin of high mass stars, the origin of the IMF, the nature of bipolar outflows and the process of star formation in embedded clusters is still far from complete. These are among the many interesting problems that remain at the frontier of star formation research as we begin the twenty-first century.

References

1. Abt, H.A. & Levy, S. 1976 *Astrophys. J. Suppl. Ser.*, **30**, 273.
2. Adams, F.C., & Shu, F. 1986 *Astrophys. J.*, **308**, 836.
3. Adams, F.C., Lada, C.J., & Shu, F. 1987 *Astrophys. J.*, **312**, 788.
4. Adams, F.C., Lada, C.J., & Shu, F. 1988 *Astrophys. J.*, **326**, 825.
5. Allen, C.W. 1973 *Astrophysical Quantities*, (Athlone Press: London).
6. Allen, L. & Hillenbrand, L. 1998 in *The Orion Complex Revisited*, eds. M.J. Mc-Caughrean and A. Burkert (ASP Conference Series), in press.
7. Andre, P. & Montmerle, T. 1994 *Astrophys. J.*, **420**, 837.
8. Andre, P. 1994 in *The Cold Universe*, eds. T. Montemerle, C.J. Lada, F. Mirabel and J. Tran Thanh Van, (Paris, 1994), 179.
9. Andre, P., Ward-Thompson, D. & Barsony, M. 1993 *Astrophys. J.*, **406**, 122.
10. Appenzeller, I., Jankovics, I & Ostreicker, R. 1984 *Astron. Astrophys.*, **141**, 108.
11. Bachiller, R. 1996 *Ann. Rev. Astron. Astrophys.*, **34**, 111.
12. Bally, J. & Lada, C.J. 1983 *Astrophys. J.*, **265**, 824.
13. Bally, J. & Lane, A.P. 1991 in *The Physics of Star Formation and Early Stellar Evolution*, eds. C.J. Lada and N.D. Kylafis, (Kluwer Academic Publishers: Dordrecht), p. 471.
14. Beckwith, S.V.W., Sargent, A.I., Chini, R. & Gusten, R. 1990 *Astron. J.*, **99**, 924.
15. Benson, P. & Myers, P. 1989 *Astrophys. J. Suppl. Ser.*, **71**, 89.
16. Bertout, C. & Basri, G. 1991 in *The Physics of Star Formation and Early Stellar Evolution*, eds. C.J. Lada and N.D. Kylafis, (Kluwer Academic Publishers: Dordrecht), p. 649.
17. Blaauw, A. 1964 *A.R.A.A.* **2**, 213.
18. Blaauw, A. 1991 in *The Physics of Star Formation and Early Stellar Evolution*, eds. C.J. Lada and N.D. Kylafis, (Kluwer Academic Publishers: Dordrecht), p. 125.
19. Blitz, L. 1980 in *Giant Molecular Clouds in the Galaxy*, eds. P Solomon and M. Edmunds, (Pergamon, Oxford), p. 1.
20. Blitz, L. 1991 in *The Physics of Star Formation and Early Stellar Evolution*, eds. C.J. Lada and N.D. Kylafis, (Kluwer Academic Publishers: Dordrecht), p. 3.
21. Bok, B.J. 1934 *Harvard Circ. No. 384*.
22. Bontemps, S., Andre, P., Tereby, S. and Cabrit, S. 1996 *Astron. Astrophys.*, **311**, 858.
23. Bouvier, J., Rigaut, F. & Nadeau, D. 1997 *Astron. Astrophys.*, **326**, 1023.
24. Calvet, N., Hartmann, L., Kenyon, S. & Whitney, B. 1994 *Astrophys. J.*, **434**, 330.
25. Casali, M.M. & Erioa, C. 1996 *Astron. Astrophys.*, **306**, 427.
26. Casali, M.M, & Matthews, H.F. 1992 *Monthly Notices Roy. Astron. Soc.*, **258**, 399.
27. Chen, H., Tokunaga, A., Strom, K. & Hodapp K. 1993 *Astrophys. J.*, **407**, 639.
28. Clemens, D. & Barvanis, R. 1988 *Astrophys. J. Suppl. Ser.*, **68**, 257.
29. D'Antona, F. & Mazzitelli, I. 1994 *Astrophys. J. Suppl. Ser.*, **41**, 467.
30. Dickman, R.L., Horvath, M.A., & Margulis, M. 1990 *Astrophys. J.*, **365**, 586.
31. Duerr, R., Imhoff, K. & Lada, C.J. 1982 *Astrophys. J.*, **261**, 135.
32. Duquennoy, A. & Mayor, M. 1991 *Astron. Astrophys.*, **248**, 485.
33. Edwards, S. *et al.*, 1987 *Astrophys. J.*, **321**, 473.
34. Elmegreen, B.G. & Lada, C.J. 1976 *Astron. J.*, **81**, 1089.
35. Elmegreen, B.G. & Lada, C.J. 1977 *Astrophys. J.*, **214**, 725.
36. Evans, N.J. *et al.* 1994 *Astrophys. J.*, **424**, 793.
37. Fletcher, A. & Stahler, S. 1994 *Astrophys. J.*, **435**, 313.
38. Fischer, D. & Marcy, G. 1992 *Astrophys. J.*, **396**, 178.
39. Friedjung, M. 1985 *Astron. Astrophys.*, **146**, 336.

40. Fukui., F. *et al.*, 1993 in *Protostars and Planets III*, eds. E. H. Levy and J. I. Lunine, (Arizona: Tucson), p. 603

41. Gehz, A. 1993 Unpublished PhD Dissertation, Caltech.

42. Goldreich, P. and Kwan, J. 1974 *Astrophys. J.*, **89**, 441.

43. Gomez, M. & Lada, C.J. 1998 *Astron. J.*, **115**, 1524.

44. Grasdalen, G., Sloan, G., Stout, N., Strom, S. & Welty, A. 1989 *Astrophys. J.*, **339**, 37.

45. Greene, T.P. *et al.* 1994 *Astrophys. J.*, **434**, 614.

46. Greene, T.P. & Lada, C.J. 1996 *Astron. J.*, **112**, 2184.

47. Greene, T.P. & Lada, C.J. 1997 *Astron. J.*, **114**, 2157.

48. Gregersen, E.M., Evans, N.J., Zhou, S., & Choi, M. 1997. *Astrophys. J.*, **484**, 256.

49. Gullbring, E., Hartmann, L., Berinco, C. & Calvet, N. 1998 *Astrophys. J.*, **492**, 323.

50. Hartmann, L. & Kenyon S. J. 1996 *Ann. Rev. Astron. Astrophys.*, **34**, 207.

51. Hayashi, C. 1966 *Ann. Rev. Astron. Astrophys.*, **4**, 171.

52. Hayashi, M., Ohashi, N. & Miyama, S.M., 1993 *Astrophys. J. Letters*, **418**, 71.

53. Heiles, C., Goodman, A.A., & Zweibel, E. 1993 in *Protostars and Planets III*, eds. G. Levy and J. Lunine, (University of Arizona Press, Tucson) p. 279.

54. Henriksen, R., Andre, P. & Bontemps, S. 1997 *Astron. Astrophys.*, **323**, 565.

55. Henyey, L.G., LeLevier, R. & Levee, R.D. 1955 *Publ. Astron. Soc. Pacific*, **67**, 154.

56. Herbig, G. 1962 *Adv.Astr.Ap* **1**, 47.

57. Herbig, G. 1977 *Astrophys. J.*, **217**, 693.

58. Hillenbrand, L. 1997 *Astron. J.*, **113**, 1733.

59. Horner, D.L., Lada, E.A., & Lada, C.J. 1997 *Astron. J.*, **113**, 1788.

60. Kenyon, S.J. & Hartmann, L. 1995 *Astrophys. J. Suppl. Ser.*, **101**, 117.

61. Kenyon, S.J. & Hartmann, L. 1987 *Astrophys. J.*, **323**, 714.

62. Kenyon, S.J., Brown, D.L., Tout, C.A., & Berlind, P. 1998, *Astron. J.*, **115**, 2491.

63. Kenyon, S.J., Hartmann, L., Strom, S.E., Strom, K.M. 1990 *Astron. J.*, **99**, 869.

64. Kroupa, P. 1995 *Monthly Notices Roy. Astron. Soc.*, **277**, 1491.

65. Kroupa, P., Tout, C. & Gilmore, G. 1990 *Monthly Notices Roy. Astron. Soc.*, **244**, 76.

66. Kroupa, P., Tout C. & Gilmore, G. 1993 *Monthly Notices Roy. Astron. Soc.*, **262**, 545.

67. Lada, C.J. *Ann. Rev. Astron. Astrophys.*, **23**, 267.

68. Lada, C.J. 1987 in *I.A.U. Symposium No. 115: Star Forming Regions*, eds. M. Peimbert and J. Jugaku, (Dordrecht, Reidel), p. 1.

69. Lada, C.J. 1991 in *The Physics of Star Formation and Early Stellar Evolution*, eds. C.J. Lada and N.D. Kylafis, (Kluwer Academic Publishers: Dordrecht), p. 329.

70. Lada, C.J. & Fich M. 1996 *Astrophys. J.*, **459**, 638.

71. Lada, C.J. & Kylafis, N.D. 1991 *The Physics of Star of Star Formation and Early Stellar Evolution*, (Dordrecht: Kluwer).

72. Lada C.J. & Lada, E.A. 1991 in *The Formation and Evolution of Star Clusters*, ed. K. Janes (ASP, San Francisco,) p. 3.

73. Lada, C.J., Margulis, M. & Dearborn, D. 1984 *Astrophys. J.*, **285**, 141.

74. Lada, C.J., Young, E.T., & Greene, T. 1993 *Astrophys. J.*, **408**, 471.

75. Lada, C.J. & Wilking, B.A. 1984 *Astrophys. J.*, **287**, 610.

76. Lada, E.A. 1992 *Astrophys. J. Letters*, **393**, 25.

77. Lada, E.A., Bally, J. & Stark, A. 1991 *Astrophys. J.*, **368**, 432.

78. Lada, E.A., DePoy, D.L., Evans, N.J. & Gatley, I. 1991 *Astrophys. J.*, **371**, 171.

79. Lada, E.A. & Lada, C.J. 1995 *Astron. J.*, **109**, 1682.

80. Lada, E.A., Lada, C.J. & Muench, G. 1998 in *The Stellar Initial Mass Function*, ASP Conference Series vol 142 eds. G Gilmore and D. Howell (ASP: San Francisco), 107.

81. Lada, E.A., Strom, K.M., & Myers, P.C. 1993 in *Protostars and Planets III*, eds. G. Levy and J. Lunine, (University of Arizona Press, Tucson) p. 245.

82. Larson, R.B. 1981 *Monthly Notices Roy. Astron. Soc.*, **194**, 271.

83. Leinert, Ch. 1993 *et al. Astron. Astrophys.*, **278**, 129.

84. Loren, R. 1989 *Astrophys. J.*, **338**, 902.

85. Luhman, K.H., Rieke, G.H., Lada, C.J. & Lada, E.A. 1998 *Astrophys. J.*, **508**, 347.

86. Lynden-Bell D. & Pringle, J. 1974 *Monthly Notices Roy. Astron. Soc.*, **168**, 168.

87. Mardones, D. *et al.* 1997 *Astrophys. J.*, **489**, 719.

88. Margulis, M., Lada, C.J. & Snell, R.N. 1989 *Astrophys. J.*, **333**, 316.

89. Mathieu, R.D. 1987 in *I.A.U. Symposium No. 115: Star Forming Regions*, eds. M. Peimbert and J. Jugaku, (Dordrecht, Reidel), p. 61.

90. Mathieu, R.D. 1994 *Ann. Rev. Astron. Astrophys.*, **32**, 465.

91. McCaughrean, M. J., Zinnecker, H., Rayner, J. Stauffer, J. 1995 in *Proceedings of the ESO Workshop: The Bottom of the Main Sequence and Beyond*, (ed. C.G. Tinney), (Heidelberg: Springer), 207.

92. McCaughrean, M.J., & O'Dell, C.R. 1996 *Astron. J.*, **111**, 1979.

93. Meyer, M.R. and Lada, E.A. 1998 in *The Orion Complex Revisited*, eds. M.J. McCaughrean and A. Burkert (ASP Conference Series), in press.

94. Mihalas, D. and Binney, J. 1981 *Galactic Astronomy*, (Freeman: San Francisco), p 217.

95. Motte, F., Andre, P. & Neri, R. 1998 *Astron. Astrophys.*, **336**, 150.

96. Mouschovias, T. 1976 *Astrophys. J.*, **207**, 141.

97. Mouschovias, T. 1991 in *The Physics of Star Formation and Early Stellar Evolution*, eds. C.J. Lada and N.D. Kylafis, (Kluwer Academic Publishers: Dordrecht), p. 61.

98. Myers, P.C. & Ladd E. F. 1993 *Astrophys. J. Letters*, **413**, 47.

99. Myers, P.C. *et al.*, 1987 *Astrophys. J.*, **319**, 340.

100. Muench, G., Lada, E.A. & Lada, C.J. 1998 *Astrophys. J.*, in preparation.

101. Murden. P.G., & Penston, M.V. 1977 *Monthly Notices Roy. Astron. Soc.*, **181**, 657.

102. Neuhauser, R. 1997 Science, **276**, 1363.

103. Palla F. & Stahler, S.W. 1993 *Astrophys. J.*, **418**, 414.

104. Parker, N.D., Padman, R., Scott, P. & Hills, R. 1991 *Monthly Notices Roy. Astron. Soc.*, **234**, 442.

105. Phelps, R.L. & Lada, E.A. 1997 *Astrophys. J.*, **477**, 176.

106. Prosser, C. *et al. Astrophys. J.*, **421**, 517.

107. Pudritz, R., Pelletier, G., Gomez de Castro, A.I. 1991 in *The Physics of Star Formation and Early Stellar Evolution*, eds. C.J. Lada and N.D. Kylafis, (Kluwer Academic Publishers: Dordrecht), p. 539.

108. Reipurth, B. 1989 *Nature* **340**, 42.

109. Roberts, M.S. 1957 *P.A.S.P.* **69**, 59.

110. Rucinski, S.M. 1985 *Astron. J.*, **90**, 2321.

111. Salpeter, E. 1955 *Astrophys. J.*, **121**, 161.

112. Scalo, J. 1983 in *Protostars and Planets II*, eds. D. Black and M. S. Matthews, (University of Arizona Press: Tucson), p. 201.

113. Scalo, J. 1986 in *Fundamentals of Cosmic Physics*, **11**, 1.

114. Scalo, J. 1990 in *Physical Processes in Fragmentation and Star Formation*, eds. R. Capuzzo-Dolcetta, C. Chiosi and A. Di Fazio (Dordrecht, Kluwer), p. 151.

115. Scalo, J. 1998 in *The Stellar Initial Mass Function*, eds. G. Gilmore and D. Howell (Astronomical Society of the Pacific, San Francisco), p.

116. Schwarzchild, M. 1958 *Structure and Evolution of Stars*, (Princeton University Press: Princeton).

117. Shu, F.H., Adams, F.C. & Lizano, S. 1987 *Ann. Rev. Astron. Astrophys.*, **25**, 23.

118. Shu, F.H., Lizano, S., Ruden, S. & Najita, J. 1988 *Astrophys. J. Letters*, **328**, 19.

119. Snell, R., Loren, R, &. Plambeck, R. 1980 *Astrophys. J. Letters*, **239**, 17.

120. Stahler, S.W. 1983 *Astrophys. J.*, **274**, 822.

121. Strom, K.M., Strom, S.E. & Merrill, M. 1993 *Astrophys. J.*, **412**, 233.

122. Tatematsu, T. *et al.*, 1993 *Astrophys. J.*, **404**, 643.
123. Terebey, S. 1998 unpublished HST observations.
124. Terebey, S., Shu, F.H. & Cassen, P. 1984 *Astrophys. J.*, **286**, 529.
125. Testi, L. & Sargent, A.I. 1998. *Astrophys. J. Letters*, (in press).
126. van Albada, T.S. 1985 in *Birth and Evolution of Massive Stars and Stellar Groups*, eds. H. van Woerden and W. Boland, (Reidel: Dordrecht), p. 89.
127. Walker, C.K., Lada, C.J., Maloney, P., Young, E.T., & Wilking, B.A. 1986 *Astrophys. J. Letters*, **309**, 47.
128. Walker, M.F. 1956 *Astrophys. J. Suppl. Ser.*, **2**, 365.
129. Walter, F. 1987 *Publ. Astron. Soc. Pacific*, **99**, 31.
130. Ward-Thompson, D. *et al.* 1996 *Monthly Notices Roy. Astron. Soc.*, **314**, 625.
131. Whitworth, A. 1979 *Monthly Notices Roy. Astron. Soc.*, **186**, 59.
132. Wilking, B.A., Lada, C.J. & Young, E.T. 1989 *Astrophys. J.*, **340**, 823.
133. Wiramihardja, S. *et al.* 1991 *Publ. Astron. Soc. Japan*, **43**, 27.
134. Wooten, A. 1989 *Astrophys. J.*, **337**, 858.
135. Yuan C. & You J. 1995 Molecular Clouds and Star Formation, (World Scientific, Singapore).
136. Zhou, S., Evans, N.J., Kompe, C. & Walmsley, C.M. 1993 *Astrophys. J.*, **404**, 232.
137. Zinnecker, H., McCaughrean, M. & Wilking, B.A. 1993 in *Protostars and Planets III*, eds. G. Levy and J. Lunine, (University of Arizona Press, Tucson) p. 429.
138. Zuckerman, B. and Evans, N. 1974 *Astrophys. J. Letters*, **187**, 67.

Charles Lada at Knossos.

Frank Shu and Robbins Bell also enjoy the first week's banquet.

LOW-MASS STAR FORMATION: THEORY

FRANK H. SHU, ANTHONY ALLEN, HSIEN SHANG,
Astronomy Department, University of California
Berkeley, CA 94720-3411, USA

EVE C. OSTRIKER
Astronomy Department, University of Maryland
College Park, MD 20742-2421, USA

AND

ZHI-YUN LI
Department of Astronomy, University of Virginia
Charlottesville, VA 22903, USA

1. Magnetic Support of Clouds

Molecular clouds, the cradles of star birth, are supported against self-gravity in part by magnetic fields (McKee, this volume). In 1956, Mestel & Spitzer (see also Mestel 1965) introduced a mass scale associated with the amount of magnetic flux Φ threaded by a self-gravitating, electrically conducting, cloud. This scale, the magnetic critical mass,

$$M_\Phi \equiv \frac{\Phi}{2\pi G^{1/2}}, \tag{1}$$

is potentially as important for our subject as the Chandrasekhar limit is for the theory of stellar evolution (see Chandrasekhar 1939 and Schwarzschild 1958). In equation (1), G is the universal gravitational constant, and the rest of the equation differs from Mestel & Spitzer's (1956) definition (see also Mouschovias & Spitzer 1976) only by the value of the numerical coefficient on the right-hand side. Our choice of the numerical coefficient $1/2\pi$ defines M_Φ to be the maximum mass that can be supported if the magnetic field acted alone to oppose the self-gravity of the cloud, so that the cloud becomes highly flattened (see below).

Consider the mechanical equilibrium of a static cloud of mass M and constant temperature T, which occupies volume V (not necessarily spheri-

C.J. Lada and N.D. Kylafis (eds.), The Origin of Stars and Planetary Systems, 193–226.

cal or highly flattened) and has a magnetic flux Φ frozen into it while it is being subjected to a slowly changing external pressure P_{ext}. The virial theorem applied to such a gas cloud yields (Strittmatter 1966; see also Chapter 24 of Shu 1992)

$$P_{\text{ext}} = \frac{a^2 M}{V} + \frac{\alpha G}{V^{4/3}} \left(M_\Phi^2 - M^2 \right),\tag{2}$$

where $a \equiv (kT/m)^{1/2}$ is the isothermal sound speed (with k being Boltzmann's constant and m being the mean molecular mass of the gas) and α is a positive numerical coefficient of order unity. The first term on the right-hand-side of equation (2) is the contribution from Boyle's law, $P_{\text{ext}} = NkT/V$ where $N \equiv M/m$, whereas the second and third terms, proportional respectively to $GM_\Phi^2/V^{4/3}$ and $GM^2/V^{4/3}$, yield the corrections associated with magnetic forces and self-gravity. Notice that the latter two forces, expressed per unit area, scale with V in the same way, as $V^{-4/3}$. Thus, on average magnetic forces act effectively to dilute self-gravity (if $M_\Phi < M$) or to reverse it (if $M_\Phi > M$). This property of magnetic forces has profound implications for the problems of gravitational collapse and fragmentation of molecular clouds. These implications can potentially nullify conclusions which are drawn from calculations that ignore its presence.

Figure 1 shows a plot of P_{ext} versus V in the two cases: $M > M_\Phi$ and $M < M_\Phi$. In either case, a cloud of large volume V compresses at first under increases of external pressure according to Boyle, because at large V the average internal thermal pressure $a^2 M/V$ is more important in balancing the external pressure than either magnetic forces or self-gravity. The situation changes dramatically at small V, where magnetic forces and self-gravity both increase more rapidly, but in opposite directions, than the internal thermal pressure.

When the cloud is *magnetically supercritical*, $M > M_\Phi$, the net force from magnetic fields and self-gravity is attractive, and less external pressure is needed to achieve a given decrease in the cloud volume than predicted by an extrapolation of Boyle's law to small V. In particular, if the external pressure increases beyond the maximum value P_{max}, the cloud can be pushed over to gravitational collapse.

When the cloud is *magnetically subcritical*, $M < M_\Phi$, the net force from magnetic fields and self-gravity in equation (2) is repulsive, and more external pressure is needed to achieve a given decrease in the cloud volume than predicted by an extrapolation of Boyle's law to small V. In particular, the external pressure required to compactify the cloud increases without bound as V approaches zero. In this case, no amount of external compression can induce the cloud to collapse. Frozen-in magnetic flux will resist the transformation of molecular clouds to stars when the cloud mass M is less than the magnetic critical mass M_Φ as much as electron degeneracy

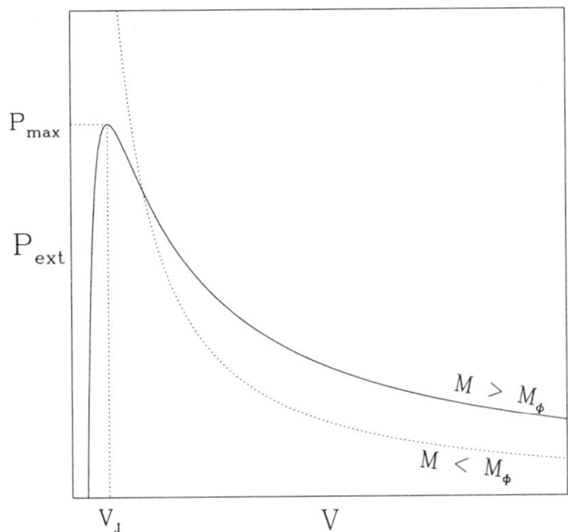

Figure 1. The external pressure P_{ext} needed to compress a magnetized, self-gravitating, molecular cloud to given volume V as estimated by the virial theorem. The two curves for the magnetically supercritical case, $M > M_\Phi$ and the magnetically subcritical case, $M < M_\Phi$ are computed for the same M_Φ, but different M.

pressure will resist the transformation of a white dwarf to a neutron star when the white dwarf's mass is less than the Chandrasekhar limit.

2. Fragmentation

For a magnetically supercritical cloud, it is a simple exercise to show from equation (2) that the condition $P_{\text{ext}} = P_{\text{max}}$ is reached at a volume $V = V_{\text{J}}$ that satisfies

$$V_{\text{J}}^{1/3} = \frac{4\alpha G M}{3a^2}\left[1 - \left(\frac{M_\Phi}{M}\right)^2\right]. \tag{3}$$

Apart from the modification factor, $[1 - (M_\Phi/M)^2]$, which has a value between 0 and 1 when $M > M_\Phi$, the length scale $V_{\text{J}}^{1/3}$ is defined so that the (negative of the) associated gravitational energy per unit mass $3\alpha G M/V_{\text{J}}^{1/3}$ equals $3/2$ of the thermal energy per unit mass, $3a^2/2$. Thus, equation (3) yields the (modified) Jeans criterion [see also Ebert (1955) and Bonnor (1956)]: A cloud of mass M and temperature T will collapse gravitationally if it is compressed to an average size less than the (magnetically modified) Jeans length $V_{\text{J}}^{1/3}$.

In the absence of magnetic fields, the Jeans mass $\propto G^{-3/2} a^3 \langle \rho \rangle^{-1/2}$ in isothermal collapse decreases as the mean density $\langle \rho \rangle$ increases upon contraction. This behavior underlies Hoyle's (1953) conjecture that a collapsing isothermal cloud will undergo hierarchical fragmention [but see the differing opinions expressed by Hunter (1962), Layzer (1964), Bodenheimer (1980), Tohline (1982), and Boss (1999)]. In the presence of frozen-in magnetic fields, where Φ remains fixed to the collapsing M, notice that the critical mass M_Φ is unaltered by contraction, completely unlike the behavior of the Jeans mass. Thus, for unstable clouds that collapse from initial states not too far removed from being critically magnetized, Mestel & Spitzer (1956) speculated that fragmentation would be suppressed by the presence of frozen-in fields. Mestel (1965, 1985) later surmised that supercritical clouds which flatten dramatically during the collapse might fragment. We shall reexamine this question in §7.

For a magnetically subcritical cloud, no upper limit exists to the external pressure that the cloud can withstand without undergoing continued gravitational contraction. In these circumstances, neither a physically relevant Jeans length nor a physically meaningful Jeans mass can be defined.

3. Empirical Measurements of the Mass-to-Flux Ratio

Given the dichotomy offered by the two choices, $M < M_\Phi$ and $M > M_\Phi$, it is clearly an important matter to measure the mass-to-flux ratio, M/Φ, in actual clouds and compare it to the critical ratio $(2\pi G^{1/2})^{-1}$ (i.e., to measure the dimensionless ratio M/M_Φ). Crutcher (1999, see also Crutcher et al. 1993) has compiled all the reliable Zeeman measurements of the magnetic field strength made to date in molecular clouds. In Table 1, we reproduce the important portion of his summary of the observational data.

In the first column, we have the region measured; in the second column, the inferred column density of molecular hydrogen, $N(H_2)$; in the third column, the measured strength of the magnetic field along the line of sight, $B_\parallel$. In the fourth column we assume a mean molecular mass $m = 2.4\, m_H$, where m_H is the mass of the hydrogen atom; and we ratio the (projected) mass per unit area, $mN(H_2)$, to the (projected) flux per unit area, $B_\parallel$, expressing the ratio in units of $(2\pi G^{1/2})^{-1}$. The fourth column has values that are typically 2 to 5. Nominally, these molecular clouds are magnetically supercritical, which is the conclusion arrived at by Crutcher, who used slightly different formulae. But as Crutcher also noted, there are two systematic effects for which we may need to apply corrections.

First, the total field strength, B, is on average *twice* as large as a component of it, $B_\parallel$, seen along some random direction. Second, if molecular clouds are highly flattened (as in the toy model in §7), then a measurement

TABLE 1. Zeeman Measurements from Crutcher (1999)

Cloud	$N(\mathrm{H_2})$ $(10^{22}\ \mathrm{cm}^{-2})$	$B_\parallel$ $(\mu\mathrm{G})$	$mN(\mathrm{H_2})/B_\parallel$ $(2\pi G^{1/2})^{-1}$	Σ/B $(2\pi G^{1/2})^{-1}$
W3OH	50.1	3100	1.1	0.3
DR21OH1	39.8	710	3.6	0.9
SgrB2	25.1	480	3.4	0.9
M17SW	12.6	450	1.8	0.5
W3	15.8	400	2.6	0.6
S106	6.3	400	1.0	0.3
DR21OH2	20.0	360	3.6	0.9
OMC1	15.8	360	2.9	0.7
NGC2024	7.9	87	5.9	1.5
S88B	2.0	69	1.9	0.5
B1	0.8	27	1.9	0.5
W49B	0.4	21	1.2	0.3
W22	1.6	18	5.7	1.4
W40	1.0	14	4.6	1.2
ρ Oph1	0.5	10	3.3	0.8
OMCN4	12.6	<280	> 2.9	> 0.7
TauG	0.4	<16	> 1.6	> 0.4
L183	0.2	<16	> 0.6	> 0.2
L1647	1.3	<15	> 5.5	> 1.4
ρ Oph2	0.4	<14	> 1.8	> 0.5
TMC1	0.8	<13	> 4.0	> 1.0
L1495W	0.4	<13	> 2.0	> 0.5
L134	0.2	<11	> 1.2	> 0.3
TMC1C	0.8	<9	> 5.7	> 1.4
L1521	0.5	<9	> 3.6	> 0.9
L889	1.0	<7	> 9.3	> 2.3
Tau16	0.5	<7	> 4.7	> 1.2

of the column density taken along some random (slant) path will typically produce a value $mN(\mathrm{H}_2)$ that is *twice* the true surface density Σ of the flattened cloud (calculated by a perpendicular projection of the mass volume density). Making corrections of two factors of 2, we replace the fourth column by Σ/B, which is 4 times smaller than $mN(\mathrm{H}_2)/B_{\parallel}$. [We remark that if the direction of the magnetic field is preferentially perpendicular to a sheetlike cloud, then the true value of Σ/B will instead be a factor 3 smaller than $mN(\mathrm{H}_2)/B_{\parallel}$.] The listed values of Σ/B in the fifth column of Table 1, measured in units of $(2\pi G^{1/2})^{-1}$, are then remarkably close to the critical value of unity.

The uncertainties in the measurements and in their corrections are too great to make definitive pronouncements at the present time. The empirical evidence does appear consistent with the oft-expressed view that interstellar magnetic fields are strong enough in molecular clouds to be dynamically important. They definitely cannot be ignored. But it is premature to decide whether most clouds are magnetically supercritical or subcritical. Indeed, an interesting case can be made that molecular clouds are nearly *marginally critical*. A speculative theoretical scenario might even explain why this is the case. Perhaps all highly supercritical molecular clouds (with $\lambda \equiv 2\pi G^{1/2}\Sigma/B \gg 1$) have collapsed to form stars long ago and are no longer around to observe in the Galaxy. Perhaps highly subcritical configurations ($\lambda \ll 1$) do not become molecular clouds, because such objects must be confined against internal expansion, not by self-gravity, but largely by external pressure. Such configurations probably form diffuse clouds rather than molecular clouds. By this line of reasoning, molecular clouds capable of star formation at the present epoch of the Galaxy are self-selected to be close to being *marginally critical* . Unfortunately, for some theoretical purposes (see below), it is insufficient to deduce $\lambda \sim 1$; rather it is quite important to determine exactly on which side of $\lambda = 1$ true clouds lie.

4. Turbulent Support of Molecular Clouds

Molecular clouds are supported against self-gravity in part by turbulent motions (McKee, this volume). If turbulence and fluctuating magnetic fields make an equally important contribution to cloud support as static magnetic fields, and if the external and thermal pressures of molecular clouds can be ignored, then λ must have a value ≈ 2 (McKee 1989, Bertoldi & McKee 1992). Unfortunately, recent numerical simulations of magneto-hydrodynamic (MHD) turbulence in self-gravitating molecular clouds may create problems for the point of view that turbulent pressure can be treated on an equivalent footing to microscopic thermal pressure.

A number of different groups have studied the rate of decay of strong

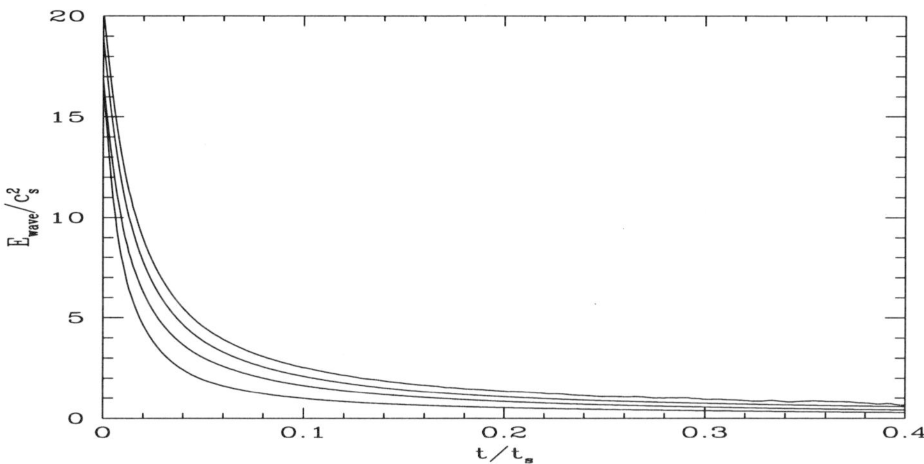

Figure 2.   Free decay of total perturbed energy (turbulent kinetic energy + turbulent magnetic energy) in non-self-gravitating 3D MHD simulations. From top-to-bottom, the curves show models with $\beta \equiv c_s^2/v_A^2 = 0.01, 0.1, 1, \infty$, where c_S and v_A are respectively the sound and Alfvén speeds. The models start with saturated turbulence and wave energy per unit mass $E_{\text{wave}}/c_s^2 = 20$, 19, 17, 16, with kinetic portions 13, 12, 13, 15, respectively. Time is given in units of the sound-crossing time $t_s \equiv L/c_s$ for the box. For details, see Stone, Ostriker, & Gammie (1998).

MHD turbulence, initially put into the simulations as a random collection of noncompressive motions – effectively nonlinear Alfvén waves. The original expectations were that nonlinear Alfvén waves should prove resilient to rapid dissipation since their primary fluid motion is perpendicular to the undisturbed magnetic field, and thus, they have no direct tendency to steepen and shock (Arons & Max 1975). MHD simulations with nontrivial variations only along one dimension showed that although Alfvén waves suffer a decay instability and do generate compressive motions, free MHD turbulence of a highly symmetric, primarily Alfvénic type could be relatively long-lived (Gammie & Ostriker 1996). However, calculations carried out in three dimensions (MacLow et al. 1998; Stone, Ostriker, & Gammie 1998; Padaon & Nordlund 1999) demonstrate that superpositions of nonlinear Alfvén waves with spatial variation in *all* directions couple more strongly to their fast and slow MHD counterparts. Fast and slow MHD waves generally involve compressive motions along the direction of the undisturbed magnetic field, and when they have nonlinear amplitudes, they readily shock. Figure 2 shows that the shock dissipation associated with nonlinear mode-coupling leads to a decay rate for free MHD turbulence that is almost as large for a strongly magnetized medium as for an unmagnetized one. The

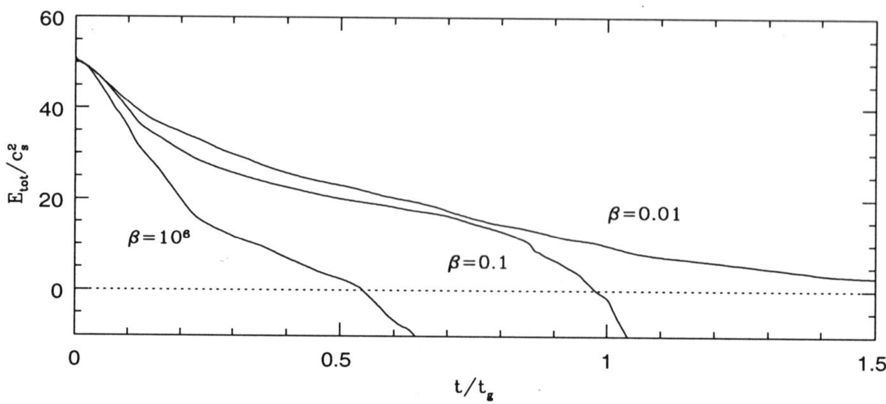

Figure 3.   Free decay of total perturbed energy (turbulent kinetic energy + turbulent magnetic energy + gravitational energy) per unit mass in self-gravitating 2.5D MHD simulations with $\beta \equiv c_s^2/v_A^2 = 0.01, 0.1, 10^6$ (essentially ∞, i.e., essentially an unmagnetized cloud). All models start with perturbed kinetic energy per unit mass $(1/2)\langle v^2/c_s^2 \rangle = 50$, and have box edge three times the initial Jeans length. Time is given in units of the gravitational time $t_g \equiv (\pi/G\langle\rho\rangle)^{1/2}$ where $\langle\rho\rangle$ is the mean density in the box. For details, see Ostriker, Gammie, & Stone (1999).

typical turbulent dissipation times are comparable to the fluid crossing times on the main energy-containing scale of the flow.

In following the evolution of self-gravitating clouds, an appropriate unit for measuring time is the gravitational time scale, $t_g \approx 3.3 t_{ff}$, where $t_{ff} \equiv (3\pi/32G\langle\rho\rangle)^{1/2}$ is the spherical free-fall time at the average density $\langle\rho\rangle$ of the cloud. Because turbulence lends support to clouds against collapse, their evolution must depend on the interplay of gravity and turbulent dissipation (and, potentially, turbulent re-excitation – see §6). What is the fate of a turbulent cloud, with total mass $\gg$ the thermal Jeans mass, but which is initially supported against self-gravity by a combination of turbulent motions and magnetic fields, if the turbulence undergoes free decay?

To answer this question, Ostriker, Gammie, & Stone (1999) directly simulated the evolution of self-gravitating, magnetized clouds containing an initial turbulent spectrum. Figure 3 shows the energy evolution in three such models with differing mean magnetization, and Figure 4 shows snapshots of the internal density, velocity, and magnetic field structure at successive intervals. Because observed self-gravitating clouds have flow crossing times comparable to $t_g/2$, the above-cited numerical results on turbulent dissipation suggest that the freely-decaying turbulence may not significantly

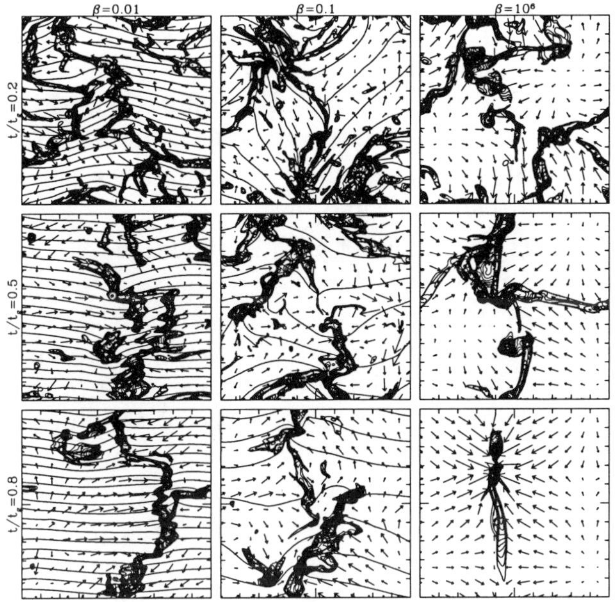

Figure 4. Snapshots of the density contours, [equally spaced by 0.1 in $\log(\rho/\langle\rho\rangle)$ starting from 0], velocity field (arrows), and magnetic field lines (light lines) at times $t/t_g = 0.2, 0.5$, and 0.8 for the self-gravitating 2.5D MHD simulations of Figure 3. For details, see Ostriker, Gammie, & Stone (1999).

prolong the support of clouds against their gravitational contraction; this is indeed the conclusion reached by direct simulations. Thus, the eventual fate of a cloud without turbulent resupply depends crucially on whether the cloud is magnetically supercritical ($\lambda > 1$) or magnetically subcritical ($\lambda < 1$). In $\lambda > 1$ clouds (e.g., the cases $\beta = 10^6$ and 0.1), Figures 3 and 4 show that the disappearance of sufficient turbulent support quickly leads to regions with overall energies that are negative and that gravitationally collapse on time scales comparable to or shorter than t_g. In $\lambda < 1$ clouds (e.g., the case $\beta = 0.01$), Figures 3 and 4 show that the removal of even all the turbulence has no such drastic effect, since the background magnetic field *by itself* suffices to prevent overall gravitational collapse. In other words, subcritical clouds can survive for times much longer than t_g, while supercritical clouds cannot, if their turbulence decays freely.

5. Filamentary Structure in Molecular Clouds

What is wrong with supposing that all molecular clouds are magnetically supercritical, and that it is the free decay of the turbulence which triggers their gravitational collapse, leading to star formation? Indeed, several recent simulations ignoring magnetic fields show highly suggestive filamentary structures to develop when clouds initially containing many thermal Jeans masses are allowed to collapse and fragment (e.g., Klessen, Burkert, & Bate 1998; Nagai, Inutsuka, & Miyama 1998). The results are similar to those obtained in cosmological studies, which provide the motivation for many of the molecular-cloud calculations.

Although filaments are observed both in the large-scale structure of the universe and in giant molecular clouds, it is unlikely that they arise from the same causes. Free gravitational collapse along the shortest available dimension (first into sheets, then into lines, then into points) is a viable option for cosmology, because the free-fall time scale t_{ff} in that problem is automatically of the same order as the age of the universe (in models having approximately the closure density). The average density in molecular clouds is 8 orders of magnitude times higher than the closure density of the universe; thus, the corresponding free-fall time $t_{ff} \propto \langle \rho \rangle^{-1/2}$ is 4 orders of magnitude lower than the age of the universe. The conversion of large pieces of molecular clouds into stars on a time scale that approaches 3×10^6 years would yield a rate of star formation in the Galaxy that exceeds the known value (3–5 $M_\odot$ yr^{-1}) by two orders of magnitude (Zuckerman & Evans 1974). Indeed, Zuckerman & Evans's original point was that the large linewidths seen in most molecular clouds could *not* have an origin in large-scale gravitational collapse, precisely because such an interpretation would lead to far too fast a turnover of molecular gas into stars.

Figure 4 contains a possible hint why filamentation may occur in molecular clouds that do not collapse but are strongly magnetized, namely, the act of dissipating MHD turbulence creates sheetlike or filamentary clouds, with the short axis aligned preferentially along the mean magnetic field. [Turbulent MHD simulations in three dimensions show filamentary structure. The simulations of Ostriker et al. (1999) depicted in Figure 4 are two-dimensional simulations. Any feature that appears in the plane of the page is assumed to extend infinitely in the direction perpendicular to the page. Thus, what appear as filaments in the page are actually sheets.] In §7, we demonstrate via a simple toy model that turbulence driven by outflows in the directions along a sheet will naturally produce true filaments.

6. Sustenance of Cloud Turbulence Through YSO Outflows

Whether or not turbulence is essential to molecular cloud support (i.e., whether or not λ is > 1), the large line-widths observed in molecular clouds do presumably have to be sustained for periods substantially longer than t_{ff}. It is a proposal of long theoretical and observational standing (see, e.g., Norman & Silk 1980, Lada 1985, Welch et al. 1985, Fukui et al. 1986) that the mechanism responsible for this sustenance is bipolar outflows from young stellar objects (YSOs) (Cudworth & Herbig 1979; Snell, Loren, & Plambeck 1980; Rodriguez et al. 1980). A simple calculation yields the required conditions. The equivalent of this calculation first appeared in a paper by McKee (1989; see also Gammie & Ostriker 1996).

Pudritz & Norman (1986) proposed that the back torques from a magnetocentrifugal wind can drive accretion through a disk, without the need for any anomalous viscosity (see also Wardle & Königl 1993). Many contemporary models invert this idea and suggest that YSO winds and jets are powered by magnetocentrifugally turning a fraction f of the accretion rate $\dot{M}_D$ of the disk (as set by Balbus-Hawley or self-gravitational instabilities) into a wind mass-loss rate $\dot{M}_w$ (e.g., Shu et al. 1988, 1994a):

$$\dot{M}_w = f \dot{M}_D. \tag{4}$$

The complementary fraction $(1 - f)$ goes to building up the central star at a rate

$$\dot{M}_* = (1 - f)\dot{M}_D. \tag{5}$$

If f is the same for all YSOs, we may temporarily take $\dot{M}_*$ to represent, not the mass-accumulation rate of any given YSO, but the rate at which molecular cloud gas is being converted by gravitational collapse into stars *throughout the whole Galaxy*. The symbol $\dot{M}_w$ would then represent the global rate at which YSO winds and jets are injecting material back to the interstellar medium (mostly molecular clouds since the heaviest mass-loss rates apply to the deeply embedded phases). The relation between $\dot{M}_w$ and $\dot{M}_*$ for a single star or for the whole Galaxy then reads

$$\dot{M}_w = \left(\frac{f}{1 - f}\right) \dot{M}_*. \tag{6}$$

We assume that these YSO winds and jets have typical terminal velocities v_w. The total momentum input per unit time into all the molecular clouds of the Galaxy from YSO winds and jets is then

$$\dot{M}_w v_w = \left(\frac{f}{1 - f}\right) \dot{M}_* v_w. \tag{7}$$

The corresponding energy injection rate into the cloud is

$$\frac{1}{2}\dot{M}_w v_w v_{sh} = \frac{1}{2}\left(\frac{f}{1 - f}\right) \dot{M}_* v_w v_{sh}, \tag{8}$$

where v_{sh} is the mean outflow shell speed. This rate must be smaller than the energy flux in the YSO outflows themselves, because the outflow-cloud interaction is probably momentum- rather than energy-conserving (see Lada 1985).

In contrast, the dissipation rate of the energy contained in cloud turbulence which is forced on a scale ℓ equals

$$\frac{M_{\mathrm{GMC}} v_{\mathrm{turb}}^2}{2 t_{\mathrm{diss}}},\tag{9}$$

where M_{GMC} is the total mass of giant molecular clouds (GMCs) in the Galaxy, v_{turb} is the typical turbulent velocity inside a GMC, and $t_{\mathrm{diss}} \approx \ell/v_{\mathrm{turb}}$ is the usual hydrodynamic time scale for turbulent dissipation and does not differ much for MHD turbulence according to the numerical simulations described in Figure 2. In steady state, we assume the energy input rate (8) is balanced by the dissipation rate (9); this yields the level of turbulence that can be maintained by YSO winds and jets:

$$v_{\mathrm{turb}} = \left(\frac{f}{1-f}\right)^{1/3} \left(\frac{\dot{M}_* \ell v_{\mathrm{sh}}}{M_{\mathrm{GMC}} v_{\mathrm{w}}^2}\right)^{1/3} v_{\mathrm{w}}.\tag{10}$$

With the substitutions, $M_{\mathrm{GMC}} \approx 10^9\ M_\odot$, $\dot{M}_* \approx 4\ M_\odot\ \mathrm{yr}^{-1}$, $v_{\mathrm{w}} \approx 200$ km s^{-1}, $\ell \approx 1$ pc, and $v_{\mathrm{sh}} \approx 10$ km s^{-1}, we obtain the numerical value,

$$v_{\mathrm{turb}} \approx \left(\frac{f}{1-f}\right)^{1/3} 2 \text{ km s}^{-1}.\tag{11}$$

Notice that the expression (10) is relatively insensitive to the product of the length and velocity scales, ℓv_{sh}, where the energy input from YSO outflows is assumed to couple into the turbulent velocity field.

The X-wind model predicts that $f \approx 1/3$ (see §12 below), which yields a value from equation (11) for v_{turb} sustainable by YSO outflows that is compatible with the 1–2 km s^{-1} actually seen in the line-widths of typical GMCs. A wind efficiency this large (where for every solar mass of material that goes into a star, roughly half a solar mass is shot back out at 200 km s^{-1}) has been claimed only for the X-wind model (cf. Shu et al. 1988, 1994a). Indeed, there is probably no source of mechanical energy in the Galaxy except for material flung from the near-vicinity of stellar surfaces that could come remotely close to replenishing the energy drain represented by the dissipation of molecular cloud turbulence.

7. A Toy Model of GMC Structure and Evolution

In the directions across magnetic fields, giant molecular clouds measure tens of parsecs, and clumps within them measure from a fraction to a few parsecs (Blitz, this volume). How thick are molecular clouds along the direction of the mean field? At first sight, the situation appears simple from a theoretical perspective. If thermal pressure is the only means of support against self-gravity in the direction of the mean field, then the density half-height of a layer with isothermal sound speed a and surface density Σ is given by Spitzer's (1942) formula:

$$z_0 = \frac{a^2}{\pi G \Sigma} \tag{12}$$

For $T = 10$ K and $m = 2.4\ m_{\mathrm{H}}$ (corresponding to $a = 0.2$ km s^{-1}), and $\Sigma = 0.01$ g cm^{-2} (corresponding to 2.5 mag of visual extinction), equation (12) yields $z_0 = 0.06$ pc. Under such circumstances, GMCs and cloud clumps should essentially be sheets (Pudritz 1986).

Internal turbulence thickens the cloud in the direction parallel to the mean magnetic field (Arons & Max 1975), whereas the extra magnetic pressure above and below the cloud resulting from field lines bending to support the cloud in the lateral direction (e.g., Li & Shu 1996) and converging turbulent flows exterior to a cloud clump (e.g., Fig. 4) would make the sheet locally even thinner than indicated by equation (12). Shu, Adams, & Lizano (1987) noted that bipolar outflows are often aligned with the ambient magnetic field (Vrba, Strom, & Strom 1976; Akeson & Carlstrom 1997), and some of these flows are still within the cloud after sweeping ambient molecular gas out to a parsec. On the other hand, Reipurth, Bally, & Devine (1997) discuss examples of parsec-long optical jets from YSOs that have clearly broken out of their local natal clouds. Thus, the question of how spatially thick are clouds in the direction parallel to the mean magnetic field remains open both theoretically and observationally.

In the rest of this section, we adopt as a toy model the hypothesis that molecular clouds are spatially flat sheets. We allow for self-gravity, magnetic fields (static or not), and fluid motions (turbulent or not as long as they take place only in the plane of the sheet), but we suppose that thermal pressure is negligible at the relatively large scales ($\gg 0.1$ pc) of primary interest here. We further suppose that the local mass-to-flux ratio has a spatially constant value:

$$\frac{\Sigma}{B_z} \equiv \Lambda = \mathrm{const}, \tag{13}$$

where z is the direction normal to the sheet and B_z equals the total B at the midplane. If field freezing holds, then the ratio Σ / B_z is also temporally constant at the value Λ as the sheet evolves in the x–y plane.

As before, it is convenient to introduce the symbol λ for the dimensionless mass-to-flux ratio:

$$\lambda = 2\pi G^{1/2} \Lambda. \tag{14}$$

Shu & Li (1997) call a magnetized sheet with a spatially and temporally constant value of λ *isopedic*. For an isopedic sheet, they showed that for arbitrary variations of $\Sigma(x, y, t)$, the magnetic forces exerted in the plane of the sheet are a (negatively) scaled version of self-gravity, with the constant of proportionality equal to $-1/\lambda^2$. An especially interesting situation occurs if the sheet is critically magnetized. *If $\lambda = 1$, then magnetic forces exactly balance gravitational forces everywhere in the sheet, independent of how complex are the motions and density (and field) distributions.* Since we have ignored the effects of thermal pressure (which requires for self-consistency that we also ignore the magnetic pressure but not the magnetic tension), this means that the equations of motion are especially simple: Small pieces of the sheet move at constant speed in a straight line until they collide with other small pieces of the sheet. (Recall that there are no other forces in the problem if we explicitly treat the turbulent motions as actual motions rather than as some sort of artificial eddy stress.) When different pieces of the sheet do collide, the shock jump conditions (for a zero temperature sheet) are also simple: The collision occurs inelastically, with the conservation of total mass and vector momentum of the various pieces and with the removal of all relative motion along the line of centers after the pieces coagulate.

Clearly, if there is continual dissipation of energy when pieces of the sheet intersect, any initial velocity field containing differential motions would tend to resolve itself into narrow dense regions ("filaments") where opposing pieces of the sheet have collided inelastically, merged, and are now traveling together in a common direction. The self-gravity in the dense filaments would be intense, but it would be balanced by an opposite and equally intense magnetic tension created by all the field lines that thread through the individual pieces and have now been bunched together tightly in the midplane while they spread out quickly above and below the sheet. The prediction that magnetic field lines thread mainly *perpendicularly* (in a cylindrical fan above and below the sheet) to the long axes of well-defined filaments distinguishes our theory from several of its competitors.

The exact balance would be lost, however, if the condition of field freezing breaks down. Realistically, we expect ambipolar diffusion and thermal pressure (relative to "turbulent pressure") to become important effects when cloud surface and volume densities rise beyond certain values (see Lizano & Shu 1989 and §8). For purposes of the simulations that follow, let us somewhat arbitrarily suppose that ambipolar diffusion becomes very efficient when the local surface density, averaged over a unit cell (representing a molecular cloud core of typical diameter 0.1 pc) in a grid-based scheme, rises to 100 times the global average value. We then suppose that the affected gas loses its magnetic support and forms stars.

In the displayed simulations, when a cell acquires 400 particles in any given time step, 32 particles are assumed to form one or more stars; 16 particles [corresponding to $f/(1-f) = 16/32 = 1/2$, i.e., to a wind fraction $f = 1/3$] are ejected in the form of a bipolar wind with a terminal velocity v_{w} equal to 230 km s^{-1}. The corresponding momentum input, 3680 particle km s^{-1}, is immediately shared by the 368 ambient particles in the cell (16 in the original wind plus 352 in the cell that did not participate in the gravitational collapse but got swept up in the outflow); thus, a molecular outflow of 10 km s^{-1} carried by these 368 particles impinges on neighboring cells as a result of the star-formation process in any dense core. Although the mass scale can be chosen arbitrarily, for definiteness we may imagine the original 400 particles to represent the typical 11 $M_\odot$ core that Onishi et al. (1998) find per star-forming event in the Taurus molecular cloud.

To keep the problem (artificially) two-dimensional, we suppose that the outflow is ejected within wedges of opening angle 45° aligned in two dia-metrically opposed directions in the plane of the sheet. On the rationale that stars are unlikely to form exactly in the midplane of the sheet, the orientation of the bipolar outflow axes is taken to be parallel to the local direction of the horizontal component of the magnetic field just above and below the midplane of the sheet. The underlying assumption is that the outflow axis lies along the spin angular momentum vector of (the inner edge of) the accretion disk (see §12), whereas magnetic braking of the pre-collapse cloud tends to align the spin vector parallel to the local direction of the interstellar magnetic field $\mathbf{B}$ (see Mouschovias & Paleologou 1980 for relative efficiencies of braking for the parallel and perpendicular rotators). Thus, YSO outflows should have a tendency to align with $\mathbf{B}$ in the ambient medium. This orientation rule yields the only reason for calculating the ge-ometries of the gravitational and magnetic fields, since, in their dynamical influence on the remaining gas, they are always assumed to balance.

The last assumption breaks down if a significant fraction of the total mass becomes locked up in stars, with the corresponding magnetic flux re-leased (rather than destroyed) back into the remaining gas. Fortunately for the meaningfulness of the simulations, the star formation rate is low (with the assumption that ambipolar diffusion becomes fast only when the local surface density rises to 100 times the global average), and the fundamental assumption of strict force-balance becomes bad only after about 50 Myr, when about 40% of the mass of the original cloud has turned into stars (giving a time scale 50 Myr/0.4 = 125 Myr for gas consumption equal to half the global average of $M_{\mathrm{GMC}}/\dot{M}_* = 10^9\ M_\odot/4\ M_\odot\ \mathrm{yr}^{-1}$).

We follow the fate of the ejected wind by the same dynamic rules as the cloud gas. Thus, the bipolar outflows sweep up ambient gas into thin moving shells that obey vector momentum conservation. This process re-supplies the "turbulence" that is lost elsewhere through inelastic collisional

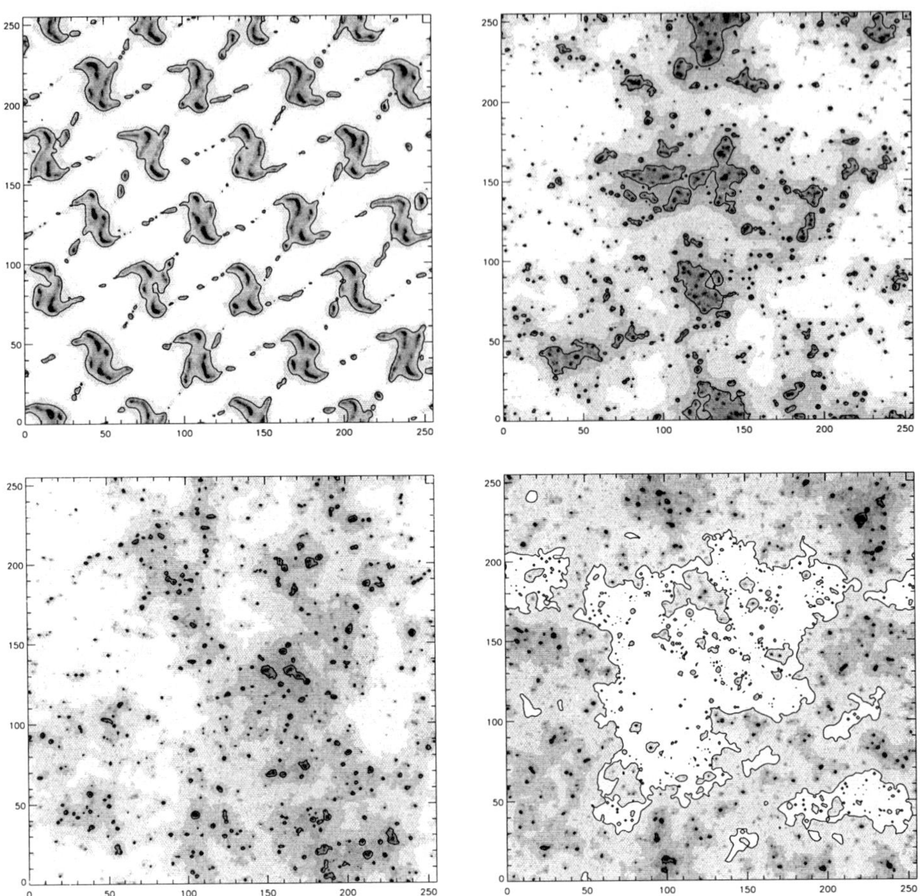

Figure 5.   Equipotential plots of toy model simulation of the evolution of a $\lambda = 1$ molecular cloud sheet. The upper left panel corresponds to a time of 2.5 Myr; the upper right panel, to a time of 15 Myr; the lower left panel, to 28 Myr; and the lower right panel, to 50 Myr. For details, consult Allen & Shu (1999a).

agglomeration. The individual moving shells also eventually merge to form new dense filaments and cores that give rise to further generations of young stars and outflows.

Figure 5 shows the evolution of a toy-model simulation with periodic boundary conditions (Allen & Shu 1999a). The computations are done on a 256×256 grid, with each cell measuring 0.1 pc along the x and y lengths. Contours of gravitational equipotential are plotted on a grey scale; iso-surface-density contours would look similar, but less smooth with the number of gas particles ($4 \times 256 \times 256$) used in the simulation. The simulations are started with a uniform and smooth surface density distribution, except that the 4 particles in each cell are given random spatial positions. They are also given small x and y velocities (of order 1–2 km s^{-1}) that vary sinusoidally with spatial position.

The upper left panel of Figure 5 yields the configuration after 2.5 million years (2.5 Myr) of evolution. The pattern looks like wallpaper, with strict periodicity destroyed only by the small random elements introduced in the initial conditions. The system quickly loses all memory of such artificialities.

After 15 Myr of evolution, the configuration looks as displayed in the upper right panel. Small dense cores have formed that have given birth to stars with outflows. The outflows replenish the "turbulent" motions, but in an anisotropic fashion, since they are directed along the x–y direction of the mean magnetic field above and below the midplane. These directions can be read off the figures by noting that the magnetic field **B** is locally parallel (or antiparallel) to the gravitational field **g** everywhere, and the components of **g** (in the x and y directions) are, of course, perpendicular to the equipotential contours that are being plotted in Figure 5.

The eye easily picks out "filaments" in all of the three later figures. The filaments are dotted by dense cores along their length. We believe that filaments form by the following "instability" process, but we do not have yet a quantitative analysis. Once there is any tendency for elongation in the x–y plane, the associated magnetic field orientations favor outflows that are directed along the short axis of that elongation. (For a vivid demonstration that nature might have reproduced this tendency, see Yu, Bally, & Devine 1997.) These bipolar outflows then sweep up ambient gas that also tends to form elongated structures with similar spatial orientations. Two adjacent parallel filaments firing outflows at each other reinforce the compressional tendency, although they also ultimately blast each other apart.

The lower left panel of Figure 5 shows the situation after 28 Myr. Long irregular filaments studded with dense cores are apparent. The overall impression is not qualitatively dissimilar to observational maps of the Taurus molecular cloud complex (cf. Myers 1995, Mizuno et al. 1995). Although cloud filaments and cloud fragments exist throughout the map, notice that the overall region is in (neutral) force balance. Gravitational collapse is not occurring anywhere except at the centers of the densest cloud cores.

The lower right panel depicts the configuration after 50 Myr, when 40% of the original mass of the cloud has been turned into stars, which are distributed more or less uniformly throughout this region (although some have migrated out of the original computational area, to be replaced by stars from neighboring areas because of the periodic boundary conditions). The gas has been pushed by outflows in the interior toward the edges of the computational box, but the periodic boundary conditions, of course, prevent any permanent loss of gas from the region.

The toy model introduced above clearly has its limitations when we attempt to apply it to the real world. Removing some of these limitations (e.g., the periodic boundary conditions) would not represent hard work; removing others (e.g., the two-dimensionality of the calculations) would require abandoning simplifying basic properties of the toy model [e.g., the Shu & Li (1997) theorems] and would add considerable computational effort. Despite its weaknesses, however, our toy model may contain some elements of essential truth. In particular, we believe that it yields a highly suggestive mechanism for the formation of filamentary clouds that does not require global free-fall gravitational collapse and fragmentation with its attendant untenably high rates of star formation.

8. Formation and Evolution of Molecular Cloud Cores

Low-mass stars form in the dense cores of molecular clouds (Myers & Benson 1983, Mizuno et al. 1994, Myers 1995). These cores have a typical size ~ 0.1 pc, a number density $\sim 3 \times 10^4$ cm^{-3} or greater, a mass ranging from a fraction of a solar mass to about 10 $M_\odot$, and an axial ratio for flattening of typically 2:1. For isolated cores, the last fact implies that agents other than isotropic thermal or turbulent pressures help to support cores against their self-gravity, although it is not yet clear observationally whether the true shapes are oblate, prolate, or triaxial (Myers 1998, personal communication). Observed cloud rotation rates are generally too small to account for the observed flattening (Goodman et al. 1993). This leaves magnetic fields, which are believed for other reasons to play a crucial role in contemporary star formation.

In one scenario, the weakening of magnetic support by ambipolar diffusion (Mestel & Spitzer 1956) leads to the continued contraction of a cloud core with ever growing central concentration (Nakano 1979, Lizano & Shu 1989, Basu & Mouschovias 1994). On a time scale given by the core scale length divided by the characteristic slip velocity between ions and neutrals, the molecular cloud core reaches a "gravomagneto catastrophe" where the central density formally tries to reach infinite values. Shu (1995) pointed out that the process has some features in common with the "gravothermal catastrophe" (Lynden-Bell & Wood 1968, Lynden-Bell & Eggleton 1980, Cohn 1980) that overtakes globular-cluster core-evolution because of the diffusion of random velocities by stellar encounters.

Myers (this volume) points out that inward fluid velocities of magnitude ~ 0.1 km s^{-1} exist at large radii in some starless cores, and that this feature is not predicted by any of the ambipolar-diffusion calculations cited above (see also Onishi, Mizuno, & Fukui 1999). Myers suggests instead that pre-existing inward motions prior to true gravitational collapse and star for-

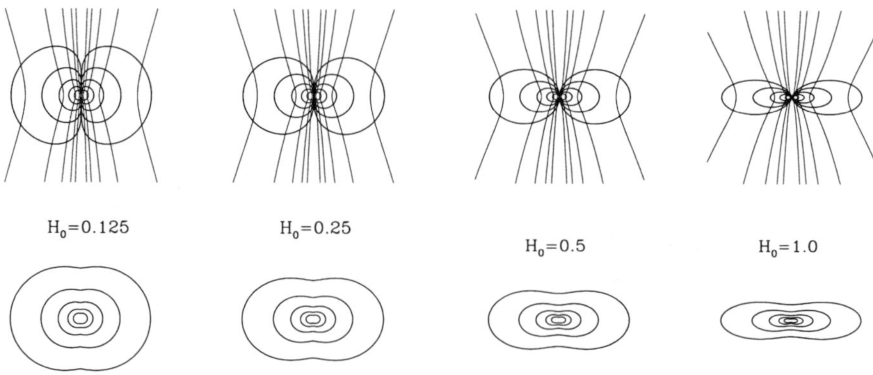

Figure 6.   Isodensity contours and magnetic field lines of isopedic singular isothermal toroids for values of the over-density parameter $H_0 = 0.125, 0.25, 0.5, 1.0$. The corresponding values of λ are $\lambda = 8.38, 4.51, 2.66, 1.79$. Below each plot is the equator-on projection of the column density. For details, consult Li & Shu (1996).

mation are better modeled as resulting from the decay of turbulence (see also Nakano 1998). Figure 2 of Lizano & Shu (1989) lends quantitative credence to this suggestion (see also Myers & Lazarian 1998). If the sequence of decaying turbulent parameter $K = 12, 10, 6, 2$ in the cited Figure were to occur over $\sim$ a few Myr, then inward motions of order $0.5a \sim 0.1$ km s^{-1} could indeed be generated at distances of order 0.1 pc from the center of the cloud core. Such subsonic motion and the generally shorter time scale may distinguish the "gravoturbo catastrophe" resulting from the decay of turbulence in a pre-existing, magnetically supercritical, core with $\lambda > 1$ from a core that reaches supercriticality via the "gravomagneto catastrophe" produced by ambipolar diffusion.

9. The Pivotal State

Li & Shu (1996) termed the cloud configuration when the central isothermal concentration first becomes formally infinite as the "pivotal" state ($t \equiv 0$). This state separates the nearly quasistatic phase of core evolution ($t < 0$) that leads to gravomagneto or gravoturbo catastrophe ($t = 0$) from the fully dynamic phase of protostellar accretion ($t > 0$). (See the first two stages depicted in Fig. 7 of Shu, Adams, & Lizano 1987.) The numerical simulations indicate that the pivotal states have a number of simplifying properties, which motivated Li & Shu to approximate them as scale-free equilibria with the power-law radial dependences for the density and magnetic flux function:

$$\rho(r,\theta) = \frac{a^2}{2\pi G r^2} R(\theta), \qquad \Phi(r,\theta) = \frac{4\pi a^2 r}{G^{1/2}} \phi(\theta). \tag{15}$$

In equation (15), we have adopted a spherical polar coordinate system (r,θ,φ), and a is the isothermal sound speed of the cloud, while $R(\theta)$ and $\phi(\theta)$ are the dimensionless angular distribution functions for the density and magnetic flux given by force balance along and across field lines. The resulting differential equations and boundary conditions yield a linear sequence of possible solutions, characterized by a single dimensionless free parameter, H_0, which represents the fractional overdensity supported by the magnetic field above that supported by thermal pressure. Comparison with the typical degree of elongation of observed cores suggests that the overdensity factor $H_0 \approx 0.5 - 1$ (corresponding to $\lambda \approx 2$).

In Figure 6, we plot isodensity contours and field lines for the cases $H_0 = 0.125, 0.25., 0.5, 1.0$. Below each case are the column density contours that would be seen by an observer in the equatorial plane of the configuration. The presence of very low-density regions at the poles of the density toroid has profound implications for the shape and kinematics of bipolar molecular outflows. [See also Torrelles et al. (1983, 1994).]

10. Self-Similar Collapse of Singular Isothermal Toroids

It is well-known that the collapse of the singular isothermal sphere (corresponding to the case $H_0 = 0$ in the previous section) occurs in a *self-similar* manner (Shu 1977). The same holds true for any member in the sequence of isopedically magnetized singular isothermal toroids. For given H_0, we have only two dimensional parameters in the problem, a and G. From a and G, we cannot form dimensions for all required physical quantities without using the independent variables of the problem themselves: r, θ, and t. The quantities r and at both have the dimensions of length, and they can be combined to form the dimensionless similarity variable

$$x = r/at, \tag{16}$$

which can then appear, along with θ, as the argument for nondimensional mathematical functions. Thus, the time-dependent solution that describes the $t > 0$ evolution of the unstable equilibrium state (15) must take the *self-similar* form:

$$\rho(r,\theta,t) = \frac{\alpha(x,\theta)}{4\pi G t^2}, \quad \mathbf{u}(r,\theta,t) = a\mathbf{v}(x,\theta), \quad \mathbf{B}(r,\theta,t) = \frac{a}{G^{1/2}t}\mathbf{b}(x,\theta), \tag{17}$$

where $\mathbf{u}$ and $\mathbf{B}$ are the poloidal fluid velocity and magnetic field, respectively. The dimensionless dependent variables $\alpha(x,\theta)$, $\mathbf{v}(x,\theta)$, and $\mathbf{b}(x,\theta)$

are called the *reduced* density, velocity, and magnetic field. Solutions with the form (17) are called "self-similar" because, apart from a simple scaling of dependent and independent variables, the solution at one instant in time looks the same as the solution at another instant in time.

For the case $H_0 = 0$ (corresponding to $\mathbf{b} = 0$), we may assume that the flow occurs spherically symmetrically. With no θ dependence and with $\mathbf{v}$ having only a radial component, the similarity solution has an easy description. At $t = 0^+$, consider initiating collapse of the densest innermost portions of a (nonrotating and unmagnetized) singular isothermal sphere to form a condensed central object (a protostar, whose physical dimensions are so small as to allow us here to approximate it as a point mass). An observer located in the outer portions of the cloud does not realize anything has happened gravitationally, because Newton's theorem for the gravitational field of spherical objects misleads the observer into thinking that all the mass interior to the observer's location equivalently lies at the center anyway. Only when a sound wave traveling outwards at the speed a reaches the observer, does the observer realize, hydrodynamically, that the bottom has dropped out. The information propagates outward, initiating a wave of falling whose location at time t has a head described by the locus $r = at$. The similarity variable description for the same locus is $x = 1$. For x larger than 1 (or $r > at$), the density profile retains its unperturbed value $\alpha = 2/x^2$ (or $\rho = a^2/2\pi G r^2$), and the fluid is at rest, $\mathbf{v} = 0$ (or $\mathbf{u} = 0$).

The solution interior to the head of the expansion wave is modified from the initial state (see Shu 1977). At $x = 1$, the solutions for α and v connect without a jump, but with a discontinuous first derivative (because of the presence of the head of the wave) to the unperturbed state corresponding to the singular isothermal sphere. Near the origin, the solutions take the asymptotic forms,

$$\alpha \to \left[\frac{m_0}{2x^3}\right]^{1/2}, \qquad v(x) \to -\left[\frac{2m_0}{x}\right]^{1/2}, \qquad \text{as} \qquad x \to 0, \qquad (18)$$

that correspond to gravitational free-fall onto a (reduced) central mass found by numerical integration to be $m_0 = 0.975$. In dimensional units, the origin contains a central mass point (star) that grows linearly with time:

$$M_* = m_0 \frac{a^3 t}{G}. \qquad (19)$$

Equation (19) constitutes the most important property of the solution – the conclusion that the gravitational collapse of a singular isothermal sphere yields, not a characteristic mass, but a characteristic *mass infall rate* given by

$$\dot{M}_{\text{in}} = m_0 \frac{a^3}{G}. \qquad (20)$$

For $a = 0.2$ km s^{-1} (Taurus region), $\dot{M}_{\mathrm{in}} \approx 2 \times 10^{-6}$ $M_\odot$ yr^{-1}. The time t required to build the typical T Tauri star of mass 0.5 $M_\odot$ is then 2.5×10^5 yr. The ratio of this time scale (for gravitational collapse before the complete reversal of the infall) to the mean age of T Tauri stars agrees reasonably well with the ratio of the numbers of embedded (Class I) sources to the numbers of the revealed (Class II) sources in the Taurus cloud. A review of more direct observational tests of theoretical collapse models is given by Evans (1999).

Inclusion of the effects of finite levels of the magnetic field (i.e., $H_0 \neq 0$) does not change the qualitative situation. The collapse still occurs self-similarly from "inside-out," but the departures from spherical symmetry implies that non-zero motions can be induced *gravitationally* ahead of the front of the expansion wave (see also Terebey, Shu, & Cassen 1984 for the case when the cause of departure from spherical symmetry is rotation). If the gravitational precursor motions are sufficiently compressive, the expansion wave, now propagating outward at the fast MHD speed, may even be preceded by a weak shock wave (Li & Shu 1997). But the main new feature introduced by the presence of the magnetic field is a flattening of the flow in the inner regions to form a "pseudodisk" (see the perturbational treatment by Galli & Shu 1993a,b of this problem).

Figure 7 is obtained by applying the ZEUS 2-D MHD code (Stone & Norman 1992) to the collapse of a singular isothermal toroid with $H_0 = 0.25$ (Allen & Shu 1999b). The computations are carried out in the usual time-dependent way, without making use of the similarity assumption. The results at time intervals of $\Delta t = 2 \times 10^{12}$ s are displayed as contour plots in the meridional plane of the similarity coordinates ϖ/at and z/at. If the numerical calculations were infinitely precise, the contours for α, $|\mathbf{v}|$, and $|\mathbf{b}|$ at different times t would all lie on top of one another, defining the exact similarity solution. Because of errors associated with finite differencing, the contours at different t do *not* lie exactly atop one another, with the errors being larger at earlier times than later ones since the nontrivial part of the flow is spanned by more grid points at later times. Thus, unlike most numerical simulations which degrade with the passage of time, these simulations get better with time, and the computed values at $t = 8 \times 10^{12}$ s (the only case for which unit vectors showing the directions of $\mathbf{v}$ and $\mathbf{b}$ are displayed) have essentially converged on the correct self-similar values.

Since unstable cloud cores probably have dimensionless mass-to-flux ratios $\lambda = 1$–3, whereas T Tauri stars have $\lambda > 5000$, the assumption of field freezing must break down at high collapse densities (Nakano & Umebayashi 1980, Li & Shu 1997). Whether flux loss in forming protostars ultimately occurs by magnetic reconnection (e.g., Mestel & Strittmatter 1967, Galli & Shu 1993b) or by C-shocks (Li & McKee 1996, Li 1998a, Ciolek & Königl

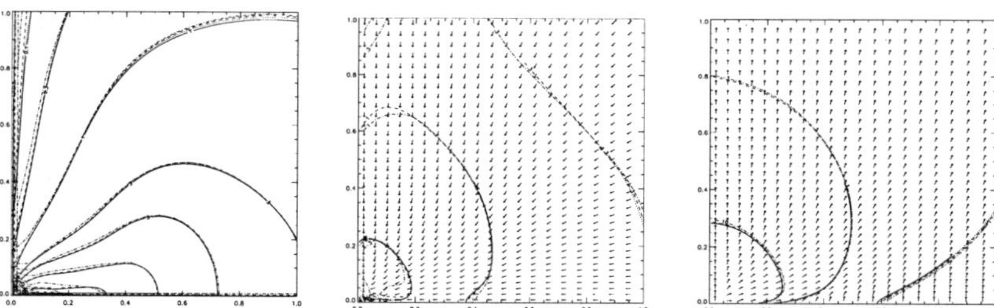

Figure 7. The reduced density α (left panel), velocity **v** (middle panel), and magnetic field **b** (right panel) plotted at time $t = 2, 4, 6, 8 \times 10^{12}$ s in the meridional plane of the similarity coordinates (ϖ/at)–(z/at). The curves give contours lines of equal density, speed, and field magnitude in intervals of 0.5 dex, while the arrows give unit vectors of the directions (on grid points) for the fluid velocity and magnetic field. For details, consult Allen & Shu (1999b).

1998) remains to be seen. Self-similar collapse calculations carried out with the assumption of strict field freezing yield important "outer solutions" for computations of the flux-loss problem in the inner regions (e.g., Li 1998b).

There is another important lesson to be learned from the self-similar collapse calculations. Contrary to earlier speculations made in the literature, magnetic tension in the envelope never suffices to hold up the envelope against the gravity of the growing protostar. When the wave of infall reaches successively larger portions of the envelope, those portions move inward to fill the hole and become part of the infalling material. In the specific case of the magnetized singular isothermal toroid, without some other influence, the mass infall rate into the central regions continues indefinitely at the constant rate given by equation (20), with the coefficient m_0 given numerically by $m_0 = 0.975(1 + H_0)^2$ when $H_0 \ll 1$ (Allen & Shu 1999b) and by $m_0 = 1.05(1 + H_0)$ when $H_0 \gg 1$ (Li & Shu 1997). In all cases, the inclusion of the effects of magnetic forces (when $H_0 > 0$) *increases* the rate of central mass accumulation in comparison with the nonmagnetized case ($m_0 = 0.975$), and does not decrease it. The main reason for the increase is that the cloud core is denser on average before it begins to collapse when supported by magnetic fields than when this support is absent. The slowing down of the infall speed by magnetic forces once collapse begins is largely offset by the increase of the signal speed at which the wave of infall propagates outward (Galli & Shu 1993a,b).

Influences other than magnetic fields can also quantitatively affect the mass infall rate, $\dot{M}_{\mathrm{in}}$, i.e., the rate at which the central YSO accumulates matter. For example, turbulence modeled by an isotropic pressure P_{turb} can also help to support a cloud core against its self-gravity. Approximated as a polytrope ($P_{\text{turb}} \propto \rho^{\Gamma}$) or as a logatrope ($P_{\text{turb}} \propto \ln \rho$), self-gravitating, turbulent, cloud cores can acquire pivotal states with density laws that depart from the $\rho \propto r^{-2}$ dependences that have become familiar for isothermal configurations (e.g., Lizano & Shu 1989, McLaughlin & Pudritz 1997, Galli et al. 1999). The gravitational collapse of such configurations will also yield mass infall rates $\dot{M}_{\text{in}}$ that are not constant in time, just as pre-existing inward motions at the pivotal instant $t = 0$ in magnetized isothermal configurations will yield non-constant mass-infall rates (Basu 1997, Ciolek & Königl 1998, Li 1998a). When added to the support provided quasistatically by thermal pressure, all of these mechanisms *increase* the mass infall rate above the base value given by the inside-out collapse solution for the singular isothermal sphere: $\dot{M}_{\text{in}} = 0.975a^3/G$. None of the considerations imply that $\dot{M}_{\text{in}}$ should *cut off* after some definite time and leave us with a star with a well-defined mass.

To obtain a stellar mass scale, one of three things might happen. First, the growing star might run out of matter to accumulate. The wave of infall might run into a vacuum because the material previously there has already fallen into some other star. This alternative requires the star formation process to be nearly 100% efficient, which we know is not the case except, possibly, in the relatively rare (at present) cases of bound cluster formation.

Second, the star might move out of its surrounding cloud core. In isolated regions of star formation, it is difficult to imagine how this occurs since the pit in the potential (where presumably the star forms) must move nearly with the center of mass of the core. In more crowded environments, a formed star might be freed from the gas dynamical forces that constrain the motion of its surrounding cloud core, and thus under the gravitational influences of the rest of the system, the star might begin to move differently from its parent core. However, the virial velocity gained by the star this way will be only comparable to, and not much larger than, the virial velocity a_{eff} (with contributions from thermal pressure, magnetic effects, and turbulence) that supports the gas against the self-gravity of the system. The resulting mass-infall rates by Bondi-Hoyle accretion will then not differ much from the canonical formula a_{eff}^3/G, unless the star-formation process has been pretty efficient, and there is not much cloud gas left in the spaces between already formed stars.

Third, the material otherwise destined to fall onto the star might be removed by some hydrodynamic, magnetohydrodynamic, or photevaporative process. Obvious candidates are mechanisms that involve energetic massive stars: HII regions, supernova explosions, etc. Observations of YSOs, however, offer an even more intriguing possibility: outflows from the central

YSO itself. Indeed, it is an empirical fact that no embedded protostar has yet been found to reveal itself as an optical object without first developing an outflow. This fact strongly suggests that stars help to define their own mass by YSO winds that reverse the continuing infall which would otherwise continue to build up the mass of the central object to ever higher masses (Shu & Terebey 1984).

The above conjecture does not mean that the interstellar medium plays *no* role in helping to define stellar masses. On empirical grounds it has been argued for some time (see, e.g., Shu, Adams, & Lizano 1987; Myers, Ladd, & Fuller 1991; Myers 1995; Shu 1995) that high-mass stars form from cores that produce high rates of infall $\dot{M}_{\mathrm{in}}$, whereas low-mass stars form from cloud cores that produce low rates of infall, consistent with the observation that the regions which form high-mass stars are more turbulent and hotter than the regions that form low-mass stars. In particular, Myers & Fuller (1993) have proposed that stars from 0.3 to 30 $M_{\odot}$, take about the same time to form, about 10^5 yr. This simple prescription corresponds to $\dot{M}_{\mathrm{in}} \propto M_*$, where M_* is the final stellar mass. Presumably, since the determination of M_* (end of collapse) comes after the determination of $\dot{M}_{\mathrm{in}}$ (beginning of collapse), a high value of $\dot{M}_{\mathrm{in}}$ *causes* a high value for M_*.

To close the logical loop, if we believe that it is outflows that sets final stellar masses, then it must also be true that stars of all masses have sufficient wind power $\dot{M}_{\mathrm{w}}$ to reverse the infall $\dot{M}_{\mathrm{in}}$ that makes them. This will require higher-mass stars to have more powerful outflows than lower-mass stars. Qualitatively, observations appear consistent with this requirement (see, e.g., Bally & Lada 1993). But quantitatively, can we make a theoretical case that $\dot{M}_{\mathrm{w}}$ increases (more or less linearly) as $\dot{M}_{\mathrm{in}}$ increases? As we shall see in the section after next, an attractive feature of X-wind theory is that it has precisely this property.

11. Mass Loss from Stars and Disks

Strong magnetic fields can considerably enhance the mass loss $\dot{M}_{\mathrm{w}}$ of thermally driven winds from the surfaces of rapidly rotating stars (Mestel, 1968). Even if the resultant flow is quite cold, Hartmann & MacGregor (1982) demonstrated that $\dot{M}_{\mathrm{w}}$ could have almost arbitrarily large values, dependent only on the ratio of the azimuthal and radial field strengths at the position where the gas is injected onto open field lines near the equator of a protostar that rotates near breakup. Shu et al. (1988) assigned the cause for the protostar to spin at breakup to a circumstellar disk that abuts against the surface of the central object and accretes onto it at a high rate $\dot{M}_D$. They also replaced Hartmann & MacGregor's arbitrary angle of injection for gas velocity and magnetic field direction with the requirement that in steady state, the wind mass-loss rate must be a definite fraction f of the disk accretion rate $\dot{M}_w = f\dot{M}_D$ (see below). [In steady state, $\dot{M}_D$ should equal $\dot{M}_{in}$. However, with the passage of time, the outflow may reduce the effective infall rate $\dot{M}_{in}$ below the undisturbed rate (20) had the outflow not been present.]

Blandford & Payne (1982) advanced an influential self-similar model of centrifugally driven winds from the surfaces of magnetized accretion disks. Pudritz & Norman (1983) applied these pure disk-wind models to bipolar outflows, and Königl (1989) investigated how the wind might smoothly join a pattern of accretion flow inside the disk. Heyvaerts and Norman (1989) studied how the winds might collimate asymptotically into jets, while Uchida & Shibata (1985) and Lovelace et al. (1991) advocated alternative driving mechanisms where magnetic pressure gradients play a bigger role in the acceleration of a disk-blown wind.

Motivated by the problem of binary X-ray sources, a parallel line of research developed concerning how magnetized stars accrete from surrounding disks. Ghosh & Lamb (1978) used order-of-magnitude arguments to show that a strongly magnetized star would truncate the surrounding accretion disk at a larger radius than the stellar radius R_* and divert the equatorial flow along closed field-line funnels toward the polar caps. Although Ghosh and Lamb thought that this inflow would spin up the central object faster than if the accretion disk had extended right up to the stellar surface, Königl (1991) made the surprising and insightful suggestion that the process might torque down the star and account for the relatively slow rate of spin of observed T Tauri stars. Observational support for magnetospheric braking and accretion was subsequently marshalled by Edwards et al. (1993) and Hartmann et al. (1994).

Arons, McKee, and Pudritz (Arons 1986) proposed that a centrifugally-driven outflow accompanies the funnel inflow [see also Camenzind (1990)]. Independently, Basri (1989, unpublished) arrived observationally at the same suggestion by extending the synthesis work by Bertout et al. (1988) on ultraviolet excesses in T Tauri stars (for a review, see Bertout 1989). Shu et al. (1994a) put these ideas together into a concrete proposal that generalized the earlier X-wind model of Shu et al. (1988). Shu et al. (1995) considered how X-winds asymptotically collimate into YSO jets.

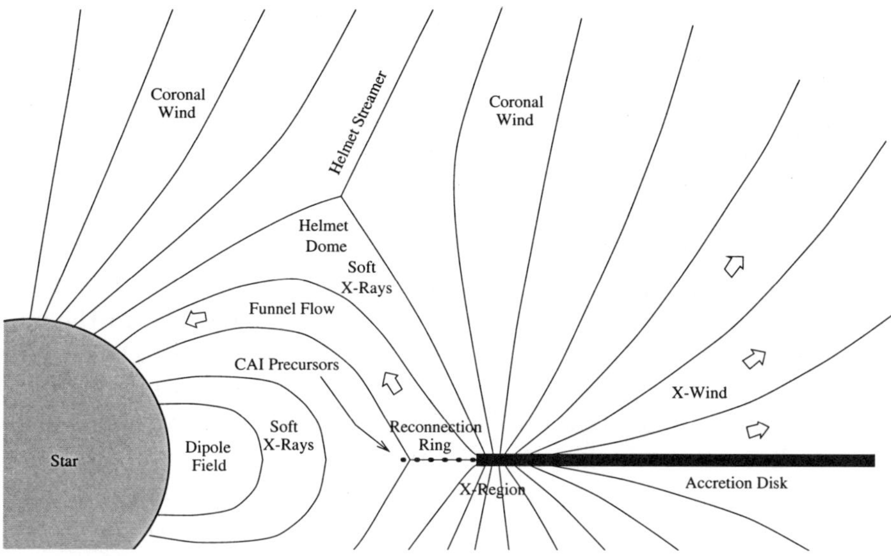

Figure 8. Schematic drawing of the X-wind model. Coronal winds from the star and the disk may help the X-wind to open field lines surrounding the helmet streamer, but this aspect of the configuration is not central to the model.

12. Generalized X-wind Model

In the generalized X-wind model, if the star has mass M_* and magnetic dipole moment μ_*, the gas disk is truncated at an inner radius,

$$R_x = \Phi_{dx}^{-4/7}\left(\frac{\mu_*^4}{GM_*\dot{M}_D^2}\right)^{1/7}. \tag{21}$$

A disk of solids may extend inward of R_x to the evaporation radius of calcium-aluminum silicates and oxides [see Shang et al. (1997) and Meyer et al. (1997)]. In equation (21), Φ_{dx} is a dimensionless number of order unity that measures the amount of magnetic dipole flux that has been pushed by the disk accretion flow to the inner edge of the disk. In the closed dead-zone model of Najita & Shu (1994) and Ostriker & Shu (1995), $\Phi_{dx} = 2\bar{\beta}_w f^{1/2}$, with $f = \dot{M}_w/\dot{M}_D$ and $\bar{\beta}_w$ defined as the streamline-averaged ratio of magnetic field to wind mass-flux. This expression for the coefficient Φ_{dx} holds when magnetic flux is swept toward R_x from both larger and smaller radii (see below), and the magnetic flux trapped in a small neighborhood of R_x is 3/2 times larger than the pure dipole value.

For a Keplerian disk, the inner edge of the disk rotates at angular speed

$$\Omega_x = \left(\frac{GM_*}{R_x^3}\right)^{1/2}. \tag{22}$$

To satisfy mass and angular momentum balance, the disk accretion divides at R_x into a wind fraction, $\dot{M}_w = f\dot{M}_D$, and a funnel-flow fraction, $\dot{M}_* = (1-f)\dot{M}_D$, where

$$f = \frac{1 - \bar{J}_* - \tau}{\bar{J}_w - \bar{J}_*}. \tag{23}$$

In equation (23) $\bar{J}_*$ and $\bar{J}_w$ are, respectively, specific angular momenta nondimensionalized in units of $R_x^2\Omega_x$ and averaged over funnel and wind streamlines, and τ is the negative of the viscous torque of the disk, $-\mathcal{T}$, acting on its inner edge and measured in units of $\dot{M}_D R_x^2\Omega_x$. In the approximation that the embedded stellar fields force the X-region to be only weakly differentially rotating (Shu et al. 1994a,b), τ is small compared to unity. Similarly, if the funnel flow does not bring so much angular momentum into the (small) star as to spin it up on a time scale much shorter than the time to accumulate the star, $\bar{J}_*$ must also be negligible in comparison to unity.

In steady state, the star is regulated to corotate with the inner disk edge,

$$\Omega_* = \Omega_x. \tag{24}$$

If corotation did not hold, the system would react to reduce the discrepancy between Ω_* and Ω_x, where Ω_x is given by equation (22) with R_x determined by how much magnetic field there is to maintain a certain average standoff distance for the stellar magnetosphere [see eq. (21)]. For example, suppose the star turns faster than the inner edge of the disk, $\Omega_* > \Omega_x$. The field lines attached to both would then continuously wrap into ever tighter trailing spirals, with the field lines adjacent to the star tugging it backward in the sense of rotation. This tug decreases the star's angular rate of rotation Ω_* to more nearly equal the rate Ω_x. Conversely, imagine that $\Omega_* < \Omega_x$. With the star turning slower than the inner edge of the disk, the field lines attached to both would continuously wrap into ever tighter leading spirals, with the field lines adjacent to the star tugging it forward in the sense of rotation. This tug increases the star's angular rate of rotation Ω_*, again to more nearly equal the rate Ω_x. In true steady state, $\Omega_* = \Omega_x$, and the funnel-flow field lines acquire just enough of a trailing spiral pattern (but without continuously wrapping up) so that the excess of material angular momentum brought toward the star by the inflowing gas is transferred outward by magnetic torques to the footpoints of the magnetic field in the disk (Shu et al. 1994a,b).

The X-wind gas gains angular momentum and the funnel gas loses angular momentum at the expense of the matter at the footpoint of the field in, respectively, the outer and inner parts of the X-region. As a consequence, this matter and the field lines across which it is diffusing pinch toward the middle of the X-region. In an idealized treatment, where the pinch is imagined to gather the trapped flux into a narrow ring (a single point in the meridional plane), 1/3 of the trapped flux yields field lines pointing in the right outward directions to launch a (cold) X-wind; another 1/3 are pointing in the right inward directions to torque down the inflowing gas into a funnel flow; and the middle 1/3 are standing too upright to induce dynamical imbalance and result in a dead zone (Ostriker & Shu 1995). We assume that the Balbus & Hawley (1992) instability occurs inside the disk on dead zone field lines. The instability allows gas near the midplane to transfer from the last of the X-wind field lines to the first of the funnel-flow field lines. Thus, of the totality of field lines in the X-region, 1/3 yield outflow via the X-wind, and 2/3 yield inflow either via the Balbus-Hawley instability or via the funnel flow. On the assumption that incoming matter from the outer disk are loaded onto the tangle of field lines in the X-region more or less randomly, we may then deduce that $\dot{M}_{\rm w} \approx \dot{M}_D/3$, i.e., that $f \approx 1/3$.

If we adopt the approximations $\bar{J}_* \approx 0$ and $\tau \approx 0$ in equation (23), the condition $f \approx 1/3$ yields $\bar{J}_{\rm w} \approx 3$ (which requires for self-consistency a field-to-wind-flux loading $\bar{\beta}_{\rm w} \approx 1$). When X-wind field lines are loaded by disk accretion so that the magnetocentrifugal fling yields an average dimensionless wind torque $\bar{J}_{\rm w}$, the formula for the terminal speed of the X-wind (averaged over streamlines) reads

$$\bar{v}_{\rm w} = \left(2\bar{J}_{\rm w} - 3\right)^{1/2} \Omega_{\rm x} R_{\rm x}. \tag{25}$$

With $\bar{J}_{\rm w} \approx 3$, equation (25) becomes $\bar{v}_{\rm w} \approx \sqrt{3}\Omega_{\rm x} R_{\rm x}$, which gives $\bar{v}_{\rm w} \approx 200$ km s^{-1} for typical estimates of the Keplerian velocity $\Omega_{\rm x} R_{\rm x}$ at the inner edges of YSO accretion disks.

Shu et al. (1997) point out that the idealized steady state of exact corotation probably cannot be maintained because of dissipative effects. Two surfaces of null poloidal field lines – labeled as "helmet streamer" and "reconnection ring" in Figure 8 – mediate the topological behavior of dipole-like field lines of the star, opened field lines of the X-wind, and trapped field lines of the funnel inflow emanating from the X-region. Across each null surface, which begin or end on "Y-points" [called "kink points" by Ostriker & Shu (1995)], the poloidal magnetic field suffers a sharp reversal of direction. By Ampére's law, large electric currents must flow out of the plane of the Figure along the null surfaces. Nonzero electrical resistivity would lead to the dissipation of these currents and to the reconnection of the oppositely directed field lines [see, e.g., Biskamp (1993)].

Thus, we anticipate that realistic YSO systems will alternate between states where the magnetic field lines of the dead zone are open (see Fig. 8) and states where they are closed (see Fig. 9). The release of magnetic energy in transforming one state to another in giant flares presumably

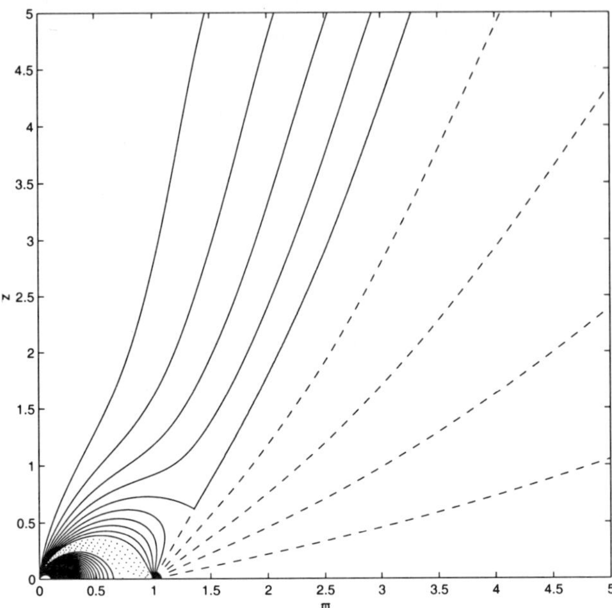

Figure 9. The poloidal magnetic field configuration resulting from the interaction of a magnetized star and a surrounding accretion disk. This Figure differs from Figure 8 in two respects. First, the dead zone is closed rather than open. Second, the field lines are not drawn schematically but result from an actual self-consistent calculation (Ostriker, Shang, and Shu, unpublished) that imposes approximate pressure balance across the interfaces of the funnel flow (dotted curves), dead zone (solid curves), and X-wind (dashed curves). Also denoted by solid curves are stellar field lines opened by the X-wind (i.e., those that connect at infinity to the dashed lines in the X-wind) and virtually unperturbed dipole field lines that are rooted in the star.

underlies a part of what observers call the "magnetic activity" of YSOs (e.g., Montmerle & André 1988; Feigelson, Carkner, & Wilking 1998). The change in magnetic topology will modify the numerical value of the coefficient Φ_{dx} in equation (21).

When R_x changes, the angular speed Ω_x of the footpoint of magnetic field lines in the X-region will vary according to equation (22). However, the considerable inertia of the star prevents its angular velocity Ω_* from changing on the time scale of magnetic reconnection at the null surfaces of the magnetosphere. The resultant shear when $\Omega_* \neq \Omega_x$ will stretch and amplify field lines attached to both the star and the disk. The poloidal field will bulge outward from the increased magnetic pressure, inserting more magnetic flux into the fan of field lines emanating from the X-region [see the simulations of Linker & Mikić (1994) and Hayashi, Shibata, & Matsumoto (1996)]. Dynamo action inside the star would presumably replace the upward rising dipole-like poloidal fields. This dynamo action would be enhanced by the wrapping of field lines between the star and disk. Averaged over long times, we envisage a secular balance, with enough dynamo-generated poloidal field being inserted into the X-region to balance the rate of field dissipation at the null surfaces.

13. Summary

We conclude with a summary. Giant molecular clouds are supported against their self-gravity in part by magnetic fields and in part by MHD turbulence. Because of nonlinear mode coupling, the decay of MHD turbulence occurs more quickly than had been previously suspected. In the absence of offsetting effects, this decay will produce highly flattened clouds. However, if the dimensionless mass-to-flux ratio is subcritical or marginally critical, $\lambda \leq 1$, the material is stable to global and local gravitational collapse. In isolated regions of star formation, ambipolar diffusion in dense cores is a process that can increase the local value of λ beyond the critical value of unity. In more crowded environments, the collisional agglomeration of cloud clumps in the direction of the mean field (which increases the mass M without increasing the flux Φ) is another process that can lead to supercritical cores.

In regions where $\lambda > 1$, the decay of turbulence can by itself lead to the formation of dense pivotal cores and the birth of stars. The inner regions of pivotal cores formed either by slow ambipolar diffusion or by faster turbulent decay are likely to assume power-law forms. The gravitational collapse of such pivotal states occurs in a self-similar fashion, from inside-out, and has no tendency to want to stop unless the region either runs out of matter (in which case the star-formation process is highly efficient), or the inflow is truncated (say, by the onset and development of a bipolar outflow). Bipolar outflows that occur in flattened sheets can produce many cloud cores strung out in long filaments. In critical or subcritical regions, the existence of such cores and filaments is not a signpost for overall gravitational collapse and hierarchical fragmentation. Rather, it simply attests to the power that YSO outflows (and other energetic events) have for sculpting the shapes and forms of molecular clouds, especially in regions of high-mass star formation (cf. Wiseman & Ho 1998).

In contrast with regions like the Taurus molecular cloud, which yield only the dispersed mode of star formation, large regions within molecular clouds which are (or become) magnetically supercritical probably give the clustered mode of star formation (Shu, Adams, & Lizano 1987, C. Lada and E. Lada, this volume). The efficiency of star formation is high in such regions, and the competition for matter among YSOs with different "feeding zones" may be the determinant process that fixes the masses of stars rather than outflows (e.g., Larson 1995). However, outflows may still play an important role (e.g., in setting the level of turbulence of such regions), although confusion limits the ability of observers to investigate such issues in crowded environments. The importance of stellar feedback for star formation in the clustered mode remains a challenge for any comprehensive theory of the birth of stars from giant molecular clouds.

Acknowledgements: This research is funded in part by a grant from the National Science Foundation and by the NASA Astrophysics Theory Program that supports a joint Center for Star Formation Studies at NASA/Ames Research Center, the University of California at Berkeley, and the University of California at Santa Cruz. The research of ECO is supported by NAG 53840.

References

Akeson, R. L., & Carlstrom, J. E. 1997, ApJ, 491, 254

Allen, A., & Shu, F. H. 1999a, ApJ, in preparation

Allen, A., & Shu, F. H. 1999b, ApJ, in preparation

Arons, J. 1986, in Plasma Penetration into Magnetospheres, ed. N. Kylafis, J. Papamastorakis, and J. Ventura (Iraklion: Crete Univ. Press), 115

Arons, J., & Max, C. E. 1975, ApJ, 196, L77

Balbus, S. A., & Hawley, J. F. 1992, ApJ, 400, 610

Bally, J., & Lada, C. J. 1983, ApJ, 265, 824

Basu, S. 1997, ApJ, 485, 240

Basu, S., & Mouschovias 1994, ApJ, 432, 720

Bertoldi, F., & McKee, C. F. 1992, ApJ, 395, 140

Bertout, C. 1989, ARAA, 27, 351

Bertout, C., Basri, G., & Bouvier, J. 1988, ApJ, 330, 350

Biskamp, D. 1993, Nonlinear Magnetohydrodynamics (Cambridge University Press)

Blandford, R. D., & Payne, D. G. 1982 MNRAS, 199, 883

Bodenheimer, P. 1980, in Fund. Prob. in Theory of Stellar Evol., IAU Symp. No. 93, ed. D. Sugimoto, D. Q. Lamb, D. N. Schramm (Dordrecht: Reidel), p. 5

Bonnor, W. B. 1956, MNRAS, 116, 351

Boss, A. 1999, in Protostars & Planets IV, ed. V. Mannings, S. Russell, & A. Boss (Tucson: Univ. Arizona Press), in press

Camenzind, M. 1990, in Reviews in Modern Astronomy 3, ed. G. Klare (Berlin: Springer), 259

Chandrasekhar, S. 1939, Stellar Structure (New York: Dover)

Ciolek, G. E., Königl, A. 1998, ApJ, 504, 257

Cohn, H. 1980, ApJ, 242, 765

Crutcher, R. M. 1999, ApJ, in press

Crutcher, R. M., Troland, T. H., Goodman, A. A., Heiles, C., Kazes, I., & Myers, P. C. 1993, ApJ, 407, 175

Cudworth, K. M., & Herbig, G. H. 1979, AJ, 84, 548

Ebert, R. 1955, Z. Ap., 37, 217

Edwards, S., Strom, S. E., Herbst, W., Attridge, J., Merrill, K. M., Probst, R., & Gatley, I. 1993, AJ, 106, 372.

Evans, N. J. 1999, ARAA, 37, in press

Feigelson, E. D., Carkner, L., & Wilking, B. A. 1998, ApJ, 494, L215

Fukui, Y., Sugitani, K., Takaba, H., Iwata, T., Mizuno, A., Ogawa, H., & Kawabata, K. 1986, ApJ, 311, L85

Galli, D., Lizano, S., Li, Z. Y., Adams, F. C., & Shu, F. H. 1999, ApJ, submitted
Galli, D., & Shu, F. H. 1993a, ApJ, 417, 220
Galli, D., & Shu, F. H. 1993b, ApJ, 417, 243
Gammie, C. F. & Ostriker, E. C. 1996, ApJ, 466, 814
Ghosh, P., & Lamb, F.K. 1978, ApJ, 223, L83
Goodman, A. A., Benson, P. J., Fuller, G. A., & Myers, P. C. 1993, ApJ, 406, 528
Hartmann, L., Hewitt, R., & Calvet, N. 1994, ApJ, 426, 669
Hartmann, L., & MacGregor, K.B. 1982, ApJ, 259, 180.
Hayashi, M. R., Shibata, K., & Matsumoto, R. 1996, ApJ, 468, L37
Heyvaerts, J., & Norman, C. 1989, ApJ, 347, 1055
Hoyle, F. 1953, ApJ, 118, 513
Hunter, C. 1962, ApJ, 135, 594
Klessen, R. S., Burkert, A., & Bate, M. R. 1998, ApJ, 501, L205
Königl, A. 1989, ApJ, 342, 208
Königl, A. 1991, ApJ, 370, L39
Lada, C. J. 1985, ARAA, 23, 267
Larson, R. B. 1995, MNRAS, 272, 213
Layzer, D. 1964, ARAA, 2, 341
Li, Z. Y. 1998a, ApJ, 493, 230
Li, Z. Y. 1998b, ApJ, 497, 850
Li, Z. Y., & McKee, C. F. 1996, ApJ, 464, 373
Li, Z. Y., & Shu, F. H. 1996, ApJ, 472, 211
Li, Z. Y., & Shu, F. H. 1997, ApJ, 475, 237
Linker, J. A., & Mikić, Z. 1994, ApJ, 430, 898
Lizano, S., & Shu, F. H. 1989, ApJ, 342, 834
Lovelace, R. V. E, Berk, H. L., & Contopoulos, J. 1991, ApJ, 379, 696
Lynden-Bell, D., & Eggleton, P. P. 1980, MNRAS, 191, 483
Lynden-Bell, D., & Wood, R. 1968, MNRAS. 138, 495
McLaughlin, D. E., & Pudritz, R. E. 1997, ApJ, 476, 750
MacLow, M. M., Klessen, R. S., Burkert, A., Smith, M. D., & Kessel, O. 1998,
    Phys.Rev.Lett. 80, 2754
McKee, C. F. 1989, ApJ, 345, 782
Mestel, L. 1965, QJRAS, 6, 161
Mestel, L. 1968, MNRAS, 138, 359
Mestel, L. 1985 in Protostars and Planets II, ed. D. C. Black & M. S. Matthews (Tucson:
    University of Arizona Press), p. 320
Mestel, L., & Spitzer, L. 1956, MNRAS, 116, 505
Mestel, L., & Strittmatter, P. A. 1967, MNRAS, 137, 95
Meyer, M. R., Calvet, N., & Hillenbrand, L. A. 1997, AJ, 114, 288
Mizuno, A., Onishi, T., Yonekura, Y., Nagahama, T, Ogawa, H., & Fukui, Y. 1995, ApJ,
    445, L161
Mizuno, A., Onishi, T., Hayashi, M., Ohashi, N., Sunada, K., Hasegawa, T., & Fukui, Y.
    1994, Nature, v. 368, no. 6473, 719
Montmerle, T., & André, P. 1988, in Formation and Evolution of Low-Mass Stars, ed. A.
    K. Dupree & M. T. V. Lago (Dordrecht: Kluwer), p. 225
Mouschovias, T. Ch., & Paleologou, E. V. 1980, ApJ, 237, 877
Mouschovias, T. Ch., & Spitzer, L. 1976, ApJ, 210, 326
Myers, P. C. 1995, in Molecular Clouds and Star Formation, ed. C. Yuan and J. H. You
    (Singapore: World Scientific)
Myers, P. C., & Benson, P. J. 1983, ApJ, 266, 309
Myers, P. C., & Fuller, G. A. 1993, ApJ, 402, 635
Myers, P. C., Ladd, E. F., & Fuller, G. A. 1991, ApJ, 372, L95
Myers, P. C., Fuller, G. A., Goodman, A. A., & Benson, P.J. 1991, ApJ, 376, 561
Myers, P. C., Lazarian, A. 1998, ApJ, 507, L157
Nagai, T., Inutsuka, S., & Miyama, S. 1998, ApJ, 506, 306

Najita, J.R., & Shu, F.H. 1994, ApJ, 429, 808
Nakano, T. 1979, PASJ, 31, 697
Nakano, T. 1998, ApJ, 494, 587
Nakano, T., & Umebayashi, T. 1980, PASJ, 32, 613
Norman, C., & Silk, J. 1980, ApJ, 238, 158
Onishi, T., Mizuno, A., & Fukui, Y. 1999, PASJ Letter, in press
Onishi, T., Mizuno, A., Kawamura, A., Ogawa, H., & Fukui, Y. 1998, ApJ, 502, 296.
Ostriker, E.C., Gammie, C.F., & Stone, J.M. 1999, ApJ, in press
Ostriker, E.C., & Shu, F.H. 1995, ApJ, 447, 813
Padoan, P., & Nordlund, A. 1999, ApJ, submitted
Pudritz, R. E. 1986, PASP, 98, 709
Pudritz, R. E., & Norman, C. A. 1983, ApJ, 274, 677
Pudritz, R. E., & Norman, C. A. 1986, ApJ, 301, 571
Reipurth, B., Bally, J., & Devine, D. 1997, AJ, 114, 2708
Rodriguez, L. F., Ho, P. T. P., & Moran, J. M. 1980, ApJ, 240, L149
Schwarzschild, M. 1958, Structure and Evoultion of the Stars (Princeton University Press)
Shang, H., Shu, F., Lee, T., & Glassgold, A.E. 1997, in Low Mass Star Formation from
    Infall to Outflow, Poster Proc. IAU Symp. No. 182, ed. F. Malbet and A. Castets
    (Grenoble: Laboratorie d'Astrophysique), 312
Shu, F. H. 1977, ApJ, 214, 488
Shu, F. H. 1992, The Physics of Astrophysics, II. Gas Dynamics (Mill Valley: University
    Science Books)
Shu, F. H. 1995, in Molecular Clouds and Star Formation, ed. C. Yuan and J. H. You
    (Singapore: World Scientific), p. 97
Shu, F. H., Adams, F. C., & Lizano, S. 1987, ARAA, 25, 23
Shu, F. H., & Li, Z. Y. 1997, ApJ, 475, 251
Shu, F.H., Lizano, S., Ruden, S., & Najita, J. 1988, ApJ, 328, L19
Shu, F.H., Najita, J., Ostriker, E. C., & Shang, H. 1995, ApJ, 455, L155
Shu, F., Najita, J., Ostriker, E., Wilkin, F., Ruden, S., & Lizano, S. 1994a, ApJ, 429,
    781
Shu, F. H., Najita, J., Ruden, S. P., & Lizano, S. 1994b, ApJ, 429, 797
Shu, F. H., Shang, H., Glassgold, A. E., & Lee, T. 1997, Science, 277, 1475
Shu, F. H., & Terebey, S. 1984, in Cool Stars, Stellar Systems, and the Sun, ed. S.
    Baliunas & L. Hartmann (Berlin: Springer-Verlag), p. 78
Spitzer, L. 1942, ApJ, 95, 329
Strittmatter, P. A. 1966, MNRAS, 132, 359
Stone, J. M., & Norman, M. L. 1992, ApJ Suppl, 80, 791
Stone, J. M, Ostriker, E. C., & Gammie, C. F. 1998, ApJ, 508, L99
Terebey, S., Shu, F. H., & Cassen, P. 1984, ApJ, 286, 529 (TSC)
Tohline, J. 1982, Fund. Cosmic Phys., 8, 1
Torrelles, J. M., Rodriguez, L. F., Canto, J., Carral, P., Marcaide, J., Moran, J. M., and
    Ho, P. T. P. 1983, ApJ 217, 214
Torrelles, J. M., Gomez, J. F. Ho, P. T. P., Rodriguez, L. F., Anglada, G., and Canto,
    J. 1994, ApJ, 435, 290
Uchida, Y., & Shibata, K. 1985, PASJ, 37, 515
Vrba, F. J., Strom, K. M., & Strom, S. E. 1976, AJ, 81, 958
Wardle, M., & Königl, A. 1993, ApJ, 410, 218
Welch, W. J., Vogel, S. N., Plambeck, R. L., Wright, M. C. H., & Bieging, J. H. 1985,
    Science, 228, 1329.
Wiseman, J., & Ho, P. T. P. 1998, ApJ, 502, 676
Yu, K. C., Bally, J., Devine, D. 1997, ApJ, 485, L45
Zuckerman, B., & Evans, N. J. 1974, ApJ, 192, L149

BIPOLAR MOLECULAR OUTFLOWS

R. BACHILLER & M. TAFALLA
IGN Observatorio Astronómico Nacional
Apartado 1143, E-28800 Alcalá de Henares, Spain

1. Introduction

Stars form by the gravitational collapse of dense gas in molecular clouds, so one would naively expect that infall motions dominate the gas kinematics in star-forming regions. Observations, however, show this not to be the case, and that outflow motions —not infall— is what prevails around young stellar objects (YSOs). This apparent contradiction was fully realized about 20 years ago, when the first systematic molecular observations of star-forming regions were done, and bipolar molecular outflows were found towards almost every region of star formation (e.g, Snell et al. 1980, Bally & Lada 1983). The discovery of outflows has deeply changed our picture of how stars are born; their observational study has dominated the field of star formation research, and their theoretical understanding has posed a serious challenge to theorists of stellar birth. After 20 years of intense outflow research, enormous progress has been done, but some basic questions still remain unanswered. In this review we attempt to briefly describe what we know about molecular outflows, and also show the challenges that lie ahead in this exciting field of star formation.

Outflows driven by young stars and protostars are observable over a wide range of wavelengths, from the ultraviolet to the radio, so it is natural that the history of outflow research started with observations in the visible, the earliest developed astronomy tool. In the early 1950's, Herbig (1951) and Haro (1952) observed a few peculiar small nebulosities with emission line spectra. These so-called Herbig-Haro (HH) objects were soon associated with stellar winds (Osterbrok 1958), and later recognized to result from the interaction of highly supersonic stellar winds with ambient material (Schwartz 1975). Measurements of proper motions (Cudworth & Herbig 1979) and the detection of highly-collimated HH jets (Mundt & Fried 1983) confirmed that the wind ejection takes place in the vicinity of a newly formed star.

227

C.J. Lada and N.D. Kylafis (eds.), The Origin of Stars and Planetary Systems, 227–266.

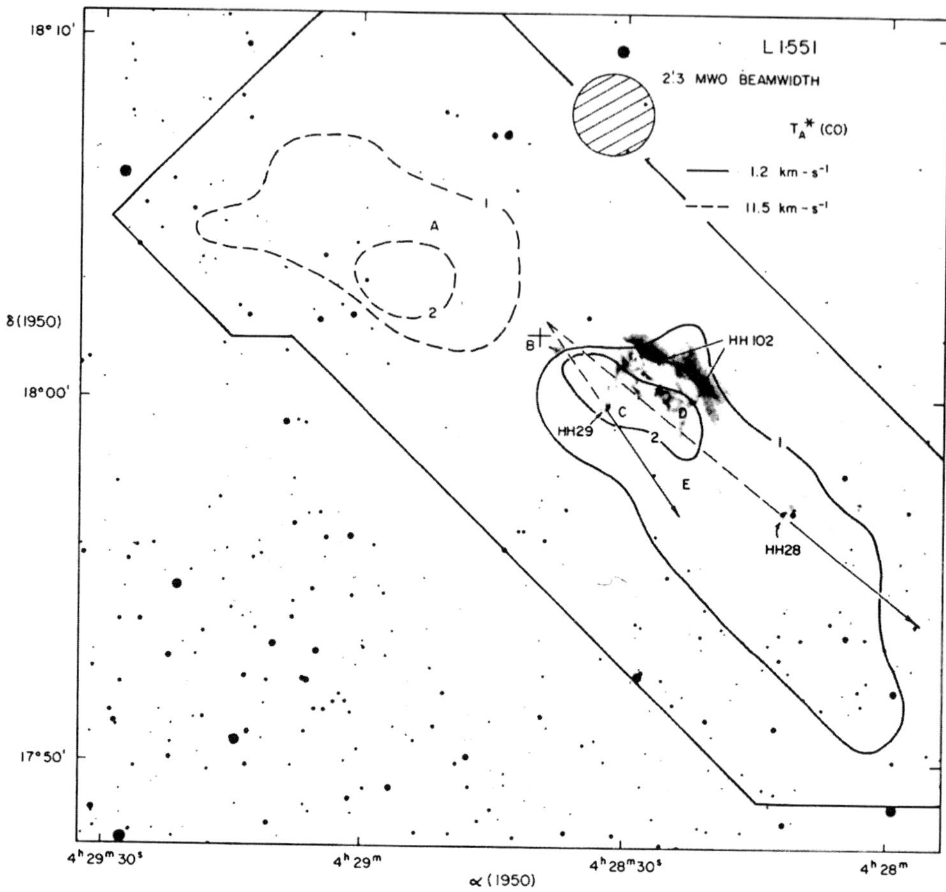

Figure 1.   Contour map of the J=1–0 CO emission from the L1551 outflow superposed on an optical photograph of the region (from Snell et al. 1980). Solid contours represent blue shifted emission and dashed contours are red shifted emission. The cross indicates the position of the driving infrared source, IRS5. Also shown are the Herbig-Haro objects in the field. The directions of the proper motions of the two compact HH objects HH28 and HH29, which are also indicated, suggest a common origin at the infrared source

The presence of winds emerging from young stars was also suggested by the profile shapes of optical spectral lines (Herbig 1960; Kuhi 1964), the presence of centimeter wavelength free-free continuum emission (Cohen et al. 1982; Rodríguez 1995), and the detection of H_2 lines in the near infrared (Beckwith et al. 1978; Hodapp & Ladd 1995). When these winds interact with the surrounding molecular cloud, they give rise to molecular outflows, which are best observed in the millimeter-wave lines of carbon monoxide (CO). The first broad CO lines, indicative of high-velocity molecular gas, were discovered toward the Orion A molecular cloud in the mid 70's (Zuckerman

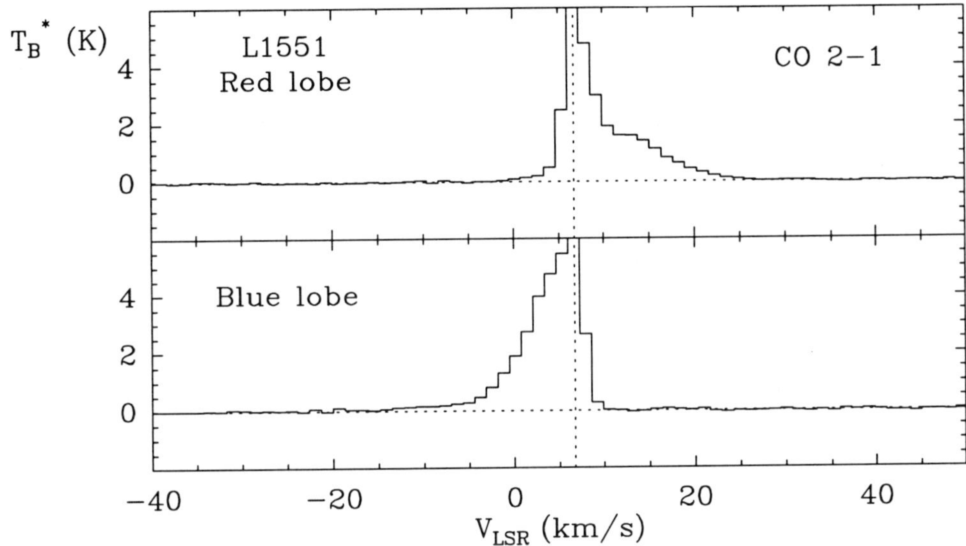

Figure 2.   CO J=2–1 spectra averaged along the main axis of the L1551 outflow. The data (from Bachiller et al. 1994) were taken with the IRAM 30-m radio telescope, whose beam size is $\sim 12''$ at 230 GHz.

et al. 1976, Kwan & Scoville 1976), and soon thereafter detailed mapping of the supersonic CO emission in different objects showed that outflowing material tends to lie in two lobes of gas, one blue shifted and the other red shifted (Snell et al. 1980; Rodríguez et al. 1980; see Figs. 1 and 2). The field of bipolar outflow research was in this way born.

The first systematic outflow searches showed that these systems are extraordinarily common around young stars, and that stars of different masses have detectable flows (Bally & Lada 1983; Edwards & Snell 1982, 1983, 1984). The number of known outflows has increased rapidly over the last 20 years. Lada (1985) compiled a first catalog containing 68 outflow sources, Fukui et al. (1993) listed 157 outflows confirmed through complete or partial mapping, and Wu et al. (1997) have cataloged around 200 outflows. It is now believed that every star, no matter its mass, goes through an outflow phase during its early life, and that the outflow phenomenon is deeply connected to the process of star formation (e.g, Shu et al. 1987).

Since the review by Bally & Lane (1991) at the Crete I Advanced Study Institute, the study of bipolar outflows has benefited enormously from the use of the new generation of millimeter and submillimeter radio telescopes and interferometers. These new observations have revealed a population of highly collimated molecular outflows with extremely high velocity and evidence for strong interaction with the surrounding gas, which seems to correspond to the early phases of the outflow life. In this chapter, we review the basic properties of the more "classical" outflows (those identified during the 80's) and those of the highly collimated flows identified in the 90's, proposing an evolutionary sequence to connect them. We refer the reader to the chapter by Shu for the theoretical modeling of the outflow phenomenon, the chapter by Churchwell for more emphasis on outflows from high mass stars, and the chapter by Reipurth & Raga for a review of HH objects and jets. Other relevant recent reviews on molecular outflows include those by Bachiller (1996), Padman et al. (1997), Cabrit et al. (1997) and Richer et al. (1999). An up to date view of the current situation in outflow research is provided by a reading of the proceedings of the recent IAU Symposium 182.

2. Determining Outflow Properties from CO Observations

Molecular outflow gas is best traced by its CO emission, as this molecule is abundant, has a very simple level structure (it is a linear rotor), and different isotopes are detectable at millimeter wavelengths (^{12}CO, ^{13}CO, $C^{18}O$). Deriving outflow properties therefore requires converting CO line observations into physical gas parameters, and this makes CO radiative transfer the most useful tool of a molecular outflow observer. As the CO emission traces ambient gas swept up by the outflow over its lifetime (see below), CO observations provide a time integrated picture of the outflow activity. It is therefore a challenge for the astronomer to read the outflow history from the observed CO lines, and from there infer the history of mass loss from the central YSO.

2.1. CO SPECTROSCOPY

The low dipole moment of CO (~ 0.1 Debye) makes this molecule easily excited and often thermalized in many astronomical situations. For this reason, a measurement of the CO excitation provides an excellent estimate of the gas kinetic temperature. To achieve this, the line ratio R between the intensity of two rotational lines (e.g., J=2–1 and J=1–0) is often used. If the CO emission is optically thin, and the filling factor and the excitation temperature are the same for both transitions, $R_{21} = 4 \exp(-11/T_{ex})$, and a measurement of R gives directly an estimate of T_{ex}. Figure 3 (left panel, upper curve) shows a plot of R_{21} as a function of T_{ex}, illustrating how the optically thin line ratio increases quickly with excitation temperature from $T_{ex} = 1$ to 50 K, and tends monotonically to a value $R=4$ when T_{ex} tends to infinity.

If the CO emission is extremely optically thick, the intensity ratio adopts the form $R = [\exp(5.5/T_{ex})-1]/[\exp(11/T_{ex})-1]$, so it remains < 1 for all values of the excitation temperature (Figure 3 left panel, lower curve); a measure of $R > 1$, therefore, indicates that the emission is not fully optically thick. In the intermediate case, the optical depth of the CO emission needs

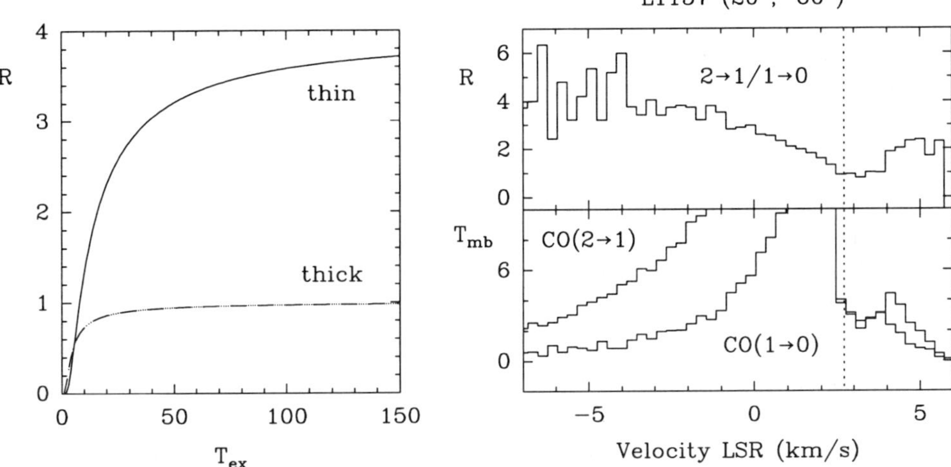

Figure 3.   Analysis of the 2–1/1–0 CO line intensity ratio, R. Left.- R as a function of excitation temperature in the optically thin and thick limits. Right.- The bottom panel shows the CO 2–1 and 1–0 spectra (convolved to the same resolution) towards a position in the L1157 outflow. The upper panel shows the line intensity ratio R across the line profile. Note that R increases from a value of ∼1 (typical of thick emission) in the inner wings to a value of ∼4 (typical of thin emission) at high velocities.

to be estimated, or a rare CO isotope used to calculate the ratio. For many outflows, the the line ratio R changes with position, and for a given position, it depends on velocity. Figure 3 (right panel) shows CO 2–1 and 1–0 spectra toward the L1157 outflow together with the ratio R_{21} as a function of velocity. R increases monotonically with velocity, indicating that the emission is thinner at higher velocities, probably because the amount of gas flowing at high velocities is much smaller than that flowing at low speeds.

To estimate the CO optical depth, lines from different isotopes should be observed. If both J=1–0 lines of ^{13}CO and ^{12}CO (CO hereafter) are optically thin, for example, the CO/^{13}CO J=1–0 intensity ratio r is given by the abundance ratio $r = x(CO)/x(^{13}CO) \sim 90$. If the 1–0 line of ^{12}CO (the main isotope) is optically thick, however, the ratio will be $r \sim 90/\tau(CO)$. Thus, if the intensity ratio r is found to be significantly lower than 90, we can conclude that the CO 1-0 line is optically thick. This is actually the case in many outflows, but again r changes with the position in the outflow and the considered velocity range.

2.2. OUTFLOW PHYSICAL PARAMETERS

From observations of two lines of CO and one line of ^{13}CO it is thus possible to estimate the excitation temperature and the opacity of the outflowing gas.

Then, it is possible to estimate the gas column densities using the standard equations of the radiation transfer and opacity (see e.g. Margulis & Lada 1985, Levreault 1988). The mass of the outflowing material can thus be calculated by assuming a CO/H_2 abundance ratio. Finally, it is possible to study the distribution of mass with the line-of-sight velocity, and estimate in this way the momentum, the kinematic energy, and the mechanical power of the outflow. All these estimates are subject to important uncertainties, like the ambiguity separating at low velocities the outflow gas from the ambient material, as outflow wings tend to merge continuously with the ambient cloud component in the spectrum.

Another source of uncertainty is our ignorance of the angle the outflow axis makes with the plane of the sky, which is necessary to convert the observed velocity radial component into the three-dimensional velocity field. One approach is to model the shape and velocity field of the outflow lobes to obtain synthetic maps for comparison with the observations. Interesting attempts in this direction have been done by using conical and paraboloidal lobe geometries (e.g. Cabrit & Bertout 1986, 1990; Cabrit et al. 1988; Meyers-Rice & Lada 1991). One important application of this technique consists of simulating the observation of a model outflow and comparing the estimated parameters with the real ones. Doing this, Cabrit & Bertout (1990) show that the most accurate methods can estimate the outflow mass within a factor of 2, the momentum within a factor of 10, the flow force within a factor of 10, and the mechanical power within a factor of 60.

Despite the large uncertainties involved, CO spectroscopy is still the best method to obtain the physical parameters of bipolar molecular outflows. From the considerable efforts devoted during these last 20 years to extensive CO mapping, we know that flow sizes range from less than 0.1 pc to several pc. The observed (unprojected) terminal velocities range from a few $km\,s^{-1}$ to about 100 $km\,s^{-1}$, the kinematical time-scale (i.e., the flow length divided by the terminal velocity) ranges from about 10^3 yr to about 10^5 yr. The total molecular mass of the outflowing material ranges from a few $10^{-2}\,M_\odot$ to several hundreds $M_\odot$.

2.3. OUTFLOW CLUMPINESS

Multiline observations of CO and ^{13}CO in a number of bipolar outflows have shown that, in many sources, the high-velocity CO emission is optically thick (e.g. Bally 1982; Plambeck et al. 1983; Snell et al. 1984). For instance, the optical depths of both the CO 1–0 and 2–1 transitions in NGC 2071 and in S 140 are greater than 3 over a velocity range of $\sim$ 30 $km\,s^{-1}$ (Snell et al. 1984). On the other hand, the excitation temperatures of these transitions are $\sim$ 12 K in both sources, and should equal the gas kinetic temperature, since the CO lines should be thermalized by collisions. As the observed antenna temperatures in the CO wings are considerably weaker than 12 K, one must conclude that the high-velocity emission arises in a clumpy gas which does not uniformly fill the telescope beam.

Plambeck et al. (1983) derived typical values of the filling factor of 0.1–0.2 in several molecular outflows, and Snell et al. (1984) found that the filling factor increases to nearly unity at the lowest outflow velocities. If the clump size distribution is independent of velocity, this means that there are many more clumps moving at low velocities than at high velocities; or, alternatively, that the smaller clumps move at higher velocities than the larger (more massive) ones.

Further evidence for the presence of clumping in the high-velocity gas has been provided by direct observations with high angular resolution. Tafalla et al. (1994) observed a striking clumped structure in the MonR2 outflow via CS observations: the flow clumps were found to be as dense as the ambient cloud (few $10^5 cm^{-3}$), and to contain several solar masses each. They concluded that most of the outflowing gas -if not all- is in the form of dense clumps. Further direct evidence for clumpiness in other outflows was provided by Wootten et al. (1984), Avery et al. (1990), Garden et al. (1991), and Kitamura et al. (1990, 1992), among others.

3. Properties of Classical CO Outflows

In sections 3 and 4 we discuss separately the properties of "classical" and "highly collimated" outflows. Although the separation of outflows into these two categories is not absolute because outflows of very different collimation have been found over many years of research, there seems to be a clear dichotomy between outflows found during the 80's, which were bipolar but poorly collimated ("classical") and some extremely collimated outflows identified in the 90's. Such a dichotomy is not the result of just resolving better already known outflows, but represents the existence of a small population of extremely well collimated outflows, only found after the advent of the latest generation of radio telescopes (e.g., IRAM 30m, JCMT).

The properties of classical outflows have been the topic of several excellent reviews (e.g., Lada 1985, Snell 1987), and here we only attempt to enumerate them briefly, referring the reader to those papers for further details. Most of these properties have been derived from CO observations using the techniques described in the previous section, although the application of different criteria by different authors (contribution of low velocity gas, different inclination correction methods, etc) makes sometimes difficult to compare outflow properties from different authors. A systematic study of the outflow phenomenon using the currently achievable high spatial resolution and applying an homogeneous method of analysis is still an overdue piece of research.

3.1. OUTFLOW GEOMETRY

The most characteristic property of outflows is their bipolarity, which arises from the presence of two separated lobes of gas, one red shifted and the other blue shifted, with a young stellar object in between (Figures 1,2). An early question in outflow research was whether bipolarity is intrinsic or it arises from external, large-scale collimation. Although extended (tenths of parsec) structures of dense gas elongated perpendicular to the flow axis ("toroids") were found in many cases (Cantó et al. 1981; Torrelles et al. 1983), it is now believed that they represent more the effect of core disruption than a collimating agent (see below). Modern high resolution observations of molecular outflows reveal that the scale at which collimation occurs can be very small (< 1000 AU for L1448, Guilloteau et al. 1992; HH211, Gueth & Gulloteau 1999), so if bipolarity is not fully intrinsic, it is associated with much smaller scales than initially thought.

To quantify outflow bipolarity, one usually defines the collimation factor R_{coll}, the ratio between the major and minor axes of the flow. This collimation factor depends on telescope angular resolution (see a nice graphic example in Snell 1987), so some care has to be exerted when using low resolution data to determine it. Figure 4 (left panel) presents a histogram of collimation factors for a sample of 26 outflows from Lada (1985), showing a wide distribution of R_{coll} values and an overabundance of systems with low collimation. This distribution implies that most outflows are poorly collimated, and that the observations cannot be explained simply from orientation effects in a sample of highly collimated outflows (Bally & Lada 1983). Different outflows must have different collimation factors, and on average, classical outflows are poorly collimated.

An important aspect of outflow morphology is the spatial distribution of the gas inside the lobes. Not many outflows have been mapped with the necessary angular resolution to study such distribution in detail, but in several cases where this has been possible, shell geometries have been found. L1551 is the classical example of a shell distribution (Moriarty-Schieven & Snell 1988, see also Figure 5), but it is by no means the only example (e.g.: NGC2071, Moriarty-Schieven et al. 1989; MonR2: Meyers-Rice & Lada 1991, Tafalla et al. 1997). In L1551, the shell surrounds a cavity devoid of molecular gas that can be seen as a hole in ^{13}CO maps (Moriarty-Schieven & Snell 1988) and as an optical reflection nebula in optical images (Graham & Heyer 1990). Such a combination of outflow shell and empty cavity strongly suggests that the outflow material represents ambient gas set into motion by an

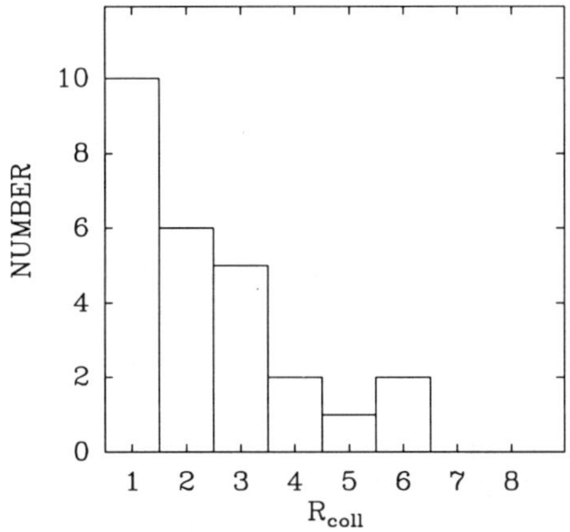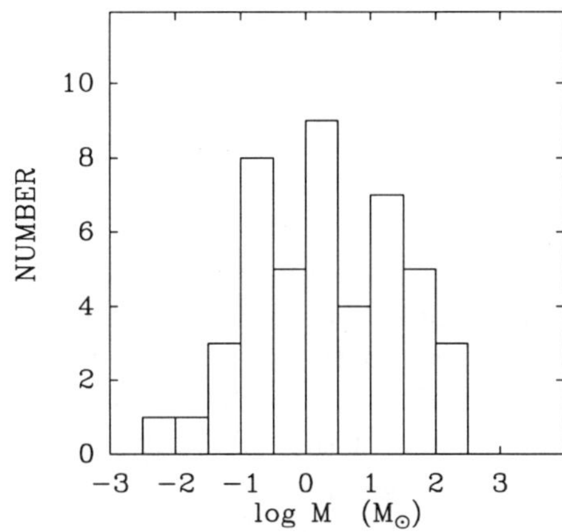

Figure 4.   Right: histogram of the outflow collimation factor R_{coll} for a sample of 26 outflows, from Lada (1985). The large number of low collimation factors argues against a population of equally collimated outflows whose only difference is the orientation in the plane of the sky. Right: histogram of outflow masses for a sample of 46 outflows, from Snell (1987). The large values at the high end of the distribution can only result from the acceleration of ambient molecular material.

(invisible) wind from the central young stellar object. Whether the situation of L1551 applies to other (classical) outflows is not clear yet, as the sample of outflows studed in high detail is small, but the presence of cometary reflection nebulae in many objects (e.g.: Staude & Elsässer 1993) suggests a common pattern. We will return to this point and its implications for outflow evolution in Section 5.

Although bipolarity is the norm in outflows, deviations from this behavior are known to exist. A case of interest is that of quadrupolar outflows, which consist of four lobes of accelerated material, two of them red and two of the blue (e.g.: L723, Moriarty-Schieven & Snell 1989, Avery et al. 1990). The most likely explanation for these systems is the that they consist of a superposition of two independent outflows, each of them bipolar itself and, in fact, multiple YSOs are often found at the center of these outflows (in the case of L723 see Anglada et al. 1991). Source multiplicity should not be surprising given that a large fraction of stars seem to be born in binaries or small groups (Mathieu 1994). A recent NIR survey by Hodapp (1994) has shown that at least one third of all known bipolar outflows are associated with clusters of young stars. The fact that simple bipolar flows are seen suggests that the outflow phase may last much less than the total star formation period in the cluster, and that outflows are lit one or just a few at a time.

3.2. PHYSICAL PROPERTIES AND KINEMATICS OF THE OUTFLOW GAS

We have seen that the distribution of CO in outflows like L1551 suggests that most of the outflow material consists of swept up ambient gas. A stronger argument in this direction comes from the estimate of outflow masses, and this is illustrated in Figure 4 (right panel) with a histogram of the distribution of those masses in a sample of 46 objects (from Snell 1987). The large values seen at the high end of the distribution (> 10 $M_\odot$) cannot arise from the material being a direct ejection from the central YSO, and the only possible source of this gas is the surrounding ambient cloud. This implies that a driving agent accelerates molecular cloud material and gives rise to the observed CO lobes. The nature of this driving agent is still a mystery, and current outflow research consists mostly of an attempt to derive properties of this agent from the observed molecular material. Current ideas on this topic will be discussed in Section 7.

Despite the acceleration to supersonic velocities that the molecular gas has suffered, the physical properties of the gas in classical outflows do not significantly differ from those of the ambient cloud. Snell et al. (1984), for example, have shown that the kinetic temperature in the outflow gas is usually similar to that of the ambient cloud, if not even slightly lower. This does not mean that shocks do not occur in classical outflows, as there is plenty of evidence for them from IR and optical emission (Hodapp 1994; Reipurth, this volume), but that most of the gas we see in CO arises from material that has have enough time to cool down after it was accelerated to supersonic speeds. Less conclusive, on the other hand, have been attempts to determine the density and chemical composition of classical outflows. For most systems, these parameters are probably similar to those of the ambient gas.

From the point of view of kinematics, one of the simplest parameters one can determine in an outflow is the so-called kinematical age, which is the ratio between the outflow length and its total velocity extent (measured to some intensity threshold). Perhaps the most complete study of this parameter is due to Levreault (1988), who calculated kinematical ages for 25 outflows selected optically (i.e, probably rather evolved). This author finds that most systems have kinematical ages in the range 2-5 $\times 10^4$ yr, which is consistent with determinations by other other authors for different types of outflows. Such low kinematical ages, however, are probably not representative of the duration of the outflow phase by almost an order of magnitude. In fact, outflow statistics suggest that the outflow phenomenon lasts about 2×10^5 yr (Parker et al. 1991).

In addition to the outflow velocity extent, the spatial distribution of the gas at different velocities offers an important clue to the acceleration mech-

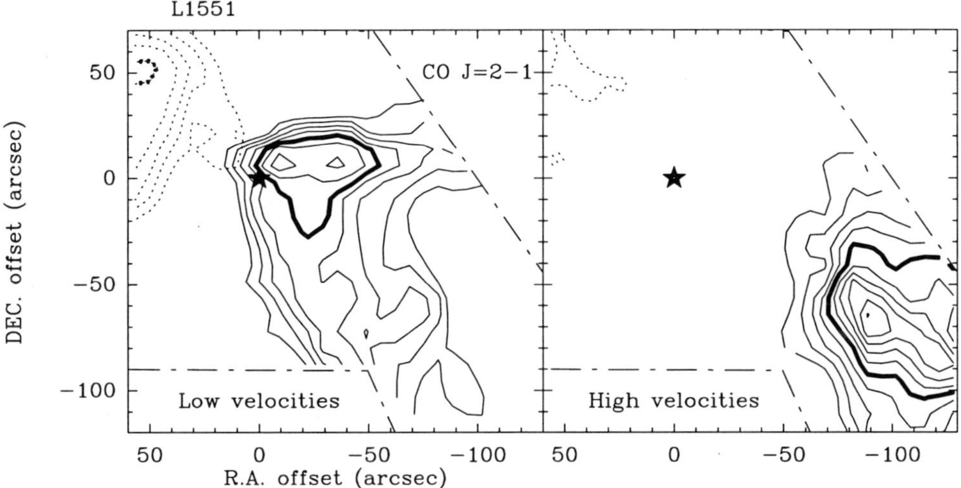

Figure 5.   CO J=2–1 emission from the central region of the L1551 outflow observed with $\sim 12''$ resolution (Bachiller & Tafalla, 1994, unpublished data). The left panel shows the lowest velocity gas and the right panel shows the highest velocity emission detected. Note the shell-like distribution at low velocities and how the fastest gas appears far away from the central source and concentrated towards the outflow axis.

anism. Snell et al. (1984) have studied the centroid of the high velocity CO emission in outflows and have found that in some cases, the highest velocities are not reached near the YSO but at some distance from it. In L1551, this behavior appears as a systematic change in the position of the shells with velocity: as the velocity increases, the shells lie further from the outflow source and closer to the outflow axis (Moriarty-Schieven & Snell 1988). More generally, position-velocity diagrams of the CO emission from outflows sometimes show a quasi-linear increase of the terminal velocity with distance to the exciting source, in a pattern which has been referred as a "Hubble-like" law (Meyers-Rice & Lada 1991). Such a pattern could arise from a real linear increase of the outflow velocity with distance (e.g., Meyers-Rice & Lada 1991), although this does not necessarily mean that the driving wind really accelerates as it moves away from the YSO. The combination of momentum conservation and a density decrease away from the outflow source can explain such a pattern (Shu et al. 1991).

A more difficult parameter to determine is the exact direction of motion of the outflow gas. Meyers-Rice & Lada (1991), from modeling outflow position-velocity diagrams, conclude that the flow has to be rather parallel to the axis if one is to reproduce the observed CO patterns. The lack of a significant component of the outflow velocity field perpendicular to the outflow axis strongly argues against energy driven outflows, which would produce large inflating bubbles. Outflow acceleration must therefore occur by momentum conservation between a stellar wind/jet and the ambient gas (see also Masson & Chernin 1993).

3.3. OUTFLOW ENERGETICS

Bipolar outflows, as we have seen, consist of large masses of ambient gas moving at supersonic velocities. They therefore represent an important deposition of energy and momentum by a newly born star on its surrounding material. Quantifying the amount of this deposition is of great interest when studying the evolution of molecular clouds after star formation has occurred.

In their classical systematic study of the outflow phenomenon, Bally & Lada (1983) found a correlation between the bolometric luminosity of the exciting object and the rate of momentum and energy injection into the outflow. Figure 6 presents a modern version of this correlation (due to Cabrit & Bertout 1992), showing a systematic behavior extending over five orders of magnitude in source luminosity. This continuity in the trend strongly suggests that a single mechanism is responsible for the production of molecular outflows from both low and high mass stars.

As mentioned before, energy is probably not conserved during outflow acceleration (it is radiated away), so the amount of this parameter in the molecular material is not necessarily a good indicator of the energy content in the driving agent. Linear momentum, on the other hand, is most likely conserved, so the momentum in the CO outflow must equal the momentum in the invisible agent. As Figure 6 shows (left panel), outflow momentum exceeds photon momentum in the stellar radiation field by about two orders of magnitude. This means that unless stellar photons are scattered hundreds of times by the outflow material (extremely unlikely), outflow acceleration cannot be due to radiation pressure (Bally & Lada 1983).

The outflow gas is accelerated ambient material, so energies and momenta measured from CO data represent amounts already deposited on molecular clouds. Even for the weakest outflows (from low mass stars), kinetic energies are of the same order of the binding energies of the cores where the exciting source is formed, so outflow-core interaction can be an important element in core dispersal (Myers et al. 1988). Direct evidence for this interaction comes from the CO maps of outflows. As mentioned before, the L1551 outflow has evacuated a large cavity in its parent cloud by setting into motion a large amount of gas (Moriarty-Schieven & Snell 1988). Similar effects have been observed in many other systems, like NGC 2071 (Moriarty-Schieven et al. 1989), L43 (Mathieu et al. 1988, Bence et al. 1998), and Mon R2 (Wolf et al. 1990; Meyers-Rice & Lada 1991, Tafalla et al. 1997). Bipolar outflows, therefore, constitute a likely mechanism by which newly born stars disperse their parental environment and make the transition from embedded object

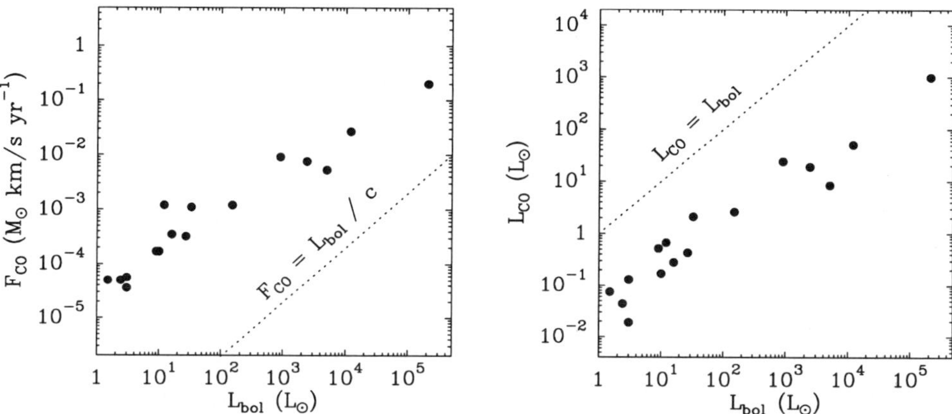

Figure 6. Correlation of outflow momentum flux (left) and mechanical luminosity (right) with source bolometric luminosity for a sample of bipolar outflows (from Cabrit & Bertout 1992). Note that the momentum flux in an outflow exceeds the YSO photon flux by almost two orders of magnitude. Bipolar outflows cannot be driven by stellar radiation pressure.

to visible star. It may even be possible that through the outflow mechanism a star is able to stop gravitational collapse and put an end to its formation phase, setting in this way its final mass (Shu et al. 1987; Velusamy & Langer 1998).

4. Properties of Highly Collimated CO Outflows

The first highly collimated outflows were identified in the early 90's thanks to the new generation of high resolution mm radio telescopes (Bachiller et al. 1990, Masson et al. 1990, André et al. 1990), and they were quickly interpreted as representing a previously missing stage of outflow life. Their discovery has been followed by a very intense search of additional examples, and during the rest of the 90's a sizable sample of collimated outflows have been built. Well known members of this outflow class include L1448 (Bachiller et al. 1990; Fig. 8), VLA 1623 (André et al. 1990), IRAS03282+3035 (Bachiller et al. 1991b), Orion B (Richer et al. 1992) HH211 (McCaughrean et al. 1994; Gueth & Guilloteau 1999; Fig. 7), NGC 2264G (Lada & Fich 1996), HH212 (Zinnecker et al. 1998), and SVS13B (Bachiller et al. 1998). In this section, we review the main properties of these highly collimated outflows, except for the heating and chemical anomalies recently found in some objects, that because of the large amount of recent results deserve a separate treatment and are discussed in Section 6. Previous reviews of the properties of highly collimated outflows can be found in Bachiller & Gómez-González (1992), Masson & Chernin (1993), and Bachiller (1996).

4.1. GEOMETRY AND COLLIMATION

High collimation is the main characteristic of this outflow class, although no clear-cut minimum value of R_{coll} has been set for an outflow to be classified as highly collimated. Values larger than 10 are easily achieved, and in some cases R_{coll} factors as high as 20 and 30 can be measured at high velocities (Richer et al. 1992, Lada & Fich 1996, Gueth & Guilloteau 1999). In most outflows collimation increases systematically with velocity, and the fastest gas lies close to the outflow axis and has the highest bipolarity. This seems a continuation of the trend found in the classical outflows (like L1551), but at speeds and R_{coll} factors that are about an order of magnitude higher.

The high bipolarity of some molecular outflows rivals the strong collimation seen in optical jets, and this has revived the idea that the relation between jets and outflows may be one of cause and effect (e.g. Masson & Chernin 1993), and not just represent two parallel phenomena having different origins (e.g. Mundt et al. 1987). To this end, it has been crucial the realization that optical jets are mostly neutral (few percent ionized, see Raga et al. 1990, Hartigan et al. 1994) and that the outflow phenomenon is much longer lived than initially thought (Parker et al. 1991), so stellar jets can potentially inject in the surrounding medium amounts of kinetic energy and linear momentum comparable to those observed in molecular outflows (Masson & Chernin 1993). In section 7 we will discuss with detail how these recent results have affected theoretical models of the outflow phenomenon.

Although the most collimated outflows look very much jet-like at the highest velocities, their behavior is much more classical at lower speeds. Figure 7 illustrates the behavior of HH211 (Gueth & Guilloteau 1999), one of the best collimated outflows known to date, which still shows a shell-like gas distribution at the lowest velocities (see top panel). Similar behavior has been observed in Orion B (Richer et al. 1992), L1448 (Bachiller et al. 1995), NGC 2264G (Lada & Fich 1996), and L1157 (Gueth et al. 1996, 1997), and may also be present in other highly collimated flows not studied yet with high enough angular resolution. In the cases of L1157 and HH211, the shells diverge away from the central source but converge at the end of the lobe, where strong H_2 (shocked) emission is observed. Bow shock models of this emission reproduce the main observed shapes (Gueth et al. 1996, Gueth & Guilloteau 1999).

4.2. KINEMATICS AND ENERGETICS

Highly collimated molecular outflows tend to have, in addition to a "classical" component, a fraction of extremely high velocity (EHV) gas moving at much higher speeds than those seen in classical outflows. In L1448, for example, terminal velocities are about 70 $\mathrm{km\,s^{-1}}$ for both lobes (Bachiller et

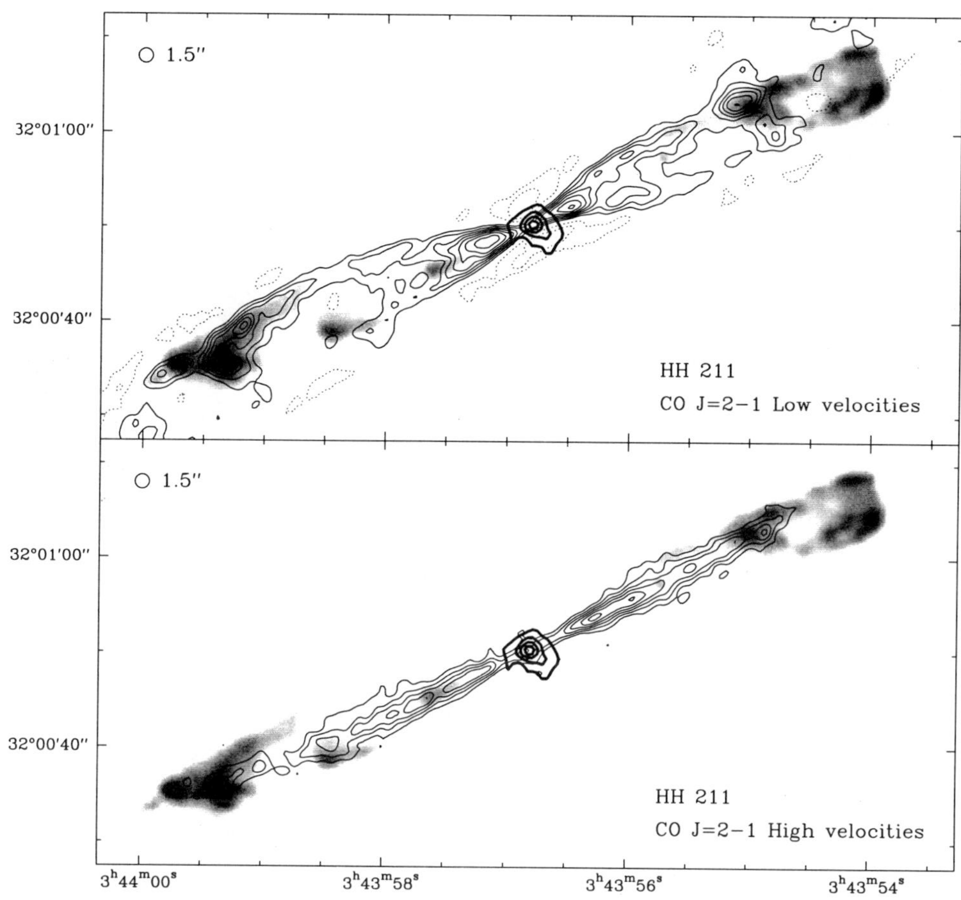

Figure 7. The highly collimated HH211 outflow in IC348 (from Gueth & Guilloteau 1999). Thin contours represent the CO 2–1 emission integrated in two different velocity intervals (low velocity on the top and high velocity on the bottom). Grey scale represents the H_2 v=1-0 S(1) emission. Thick contours are λ 1.3 mm continuum emission from the protostellar condensation HH211-mm.

al. 1990), and in HH7-11 (probably the record holder), molecular gas up to 150 km s^{-1} (with respect to the ambient cloud) has been found (Bachiller & Cernicharo 1990, Masson et al. 1990, Koo 1990). These observed velocities have to be corrected for projection effects, which are likely to be important given the the small angle with the plane of the sky that some highly collimated outflows seem to make. The resulting corrected velocities are among the highest observed in molecular gas.

In many highly collimated outflows, the EHV component appears forming discrete condensations often referred as molecular "bullets." Figure 8 illustrates the case of L1448, where the "bullets" are very well defined in position and velocity, and lie almost symmetrically located at each side of the exciting source (see also Guilloteau et al. 1992; Dutrey et al. 1997). Other sources with bullets include HH7-11 (Masson et al. 1990, Bachiller & Cernicharo 1990), VLA16293 (André et al. 1990), Orion B (Richer et al. 1992), IRAS03282+3035 (Bachiller et al. 1991b), IRAS2005+2720 (Bachiller et al. 1995), and HH111 (Cernicharo & Reipurth 1996, Nagar et al. 1997). From CO observations, the bullets are estimated to have typical sizes of a few 10^{-2} pc and typical masses of a few $10^{-4} M_\odot$ (Bachiller et al. 1990). Their kinematic time scales range from a few 10^2 to a few 10^3 yr.

In cases like L1448 and IRAS03282+3035, the terminal velocity of the EHV gas (bullets) decreases with distance from the outflow origin, while the terminal velocity of the low-velocity outflow increases according to an approximate "Hubble-like" law. This behavior suggests that the EHV jet-like component could be injecting momentum into the ambient gas to produce the lower velocity outflow, and in fact, the EHV gas sometimes contains amounts of energy and momentum comparable to those of the lower velocity outflow (Bachiller et al. 1990, 1991b). Whether the EHV gas is in fact the driving agent of the outflow or a manifestation of a hidden wind will be further discussed in Section 7 in the context of outflow models.

The physical origin of the molecular bullets is still subject to debate. Bachiller et al. (1990) interpreted them as coherently moving pieces of gas ejected from the immediate vicinity of the central YSO. Richer et al. (1992) have argued that given the internal velocity dispersion of the bullets ($\sim$ 6 km s^{-1} in L1448), if they were ballistic pieces of gas, they should expand and dissolve as they move away from the YSO, so either they are somehow confined or they are produced in situ, most likely through some type of flow instability. Chernin et al. (1994), from numerical simulations, produce features similar to the observed bullets by reformation of molecules in post-shock gas. Although it is not clear which of these interpretations is correct, an interesting clue to their origin comes from the symmetrical location of bullets in both position and velocity with respect to the central source (at least in L1448, Bachiller et al. 1990; IRAS03282+3035, Bachiller et al. 1991b; HH111, Cernicharo & Reipurth 1996). This symmetry implies that there are pairs of bullets, one of them red and the other one blue, which have the same kinematical age and probably a related origin. This is best understood if there is some type of instability in the central source that gives rise to quasi-periodic ejections every $10^2 - 10^3$ yr. Whether all outflow material is ejected during these outbursts and whether the bullets represent ejected component or ambient gas excited by the ejection are still mysteries that need to be solved. In any case, the bullets illustrate the violent and non steady nature

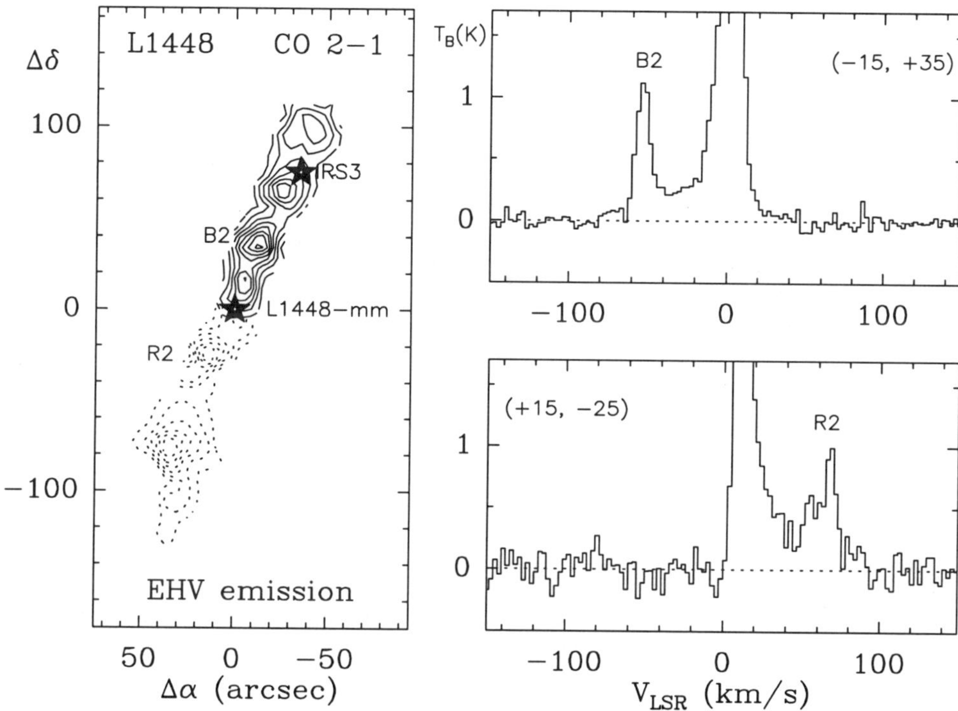

Figure 8. CO 2–1 emission in the L1448 outflow (from Bachiller et al. 1990). The left panel shows the EHV component of the outflow. Solid contours represent blue shifted emission in an interval of 30 $km\,s^{-1}$ wide centered at –50 $km\,s^{-1}$ from the ambient velocity. Dashed contours are for red shifted emission in a similar interval centered at +50 $km\,s^{-1}$ from the ambient velocity. Note the presence of high-velocity molecular "bullets" along the jet axis. The CO spectra obtained toward the positions of bullets B2 and R2 are shown in the right panels. Position offsets are in arcseconds with respect to the position of the Class 0 source L1448-mm.

of the outflow phenomenon.

Overall, highly collimated outflows are more energetic than classical flows from sources of comparable luminosity. This is not just due to the presence of an EHV component, although the EHV gas makes a substantial contribution to the outflow energy and momentum (40% and 60%, respectively for IRAS03282+3035) despite constituting a small fraction of the total mass (20% in IRAS03282+3035). This systematic higher energy and momentum content of the highly collimated outflows is nicely illustrated in the recent study by Bontemps et al. (1996). These authors have analyzed a sample of 45 low-mass, nearby outflows both from Class 0 sources (highly collimated) and from Class I YSOs (classical), and in Figure 9 we present their plot of outflow momentum flux versus source bolometric luminosity in a similar manner than was presented for classical outflows in Figure 6. In

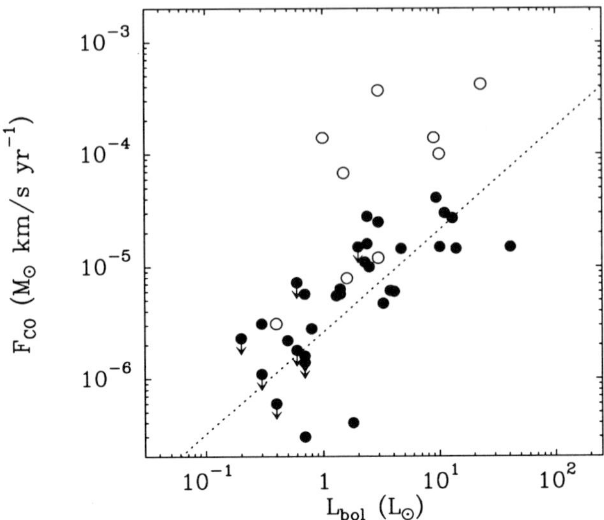

Figure 9. CO momentum flux F_{CO} versus bolometric luminosity L_{bol} for 41 young stellar objects (Bontemps et al. 1996 data). Class 0 and Class I sources are represented as open and filled circles, respectively. The dashed line represents the "best fit" correlation found only for Class I sources. Note that all Class 0 sources lie well above this line.

agreement with Bally & Lada (1983) (also Cabrit & Bertout 1992), classical outflows follow a linear relation between momentum flux and source luminosity. Highly collimated outflows, on the other hand, have systematically an order of magnitude more momentum than classical sources of equal luminosity, and this suggests that the exciting sources of these outflows are losing mass in a more violent manner than the sources of the classical flows. The origin of this energy excess, and of the other extreme properties of the highly collimated outflows, constitutes one of the most intriguing aspects of outflow research, and in the next section we explore the possibility that it is related to the extreme youth of the outflow exciting sources.

5. Outflow Evolution

After reviewing in the last two sections the properties of classical and highly collimated outflows as derived from observations, we now take a more speculative approach and explore how the differences between these two types of outflows may be understood in terms of outflow evolution. This is a topic that has only begun to receive the attention it deserves (e.g., Padman 1997, André 1997), so many of the ideas we present here are still tentative and may be subject to change. Still, understanding how outflows evolve with time and how the observed properties depend on the outflow evolutionary status appears as the best strategy to organize a rather diverse and some-

times seemingly conflicting collection of observational results. It also offers
the hope to clarify the still mysterious origin of bipolar flows.

One of the main obstacles when studying outflow evolution is the identification of a reliable time indicator, an internal clock that allows one to order
outflows by age. Using outflow properties to set this clock may produce a
circular argument when studying outflow evolution, so it seems preferable
to use the properties of the central YSO as an age indicator. This has the
additional advantage of having already a time sequence of YSO evolution in
terms of discrete classes (Adams et al. 1987; Lada 1991; André et al. 1993),
with an even finer subdivision in terms of the bolometric temperature (Myers
& Ladd 1993). For the purposes of our discussion, the division into classes
will suffice, as our arguments are still rather qualitative.

The first evidence for age being at the root of the differences between
classical and highly collimated outflows comes from the different classes of
the sources powering these two types of systems. Highly collimated outflows
like L1448 (Fig. 8) and HH211 (Fig. 7) are powered by Class 0 objects, which
are considered the youngest protostars known (André et al. 1993), while
classical outflows like L1551 are excited by Class I YSOs, which represent a
more evolved state.

5.1. CHANGES IN OUTFLOW COLLIMATION

Outflows from Class 0 YSOs like HH211 (Fig. 7) and L1448 (Fig 8) appear
highly collimated, while outflows from Class I objects tend to be much less
collimated. Even L1551 (Fig. 1), one of the most collimated classical outflows, is still different from the almost jet-like systems HH211 and L1448,
both powered by Class 0 YSOs. This suggests that by the time a YSO reaches
the Class I stage, its bipolar outflow has lost an appreciable part of its collimation. Molecular outflows, therefore, decollimate as they evolve, starting
like highly bipolar systems and changing to classical flows.

Outflow decollimation seems accompanied by the opening of a cavity in
the parent cloud. In HH211 (Fig. 7) and L1448 (Bachiller et al. 1995), this
cavity is narrow and has straight walls, with an opening angle of about 30°,
while in L1551 the cavity is much broader ($\sim$ 90°) and has almost para-
bolic walls (Figs. 1 and 5). This is consistent with the outflow eroding the
parent cloud, and opening a broader and broader channel as time passes.
How exactly this process occurs is still a matter of debate. In the jet model
of Masson & Chernin (1993), the jet "wanders" in angle, sweeping up different portions of the ambient cloud and therefore broadening with time.
Raga & Cabrit (1993) have proposed that bow shocks give outflows their
transversal dimension, so in this picture outflow broadening will imply bow
shock broadening. Another possibility is that in addition to a highly collim-

ated component, outflows also contain a broader wind. In this case, the jet component would dominate at early times because of its larger directivity, and the wider angle component would become apparent only later, perhaps because a central channel along the outflow has been opened.

5.2. CHANGES IN THE KINEMATICS AND ENERGETICS

Outflows from Class 0 sources have faster gas than outflows from Class I YSOs, so evidence points to a slowing down with time of the outflow material. This seems to occur by the weakening of the EHV component, which is already undetectable by the end of the Class 0 stage, while outflows from both classes have comparable amounts of slow moving gas. Outflows therefore evolve in time by increasing the relative fraction slow material with respect to fast moving gas, an idea previously proposed by Bally (1986) in the context of classical high mass outflows.

The above evolutionary scenario, however, seems to conflict with the work from Masson & Chernin (1992), who have studied the distribution of mass with velocity in several outflows. These authors find distributions with similar slopes, suggesting that the ratio between high and low velocity gas is common to all outflows. The contradiction between this result and our proposed evolutionary scenario is only apparent, and a more careful study of the distribution of mass with velocity shows so. In Figure 10 (adapted from Tafalla 1993) we present the distributions of outflow wing intensity with velocity for a sample of 6 outflows, including the data for L1551 and NGC2071 that Masson & Chernin used in their study. As the figure shows, all outflows have similar slopes at low velocities, with an average value of about -1.5, in agreement with Masson & Chernin (1992). At higher speeds, however, the slope of the distribution changes, becoming steeper and reaching values of the order of -4. Such a steepening in the gas distribution was first noticed by Kuiper et al. (1981) for the Orion outflow, and a nice recent example was presented by Lada & Fich (1996) for NGC 2264G. The outflows in Figure 10 have been ordered vertically by the velocity of the break point, which is largest in L1448 (≈ 30 km s^{-1}) and smaller in Mon R2 (≈ 6 km s^{-1}). This ordering, in addition, matches the ordering by kinematical age (Tafalla 1993) and agrees intuitively with the expected ordering of YSO age (note however, mixture of low and high mass objects). The solution to the kinematical evolution of outflows may therefore arise from the combination of a constant slope at low velocities and a systematic decrease in the fraction of high velocity gas, which makes the break point in the mass distribution move with time toward lower and lower velocities.

Additional evidence for the slowing down and weakening of outflows with time comes from the mentioned fact that young, highly collimated outflows

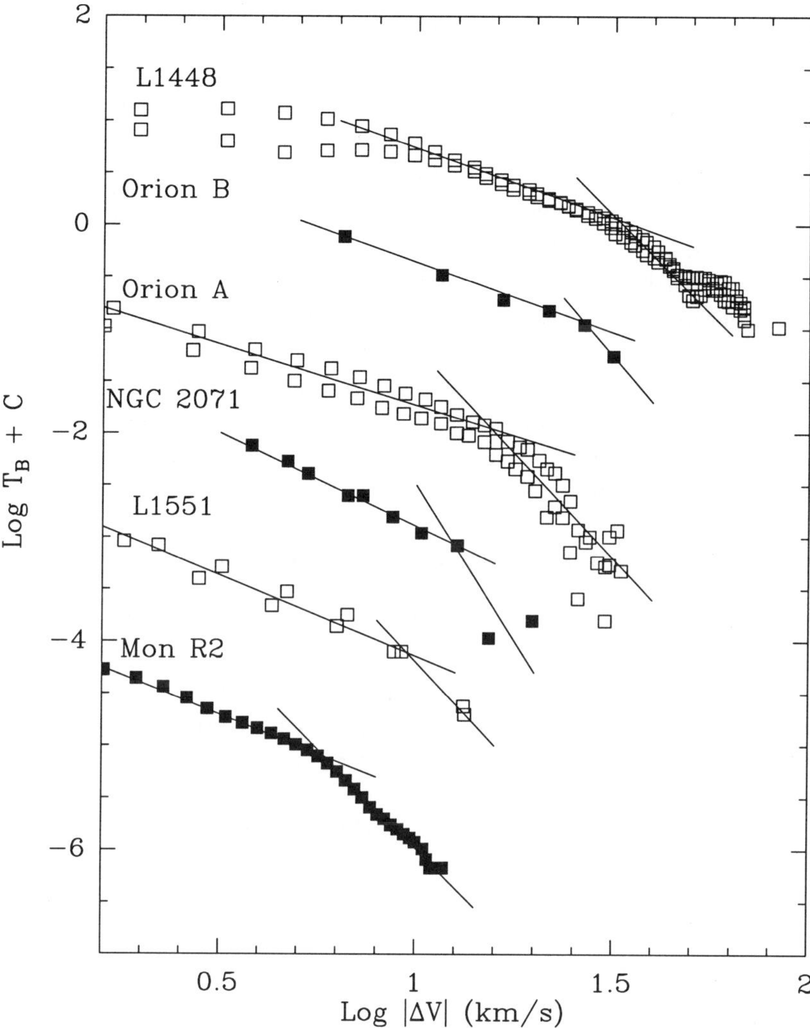

Figure 10. Log-log brightness distribution with velocity for a series of outflows of different ages. Note that the point where the distributions changes slope moves to lower velocities from top to bottom of the figure. This change seems associated with an increase of the outflow age. Data from different authors compiled by Tafalla (1993).

are more energetic than classical flows (Section 4.2). The work of Bontemps et al. (1996) shows that there is a very good correlation between the outflow momentum flux and the amount of circumstellar mass, and this is illustrated in Figure 11 (see also Saraceno et al. 1996). As circumstellar mass is a likely indicator of age (mass decreases as age increases), the correlation in the Figure suggests again that outflow power decreases with time.

5.3. CHANGES IN THE PHYSICAL AND CHEMICAL CONDITIONS OF THE OUTFLOW GAS

As was mentioned in Section 4 and will be discussed in the next section, high velocity outflows present evidence for warm molecular gas and spectacular chemical anomalies. These features can be explained as the result of strong shocks in the outflow gas, something that is not found, at least with that magnitude in classical flows. High velocity outflows, therefore, seem to interact more strongly with the surrounding environment, and most of the signatures of such an interaction (like chemical anomalies) have vanished by the time an outflow appears as a classical flow. Whether such a decrease in the interaction occurs because of the gradual clearing of the outflow path or because of an intrinsic weakening of the driving wind is not clear yet, but the strength of these shock signatures seems a good indicator of the outflow age (Codella et al. 1999).

5.4. AN EMPIRICAL TIME SEQUENCE OF LOW-MASS OUTFLOWS

After reviewing the fragmentary evidence for outflow evolution and discussing how observations suggest a transition between different stages, we conclude this Section proposing an empirical time sequence of outflows. Given that most observations of extremely young outflows (Class 0) are restricted to low mass systems, we will also restrict ourselves to this type of outflows. Our purpose here is to present a time sequence based on well known outflows, with the implicit assumption that differences between them are due to their different evolutionary status.

- First stage, L1448 (also IRAS03282+3035). The molecular outflow is highly collimated, almost jet-like, with a large fraction of extremely high velocity gas. Molecular "bullets" appear as secondary components in the spectra at extreme velocities and in the maps they look like high velocity clumps. SiO abundance is anomalous by a large factor ($> 10^4$) with respect to quiescent gas. The powering source is an extremely embedded YSO not visible in the near IR (Class 0).

- Second stage, L1157 (also BHR 71). The outflow appears less collimated and its terminal velocity is lower than in the phase before. There is no evidence for "bullets" in the spectra, but the maps show regions with large amounts of warm (~ 100 K) gas (hot spots). Very strong abundance anomalies: molecules like SiO and CH_3OH produce lines with prominent wing components, characteristic of high excitation and high molecular abundance. Exciting source still invisible in the NIR (Class 0).

- Third stage, L1551 (also NGC 2071). The systematic decrease of collimation and terminal velocity continues. Clear shell structure surrounding an evacuated cavity with a reflection nebula. Optical Herbig-Haro objects

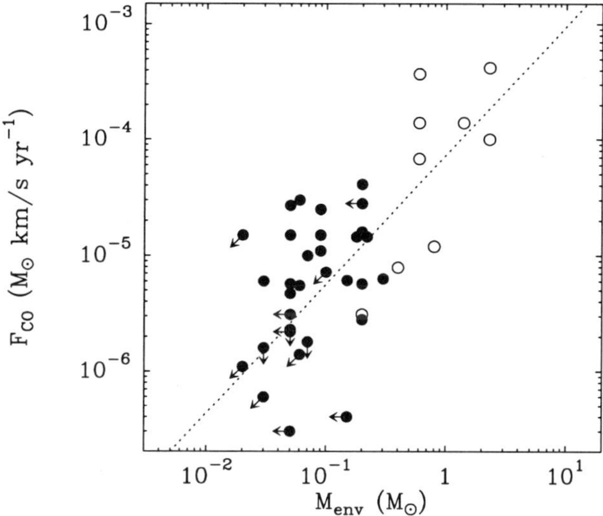

Figure 11. CO momentum flux F_{CO} versus circumstellar envelope mass M_{env} for 41 young stellar objects (Bontemps et al. 1996 data). Class 0 and Class I sources are represented as open and filled circles, respectively. The dashed line represents the "best fit" correlation for entire sample. Note that all Class 0 and Class I sources seem to follow the same correlation.

are visible. Large fraction of low velocity gas, which may appear in extreme cases forming detached wings in the CO spectra. Evidence for perturbation in the dense core and no chemical anomalies. Exciting source is Class I.

6. Shock Effects

Protostellar outflows are strongly supersonic, so they are expected to shock the surrounding cloud giving rise to rapid heating and chemical processing. In the more violent outflows, those from Class 0 protostars, these processes tend to be enhanced.

6.1. SHOCK HEATING

The molecular gas subjected to the passage of shocks is rapidly heated and then cools very rapidly (< 100 yr) due to line radiation (e.g. Kaufman & Neufeld 1996). Different spectral lines can dominate the cooling process, depending on the temperature regime, and such lines will indeed be the best tracers to diagnose the cooling region. The importance of the dust emission is uncertain. For instance, IRAS detected diffuse far-IR emission from the shocked region in L1551 to be 18% of the bolometric luminosity of the central source (Edwards et al. 1986; Clark & Laurejis 1986), but it is not possible to estimate the possible contribution from spectral lines (e.g. H_2, CO) only with the IRAS data.

At temperatures up to $\sim$5000 K, attained in regions subjected to dissociative J-shocks, the main coolant is the 63 μm line of [OI] (Hollenbach 1985). Observations with the KAO and ISO have allowed the measurement of this line in a handful of HH objects and outflows from Class 0 objects (Cohen et al. 1988, Ceccarelli et al. 1997, Saraceno et al. 1998). The [OI] line intensity, which is proportional to the mass flux into the J-shock, is correlated with the force of the CO flow F(CO) (Saraceno et al. 1998).

At temperatures around 2000 K the main coolants of the dense shocked regions are the vibrational near-IR lines of H_2 (Neufeld & Kaufman 1993). This temperature is typical of shocks with velocities 10–30 km s^{-1} (Shull & Beckwith 1982, Draine et al. 1983, Smith 1994). Indeed the v=1–0 S(1) line of H_2 at 2.122 μm is detected toward a high number of HH and CO outflows. The comprehensive survey by Hoddap (1994) shows that the H_2 emission adopts a variety of complex morphologies, with H_2 jet-like filaments and bow-shaped working surfaces being observed in a number of sources. The ratio of H_2 to the outflow power (estimated from CO) is observed to vary strongly from source to source (e.g. Davis & Eislöffel 1995), probably because of the combination of several factors such as extinction, H_2 emission decline due to cooling, and outflow power time variations.

At temperatures from 1000 K down to a few 100 K, the pure rotational lines of H_2 in the mid-IR, and the high-J (up to J$\sim$30) lines of CO in the far-IR are observed to be strong (e.g. Cabrit et al. 1998, Nisini et al. 1998). These temperatures are typical of slow shocks with velocities 10–25 km s^{-1} and densities around 10^5 cm^{-3}.

Molecular lines of shock-enhanced species (e.g. NH_3, CH_3OH) trace chemically processed gas at temperatures as low as 60– 100 K (Bachiller et al. 1993, 1995; Tafalla & Bachiller 1995), which is nevertheless significantly hotter than the ambient gas (usually at 10– 20 K). This temperature regime seems to be particularly interesting for the chemical studies which are carried out at mm wavelengths.

6.2. SHOCK CHEMISTRY

The heating and compression caused by shocks often gives rise to dramatic effects in the chemical composition of the surrounding molecular cloud. Processes such as dissociation, endothermic reactions, sublimation of ices, and disruption of dust grains lead to a "shock-chemistry" which is radically different from that operating in the quiescent medium. Some of these processes are fast, and the cooling times of the shocked region are short, making the chemistry strongly time dependent.

Dissociative J-shocks destroy molecules in the resulting hot ($\sim10^5$ K) layer of shocked gas, but molecules can reform when the gas cools, or within the warm region around the hot layer. Such violent shocks can also destroy dust grains and produce thermal sputtering, resulting in the injection of refractory elements (e.g. Si and Fe) into the gas phase.

C-shocks, on the other hand, are particularly efficient at triggering a rich molecular chemistry, since they increase the temperature to moderate values ($\sim$2000 K) in a thick layer where the molecules are preserved. Reactions that overcome energy barriers of a few 1000 K can proceed in such conditions, and in particular the very chemically active OH radical is formed by $O+H_2 \rightarrow OH+H$ (energy barrier of 3160 K). OH contributes to the formation of water by further reaction with H_2: $OH+H_2 \rightarrow H_2O+H$ (energy barrier of 1660 K). HS and H_2S are formed by similar reactions. Additionally, in C-shocks, non-thermal sputtering due to dust-molecule or to dust-dust collisions (Schilke et al. 1997, Caselli et al. 1997, Pineau des Forêts et al. 1997) results in the injection of refractory and volatile species, mainly from grain mantles, into the gas phase. The entrance of this material together with the presence of a high abundance of OH in the gas phase will produce oxides such as SO and SiO (see e.g. Bachiller 1997, van Dishoeck & Blake 1998, and references therein).

As the gas cools, reactions with high energy barriers will not be operative, and the chemistry will be dominated again by the usual low-temperature reactions. Depletion of molecules onto dust grains starts again to be important, reducing the abundances of some of the newly formed molecules (e.g. see the case of water, Bergin et al. 1998). Nevertheless, the chemical composition of both the gas and the solid phases will remain altered, keeping a memory of the passage of the shock wave.

Recent observations with high angular resolution and sensitivity are providing important insights in the shock-chemistry phenomena. Observations of the IRc2 outflow in Orion A have provided valuable information about the chemical processes induced in the surrounding medium by the formation of a high-mass star (Blake et al. 1987). But the cases of high-mass stars are particularly complex due to the confusion among different gas components along the line of sight.

The chemical effects of shocks are better studied in outflows from Class 0 sources because these are particularly energetic and highly collimated, so they contain shocked regions well separated spatially from the quiescent protostellar envelope. Recent studies of this type of outflows include NGC1333/IRAS4 (Blake et al. 1995), NGC1333/IRAS2 (Blake 1996, Langer et al. 1996, Bachiller et al. 1998), L1641-N (McMullin et al. 1994), HH25-MMS (Gibb & Davis 1998), and BHR-71 (Bourke et al. 1997; Garay et al. 1998). IRAS16293-2422 was studied in detail by Blake et al. (1994) and van Dishoeck et al. (1995), but only toward the position of the driving object.

6.3. OBSERVATIONS OF SHOCK CHEMISTRY IN L1157

A comprehensive chemical study including many different species has been recently carried out on the outflow from the Class 0 source L1157-mm (Bachiller & Pérez Gutierrez 1997). Due to its proximity (440 pc), high collimation, favorable orientation in the sky, and high column densities in the shocked region, L1157 appears as one of the best objects to study the rich phenomenology associated with young bipolar outflows.

The CO outflow in L 1157 (Umemoto et al. 1992; Bachiller & Pérez Gutiérrez 1997, Fig. 12) is driven by IRAS 20386+6751 (L1157-mm), a very cold source of about 11 $L_\odot$ embedded in a dense NH_3 core (Bachiller et al. 1993). The central source presents all the attributes of a Class 0 source, including a relatively strong millimeter flux due to thermal dust emission (Gueth et al. 1997). The mm-wave continuum emission consists of a compact ($\leq 1''$) source surrounded by spatially-extended, low-level emission which arises from the heated edges of the cavity excavated by the outflow. High-angular resolution images in the J=1–0 line of CO, ^{13}CO, and $C^{18}O$ (Gueth et al. 1997) show that the $C^{18}O$ emission is associated with the protostellar condensation, whereas the ^{13}CO line emission mostly originates from a extended envelope and the limb-brightened edges of the outflow; the CO line wings trace the bulk of the outflowing material. Possible gravitational infall is suggested by the self-reversed ^{13}CO profiles and the narrow $C^{18}O$ lines.

The sharp increase of the CO brightness temperature towards the blue shifted lobe in L 1157 (Fig. 12) is attributed to gas heating produced by the shock. Multiline observations of ammonia, the best interstellar thermometer, confirm this interpretation, and provide a good estimate of the gas kinetic temperature in the shock, which results to be in the range 60– 100 K (Bachiller et al. 1993). The bow shocks at the head of the cavities are particularly bright in NH_3 (3,3), and VLA images from Tafalla & Bachiller (1995) reveal the hot regions of the bow shock structures seen in the high-velocity CO. The ammonia abundance can be estimated in both the quiescent gas and in the high-velocity outflow, and it results to be enhanced by at least a factor of 10 in the shocked region.

Bachiller & Pérez Gutiérrez (1997) have carried out a survey of the blue shifted lobe in L 1157 to study the chemical composition of the gas affected by the passage of the shock. About 50 different lines of 27 molecules (including rare isotopic species) were observed toward the central source and toward several positions with evidence for shock, in particular the two prominent NH_3 peaks of the blue shifted lobe: B1 at $(20'', -60'')$, and B2 at $(35'', -95'')$.

An immediate result from these observations is that some molecular species such as C_3H_2, N_2H^+, $H^{13}CO^+$, and DCO^+ are only observed toward the cold gas condensation around the exciting source, whereas some other

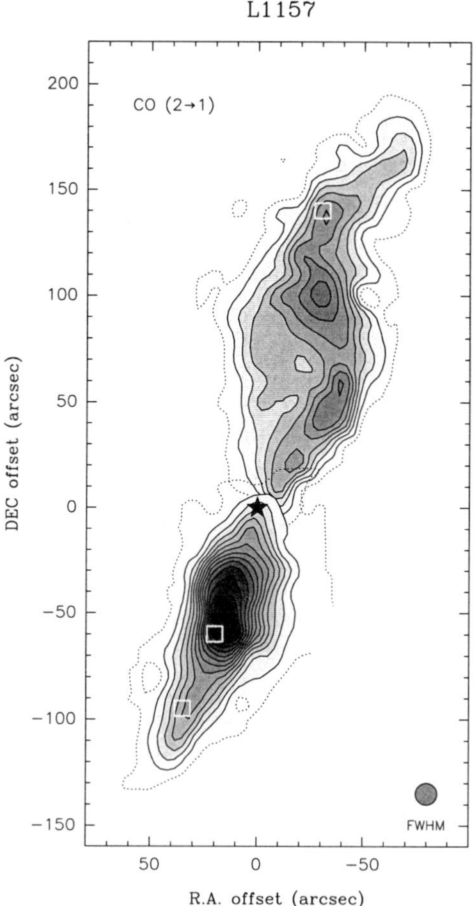

Figure 12. CO 2–1 emission from the L1157 outflow integrated over velocity intervals from −20 to 3 km s^{-1} (blue shifted gas to the south of the driving source) and from 3 to 30 km s^{-1} (red shifted gas toward the north). Position offsets are in arcsec with respect L1157-mm, the outflow exciting source. First contour and step are 11 K km s^{-1}. The squares mark the positions of SiO emission peaks in which Bachiller & Pérez Gutiérrez (1997) carried out a survey of molecular lines.

molecules (such as SiO) only trace the warm gas in the shock. This result is important for studies of the structure of the central infalling condensation, since one can avoid the outflow contribution to the line profiles by observing the source vicinity in lines of C_3H_2, N_2H^+, $H^{13}CO^+$, and DCO^+. This does not mean that the outflowing gas is fully free of these molecules, but that the column densities are insufficient to produce significant wings in the profiles.

All the other detected molecules (SiO, CH_3OH, H_2CO, HCO^+, HCN, HNC, CN, SO, SO_2, CS) present emission in the outflow. The extreme case is SiO which is enhanced by a factor of $\sim 10^6$ (see also Mikami et al. 1992,

Zhang et al. 1996, Avery & Chiao 1996). This is similar to what it is observed at the heads and along the axes of some other highly-collimated outflows (Bachiller et al. 1991a; Martín-Pintado et al. 1992; McMullin et al. 1994). The other molecular species with broad emission are observed to be enhanced by factors ranging from a few to a few hundred.

Figure 13 shows molecular line maps of the blueshifted lobe of L1157. There are important differences in the spatial distribution among the different species: some molecules such as HCO^+ and CN present their maxima close to the central source, while SO and SO_2 are brightest (in peak intensity) around the most distant shocks, with OCS presenting the most distant peak. Other molecules such as SiO, CS, CH_3OH, and H_2CO present an intermediate behavior. Such differences cannot be attributed solely to excitation conditions: an important gradient in chemical composition is observed along the outflow. It is very likely that this gradient is related to the strong time dependence of shock chemistry.

As an example of time dependence, consider the chemistry of SO, SO_2, and OCS, which has been recently modeled by Charnley (1997). It is believed that sulfur is released from grains in the form of H_2S, and that it is then oxidized to SO and SO_2 in a few 10^3 yr; the formation of OCS needs a few 10^4 yr. This is in agreement with the observations of L 1157, since the SO/H_2S and SO_2/H_2S ratios do increase with distance from the source (i.e. with time), and OCS emission in only observed in the most distant position (i.e the oldest shock).

Methanol (CH_3OH), with an abundance of a few percent, is thought to be the most abundant molecule in ices after water (Allamandola et al. 1992). Indeed, millimeter-wave observations of L 1157 show that the abundance of methanol is enhanced by a factor of 400 in the shock (Bachiller et al. 1995, Avery & Chiao 1996). There is increasing evidence that high abundances of methanol in gas phase are able to drive a chemistry completely different from the usual chemistry operating in dark clouds. In particular, laboratory experiments show that a significant fraction of the methanol released into the gas phase could produce H_2CO in a relatively short time (e.g. Bernstein et al. 1995). Solid formaldehyde (H_2CO) has also been recently discovered in the vicinity of protostars (Schutte et al. 1996), and our observations of L 1157 indicate that H_2CO in enhanced by a factor close to 100 in the shocked region. The chemistry of formaldehyde is probably closely related to that of methanol in both solid and gas phases.

Chemical studies provide a powerful tool to investigate the physical conditions of the shocked gas. As discussed above, NH_3 provides good estimates of the kinetic temperature. Once the temperature is known, multi line studies of SiO and CH_3OH provide reliable estimates of the volume densities. In the case of the L 1157 bow shocks, the density is found to be in excess of 10^6

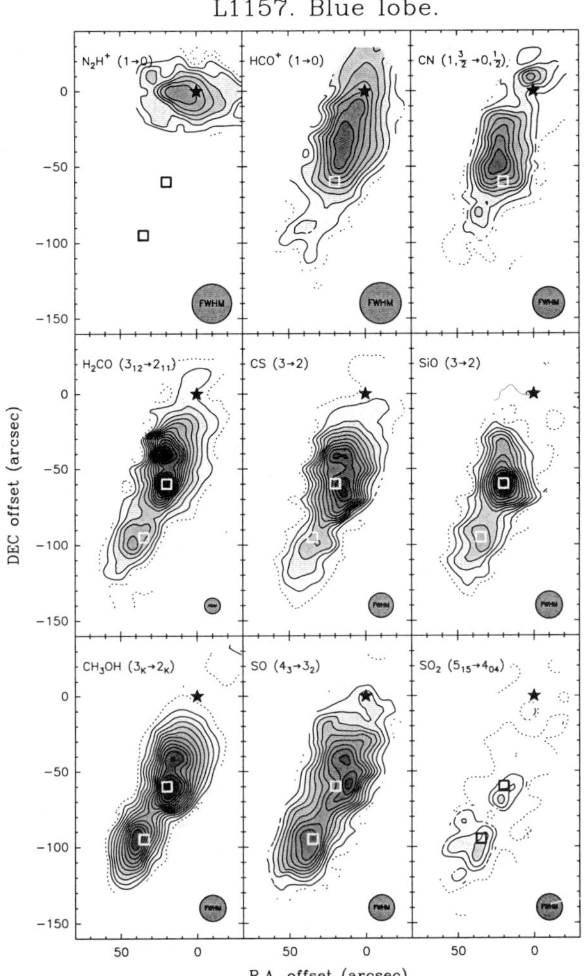

Figure 13. Maps of integrated intensity for different molecular lines in the blue shifted lobe of the L 1157 outflow obtained with the IRAM 30-m telescope. The white squares indicate the positions of the SiO emission peaks. Note the large differences observed in the distribution of the different molecules (see discussion in the text).

cm^{-3} (Bachiller et al. 1993, 1995).

7. Properties of the Primary Wind

If radiation pressure from the central YSO lacks enough momentum to drive a molecular outflow (Bally & Lada 1983, see also Section 3), a mechanical agent like a stellar wind seems the most likely responsible for this phenomenon. Despite many years of both observational and theoretical work, however, basic characteristics of this agent, like its composition and collimation, are still subject to strong debate, and different theories for the launching mechanism have been proposed. In this section, we review the status of this important topic, discussing both observational attempts to detect the outflow agent and theoretical models of outflow acceleration.

7.1. OBSERVATIONS

The driving wind could be either neutral or ionized, and if neutral, it could be either molecular or atomic. Depending on this state, its detection could be done at radio, infrared, or optical wavelengths.

If the driving agent is molecular, it could probably be detected using the same tracers used to detect molecular clouds: line emission from abundant species like CO. However, the molecular emission from an outflow comes mostly from swept-up ambient material (Section 3), so any possible driving wind contribution should be separated from the accelerated ambient component. As the driving wind should move faster than the swept up ambient gas, there is the possibility that the EHV component seen in highly collimated outflows represents the driving agent. If this were the case, however, classical outflows, which lack EHV component, would be either driven by an additional (invisible) agent, or they would have an undetected EHV component. Deep searches for EHV gas towards classical outflows have been carried out (Margulis & Snell 1989; Choi, Evans, & Jaffe 1993), but they have tended to concentrate on high mass outflows. For the case of L1551 (see Koo 1989), no clear evidence for a EHV component, despite the strong outflow signal at low velocities. It seems, therefore, that the EHV gas, although closely connected to the driving wind (may be recently accelerated), is not the agent of outflow acceleration. Deeper searches for a molecular driving wind component should be made to further explore this possibility.

If the driving wind is atomic, 21-cm HI emission could reveal its presence. In fact, HI observations have allowed the detection of a neutral component towards a few outflows, like NGC 2071 (Bally & Stark 1983), HH7-11 (Lizano et al. 1988, Rodríguez et al. 1990), and L1551 (Lizano et al. 1988, Giovanardi et al. 1992). The origin of this atomic component, however, has been interpreted differently by different authors: Bally & Snell (1983) propose it arises from the dissociation of ambient molecular hydrogen due to outflow shocks, while Lizano et al. (1988) propose the HI emission traces a wide-angle neutral wind from the central protostar. Unfortunately, the extreme weakness of the HI line, together with the lack of spatial resolution at low frequencies, makes it very difficult to fully characterize this component. If it represents a truly driving wind, the data from L1551 and HH7-11 suggest velocities of about 200 $km\,s^{-1}$ and YSO mass loss rates of several 10^{-6} $M_\odot$ yr^{-1}

(Giovanardi et al. 1992)

For the case of an ionized driving wind, a component of potential interest is that detected at cm wavelengths (see recent review by Rodríguez 1997). The spectral index of this emission is usually consistent with models of free-free emission from an ionized wind, and in cases where the emission has been spatially resolved, it is elongated along the outflow direction (e.g., Anglada 1995). The momentum content of this component, however, is typically one or two orders of magnitude lower than the momentum in molecular outflows (Snell & Bally 1986), so it cannot by itself drive molecular flows. This component probably represents gas shock ionized by the outflow driving wind, so it constitutes a consequence, not a cause, of the bipolar outflow.

A final candidate for the outflow driving wind is the gas seen at optical and IR wavelengths in the form of Herbig-Haro objects and jets (see chapter by Reipurth & Raga in this volume for a more comprehensive discussion). As in the cm-wave emitting component, the momentum in the optically visible component is much too small to power molecular outflows (Mundt, Brugel, & Buhrke 1987), although this component is probably accompanied by a large amount of neutral gas, as jets seem lightly (few percent) ionized (Raga et al. 1990, Hartigan et al. 1994). This neutral component (invisible jet) has enough energy and momentum to drive in principle a weak molecular outflow (Hartigan et al. 1994), so it cannot be ruled as the outflow driving wind. Unfortunately, the lack of direct observability of this neutral component, together with the short radiative time of the HH objects and jets, makes it only possible to study the location of the most recent interaction, and the optical emission does not provide a complete history of the wind. The recent identification of large scale (super) jets (Bally & Devine 1997; Reipurth et al. 1997; see also the chapter by Reipurth & Raga in this volume) may help to circumvent this problem.

In summary, we do not have a conclusive direct observation of the primary agent which drives molecular outflows. Jets with large amounts of energy and momentum are present, and they seem to be largely neutral, but we cannot rule out that there exists a wide angle atomic component coexisting with the neutral jet and contributing to the outflow acceleration. Determining the nature of this agent, its collimation and velocity, remains the most critical problem in outflow research.

7.2. MODELS

It is not possible to make justice in the little space of this section to the enormous theoretical effort dedicated to modeling the outflow phenomenon and its many features. Here we will limit ourselves to reviewing a small sample of this work, distinguishing two types of models: those which attempt to explain the molecular gas acceleration starting with a given driving agent (jet, wide angle wind), and those which attempt to explain the outflow launching by the central star/disk system. Models from the first type are necessarily incomplete, as their goal is to *reproduce* outflow observations, leaving unanswered the question of how the primary wind is produced in first place. However, they permit an easy comparison with observations and therefore allow one to infer properties of the driving wind (collimation, velocity, etc). Models from the second type attempt to *produce* the outflow from first principles, and their goal is to offer the complete picture of the outflow phenomenon, from the launching of the wind to the acceleration the ambient cloud. The complexity of the whole process makes them harder to compare with observational data.

7.2.1. *Models of Molecular Gas Acceleration*

We first discuss models that start with some form of wind and study its propagation through the cloud, giving rise to a molecular outflow. For further details, we refer the reader to the excellent and up-to-date comparison between models and observations by Cabrit, Raga, & Gueth (1997). Models can be divided according to the agent collimation in jets and wide angle winds, a dichotomy that reflects the distinction between classical and highly collimated outflows discussed in the previous sections. Not surprisingly, each type of model reproduces well one of the two types of outflows, but has more difficulty explaining the other. Achieving a model which can reproduce both outflow types and can explain the transition between them remains an unfulfilled challenge.

Wide angle winds.   Among wide angle models, the one by Shu et al. (1991) is probably the best studied. In this model, a YSO is embedded in a dense core with a density distribution that radially decreases as r^{-2} (Shu 1977) and has a flattening term $Q(\theta)$, where θ is the polar coordinate. A wind from the YSO emerges with an angular distribution which depends on the same polar angle θ through a function $P(\theta)$. As the wind propagates, it sweeps ambient medium forming a thin shell which constitutes the molecular outflow. Momentum conservation in every direction θ gives rise to a velocity law that depends angularly as $\sqrt{P(\theta)/Q(\theta)}$, which therefore defines the shape of the shell.

In their original paper, Shu et al. (1991) favored smooth forms for both $P(\theta)$ and $Q(\theta)$ at all angles, in order to reproduce the shapes of known outflows. Masson & Chernin (1992), however, pointed out that such choices of P and Q give rise to distributions of mass with velocity (and therefore to spectral profiles) that have too much gas at high velocities, in contradiction with observations. In a more recent elaboration of the wide-angle wind model,

Li & Shu (1996) favor density distributions with little gas along the polar axis ("toroids") and winds with a stronger axial momentum component, avoiding the problem of the mass distribution while still producing reasonably-looking shells. Although the Li & Shu (1996) model seems to reproduce the main characteristics of classical outflows, it is not clear whether it can do the same with highly collimated molecular flows like HH211 (it does reproduce optical images from jets, see Shang et al. 1998). Also, if outflows evolve self similarly (Shu et al. 1991), it seems difficult to explain both the classical and highly collimated outflows. Still, it is possible that further refinements of this model (including time variability, for example) can reproduce these and other aspects, like EHV bullets.

Jet models. Jet-driven outflow models have attracted increasing interest after the discovery of the highly collimated outflows in the early 90's. As one could expect, this family of models reproduces best the highly collimated flows, but has more difficulty explaining classical outflows. Masson & Chernin (1993) have nicely discussed the basic physics involved in the driving of a molecular outflow by a jet. They argue that for molecular cloud densities ($n \sim 10^3 - 10^4$ cm^{-3}) and typical jet velocities (~ 300 km s^{-1}), the jet-cloud interaction is highly dissipative and a narrow bow shock forms at the jet end; the swept up material is accelerated mostly forward. Parameterizing the complicated bow shock physics in simple terms and adjusting the parameters from HH jet observations, Masson & Chernin (1993) construct a jet driven outflow model which they compare with observations. They succeed reproducing several observed features like the forward momentum, "Hubble-law" behavior, and distribution of mass with velocity, but their model produces jets that are too long and thin if they are allowed to evolve for several 10^5 yr (expected outflow lifetime, Parker et al. 1991). Masson & Chernin argue that this problem can be circumvented if the jet is allowed to wonder in angle, so it accelerates different pieces of gas with time, increasing the lobes width but limiting their length.

Another analytical model to reproduce a jet driven outflow is due to Raga & Cabrit (1993), who consider the situation of an internal jet working surface caused by time dependence in the flow velocity field. In this model, a bow shock compresses and sweeps ambient material, which later re-expands into the cavity left behind, forming a turbulent wake identified with the molecular flow. With this model, and assuming the size of the bow shock increases with distance (due to the cloud density decreasing away from the YSO), Raga & Cabrit reasonably reproduce observed shapes of outflows. Their distribution of mass decreases with velocity, as observed in molecular outflows, although the decrease seems too rapid, producing triangular spectra.

In addition to the above analytical models (and others due to Stahler 1993, Wilkin 1996, Zhang & Zheng 1996), numerical simulations of the propagation of a jet through a cloud have been presented by a number of authors (e.g. Chernin et al. 1994, Stone & Norman 1994, Raga et al. 1995, Frank & Mellema 1996, Smith et al. 1997, Suttner et al. 1997). Unfortunately, it is difficult to follow the evolution of a high velocity jet moving at several hundred $km\,s^{-1}$ over the time scales typical of outflows (several 10^5 yr), so only the early stages of the outflow production can be investigated in this way. Hopefully, longer calculations will be possible in the next future, opening the possibility of comparing numerical simulations with CO observations of mature outflows.

7.2.2. *Wind Launching Models*

Although a variety of theories was proposed soon after the discovery of bipolar outflows (e.g., Barral & Cantó 1981, Königl 1982, Torbett 1984), only the family of magnetohydrodynamic (MHD) models seems to have survived the test of time. As pointed out by De Campli (1981), thermal expansion and line radiation pressure are not efficient enough driving the winds observed in some TTauri stars, which are most likely weaker than those at the origin of bipolar flows. MHD models seem capable to do the job and to meet the energy and momentum requirements by converting a fraction of the gravitational energy released by mass accretion into outflow motion. This is achieved by the coupling of gas rotation with the magnetic field, in a mechanism analogous to the behavior of a bead on a rotation, rigid wire (Henriksen & Rayburn 1971). In this situation, if the magnetic field is coupled to the rotating matter (through flux freezing) and if it makes a sufficiently large angle with the totation axis ($> 30°$), corrotation accelerates and ejects the material along the field lines. Achieving a geometry where the magnetic field makes such an angle with the rotation vector is the crucial point in launching the outflow.

The above situation can be obtained in at least two configurations. In one, the outflow arises from a disk which is threaded by an inclined (concave) magnetic field (Blandford & Payne 1982). In the other, the outflow arises from the surface of the star, taking advantage of the natural curvature of the stellar surface near the equator (Belcher & MacGregor 1976).

The original disk outflow models (Pudritz & Norman 1983, Pudritz 1985, Uchida & Shibata 1985) required very massive disks (several $M_\odot$), which are not supported by observations. More modern versions of this model (Königl 1989, 1995, Lovelace et al. 1991, Ferreira & Pelletier 1993) use much more reasonable parameters. One nice feature of a wind model is that the wind can carry away part of the angular momentum, allowing disk accretion without the need for an anomalous viscosity (Pudritz 1985). The main criticism to this model (Shu et al. 1987) is the difficulty of producing a strong bend in the magnetic field, necessary to launch the wind, against the natural trend of the magnetic lines to straighten up.

The alternative situation was proposed by Shu et al. (1988) (see also Hartmann & MacGregor 1982), with the outflow being launched from the stellar surface, near the equator. This so-called X-wind mechanism is explained with detail in the chapter by Shu in this book. In this model, a magnetic field, emerging almost perpendicular to the star surface, makes the adequate angle to centrifugally drive a wind. For this to happen, however, the star should be rotating near break up velocity, which is much faster than observed in TTauri stars (Bouvier et al. 1986, Hartmann et al. 1986). A more recent elaboration of this model (Shu et al. 1994a, b; Najita & Shu 1994) avoids this problem by having the strong stellar magnetic field truncate the disk at some distance from the stellar surface. This point, located where the disk material corrotates with a slower rotating star, provides a strong lever arm for launching energetic winds; both the bipolar outflow and an infall funnel flow emerge from this X-radius.

Acknowledgements. It is a pleasure to acknowledge Drs. André, Bontemps, Cabrit, Cernicharo, Codella, Colomer, Dutrey, Fuente, Gueth, Guilloteau, Martín-Pintado, Myers, Pérez Gutiérrez, Planesas, Reipurth, Richer, Shu, and Walmsley for interesting discussions and in many cases for fruitful collaborative work on outflows over the past several years. Funding support from Spanish DGES grant PB96-104 is also gratefully acknowledged.

References

Adams, F.C., Lada, C.J., Shu, F.H. 1987. ApJ 312, 788

Allamandola, L.J., Sandford, S.A., Tielens, A.G.G.M., Herbst, T. 1992, ApJ 399, 134

André, P. 1997 in *IAU Symposium No. 182*, eds. B. Reipurth and C. Bertout (Kluwer Academic Publishers), p. 483

André, P., Martín-Pintado, J., Despois, D., Montmerle, T. 1990, A&A 236, 180

André, P., Ward-Thompson, D., Barsony. M. 1993, ApJ 406, 122

Anglada, G. 1995, Rev. Mex. Astron. Astrofís. Ser. de Conf. 1, 67

Anglada, G., Estalella, R., Rodríguez, L.F., Torrelles, J.M, López, R., Cantó, J. 1991, ApJ 376, 615

Avery, L.W., Hayashi, S.S., White, G.J. 1990, ApJ 357, 524

Avery, L.W., Chiao, M. 1996, ApJ 463, 642

Bachiller, R. 1996, ARAA 34, 111

Bachiller, R., Cernicharo, J. 1990 A&A 239, 276

Bachiller, R., Gómez-González, J. 1992 A&A Review 3, 257

Bachiller, R., Cernicharo, J., Martín-Pintado, J., Tafalla, M., Lazareff, B. 1990 A&A 231, 174

Bachiller, R., Codella, C., Colomer, F., Liechti, S., Walmsley, C.M. 1998, A&A 335, 266

Bachiller, R., Fuente, A., Tafalla, M. 1995a, ApJ 445, L51

Bachiller, R., Guilloteau, S., Dutrey, A., Planesas, P., Martín-Pintado, J. 1995, A&A 299, 857

Bachiller, R., Guilloteau, S., Gueth, F., Tafalla, M., Dutrey, A., Codella, C., Castets, A.

1998, A&A, in press

Bachiller, R., Liechti, S., Walmsley, C.M., Colomer, F. 1995, A&A 295, L51

Bachiller, R., Martín-Pintado, J., Fuente, A. 1991a, A&A 243, L21

Bachiller, R., Martín-Pintado, J., Fuente, A. 1993, ApJ 417, L45

Bachiller, R., Martín-Pintado, J., Planesas, P. 1991b, A&A 251, 639

Bachiller, R., Pérez Gutiérrez, M. 1997, ApJ 487, L93

Bachiller, R., Tafalla, M., Cernicharo, J. 1994, ApJ 425, L93

Bally, J. 1982, ApJ 262, 558

Bally, J. 1986, Can. J. Phys. 64, 383

Bally, J., Lada C.J. 1983. ApJ 265, 824

Bally, J., Lane A.P. 1991, in *The Physics of Star Formation and Early Stellar Evolution*, ed. C.J. Lada and N.D. Kylafis (Kluwer, Dordrecht), pag. 471

Bally, J., Stark, A.A. 1983, ApJ 266, L61

Bally, J., Devine, D. 1997, in IAU Symp. 182 *Herbig-Haro Flows and the Birth of Low Mass Stars* (Eds. B. Reipurth & C. Bertout). Kluwer, Dordrecht, pag. 29

Barral, P., Cantó, J. 1981 Rev. Mex. Astron. Astrofís. 5, 101

Belcher, J. W., MacGregor, K.B. 1976, ApJ 210, 498

Bence, S.J, Padman, R., Isaak, K.G., Wiedner, M.C., Wright, G.S. 1998, MNRAS 299, 965

Bergin, E.A., Neufeld, D. A., Melnick, G.J. 1998, ApJ 499, 777

Bernstein, M.P., Sandford, S.A., Allamandola, L.J., Chang, S., Scharberg, M.A. 1995, ApJ 454, 327

Blake, G.A. 1996, in IAU Symp. 178, Molecules in Astrophysics: Probes and Processes (ed. E.F. van Dishoeck). Kluwer, Dordrecht, pag. 31

Blake, G.A., Sutton, E.C., Masson, C.R., Phillips, T.G. 1987, ApJ 315, 621

Blake, G.A., van Dishoeck, E.F., Jansen, D.J., Groesbeck, T.D., Mundy, L.G. 1994, ApJ 428, 680

Blake, G.A., Sandell, G., van Dishoeck E.F., Groesbeck, T.D., Mundy, L.G., Aspin, C. 1995, ApJ 441, 689

Blandford, R.D., Payne, D.G. 1982, MNRAS 199, 883

Bontemps, S., André, P., Terebey, S., Cabrit, S. 1996. A&A 297, 98

Bouvier, J., Bertout, C., Benz, W., Mayor, M. 1986, A&A 165, 110

Bourke, T.L, Garay, G., Lehtinen, K.K., Koehnenkamp, I., Launhardt, R., Nyman, L.A., May, J., Robinson, G., Hyland, A.R. 1997, ApJ 476, 781

Cabrit, S., Bertout, C. 1986. ApJ 307, 313

Cabrit, S., Bertout, C. 1990. ApJ 348, 530

Cabrit, S., Bertout, C. 1992. A&A 261, 274

Cabrit, S., Goldsmith, P.F., Snell, R.L. 1988. ApJ 334, 196

Cabrit, S., Raga, A.C., Gueth, F. 1997, in IAU Symp. 182 *Herbig-Haro Flows and the Birth of Low Mass Stars* (Eds. B. Reipurth & C. Bertout). Kluwer, Dordrecht, pag. 163

Cabrit, S., Couturier, Andre, P., Boulade, O., Cesarsky, C. J. and Lagage, P. O., Sauvage, M., Bontemps, S., Nordh, L. and Olofsson, G., Boulanger, F., Sibille, F., Siebenmorgen, R. 1998. In *ASP Conf. Ser. 132: Star Formation with the Infrared Space Observatory*, eds. Yun, J. and Liseau, R. (ASP), pag. 326

Cantó, J., Rodriguez, L.F., Barral, J.F., Carral, P. 1981, Apj 244, 102

Caselli, P., Hartquist, T.W., Havnes, O. 1997, A&A 322, 296

Ceccarelli, C., Haas, M. R., Hollenbach, D. J., Rudolph, A. L. 1997. ApJ 476, 771

Cernicharo, J., Reipurth, B.1996. ApJ 460, L57

Charnley, S.B. 1997, ApJ 481, 396

Chernin, L.M., Masson, C.R., Fuller, G. 1994, ApJ 436, 741

Choi, M., Evans, N.J. II, Jaffe, D.T. 1993, ApJ 417, 624

Clark, F.O., Laurejis, R.J. 1986, A&A 154, L26

Codella, C., Bachiller, R., Reipurth, B. 1999, A&A, in press

Cohen, M., Bieging, J.H., Schwartz, P.R. 1982, ApJ 253, 707

Cohen, M., Hollenbach, D. J., Haas, M. R. and Erickson, E. F. 1988, ApJ 329, 863
Cudworth, K.M., Herbig, G.H. 1979, Astron. J. 84, 548
Davis, J.J., Eisloeffel, J. 1995, A&A 300, 851
De Campli, W.M. 1981, ApJ 244, 124
Draine, B.T., Roberge, W.G., Dalgarno, A. 1983, ApJ 264, 485
Dutrey, A., Guilloteau, S., Bachiller, R., 1997 A&A 325, 758
Edwards, S., Snell, R.L. 1982, ApJ 261, 151
Edwards, S., Snell, R.L. 1983, ApJ 270, 605
Edwards, S., Snell, R.L. 1984, ApJ 281, 237
Edwards, S., Strom, S.E., Snell, R.L., Jarrett, T.H., Beichman, C.A., Strom, K.M. 1986,
    ApJ 307, L65
Frank, A., Mellema, G. 1996, ApJ 472, 684
Ferreira, J., Pelletier, G. 1993, A&A 276, 625
Fukui, Y., Iwata, T., Mizuno, A., Bally, J., Lane, A.P. 1993, in *Protostars and Planets
    III*, ed. EH Levy, JI Lunine. Tucson: Univ. Ariz. Press
Garay, G., Köhnenkamp, I., Bourke, T.L., Rodríguez, L.F., Lehtinen, K.K. 1998, ApJ, in
    press
Garden, R.P., Hayashi, M., Gatley, I., Hasegawa, T., Kaifu, N. 1991, ApJ 374, 540
Gibb, A.G., Davis, C.J., 1998, MNRAS 298, 644
Giovanardi, C., Lizano, S., Natta, A., Evans, N.J. II, Heiles, C. 1992, ApJ 397, 214
Graham, J.A., Heyer, M.H. 1990, PASP 102, 972
Gueth, F., Guilloteau, S. 1999, A&A, in press
Gueth, F., Guilloteau, S., Bachiller, R. 1996, A&A 307, 891
Gueth, F., Guilloteau, S., Dutrey, A., Bachiller, R. 1997, A&A 323, 943
Guilloteau, S., Bachiller, R., Lucas, R., Fuente, A. 1992, A&A 265, L49
Haro, G. 1952, ApJ 115, 572
Hartigan, P., Morse, J. and Raymond, J. 1994, ApJ 436, 125
Hartmann, L., Hewett, R., Stahler, S., Mathieu, R.D. 1986, ApJ 309, 275
Hartmann, L., MacGregor, K.B. 1982, ApJ 259, 180
Henriksen, R.N., Rayburn, D.R. 1971, MNRAS 152, 152
Herbig, G. 1951, ApJ 113, 697
Herbig, G. 1962, Avd. A&A 1, 47
Hodapp, K.W. 1994, ApJSS 94, 615
Hodapp, K.W., Ladd, E.F. 1995, ApJ 453, 715
Hollenbach, D. 1985, Icarus 61, 36
Kaufman, M. J., Neufeld, D.A. 1996, ApJ 456, 611
Kitamura, Y., Kawabe, R., Yamashita, T., Hayashi, M. 1990, ApJ 363, 180
Kitamura, Y., Kawabe, R., Ishiguro, M. 1992, PASJ 44, 407
Königl, A. 1982, ApJ 261, 115
Königl, A. 1989, ApJ 342, 208
Königl, A. 1995, Rev. Mex. Astron. Astrofís. (Ser. de Conf.) 1, 275
Koo, B.-C. 1989, ApJ 337, 318
Koo, B.-C. 1990, ApJ 361, 145
Kuhi, L.V. 1964 ApJ 140, 1409
Kuiper, T.B.H., Zuckerman, B., Rodríguez-Kuiper, E.N. 1981, ApJ 251, 88
Kwan, J., Scoville, N. 1976, ApJ 210, L39
Lada, C.J. 1985, ARAA 23, 267
Lada, C.J. 1991, in *The Physics of Star Formation and Early Evolution*, eds. CJ Lada,
    ND Kylafis. Kluwer: Dordrecht
Lada, C.J., Fich, M. 1996, ApJ 459, 638
Langer, W.D., Castets, A., Lefloch, B. 1996, ApJ 471, 111
Levreault, R.M. 1988, ApJSS 67, 283
Li, Z.-Y., Shu, F.H. 1996, ApJ 468, 261
Lizano, S., Heiles, C., Rodriguez, L.F., Koo, B., Shu, F.H., Hasegawa, T., Hayashi, S.,
    Mirabel, I.F. 1988, ApJ 328, 763

Lovelace, R.V.E., Berk, H.L., Contopoulos, J. ApJ 376, 696
Margulis, M., Lada, C.J. 1985, ApJ 299, 925
Margulis, M., Snell, R.L. 1989, ApJ, 343, 779
Martín-Pintado, J., Bachiller, R., Fuente, A. 1992, A&A 254, 315
Masson, C.R., Chernin, L.M. 1992, ApJ 387, L47
Masson, C.R., Chernin, L.M. 1993, ApJ 414, 230
Masson, C.R., Mundy, L.G., Keene, J. 1990, ApJ 357, L25
Mathieu, R.D. 1994, ARAA 32, 465
Mathieu, R.D., Benson, P.J., Fuller, G.A., Myers, P.C., Schild, R.E. 1988, ApJ 330, 385
McCaughrean, M.J., Rayner, J.T., Zinnecker, H. 1994, ApJ 436, L189
Meyers-Rice, B.A., Lada, C.J. 1991, ApJ 368, 445
McMullin, J.P., Mundy, L.G., Blake, G.A. 1994, ApJ 437, 305
Mikami, H., Umemoto, T., Yamamoto, S., Saito, S. 1992, ApJ 392, L87
Moriarty-Schieven, G.H., Snell, R.L. 1988, ApJ 332, 364
Moriarty-Schieven, G.H., Snell, R.L. 1989, ApJ 338, 952
Moriarty-Schieven, G.H., Snell, R.L., Hughes, V.A. 1989, ApJ 347, 358
Mundt, R., Brugel, E.W., Bührke, T. 1987, ApJ 319, 275
Mundt, R, Fried, J.W. 1983, ApJ 274, L83
Myers, P.C., Heyer, M., Snell, R.L., Goldsmith, P.F. 1988, ApJ 324, 907
Myers, P.C., Ladd, E.F. 1993, ApJ 413, L47
Nagar, N.M., Vogel, S.N., Stone, J.M., Ostriker, E.C. 1997, ApJ 482, 195
Najita, J.R., Shu, F.H. 1994, ApJ 429, 808
Neufeld, D.A., Kaufman, M.J. 1993, ApJ 418, 263
Nisini, B., Giannini, T., Molinari, S., Saraceno, P., Caux, E., Ceccarelli, C., Liseau,
    R., Lorenzetti, D., Tommasi, E., White, G. J. 1998, in *ASP Conf. Ser. 132: Star
    Formation with the Infrared Space Observatory*, eds. Yun, J. and Liseau, R. (ASP),
    p. 256
Osterbrock, D.E. 1958, PASP 70, 399
Padman, R., Bence, S., Richer, J. 1997, in *IAU Symposium No. 182*, eds. B. Reipurth
    and C. Bertout (Kluwer Academic Publishers), p. 123
Parker, N.D., Padman, R., Scott, P.F. 1991, MNRAS 252, 442
Pineau des Forêts, G., Flower, D.R., Chièze, J.P. 1997, in IAU Symp. 182 *Herbig-Haro
    Flows and the Birth of Low Mass Stars* (Eds. B. Reipurth & C. Bertout). Kluwer,
    Dordrecht, pag. 199
Plambeck, R.L., Snell, R.L., Loren, R.B. 1983. ApJ 266, 321
Pudritz, R.E. 1985, ApJ 293, 216
Pudritz, R.E., Norman, C.A. 1983, ApJ 274, 677
Raga, A.C., Cabrit, S. 1993, A&A 278, 267
Raga, A.C., Binette, L., Cantó, J. 1990, ApJ 360, 612
Raga, A.C., Taylor, S.D., Cabrit, S., Biro, S. 1995, A&A 296, 833
Reipurth, B., Bally, J., Devine, D. 1997, AJ 114, 2708
Richer, J.S., Hills, R.E., Padman, R. 1992, MNRAS, 254, 525
Richer, J., Shepherd, D., Cabrit, S., Bachiller, R., Churchwell, E. 1999, in *Protostars &
    Planets IV*, in press
Rodríguez, L.F. 1995, Rev. Mex. Astron. Astrofís. (Ser. de Conf.) 1, 1
Rodríguez, L.F. 1997, in IAU Symp. 182 *Herbig-Haro Flows and the Birth of Low Mass
    Stars* (Eds. B. Reipurth & C. Bertout). Kluwer, Dordrecht, pag. 83
Rodríguez, L.F., Ho, P.T.P., Moran, J.M. 1980, ApJ 240, L149
Rodríguez, L.F., Lizano, S., Cantó, J., Escalante, V., Mirabel, I.F. 1990, ApJ 365, 261
Saraceno, P., André, P., Ceccarelli, C., Griffin, M., Molinari, S. 1996, A&A 309, 827
Saraceno, P., Nisini, B., Benedettini, M., Ceccarelli, C., Di Giorgio, A. M., Giannini, T.,
    Molinari, S., Spinogl, L., Clegg, P. E., Correia, J., Griffin, M. J., Leeks, S. J., White,
    G. J., Caux, E., Lorenzetti, D., Tommasi, E., Liseau, R., Smith, H. A. 1998, in *ASP
    Conf. Ser. 132: Star Formation with the Infrared Space Observatory*, eds. Yun, J. and
    Liseau, R. (ASP), p. 233

Schilke, P., Walmsley, C.M, Pineau des Forêts, G., Flower, D.R. 1997, A&A 321, 293

Schutte, W.A., Gerakines, P.A., Geballe, T.R., van Dishoeck, E.F., Greenberg, J.M. 1996, A&A 309, 633

Schwartz, R.D. 1975, ApJ 195, 631

Shang, H., Shu, F.H., Glassgold, A.E. 1998, ApJ 493, L91

Shu, F.H. 1977, ApJ 214, 488

Shu, F.H., Adams, F.C., Lizano, S. 1987, ARAA 25, 23

Shu, F.H., Lizano, S., Ruden, S.P., Najita, J. 1988, ApJ 328, L19

Shu, F.H., Najita, J., Ostriker, E., Wilkin, F., Ruden, S.P., Lizano, S. 1994a, ApJ 429, 781

Shu, F.H., Najita, Ruden, S.P., Lizano, S. 1994b, ApJ 429, 797

Shu, F.H., Najita, J.R., Shang, H., Li, Z.-Y. 1999, in *Protostars & Planets IV*, in press

Shu, F.H., Ruden, S.P., Lada, C.J., Lizano, S. 1991, ApJ 370, L31

Shull, J.M., Beckwith, S. 1982, ARAA 20, 163

Smith, M.D. 1994, MNRAS 266, 238

Smith, M.D., Suttner, G., Yorke, H.W. 1997, A&A 323, 223

Snell, R.L.: 1987, in *IAU Symposium No. 115: Star Forming Regions*, eds. M. Peimbert and J. Jugaku (Reidel, Dordrecht), p. 213

Snell, R.L., Bally, J. 1986, ApJ 303, 683

Snell, R.L., Loren, R.B., Plambeck, R.L. 1980, ApJ 239, L17

Snell, R.L., Scoville, N.Z., Sanders, D.B., Erickson, N.R. 1984, ApJ 284, 176

Stahler, S.W. 1993, in *Astrophysical Jets*, ed. M Livio, C O'Dea, D. Burgarella. Cambridge: Cambridge University Press

Staude, H.J., Elsässer, H. 1993, A&A Rev. 5, 165

Stone, J.M., Norman, M.L. 1994, ApJ 420, 237

Suttner, G., Smith, M.D., Yorke, H.W., Zinnecker, H. 1997, A&A 318, 595

Tafalla, M., 1993, Ph. D., Univ. of California, Berkeley

Tafalla, M., Bachiller, R. 1995, ApJ 443, L37

Tafalla, M., Bachiller, R., Wright, M.C.H. 1994, ApJ 432, L127

Tafalla, M., Bachiller, R., Wright, M.C.H., Welch, W.J. 1997, ApJ 474, 329

Torbett, M.V. 1984, 278, 318

Torrelles, J.M., Rodríguez, L.F., Cantó, J., Carral, P., Marcaide, J., Moran, J.M., Ho, P.T.P., 1983, ApJ 274, 214

Uchida, Y., Shibata, K. 1985, PASJ 37, 515

Umemoto, T., Iwata, T., Fukui, Y., Mikami, H., Yamamoto, S., Kamaya, O., Hirano, N. 1992, ApJ 392, L83

van Dishoeck, E.F., Blake, G.A. 1998, ARAA, in press

van Dishoeck, E.F., Blake, G.A., Jansen, D.J., Groesbeck, T.D. 1995, ApJ 447, 760

Velusamy, T., Langer, W.D. 1998, Nature 392, 685

Wilkin, F.P. 1996, ApJ 459, 31

Wolf, G.A., Lada, C.J., Bally, J. 1990, AJ 100, 1892

Wootten, A., Loren, R.B., Sandqvist, A., Friberg, P., Hjalmarson, A. 1984, ApJ 279, 633

Wu, Y., Huang, M., He, J. 1996, A&ASS 115, 283

Zhang, Q., Ho, P.T.P., Wright, M.C.H., Wilner, D.J. 1995, ApJ 451, L71

Zhang, Q., Zheng, X. 1997, ApJ 474, 719

Zinnecker, H., McCaughrean, M.J., Rayner, J.T. 1998, Nature 394, 862

Bo Reipurth

HERBIG-HARO FLOWS

BO REIPURTH
Center for Astronomy and Space Astrophysics
University of Colorado
Boulder, CO 80309, USA

AND

A.C. RAGA
Instituto de Astronomía
UNAM, Ap. 70-264, 04510 DF
México

1. Introduction

In the course of objective prism surveys of dark clouds, Herbig (1950, 1951, 1952) and Haro (1952, 1953) discovered a number of small objects, which display characteristic emission line spectra of hydrogen and forbidden low excitation lines of mainly [OI], [NII] and [SII], with at most very weak continua. On direct photographic plates, Herbig and Haro found these objects to be tiny, almost semi-stellar nebulae, with a tendency to be located in groups, often aligned in strings of objects. Ambartsumian (1954) subsequently coined the term Herbig-Haro objects for these tiny nebulae. From their location in dark clouds often containing nebulous stars or the newly recognized T Tauri stars (cf. Joy 1942, Ambartsumian 1947), early researchers concluded that the HH objects were somehow related to the process of star birth.

2. Herbig-Haro Objects at Different Wavelengths

2.1. OPTICAL SPECTRA

One of the brightest known HH objects, HH 1, has been the subject of the most detailed spectroscopic studies up to now. Its spectrum was first described by Herbig (1951), and subsequent spectrophotometric studies (e.g. Böhm 1956, Osterbrock 1958, Haro & Minkowski 1960, Böhm *et al.* 1973)

C.J. Lada and N.D. Kylafis (eds.), The Origin of Stars and Planetary Systems, 267–304.

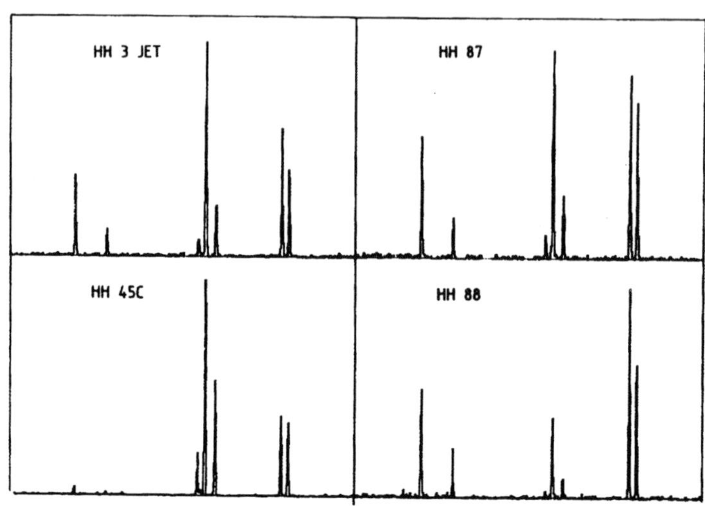

Figure 1.   Low resolution spectra of four HH objects, showing the large variation of emission line strengths. The lines seen are [OI] 6300, [OI] 6363, [NII] 6548, Hα, [NII] 6584, [SII] 6717 and [SII] 6731. From Reipurth (1989b).

further established its peculiar nature. In an HH object, permitted and forbidden emission from neutral atoms (e.g. [OI], OI, [CI], [NI]) and from ions of low ionization energy (e.g. CaII, [CaII], [FeII], MgII, [SII]) are much stronger than in common photo-ionized nebulae. However, line-ratios can vary greatly from one HH object to another, and HH objects are often classified according to their level of "excitation" (e.g. Böhm 1983). HH 1 is the prototype of a high excitation object: for example, its [SII]6717/Hα ratio is very low ($\sim$0.2), its [OIII]5007/Hβ ratio is high ($\sim$0.6), its [NI]5198/5200 to Hβ ratio is low ($\sim$0.2), and its [CI]9848/Hα ratio is very low ($\sim$0.02) (Brugel *et al.* 1981a, Solf *et al.* 1988). HH 7, on the other hand, is a typical low excitation object, with a very high [SII]6717/Hα ratio ($\sim$2.8), no detection of [OIII]5007, a very high [NI]5198/5200 to Hβ ratio ($\sim$1.8) and a very high [CI]9848/Hα ratio of $\sim$1.3 (Böhm *et al.* 1980, 1983). Most HH objects have line-ratios which lie within these values. Fig. 1 illustrates a variety of HH spectra in the region around Hα.

For a derivation of line-ratios, and a meaningful interpretation of them, HH spectra must be de-reddened. This correction can be significant, because HH objects mostly are located in regions of rather high extinction. The best method to correctly determine the reddening of an HH object is the one of Miller (1968). In this method, the sum of the auroral lines of [SII] at $\lambda\lambda$ 4068/4076 are compared with the sum of the transauroral lines of [SII] at $\lambda\lambda$ 10318/10337. These lines appear in very different parts of the spectrum, but originate from the same upper energy level. Their theoretical ratio is therefore fully determined by the ratios of their transition probabilities and their frequencies; but the observed ratio is additionally affected by reddening. Adopting an extinction curve then leads directly to an E(B–V) for the observed object. While easy in principle, in practice it is not trivial to get accurate, good signal-to-noise observations of the [SII] lines beyond 1 μm. Another similar method is to compare Balmer and Paschen lines which come from the same upper level; for a careful discussion of these reddening methods see Solf $et\ al.$ (1988). For spectra that do not have sufficient spectral coverage to include both sets of lines, more indirect methods exist (e.g. Dopita 1978). In the near-infrared, the [FeII] lines at 1.64 and 1.25 μm also arise in the same upper level and have been used to infer reddenings in HH objects (e.g. Gredel 1994).

In addition to emission lines, HH objects have very weak continua (Herbig 1951, Böhm 1956, Böhm $et\ al.$ 1974, Brugel $et\ al.$ 1981b). Although in a few HH objects one observes red continua due to reflected light from the stellar energy source, the majority of continua are blue, and indeed keep rising into the ultraviolet. These blue continua were interpreted as the result of a collisionally enhanced two photon decay mechanism in neutral atomic hydrogen by Dopita $et\ al.$ (1982a) and Brugel $et\ al.$ (1982). High signal-to-noise data for HH 1 corroborate this suggestion (Solf $et\ al.$ 1988).

An important step towards interpreting the spectra of HH objects came with the realization that their line spectra could originate in the cooling regions behind shock waves (Schwartz 1975, 1978). This was borne out in more detailed comparisons between observed line ratios and plane-parallel shock wave models (Dopita 1978, Raymond 1979). The best fits to the observed data came for shocks with velocities of typically $70 - 100$ km s^{-1} moving into a rather low density, almost neutral medium. However, none of the models could reconcile the simultaneous presence of low excitation lines of [OI] and [SII] with highly ionized species like CIV. The introduction of bow shock models has resolved many of the inconsistencies inherent in the plane-parallel shock geometry. A bow shock produces a range of shock velocities, because only the component of the flow velocity perpendicular to the bow shock is thermalized. A single bow shock can therefore produce high excitation near its apex and simultaneously much lower excitation lines from the oblique shocks along its wings (e.g. Hartmann & Raymond 1984, Choe $et\ al.$ 1985, Raga & Böhm 1985, 1986, Hartigan $et\ al.$ 1987, Raymond $et\ al.$ 1988).

The relevance of bow shock models for HH objects is supported not only by the line ratios they produce, but also by the line shapes they predict. The kinematics of the flow in the recombination region behind the shock can be

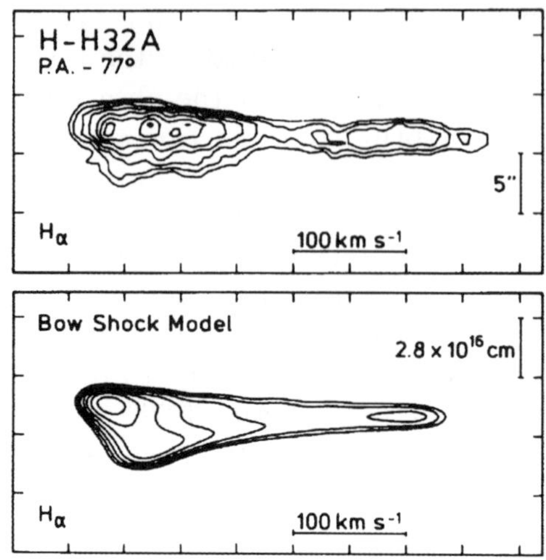

Figure 2.   An observed Hα long slit spectrum of HH 32A and the predicted position-velocity diagram from a bow shock model.   From Solf et al. (1986).

accurately predicted, and by integrating the emission over an appropriate volume (e.g. corresponding to what is seen through a spectrograph slit) a direct comparison with observations can be made. The results can be presented either as line profiles (e.g. Hartigan *et al.* 1987, Raymond *et al.* 1988) or also including the spatial dimension as position-velocity diagrams (e.g. Choe *et al.* 1985, Raga & Böhm 1985, 1986, Hartigan *et al.* 1990).

Fig. 2 shows a comparison of the observed long-slit spectrum of the Hα line in HH 32A with a position-velocity diagram calculated to reproduce the observed features, from a detailed study of the HH 32 complex by Solf *et al.* (1986). On the spatial (vertical) axis, the source is located towards the bottom, and on the velocity (horizontal) axis lower (more negative) velocities are towards the left, and higher (more positive) velocities towards the right. The immediately striking feature of the observed line profile is its very large width, of more than 350 km s⁻¹. This can readily be reproduced by a bow shock because it, as distinct from a plane-parallel shock, contains a large range of velocities projected along the line of sight, in addition to the thermal broadening of the emission from the hot post-shock gas. The observed emission clearly has two peaks, which are rather well reproduced in the model; they arise from regions near the bow apex and from the bow wings, respectively. Finally, the line has a distinct triangular shape. This asymmetry is related to the angle under which the bow shock is observed: for a bow shock moving in the plane of the sky the line is completely symmetric, but for bow shocks moving at increasingly small angles to the line of sight the asymmetry becomes more and more pronounced (Raga & Böhm 1986, Hartigan *et al.* 1987). The angle of the flow is a parameter to which the models are very sensitive. The parameters which Solf *et al.* (1986) found best match the observation in Fig. 2 are: bow shock velocity 300 km s^{-1} at the stagnation point, and an angle to the line of sight of 20°, with the object moving away from the observer. Hartigan *et al.* (1987) present simple analytic formulae that can be used to estimate the shock velocity and bow shock orientation from a single high resolution spectrum of a low excitation line.

Ideally, spectroscopic observations should provide both the large spectral range needed to yield numerous line-ratios, and the high spectral and spatial resolution necessary to map the kinematics of the flow in each emission line (e.g. Solf *et al.* 1988, Böhm & Solf 1990, Solf & Böhm 1991). For further details on the optical emission line spectra of HH objects, see the reviews of Böhm (1990) and Böhm & Goodson (1997).

2.2. INFRARED SPECTRA AND IMAGES

The near-infrared spectral region contains a number of emission lines of molecular hydrogen, which have proven to be important diagnostics for the molecular component of the gas in HH objects. The strongest and most commonly studied line is the 2.12 μm $v = 1 - 0$ S(1) line of H_2. This line was first detected in HH objects around T Tauri by Beckwith *et al.* (1978). Elias (1980) detected the line in a number of southern HH objects, and from a study of line-ratios concluded that the molecular gas was dense ($\sim 10^4$ cm^{-3}) and heated by shock waves. It is now clear that molecular hydrogen can contribute sometimes significantly to postshock cooling, and that shock-codes should incorporate both atomic and molecular processes. A considerable fraction of the HH objects detected have low-excitation optical spectra formed in low velocity or oblique shocks.

Near-infrared imagers and spectrographs have now reached such a level of development that it is possible to study the morphology and line strengths of the infrared emission from a large number of HH flows, particularly the strong and abundant infrared lines of H_2. Additionally, spectra in the H-band will record the [FeII] lines at 1.64 μm and 1.25 μm, which both arise from the same upper level, and therefore can be used to derive the reddening in an HH flow. [FeII] lines are also useful indicators of electron density.

The ground state of H_2 has many vibration-rotation levels with excitation energies from less than 1000 K to several times 10000 K. The resulting lines are likely to be collisionally excited in shocked gas with temperatures of a few thousand degrees, as originally discussed by Elias (1980). Such thermal excitation will populate only the lowest vibrational levels (v≤3). Other non-thermal mechanisms have been discussed as well, particularly excitation due to fluorescence pumped by UV photons. Such processes will populate also high (v>3) vibrational levels. High signal-to-noise spectra of spectral regions where higher vibrational level lines should appear are therefore important in order to distinguish between thermal and non-thermal excitation mechanisms.

A convenient way of presenting the information of molecular hydrogen spectra is the H_2 excitation diagram, in which a measure of population density is plotted against excitation energy of the upper level for each transition. Gredel (1994) discusses the underlying physics, here it suffices to say that if the emitting gas can be characterized by a single temperature then the data points will fall along a straight line, with a slope proportional to the inverse of the excitation temperature. But if the emission comes from gas at a range of temperatures, then the line will be curved.

Gredel (1994) has obtained relatively high resolution infrared spectra of a large number of HH flows. For all of the objects, he finds a thermal population distribution which can be described by single excitation temperatures in the range from 2000 K to 2700 K. Fig. 3 shows in panel A the K-band spectrum of HH 43, one of the brightest H_2 emitters among HH objects, in panel B the H-band spectrum, and in panel C the resulting excitation diagram. Gredel derives an excitation temperature for HH 43 of 2200 K. In the H-band spectrum there should be five (6,4) Q(1)-Q(5) lines in the 1.60 to 1.64 μm region if any non-thermal emission was present. None are seen. There *is* in fact fluorescently excited H_2 in HH 43, but it has so far only been detected in the UV, where Schwartz (1983) observed eight Lyman band emission lines between 1250 Å and 1600 Å.

Both C-type (continuous) and J-type (jump shocks) have been invoked to explain the infrared line spectra in HH flows. C-shocks require a magnetized medium of low ionization (e.g. Smith 1993). J-shocks require a medium of higher ionization that is not so strongly magnetized (Smith 1994a). The intensities in the (1,0) S(1) H_2 line that Gredel (1994) observed in various HH objects can be produced by slow non-dissociative J-shocks. The observed intensities of the [FeII] 1.64 μm line are, on the other hand, well modelled by fast, dissociative J-shocks in the 40-60 kms^{-1} range. This can be reconciled only if the H_2 and [FeII] emission originate in separate regions, and Gredel (1994) suggests that curved bow shocks of J-type could explain the data, with the [FeII] emission arising in the fast, dissociative region near the apex of the bow shock, and the H_2 emission in the slower, non-dissociative wings. Schwartz et al. (1995) arrive at similar conclusions.

The similarities and differences between distributions of optical and molecular shock regions may also provide essential clues to the flow struc-

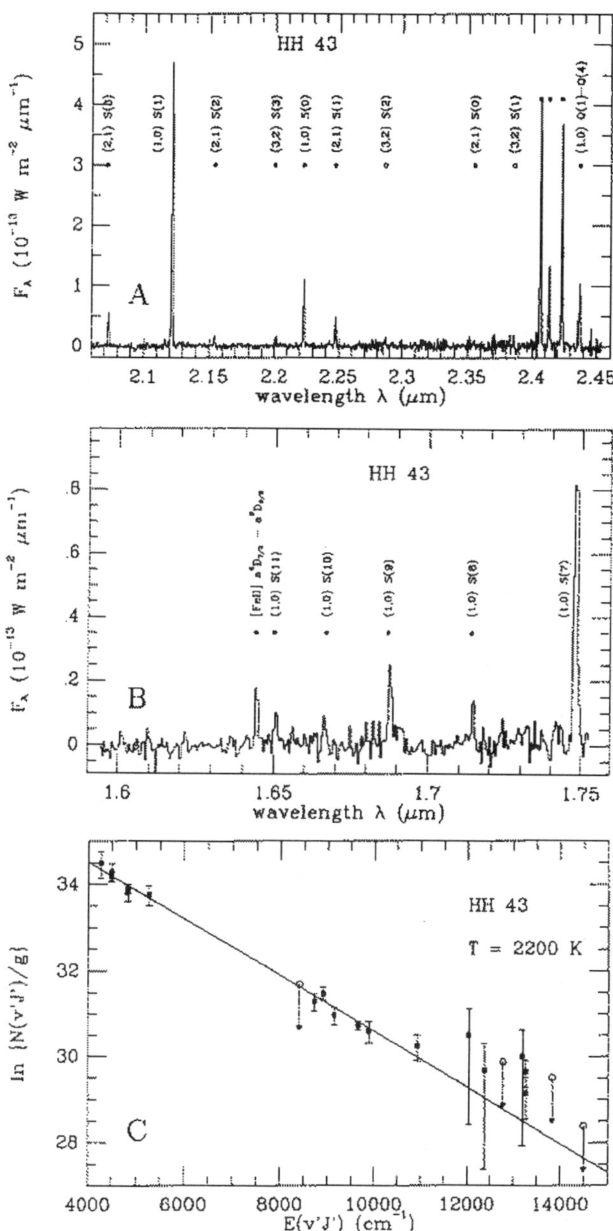

Figure 3. *Infrared spectroscopy of HH 43. Panel A shows a K-band spectrum, and Panel B an H-band spectrum. Panel C shows the resulting H_2 excitation diagram. From Gredel (1994).*

ture in HH objects. Such studies have been performed by e.g. Schwartz *et al.* (1988), Hartigan *et al.* (1989), Garden *et al.* (1990), Wilking *et al.* (1990), Zealey *et al.* (1992), Carr (1993), Eislöffel *et al.* (1994b), Davis *et al.* (1994), Gredel & Reipurth (1994). It appears that the H_2 emission occurs mainly in the wings of the bow shocks, where the shocks are oblique and thus much weaker.

2.3. ULTRAVIOLET SPECTRA

It came as a surprise when Ortolani & D'Odorico (1980) discovered a strong continuum and emission lines in the ultraviolet spectrum of HH 1, considering its not insignificant reddening. Subsequently, more than a dozen studies of ultraviolet spectra of HH objects have confirmed the importance of UV radiation from these objects.

The UV spectra of high-excitation objects are dominated by emission lines of [CIV] 1548/1551 and [CIII] 1907/1909; low-excitation objects, on the other hand, almost exclusively show fluorescent lines of the Lyman bands of the H_2 molecule (Schwartz 1983). The UV continuum rises towards shorter wavelengths, particularly rapidly between 1700Å and 1620Å, with a peak around 1575Å. This continuum is likely to be a continuation of the observed blue continuum, and is also believed to be due to a collisionally enhanced two photon decay mechanism in neutral atomic hydrogen (Brugel *et al.* 1982), but an additional component due to fluorescent H_2 appears to play a role at the shortest wavelengths (Böhm *et al.* 1987).

Interpretations of UV observations are extremely sensitive to the reddening corrections applied. Although the amount of extinction can be reasonably determined using the Miller (1968) method, the shape of the extinction curve may differ from region to region. Böhm-Vitense *et al.* (1982) argued that flatter than normal extinction curves (like the θ Ori curve) may be appropriate, though this result was partially called into question by Böhm et al. (1991).

All the brightest HH objects have now been observed in the UV (Brown *et al.* 1981; Böhm *et al.* 1981, Böhm & Böhm-Vitense 1984; Brugel *et al.* 1985; Schwartz *et al.* 1985; Noriega-Crespo *et al.* 1989, Cameron & Liseau 1990, Böhm *et al.* 1993, Liseau *et al.* 1996 and Hartigan *et al.* 1999; and above mentioned references).

3. Morphological and Kinematic Properties of Jets

A particularly intriguing subset of Herbig-Haro objects is formed by the highly collimated HH jets (Fig. 4 and Fig. 5). The first jet recognized as such was the HH 47 jet (Dopita *et al.* 1982b), which remains one of the finest cases known. Mundt & Fried (1983) and Mundt *et al.* (1983, 1984) did

CCD imaging of young stars and known HH objects and discovered several more HH jets. A large number of detailed studies of HH jets now exists in the literature, e.g. Graham & Elias (1983), Reipurth *et al.* (1986), Mundt *et al.* (1990, 1991), Reipurth & Heathcote (1991), Eislöffel *et al.* (1994a), Bally *et al.* (1995), Heathcote *et al.* (1996), Reipurth *et al.* (1997a).

The opening angle of jets are in general small, of the order of a few degrees. Length-to-width ratios vary considerably, from fairly poorly collimated jets with a ratio of 5 or so to extremely high ratios exceeding 100 (e.g. HH 333; Bally *et al.* 1996). The widths of jets are mostly poorly determined, but are in a few cases known well; the HH 34 jet for example has a width of 0.7 arcseconds or 300 AU (Raga & Mateo 1988, Bührke *et al.* 1988, Raga *et al.* 1991b) and a width of about 0.8 arcseconds was found for HH 111 (Reipurth *et al.* 1992, 1997a).

All jets have a knotty sub-structure, with typical semi-regular spacings of one to a few arcseconds. These knots do not always trace a perfectly straight line, in fact it is common for jets to show considerable wiggling. The knots provide a means to measure transverse motions of jets; proper motions have been measured quite accurately for many jets by now, and typical transverse motions are around $200-300$ km s^{-1} (e.g. Herbig & Jones 1981, 1983, Heathcote & Reipurth 1992, Eislöffel *et al.* 1994a).

In many cases HH jets are bipolar, tracing an approaching and a receding lobe. The two lobes are generally very asymmetric, usually due to uneven obscuration. However, cases exist where the asymmetries appear to be real, either because the jet-flow is intrinsically asymmetric, or because of differences in the ambient medium into which the jets flow (e.g. Mundt *et al.* 1987, Hirth *et al.* 1994, Reipurth *et al.* 1998a).

Long slit spectra along a jet axis give information on the variation of the radial velocity, line width, excitation and electron density in the flow. Such studies have now been made for many jets, see for example Solf (1987), Meaburn & Dyson (1987), Solf & Böhm (1987), Mundt *et al.* (1987), Bührke *et al.* (1988), Reipurth (1989a,b), Mundt *et al.* (1990), Hartigan *et al.* (1990), Heathcote & Reipurth (1992), Solf & Böhm (1993). Typical radial velocities are high, around $100 - 200$ km s^{-1}. Jets show velocity variations, which are often complex but rarely major. When an ordered motion is found it is mostly a decrease in the radial velocity; in one case, HH 83, a major increase in radial velocity is measured as one moves away from the source (Reipurth 1989b).

Electron densities as derived from the [SII] 6717/6731 ratio vary considerably, with values ranging typically from a few hundred to a few thousand electrons per cm^3. When changing systematically along the jet, the electron density more often decreases, which in the simplest interpretation suggests diverging flows (e.g. Mundt *et al.* 1987). It is not so easy to determine the

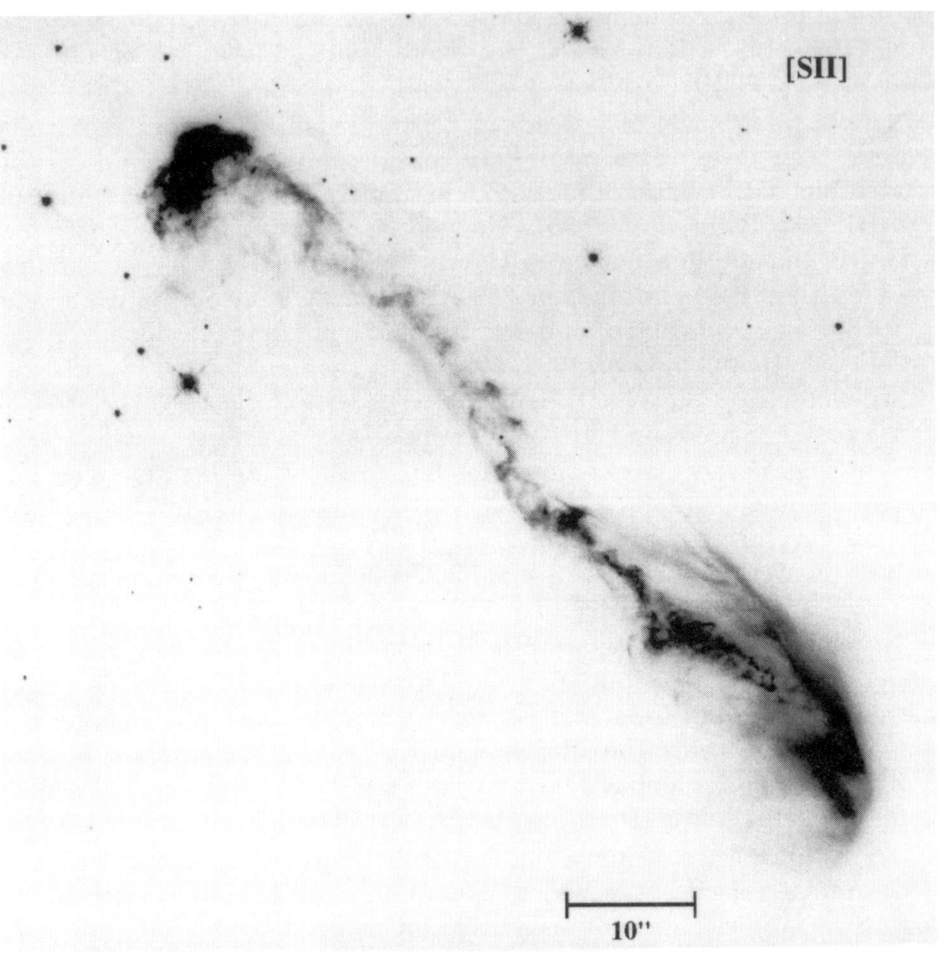

Figure 4.   A [SII] image of the HH 47 jet obtained with WFPC2 at HST. The jet emerges from a Bok globule in the Gum Nebula, and is powered by an embedded IRAS source located near the lower right hand corner of the image. The source illuminates the walls of a cavity through which the jet emerges, producing a bright reflection nebula, which curves around and above the beginning of the jet. The body of the jet, which is very bright in [SII], consists of a sinuous chain of knots. The jet ends in a bright, extended working surface, HH 47A, surrounded by a faint veil that sweeps backwards. At this point, jet material is shocked and decelerated, while ambient material is shocked and accelerated. HH 47A is not the terminal working surface, as at least one more bow shock is known further out along the jet axis. The jet is not flowing along the complicated path seen here, rather, the meandering structure most probably reflects a wandering of the ejection axis near the source. From Heathcote et al. (1996).

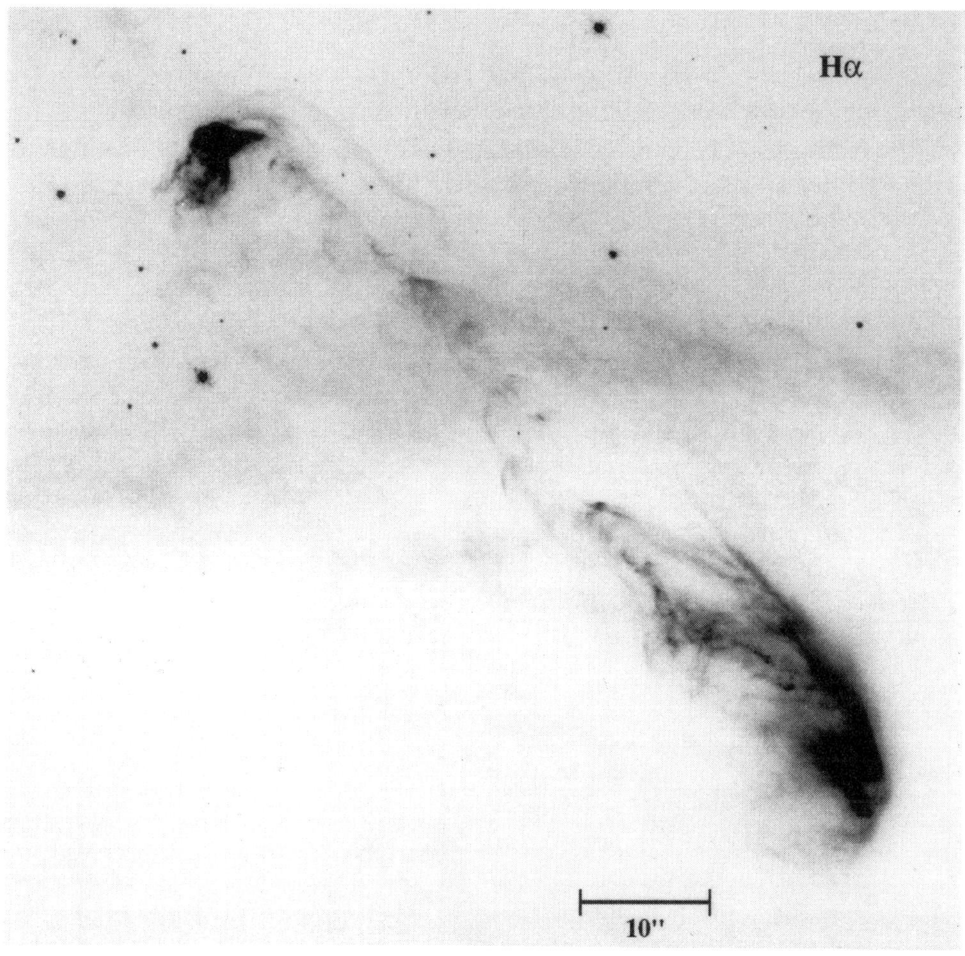

Figure 5. An Hα image of the HH 47 jet obtained with WFPC2 at HST. The winding chain of knots seen in [SII] is here replaced with a delicate tracery of narrow filaments and wisps, with trailing wings swept back at an oblique angle to the flow direction. It turns out that the [SII]-bright knots delineate the "core" of the jet, while the Hα emission comes predominantly from the zone along the edges of the jet. The Hα wisps appear to resemble one-sided bow shocks that extend into and accelerate the ambient medium. This is consistent with kinematic studies which show that the core of the jet moves faster, and the edges slower. Entrainment and acceleration of the medium through which the jet flows thus occur both at the working surface of the jet and also along the body of the jet. The result is a molecular outflow which can be detected at millimeter wavelengths. From Heathcote et al. (1996).

actual gas density of jets, because they have rather low ionization fractions, so that the gas number density is normally considerably larger than the observed electron density. In the model calculations of Raga *et al.* (1990b), the hydrogen ionization fraction is found to be at most 10%. Bacciotti et al. (1995) have suggested a line ratio analysis useful for determining the ionization fraction of HH objects, but the uncertainties involved are quite large (due to the strong temperature dependence of the line ratios).

The excitation of jets is generally low, and jets are normally prominent in CCD images obtained through a [SII] filter, and much weaker in Hα images. The low excitation character of HH jets and their very high space motions would appear contradictory, but could be reconciled either if the shocks in jets are very oblique (see e.g. Raga 1989), or, as we believe now, are due to small velocity variations along the flow axes (Raga *et al.* 1990a).

The velocity dispersions in HH jets are, with some exceptions, generally low or moderately low. In HH 34, for example, the widths of the [SII] 6717/6731 lines are less than 30 km s^{-1} (Bührke *et al.* 1988). In several jets, one sees a splitting of the lines, e.g. in HH 24 (Solf 1987), in HH 46/47 (Meaburn & Dyson 1987), in HH 34 (Bührke *et al.* 1988, Heathcote & Reipurth 1992) and in several jets in the HL Tau region (Mundt *et al.* 1990). In HH 24, Solf (1987) found two velocity components close to the source, a high velocity component displaying a gentle velocity decrease away from the source, and a weaker lower velocity component with a steep acceleration. ˙Solf (1987) interprets this in terms of a jet model with an inner cylindrical core region and a co-axial envelope. The inner high velocity core is the jet itself and the envelope is ambient material dragged along with the jet; their mutual interaction will slightly slow down the jet and accelerate the entrained material. Using Fabry-Perot observations, Hartigan *et al.* (1993) directly showed how low-velocity gas in the HH 47 jet is surrounding the inner high-velocity gas, similar to a viscous flow in a pipe. The interaction of jets with their ambient medium is generally assumed to be the driving mechanism of the ubiquitous molecular outflows (see the chapter by Bachiller in these proceedings).

4. The HH 111 Jet Complex

One of the finest collimated HH jets is the HH 111 jet in Orion. It emerges from a dense cloud core in the L1617 cloud, which harbors a 25 L$_\odot$ infrared source (Reipurth 1989a). Fig. 6 shows the complex in a large field CCD image. The energy source, a class I source in the classification scheme of Lada (1987), is located at the center of the cloud core, which is outlined by faint Hα emission caused by the OB stars in the Ori Ib association. Light from the source illuminates a conical reflection nebula through which the jet emerges. The flow terminates in a bow shock (V) 142 arcseconds from the source, corresponding to 0.32 pc in projection. On the opposite

Figure 6. *A red broadband CCD mosaic of the HH 111 jet complex, from the ESO 3.6m telescope. The jet emerges from an IRAS source embedded in a cloud core outlined by faint bright rims. Several bow shocks (V, L, X, Y) can be seen on either side of the cloud core. A bright knot (Z) in a very faint counter jet is also visible. From Reipurth et al. (1992).*

side of the source, at a separation of 200 and 226 arcsec, respectively, two other bow shocks are found. The total extent of this complex is more than 0.8 pc in projection (as discussed later, the HH 111 flow actually extends much further than this). Spectroscopic observations show moderate radial velocities, with the western lobe containing the jet which approaches us, while the eastern lobe is receding. Proper motions have been determined for the flow, they are large, of the order of 300 km s^{-1} and directed away from the source (Reipurth *et al.* 1992). Fig. 7 shows the intensity profile of the jet in two Hα images, from 1987 and 1990; the shift due to the proper motion is obvious. From the proper motion and radial velocities one can derive an inclination of the jet to the plane of the sky of only about 10°.

The jet is very bright in [SII], suggesting that the shocks are weak. From the [SII] lines the electron density can be determined, and it is found to decrease steeply with distance from the source, which in the simplest interpretation suggests a constant mass flow through a slightly widening channel. But the electron density derived from the [SII] lines refers to the recombination region in the post-shock gas. Although a simple relation exists between pre-shock and post-shock density as a function of shock velocity (Dopita 1978), a jet consists of many shocks, and so the density may become a complex function of the flow geometry.

The HH 111 jet has been the subject of a very detailed study based on deep images in Hα and in [SII] obtained with WFPC2 onboard the Hubble Space Telescope (Reipurth et al. 1997a). The HST images of the almost perfectly straight HH 111 jet offer an unprecedented opportunity for studying the HH jet phenomenon. At the distance of Orion, a WFPC2

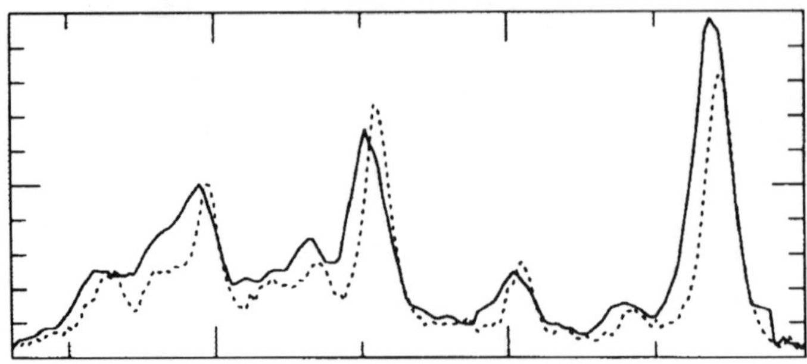

Figure 7.   Intensity profile of the HH 111 jet in Hα, observed in 1987 (full line) and in 1990 (dotted line).   The driving source is towards the left.   The proper motion of individual knots is clearly visible. From Reipurth et al. (1992).

pixel of 0.1 arcsecond corresponds to 46 AU.

Figure 8 shows the HST images in Hα, [SII] and Hα−[SII] of the leading working surface V. This working surface is composed of two brightly emitting regions in Hα and in [SII] (Figure 8a and b), which are displaced from each other, forming a bow shock (forward shock) and Mach disk (reverse shock or jet shock), as can be seen in the Hα−[SII] image in Figure 8c. Closer examination of the Hα image reveals that the brightest emission comes from a very thin filament, unresolved even with HST, which curves around the very front of the bow shock. The filament itself does not emit in [SII], which only appears about 0.5″ upstream. This feature is interpreted as a layer where neutral hydrogen atoms are collisionally excited thus giving rise to strong Balmer emission (Chevalier & Raymond 1978). The [SII] emission arc immediately behind the collisionally excited Hα filament arises in the resolved cooling zone, where gas heated in the forward shock cools and recombines, emitting forbidden metallic lines and Balmer emission.

Contour plots of the Hα and [SII] HST images of the HH 111 jet body are shown in Figure 9. These images provide a wealth of new information on the jet structure. The jet appears to be fully resolved perpendicular to the flow direction, and the knotty substructure is rich and complex. Knot L has the morphology of a well-defined bow shock, best seen in the Hα image. In the body of the jet, most of the brighter knots also have bow shock shapes, with a compression in the contours in the flow direction. Bow shock wings sweep backwards on either one or both sides of the bright knots. The fainter knots, on the other hand, have no characteristic structure.

In most places the jet is much brighter in [SII] than in Hα. Morphologically it is also different in the two lines: seen at lower light levels the jet knots are more distinct in Hα, whereas in [SII] the knots appear to be surrounded by a sheath of faint, diffuse emission. The brightest portion of

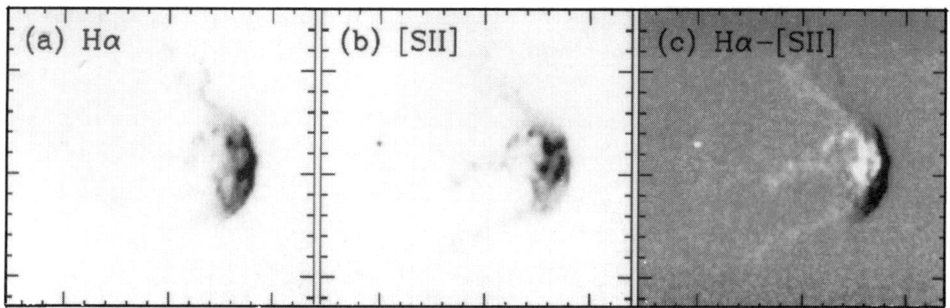

Figure 8.   The HH 111V working surface as seen with HST: (a) Hα; (b) [SII]; and (c) an Hα−[SII] difference image in which Hα-bright regions are black and [SII]-bright regions are white. From Reipurth & Heathcote (1997).

this [SII] tube wavers from side to side as it follows behind the Hα emission, but, as discussed in more detail by Reipurth et al. (1997a), there is no evidence that the jet flow actually spirals about its axis.

Apart from the bright bow shocks L and E, the jet maintains a fairly constant width along its axis. The width of the jet body, measured by collapsing the jet between (and excluding) L and E and measuring the FWHM of the result, is 0.8 arcsec, both in [SII] and in Hα. However, it should be kept in mind that it is difficult to characterize a structure as complex as the HH 111 jet body with a single number. Figure 9 gives the impression that the jet becomes narrower between knots H and L, but this appearance simply results from the jet being fainter in this region.

Even higher resolution observations have been performed of the region of the embedded energy source, using the VLA in the A configuration at 3.6 cm (Rodríguez & Reipurth 1994). The source is detected as a relatively bright (0.9 mJy) object at 3.6 cm, and is shown in Figure 10 with a resolution of about 0.2 arcsec. The source clearly consists of a core with a one-sided radio jet extending towards the West at exactly the position angle of the optical jet. The angular "width" of the radio jet, 0.1 arcsec, is much smaller than the width of the optical jet, ∼0.8 arcsec. The ionized gas that emanates from the exciting star is very young: assuming a jet velocity of 320 kms^{-1} from the optical observations suggests an age of only a few years. The HH 111 energy source is thus at the very moment actively forming the jet.

5. Giant Herbig-Haro Jets

Until recently, the dimensions of HH flows have generally been assumed to be about a few arcminutes, which at a distance of, say, Orion corresponds to a fraction of a parsec, typically about 0.3 pc. A few flows were known with

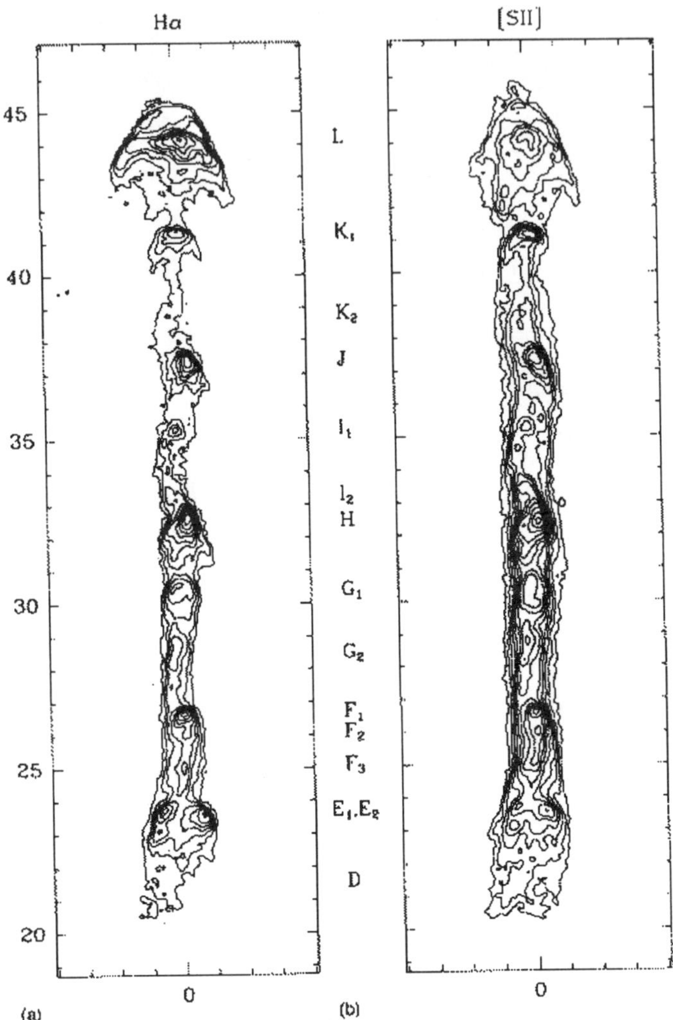

Figure 9.   Contour plots of HST images of the HH 111 jet in Hα (left) and [SII] (right). From Reipurth et al. (1997a).

larger dimensions, but they were associated with more luminous stars. It is now of interest to note that the dimensions of a few arcminutes correspond well to the field size of the first few generations of CCD detectors. When Bally & Devine (1994) used a new wide field CCD imager in the HH 34 region, they saw faint emission features connecting the HH 34 complex with other HH regions previously assumed to be independent flows and all lying along a single well defined axis, and they realized that the HH

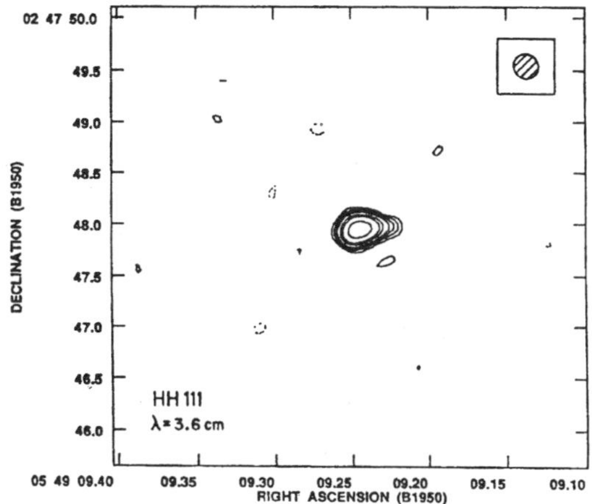

Figure 10. Cleaned natural-weight VLA-A map of the VLA source at HH 111 at 3.6 cm. Contours are -3, 3, 4, 5, 6, 8, 10, and 15 times 20 μJy beam^{-1}, the rms noise of the map. The half power contour of the beam is also shown. A small jet is seen to emanate in the direction towards the optical jet. From Rodríguez & Reipurth (1994).

34 flow stretches to HH 33/40 to the North and to HH 86-87-88 to the South, spanning 2.8 pc. In a detailed follow-up study, Devine *et al.* (1997) showed that all HH objects South of the HH 34 source are blueshifted and have proper motions to the South, while all HH objects to the North are redshifted and have proper motions to the North, thus establishing not only a positional and morphological link to the HH 34 source, but also a kinematical connection.

Following the recognition that HH flows from low mass stars can attain very large dimensions, a systematic survey for such giant flows was undertaken using large-field Schmidt films from the ESO Schmidt Telescope and large-field CCD images from CTIO and Kitt Peak. As a result, a large number of HH flows of gigantic proportions were discovered (Reipurth *et al.* 1997b, 1998b).

5.1. THE GIANT HH 111 JET COMPLEX

The HH 111 jet is the object which comes closest to the text book idealization of how a jet should look (see sect.4). Reipurth *et al.* (1997b) have demonstrated, on structural and kinematic grounds, that the HH 111 complex contains even more distant bow shocks, spanning 57 arcminutes, that is, almost a full degree on the sky; at a distance of 460 pc and inclined 10° to the plane of the sky, this corresponds to a physical extent of 7.7 pc.

The central panel in Figure 11 shows a [SII] CCD image of what we have

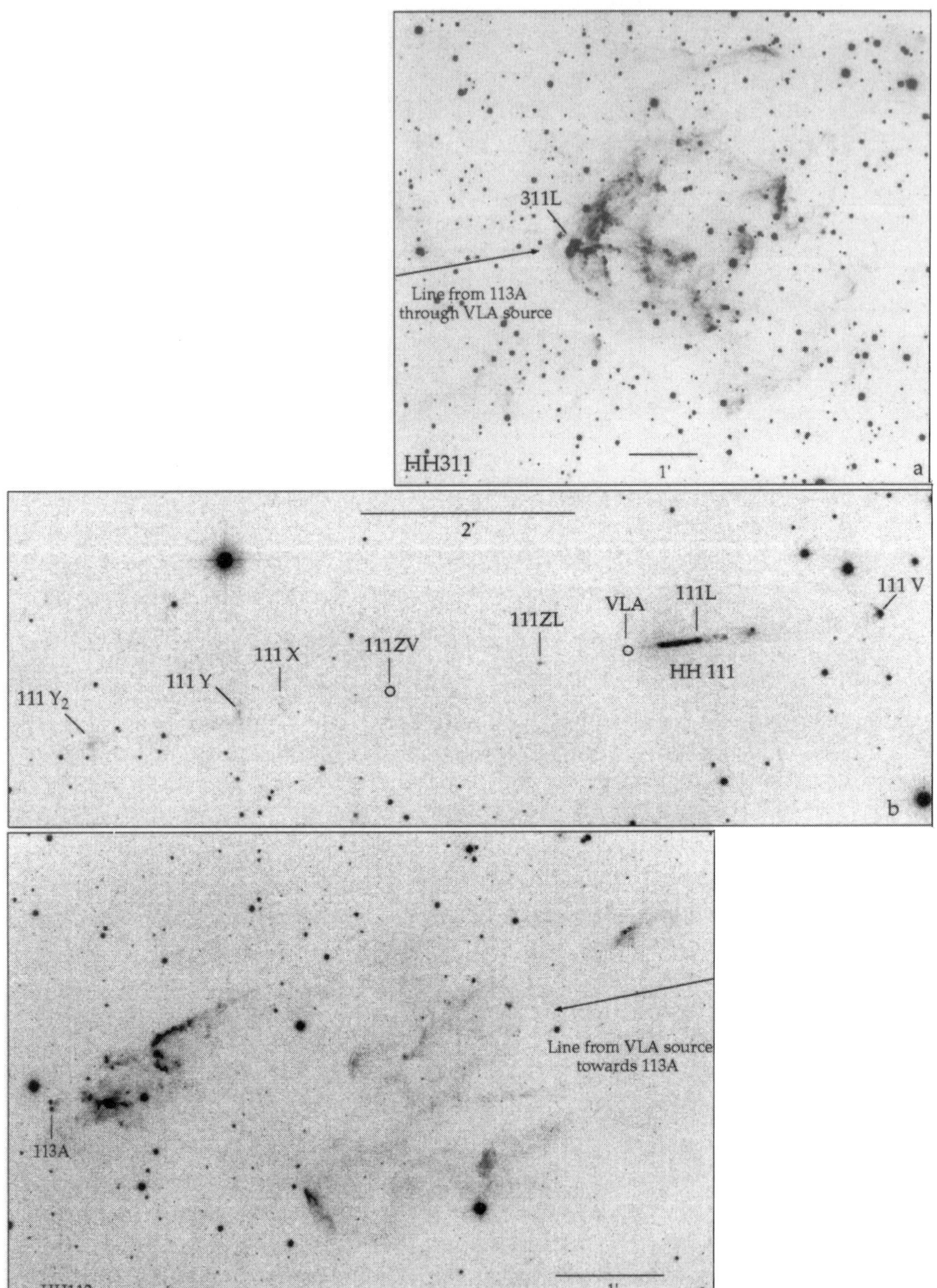

Figure 11.   A mosaic of the principal components of the giant HH 111 flow. From Reipurth et al. (1997b).

up to now thought of as the HH 111 complex, from V to the West and to Y in the East. Another new knot, $111Y_2$, is also identified. If one moves further eastward, about 15 arcminutes from the VLA source, one finds another HH knot, $111Y_3$, on the well defined flow axis. Continuing another about 12 arcminutes to the East, one encounters the HH 113 complex. Based on small-field CCD images, Reipurth & Olberg (1991) considered this to be a separate HH flow, although they were puzzled about the absence of an energy source.

In the lower panel of Figure 11 we see new wide field $H\alpha+[SII]$ images from the ESO New Technology Telescope. The HH complex appears to be a very large cone-shaped bow shock, with HH 113A at its apex. A line has been drawn pointing from the VLA source to the apex. The position angle of the line between the VLA source and the apex A is 101 degrees, only a few degrees larger than the values determined from the knots much closer to the source. It is remarkable that the flow can maintain such a directional stability over a length of 3.8 pc, which is the distance between the VLA source and knot A.

Precisely symmetrical with HH 113 around the VLA source, we find HH 311. The upper panel of Figure 11 shows the sum of $H\alpha$ and [SII] images of HH 311 taken at the ESO NTT. The object is large, with a width and length of about 5 arcmin, corresponding to a projected size of 0.7×0.7 pc.

Multiple epoch CCD images of HH 113 and HH 311 have been obtained, from which proper motions have been derived. The principal knots in HH 113 are all moving towards the Southeast with velocities in the plane of the sky of order 50-150 $km\,s^{-1}$, while knots in HH 311 are moving towards the Northwest with similar velocities. These proper motions provide very strong support for the idea that HH 113 and HH 311 form distant bow shocks in the HH 111 complex.

5.2. THE IMPORTANCE OF GIANT HH FLOWS

In their study, Reipurth et al. (1997b) discovered 9 HH flows with parsec-scale dimensions, and listed another 13 likely giant HH flows. The ubiquity of HH flows with such gigantic proportions has important consequences for our understanding of the HH phenomenon, as summarized in the following.

1. As the dimensions of HH flows increase by a factor of 5 or so, their dynamical ages increase by at least as much. In their detailed kinematical study of HH 34, Devine *et al.* (1997) demonstrated that the different components of the HH 34 giant flow decrease in space velocity linearly with distance from the source, as momentum is expended in pushing the flow through the interstellar medium. Alternatively, the observed velocity de-

crease with increasing distance from the outflow source could be the result of a gradual "turning on" of the jet flow at the beginning of its evolution. For the newly discovered giant flows, the dynamical ages of the most distant components are of the order of 10^4 - 10^5 years. The longer timescale is comparable to the stellar accretion phase and also to the lifetimes of molecular outflows estimated from millimeter observations.

2. Virtually all well developed jets have multiple working surfaces. This is also true on the larger scales of the giant HH flows, and it provides strong evidence for significant and periodic variability of the source on time scales of many hundred to several thousand years during the dynamical life time of the flow. When known, the space motions of the individual working surfaces exceed their shock velocities, implying that they are faster fluid elements overtaking slower moving material. It follows that most HH objects are internal shocks that define zones of interaction between material ejected at different times with different velocities.

3. Giant HH flows provide a record of the activity of the driving source on timescales significant compared to collapse time scales. Many parsec scale flows show s-shaped point symmetry around the source. This tells us that the axis of ejection is not stable, either because it precesses, or more likely because it wanders or wobbles under the erratic gravitational influence of forming and dissolving protoplanetary or protostellar condensations in its disk. Little more can be said about this influence until a clearer picture emerges of how jets are formed.

4. The behaviour of HH objects as they move away from their source is poorly understood, but from the currently available data it appears that HH objects gradually slow down with time, while at the same time their physical size increases, and they take increasingly chaotic shapes.

5. With sizes of several parsecs, HH flows can completely leave the parental cloud in which they formed and travel into the intercloud medium. One aspect of this is that newborn stars can inject energy and momentum into the interstellar medium far from their location. Another aspect is that terminal working surfaces of giant HH flows can be used as probes of the physical conditions in the surroundings of molecular clouds, because their physical and kinematical properties also depend on their environment.

6. When supersonic mass ejectae of HH flows encounter molecules in the ambient medium, the molecules can be dissociated, and thus the chemical abundances can be reset to values closer to the ones present at an earlier stage of the chemical evolution of the cloud (see, e. g., Charnley *et al.* 1988). They may have an effect on their environment which would be unexpected in the absence of UV radiation from massive stars. Finally they may contribute to the generation of turbulence in molecular clouds.

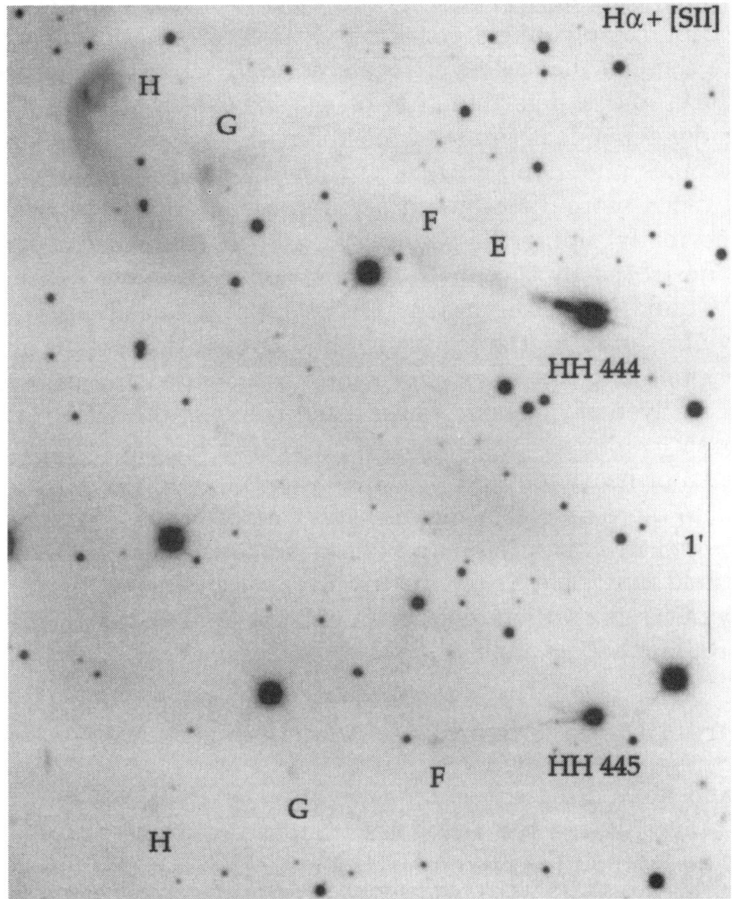

Figure 12.   Two irradiated jets, HH 444 and HH 445, located just East of the massive O-star σ Orionis. Both jets are subjected to and partially ionized by the strong UV radiation field of the O-star. From Reipurth et al. (1998a).

6. Irradiated Jets

Most low-mass stars are born not in tranquil environments like the Taurus clouds or other T Tauri associations, but in dense clusters which include one or more massive OB stars. As a result, the young low-mass stars sooner or later become subjected to an intense UV irradiation, depending on when they were born relative to the appearance of the OB stars and on how soon their placental material is being eroded and swept away. This process may have important effects on the amount of material these stars have at their disposal for accretion, and thus it may affect the shape of the initial mass function in the cluster. It also has two effects related to the HH jets formed by these young stars. Firstly, the UV radiation may partly or fully ionize the jets themselves. HH jets normally emit through the shocks they produce, and a determination of the mass loss rate must therefore involve model dependent assumptions about the nature of the shocks. But if a jet becomes photoionized, it is possible to determine the density and mass flow much more directly and accurately. Secondly, the circumstellar material normally hiding the young star and its immediate surroundings is rapidly destroyed by the photo evaporation and photo ablation caused by the UV radiation of the nearby OB stars, and this opens the exciting possibility of examining the jet launch and collimation region of newborn stars, a region that has so far been largely excluded from observational scrutiny. Finally, a curious and unexpected result of the study of the more than a dozen irradiated jets known so far is that virtually all of these jets are one-sided or nearly one-sided. Whereas extinction is usually hiding one side of jets in more quiescent regions, this cannot easily be invoked for the highly exposed irradiated jets, and one is left to wonder if the strong UV radiation field may paly a role in this strong asymmetry. Altogether, the new study of irradiated jets may open up new approaches to the study of HH jets (Reipurth *et al.* 1998a).

7. Jets from Massive Stars

The jets and flows discussed so far have all emanated from low-luminosity low-mass stars. But a few HH flows from intermediate and higher mass stars have been found (e.g. Axon & Taylor 1984, Reipurth & Graham 1988, Ray *et al.* 1990, Poetzel *et al.* 1992, Ogura & Walsh 1992, Corcoran & Ray 1998, Devine *et al.* 1999).

HH 80/81 are the intrinsically brightest HH objects known, emanating from an embedded high-luminosity IRAS source (Reipurth & Graham 1988). The source is associated with a reflection nebula, GGD 27, and a large bipolar outflow, and is embedded in a dense flattened CS core (Yamashita *et al.* 1989). Compact radio continuum and H_2O maser emission was detected by Rodríguez *et al.* (1980), who suggested a distance of 1.7 kpc. At this distance HH 80/81 lie at a projected distance from the source of 2.3 pc.

Continuum observations of the region were made at 6 cm with the VLA–D array (Rodríguez & Reipurth 1989); both HH 80 and 81 were detected, and the central source was found to be significantly extended in the direction towards the HH objects. In three more recent studies Martí *et al.* (1993,1995,1998) used the VLA–A array to achieve maps of the region with a resolution of better than 1 arcsec. They found that the central source

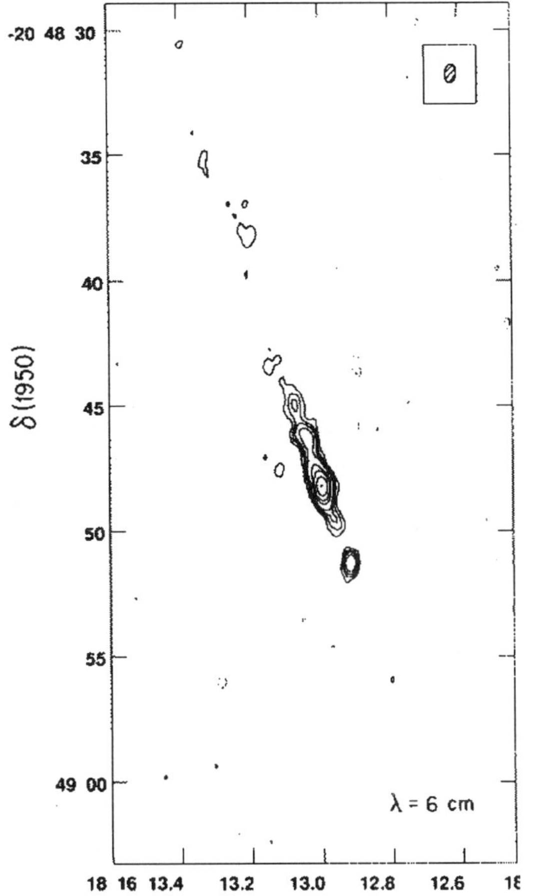

Figure 13.   The highly collimated jet in the HH 80/81 complex.   The central source is an 18000 $L_\odot$ B star, located in the bright central knot.   The map is a 6 cm VLA map. At a distance of 1700 pc, 20 arcsecond correspond to 0.17 pc. From Martí et al. (1993).

powers a highly collimated bipolar radio jet, which shows high proper motions away from the source. The central region of the jet is shown in Fig. 13, it is unresolved perpendicular to the flow direction. The source is located at the brightest point of the jet. Like jets from low luminosity young stars, the HH 80/81 jet consists of a series of knots. The knots extend beyond the central region shown here, and the south-western lobe points directly towards HH 80 and 81. In the opposite direction, along the well defined flow axis, a radio counterpart to the visible HH objects is found; this object is hidden from view by a large molecular cloud. The total projected extent of this bipolar jet system exceeds 5 pc. The dynamical timescale for the system is likely to be short, because extremely high radial velocities, up to 700 km s^{-1}, were detected in HH 80/81 (Heathcote *et al.* 1998).

The case of the HH 80/81 jet system demonstrates that even massive young stars can produce highly collimated bipolar jets, which appear to be rather similar to those from low luminosity young stars, except for being much larger and faster, and probably also considerably more massive.

8. Theoretical models of HH objects

8.1. BASIC THEORETICAL IDEAS

A variety of mechanisms have over the years been proposed to explain the shock waves responsible for HH emission. Early models for the formation of shock waves in HH objects include the "spherical explosion" scenario (Böhm 1978) and the wind focussing mechanisms of Cantó (1980) and Königl (1982) (also studied by Cantó & Rodríguez 1980; Barral & Cantó 1981; Smith 1986; Cantó *et al.* 1986). In these models, the HH emission was assumed to be associated with the walls and tips of the elongated cavities produced in the wind focussing process.

These wind focussing models were also proposed as possible jet collimation mechanisms, the HH objects then being associated with the jet rather than with the wind cavity walls (Königl 1982; Cantó *et al.* 1988; Raga & Cantó 1989a,b; Kim & Raga 1991; somewhat related calculations also being carried out by Peter & Eichler 1996). The effect on the wind collimation due to the presence of an infalling envelope was also studied (Frank & Noriega-Crespo 1994; Frank & Mellema 1996; Mellema & Frank 1997; Wilkin & Stahler 1998). However, these wind focussing mechanisms are now less generally accepted than the MHD wind production/collimation models (see the review of Shu in these Proceedings).

One of the initial suggestions was that the HH knots were slowly moving condensations, embedded in a supersonic wind ejected from a young star. This "shocked cloudlet" model was initially suggested by Schwartz (1978), and then studied further by Schwartz & Dopita (1980), Raga & Wang (1994) and Raga *et al.* (1998a). An alternative "interstellar bullet" scenario (in which HH objects correspond to fast moving cloudlets embedded in a more or less stationary medium) was suggested by Norman & Silk (1979).

More recent models are based on the idea that the HH emission is associated with shock waves produced in a jet-like flow. Dopita *et al.* (1982b) made the initial suggestion that the emission of some HH objects could be the result of shock waves produced at the "head" or "working surface" of a jet-like flow. Following this idea, a considerable effort has been made to model HH objects in terms of a jet beam interacting with the surrounding environment. In these models, the production of the jet itself is not considered, assuming that it is a decoupled problem (see the paper of Shu in these Proceedings),

In the following subsections, we describe the efforts that have been made to model HH objects as plane or bow shocks (section 8.2), to describe them as working surfaces at the heads of jet (section 8.3), and the attempts to model jet-like HH objects in terms of stationary (section 8.4) or travelling (section 8.5) "crossing shocks". The possibly most favoured current model of knots as "internal working surfaces" is described in section 8.6. Finally, in section 8.7 we discuss the implications of current observations on the models of HH objects.

8.2. PLANE-PARALLEL AND BOW SHOCKS

Following the initial suggestion of Schwartz (1978), the emission of HH objects has since been modelled as the result of the excitation of gas by passing shock waves. The most simple problem that can be treated is of course the one of a stationary, plane-parallel shock wave, and quite detailed calculations of this problem (including detailed microphysical processes, as well as the transport of the diffuse, ionizing radiation) have been made in the context of HH objects.

Several papers give predictions of the atomic/ionic emission line fluxes of plane-parallel, stationary HH shocks (Dopita 1978; Raymond 1979; Shull & McKee 1979; Dopita *et al.* 1982a; Binette *et al.* 1985; Hartigan *et al.* 1987, 1994, 1995). The radio free-free continuum of such shocks was modelled by Curiel *et al.* (1987) and by Ghavamian & Hartigan (1998). Also, predictions of observed line fluxes were carried out for the plane-parallel, time-dependent case (Innes *et al.* 1987a,b,c; Gaetz *et al.* 1988; Gaetz 1990).

Given the fact that the "shocked cloudlet", "interstellar bullet" and "jet" models for HH objects (see section 8.1) all predict that at least part of the emission is formed in a bow shock, Hartmann & Raymond (1984) calculated the emission of such a bow shock using a superposition of plane-parallel shocks. This "quasi-1D" bow shock model approach was successfully used to obtain predictions of emission line ratios, line profiles, spatially resolved line profiles and narrow-band images of HH objects (Choe *et al.* 1985; Raga & Böhm 1985, 1986; Raga 1986; Raga *et al.* 1986; Hartigan *et al.* 1987, 1990; Noriega-Crespo *et al.* 1989, 1990; Morse *et al.* 1994; Indebetouw & Noriega-Crespo 1995). Such "quasi-1D" models have also been used to study the non-axisymmetric structures that result from the interaction of a jet with a non-uniform medium (Henney 1996), and to describe the kinematical properties of the proper motions of condensations in a bow shock flow (Raga *et al.* 1997a). Regrettably, the topic of preionization in a bow shock has received relatively little attention in the literature of HH objects (Raymond *et al.* 1988; Raga & Wang 1994).

The recent explosion in observational studies of H_2 emission in HH objects has motivated models of the molecular excitation in plane-parallel and bow shocks (Hartigan *et al.* 1989; Smith & Brand 1990a,b,c,d; Smith 1991; Smith *et al.* 1991a,b; Wolfire & Königl 1991, 1993). Interestingly, the issue of molecular excitation in HH objects is far from settled, and important questions remain regarding whether the molecular emission is produced in a magnetic or radiative precursor, or alternatively is the result of the excitation by a (very !) highly magnetized, sub-Alfvénic "C-shock".

It was already seen several years ago (and clearly more so in the recent HST results) that the "quasi-1D" bow shock models were insufficient for modelling spatially resolved observations of HH objects (Raga & Binette 1991). This result motivated the calculation of fully 2D, axisymmetric bow shocks (Raga & Böhm 1987; Raga *et al.* 1988, 1997b; Cantó & Raga 1998) produced by the interaction of a supersonic wind with a rigid obstacle. These models provided the first available dynamical explanation of the appearance of the condensations observed in HH bow shocks.

Finally, an interesting sideline in modelling the emission line profiles and images of HH bow shocks has been to consider the scattering of the direct emission on the surrounding, dusty environment (Calvet *et al.* 1992; Noriega-Crespo *et al.* 1991; Feldman & Raga 1991; Henney *et al.* 1994, 1995). These models have been successfully applied to modelling the line profiles of HH 1, and might have more important applications in the future, when more extended spectropolarimetric observations of HH objects become available. A second interesting topic that is becoming very important from the observational point of view is the formation of masers in outflows from young stars, and we are likely to see more theoretical work on this subject in the near future (see, e.g., Mac Low *et al.* 1994).

8.3. WORKING SURFACES

The next step in modelling HH objects has been to consider the real, two-shock structure of the working surface at the head of a jet. This structure is schematically shown in Fig. 14.

In the context of the highly radiative HH jets, the first numerical models of working surfaces in jets were the ones of Raga (1988) and Blondin *et al.* (1989, 1990). These models were used to obtain predictions of narrow-band, emission line maps. Also, Hartigan (1989) and Raga & Noriega-Crespo (1993) obtained predictions of the relative contributions to the observed line fluxes of the bow shock and Mach disk from plane-parallel shock models.

More sophisticated numerical simulations of HH jets were later carried out including a 3D description and/or a more realistic treatment of the

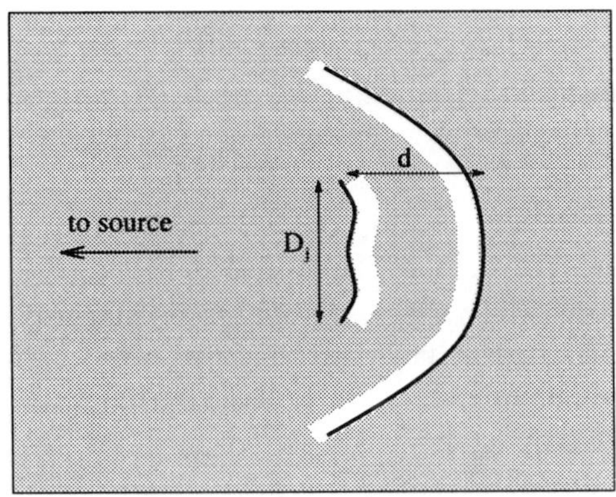

Figure 14. Schematic diagram of a working surface in a jet flow. Such a shock structure is formed at the "head" of a jet (where the jet beam interacts directly with the undisturbed environment). Also, "internal working surfaces" can be formed in a jet ejected from a variable source as a result of the "catching up" of slower moving material by faster material ejected at later times. The working surface has two shocks : a Mach disk or "jet shock" (of diameter D_j) and a bow shock. For the case of strongly radiative HH jets, the expectation is that the separation d between the jet and bow shocks would satisfy the condition $d \ll D_j$. However, there are different possible alternatives for a working surface to violate this condition (see Cerqueira et al. 1997; Frank et al. 1998; Raga et al. 1998b), which can be invoked to model HH objects with large separations between the Mach disk and the bow shock.

relevant microphysical processes (de Gouveia dal Pino & Benz 1993; Stone & Norman 1993). The effects of a stratified medium (Stone & Norman 1994; de Gouveia dal Pino *et al.* 1996; de Gouveia dal Pino & Birkinshaw 1996) and of a supersonic sidewind (Lim & Raga 1998) on radiative HH jets have been studied with 3D numerical simulations. Also, both analytic and numerical working surface models have been proposed to model the coupling between HH jets and the surrounding molecular outflows (Masson & Chernin 1993; Raga & Cabrit 1993; Chernin *et al.* 1994; Raga *et al.* 1995; Suttner *et al.* 1997).

In recent papers it has been shown that jet cross sections with centrally peaked ram pressure distributions (Raga *et al.* 1998b) or the presence of a toroidal magnetic field (Cerqueira *et al.* 1997; Frank *et al.* 1998) result in the formation of an axially extended cool "nose cone" separating the bow shock and the Mach disk. This prediction appears to have a clear application to the structures observed in the HST images of HH 47A.

Finally, an interesting aspect of HH jets has been the study of collisions of jets with dense clouds (Reipurth *et al.* 1996; Raga & Cantó 1995; Raga *et al.* 1996a). However, these collisions appear to be rare, and the calculations currently only have applications to HH 110 and possibly to HH 7-11.

8.4. STEADY JETS

It has been suggested that the knots along HH jets could be the result of stationary recollimation shocks (Falle *et al.* 1987; Cantó *et al.* 1989; Raga 1991; Raga *et al.* 1990b, 1991a; Biro *et al.* 1993) or MHD compression waves (Pelletier & Pudritz 1992; Bacciotti *et al.* 1997; Ouyed & Pudritz 1997a) formed within the jet beam. The theory of such stationary structures has been studied in detail, and both analytical and numerical descriptions have been obtained for the radiative, HH jet case, and detailed predictions of observable quantities have been made. Detailed studies have been carried out of the effects of the gravity of the central object (Fiege & Henriksen 1996a,b), of an environmental pressure gradient (Smith 1994b; Cantó & Raga 1996; Raga & Cantó 1996) and of the presence of a supersonic sidewind (Cantó & Raga 1995). Also, predictions of the observational properties of a steady jet launched by an MHD wind have been made (Shang *et al.* 1998).

The stationary recollimation shock models reproduce very well the knot structures that are produced in laboratory jets. However, the observation of high proper motions in many HH jets clearly indicates that the knots in HH objects are of a different origin from the knots seen in laboratory jets, and that they do not correspond to steady recollimation shocks or compression waves. Because of this fact, recent models of knots in HH jets have focussed on other mechanisms that do produce travelling knots.

8.5. INSTABILITIES IN JET BEAMS

There are several studies in which condensations in HH flows are modelled as being the result of thermal (de Gouveia dal Pino & Opher 1990), hydrodynamic (Silvestro *et al.* 1987; Massaglia *et al.* 1992; Stone *et al.* 1995, 1997; Hardee & Stone 1997; Rossi *et al.* 1997; Micono *et al.* 1998a,b; Downes 1998) or magnetohydrodynamic instabilities (Ouyed & Pudritz 1997b; Goodson *et al.* 1997; Frank *et al.* 1998) in a jet beam, which all produce travelling shock structures. The most detailed predictions of the observable properties of such instabilities are the hydrodynamic simulations of Micono *et al.* (1998b), who compute predicted emission line maps.

The work discussed above has focussed on the production of knot structures in jets as a result of instabilities. It has also been argued that hydrodynamic K-H instabilities in the jet beam might lead to the formation of a turbulent mixing layer (at the outer boundary of the beam) or to a fully turbulent jet (see Kahn 1980), both possibilities being important either for producing the optical/IR emission of HH jets or as coupling mechanisms between the jet and the surrounding molecular outflow (see the review of Bachiller in these Proceedings). Both the turbulent mixing layers and the fully turbulent regime have been modelled in terms of a "turbulent viscosity approach" (Cantó & Raga 1991; Richer *et al.* 1992; Raga *et al.* 1993b; Raymond *et al.* 1994; Stahler 1994; Taylor & Raga 1995; Raga *et al.* 1995; Noriega-Crespo *et al.* 1996), which leads to predictions of the observable properties of a spatially averaged "mean flow". The detailed spatial structure of the turbulent eddies is not modelled with this approach.

The remaining step in the study of hydrodynamic instabilities is to study the transition from the "coherent eddie" regime described by the numerical simulations to the "fully turbulent" regime described by the turbulent viscosity approach. In order to do this, 3D numerical simulations are necessary (as the 2D and 3D turbulent cascades have very different properties). We now see the first attempts to carry out this step for the case of adiabatic, extragalactic jets, and studies for the radiative, HH jet case will probably appear in the literature in the near future.

8.6. JETS FROM VARIABLE SOURCES

It was first suggested by Rees (1978), that a time-variability in the velocity with which the jet is ejected would result in the formation of shock waves travelling down the jet flow. This was studied numerically by Wilson (1984), who tried to model double-lobed radio sources in terms of jets resulting from two "outflow events". The first theoretical study of HH jets from variable sources appears to be the "pulsed jet" model of Choe & Henriksen (1986).

Raga *et al.* (1990a) argued that velocity time-variabilities of supersonic amplitude would result in the formation of two-shock "internal working surfaces" travelling down the jet beam, and that at least part of the emission observed in HH jets could correspond to such structures. This idea has been studied quite extensively both analytically (Kofman & Raga 1992; Raga & Kofman 1992; Raga & Noriega-Crespo 1992; Raga 1992; Raga & Cabrit 1993; Smith *et al.* 1997a) and numerically (Hartigan & Raymond 1993; Stone & Norman 1993; Falle & Raga 1993, 1995; de Gouveia dal Pino & Benz 1994; Biro & Raga 1994; Biro 1996; Suttner *et al.* 1997; Smith *et al.* 1997b) with specific application to HH jets and molecular outflows.

The formation of working surfaces in a variable ejection velocity jet is an intuitively simple process. During the periods in which the ejection velocity is growing as a function of time, one produces a situation in which the faster material ejected at later times "catches up" with the slower material ejected earlier. This "catching up" occurs at a finite distance x_c from the source, which is given by the expression :

$$x_c = \left(\frac{u_0{}^2}{du_0/d\tau} \right)_{min} , \qquad (1)$$

where $u_0(\tau)$ is the ejection velocity (as a function of the time τ at which the flow is ejected), and the minimum is obtained over the period of the ejection velocity variability.

For distances larger than x_c, an internal working surface is formed (see Fig. 14). This working surface then travels down the jet flow. Initially, the strength of the shocks of the working surface grow with increasing x, then reach a maximum, and eventually decay monotonically as the working surface travels away from the source.

If one allows the source to have an ejection velocity variability that can be described by a superposition of sinusoidal modes, it is possible to obtain complex morphologies. In order to understand the effects of such a multi-mode variability, one can rewrite eq. (1) in the approximate way :

$$x_{c,a} \approx \frac{\bar{v}^2 \tau_a}{2v_a}, \qquad (2)$$

where $x_{c,a}$ is the distance from the source at which the working surfaces of the mode with period τ_a and (full) amplitude v_a are formed. In this equation, $\bar{v}$ is the mean velocity of the time-dependent ejection velocity. From equation (2), we then see that for a fixed amplitude, the larger period modes form working surfaces at increasing distances from the source.

The working surfaces for the different modes are formed at a distance $x_{c,a}$ given (in an approximate way) by equation (2). The shocks associated with the working surface first grow in strength, peak at a distance $x_{m,a} \approx \pi x_{c,a}/2$, and then decay for larger distances from the source (with shock velocities following a $v_s \propto 1/x$ law, see Raga & Kofman 1992).

The different modes of a variable ejection velocity have an interesting interaction. The long period modes produce a modulation of the amplitude of the short period modes, resulting in the production of "trains" of short period knots travelling down the jet beam. This mechanism has been proposed by Raga & Noriega-Crespo (1998) to explain the chains of knots observed in some HH jets. In the following section, this work is described in more detail.

Finally, the effect of a time-variability in the ejection direction of HH jets has also been studied (Raga *et al.* 1993a; Biro *et al.* 1995; Cliffe *et al.* 1996). Furthermore, a general velocity+direction time variability has been suggested as a mechanism to convert an HH jet into a series of disconnected "bullets" (Raga & Biro 1993; Suttner *et al.* 1997). A detailed study of the application of these models to actual HH jets still remains to be done.

8.7. COMPARING MODELS AND OBSERVATIONS OF HH JETS

Most of the comparisons between observations of HH objects and theoretical models have been limited to analyses based either on plane-parallel shock or on bow shock models (see, e.g., Solf *et al.* 1986; Hartigan *et al.* 1987; Raga *et al.* 1986). Comparisons with bow shock models of observed line ratios, line profiles, spatially resolved line profiles and narrow band images have been carried out.

Interestingly, more advanced models have mostly been used only for carrying out purely qualitative comparisons with observations. For example, it has been shown that both models from variable sources as well as models of K-H instability modes can produce knots that qualitatively resemble the knots along HH jets (in the sense that predicted Hα emission line maps do show a series of aligned knots).

It has been pointed out that objects that show knots with a strong jet to counter-jet symmetry probably are the result of a source variability, since there is no reason to expect that Kelvin-Helmholtz modes (arising from instabilities in the jet beam/environment boundary) would show such a symmetry (Gredel & Reipurth 1994, Zinnecker *et al.* 1998). However, the converse argument (i.e., that asymmetric jet/counterjet systems have to be the result of Kelvin-Helmholtz modes) does not necessarily follow, since the propagation of internal working surfaces could be strongly affected by the existence of differences in the properties of the environments surrounding the jet and the counterjet. This, however, is a topic that has not yet been explored in a detailed way.

Recently, Raga & Noriega-Crespo (1998) have used analytic considerations to derive parameters for a multimode variability that could explain the observed proper motions of the knots along the HH 34 jet. The Hα intensity maps predicted from a numerical simulation (computed with the derived parameters) are shown in Fig. 15. It is clear that the predicted intensity maps do show a strong resemblance to the morphology and intensities observed for the knots along the HH 34 jet.

In the future, more efforts will have to be made to attempt to reproduce the kinematical properties, line ratios, and morphologies (particularly, the high resolution maps obtained with HST) of different HH jets. The best candidates for such studies are of course the objects that show the more organized spatial structures (e.g., HH 34, 111, 211, 212, 228), as these structures might be easier to model than the more complex and chaotic structures observed in other HH objects.

Even though the more promising model for HH jets at this time appears to be the "internal working surface" model, important questions still need to be answered. For example :

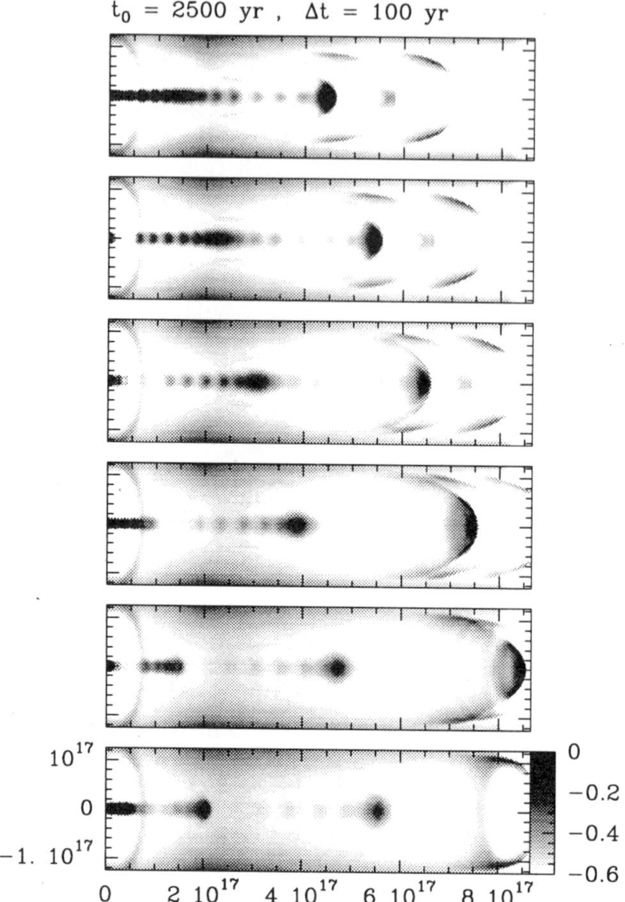

Figure 15.   *Time sequence of Hα intensity maps obtained from a three-mode, variable ejection velocity model of HH 34 (from Raga & Noriega-Crespo 1998). The jet is ejected with a velocity variability with three sinusoidal modes (with half-amplitudes of 70, 40 and 15 km/s, and periods of 1200, 310 and 27 years, respectively) chosen so as to reproduce the present-day structure of the HH 34 jet. The Hα maps have been calculated assuming an angle of 30° between the outflow axis and the plane of the sky. The top frame corresponds to an integration time of 2500 yr, and the following frames correspond to time-intervals of 100 yr. The greyscale is a logarithmic representation of the Hα intensity (the relative scale, normalized to the peak of the t=2900 yr panel, is shown in the wedge to the right of the lower frame).*

    - can variable velocity jet models explain the spatial distribution of emitting knots in different HH jets (this has only been attempted for HH 34, see above) ?

    - can the interaction with the surrounding environment modify the properties of internal working surfaces, or do knots in asymmetric jet/counterjets require a different interpretation (e.g., K-H modes) ?

- what are the detailed consequences of the presence of a stratified environment and/or motion of the source ?

Finally, an important problem in modelling HH objects is the interpretation of the emission line ratios. Though the observed line ratios were the inspiration that initially led Schwartz (1978) to suggest that the emission of HH objects was excited by the passage of shock waves, the detailed modelling of these line ratios has proved to be more difficult than expected. Some of the problems encountered are summarized by Raga *et al.* (1996b), and are described in many papers in the literature.

9. References

Ambartsumian, V.A. 1947, *Stellar Evolution and Astrophysics* (Yerevan, Acad. Sci. Armenian SSR.)

Ambartsumian, V.A. 1954, *Comm. Byurakan Obs.* No. 13

Axon, D.J., Taylor, K. 1984, *MNRAS* 207, 241

Bacciotti, F., Chiuderi, C., Oliva, E. 1995, *A&A* 296, 185

Bacciotti, F., Chiuderi, C., Pouquet, A. 1997, *ApJ* 478, 594

Bally, J., Devine, D. 1994, *ApJ* 428, L65

Bally, J., Devine, D., Fesen, R., Lane, A. 1995, *ApJ* 454, 345

Bally, J., Devine, D., Reipurth, B. 1996, *ApJ* 473, L49

Barral, J. F., Cantó, J. 1981, *RMxAA* 5, 101

Beckwith, S., Gatley, I., Matthews, K., Neugebauer, G. 1978, *ApJ* 223, L41

Binette, L., Dopita, M. A., Tuohy, I. R. 1985, *ApJ* 297, 476

Biro, S. 1996, *MNRAS* 278, 990

Biro, S., Raga, A. C. 1994, *ApJ* 434, 221

Biro, S., Cantó, J., Raga, A. C., Binette, L. 1993, *RMxAA* 25, 95

Biro, S., Raga, A. C., Cantó, J. 1995, *MNRAS* 275, 557

Blondin, J. M., Fryxell, B. A., Königl, A. 1990, *ApJ* 360, 370

Blondin, J. M., Königl, A., Fryxell, B. A. 1989, *ApJ* 337, L37

Böhm, K.H. 1956, *ApJ* 123, 379

Böhm, K. H. 1978, *A&A* 64, 115

Böhm, K.H. 1983, *Rev. Mex. Astron. Astrofis.* 7, 55

Böhm, K.H. 1990, in *Structure and Dynamics of the Interstellar Medium*, eds. G. Tenorio-Tagle, M. Moles, J. Melnick, Springer, p. 282

Böhm, K.H., Böhm-Vitense, E. 1984, *ApJ* 277, 216

Böhm, K.H., Raga, A.C. 1987, *PASP* 99, 265

Böhm, K.H., Solf, J. 1990, *ApJ* 348, 297

Böhm, K.H., Goodson, A.P. 1997, in *Herbig-Haro Flows and the Birth of Low Mass Stars*, IAU Symp. No. 182, Eds. Bo Reipurth & C. Bertout, p.47

Böhm, K.H., Perry, J.F., Schwartz, R.D. 1973, *ApJ* 179, 149

Böhm, K.H., Schwartz, R.D., Siegmund, W.A. 1974, *ApJ* 193, 353

Böhm, K.H., Brugel, E.W., Mannery, E. 1980, *ApJ* 235, L137

Böhm, K.H., Böhm-Vitense, E., Brugel, E.W. 1981, *ApJ* 245, L113

Böhm, K.H., Brugel, E.W., Olmsted, E. 1983, *A&A* 125, 23

Böhm, K.H., Bührke, Th., Raga, A.C., Brugel, E.W., Witt, A.N., Mundt, R. 1987, *ApJ* 316, 349

Böhm, K.H., Raga, A.C., Binette, L. 1991, *PASP* 103, 85
Böhm, K.H., Noriega-Crespo, A., Solf, J. 1993, *ApJ* 416, 647
Böhm-Vitense, E., Böhm, K.H., Cardelli, J.A., Nemec, J.M. 1982, *ApJ* 262, 224
Brown, A., Jordan, C., Millar, T.J., Gondhalekar, P., Wilson, R. 1981, *Nature* 290, 34
Brugel, E.W., Böhm, K.H., Mannery, E. 1981a, *ApJS* 47, 117
Brugel, E.W., Böhm, K.H., Mannery, E. 1981b, *ApJ* 243, 874
Brugel, E.W., Shull, J.M., Seab, C.G. 1982, *ApJ* 262, L35
Brugel, E.W., Böhm, K.H., Shull, J.M., Böhm-Vitense, E. 1985, *ApJ* 292, L75
Bührke, T., Mundt, R., Ray, T.P. 1988, *A&A* 200, 99
Calvet, N., Cantó, J., Binette, L., Raga, A. C. 1992, *RMxAA* 24, 81
Cameron, M., Liseau, R. 1990, *A&A* 240, 409
Cantó, J. 1980, *A&A* 86, 327
Cantó, J., Rodríguez, L. F. 1980, *ApJ* 239, 982
Cantó, J., Raga, A. C. 1991, *ApJ* 372, 646
Cantó, J., Raga, A. C. 1995, *MNRAS* 227, 1120
Cantó, J., Raga, A. C. 1996, *MNRAS* 280, 559
Cantó, J., Raga, A. C. 1998, *MNRAS* 297, 383
Cantó, J., Raga, A. C., Binette, L. 1989, *RMxAA* 17, 65
Cantó, J., Sarmiento, A., Rodríguez, L. F. 1986, *RMxAA* 13, 107
Cantó, J., Tenorio-Tagle, G. Róźyczka, M. 1988, *A&A* 192, 287
Carr, J.S. 1993, *ApJ* 406, 553
Cerqueira, A. H., de Gouveia dal Pino, E. M., Herant, M. 1997, *ApJ* 489, L185
Charnley, S. B., Dyson, J. E., Hartquist, T. W., Williams, D.A. 1988, *MNRAS* 235, 1257
Chernin, L., Masson, C., de Gouveia dal Pino, E., Benz, W. 1994, *ApJ* 426, 204
Chevalier, R.A., Raymond, J.C. 1978, *ApJ* 225, L27
Choe, S.U., Böhm, K.H., Solf, J. 1985, *ApJ* 288, 338
Choe, S. U., Henriksen, R. N, 1986, *ApJ* 305, 131
Cliffe, J. A., Frank, A., Jones, T. W. 1996, *MNRAS* 282, 1114
Corcoran, M., Ray, T.P. 1998, *A&A* 336, 535
Curiel, S., Cantó, J., Rodríguez, L. F. 1987, *RMxAA* 14, 595
Davis, C.J., Eislöffel, J., Ray, T.P. 1994, *ApJ* 426, L93
Devine, D., Bally, J., Reipurth, B., Heathcote, S. 1997, *AJ* 114, 2095
Devine, D., Bally, J., Reipurth, B., Shepherd, D., Watson, A. 1999, *AJ* in press
Dopita, M.A. 1978, *ApJS* 37, 117
Dopita, M.A., Binette, L., Schwartz, R.D. 1982a, *ApJ* 261, 183
Dopita, M.A., Schwartz, R.D., Evans, I. 1982b, *ApJ* 263, L73
Downes, T. P. 1998, *A&A* 331, 1130
Eislöffel, J., Mundt, R. 1994, *A&A* 284, 530
Eislöffel, J., Mundt, R., Böhm, K.-H. 1994a, *AJ* 108, 1042
Eislöffel, J., Davis, C.J., Ray, T.P., Mundt, R. 1994b, *ApJ* 422, L91
Elias, J.H. 1980, *ApJ* 241, 728
Falle, S. A. E. G., Raga, A. C. 1993, *MNRAS* 261, 573
Falle, S. A. E. G., Raga, A. C. 1995, *MNRAS* 272, 785
Falle, S. A. E. G., Innes, D. E., Wilson, M. J. 1987, *MNRAS* 225, 741
Feldman, H. A., Raga, A. C. 1991, *AJ* 102, 2049
Fiege, J. D., Henriksen, R. N. 1996s, *MNRAS* 281, 1038
Fiege, J. D., Henriksen, R. N. 1996b, *MNRAS* 281, 1055
Frank, A., Noriega-Crespo, A. 1994, *A&A* 290, 643

Frank, A., Mellema, G. 1996, *ApJ* 472, 684
Frank, A., Ryu, D., Jones, T. W., Noriega-Crespo, A. 1998, *ApJ* 494, L79
Gaetz, T. J. 1990, *ApJ* 353, 357
Gaetz, T. J., Edgar, R. J., Chevalier, R. A. 1988, *ApJ* 329, 927
Ghavamian, P., Hartigan, P. 1998, *ApJ* 501, 687
Goodson, A. P., Winglee, R. M., Böhm, K. H. 1997, *ApJ* 489, 199
de Gouveia dal Pino, E. M., Opher, R. 1990, *A&A* 231, 571
de Gouveia dal Pino, E. M., Benz, W. 1993, *ApJ* 410, 686
de Gouveia dal Pino, E. M., Benz, W. 1994, *ApJ* 435, 261
de Gouveia dal Pino, E. M., Birkinshaw, M. 1996, *ApJ* 471, 832
de Gouveia dal Pino, E. M., Birkinshaw, M., Benz, W. 1996, *ApJ* 460, L111
Graham, J.A., Elias, J.H. 1983, *ApJ* 272, 615
Gredel, R. 1994, *A&A* 292, 580
Gredel, R., Reipurth, B. 1994, *A&A* 289, L19
Hardee, P. E., Stone, J. M. 1997, *ApJ* 483, 136
Haro, G. 1952, *ApJ* 115, 572
Haro, G. 1953, *ApJ* 117, 73
Haro, G., Minkowski, R. 1960, *AJ* 65, 490
Hartigan, P. 1989, *ApJ* 339, 987
Hartigan, P., Raymond, J. C. 1993, *ApJ* 409, 705
Hartigan, P., Raymond, J., Hartmann, L. 1987, *ApJ* 316, 323
Hartigan, P., Curiel, S., Raymond, J. 1989, *ApJ* 347, L31
Hartigan, P., Raymond, J., Meaburn, J. 1990, *ApJ* 362, 624
Hartigan, P., Morse, J. A., Raymond, J. C. 1994, *ApJ* 436, 125
Hartigan, P., Morse, J. A., Raymond, J. C. 1995, *ApJ* 444, 943
Hartigan, P., Morse, J., Tumlinson, J., Raymond, J., Heathcote, S. 1999, *ApJ* in
    press
Hartmann, L., Raymond, J. 1984, *ApJ* 276, 560
Heathcote, S., Reipurth, B. 1992, *AJ* 104, 2193
Heathcote, S., Morse, J.A., Hartigan, P., Reipurth, B., Schwartz, R.D., Bally, J.,
    Stone, J.M. 1996, *AJ* 112, 1141
Heathcote, S., Reipurth, B., Raga, A.C. 1998, *AJ* 116, 1940
Henney, W. J. 1996, *RMxAA* 32, 3
Henney, W. J., Raga, A. C., Axon, D. J. 1994, *ApJ* 427, 305
Henney, W. J., Axon, D. J. 1995, *ApJ* 454, 233
Herbig, G.H. 1950, *ApJ* 111, 11
Herbig, G.H. 1951, *ApJ* 113, 697
Herbig, G.H. 1952, *J.R. Astron. Soc. Can.* 46, 222
Herbig, G.H., Jones, B.F. 1981, *AJ* 86, 1232
Herbig, G.H., Jones, B.F. 1983, *AJ* 88, 1040
Hirth, G.A., Mundt, R., Solf, J. 1994, *ApJ* 427, L99
Indebetouw, R., Noriega-Crespo, A. 1995, *AJ* 109, 752
Innes, D. E., Giddings, J. R., Falle, S. A. E. G. 1987a, *MNRAS* 224, 179
Innes, D. E., Giddings, J. R., Falle, S. A. E. G., 1987b, *MNRAS* 226, 67
Innes, D. E., Giddings, J. R., Falle, S. A. E. G., 1987c, *MNRAS* 227, 1021
Joy, A.H. 1942, *PASP* 54, 15
Kahn, F. D. 1980, *A&A* 83, 303
Kim, S.-H., Raga, A. C. 1991, *ApJ* 379, 689
Kofman, A. C., Raga, A. C. 1992, *ApJ* 390, 359
Königl, A. 1982, *ApJ* 261, 115

Lada, C.J. 1987, in IAU Symp. no. 115 *Star Forming Regions*, eds. M. Peimbert, J. Jugaku (Reidel), p. 1

Lim, A. J., Raga, A. C. 1998, *MNRAS* 298, 871

Liseau, R., Huldtgren, M., Fridlund, C.V.M., Cameron, M. 1996, *A&A* 306, 255

Mac Low, M. M., Elitzur, M., Stone, J. M., Königl, A. 1994, *ApJ* 427, 914

Martí, J., Rodríguez, L.F., Reipurth, B. 1993, *ApJ* 416, 208

Martí, J., Rodríguez, L.F., Reipurth, B. 1995, *ApJ* 449, 268

Martí, J., Rodríguez, L.F., Reipurth, B. 1998, *ApJ* 502, 337

Massaglia, S., Trussoni, E., Bodo, G., Rossi, P., Ferrari, A. 1992, *A&A* 260, 243

Masson, C., Chernin, L. 1993, *ApJ* 414, 230

Meaburn, J., Dyson, J. 1987, *MNRAS* 225, 863

Mellema, G., Frank, A. 1997, *MNRAS* 292, 795

Micono, M., Massaglia, S., Bodo, G., Rossi, P., Ferrari, A. 1998a, *A&A* 333, 989

Micono, M., Massaglia, S., Bodo, G., Rossi, P., Ferrari, A. 1998b, *A&A* 333, 1001

Miller, J.S. 1968, *ApJ* 154, L57

Morse, J. A., Hartigan, P., Heathcote, S., Raymond, J. C., Cecil, G. 1994, *ApJ* 425, 738

Mundt, R., Fried, J.W. 1983, *ApJ* 274, L83

Mundt, R., Stocke, J., Stockman, S. 1983, *ApJ* 265, L71

Mundt, R., Bührke, T., Fried, J.W., Neckel, T., Sarcander, M., Stocke, J. 1984, *A&A* 140, 17

Mundt, R., Brugel, E.W., Bührke, T. 1987, *ApJ* 319, 275

Mundt, R., Ray, T.P., Bührke, T., Raga, A.C., Solf, J. 1990, *A&A* 232, 37

Mundt, R., Ray, T.P., Raga, A.C. 1991, *A&A* 252, 740

Noriega-Crespo, A., Böhm, K. H., Raga, A. C. 1989, *AJ* 98, 1388

Noriega-Crespo, A., Böhm, K. H., Raga, A. C. 1990, *AJ* 99, 1918

Noriega-Crespo, A., Böhm, K. H., Calvet, N. 1991, *ApJ* 379, 676

Noriega-Crespo, A., Garnavich, P. M., Raga, A. C., Cantó, J., Böhm, K. H. 1996, *ApJ* 462, 804

Norman, C., Silk, J. 1979, *ApJ* 228, 197

Ogura, K., Walsh, J. 1992, *ApJ* 400, 248

Ortolani, S., D'Odorico, S. 1980, *A&A* 83, L8

Osterbrock, D.E. 1958, *PASP* 70, 399

Ouyed, R., Pudritz, R. E. 1997a, *ApJ* 482, 712

Ouyed, R., Pudritz, R. E. 1997b, *ApJ* 484, 794

Pelletier, G., Pudritz, R. E. 1992, *ApJ* 394, 117

Peter, W., Eichler, D. 1996, *ApJ* 466, 840

Poetzel, R., Mundt, R., Ray, T.P. 1992, *A&A* 262, 229

Raga, A. C. 1986, *AJ* 92, 637

Raga, A. C. 1988, *ApJ* 335, 820

Raga, A. C. 1991, *AJ* 101, 1472

Raga, A. C. 1992, *MNRAS* 258, 301

Raga, A.C., Böhm, K.H. 1985, *ApJS* 58, 201

Raga, A.C., Böhm, K.H. 1986, *ApJ* 308, 829

Raga, A.C., Böhm, K.H. 1987, *ApJ* 323, 193

Raga, A.C., Mateo, M. 1987, *AJ* 94, 684

Raga, A.C., Mateo, M. 1988, *AJ* 95, 543

Raga, A. C., Cantó, J. 1989a, *ApJ* 344, 404

Raga, A. C., Cantó, J. 1989b, *PASP* 101, 1151

Raga, A. C., Binette, L. 1991, *RMxAA* 22, 265

Raga, A. C., Kofman, L. 1992, *ApJ* 386, 222
Raga, A. C., Noriega-Crespo, A. 1992, *RMxAA* 24, 9
Raga, A. C., Noriega-Crespo, A. 1993, *RMxAA* 25, 149
Raga, A. C., Biro, S. 1993, *MNRAS* 264, 758
Raga, A.C., Cabrit, S. 1993, *A&A* 278, 267
Raga, A. C., Wang, L. 1994, *MNRAS* 268, 354.
Raga, A. C., Cantó, J. 1995, *RMxAA* 31, 51
Raga, A. C., Cantó, J. 1996, *MNRAS* 280, 567
Raga, A. C., Noriega-Crespo, A. 1998, *AJ* 116, 2943
Raga, A.C., Böhm, K.H., Solf, J. 1986, *AJ* 92, 119
Raga, A.C., Mateo, M., Böhm, K.H., Solf, J. 1988, *AJ* 95, 1783
Raga, A.C., Cantó, J., Binette, L., Calvet, N. 1990a, *ApJ* 364, 601
Raga, A.C., Binette, L., Cantó, J. 1990b, *ApJ* 360, 612
Raga, A. C., Biro, S., Cantó, J., Binette, L. 1991a, *RMxAA* 22, 243
Raga, A.C., Mundt, R., Ray, T.P. 1991b, *A&A* 252, 733
Raga, A. C., Cantó, J., Biro, S. 1993a, *MNRAS* 260, 163
Raga, A. C., Cantó, J., Calvet, N., Rodríguez, L. F., Torrelles, J. M. 1993b, *A&A* 276, 539
Raga, A. C., Cabrit, S., Cantó, J. 1995a, *MNRAS* 273, 422
Raga, A. C., Taylor, S. D., Cabrit, S., Biro, S. 1995b, *A&A* 296, 833
Raga, A. C., Böhm, K. H., Cantó, J. 1996a, *RMxAA* 32, 161
Raga, A. C., Cantó, J., Steffen, W. 1996b, *QJRAS* 37, 493
Raga, A. C., Cantó, J., Curiel, S., Noriega-Crespo, A., Raymond, J. C. 1997a, *RMxAA* 33, 157
Raga, A. C., Mellema, G., Lundqvist, P. 1997b, *ApJS* 109, 517
Raga, A. C., Cantó, J., Curiel, S., Taylor, S. 1998a, *MNRAS* 295, 738
Raga, A. C., Cantó, J., Cabrit, S. 1998b, *A&A* 332, 714
Ray, T.P., Poetzel, R., Solf, J., Mundt, R. 1990, *ApJ* 357, L45
Raymond, J.C. 1979, *ApJS* 39, 1
Raymond, J. C., Hartmann, L., Hartigan, P. 1988, *ApJ* 326, 323
Raymond, J. C., Morse, J. A., Hartigan, P., Curiel, S., Heathcote, S. 1994, *ApJ* 434, 232
Rees, M. J. 1978, *MNRAS* 184, 61
Reipurth, B. 1989a, *Nature* 340, 42
Reipurth, B. 1989b, *A&A* 220, 249
Reipurth, B., Graham, J.A. 1988, *A&A* 202, 219
Reipurth, B., Heathcote, S. 1991, *A&A* 246, 511
Reipurth, B., Olberg, M. 1991, *A&A* 246, 535
Reipurth, B., Heathcote, S. 1997, in IAU Symp. No. 182 *Herbig-Haro Flows and the Birth of Low Mass Stars*, Eds. Bo Reipurth & C. Bertout, p. 3
Reipurth, B., Bally, J., Graham, J., Lane, A., Zealey, W. 1986, *A&A* 164, 51
Reipurth, B., Raga, A.C., Heathcote, S. 1992, *ApJ* 392, 145
Reipurth, B., Raga, A.C., Heathcote, S. 1996, *A&A* 311, 989
Reipurth, B., Hartigan, P., Heathcote, S., Morse, J., Bally, J. 1997a, *AJ* 114, 757
Reipurth, B., Bally, J., Devine, D. 1997b, *AJ* 114, 2708
Reipurth, B, Bally, J., Fesen, R., Devine, D. 1998a, *Nature* 396, 343
Reipurth, B., Devine, D., Bally, J. 1998b, *AJ* 116, 1396
Richer, J. S., Hills, R. E., Padman, R. 1992, *MNRAS* 254, 525
Rodríguez, L.F., Reipurth, B. 1989, *Rev. Mex. Astron. Astrofis.* 17, 59
Rodríguez, L.F., Reipurth, B. 1994, *A&A* 281, 882

Rodríguez, L.F., Moran, J.M., Ho, P.T.P., Gottlieb, E.W. 1980, *ApJ* 235, 845
Rossi, P., Bodo, G., Massaglia, S., Ferrari, A. 1997, *A&A* 321, 672
Schwartz, R.D. 1975, *ApJ* 195, 631
Schwartz, R.D. 1978, *ApJ* 223, 884
Schwartz, R.D. 1983, *ApJ* 268, L37
Schwartz, R. D., Dopita, M. A. 1980, *ApJ* 236, 543
Schwartz, R.D., Dopita, M.A., Cohen, M. 1985, *AJ* 90, 1820
Schwartz, R.D., Williams, P.M., Cohen, M., Jennings, D.G. 1988, *ApJ* 334, L99
Schwartz, R.D., Schultz, A.S.B., Cohen, M., Williams, P.M. 1995, *ApJ* 446, 318
Shang, H., Shu, F. H., Glassgold, A. E. 1998, *ApJ* 439, L91
Shull, J. M., McKee, C. F. 1979, *ApJ* 227, 131
Silvestro, G., Ferrari, A., Rosner, R., Trussoni, E., Tsinganos, K. 1987, *Nature* 325, 228
Smith, M. D. 1986, *MNRAS* 223, 57
Smith, M. D. 1991, *MNRAS* 252, 378
Smith, M.D. 1993, *ApJ* 406, 520
Smith, M.D. 1994a, *MNRAS* 266, 238
Smith, M. D. 1994b, *ApJ* 421, 400
Smith, M. D., Brand, P., 1990a, *MNRAS* 242, 495
Smith, M. D., Brand, P., 1990b, *MNRAS* 243, 498
Smith, M. D., Brand, P., 1990c, *MNRAS* 244, 384
Smith, M. D., Brand, P., 1990d, *MNRAS* 245, 108
Smith, M. D., Brand, P., Moorhouse, A. 1991a, *MNRAS* 248, 451
Smith, M. D., Brand, P., Moorhouse, A. 1991b, *MNRAS* 248, 730
Smith, M. D., Suttner, G., Zinnecker, H. 1997a, *A&A* 320, 325
Smith, M. D., Suttner, G., Yorke, H. W. 1997b, *A&A* 323, 223
Solf, J.   1987, *A&A* 184, 322
Solf, J., Böhm, K.H.   1987, *AJ* 93, 1172
Solf, J., Böhm, K.H.   1991, *ApJ* 375, 618
Solf, J., Böhm, K.H.   1993, *ApJ* 410, L31
Solf, J., Böhm, K.H., Raga, A.C. 1986, *ApJ* 305, 795
Solf, J., Böhm, K.H., Raga, A.C. 1988, *ApJ* 334, 229
Stahler, S. W. 1994, *ApJ* 422, 616
Stone, J. M., Norman, M. L. 1993a, *ApJ* 413, 198
Stone, J. M., Norman, M. L. 1993b, *ApJ* 413, 210
Stone, J. M., Norman, M. L. 1994, *ApJ* 420, 237
Stone, J. M., Xu, J., Mundy, L. G. 1995, *Nature* 377, 315
Stone, J. M., Xu, J., Hardee, P. E. 1997, *ApJ* 483, 136
Suttner, G., Smith, M. D., Yorke, H. W., Zinnecker, H. 1997, *A&A* 318, 595
Taylor, S. D., Raga, A. C. 1995, *A&A* 296, 823
Wilkin, F. P., Stahler, S. W. 1998, *ApJ* 502, 661
Wilson, M. J. 1984, *MNRAS* 209, 923
Wolfire, M. G., Königl, A. 1993, *ApJ* 415, 204
Wolfire, M. G., Königl, A. 1991, *ApJ* 383, 205
Yamashita, T., Suzuki, H., Kaifu, N., Tamura, M., Mountain, C.M., Moore, T.J.T. 1989, *ApJ* 347, 894
Zealey, W., Williams, P., Sandell, G., Taylor, K., Ray, T. 1992, *A&A* 262, 570
Zinnecker, H., Mundt, R., Geballe, T.R., Zealey, W.J. 1989, *ApJ* 342, 337
Zinnecker, H., McCaughrean, M., Rayner, J. 1998, *Nature* 394, 862

MAGNETIC FIELDS AND STAR FORMATION: A THEORY
REACHING ADULTHOOD

TELEMACHOS CH. MOUSCHOVIAS
University of Illinois at Urbana-Champaign
Departments of Physics and Astronomy
1002 West Green Street
Urbana, IL 61801, U.S.A.

E-mail: tchm@astro.uiuc.edu

and

GLENN E. CIOLEK
NY Center for Studies on the Origins of Life (NSCORT)
Rensselaer Polytechnic Institute
110 8th Street, Troy, NY 12180, U.S.A.

E-mail: ciolek@charon.phys.rpi.edu

1. Introduction – A Modern Theory of Star Formation

The correctness of any theory rests with the lack of mathematical or numerical errors, so that its conclusions would follow from its assumptions and known physical laws. A correct theory, however, can be irrelevant unless its assumptions are consistent with observations or experiments. It is therefore essential that the physical properties of star-forming interstellar clouds be understood quantitatively before a theory of star formation becomes both correct and relevant.

Typical, massive ($M \sim 10^2 - 10^5 \ M_\odot$) interstellar molecular clouds, in which most star formation activity is observed, are relatively dense, cold, nearly isothermal objects (number density of neutral particles $n_n \gtrsim 10^3$ cm^{-3}, and temperature $T \simeq 10$ K). High-energy (> 100 MeV) cosmic rays maintain a degree of ionization $x_i \equiv n_i/n_n \lesssim 10^{-7}$ in their deep interiors, and ultraviolet (UV) radiation is responsible for the much greater degree of ionization ($x_i \sim 10^{-4}$) in the outer envelopes. Approximately 1% of the mass of a cloud is in the form of dust particles (or grains), 90% of which carry a single electronic charge $-e$ at any one time, the remaining 10% being electrically neutral. In the density range of main interest ($\sim 10^3 - 10^{10}$ cm^{-3}), the dominant ions in dense cores are HCO$^+$, Mg$^+$, Na$^+$, and Fe$^+$, and in cloud envelopes C$^+$, S$^+$, Mg$^+$, Si$^+$, Na$^+$, Fe$^+$, and HCO$^+$. Zeeman measurements yield magnetic field strengths $B \simeq 10 - 150$ μG, with 30 μG being typical, while polarization observations reveal nicely ordered magnetic fields inside and in the neighborhoods of clouds. Molecular clouds rarely exhibit rotation significantly higher than that of the background medium and, when they do, their measured angular velocities are much smaller than those expected from angular-momentum conservation from an initial angular velocity $\Omega_0 \simeq 10^{-15}$ s^{-1} and density $n \simeq 1$ cm^{-3} (the approximate local Galactic rotation rate and mean interstellar density). The observed spectral linewidths are almost always supersonic but subAlfvénic.

Despite the fact that the critical mass for collapse of a pressure-bounded isothermal sphere is only 5.8 $M_\odot$ at a density $n_n = 10^3$ cm^{-3} and temperature $T = 10$ K (Bonnor 1956; Ebert 1955, 1957), *molecular clouds are not collapsing; they are magnetically supported* (Mouschovias 1976a, b; Mouschovias and Spitzer 1976; Spitzer 1978; Mouschovias 1978). The need for multifluid MHD as the proper tool for the theoretical description of star

305

C.J. Lada and N.D. Kylafis (eds.), The Origin of Stars and Planetary Systems, 305–340.

formation arises from the fact that *ambipolar diffusion*[1] is an inevitable process in magnetically supported,[2] self-gravitating clouds. Its timescale (with τ_{ff} denoting the free-fall and τ_{ni} the neutral-ion collision time)

$$\tau_{AD} \simeq \frac{\tau_{ff}^2}{\tau_{ni}} \simeq 2 \times 10^6 \left[\frac{n_i/n_{H_2}}{10^{-7}}\right] \quad \text{yr} \tag{1}$$

was calculated and shown to be typically three orders of magnitude smaller in the deep interior of a cloud than in the outermost envelope, thus suggesting that fragmentation and star formation can be *self-initiated* (Mouschovias 1977, 1979b): Ambipolar diffusion allows fragments, each having mass somewhat greater than the critical thermal mass ($\simeq$ Jeans mass) to form and contract *quasistatically* (i.e., with negligible acceleration) through essentially stationary plasma and magnetic field lines. The presence of the magnetic field does not prevent this process. It simply regulates it. That is, it slows down the contraction through plasma-neutral drag and converts the would-be violent free fall into a quasistatic process. The fragmentation (or core-formation) timescale is the mass-redistribution timescale among the central flux tubes of a cloud. No magnetic flux is lost by the cloud as a whole. The essence of ambipolar diffusion in this modern picture of its role in star formation is precisely this kind of redistribution of mass or, equivalently, the increase of the central mass-to-flux ratio $(dm/d\Phi_B)_c$ (Mouschovias 1978, 1979b). Eventually, $(dm/d\Phi_B)_c$ exceeds the critical (central) value for collapse

$$\left[\frac{dm}{d\Phi_B}\right]_{c,\,crit} = 1.5 \left[\frac{1}{63G}\right]^{1/2} \tag{2}$$

(Mouschovias and Spitzer 1976) and dynamical contraction ensues. For a cloud as a whole and for a core during its quasistatic (or subcritical) phase of contraction, *magnetic braking* is very effective and keeps the cores and the cloud essentially corotating with the background (Mouschovias 1977, 1979a; Mouschovias and Paleologou 1979, 1980; Basu and Mouschovias 1994). Hence, the centrifugal forces have a negligible effect on the evolution. In the case of an *aligned disk rotator* (cloud or fragment) of density ρ_{cl} and half-thickness Z, with field lines "fanning out" away from it, surrounded by an "external" medium (or envelope) of density ρ_{ext}, the magnetic-braking timescale is given by (Mouschovias 1983)

$$\tau_{\parallel\,fan} = \frac{\rho_{cl}}{\rho_{ext}} \frac{Z}{v_{A,ext}} \left[\frac{R_{cl}}{R_0}\right]^4 \equiv \frac{\sigma_{n,cl}}{2\rho_{ext}\,v_{A,ext}} \left[\frac{R_{cl}}{R_0}\right]^4 , \tag{3a, b}$$

$$\equiv \left[\frac{\pi}{\rho_{ext}}\right]^{1/2} \frac{M}{\Phi_B} \left[\frac{R_{cl}}{R_0}\right]^2 , \tag{3c}$$

where $v_{A,ext}$ is the Alfvén speed in the external medium, $\sigma_{n,cl} \equiv M/\pi R^2$ the column density of matter along field lines, R the equatorial radius, Φ_B the magnetic flux threading the cloud

---

[1]Ambipolar diffusion refers to the magnetically driven drift of the plasma relative to the neutrals. It was proposed by Mestel and Spitzer (1956) as a way for a cloud to lose its magnetic flux and thus collapse and fragment to form stars. With the magnetic field frozen in the matter, the critical mass would be $M_{crit} \propto B^3/n^2 = const$ during spherical, isotropic contraction, thought at the time to be the collapse mode; hence, fragmentation would be impossible.

(or fragment), R_0 the original equatorial radius of the flattened cloud, when its density was ρ_0 ($> \rho_{\text{ext}}$), because of its formation by motions along field lines, and its magnetic field was equal to that of the external medium, B_{ext} (see Mouschovias and Morton 1985b, Fig. 2, which shows this geometry, but for several fragments). Note that $\tau_{\parallel\,\text{fan}}$ is equal to the ratio of the moment of inertia of the cloud and the rate of increase of the moment of inertia of that part of the external medium affected by the outward-propagating torsional Alfvén waves. For a *perpendicular rotator* with $\rho_{\text{cl}}/\rho_{\text{ext}} \gg 1$, the magnetic-braking timescale is

$$\tau_{\perp} = \frac{1}{2} \left[\frac{\rho_{\text{cl}}}{\rho_{\text{ext}}} \right]^{1/2} \frac{R}{v_A(R)} \equiv 2 \left[\frac{\pi}{\rho_{\text{cl}}} \right]^{1/2} \frac{M}{\Phi_B} , \qquad \text{(4a, b)}$$

(Mouschovias and Paleologou 1979; Mouschovias 1985), where $v_A(R)$ is the Alfvén speed just outside the cloud (or core) surface. It follows from equation (3c) that, as a cloud (or core) contracts at constant M/Φ_B in an environment whose properties (ρ_{ext}) do not change much as a result of the cloud's (or core's) contraction, $\tau_{\parallel\,\text{fan}}$ decreases as R_{cl}^2 ($\propto Z \propto \rho_{\text{cl}}^{-1/2}$ for isothermal contraction with balance of forces maintained along field lines). A similar contraction for a perpendicular rotator yields that $\tau_{\perp}$ also decreases as $\rho_{\text{cl}}^{-1/2}$ and, moreover, does not depend on ρ_{ext}.

During the collapse phase, force balance is maintained *along* field lines, and both magnetic flux and angular momentum tend to get trapped inside the magnetically and thermally supercritical protostellar cores, with a consequent increase of the magnetic field and angular velocity with gas density as $B_c \propto \rho_c^{1/2}$ and $\Omega_c \propto \rho_c^{1/2}$ (Mouschovias 1976b, 1979a; 1989; Fiedler and Mouschovias 1993; Ciolek and Mouschovias 1994; Basu and Mouschovias 1994). Hence, the leftover angular momentum and magnetic flux can be calculated from a knowledge of the density at which dynamical contraction sets in. For oblate clouds maintaining force balance along field lines, the central density enhancement at which the central mass-to-flux ratio becomes critical is given by

$$\frac{n_{\text{n,c,crit}}}{n_{\text{n,c0}}} \simeq \left[\frac{\sigma_{\text{n,c,crit}}}{\sigma_{\text{n,c0}}} \right]^2 \equiv \left[\frac{B_{\text{c,crit}}}{B_{\text{c0}}} \right]^2 \left[\frac{1}{\mu_{\text{c0}}} \right]^2 \qquad (5)$$

(Mouschovias 1991c, eqs. [9], [10], and § 2.4). The quantity σ_n denotes the neutral column density along field lines; the ratio $B_{\text{c,crit}}/B_{\text{c0}}$ is the total enhancement of the central field strength during the subcritical phase of contraction (it is typically $\simeq 1.3 - 1.7$); and μ_{c0} is the initial central mass-to-flux ratio in units of its critical value for collapse. If the mass-to-flux ratio in the envelope of a molecular cloud is measured, one may compare it to the critical value (eq. [2]) to obtain μ_{c0}, and then use equation (5) with $(B_{\text{c,crit}}/B_{\text{c0}})^2 \simeq 2 - 3$ to obtain an approximate but reasonable value for the factor by which the initial central density will need to increase for critical conditions to be achieved in a core.

Axisymmetric numerical simulations have shown the remarkable result that, even during the dynamical phase of contraction, the infall of a core does not evolve into free fall (see review by Mouschovias 1995). Even at a central density as high as 10^{10} cm^{-3}, the *maximum* infall acceleration does not exceed 30% that of gravity, and the magnetic force dominates the thermal-pressure force everywhere in the supercritical core, except in a small region of

---

[2]Magnetic support can include a contribution from hydromagnetic (HM) waves. In fact, it was suggested that it is precisely the damping of HM waves by ambipolar diffusion in magnetically supported molecular clouds, rather than the growth of perturbations in collapsing clouds, that initiates fragmentation and star formation (Mouschovias 1987a).

size $\simeq$ 30 AU, in which the two forces are comparable.

The leftover magnetic flux at the end of these, relatively early, isothermal stages of contraction ($n_n \lesssim 10^{10}$ cm^{-3}) will not by any means survive to main-sequence densities. Pneuman and Mitchell (1965) had shown that Ohmic dissipation is effective during the opaque stages of contraction. Yet, calculations accounting for magnetic effects applicable to the opaque phases have so far been confined to estimates and comparisons of timescales (i.e., τ_{ff}, τ_{AD}, and τ_{ohmic}; e.g., see Nakano and Umebayashi 1986a, b; Nishi, Nakano, and Umebayashi 1991). Recent dynamical calculations by Desch and Mouschovias (1999), which follow the evolution up to densities $n_n = 10^{12}$ cm^{-3}, show that *ambipolar diffusion is much more effective than Ohmic dissipation in decoupling the magnetic field from the neutral matter*. They predict a magnetic field $\simeq$ 0.1 G in the innermost 20 AU of the protoplanetary disk. This is in excellent agreement with measurements of remanent magnetization in stony chondrites and in very good agreement with measurements in iron-nickel meteorites. As for the angular momenta of main-sequence stars, although we have shown that magnetic braking can resolve the thorny angular momentum problem of star formation during the isothermal phase of contraction and that it can account for the periods of binary stars from 10 hr to 100 yr as well as for the observed rotation of O5 to F5 main-sequence stars, refinements are still needed before the claim can be made that the stellar rotation-mass relation has been understood.

We first summarize in § 2 the five-fluid (neutral molecules, electrons, ions, charged grains and neutral grains) description of the evolution of isothermal, self-gravitating molecular clouds as it relates to star formation. The emphasis is on the physical origin and interpretation of the equations. The conditions for charged-particle attachment to the magnetic field are elucidated, including those for the effect of partial, indirect attachment of negatively-charged grains due to electrostatic attraction by well-attached ion "quasiparticles", and the detachment of ions due to electrostatic coupling to (at least partially detached) negatively charged grains. The two-fluid (plasma, i.e., electrons and ions, plus neutral molecules) approximation is described in § 3 for both nonrotating and rotating model clouds; key results are explained physically in § 4. In § 5 we discuss the effects introduced by the presence of grains. For most of the density range ($10^3 - 10^{10}$ cm^{-3}) of primary concern in this paper, a four-fluid (neutrals, plasma, charged and neutral grains) description is sufficient. A summary of the results of a detailed, quantitative comparison of an evolutionary model with observations is given is § 5.2, while § 5.3 summarizes the effects of UV ionization on the evolution. Some consequences of the presence of hydromagnetic waves are discussed briefly in § 6. The main conclusions are stated in § 7.

2. Five-Fluid MHD Description of Protostar Formation

The three-fluid equations, including their validity conditions, have been reviewed at some length (Mouschovias 1991b), and the five-fluid equations have been developed in Ciolek and Mouschovias (1993). Here we summarize the results (see also Mouschovias 1996a).

2.1. Physical Origin of the Five-Fluid Equations

As explained in § 1, in the density range of interest ($\sim 10^3 - 10^{10}$ cm^{-3}), a molecular cloud is a nearly isothermal, weakly ionized physical system consisting of neutral molecules (mainly H_2, with a He abundance 20% by number), electrons, ions (each of charge $+e$), singly negatively charged grains, and neutral grains. Electric (E) and magnetic (B) fields affect the charged particles, and gravitational ($g = - \nabla\psi$) fields affect, at least in principle, all species. Elastic collisions transfer momentum primarily between the dominant neutral molecules and each of the other species. However, (inelastic) electron capture by neutral grains and ion capture (and neutralization) by charged grains represent a momentum transfer between the charged- and neutral-grain fluids whose effect on at least the less abundant neutral grains cannot be ignored. For the low-frequency phenomena of interest here, we may assume charge neutrality, i.e.,

$$n_i = n_e + n_{g_-} , \tag{6a}$$

so as to suppress or average over (high-frequency) plasma phenomena, and we may ignore the displacement current in Ampere's law, which is then written as

$$\nabla \times B = (4\pi/c)\, j . \tag{6b}$$

The electric current density is defined by

$$j = e(n_i v_i - n_e v_e - n_{g_-} v_{g_-}) , \tag{6c}$$

where n_s and v_s denote the number density and fluid velocity of species s. The time evolution of the magnetic field is given by Faraday's law of induction,

$$\frac{\partial B}{\partial t} = - c\,(\nabla \times E) , \tag{6d}$$

with B satisfying the condition $\nabla \cdot B = 0$ at all times. In quasineutral plasmas the electric field E is not calculated from Maxwell's equation $\nabla \cdot E = 4\pi \rho_q$, where ρ_q is the net charge density. It is obtained instead from the electron or ion force equations in the form of a generalized Ohm's law (e.g., see review by Mouschovias 1991b, § 2.2).

Keeping only the dominant terms, the five momentum (or force) equations are

$$\frac{\partial}{\partial t}(\rho_n v_n) + \nabla \cdot (\rho_n v_n v_n) = - \nabla P_n - \rho_n \nabla \psi + F_{ni} + F_{ne} + F_{ng_-} + F_{ng_0} , \tag{6e}$$

$$0 = - n_e e \left[E + \frac{v_e}{c} \times B \right] + F_{en} , \tag{6f}$$

$$0 = + n_i e \left[E + \frac{v_i}{c} \times B \right] + F_{in} , \tag{6g}$$

$$0 = - n_{g_-} e \left[E + \frac{v_{g_-}}{c} \times B \right] + F_{g_- n} + F_{g_- g_0, \mathrm{inel}} , \tag{6h}$$

$$0 = + F_{g_0 n} + F_{g_0 g_-, \mathrm{inel}} , \tag{6i}$$

where the frictional force on species s due to collisions with species l is denoted by F_{sl}. All these forces arise because of elastic collisions, except $F_{g_- g_0, \mathrm{inel}}$ (and the equal but opposite force $F_{g_0 g_-, \mathrm{inel}}$), which represents the transfer of momentum from the (negatively) charged-grain fluid, because of ion attachment and neutralization, to the neutral-grain fluid (and vice versa for $F_{g_0 g_-, \mathrm{inel}}$). Expressions for these forces are given below.

The thermal pressure and mass density of the neutrals, P_n and ρ_n respectively, are related through the equation of state

$$P_n = \rho_n C_n^2 , \tag{6j}$$

where $C_n = (k_B T / \mu m_H)^{1/2}$ is the isothermal speed of sound in the neutrals, with k_B being the Boltzmann constant and μ the mean mass per neutral particle in units of the atomic-hydrogen mass, m_H. Because the grains contribute only 1% to the mass of a cloud, while the ions and electrons contribute much less than that, the gravitational potential ψ appearing in equation (6e) is calculated from Poisson's equation using only the density of neutrals as a source term,

$$\nabla^2 \psi = 4\pi G \rho_n \ . \tag{6k}$$

Moreover, ionizations and recombinations have a negligible effect on ρ_n, which permits us to write the continuity equation

$$\frac{\partial \rho_n}{\partial t} + \nabla \cdot (\rho_n \mathbf{v}_n) = 0 \ . \tag{6l}$$

It is also a reasonable approximation to use a mass continuity equation for the grains (charged + neutral) at least in the case of clouds (the ones of interest here) which have not yet given birth to stars:

$$\frac{\partial(\rho_{g_-} + \rho_{g_0})}{\partial t} + \nabla \cdot (\rho_{g_-} \mathbf{v}_{g_-} + \rho_{g_0} \mathbf{v}_{g_0}) = 0 \ . \tag{6m}$$

The system of 13 equations (6a) - (6m) contains 15 unknowns. It will be closed once a prescription is given for calculating the number densities of electrons and ions. (The four elastic-collision forces appearing in eqs. [6e] - [6i] will be expressed in terms of known quantities and dependent variables already appearing in the equations, while elimination of the inelastic forces in eqs. [6h] and [6i] will introduce only the velocity of neutral grains as a new unknown, which has already been counted as part of the set of fifteen. The quantities ρ_{g_-} and n_{g_-} are, of course related by $\rho_{g_-} = n_{g_-} m_g$, where m_g is the mass of an individual dust particle; a similar relation exists between ρ_{g_0} and n_{g_0}.)

In our numerical simulations of the formation and evolution of protostellar cores (from 3×10^3 cm^{-3} to 3×10^9 cm^{-3}) we find that the timescales of the microscopic processes (ionization, charge transfer, and recombination, including grain effects) that determine the abundances of charged species are much smaller than the evolutionary timescales of the cores. We may therefore assume that a chemical equilibrium is established and maintained among the charged particles. In such a case, one may adopt the chemistry designed for static clouds (e.g., see Elmegreen 1979). We consider two kinds of ions, molecular (denoted by the subscript m_+), such as HCO$^+$, and metallic or atomic (i.e., elements heavier than carbon; subscript a_+), such as Mg$^+$ and Na$^+$. The creation/destruction rate equations for electrons, molecular ions, and atomic ions are, respectively,

$$\zeta_{CR} n_n = n_e (\alpha_{dr} n_{m_+} + \alpha_{rr} n_{a_+} + \alpha_{eg_0} n_{g_0}) \ , \tag{7a}$$

$$\zeta_{CR} n_n = n_{m_+} (\alpha_{dr} n_e + \beta n_{a_0} + \alpha_{m_+ g_-} n_{g_-}) \ , \tag{7b}$$

$$\beta n_{a_0} n_{m_+} = n_{a_+} (\alpha_{rr} n_e + \alpha_{a_+ g_-} n_{g_-}) \ . \tag{7c}$$

The quantity ζ_{CR} is the cosmic-ray ionization rate ($\simeq 10^{-17} - 10^{-16}$ s^{-1}, Spitzer and Tomasko 1968; Payne, Salpeter, and Terzian 1984); α_{dr} the dissociative recombination rate ($\simeq 10^{-6}$ cm^3 s^{-1}, Langer 1985; Dalgarno 1987); α_{rr} the radiative recombination rate ($\simeq 10^{-11}$ cm^3 s^{-1}, Oppenheimer and Dalgarno 1974); α_{eg_0} the electron capture rate by neutral grains (Spitzer 1948; Hollenbach and Salpeter 1970; Watson and Salpeter 1972); β the charge transfer rate from molecular ions to metallic atoms ($\simeq 0.5 - 8 \times 10^{-9}$ cm^3 s^{-1}, Watson 1976); $\alpha_{m_+ g_-}$ the capture rate of molecular ions by negatively charged grains; and $\alpha_{a_+ g_-}$ the atomic-ion capture rate by negative grains. The above cited values of the reaction rates refer to a $T \simeq 10$ K plasma. The electron and ion capture rates by neutral and negative grains, respectively, are given by Spitzer (1941, 1948) and modified by Draine and Sutin (1987) to account for grain polarizability (see Ciolek and Mouschovias 1993, eqs. [57a] - [57c]). UV ionization has been neglected in equations (7a) - (7c) only for simplicity in presentation (see § 5.3). Also, in equation (7b), cosmic-ray ionization of the dominant neutral molecule (H_2) does not directly lead to the formation of HCO^+. The relevant chain of reactions is: $H_2 + CR \rightarrow H_2^+ + e$, $H_2^+ + H_2 \rightarrow H_3^+ + H$, $H_3^+ + CO \rightarrow HCO^+ + H_2$, but equation (7b) gives a sufficiently accurate molecular-ion abundance for our present purposes because each cosmic-ray ionization of H_2 leads to the formation of a HCO^+ molecule relatively rapidly.[3] (The detailed work of Ciolek and Mouschovias 1995 includes all the relevant intermediate reactions in the chemical model, which also accounts for UV ionization.) The total ion density appearing in equation (6g) is the sum of the molecular and metallic ion densities,

$$n_i = n_{m_+} + n_{a_+} . \tag{8}$$

Although equations (7a) - (7c) refer to electrons and ions, they imply an equilibrium abundance of charged grains at each time during the evolution. By subtracting equations (7b) and (7c) from (7a), we find that

$$\alpha_{eg_0} n_e n_{g_0} = n_{g_-} (\alpha_{a_+g_-} n_{a_+} + \alpha_{m_+g_-} n_{m_+}) . \tag{9}$$

This equation states that the rate at which charged grains are produced by electron capture onto neutral grains is equal to the rate at which they are neutralized by metallic and molecular ion captures. Typically, 90% of the grains are charged and 10% are neutral.

It remains to specify the frictional forces. The force (per unit volume) on species s due to collisions with neutrals is given by

$$F_{sn} = \frac{\rho_s}{\tau_{sn}} (v_n - v_s) , \qquad s = i, e, g_-, g_0 , \tag{10a}$$

where the mean (momentum exchange) collision times, accounting for both s-H_2 and s-He collisions, are

$$\tau_{sn} = \frac{\tau_{sH_2}}{a_{sHe}} = \frac{1}{a_{sHe}} \frac{m_{H_2} + m_s}{\rho_{H_2} \langle \sigma w \rangle_{sH_2}} , \qquad s = i, e, g_-, g_0 , \tag{10b}$$

and the quantity

$$a_{sHe} = 1.14 \quad \text{for } s = i, \tag{10c}$$

$$= 1.16 \quad \text{for } s = e, \tag{10d}$$

$$= 1.28 \quad \text{for } s = g_- \text{ or } g_0 \tag{10e}$$

is the factor by which the presence of He reduces the slowing-down time of species s due to s-H_2 collisions alone. The quantity $\langle \sigma w \rangle_{iH2}$ is the mean collisional rate between ions of mass m_i and hydrogen molecules of mass m_{H2}; it is equal to 1.69×10^{-9} cm^3 s^{-1} for HCO^+ - H_2 collisions, and to a similar value for Na^+ - H_2 and Mg^+ - H_2 collisions (McDaniel and

---

[3] Dissociative recombination of H_3^+, which in principle would compete with the $H_3^+ + CO$ reaction, is negligible despite the measured (see Larsson *et al.* 1993) relatively large rate coefficient. This is so because in the core the electron abundance is small relative to that of CO, and in the envelope atomic ions, such as C^+ and Mg^+, dominate HCO^+ despite the relatively large electron abundance there.

Mason 1973). In calculating a_{sHe} in equations (10c) - (10e) we have made use of the following facts. The ratio of the ion-He and ion-H_2 collisional rates is $\langle \sigma w \rangle_{iHe}/\langle \sigma w \rangle_{iH2} = 0.3679$. The measured value of the e-H_2 cross section at low energies is 6.4×10^{-16} cm^2, and the electron speed $(8k_B T/\pi m_e)^{1/2} = 2.0 \times 10^6$ cm s^{-1} at $T = 10$ K; hence, $\langle \sigma w \rangle_{eH2} = 1.3 \times 10^{-9}$ cm^3 s^{-1}.[4] The grain-H_2 collisional rate is given by

$$\langle \sigma w \rangle_{gh} = \pi a^2 (8k_B T/\pi m_h)^{1/2} , \qquad h = H_2, He, \tag{10f}$$

for the case $|v_n - v_g| < C_n$, a condition fulfilled in the density range of interest here; this rate is the same for both charged and neutral grains of radius $a \geq 10^{-6}$ cm (the kind of grains we consider). The quantity in parentheses on the right-hand side of equation (10f) is the most probable (Maxwellian) speed of a neutral particle s (H_2 or He), and πa^2 is the geometric cross section of a grain.

The collisional forces F_{ns} in equation (6e) are obtained from F_{sn}, specified above, by using Newton's third law, i.e., $F_{ns} = - F_{sn}$, which implies that

$$\tau_{ns} = (\rho_n/\rho_s)\, \tau_{sn} , \qquad s = i, e, g_-, g_0; \tag{11a}$$

$$= a_{He\text{-}s}\, \tau_{H_2 s} = a_{He\text{-}s}\, \frac{m_s + m_{H_2}}{\rho_s \langle \sigma w \rangle_{sH_2}} , \qquad s = i, e, g_-, g_0; \tag{11b}$$

$$\simeq a_{He\text{-}s}\, \frac{1}{n_\alpha \langle \sigma w \rangle_{\alpha H_2}} , \qquad \alpha = i, g_-, g_0 . \tag{11c}$$

The quantity $a_{He\text{-}s}$ is the factor by which the presence of He lengthens the slowing-down time relative to the value it would have if only H_2-s collisions were considered. We find

$$a_{He\text{-}s} = 1.23 \quad \text{for } s = i, \tag{11d}$$

$$= 1.21 \quad \text{for } s = e, \tag{11e}$$

$$= 1.09 \quad \text{for } s = g. \tag{11f}$$

The approximate form of τ_{ns} in equation (11c) follows from the fact that $m_g \gg m_i \gg m_{H_2}$. (Note that eq. [11c] is not valid for H_2-e collisions because $m_e \ll m_{H2}$.) If He-s collisions are ignored but the contribution of He to the inertia of the H_2 fluid is accounted for, then $a_{iHe} = a_{eHe} = a_{gHe} = 1$ and, consequently, $a_{He\text{-}i} = a_{He\text{-}e} = a_{He\text{-}g} = 1.4$.

The inelastic forces appearing in equations (6h) and (6i) are written as

$$F_{g_- g_0, \text{inel}} = \frac{\rho_{g_-}}{\tau_{g_-\, i, \text{inel}}} (v_{g_0} - v_{g_-}) , \qquad F_{g_0 g_-, \text{inel}} = \frac{\rho_{g_0}}{\tau_{g_0\, e, \text{inel}}} (v_{g_-} - v_{g_0}) , \tag{12a, b}$$

where the ion- and electron-capture timescales by charged and neutral grains, respectively, are given in terms of the corresponding capture rates, introduced in connection with equations (7a) - (7c), by

---

[4]If the Langevin approximation were valid not only for ion-neutral but for electron-neutral collisions as well, one would find that $\langle \sigma w \rangle_{eH2} \simeq (m_{H2}/m_e)^{1/2} \langle \sigma w \rangle_{iH2}$ and that, therefore, $\tau_{ni}/\tau_{ne} \simeq (m_e/m_{H2})^{1/2} (n_e/n_i) = 1.65 \times 10^{-2} (n_e/n_i) \ll 1$; i.e., F_{ne} could be neglected compared to F_{ni}. Actually, the Langevin approximation is not valid for e-H_2 collisions (see Mott and Massey 1971). With the values of $\langle \sigma w \rangle_{iH2}$ and $\langle \sigma w \rangle_{eH2}$ given above, we find that $\tau_{ni}/\tau_{ne} \simeq 2.0 \times 10^{-4} (n_e/n_i)$, a value $\simeq 100$ times smaller than that found from the Langevin approximation. Hence, electrons are completely insignificant, relative to the ions, in transmitting the magnetic forces to the neutrals.

$$\tau_{g\_i,inel} = 1/n_i \alpha_{ig\_} \; , \qquad \tau_{g_0 e, inel} = 1/n_e \alpha_{eg_0} \; . \qquad (12c, d)$$

2.2. Reduction of the Five-Fluid Equations: Flux-Freezing in the Plasma

The sum of the four drag forces appearing on the right-hand side of equation (6e) is eliminated in favor of the magnetic field by adding equations (6f) – (6i) and using the charge-neutrality equation (6a), Newton's third law, and Ampere's law (eq. [6b]) to find that

$$F_{ni} + F_{ne} + F_{ng\_} + F_{ng_0} = \frac{1}{4\pi}(\nabla \times B) \times B \; . \qquad (13)$$

Then the force equation (6e) for the neutrals becomes

$$\frac{\partial}{\partial t}(\rho_n v_n) + \nabla \cdot (\rho_n v_n v_n) = - C^2 \nabla \rho_n - \rho_n \nabla \psi + \frac{1}{4\pi}(\nabla \times B) \times B \; , \qquad (14)$$

where the equation of state (6j) has also been used to eliminate P_n in favor of ρ_n.

A further simplification occurs by noting that the third term, F_{en}, in the force equation (6f) for the electrons is negligible compared to the magnetic force (the second term) for the physical conditions characterizing the isothermal stage of evolution. The ratio of the magnitudes of the two terms can be written as

$$\frac{c |F_{en}|}{n_e e |v_e \times B|} = \frac{1}{\omega_{c,e}\tau_{en}} \frac{|v_e - v_n|}{|v_e|} \; , \qquad (15a)$$

where $\omega_{c,e} = eB/m_e c$ is the electron cyclotron frequency. We evaluate $\omega_{c,e}\tau_{en}$ to find that

$$\omega_{c,e}\tau_{en} = 1.4 \times 10^8 \left(\frac{B}{30 \; \mu G}\right) \left(\frac{3 \times 10^3 \; cm^{-3}}{n_n}\right) \left(\frac{10 \; K}{T}\right)^{1/2} \; . \qquad (15b)$$

The magnetic field and gas density on the right-hand side have been normalized to values typical of initial conditions at the cloud center, before a protostellar core forms. Our numerical simulations show that, during the quasistatic phase of contraction, B in the core increases by less than a factor of 2, while n_n increases by a factor 10^1 - 10^2 and $|v_e - v_n|/|v_e| \simeq 20$ (because the electrons are essentially held in place by magnetic forces as the neutrals contract through them). So, the ratio of forces in equation (15a) is negligible. By the end of a typical run, the neutral density and magnetic field strength in the core increase by a factor of 10^6 and 10^2, respectively, while $|v_e - v_n|/|v_e| \simeq 1/3$. Hence, $|F_{en}|$ is still nearly 10^4 times smaller than that required to detach the electrons from the magnetic field. For the ions, we have that

$$\omega_{c,i}\tau_{in} = 3.2 \times 10^4 \left(\frac{B}{30 \; \mu G}\right) \left(\frac{3 \times 10^3 \; cm^{-3}}{n_n}\right) \; . \qquad (16)$$

A similar argument shows, therefore, that the neutral drag on the ions falls short, by a factor $> 10^3$ initially, of the magnitude needed to dislodge the ions from the field lines, but that by a central density of 3×10^9 cm^{-3} the ions are on the verge of becoming detached from the field (cf. § 2.4). We may thus obtain the electric field from equations (6f) and (6g),

$$E = - \frac{v_e}{c} \times B = - \frac{v_i}{c} \times B \; , \qquad (17a, b)$$

which is then used to express Faraday's law (6d) as

$$\frac{\partial B}{\partial t} = \nabla \times (v_e \times B) = \nabla \times (v_i \times B) , \qquad (18a, b)$$

with equations (17b) and (18b) being valid only for densities $\leq 10^9$ cm^{-3}. In summary, *the magnetic flux is frozen in the plasma during the isothermal phase of star formation.* The electron and ion fluids may thus be treated as a single fluid, the plasma.

2.3. Grain Motion and Attachment to the Magnetic Field

The physics of charged-grain coupling to the magnetic field can be most easily understood if we ignore for the moment the presence of neutral grains. The *direct attachment parameter* of grains to the field, defined by $\Gamma_{dir} = \omega_{c,g} \tau_{gn}$, is evaluated to find that

$$\Gamma_{dir} \equiv \omega_{c,g} \tau_{gn} = 4.4 \times 10^2 \left[\frac{B}{30 \ \mu G} \right] \left[\frac{3 \times 10^3 \ \text{cm}^{-3}}{n_n} \right] \left[\frac{10 \ \text{K}}{T} \right]^{1/2} \left[\frac{10^{-6} \ \text{cm}}{a} \right]^2 , \quad (19)$$

with the typical grain radius used in our numerical simulations being $a = 3.75 \times 10^{-6}$ cm. In a typical model cloud, the subcritical phase of core formation and contraction ends by a central density enhancement $\leq 10^2$ and a corresponding increase of the central field strength by a factor < 2. Hence, $\Gamma_{dir} \simeq 1$ at the end of this phase of evolution, and the grains are expected to decouple from the field. This, however, is only part of the story. As the (negatively charged) grains begin to fall in (because of drag by the contracting neutrals) faster than the ions, which are still very well attached to the magnetic field, electrostatic attraction by the ions tends to keep the grains at least partially attached to the field even at densities at which $\Gamma_{dir} \ll 1$. This is an *indirect attachment* mechanism. Because, as we have seen, the electrons are extremely well attached to the field at this stage, they tend to shield the ionic charge, thus reducing the grain-ion electrostatic attraction. (Recall that $n_e = n_i - n_{g_-}$, so that the shielding is never complete.) The electron-shielded ions, to which we refer as "*quasiparticles*", thereby behave as if they carried a fractional electronic charge equal to $[(n_i - n_e)/n_i]e$. Partial coupling occurs if the *indirect attachment parameter*

$$\Gamma_{indir} \equiv [(n_i - n_e)/n_i] \, \omega_{c,i} \tau_{in} = (n_{g_-}/n_i) \, \omega_{c,i} \tau_{in} \equiv \omega_{c,q} \tau_{in} , \qquad (20)$$

is ≥ 1. The quantity $\omega_{c,q}$ is the cyclotron frequency of the ion quasiparticles.

The above physical argument is put on a rigorous mathematical footing as follows. For simplicity, we consider a cylindrically symmetric model cloud with the magnetic field lines parallel to the $(z-)$ axis of symmetry. The force equation (6h) for the grains without the last term can be solved for the radial component of the grain velocity (denoted simply by v_{g_-}), by using equation (17b) to eliminate the electric field, equation (10a) and Newton's third law, and the facts that $\omega_{c,g} \tau_{gn} \ll \omega_{c,i} \tau_{in} \ll \omega_{c,e} \tau_{en}$ and $\tau_{ni}/\tau_{ne} \ll 1$:

$$v_{g_-} = \frac{\Theta_g}{\Theta_g + 1} v_i + \frac{1}{\Theta_g + 1} v_n , \qquad (21)$$

where the function Θ_g is given by

$$\Theta_g = (1 + \tau_{ni}/\tau_{ng_-}) (\omega_{c,g_-} \tau_{g_- n})^2 , \qquad (22a)$$

$$= (\omega_{c,g_-} \tau_{g_- n})^2 + (n_{g_-}/n_i) (\omega_{c,i} \tau_{in}) (\omega_{c,g_-} \tau_{g_- n}) . \qquad (22b)$$

It is clear from equation (21) that, if $\Theta_g \gg 1$, then $v_{g_-} = v_i$, i.e., the grains are well attached to the magnetic field. If $\Theta_g \ll 1$, then $v_{g_-} = v_n$, and the grains fall in with the neutrals, i.e., they are decoupled from the field. *Partial attachment to the field can still take place if*

$$\Theta_g \equiv (\omega_{c,g_-} \tau_{g_- n})^2 + (n_{g_-} /n_i) (\omega_{c,i} \tau_{in}) (\omega_{c,g_-} \tau_{g_- n}) \gtrsim 1 . \qquad (23a)$$

For example, if $\Theta_g = 1$, $v_{g_-} = (v_i + v_n)/2$, the arithmetic mean of the ion and neutral infall speeds. Condition (23a) can be satisfied even if $\omega_{c,g_-} \tau_{g-n} \leq 1$, in which case the grains would not be *directly* attached to the field, provided only that $(n_{g_-} /n_i)(\omega_{c,i} \tau_{in})(\omega_{c,g_-} \tau_{g-n}) \gtrsim 1$. This means that partial attachment can be achieved *indirectly*, through grain-quasiparticle attraction, even if $n_{g_-} /n_i \ll 1$ as long as $\Gamma_{indir} \equiv (n_{g_-} /n_i)(\omega_{c,i} \tau_{in}) > 1$. We find in our numerical simulations in axisymmetric geometry that the last condition is satisfied (because $\omega_{c,i} \tau_{in} \gg 1$) even at densities high enough that the direct attachment parameter (see eq. [19]) is $\Gamma_{dir} \ll 1$ (see § 5.1, Fig. 7e).

Neutral grains behave as if they are attached to the field if the electron-capture timescale $\tau_{g_0 e,inel}$ is much smaller than $\tau_{g_0 n}$, the momentum-transfer timescale to the neutral grains by collisions with neutral molecules. If the opposite is true, i.e., if $\tau_{g_0 e,inel} \gg \tau_{g_0 n}$, the neutral grains fall in with the neutrals. Equation (6i) can be solved for v_{g_0} in terms of v_{g_-} and v_n to obtain an equation analogous to (21), and thus quantify the degree to which neutral grains are attached to the field (see Ciolek and Mouschovias 1993, eq. [44]).

2.4. Modification of the Ion Attachment Parameter by Grains

The same electrostatic forces (between negatively charged grains and positive ions) which keep the grains partially attached to the magnetic field lines at higher neutral densities than direct attachment alone would imply have an important effect on the ions: They tend to detach the ions from the field lines at *lower* densities than expected from direct ion attachment given by equation (16). Desch and Mouschovias (1999) studied the evolution of protostellar fragments above densities 10^9 cm^{-3}, at which ions are only marginally directly attached to the field lines and both negative and positive grains are abundant. They find an *ion attachment parameter* Θ_i, corresponding to Θ_g in equation (22) or (23a) for the grains, given by

$$\Theta_i = (\omega_{c,g} \tau_{gn}) (\omega_{c,i} \tau_{in}) \frac{n_{g_-} + n_{g_+}}{n_{g_-} - n_{g_+}} + \frac{n_e}{n_{g_-} - n_{g_+}} . \qquad (23b)$$

Detachment of ions from field lines (i.e., $\Theta_i < 1$) becomes significant at densities as low as 3×10^8 cm^{-3}. This implies that Faraday's law in the form of equation (18b) cannot be used above these densities. One must use equation (18a), at least until the electrons themselves become detached from the field lines or disappear by attachment onto grains.

3. Two-Fluid Description of Protostar Formation

3.1. Basic Equations

The modern theory of star formation outlined in § 1 is mainly based on analytical and numerical results obtained over the last twenty three years on the basis of a two-fluid (neutrals and electron-ion plasma) MHD description of the process of ambipolar-diffusion-induced fragmentation and star formation. With the macroscopic effects of grains ignored, but their role in the microscopic processes that determine the degree of ionization retained, the five-fluid equations reduce to

$$\frac{\partial \rho_n}{\partial t} + \nabla \cdot (\rho_n v_n) = 0 , \qquad (24a)$$

$$\frac{\partial}{\partial t}(\rho_n \mathbf{v}_n) + \nabla \cdot (\rho_n \mathbf{v}_n \mathbf{v}_n) = - C_n^2 \nabla \rho_n - \rho_n \nabla \psi + \frac{1}{4\pi}(\nabla \times \mathbf{B}) \times \mathbf{B} , \qquad (24b)$$

$$\mathbf{v}_i = \mathbf{v}_n + \frac{\tau_{ni}}{\rho_n} \frac{1}{4\pi}(\nabla \times \mathbf{B}) \times \mathbf{B} , \qquad (24c)$$

$$\tau_{ni} = a_{\text{He-i}} \frac{m_i + m_{H_2}}{\rho_i \langle \sigma w \rangle_{iH_2}} , \qquad (24d)$$

$$\frac{\partial \mathbf{B}}{\partial t} = \nabla \times (\mathbf{v}_i \times \mathbf{B}) , \qquad (24e)$$

$$\nabla^2 \psi = 4\pi G \rho_n . \qquad (24f)$$

As shown in § 2.1, $a_{\text{He-i}}$ is equal to 1.23 if He-ion collisions are accounted for, and to 1.4 if they are ignored but the contribution of the He inertia to that of H_2 is included. These are six equations for seven unknowns (ρ_n, $\mathbf{v}_n$, ρ_i, $\mathbf{v}_i$, τ_{ni}, $\mathbf{B}$, and ψ). Equations (7a) - (7c) can be used to determine the ion density and thus close the system. In the past, calculations of ionization equilibrium in *static* clouds (Elmegreen 1979; Nakano 1979) have been used to parametrize the ion density in terms of the neutral density as

$$\rho_i = m_i K \left[\frac{\rho_n / m_n}{10^5 \text{ cm}^{-3}} \right]^k , \qquad (25)$$

in the density range $10^3 \leq n_n \leq 10^9$ cm^{-3}. The constant $K \simeq 3 \times 10^{-3}$ cm^{-3} and the exponent $k \simeq 0.5$. Although these "canonical" values of K and k are routinely adopted (e.g., Lizano and Shu 1989; Tomisaka, Ikeuchi, and Nakamura 1990; Galli and Shu 1993a, b; Li and McKee 1996; Safier, McKee, and Stahler 1997), it should be borne in mind that K can easily differ from this value by a factor of 10 (or more) due to uncertainties in the rate of cosmic-ray ionization and metal depletion onto grains, and k is uncertain by about 20% - 30%. In fact, our recent determination of the charged-species abundances in a fashion consistent with the time evolution of model clouds has shown that no constant k can accurately represent the ion density in the above range of n_n (see § 5, Fig. 7g). We may nevertheless use equation (25) in a first approach to the problem, to avoid adopting a detailed chemical model, but K and k should then be regarded as free parameters, the former allowed to vary by about a factor of 10 and the latter restricted in the range 0.6 - 0.2. Fiedler and Mouschovias (1992, 1993) and Basu and Mouschovias (1994, 1995a, b) have taken this approach in studying the ambipolar-diffusion-induced protostar formation in axisymmetric geometry in the absence and presence of rotation, respectively, while Ciolek and Mouschovias (1993, 1994, 1995), in their five-fluid simulations, have calculated the ion density through detailed chemical models in a fashion consistent with the time evolution.

The continuity equation (24a) can be used in the left-hand side of the force equation (24b) to write it in the more familiar form $\rho_n(d\mathbf{v}_n/dt)$, where $d/dt = \partial/\partial t + \mathbf{v}_n \cdot \nabla$ is the (Lagrangian) time derivative comoving with the neutrals.

3.2. Ambipolar Diffusion: General Considerations, and Timescales

With the ion-neutral drift velocity denoted by $\mathbf{v}_D$ ($\equiv \mathbf{v}_i - \mathbf{v}_n$), equation (24c) implies that

$$v_D \simeq \frac{v_A^2 \tau_{ni}}{L} , \qquad (26a)$$

or

$$v_D \simeq 0.1 \left[\frac{v_A}{1\ \mathrm{km\ s^{-1}}}\right]^2 \left[\frac{\tau_{ni}}{10^4\ \mathrm{yr}}\right] \left[\frac{0.1\ \mathrm{pc}}{L}\right], \quad \mathrm{km\ s^{-1}} \tag{26b}$$

where $|\nabla| \sim 1/L$, $v_A = B/(4\pi\rho_n)^{1/2}$ is the Alfvén speed in the neutrals, and in equation (26b) all quantities have been normalized to values typical of the initial stages of core formation in molecular clouds. Equation (26a) suggests a natural ambipolar-diffusion timescale $L/v_D \simeq \tau_A^2/\tau_{ni}$, so that

$$\tau_{AD} \simeq \frac{\tau_A^2}{\tau_{ni}}, \tag{26c}$$

$$\simeq 10^6 \left[\frac{1\ \mathrm{km\ s^{-1}}}{v_A}\right]^2 \left[\frac{10^4\ \mathrm{yr}}{\tau_{ni}}\right] \left[\frac{L}{0.1\ \mathrm{pc}}\right]^2 \quad \mathrm{yr}, \tag{26d}$$

where $\tau_A \equiv L/v_A$ is the Alfvén crossing time. The time evolution (due to ion-neutral drift) of the magnetic field threading a predominantly neutral fluid element can be obtained from the plasma flux-freezing equation (24e) by eliminating v_i in favor of v_D and v_n (from the definition of the drift velocity), and using equation (24c) to find

$$\frac{\partial B}{\partial t} - \nabla \times (v_n \times B) = \nabla \times (v_D \times B), \tag{27a}$$

$$= \nabla \times \left\{ \frac{\tau_{ni}}{\rho_n} \left[\frac{1}{4\pi}(\nabla \times B) \times B \right] \times B \right\}, \tag{27b}$$

$$= \nabla \times \left\{ \frac{\tau_{ni}}{\rho_n} \left[-\nabla\left(\frac{B^2}{8\pi}\right) + \frac{1}{4\pi}(B \cdot \nabla)B \right] \times B \right\}. \tag{27c}$$

This is a nonlinear diffusion equation, with a diffusion coefficient $\mathscr{D} \simeq v_A^2\tau_{ni}$. Hence, the diffusion timescale is $\tau_{\mathrm{diff}} \simeq L^2/\mathscr{D} \simeq \tau_A^2/\tau_{ni}$, which is identical with τ_{AD} (see eq. [26c]).

The above considerations and estimate of τ_{AD} do not account for self-gravity. In order to do so, we consider the early, subcritical phase of ambipolar-diffusion-induced core formation and evolution, during which the acceleration of the neutrals is negligible (i.e., the contraction is quasistatic) and the magnetic forces dominate the thermal-pressure forces in opposing gravity. (During dynamical contraction, magnetic flux tends to get trapped inside the collapsing core; see eq. [31] below, and Fiedler and Mouschovias 1993, Fig. 1b.) Combining the neutral-particle and plasma force equations ([24b] and [24c], respectively), we find that the drift velocity can be expressed as

$$v_D = -\tau_{ni} g_\perp, \tag{28}$$

where $g \equiv -\nabla\psi$, and the subscript $\perp$ denotes a direction perpendicular to the field lines. In the equatorial plane of the forming core, we have that $|g_\perp| \propto G\rho_n r \propto r/\tau_{ff}^2$. Substituting this in equation (28) and using the resulting expression for v_D in the definition of the ambipolar-diffusion timescale in the core, $\tau_{AD} \equiv r/|v_D|$, we find that

$$\tau_{AD} = const \times \frac{\tau_{ff}^2}{\tau_{ni}}, \tag{29a}$$

$$\propto n_i/n_n, \tag{29b}$$

where the definition of τ_{ni} (eq. [11b]) has been used to obtain (29b) from (29a). (This is the origin of eq. [1].) The *const* on the right-hand side of equation (29a) is near 1, independent of geometry (Mouschovias 1987b), and *the expression* (29a) *for* τ_{AD} *is seen to be a general property of the equations describing ambipolar diffusion in quasistatically contracting, magnetically supported cores.*

This equation can be used to define a parameter ν_{ff} by

$$\nu_{ff} \equiv \tau_{ff}/\tau_{ni} \simeq \tau_{AD}/\tau_{ff} \, , \tag{30}$$

which is clearly the factor by which ambipolar diffusion retards the contraction relative to free fall. We therefore refer to it as the *collapse retardation factor.*

After dynamical contraction begins and equation (29a) for τ_{AD} is no longer valid, one may still calculate τ_{AD} analytically from first principles (see Mouschovias 1989, § IIIb). The resulting expression, accounting for the He inertia but ignoring He-ion collisions, is

$$\tau_{AD} = 0.4 \frac{\tau_{ff}^2}{\tau_{ni}} \left[\frac{\Phi_{B,crit}}{\Phi_B} \right]^2 , \qquad \text{for} \quad \Phi_B \le \Phi_{B,crit}; \tag{31a}$$

$$\simeq 1 \times 10^3 \left[\frac{n_i/n_{H_2}}{10^{-10}} \right] \left[\frac{\Phi_{B,crit}}{\Phi_B} \right]^2 \quad \text{yr}, \tag{31b}$$

where Φ_B is the actual flux of the core at any stage past the onset of dynamical contraction, and $\Phi_{B,crit}$ is its total flux at the onset of dynamical contraction and is uniquely determined by the mass of the core as $\Phi_{B,crit} = (63G)^{1/2}M$. The normalization of n_i/n_{H2} in equation (31b) refers to a neutral density $\sim 10^9$ cm^{-3}, above which n_i no longer increases significantly with n_n (Spitzer 1963; Elmegreen 1979; Nakano 1979). Canonical values of K and k have been assumed (see eq. [25]). The free-fall time for a spherical core is

$$\tau_{ff} = (3\pi/32G\rho)^{1/2} . \tag{32}$$

The presence of the factor Φ_B^{-2} on the right-hand side of equation (31) shows clearly the tendency of magnetic flux to get trapped inside a collapsing core, unless the degree of ionization were to decrease very rapidly.

In the presence of grains, if they are well attached to the magnetic field, one can easily show that τ_{ni} in equations (30) and (31a) is replaced by the mean collision time τ_{coll} of neutrals with both ions and charged grains, where $(\tau_{coll})^{-1} = (\tau_{ni})^{-1} + (\tau_{ng-})^{-1}$. (Physically, collisions of neutrals with ions and grains are processes occurring in parallel, hence the combined timescale is obtained in the same fashion that the net resistance is found in an electrical circuit consisting of a parallel combination of resistances.) A general expression for ν_{ff} is given by Ciolek and Mouschovias (1993) accounting for partial attachment of the charged grains to the field and for the presence of neutral grains. The typical value of ν_{ff} in molecular clouds is near 10, with the range $\simeq 3 - 30$ permissible by observations (see review by Mouschovias 1995).

3.3. Magnetic Braking

The five-fluid equations were presented in §§ 2.1 and 2.2 in vectorial form, and so were the two-fluid equations in § 3.1. Specifically, no component of the velocities or of the magnetic field was assumed to vanish. Rotating model clouds and magnetic braking may therefore be investigated using these equations.

Magnetic braking of quasistatically contracting clouds (or fragments) was studied analytically for both aligned and perpendicular rotators (Mouschovias and Paleologou 1979,

1980), and for magnetically linked, aligned rotators (Mouschovias and Morton 1985a, b). The propagation of the torsional Alfvén waves, including their partial reflection and transmission at cloud surfaces, was followed and interpreted physically using exact analogies with transverse waves on a string of varying density. For typical cloud (or fragment) parameters, magnetic braking was found to be very effective and to wipe out differential rotation on a timescale much shorter than even the free-fall timescale, which is a strict lower limit on the contraction timescale. Hence, if one is interested in the loss of angular momentum by a cloud as a whole or by a protostellar fragment within a cloud, one may ignore (the important for other purposes) transient phenomena involving torsional Alfvén waves and treat the cloud or fragment as a rigid rotator. This approximation, however, would break down during the dynamical phase of contraction of a protostellar core because of the rapid and nonhomologous nature of the contraction. Angular momentum would tend to get trapped inside, and differential rotation would be unavoidable. A simplification, nevertheless, can be introduced even in this case. Two-fluid calculations ignoring rotation have shown that, because of flattening along field lines, balance of forces is rapidly established and maintained *along* field lines even during the dynamical-contraction phase perpendicular to the field lines. Under these conditions magnetic braking can establish isorotation along each magnetic flux tube, while differential rotation can be generated and maintained perpendicular to the field lines. Basu and Mouschovias (1994, 1995a, b) followed this approach to study the combined effects of ambipolar diffusion and magnetic braking on the formation and evolution of protostellar fragments (or cores) in model molecular clouds, up to $n_n \simeq 10^{10}$ cm^{-3}. The dependence of the solution on the free parameters of the problem was also investigated. Some of the results are summarized in § 4.2.

Mouschovias and Paleologou (1986) showed that ambipolar diffusion during the quasistatic phase of contraction lengthens the magnetic-braking timescale by only a few percent. Hence, *it is the onset of dynamical contraction, rather than ambipolar diffusion, that brings an end to redistribution of angular momentum by magnetic braking and leads to angular-momentum trapping in a collapsing core* (see also Mouschovias 1979a). By the time dynamical contraction sets in ($n_n \gtrsim 10^5$ cm^{-3}), magnetic braking has removed enough angular momentum for binary stars and even solar systems to become dynamically possible; i.e., they can form without interference from centrifugal forces (see § 4.2).

4. Results of Two-Fluid Description of Protostar Formation

We have studied the self-initiated (due to ambipolar diffusion) formation and contraction of protostellar cores in self-gravitating, isothermal, primarily magnetically supported model molecular clouds, by developing and using fully implicit, multifluid, second-order accurate, adaptive-grid codes. A recent review of the main results is found in Mouschovias (1995). The formulation of the problems, the mathematical and numerical techniques, and detailed descriptions of the results obtained from the two-fluid approximation are given in Fiedler and Mouschovias (1992, 1993) for nonrotating clouds, and in Basu and Mouschovias (1994, 1995a, b) for rotating models.

A theory of protostar formation must ultimately explain the observed mass spectrum of protostellar cores (see Fig. 1a) and the observed relation between the magnetic field strength and gas density (see Fig. 1b). It is clear from Figure 1a that the most common core mass is almost 10 $M_\odot$. While it is hardly legitimate to refer to objects with masses greater than about 10^2 $M_\odot$ as cores (they are clouds in their own right), this should not obscure the fact that structure exists in clouds and cloud complexes ranging in mass from about 1 $M_\odot$ to more than 10^3 $M_\odot$. The low-density low-field points in Figure 1b refer to HI Zeeman measurements, as do the boxes, each of which contains several measurements (see Troland

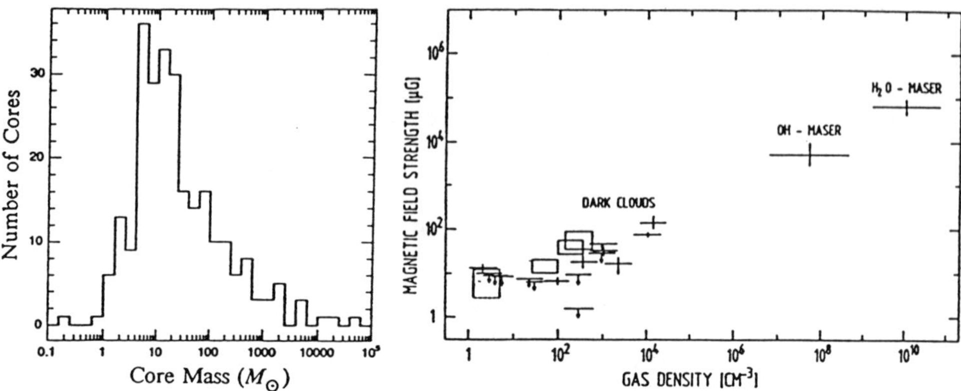

Figure 1. (**a**, *left*) *Number of Cores as a function of their mass, in $M_\odot$* (from Wood *et al.* 1994). (**b**, *right*) *Observed magnetic field strength as a function of gas density* (courtesy of D. Fiebig).

and Heiles 1986, Fig. 1). The points labeled as "dark clouds" represent several OH Zeeman measurements. The "error bars" shown for OH and H_2O masers are meant to reflect the range of observed densities and fields, not the errors in any one measurement. One may summarize Figure 1b as showing very little, if any, enhancement of the field strength at densities below 10^4 cm^{-3}, and a relation $B \propto n^{1/2}$ above this density. The calculations about to be summarized explain Figure 1b and give as typical protostellar core masses somewhat less than 10 $M_\odot$, which is consistent with the peak in Figure 1a.

4.1. Ambipolar Diffusion in Nonrotating Models

Choosing as units of density, velocity, time, and magnetic field the initial central values of the neutral density $\rho_{n,c0}$, Alfvén speed in the neutrals $v_{A,c0}$, neutral-ion collision time $\tau_{ni,c0}$, and magnetic field strength B_{c0}, it is easily shown that the two-fluid MHD equations (24a) – (25) contain three dimensionless free parameters:

$$\nu_{ff,c0} \equiv \frac{\tau_{ff,c0}}{\pi^{1/2}\tau_{ni,c0}}, \qquad \alpha_{c0} \equiv \frac{B_{c0}^2}{8\pi\rho_{n,c0}C^2}, \qquad k \equiv \frac{d\ln n_i}{d\ln n_n}. \qquad (33a, b, c)$$

The quantity $\nu_{ff,c0}$ is essentially the initial central collapse retardation factor defined in equation (30). The second parameter α_{c0} is the initial value of the ratio of magnetic and thermal pressures at the center; and k is the exponent in the parametrization of the ion density in terms of the neutral density, $n_i \propto n_n^k$ (see eq. [25]). The boundary conditions introduce two free parameters, to which the solution is very insensitive, provided that the model clouds are sufficiently thermally supercritical (as suggested by observations). The initial conditions introduce the central mass-to-flux ratio relative to its critical value for collapse as a fourth free parameter, i.e.,

$$\mu_{c0} \equiv \frac{(dm/d\Phi_B)_{c0}}{(dm/d\Phi_B)_{c,crit}}, \qquad (33d)$$

where the critical value $(dm/d\Phi_B)_{c,crit}$ was calculated from a sequence of exact equilibrium states, and is given by equation (2). [For a uniform (hence, nonequilibrium) thin disk of infinite radius, Nakano and Nakamura (1978) find a smaller critical value by the factor 1.19.] If the initial states are taken to be exact magnetohydrostatic (MHS) equilibrium states with the magnetic field frozen in the matter (i.e., in the absence of ambipolar diffusion), α_{c0} and μ_{c0} are related by

$$(\alpha_{c0}\mu_{c0}{}^2)_{equil} = const \simeq 1 \tag{33e}$$

(Mouschovias 1991c). Hence, there are only three significant free parameters in the problem. Observations suggest the following set of typical values: $\nu_{ff,c0} = 8.3$, $\alpha_{c0} = 26$, $\mu_{c0} = 1/4$, and $k = 1/2$. Observations and theoretical considerations limit the range of values to be investigated to $1 \le \nu_{ff,c0} \le 10$, $1 < \alpha_{c0} < 100$, $0.1 \le \mu_{c0} \le 1.0$, and $1/3 \le k \le 1/2$.

In this investigation, model clouds are initially in uniform states, spherical or cylindrical in shape, threaded by a uniform magnetic field aligned with the $(z-)$axis of symmetry of a cylindrical polar coordinate system (r, ϕ, z) --equilibrium initial states are considered in §§ 4.2 - 5.3. We study first the evolution of a typical model cloud, having $\nu_{ff,c0} = 8.21$, $\mu_{c0} = 0.246$, $\alpha_{c0} = 86.5$, and $k = 1/2$. These values could, for example, represent a region $Z = R = 0.75$ pc in a molecular cloud, of neutral density $n_{n,0} = 300$ cm^{-3}, ion density 5.36×10^{-3} cm^{-3}, temperature $T = 10$ K, mass 45.5 $M_\odot$, threaded by a magnetic field $B_0 = 30$ μG.

Figure 2a exhibits the central density $n_c \equiv n_{n,c}$ (in cm^{-3}) as a function of time (in Myr). *The evolution occurs in three distinct phases.* (1) *The cloud relaxes along field lines to a quasi-equilibrium state*, and the central density increases by a factor $\simeq 8$ in about 6 Myr. (If ambipolar diffusion were not present, an equilibrium state would be reached having a central density greater than its initial value by a factor of 3.4, instead of 8.) (2) *Quasistatic, ambipolar-diffusion-controlled contraction follows.* It takes place on the central flux-loss timescale,

$$\tau_{\Phi,c0} = \left[\frac{\Phi_B}{|d\Phi_B/dt|}\right]_{c0} = \frac{1}{2}\tau_{AD,c0} \; , \tag{34a, b}$$

of the quasi-equilibrium state, which is $\simeq 9$ Myr. This phase lasts until $t \simeq 15$ Myr, at which time the central density has increased to about 3.3×10^4 cm^{-3} and the central mass-to-flux ratio has reached the critical value for collapse. Soon thereafter, (3) *dynamic contraction in the radial direction sets in.* Even at the end of the run, however, at a central density greater than 3×10^8 cm^{-3}, the central acceleration is only 30% of the local gravitational field; i.e., the dynamic contraction is by no means a free fall (see Fig. 2b). At this time, the r-components of the central magnetic and thermal-pressure forces are 40% and 30% of the r-component of the gravitational force, respectively. *Along field lines*, once balance of thermal-pressure and gravitational forces is established at the end of the relaxation phase, it is maintained even during the lateral dynamic-contraction phase (see Fig. 2c).

Isodensity contours, field lines, and velocity vectors are displayed in Figure 3a at time $t = 16.135$ Myr, near the end of the run, when the central density is $n_c = 2.9 \times 10^8$ cm^{-3}. The equatorial radius of the supercritical core is $R_{core} = 0.14R = 0.10$ pc, and has been at this value for the last four orders of magnitude enhancement of the central density. During this time, its mass has been $M_{core} \simeq 5.0$ $M_\odot$. Magnetic forces provide very effective support of the envelope beyond $r \gtrsim 0.3R$. *The isodensity contours reveal the formation of a protostellar disk.* The field lines have deformed appreciably only in the innermost $\simeq 15\%$ of the cloud. The innermost 3% is shown in Figure 3b; to better represent the magnetic field, six field lines are displayed for each one shown in Figure 3a. The velocity vectors are normalized to the magnitude of the maximum infall velocity $|v_n|_{max} = 0.49$ km s^{-1}, which occurs on the

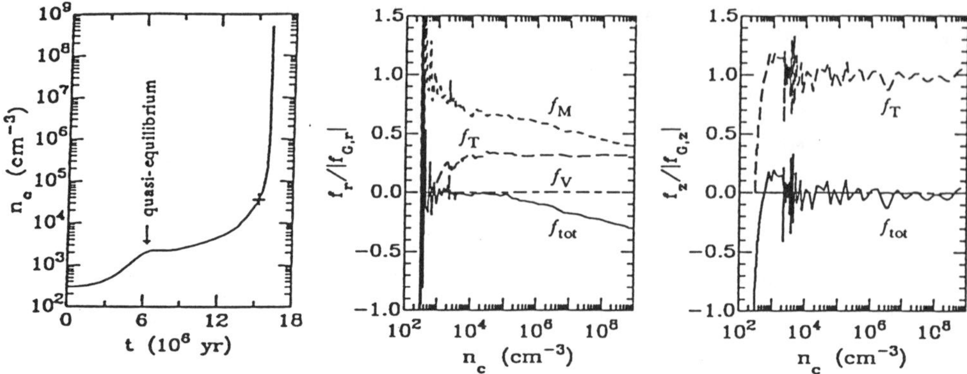

Figure 2. *Evolution of central values.* A + marks the point at which the central mass-to-flux ratio becomes critical. (**a**, *left*) Neutral density as a function of time. (**b**, *middle*) Radial components of forces (per unit volume), normalized to the magnitude of the r-component of the gravitational force $f_{G,r}$, as functions of density. Total force f_{tot}, magnetic force f_M, thermal-pressure force f_T, and artificial-viscosity force f_V. (**c**, *right*) z-components of f_{tot}, f_T, and the other two forces, which are negligible, normalized to the magnitude of the z-component of the gravitational force $f_{G,z}$, as functions of density.

axis of symmetry at $z \simeq 3 \times 10^{-3} R \simeq 2.3 \times 10^{-3}$ pc ($\simeq 464$ AU). A shock is beginning to form in this region. (The shock evident in Fig. 3a at $z \simeq 0.3R$, near the z-axis, is a result of the early relaxation along field lines; it is not present if equilibrium states are used as initial states.) The mass inside each flux tube visible in Figure 3b (found by integrating from $r = 0$ to the field line and from $-Z$ to $+Z$) is 0.0525, 0.199, 0.419, 0.695, 1.02, and 1.38 $M_\odot$. The corresponding initial masses were 5.61×10^{-3}, 0.0225, 0.0505, 0.0899, 0.140, and 0.202 $M_\odot$. More than 77% of the mass in each of these flux tubes lies inside the isodensity contour labeled "10,000".

Figure 4a shows B_c as a function of n_c for three models (one being that of Figs. 2 - 3) differing only in the initial value of the dimensionless mass-to-flux ratio μ_{c0}, whose values label the three curves. The more subcritical a cloud is, the greater the factor ($\simeq \mu_{c0}^{-2}$; see eq. [5]) by which the central density increases (with hardly any increase in the field strength) before critical conditions are reached and the core begins to contract dynamically. Only then does B_c increase (as n_c^κ, with $\kappa = 0.47$, a value near the flux-freezing limit of 1/2). Once a supercritical core forms, however, memory of the initial μ_{c0} is lost, and the value of the field is typically in the milligauss range for densities 10^{8-9} cm^{-3}. Herein may lie the explanation for the magnetic-field observations shown in Figure 1b; i.e., for (1) the near constancy of the observed field strength below densities $\simeq 10^4$ cm^{-3}, and (2) the observed $B \simeq$ a few mG - several $\times$ 10 mG in OH and H_2O masers, which typically have densities $\sim 10^8$ cm^{-3} and $> 10^9$ cm^{-3}, respectively. (This is not to imply, however, that OH and H_2O masers are protostars. Simply, the physical conditions of density and associated field strength predicted by our calculations require only a pump for maser action consistent with observations to take place.)

The central mass-to-flux ratio, normalized to the critical value for collapse, is shown in Figure 4b as a function of central density for the three models of Figure 3a. It increases by a factor $\simeq \mu_{c0}^{-1}$) during the subcritical phase of evolution, and by an additional factor of

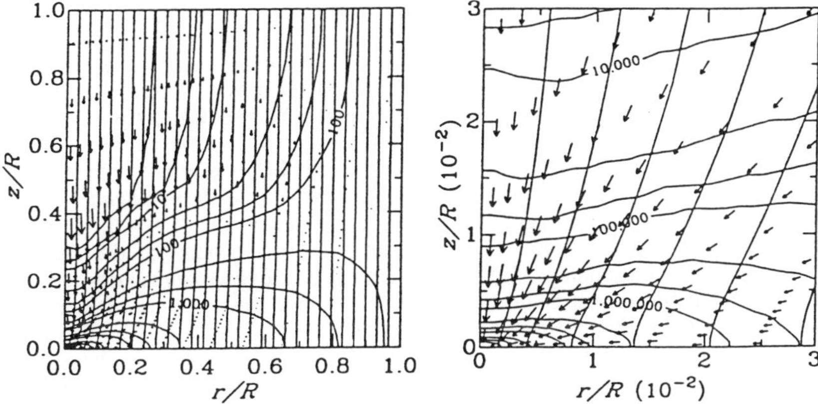

Figure 3. *Isodensity contours, field lines, and velocity vectors*, normalized to the magnitude of the maximum infall velocity (= 0.49 km s^{-1}), near the end of the run (when t = 16.135 Myr, and n_c = 2.9 × 10^8 cm^{-3}). The two unlabeled contour levels between labeled ones have density values 3 and 6 times greater than the previous label. The r and z coordinates are normalized to the cloud radius R. (**a**, *left*) The whole cloud. (**b**, *right*) The innermost 3% of the cloud.

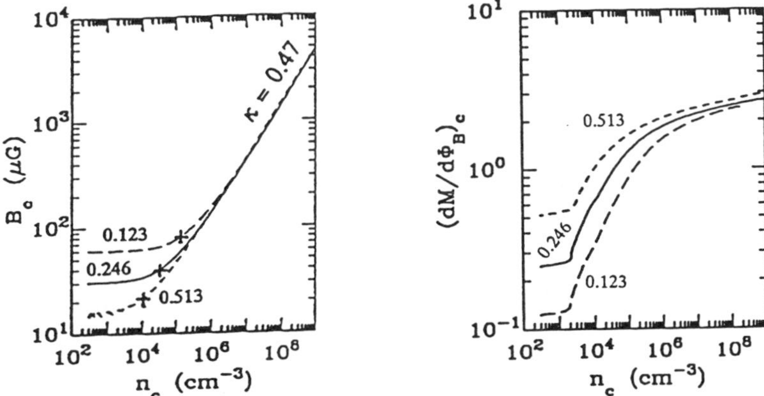

Figure 4. (**a**, *left*) *Dependence of the B_c-n_c relation and* (**b**, *right*) *the central mass-to-flux ratio on the free parameter* μ_{c0}, whose values label the three curves.

about 2.6, almost independent of its initial value, during the dynamic phase. The factor by which the mass-to-flux ratio increases during the dynamic phase depends on the values of the other two free parameters, $\nu_{ff,c0}$ and k (see Fiedler and Mouschovias 1993, Figs. 5b and 6b). The smaller these values are, the greater its increase. Physically, the smaller $\nu_{ff,c0}$ is, the smaller the degree of ionization (hence, the more effective the ambipolar diffusion) *at all densities*. The smaller k is, the smaller the degree of ionization (and the more rapid the ambipolar diffusion) becomes *as the neutral density increases*.

4.2. Rotating Models: Ambipolar Diffusion and Magnetic Braking

In order to account for rotation (and, later, other physical phenomena), we take advantage of the result found above, namely, that relatively rapid relaxation along field lines leads to the formation of a thin disk, and establishes and maintains balance of forces along field lines at all times (certainly up to $n_{n,c} \lesssim 10^{10}$ cm^{-3}). We start the calculation from exact MHS equilibrium states at time $t = 0$. Thermal-pressure forces balance gravity along field lines, while magnetic, centrifugal, and thermal-pressure forces do so perpendicular to the field lines. Ambipolar diffusion and magnetic braking are switched on at $t = 0$. The two-fluid MHD equations are integrated over z (i.e., perpendicular to the plane of the disk), thereby reducing the spatial dimensionality of the problem and improving enormously the computational efficiency. For the model cloud described below, the mass-to-flux ratio in the initial equilibrium state is the same as that of a "reference" state threaded by a uniform magnetic field B_{ref}, but with a column density profile $\sigma_n(r) = \sigma_{c,ref}[1 + (r/l_{ref})^2]^{-3/2}$. (In galactic dynamics, this column density distribution is known as "Toomre's model 1"; Toomre 1963.) The quantity $\sigma_{c,ref}$ is the central column density, and l_{ref} is the characteristic length of its distribution; as suggested by observations, l_{ref} must be significantly greater than the critical thermal lengthscale $\lambda_{T,cr} \sim C\tau_{ff,c}$ (see Mouschovias 1991a), where C is the isothermal sound speed and $\tau_{ff,c}$ the central free-fall time for a thin disk. For the initial angular velocity profile, we use that obtained from a linear dependence of the specific angular momentum ($\ell = \Omega r^2$) on mass; this corresponds to the angular momentum of a uniformly rotating disk of uniform column density. The initial equilibrium state is calculated from the corresponding reference state by allowing the model cloud to contract under magnetic-flux and angular-momentum conservation until force balance is achieved. Because the model clouds are magnetically subcritical, the equilibrium central values $n_{n,c0}$, $B_{z,c0}$, and Ω_{c0} are typically greater than their values in the corresponding reference state by only 2% - 4%. Given the column density $\sigma_n(r, t)$ and a fixed external pressure P_{ext}, the mass density $\rho_n(r, t)$ is uniquely determined analytically from the condition of force balance along field lines. The half-thickness is then given by $Z(r, t) = \sigma_n(r, t)/2\rho_n(r, t)$. An important new free parameter is now present in the problem, namely, the ratio of the "external" and central neutral densities in the reference state:

$$\tilde{\rho} = \frac{\rho_{ext}}{\rho_{n,c,ref}};$$

(35)

its typical value is 0.01. If $\tilde{\rho} = 0$, no magnetic braking can take place, and angular momentum is conserved.

The evolution of a typical model cloud ($n_{n,c,ref} = 3 \times 10^3$ cm^{-3}, $T = 10$ K, $\Omega_{c,ref} = 0.48$ km s^{-1} pc^{-1} = 1.55×10^{-14} rad s^{-1}, $B_{ref} = 30$ μG, degree of ionization $x_{i,c} \equiv n_{i,c}/n_{n,c} = 1.72 \times 10^{-7}$, $l_{ref} = 1.15$ pc, $R_0 = 5l_{ref}$, total mass $M_{cl} = 191$ $M_\odot$) is followed to an enhancement of the central density by a factor of 10^6. These dimensional quantities imply that $\mu_{c0} = 0.324$ (using the disk critical mass-to-flux ratio) and $\nu_{ff,c0} = 21.6$; we still take $k = 1/2$. [The greater value of $\nu_{ff,c0}$ in the present model compared to that of the nonrotating model of § 4.1 is almost entirely due to the fact that the central free-fall time of the reference disk exceeds that of a uniform sphere of the same density by the factor 2.4 --the gravitational field of a disk at r is weakened by matter outside the radius r.]

Figure 5a shows the central density and angular velocity as functions of time. [Note: 1 km s^{-1} pc^{-1} = 3.24×10^{-14} rad s^{-1}.] As in the absence of magnetic braking, the density increases slowly during the ambipolar-diffusion-controlled, quasistatic phase, by a factor $\simeq$ 20. Once a supercritical core forms, rapid contraction ensues. *The angular velocity evolves in three distinct phases.* (1) *It decreases exponentially*, to almost its background value, on the magnetic-braking timescale $\tau_J = \tau_\parallel \equiv \tau_{\parallel,fan}(R_{cl} = R_0)$ (see eq. [3a]) for an aligned disk rotator with straight-parallel field lines --the quantity $v_{A,ext}$ is the Alfvén speed in the external medium near the axis of symmetry. (2) *It remains essentially constant*, because of very effective magnetic braking, during the entire quasistatic phase of contraction. (3) *It increases almost as* $n_{n,c}^{1/2}$ during the subsequent collapse phase, implying that angular momentum is nearly conserved in the rapidly contracting core. During this stage of evolution, $\tau_J \simeq \tau_\Phi \simeq 10\tau_\sigma$ near $r = 0$, and τ_J is well approximated by $\tau_{\parallel\,fan}$ (see eq. [3a]).

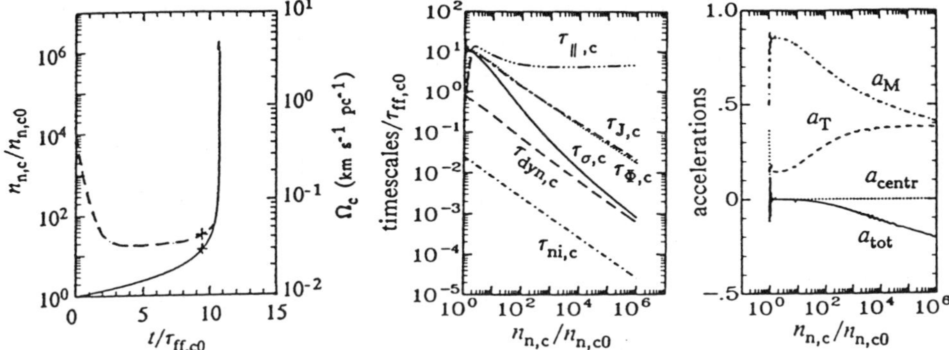

Figure 5. *Evolution in the presence of magnetic braking.* (**a,** *left*) Central neutral density (*solid line*), normalized to its initial value $n_{n,c0} = 3.12 \times 10^3$ cm^{-3}, as a function of time, normalized to the initial central free-fall time $\tau_{ff,c0} = 1.48 \times 10^6$ yr. Also, central angular velocity (in km s^{-1} pc^{-1}; *dashed line*; scale on right side of the frame). Meaning of + is as in Fig. 2a. (**b,** *middle*) Central timescales, normalized to $\tau_{ff,c0}$, as functions of central density enhancement $n_{n,c}/n_{n,c0}$. Magnetic-braking timescale $\tau_{J,c}$ (*long-dash, short-dash line*), magnetic-braking time for rotator with straight-parallel field lines $\tau_{\parallel,c}$ (*dash, triple-dot line*), flux-loss time $\tau_{\Phi,c}$ (*dotted line*), column-density e-folding time $\tau_{\sigma,c}$ (*solid line*), dynamical time $\tau_{dyn,c}$ (*dashed line*), and neutral-ion collision time $\tau_{ni,c}$ (*dash-dot line*). (**c,** *right*) Forces (per gram of neutral matter), normalized to $|g_r|$, in the limit $r \rightarrow 0$. Magnetic force a_M (*dash-dot line*), thermal-pressure force a_T (*dashed line*), centrifugal force a_{centr} (*dotted line*), and total force a_{tot} (*solid line*).

The various timescales evaluated at the center and normalized to the initial central free-fall time of the disk, $\tau_{ff,c0} = 1.48 \times 10^6$ yr, are shown in Figure 5b as functions of the central density enhancement, $n_{n,c}/n_{n,c0}$. The central "dynamical timescale" is operationally defined by

$$\tau_{dyn,c} \equiv \lim_{r \rightarrow 0} \left(\frac{r}{|g_r|}\right)^{1/2} , \tag{36}$$

and is approximately (but not exactly) equal to the central free-fall time. The timescale $\tau_{J,c}$ (*long-dash short-dash line*) is only slightly smaller than $\tau_{\parallel,c}$ initially, since the field lines are only slightly compressed in the central part of the initial equilibrium state. However, $\tau_{J,c}$ rises quickly in the early evolution as the angular velocity decreases toward the background value; i.e., magnetic braking renders itself ineffective after it achieves corotation with the background. The rise in $\tau_{J,c}$ is halted when it becomes approximately equal to the column-density e-folding time, $\tau_{\sigma,c}$, as a balance is established between the rates of inward lateral advection of angular momentum (perpendicular to the field lines) and its transport away from the core along magnetic field lines. The relation $\tau_{J,c} \simeq \tau_{\sigma,c}$ is maintained during the quasistatic phase of contraction, because magnetic braking can remove angular momentum just as rapidly as it is advected inward. Since the inward advection of mass and angular momentum is governed by ambipolar diffusion during the quasistatic contraction phase, Figure 5b also shows that $\tau_{\Phi,c} \simeq \tau_{J,c} \simeq \tau_{\sigma,c}$. After the onset of dynamic contraction of the core, $\tau_{J,c}$ is found to decrease as $n_{n,c}^{-1/2}$, in accordance with the prediction of equation (3c) --see discussion following equation (4). Although the relation $\tau_{J,c} \simeq \tau_{\Phi,c}$ is maintained even during the dynamic phase, these timescales become significantly greater than $\tau_{\sigma,c}$ as this phase progresses; in fact, $\tau_{\sigma,c}$ approaches $\tau_{dyn,c}$ by the end of the run. The timescale $\tau_{\parallel}$ becomes the longest of all timescales, but is also irrelevant to the process of angular momentum loss by the core, because of the significant deformation of the field lines. Note that $\tau_{\sigma,c}$ is only 4×10^5 yr at a density enhancement of 10^2 (i.e., for typical parameters, at $n_{n,c} = 3 \times 10^5$ cm^{-3}), and only 1.2×10^3 yr at the end of the run (at $n_{n,c} = 3 \times 10^9$ cm^{-3}).

Figure 5c shows the central forces (magnetic, thermal-pressure, centrifugal, and total), normalized to the magnitude of the local gravitational force, as functions of central density enhancement. Magnetic braking is so effective initially that Ω_c decreases very rapidly, which in turn results in a dramatic decrease in rotational support (from 36% to 0.1%) with hardly any increase in central density. However, the magnetic force rises just as rapidly and compensates for the loss of centrifugal support (it increases its share of support against gravity from 50% to 85%) while the share of the thermal-pressure support remains fixed at 14%. This is so because, in a subcritical cloud, only a slight contraction is required for the magnetic-pressure and tension forces to make up for the loss of centrifugal support. After its precipitous initial drop, the centrifugal force never again becomes a significant source of support in the core. It is also clear from Figure 5c that magnetic opposition to gravity decreases and thermal-pressure opposition increases as the evolution progresses into the supercritical phase. However, even at the end of the run, the central magnetic force remains somewhat greater than the thermal-pressure force, and contributes slightly more than 40% to the support of the cloud; it is much greater than the thermal-pressure force everywhere else in the cloud. The magnitude of the total central (infall) acceleration rises to 20% of the gravitational acceleration by the end of the run. This is dynamic contraction, but by no means free fall --even at a central density of 3×10^9 cm^{-3}.

The angular velocity as a function of radius is shown in Figure 6a at seven different times t_j ($j = 0, 1, ..., 6$) chosen such that the central density enhancement is 10^j at time t_j. It is clear that *the initial angular momentum distribution* (curve labeled t_0) *is "forgotten" very quickly because of effective magnetic braking.* Only in the supercritical core does Ω increase with time. Centrifugal forces are rapidly reduced to an insignificant level *everywhere* in the cloud; they remain negligible at subsequent times. Uniform rotation is maintained in the central region and in the envelope, which contains most of the cloud's mass, while most of the supercritical core exhibits differential rotation. In fact, in most of the core at the end of the run, v_ϕ is nearly constant (see Fig. 6b). Significant infall velocities develop only in the supercritical core. Even these, however, remain subsonic (see Fig. 6c) at all times.

The central magnetic field increases by a factor of 1.6 during the subcritical phase of contraction, and by a factor of $\simeq 180$ by the end of the run. The exponent $\kappa = 0.48$ at late times. The maximum value of B_r at the upper surface of the disk ($z = +Z$) is $(B_{r,Z})_{max} = B_z/4$ and occurs at $r \simeq 7 \times 10^{-5} R_0 = 4.0 \times 10^{-4}$ pc $\simeq 83$ AU. Elsewhere, $B_{r,Z} \ll B_z$. It is always the case that $|B_{\phi,Z}| \ll B_z$. Magnetic forces support the envelope very well against gravity. The mass and radius of the supercritical core are $M_{core} \simeq 10\ M_\odot$ (in excellent agreement with the peak in Fig. 1a) and $R_{core} \simeq 0.14$ pc, respectively. Its specific angular momentum has decreased by more than two orders of magnitude, to $\ell = 10^{20}$ cm^2 s^{-1}, which is comparable to that of wide visual binaries. The innermost $1\ M_\odot$ has $\ell = 10^{19}$ cm^2 s^{-1}, which is comparable to that of the protosolar nebula.

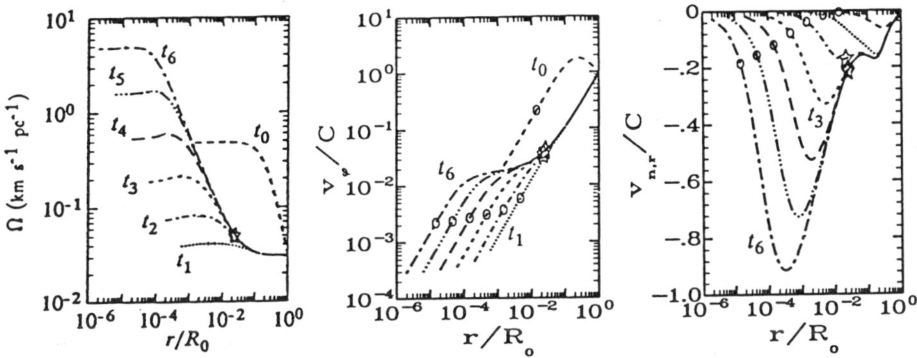

Figure 6. *Angular velocity and components of the linear velocity as functions of radius at seven different times* (0, 13.379, 15.560, 15.846, 15.896, 15.907, and 15.910 Myr); R_0 = 5.75 pc. A *star* on a curve marks the instantaneous radius of the supercritical core. (**a**, *left*) Angular velocity. (**b**, *middle*) Azimuthal v_ϕ and (**c**, *right*) radial v_r components of the velocity, normalized to the isothermal sound speed C (= 0.188 km s^{-1} at T = 10 K).

As part of an extensive parameter study, the dependence of the physical properties of rotating protostellar cores on the initial mass-to-flux ratio μ_{c0} (in units of its thin-disk critical value) was determined (Basu and Mouschovias 1995b). For the range $0.1 \leq \mu_{c0} \leq 1.0$, suggested by observations, the core mass is found to scale as $M_{\text{core}} \propto \mu_{c0}$, while its specific angular momentum scales as $(J/M)_{\text{core}} \propto \mu_{c0}^2$. Hence, $(J/M)_{\text{core}} \propto M_{\text{core}}^2$.

5. Protostar Formation in the Four-Fluid Approximation

As shown in § 2.2, for as long as the electrons and the ions are well attached to the magnetic field, they can be treated as a single fluid (the plasma), leading to a four-fluid (neutrals, plasma, charged and neutral grains) description of protostar formation. The details of the formulation of the problem and the results are found in Ciolek and Mouschovias (1993, 1994, 1995), and the numerical techniques are described in Morton, Mouschovias, and Ciolek (1994). Summaries of the results from different perspectives are given by Mouschovias (1995, 1996b), and Ciolek (1996).

5.1. Main Effects of Grains

Microscopically, it is well known that grains play an important role in the chemical reactions that determine the degree of ionization. Macroscopically, the charged grains, if attached to the field, can exert a drag on the neutrals in addition to that exerted by ions, as originally suggested by Baker (1979) and Elmegreen (1979). We use the proper creation/destruction rate equations (7a) - (7c) to calculate in a self-consistent manner the abundances of charged species, i.e., electrons, molecular ions, atomic (metallic) ions, and charged grains, at each time step of the evolution; this includes the conversion of neutral grains to charged grains and *vice versa*. The high-energy ($\gtrsim$ 100 MeV) cosmic-ray ionization rate is taken to be ζ_{CR} = 5 × 10^{-17} s^{-1} (Spitzer and Tomasko 1968; Payne, Salpeter, and Terzian 1984). Grain-neutral frictional forces are also properly accounted for in the momentum equations (see

§ 2.1). *The quantities* $v_{ff,c0}$ *and* k (see eqs. [33a] and [33c]) *are no longer free parameters*, since the ion density is uniquely determined as a function of (the time-dependent) neutral density everywhere in a model cloud. As in the investigations ignoring grains, the initial central mass-to-flux ratio (μ_{c0}) is an important free parameter in the problem. The most significant new free parameters introduced by the physics of grains are (1) the dimensionless grain-neutral collision rate, and (2) the rate of (inelastic) electron attachment onto neutral grains relative to the rate of elastic grain-neutral collisions. They are most sensitive to the value of the grain radius a (see Ciolek and Mouschovias 1993, 1994), which, as is intuitively clear, affects both the rate of chemical reactions involving grains and the grain-neutral drag forces. In what follows we use the "standard" value $a = 3.75 \times 10^{-6}$ cm.

We use a reference state to calculate the mass-to-flux ratio and determine an initial equilibrium state as described in § 4.2. The initial central mass-to-flux ratio, in units of its thin-disk critical value, is now $\mu_{c0} = 0.256$. The initial equilibrium state could represent a cloud with $n_{n,c0} = 2.60 \times 10^3$ cm^{-3}, $B_{z,c0} = 35.3$ μG, $T = 10$ K, $l_{ref} = 0.858$ pc, $R_0 = 5l_{ref} = 4.29$ pc, $M_{cl} = 98.3$ M$_\odot$, and a grain mass fraction $\chi_{g,0} = 0.01$. The main difference between this equilibrium state and that of § 4.2, except for the presence of grains and absence of rotation, is that l_{ref} has now been taken to be smaller than that of the rotating cloud by the factor 1.34. As explained earlier, the results relating to the formation and collapse of protostellar cores are not affected by the choice of l_{ref}, provided that it is significantly greater than the critical thermal lengthscale; in this case, we have that $l_{ref} = 5.5\lambda_{T,cr,ref}$.

Ambipolar diffusion is turned on at time $t = 0$ and the evolution is followed, as usual, to six orders of magnitude enhancement of $n_{n,c}$. Figure 5a describes the evolution of $n_{n,c}$ of the present model cloud as well, except for the fact that, instead of taking $\simeq 10\tau_{ff,c0}$ to achieve supercritical conditions in the central flux tube, it now takes $\simeq 13\tau_{ff,c0}$.

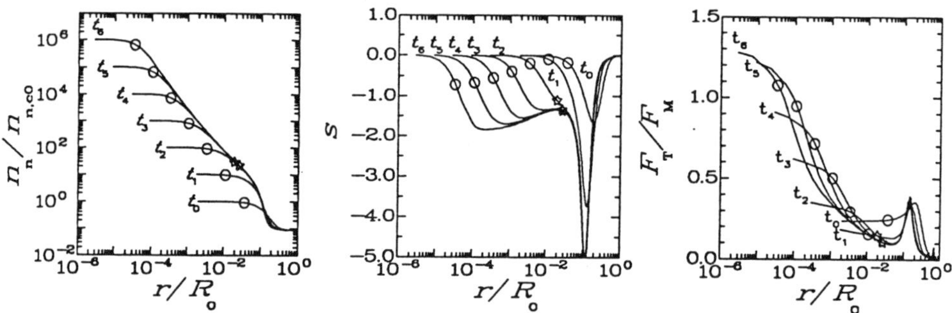

Figure 7. *Evolution in the presence of grain-neutral drag.* Spatial profiles of physical quantities are shown at seven different times ($t = 0$, 10.667, 13.279, 13.614, 13.663, 13.673, and 13.675$\tau_{ff,c0}$; $\tau_{ff,c0} = 1.2437\tau_{dyn,c0} = 1.3295$ Myr). The radial coordinate r is normalized to $R_0 = 4.29$ pc. See text for meaning of *stars* and *open circles*. (**a**, *left*) Neutral density, normalized to its initial central value $n_{n,c0} = 2.60 \times 10^3$ cm^{-3}. (**b**, *middle*) Slope $s \equiv \partial\ln n_n/\partial\ln r$. (**c**, *right*) Ratio of thermal-pressure and magnetic forces.

Figure 7a shows the neutral density n_n as a function of radius r at seven different times t_j ($j = 0, 1, ..., 6$) chosen such that the central density increases by a factor 10^j in the time t_j. A "*star*" on a curve, present only after a supercritical core forms, marks the instantaneous radius of the core. Once such a core forms, it is characterized by a compact, nearly uniform-density central region [whose size is determined by the instantaneous value of the critical thermal lengthscale $\lambda_{T,cr} = C^2/2G\sigma_{n,c}(t)$, marked by an *open circle* on each curve], and a "tail" of infalling matter left behind by the shrinking (both in size and in mass) central region. A near power-law density profile is established in the tail, $n_n \propto r^s$, with $-1.84 \lesssim s \lesssim -1.50$ in the region $1.0 \times 10^{-4} \lesssim r/R_0 \lesssim 8.0 \times 10^{-3}$ at time t_6 (see Fig. 7b). *A break in the slope of the density profile occurs almost exactly at the boundary of the supercritical core*, reflecting the different physics governing the evolution of the core and that of the subcritical, magnetically well-supported envelope. At the end of the run, the mass and radius of the core are $M_{core} = 8.21 \times 10^{-2} M_{cl} \simeq 8.1\ M_\odot$ (in excellent agreement, once again, with the observations shown in Fig. 1a), and $R_{core} = 2.67 \times 10^{-2} R_0 \simeq 0.11$ pc (which is also a typically observed core radius). The radius of the uniform-density central region is $\simeq \lambda_{T,cr}(t_6) = 3.38 \times 10^{-5} R_0 \simeq 30$ AU. Although the central density at this time is 2.6×10^9 cm^{-3}, the *mean* density of the supercritical core is only $\langle n_n \rangle_{core} = 39.5 n_{n,c0} = 1.0 \times 10^5$ cm^{-3}, comparable to the density of observed protostellar cores using NH_3 as a tracer molecule (e.g., see Bachiller *et al.* 1990; Crutcher *et al.* 1994). Similarly, the central field is 4 mG, but its mean value in the core is only $\simeq 65\ \mu$G.

Perpendicular to the field lines, the magnetic force F_M dominates the thermal-pressure force F_T in opposing gravity at all times and everywhere in the cloud, including the supercritical core, except in the uniform-density central region (within $r < 10^{-4} R_0$) at the end of the run, where $F_T/F_M \simeq 1.0 - 1.3$ (see Fig. 7c).[5] The maximum infall speed occurs at the end of the run and is only $\simeq 0.6C$, i.e., the infall is subsonic and far from being free fall. Its location is $r \simeq 2 \times 10^{-4} R_0$; it decreases rapidly on either side of this radius, and is negligible farther out in the subcritical envelope. Qualitatively, it behaves as shown in Figure 6c. The magnitude of the maximum infall acceleration occurs at a slightly smaller radius than the infall-speed maximum, and is only $\simeq 16\%$ that of gravity.

The exponent $\kappa \equiv d\ln B_c/d\ln n_{n,c}$ as a function of central density is shown in Figure 7d. Ion depletion onto grains would by itself tend to allow ambipolar diffusion to operate effectively even at high densities, during the collapse phase. However, grain-neutral drag almost compensates for the effect of ion depletion, and κ increases to 0.48 by the end of the run, a value near the flux-freezing limit of 1/2. During the early, quasistatic phase of contraction κ is very small, revealing that the neutrals are contracting through essentially stationary field lines (and plasma).

The direct (Γ_{dir}; see eq. [19]) and indirect (Γ_{indir}; see eq. [20]) attachment parameters (of grains to the magnetic field) are shown in Figure 7e as functions of central density. It is clear that, if it were not for the grain-quasiparticle electrostatic attraction, the grains would detach from the field at density enhancements $\gtrsim 10^2$. In reality, they remain partially coupled to the field even at the end of the run, at a central density of 2.6×10^9 cm^{-3}.

Customarily, it is thought that the microscopic physical processes that determine the degree of ionization affect the rate at which ambipolar diffusion progresses and, therefore, the evolution of a cloud. Not much attention has been paid to the effect of the evolution on the microscopic physics. Figure 7f shows the central number densities, relative to that of the neutrals, of ions ($x_{i,c}$), electrons ($x_{e,c}$), charged grains ($x_{g_-,c}$), and neutral grains ($x_{g_0,c}$) as functions of neutral density enhancement $n_{n,c}/n_{n,c0}$. It is clear that, during the quasistatic phase of contraction, the dust-to-gas ratio *decreases* in the central flux tubes of a cloud because the grains are well attached to the magnetic field and are thus left behind by the

---

[5]Hence, calculations that ignore the magnetic field or treat it as a perturbation on the zeroth-order flow of a singular, nonmagnetic, isothermal sphere (e.g., Galli and Shu 1993a, b) are based on a fundamentally incorrect assumption, which is responsible for erroneous conclusions and speculations, such as large magnetic pinching forces and magnetic reconnection during the isothermal phases of star formation.

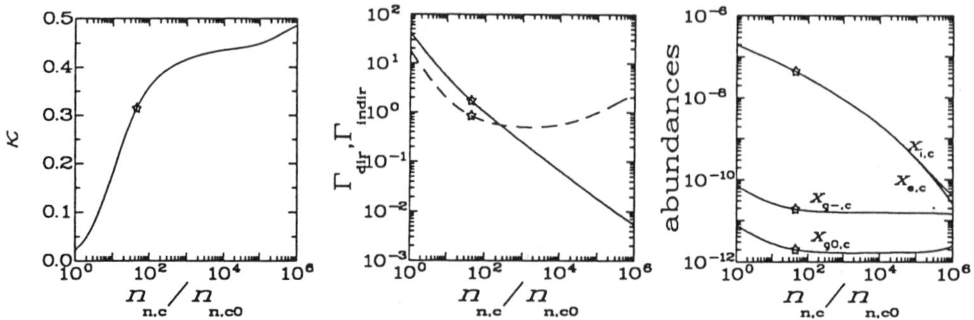

Figure 7, cont'd. *Central values of parameters as functions of central density* $n_{n,c}$, *normalized to its initial value as in Fig. 7a.* (**d**, *left*) *Exponent* $\kappa \equiv d\ln B_c/d\ln n_{n,c}$. (**e**, *middle*) *Charged-grain direct* (Γ_{dir}; *solid line*) *and indirect* (Γ_{indir}; *dashed line*) *magnetic attachment parameters.* (**f**, *right*) *Abundances of ions* ($x_{i,c}$), *electrons* ($x_{e,c}$), *charged grains* ($x_{g-,c}$), *and neutral grains* ($x_{g0,c}$) *relative to neutral molecules.*

contracting neutrals. This reduces the rate of electron and ion capture by grains. If we represent the relation between the ion and neutral densities in the usual manner, $n_i \propto n_n^{\,k}$, *no constant k can accurately describe the ion density at all times.* Figure 7g exhibits the central value of the exponent k as a function of central density enhancement. It is seen (1) that $k_c > 1/2$ during the quasistatic phase (i.e., there are more ions than one would deduce from chemistry in static cloud models); and (2) that k_c decreases to $\simeq 0.1$ at the end of the run. *Hence, the magnetic field, through its effects on the evolution, affects significantly the chemistry in contracting cores.*

Figure 7g. *Exponent* $k_c \equiv d\ln n_{i,c}/d\ln n_{n,c}$ *as a function of central density enhancement* $n_{n,c}/n_{n,c0}$. *It is greater than 1/2 (the value predicted by chemical models of static clouds) during the quasistatic phase, and becomes significantly smaller than 1/2 during the dynamic phase of core contraction. The decrease of grain abundance in the core during quasistatic contraction (see Fig. 7f) allows more ions and electrons to remain in the gas phase. The constant relative abundance of grains during collapse leads to effective attachment of electrons and ions onto grains and, therefore, reduction of* x_e *and* x_i *compared to their expected values in the absence of grain chemistry.*

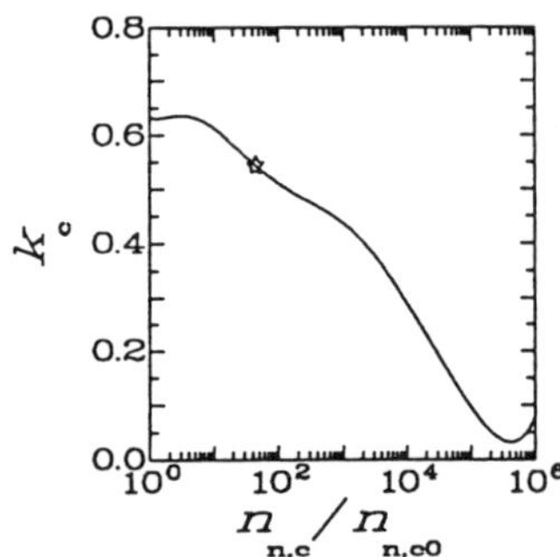

The precise factor by which the dust abundance decreases in a supercritical core relative to its value in the envelope depends on the degree to which grains are attached to the field. For well-attached grains (i.e., $a \lesssim 7 \times 10^{-6}$ cm), this factor is equal to $1/\mu_{c0}$, the initial central mass-to-flux ratio in units of its critical value for collapse. In practice, the grain abundance in a core relative to that of the parent envelope may be easier to determine observationally than the *initial* mass-to-flux ratio of the core, which is essentially that of the envelope in the vicinity of the core. Hence, one may turn the problem around and, *from the observable reduction of the grain abundance in a core, one may straightforwardly infer the initial mass-to-flux ratio relative to its critical value.* Only after a core becomes magnetically and thermally supercritical, and dynamical contraction ensues, does its grain mass fraction become "frozen" in the matter. Since small grains are better attached to the magnetic field lines than more massive grains, the dynamical effect described above implies that the larger grains will tend to be more abundant in high-density cores. To observations this would appear as grain growth with increasing density (see Vrba *et al.* 1981, 1993), when it may actually be a simple and direct consequence of the fact that the small grains, being better attached to the field lines, are preferentially left behind during the formation and contraction of protostellar cores.

Ciolek and Mouschovias (1996) derived analytical expressions that relate the predicted grain abundances in dense cores to physical quantities such as the grain radii and the initial central mass-to-flux ratios of the parent molecular clouds. They first considered grains of a single radius, and they then extended their results to a grain population having a MRN (Mathis, Rumpl, and Nordsieck 1977) distribution of radii ($a^{-3.5}$). Their expressions yield predictions in excellent agreement with the results of the numerical calculations summarized above as well as cover cases not yet considered by numerical calculations.

Jenniskens and Greenberg (1993) suggested that the lack of increase of "the linear rise" region in interstellar extinction curves for dense regions of molecular clouds results from reduction of the abundances of small grains at large optical depth. They argue that this is consistent with models of accretion of small grains onto larger grains. However, as can be seen in Figure 7f, ambipolar diffusion leads naturally to significant reduction of the abundance of small grains in dense cores. Growth of large grains has also been invoked to interpret optical and near-infrared (NIR) polarimetric observations, which find an increase in the wavelength of maximum polarization (λ_{max}) with increasing extinction (e.g., Vrba, Coyne, and Tapia 1981; Whittet *et al.* 1994). Moreover, Vrba *et al.* (1993) suggest that their data for ρ Ophiuchi, when compared to other dark clouds, indicate that large grains are formed predominantly in the very early stages of cloud evolution and that, after 10^6 yr, grain evolution is mainly a process of destroying larger grains in the outer several magnitudes of extinction (i.e., in the envelope of a cloud). This physical picture of *segregation* (or, concentration) of large grains at high extinction is also consistent with reduction of the abundances of small grains by ambipolar diffusion in the inner flux tubes of molecular clouds, as described above. An increase in λ_{max} in a core may occur because of a significant decrease in the abundance of the smaller polarizing grains ($a \simeq 0.7 - 3 \times 10^{-5}$ cm), which is exactly what occurs in the cores of models 2 - 4 of Ciolek and Mouschovias (1996; see Figs. 2 and 3 and Tables 2 and 3 therein).

Reduction of grain abundances in cores would decrease the metallicity of the stars that form in them. This, in turn, would affect the measurement of interstellar abundances that use the atmospheres of young stars to investigate the metallicity of the interstellar medium from which these stars formed. Recent observations of the abundance of carbon based on the analysis of photospheres of young stellar objects (YSO) reveal the abundance of carbon (and other heavy elements) in the interstellar medium to be only $\simeq 60\%$ of the solar value. Dwek (1997) suggests as one possibility that this "interstellar carbon crisis" may reflect depletion of or "straining" of the grains by the magnetic field and the ambipolar diffusion mechanism proposed by Ciolek and Mouschovias (1996), summarized here.

Ciolek and Mouschovias (1998) incorporated the dynamical effect of the contraction of the neutrals and the reduction of the dust-to-gas ratio in the chemical rate equations for simplified model clouds. The resulting semi-analytical expressions derived to calculate the ion abundances in contracting cores for densities $n_{n,c} \lesssim 10^9$ cm^{-3} were found to be in excellent agreement with the detailed numerical simulations of Ciolek and Mouschovias

(1994, 1995) and, specifically, with the variation of the exponent k_c with central neutral density shown in Figure 7g. It follows that *measurements of ion abundances during the early stages of core formation ($n_{n,c} \lesssim 10^5$ cm^{-3}) can be used, in principle, to observationally determine whether ambipolar diffusion is responsible for initiating the formation and contraction of protostellar cores.* Ideally, this should be done by separately mapping the radial distribution of the ion and neutral densities in a young core that has not yet collapsed to form stars (i.e., the "prestellar" or "starless" cores studied by Ward-Thompson *et al.* 1994, and by André *et al.* 1996). From these radial profiles the relation $k_c(n_{n,c}) = (d\ln n_{i,c}/d\ln r)/(d\ln n_{n,c}/d\ln r)$ could then be deduced for that particular core. Note that $n_{i,c}$ is the total ion density. The model calculations show that the abundance of atomic ions is much greater than that of molecular ions. Hence, measurement of the density of atomic ions (predominantly Mg$^+$ in our models) would effectively determine the total ion density. Detection of values of $k_c > 1/2$ should correlate with measurements of reduced grain abundances in cores.

5.2. Detailed Comparison with Observations

The results of these calculations have been used to construct a detailed evolutionary model for the Barnard 1 cloud (Crutcher, Mouschovias, Troland, and Ciolek 1994). We first showed that the profile of the column density (as revealed by C^{18}O and ^{13}CO observations) of the "internal envelope" of the cloud (see Bachiller *et al.* 1990), of mass 600 $M_\odot$, can be well represented by an exact MHS equilibrium model of a disklike cloud, whose axis of symmetry (the z-axis) is inclined by 20° with respect to the plane of the sky so as to explain the observed 3:1 axes ratio. We then turned on ambipolar diffusion. A magnetically supercritical core formed in about $\simeq 10^7$ yr, but, once the core formed, it collapsed in less than 1 Myr. The predicted properties of the protostellar core include a mass = 13.4 $M_\odot$, radius = 0.13 pc, mean density = 1.3×10^5 cm^{-3}, and mean magnetic field strength along the line of sight = 29.1 μG. These are in excellent agreement with the corresponding observed values of the NH$_3$ core; namely, mass $\simeq 13$ $M_\odot$, radius $\simeq 0.15$ pc, mean density > 8×10^4 cm^{-3}, and $B_{los} = 30 \pm 4$ μG. Moreover, the calculated spatial profile of the Alfvén speed compares well with observations of the linewidth as a function of radius along the major axis. The model makes further predictions concerning the structure of the protostellar core that can be tested by higher spatial resolution observations.

Benson, Caselli, and Myers (1998) have carried out high spatial and spectral resolution observations of lines of N$_2$H$^+$ and C$_3$H$_2$ in sixty dense molecular cloud cores. They find a good correlation between the velocities and the linewidths of the ionized and neutral molecules, with an upper limit on the relative drift speed $v_{D,max} = 0.03$ km s^{-1}. Although the authors do not regard this as a definite detection of ambipolar diffusion, they do point out that this result is completely consistent with the ambipolar-diffusion-induced drift speeds predicted by Ciolek and Mouschovias (1995). Moreover, the implied ambipolar-diffusion timescale is an order of magnitude greater than the dynamical timescale in these cores, which is in excellent agreement with equation (1) and with the early work by Mouschovias (1983) --see, also, Figure 5b above, curves labeled $\tau_{\Phi,c}$ and $\tau_{dyn,c}$.

5.3. Effect of Ultraviolet Radiation

In addition to cosmic-ray ionization and the microscopic and macroscopic effects of grains described in § 5.1, Ciolek and Mouschovias (1995) have included the effect of UV ionization. We take $\zeta_{uv} = 1.2 \times 10^{-11}$ s^{-1} at the cloud surface, and we account for the attenuation due to grain absorption and carbon ionization. In addition to the ions HCO$^+$,

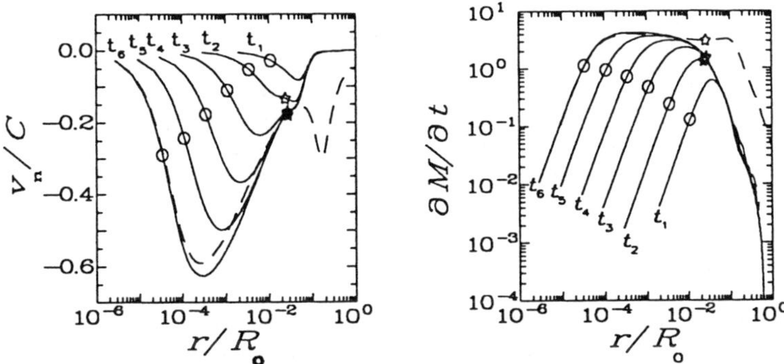

Figure 8. *Effect of UV ionization.* Spatial profiles of (**a**, *left*) the neutral velocity v_n in the equatorial plane, normalized to the isothermal sound speed $C = 0.188$ km s^{-1}, and (**b**, *right*) the mass infall rate (in $M_\odot$/Myr), at six times t_1, t_2, ..., t_6 for a model identical to that of Figs. 7a–7g, except that UV ionization is now accounted for. The times are chosen such that the central density at time t_i exceeds its initial value by the factor 10^j; they are equal to 8.614, 10.507, 10.774, 10.818, 10.827, and 10.829$\tau_{ff,c0}$; $\tau_{ff,c0} = 1.3295$ Myr. The *dashed line* refers to the model of § 5.1, which ignores UV ionization, at the end of the run. The "stars" and open circles have the same meaning as in Fig. 7.

Mg$^+$, and Na$^+$, other ions (e.g., S$^+$, Si$^+$, and C$^+$) now play a significant role (see Appendix). Figure 8a shows the infall speed of the neutrals as a function of radius at six different times. The dashed line represents the velocity profile at the end of the run in the absence of UV ionization (the model cloud of Figs. 7a – 7g). Clearly, although the infall speed within the core is virtually unaffected by UV ionization, infall is drastically reduced in the envelope. Figure 8b exhibits the mass infall rate dM/dt (in $M_\odot$/Myr) as a function of r/R_0 at six different times. The dashed line represents the infall rate at the end of the run in the absence of UV ionization, as in Figure 8a. Although the formation and evolution of the supercritical protostellar core is unaffected (except for the fact that it now takes 30% *less* time to form[6]), *infall beyond the supercritical core is effectively cut off by the higher degree of ionization in the envelope.* The maximum mass that may accrete onto the forming star is therefore that of the supercritical core. Some of the accreting mass may be subsequently ejected in the form of a wind during the opaque (nonisothermal) phases of star formation.

6. Hydromagnetic Waves, Linewidths, Ambipolar Diffusion, and Star Formation

The presence of HM waves in molecular clouds, and the role they may play in supporting the clouds certainly along field lines (Mouschovias and Morton 1985a, b; Mouschovias 1987a), does not alter in any significant way the ambipolar-diffusion-induced formation and

---

[6]The now better magnetic support of the envelope keeps matter farther away from the axis of symmetry of the protostellar *disk*. Hence, there is less dilution of the core's gravitational field by the mass in the magnetically supported envelope. This leads to a faster formation and evolution of a supercritical core. Arguments that UV ionization increases τ_{AD} too much to account for protostar formation are thus seen to be in error.

collapse of protostellar fragments (or cores) described above. This is so because, for typical molecular cloud parameters ($n_n \sim 10^3$ cm^{-3}, $T \simeq 10$ K, $B \simeq 30$ μG), the size of the region that can just become gravitationally unstable because of ambipolar diffusion happens to be essentially equal to the Alfvén lengthscale λ_A (Alfvén waves with wavelengths $\lambda \leq \lambda_A$ cannot propagate in the neutrals because of damping by ambipolar diffusion) --see Mouschovias 1987a; or 1991a, eqs. (18a, b). In fact, it is precisely the decay of HM waves due to ambipolar diffusion that removes part of the support against gravity over the critical thermal lengthscale and thus initiates fragmentation (or core formation) in molecular clouds (Mouschovias 1987, § 2.2.6[b]). One may therefore consider the issue of whether HM waves can explain the observed, usually supersonic but subAlfvénic linewidths without concern that attributing the linewidths to such waves might modify substantially our current understanding of the role of magnetic fields and ambipolar diffusion in star formation (see Mouschovias and Psaltis 1995; review by Mouschovias 1996a, § 6).

If the observed linewidths are due to large-amplitude, long-wavelength ($\lambda \gtrsim 0.1$ pc), hence long-lived, Alfvén waves, they should be related to the magnetic field strength B and the cloud (or core) radius R by

$$(\Delta v)_{\text{wave}} \simeq 1.4 \left[\frac{B}{30 \ \mu G} \right]^{1/2} \left[\frac{R}{1 \ \text{pc}} \right]^{1/2} \quad \text{km s}^{-1} \tag{37}$$

(Mouschovias 1987a, § 2.2.1; Mouschovias and Psaltis 1995). Thus, the "turbulence law" found in molecular clouds and cloud cores (Larson 1981; Leung, Kutner, and Mead 1982; Myers 1983, 1985; Falgarone and Pérault 1987; Myers and Goodman 1988; Fuller and Myers 1992) is a special case of this more general result.

7. Summary

A five-fluid (neutral molecules, ions, electrons, charged and neutral grains) MHD description of protostar formation has been physically motivated and justified. For the early isothermal phase (neutral density $n_n \leq 10^9$ cm^{-3}), the electrons and ions behave as a single fluid (the plasma) well coupled to the magnetic field; hence, a four-fluid description becomes appropriate. The charged grains (typical radius $a \simeq 4 \times 10^{-6}$ cm) are directly attached to the field only up to densities $\sim 10^5$ cm^{-3}. Indirect, at least partial attachment, due to electrostatic attraction by well-attached electron-shielded ion quasiparticles, dominates at higher densities. The same electric field which is responsible for the partial attachment of grains to the magnetic field lines also causes detachment of the ions from the field lines at smaller densities ($\lesssim 3 \times 10^8$ cm^{-3}) than previously thought possible. Only the electrons remain attached to the field lines at higher densities.

Theoretical calculations have predicted and observations have shown that star formation in a typical, self-gravitating molecular cloud is not the result of collapse of the cloud as a whole, despite the fact that its observed mass exceeds the critical thermal ($\simeq$ Jeans) mass by a factor $\sim 10^2$ - 10^5. Clouds are primarily magnetically supported, with hydromagnetic (HM) waves probably providing some support at least along field lines. In such magnetically subcritical clouds, ambipolar diffusion is an unavoidable process. Analytical and numerical two-fluid calculations first, and four-fluid axisymmetric simulations later (accounting for both cosmic-ray and UV ionization), have shown that star formation is a self-initiated process, and have followed the formation and contraction of protostellar fragments (or cores) from typical densities $\sim 10^3$ cm^{-3} to central densities $\sim 10^{10}$ cm^{-3}, by which the ions are no longer well attached to the field.

Ambipolar diffusion damps the HM waves and allows gravity to initiate quasistatic contraction in the deep interior of a cloud, where the degree of ionization is $x_i \lesssim 10^{-7}$, over a region containing approximately one critical thermal mass (= a few $M_\odot$ at $n_n \sim 10^3$ cm^{-3} and $T \simeq 10$ K), while the field lines and the attached plasma and grains remain essentially stationary. A disklike core begins to form and to contract on a timescale equal to the central magnetic-flux redistribution timescale $\tau_{\Phi,c}$ (= $\tau_{AD,c}/2$) $\lesssim 10^7$ yr, which exceeds the central free-fall timescale $\tau_{ff,c}$ typically by a factor $\simeq 10$. Magnetic braking is very effective during this phase of evolution, and keeps the entire cloud essentially corotating with the background. Centrifugal forces are reduced to a negligible level, never to play a significant role in the evolution during the isothermal phase of star formation ($n_n \lesssim 10^{10}$ cm^{-3}).

The ambipolar-diffusion-induced and controlled redistribution of mass in the central flux tubes eventually increases the central mass-to-flux ratio to the critical value for collapse [=1.5(63G)$^{-1/2}$], and dynamic contraction of the thermally and magnetically supercritical core ensues. The density enhancement at which critical conditions are achieved is essentially equal to μ_{c0}^{-2}, where μ_{c0} (typically 0.1 - 1.0) is the initial central mass-to-flux ratio is units of its critical value for collapse. Even though the lateral contraction (i.e., perpendicular to the field lines) is dynamic, the remarkable result is found that force balance is maintained *along* field lines. Magnetic forces represent the dominant opposition to gravity (in the lateral direction), even in the supercritical core, up to central densities $\simeq 3 \times 10^9$ cm^{-3}. Only in a small central region, of radius approximately equal to the instantaneous value of the critical thermal lengthscale $\lambda_{T,crit}(t) \simeq C\tau_{ff,c}(t)$ ($\simeq 30$ AU; C is the isothermal sound speed), does the thermal-pressure force become comparable to the magnetic force near the end of a typical run, at $n_{n,c} = 3 \times 10^9$ cm^{-3}. Thermal pressure is responsible for maintaining a uniform-density central region of radius $\simeq \lambda_{T,crit}(t)$ during the evolution. This region shrinks in both size and mass, leaving behind a "tail" of matter (all within the supercritical core), in which a near power-law density profile is established $n_n \propto r^s$, with $-1.8 \lesssim s \lesssim -1.5$. Both angular momentum and magnetic flux tend to get trapped inside the core during the collapse phase, so that the magnetic field and the angular velocity vary almost as $B_c \propto n_{n,c}^{1/2}$ and $\Omega_c \propto n_{n,c}^{1/2}$, the magnetic-flux and angular-momentum conservation laws for a disk. The predicted typical physical properties of protostellar cores are in excellent agreement with observations: mass $\simeq 1 - 15\ M_\odot$; radius $\simeq 0.1 - 0.2$ pc; *mean* field $\langle B \rangle$ greater than that of the parent envelope by less than a factor of 2 (although the *central* field is typically $\simeq$ a few mG at $n_n \simeq 10^8 - 10^9$ cm^{-3}); *mean* density $\langle n_n \rangle \simeq 10^5$ cm^{-3}; specific angular momentum $J/M \simeq 10^{19} - 10^{21}$ cm^2 s^{-1}, which is comparable to that of wide binary systems. The innermost 1 $M_\odot$ has a $J/M \simeq 10^{19}$ cm^2 s^{-1}, which is the inferred value for the protosolar nebula (see Bodenheimer et al. 1993). Hence, magnetic braking resolves the angular momentum problem both for binary and solar-type stars. During the formation and contraction of protostellar cores, the cloud envelopes, which typically contain $\simeq 90\%$ of the mass, hardly suffer any evolution; they remain magnetically supported.

The theory also predicts that the presence of large-amplitude, long-wavelength Alfvén waves in self-gravitating clouds implies a linewidth-size relation which is in excellent quantitative agreement with observations, namely, $\Delta v \simeq 1.4\ (B/30\ \mu G)^{0.5}(R/1\ pc)^{0.5}$ km s^{-1}.

The process of core formation itself affects the relative abundance of grains (hence, the chemistry) in dense cores in a profound way: grains attached to the field are left behind during the quasistatic phase of contraction, thus reducing the dust-to-gas relative abundance by the factor $1/\mu_{c0}$. Aside from its evident implications on the chemistry in dense cores, this result can be used to deduce the initial mass-to-flux ratio of a cloud if the relative abundances of grains in a core and the envelope are observed. Also, since small grains are better attached to the field than larger grains, we predict a stratification of grains according to size during core formation. This effect gives the *impression* of grain growth with optical depth. A number of commonly used assumptions (e.g., a fixed dust-to-gas mass ratio as a function of density, and the parametrization of the ion density in terms of the neutral density as a power law with a single, constant exponent) are found to be incorrect. We have described a modern theory of star formation incorporating these new results.

There is little doubt that the multifluid–MHD description of star formation will uncover many more surprises when the later stages of evolution ($n_n > 10^{10}$ cm^{-3}) are explored. Desch and Mouschovias (1999) have already shown that (1) the isothermality assumption for protostellar fragments (or cores) does not necessarily break down at central densities $n_{n,c} \simeq 10^{10}$ cm^{-3}; (2) ambipolar diffusion is much more effective than Ohmic dissipation (by almost two orders of magnitude) in decoupling the magnetic field from the neutral matter in the density range 10^9 - 10^{12} cm^{-3}, contrary to current expectations; (3) the predicted magnetic field strength in the innermost 20 AU of the protoplanetary disk is 0.1 G. This is in excellent agreement with measurements of the remanent magnetization in relevant meteorites (mainly stony chondrites, but also iron–nickel meteorites).

Acknowledgements. This work was supported by the National Science Foundation under grant AST 93-20250 to the University of Illinois. Discussions on chemistry issues with Tom Hartquist are gratefully acknowledged. GC gratefully acknowledges support by the New York Origins of Life Center (NSCORT) and by the Department of Physics at RPI, under NASA grant NAG5-7598.

Appendix: Relevant Chemical Reactions in the Presence of UV Ionization

Detailed chemical models for molecular clouds have been presented by Solomon and Klemperer (1972), Herbst and Klemperer (1973), Oppenheimer and Dalgarno (1974), Glassgold and Langer (1974, 1976), Black and Dalgarno (1977), Mitchell et al. (1978), Prasad and Huntress (1980), and Graedel et al. (1982). Here we consider only the reactions affecting the equilibrium abundances of the relevant charged species.

Cosmic-ray ionization: $\qquad\qquad\qquad H_2 + CR \rightarrow H_2^+ + e$. $\hfill$ (A1)

UV ionization: $\quad \mathcal{A} + UV \rightarrow \mathcal{A}^+ + e$, $\qquad \mathcal{A} = $ C, Mg, Na, Fe, S, Si. $\hfill$ (A2)

HCO$^+$ formation: $\ H_2^+ + H_2 \rightarrow H_3^+ + H$, $\qquad H_3^+ + CO \rightarrow HCO^+ + H_2$. $\hfill$ (A3a, b)

$$C^+ + OH \rightarrow CO^+ + H , \qquad CO^+ + H_2 \rightarrow HCO^+ + H .$$ $\hfill$ (A3c, d)

$$C^+ + H_2O \rightarrow HCO^+ + H .$$ $\hfill$ (A3e)

Dissociative recombination: $\qquad HCO^+ + e \rightarrow H + CO$. $\hfill$ (A4)

Radiative recombination: $\qquad\quad \mathcal{A}^+ + e \rightarrow \mathcal{A} + h\nu$. $\hfill$ (A5)

Charge transfer: $\quad A + HCO^+ \rightarrow A^+ + HCO$, $\qquad A = $ Mg, Na, Fe; $\hfill$ (A6a)

$$A + M^+ \rightarrow A^+ + M , \qquad M = \text{C, S, Si;}$$ $\hfill$ (A6b)

$$s + C^+ \rightarrow s^+ + C , \qquad s = \text{S, Si;}$$ $\hfill$ (A6c)

$$Si + S^+ \rightarrow Si^+ + S .$$ $\hfill$ (A6d)

Electron and ion capture by grains:

$$e + g_0 \rightarrow g_- , \qquad HCO^+ + g_- \rightarrow HCO + g_0 , \qquad \mathcal{A}^+ + g_- \rightarrow \mathcal{A} + g_0 .$$ $\hfill$ (A7a, b, c)

As explained in § 2.1, footnote #3, dissociative recombination of H_3^+, which would compete with reaction (A3b), can be neglected.

References

1. André, P., Ward-Thompson, D., and Motte, F. 1996, *A&A*, **314**, 625
2. Bachiller, R., Menten, K. M., and del Rio-Alvarez, S. 1990, *A&A*, **236**, 461
3. Baker, P. L. 1979, *A&A*, **75**, 54
4. Basu, S., and Mouschovias, T. Ch. 1994, *ApJ*, **432**, 720
5. ____. 1995a, *ApJ*, **452**, 386
6. ____. 1995b, *ApJ*, **453**, 271
7. Benson, P. J., Caselli, P., and Myers, P. C. 1998, *ApJ*, **506**, 743
8. Black, J. H., and Dalgarno, A. 1977, *ApJS*, **34**, 405
9. Bodenheimer, P., Ruzmaikina, T., and Mathieu, R. D. 1993, in *Protostars and Planets III*, ed. E. H. Levy and J. Lunine (Tucson: Univ. of Arizona), 367
10. Bonnor, W. B. 1956, *MNRAS*, **116**, 351
11. Ciolek, G. E. 1996, in *The Role of Dust in the Formation of Stars*, ed. H. U. Käufl (Berlin: Springer), 367
12. Ciolek, G. E., and Mouschovias, T. Ch. 1993, *ApJ*, **418**, 774
13. ____. 1994, *ApJ*, **425**, 142
14. ____. 1995, *ApJ*, **454**, 194
15. ____. 1996, *ApJ*, **468**, 749
16. ____. 1998, *ApJ*, **504**, 280
17. Crutcher, R. M., Mouschovias, T. Ch., Troland, T., and Ciolek, G. E. 1994, *ApJ*, **427**, 839
18. Dalgarno, A. 1987, in *Physical Processes in Interstellar Clouds*, ed. G. E. Morfill and M. Scholer (Dordrecht: Reidel), 219
19. Desch, S. J, and Mouschovias, T. Ch. 1999, *ApJ*, submitted
20. Draine, B. T., and Sutin, B. 1987, *ApJ*, **320**, 803
21. Dwek, E. 1997, *ApJ*, **484**, 779
22. Ebert, R. 1955, *Zs. Ap.*, **37**, 217
23. ____. 1957, *Zs. Ap.*, **42**, 263
24. Elmegreen, B. G. 1979, *ApJ*, **232**, 729
25. Falgarone, E., and Pérault, M. 1987, in *Physical Processes in Interstellar Clouds*, ed. G. E. Morfill and M. Scholer (Dordrecht: Reidel), 59
26. Fiedler, R. A., and Mouschovias, T. Ch. 1992, *ApJ*, **391**, 199
27. ____. 1993, *ApJ*, **415**, 680
28. Fuller, G. A., and Myers, P. C. 1992, *ApJ*, **384**, 523
29. Galli, D., and Shu, F. H. 1993a, *ApJ*, **417**, 220
30. ____. 1993b, *ApJ*, **417**, 243
31. Glassgold, A. E., and Langer, W. D. 1974, *ApJ*, **193**, 73
32. Graedel, T. E., Langer, W. D., and Frerking, M. A. 1982, *ApJS*, **48**, 321
33. Herbst, E., and Klemperer, W. 1973, *ApJ*, **185**, 505
34. Hollenbach, D., and Salpeter, E. E. 1970, *J. Chem. Phys.*, **53**, 79
35. Jeans, J. H. 1928, *Astronomy and Cosmogony* (Cambridge: Cambridge Univ.)
36. Jenniskens, P., and Greenberg, J. M. 1993, *A&A*, **274**, 439
37. Langer, W. D. 1985, in *Protostars and Planets II*, ed. D. C. Black and M. S. Mathews (Tucson: Univ. of Arizona), 650
38. Larson, R. B. 1981, *MNRAS*, **194**, 809
39. Larsson, M. et al. 1993, *Phys. Rev. Lett.*, **70**, 430
40. Leung, C. M., Kutner, M. L., and Mead, K. N. 1982, *ApJ*, **262**, 583
41. Lizano, S., and Shu, F. H. 1989, *ApJ*, **342**, 834
42. Li, Z.-Y., and McKee, C. F. 1996, *ApJ*, **464**, 373
43. Mathis, J. S., Rumpl, W., and Nordsieck, K. H. 1977, *ApJ*, **217**, 425 (MRN)

44. McDaniel, E. W., and Mason, E. A. 1973, in *The Mobility and Diffusion of Ions and Gases* (New York: Wiley)
45. Mestel, L., and Spitzer, L., Jr. 1956, *MNRAS*, **116**, 503
46. Mitchell, G. F., Ginsburg, J. L., and Kuntz, P. J. 1978, *ApJS*, **38**, 39
47. Morton, S. A., Mouschovias, T. Ch., and Ciolek, G. E. 1994, *ApJ*, **421**, 561
48. Mott, N. F., and Massey, H. S. W 1971, *The Theory of Atomic Collisions*, 3rd ed. (Oxford: Oxford Univ.)
49. Mouschovias, T. Ch. 1976a, *ApJ*, **206**, 753
50. _____ . 1976b, *ApJ*, **207**, 141
51. _____ . 1977, *ApJ*, **211**, 147
52. _____ . 1978, in *Protostars & Planets*, ed. T. Gehrels (Tucson: Univ. of Arizona), 209
53. _____ . 1979a, *ApJ*, **228**, 159
54. _____ . 1979b, *ApJ*, **228**, 475
55. _____ . 1983, in *Solar and Stellar Magnetic Fields: Origins and Coronal Effects*, ed. J. O. Stenflo (Dordrecht: Reidel), 479
56. _____ . 1985, *A&A*, **142**, 41
57. _____ . 1987a, in *Physical Processes in Interstellar Clouds*, ed. G. E. Morfill and M. Scholer (Dordrecht: Reidel), 453
58. _____ . 1987b, in *Physical Processes in Interstellar Clouds*, ed. G. E. Morfill and M. Scholer (Dordrecht: Reidel), 491
59. _____ . 1989, in *The Physics and Chemistry of Interstellar Molecular Clouds*, ed. G. Winnewisser and J. T. Armstrong (Berlin: Springer), 297
60. _____ . 1991a, *ApJ*, **373**, 169
61. _____ . 1991b, in *The Physics of Star Formation and Early Stellar Evolution*, ed. C. J. Lada and N. D. Kylafis (Dordrecht: Kluwer), 61
62. _____ . 1991c, in *The Physics of Star Formation and Early Stellar Evolution*, ed. C. J. Lada and N. D. Kylafis (Dordrecht: Kluwer), 449
63. _____ . 1995, in *The Physics of the Interstellar Medium and Intergalactic Medium*, ed. A. Ferrara, C. F. McKee, C. Heiles, and P. R. Shapiro (San Francisco: ASP), **80**, 184
64. _____ . 1996a, in *Solar and Astrophysical Magnetohydrodynamic Flows*, ed. K. Tsinganos (Dordrecht: Kluwer), 505
65. _____ . 1996b, in *The Role of Dust in the Formation of Stars*, ed. H. U. Käufl (Berlin: Springer), 382
66. Mouschovias, T. Ch., and Morton, S. A. 1985a, *ApJ*, **298**, 190
67. _____ . 1985b, *ApJ*, **298**, 205
68. Mouschovias, T. Ch., and Paleologou, E. V. 1979, *ApJ*, **230**, 204
69. _____ . 1980, *ApJ*, **237**, 877
70. _____ . 1986, *ApJ*, **308**, 781
71. Mouschovias, T. Ch., and Psaltis, D. 1995, *ApJ*, **444**, L105
72. Mouschovias, T. Ch., and Spitzer, L., Jr. 1976, *ApJ*, **210**, 326
73. Myers, P. C. 1983, *ApJ*, **270**, 105
74. _____ . 1985, in *Protostars & Planets II*, ed. D. C. Black and M. S. Matthews (Tucson: Univ. of Arizona), 81
75. Myers, P. C., and Goodman, A. A. 1988, *ApJ*, **326**, L27
76. Nakano, T. 1979, *PASJ*, **31**, 697
77. Nakano, T., and Nakamura, T. 1978, *PASJ*, **30**, 671
78. Nakano, T., and Umebayashi, T. 1986a, *MNRAS*, **218**, 663
79. _____ . 1986b, *MNRAS*, **221**, 319
80. Nishi, R., Nakano, T., and Umebayashi, T. 1991, *ApJ*, **368**, 181
81. Oppenheimer, M., and Dalgarno, A. 1974, *ApJ*, **192**, 29

82. Payne, H. E., Salpeter, E. E., and Terzian, Y. 1984, *AJ*, **89**, 668
83. Pneuman, G. W., and Mitchell, T. P. 1965, *Icarus*, **4**, 494
84. Prasad, S. S., and Huntress, W. T., Jr. 1980, *ApJS*, **43**, 1
85. Safier, P. N., McKee, C. F., and Stahler, S. W. 1997, *ApJ*, **485**, 660
86. Solomon, P. M, and Klemperer, W. 1972, *ApJ*, **178**, 389
87. Spitzer, L., Jr. 1941, *ApJ*, **93**, 369
88. ____. 1948, *ApJ*, **107**, 6
89. ____. 1963, in *Origin of the Solar System*, ed. R. Jastrow and A. G. W. Cameron (New York: Academic), 39
90. ____. 1978, *Physical Processes in the Interstellar Medium* (New York: Wiley-Interscience)
91. Spitzer, L., Jr., and Tomasko, M. G. 1968, *ApJ*, **152**, 971
92. Tomisaka, K., Ikeuchi, S., and Nakamura, T. 1990, *ApJ*, **362**, 202
93. Toomre, A. 1963, *ApJ*, **138**, 385
94. Troland, T. H., and Heiles, C. 1986, *ApJ*, **301**, 339
95. Vrba, F. J., Coyne, G. V., and Tapia, S. 1981, *ApJ*, **243**, 489
96. ____. 1993, *AJ*, **105**, 1010
97. Ward-Thompson, D., Scott, P. F., Hills, R. E., and André, P. 1994, *MNRAS*, **268**, 276
98. Watson, W. D. 1976, *Rev. Mod. Phys.*, **48**, 513
99. Watson, W. D., and Salpeter, E. E. 1972, *ApJ*, **174**, 32
100. Whittet, D. C. B., Gerakines, P. A., Carkner, A. L., Hough, J. H., Martin, P. G., Prusti, T., and Kilkenny, D. 1994, *MNRAS*, **268**, 1
101. Wood, D. O. S., Myers, P. C., and Daugherty, D. A. 1994, *ApJS*, **95**, 457

Telemachos Mouschovias and Nick Kylafis show how it's done at the first week's banquet.

Rafael Bachiller, Claudia Lavalley and Francois Menard at the afternoon
coffee break.

THE NATURE OF YOUNG SOLAR-TYPE STARS

FRANCOIS MÉNARD

*Canada-France-Hawaii Telescope Corporation, PO Box 1597,
Kamuela, HI 96743, USA*
*Laboratoire d'Astrophysique, Observatoire de Grenoble, UMR
5571 CNRS/UJF, BP 53, F-38041 Grenoble cedex 9, France*

AND

CLAUDE BERTOUT

*Institut d'Astrophysique de Paris, 98bis Boulevard Arago,
F-75014 Paris, France*

1. Introduction

We focus in this review on the properties of optically visible, young stellar objects (YSOs). We hope we will not be accused of undue anthropocentric reasoning if we say that these stars have an activity level compatible with their youth. Although each of these stars may display its own spectroscopic and photometric variability pattern at various wavelengths – which makes them fascinating objects to observe – they share enough basic properties to appear as classes of stars. We go through these various classes and their historical definitions in the following section. We then briefly address the question of their age; i.e., how do we know that they are young pre-main sequence stars? The rest of the review is devoted to the physical picture that has emerged over the last decade, focusing mainly on the physics of light variability, disk properties, and accretion. In each of these topics, we emphasize areas where we expect rapid progress in the next few years.

1.1. YOUNG SOLAR-TYPE STARS: THE DIFFERENT FLAVORS

The main reference catalog for young stellar objects is the unpublished Herbig and Bell catalog [86], which provides a census of Orion population emission-line objects in nearby star-forming regions. In the following, we focus on the various classes of solar-type stars in this sample.

C.J. Lada and N.D. Kylafis (eds.), The Origin of Stars and Planetary Systems, 341–374.
© 1999 *Kluwer Academic Publishers. Printed in the Netherlands.*

1.1.1. *T Tauri Stars*

Classical T Tauri Stars (CTTSs). Alfred H. Joy discovered "a new type of variable stars whose prototype is T Tauri" [94], based on the following criteria: irregular light variations of about three magnitudes, spectral type F5-G5 with emission lines resembling the solar chromosphere, low luminosity, and association with dark or bright nebulosity. Joy gave a list of eleven stars and added "These stars differ from other known variables, especially in their low luminosity and the high intensity of bright H and K [Ca lines] in their spectra". None of these original stars has actual spectral type in the range F5-G5; instead they range from K to M. This illustrates the difficulties involved in classifying T Tauri stars (TTSs), especially when using low-resolution spectrograms. Selection criteria for TTSs were refined by Herbig [84] who defined the class of TTSs in terms of optical spectroscopic properties, first of all hydrogen and calcium line emission. Herbig also noted that stars with the above spectral properties are all irregular variables, and are all associated with nebulosity, which makes two of Joy's original criteria redundant. The youth of TTSs was recognized first by Ambartsumian [3], and it took a few years before this idea met with recognition in the rest of the world.

YY Orionis Stars. The term YY Orionis star was introduced by Walker [157], [158] to define the subclass of CTTSs that display inverse P Cygni absorption components at the edge of their Balmer and CaII emission lines. These absorption components are red-displaced by typically $300 \mathrm{km\ s^{-1}}$, which is a direct indication that circumstellar matter is accreting onto the star at close to free-fall velocity. The YY Orionis phenomenon seemed at first restricted to CTTSs which have a strong UV excess in their spectral energy distribution. Appenzeller [5] then found that about 75% of the YY Orionis stars show a sizable UV excess while the same is true for only around 50% of all CTTSs. More recently, high-resolution spectroscopy of a large CTTS sample revealed that inverse P Cygni profiles are present in most CTTSs in at least one of their Balmer lines, though they appear much shallower than in YY Orionis stars ([50], [120], [118]). Thus the phenomenon of inverse P Cygni profiles seems to be rather ubiquitous in CTTSs, and one may doubt that YY Orionis characteristics are defining a peculiar class of young stellar objects. Nevertheless the question of why YY Orionis red-displaced absorption lines are much deeper than in CTTSs still remains.

Weak-Emission Line T Tauri Stars (WTTSs). Stars which obey Herbig's spectroscopic selection criteria are now called "classical" TTSs. They were discovered mainly during Hα surveys and they harbor (among other characteristics mentioned above) strong Balmer line emission, with Hα equivalent

width (EW) larger than 10Å. In more recent years, another population of
visible solar-type young stars that had escaped detection so far was discovered because of their strong X-ray emission ([159], [54], [160]). The basic
criteria for selection of these objects are: late-type spectrum, Hα emission
with EW smaller than about 10Å, LiI λ6707 absorption with EW > 0.1Å,
and association with nebulosity. Because they display little activity in the
optical range, and notably no line emission except for weak Hα emission,
these objects are named Weak-emission line T Tauri Stars (WTTSs). The
relative abundances of CTTSs and WTTSs in nearby star-forming regions
is still a matter of controversy, caused in good part by the recent discovery
by ROSAT of a population of X-ray active late-type stars that often have
the properties given above but are found outside of star-forming regions,
i.e., they are not associated with nebulosity (see §1.2.1).

The Current Paradigm for TTSs.   A consensual physical picture of TTSs
emerged in the late 80s. The activity of WTTSs is believed to be entirely
of magnetic nature. Their rotation rates are quite large compared to the
Sun ($v_{eq} \sim 10\text{-}20$ km/s), presumably generating a strong dynamo which
amplifies the stellar magnetic field, leading to strong and variable X-ray
emission, flaring at short optical wavelengths, and formation of large, cool
spots. All these properties - which are also seen in other fast-rotating late-
type stars such as the RS CVn - will be discussed in the following sections,
but we want to emphasize here that there is no apparent need to assume any
other physical process for explaining the properties of WTTSs. CTTSs, on
the other hand, are also magnetically active stars but they are surrounded
by, and interacting with, a circumstellar, dusty accretion disk that results,
together with the star, from the gravitational collapse of a rotating cloud
core. The interaction region is currently thought to be the interface between disk and stellar magnetosphere, and accretion of disk matter onto
the star is likely chanelled by the stellar magnetic field, resulting in localized
(hot) accretion spots at the stellar surface. The interaction region is also
thought to be the base of the observed collimated T Tauri wind, although
the exact mechanism driving this wind is still a matter of speculation (see
the reviews by Shu and Reipurth in this volume). Some CTTSs, presumably the youngest, are also surrounded by a remnant protostellar envelope,
which feeds the circumstellar accretion disk. While the IR and sub-mm
spectral energy distribution gives indications about physical conditions in
the circumstellar disk and envelope, the main accretion diagnostics are in
the optical and UV range: the spectral "veiling" - an extra-photospheric
blue/UV continuum superimposed on the optical spectrum - is a measure
of the accretion power, as is the Balmer jump, seen in emission in accreting
CTTSs.

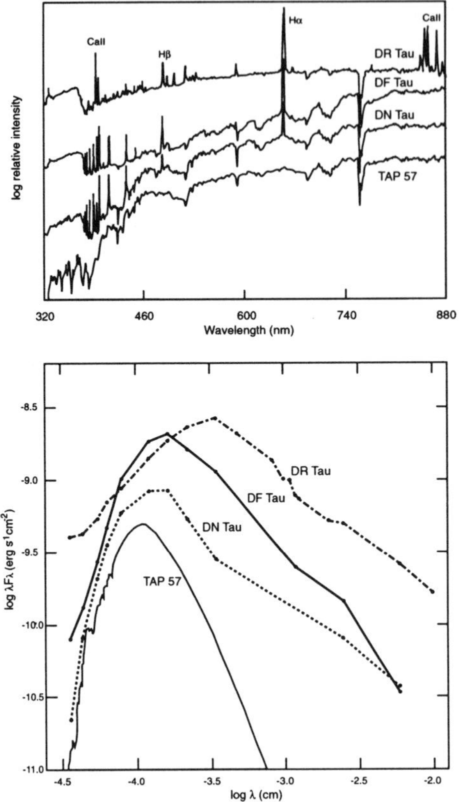

Figure 1. **Top Panel:** Medium resolution spectrograms that illustrate the variety of spectra that TTS can exhibit, ranging from the basically normal late-type spectrum of TAP57, a WTTS, through moderate CTTSs, such as DN Tauri and DF Tauri that show Balmer and Ca II emission and varying amounts of veiling , to extreme CTTSs – here DR Tauri – that displays no photospheric spectrum but many emission lines of HeI, CaII, TiII, and FeII. **Bottom Panel:** Spectral Energy Distributions of the same stars, showing the increasing infrared excess from WTTS to extreme CTTS. Note that these data were not corrected for extinction.

1.1.2. *FU Ori Stars*

This class of eruptive young objects was introduced by Herbig [85]. On the basis of one pre-outburst spectrogram of V1057 Cygni showing a spectrum characteristic of strong emission-line CTTSs, it is assumed that FU Orionis stars are in fact CTTSs undergoing a strong luminosity outburst (see also Section 2.2). Their post-outburst spectral type appears to depend on the wavelength range used for the spectral classification. For V1057 Cyg, an early A type of fairly high luminosity is derived from the blue spectral range, whereas the red range shows an approximate F5 II spectrum. Spectrograms of V1057 Cyg obtained during subsequent years show a slow decline of effective temperature and luminosity. Five years after the outburst, Herbig classified V1057 Cyg as G2-5 Ib,II in the red spectral region. Herbig [85] pointed out that each T Tauri star must undergo about one hundred FU Ori events in its lifetime of about one million years, based on the (conservative) assumption that the three observed outbursts (in FU Orionis, V1057 Cygni, and V1515 Cygni) represent all FU Ori events which occurred within 1 kpc of the Sun in the last eighty years. Properties of FU Orionis stars and their relevance to the formation of low-mass stars are discussed elsewhere in this volume by Kenyon. There have been many attempts to model FU Orionis eruptions, and a consensual picture has emerged in which the eruptive events are caused by instabilities in the circumstellar disk, leading to phases of increased accretion. Because of the high mass accretion rate, the disk brightness overwhelms the star's and the spectral signatures mentioned above trace the temperature gradient in the accretion disk.

1.2. THE CASE FOR STELLAR YOUTH

Star formation is widely thought to occur in molecular clouds, which provide the raw material from which stars can accumulate during the course of the gravitational collapse. While many details of the star formation process remain topics of controversy, the idea that most YSOs should be physically associated with molecular clouds remained unchallenged until recently.

Four main arguments are used to infer pre-main sequence status of a solar-type star: association with OB associations (e.g., in the Orion Trapezium region); association with dark or bright nebulosity (e.g., in the Taurus-Auriga region); location above the main-sequence in the Herzsprung-Russell Diagram (HRD); presence in the spectrum of the $\lambda 6707\text{Å}$ LiI resonance line, since this element is destroyed early in the star history.

1.2.1. *Association with OB Associations and Molecular Clouds*

The first two criteria for youth are straightforward: OB associations are short-lived, which guarantees the youth of associated low-mass objects, and physical association of a star with a cloud is either seen at the telescope when close reflection and emission nebulae are present – as is the case in the vicinity of many CTTSs – or inferred from kinematic studies (e.g., [93]).

However, it was recently argued from ROSAT data that a population of X-ray active pre-main sequence stars – many of which were identified as WTTSs – extend far beyond the boundaries of known star-forming regions (e.g.,[122],[123]). This result was at first taken as an indication that young stellar objects could migrate far away from the region of their putative formation. Propositions to explain this fact included the formation of stars in fast-moving molecular cloudlets [53] or tidally-induced escapes in multiple system members [8]. However, the recent finding that many of these PMS objects are likely the low-mass counterpart of the Gould Belt OB associations ([69],[70]) reconciles the ROSAT results with the conventional idea that these stars formed in now dispersed molecular clouds.

Another observation which may contradict the hypothesis that YSOs and molecular clouds are associated was reported by Favata et al. [52]. These authors assert, on the basis of the Hipparcos satellite astrometric

TABLE 1. Distances to the Taurus cloud and to associated young stars

Taurus	π_w	σ_w	distance (pc)
Distance indicators	6.78	0.46	147^{+11}_{-9}
Young stars	7.21	0.57	139^{+12}_{-10}
Young stars (constrained)	6.94	0.58	144^{+13}_{-11}

measurements [90], that some CTTSs of the Taurus-Auriga star-forming region are apparently much closer than previously thought and may not belong to the Taurus clouds. It was shown recently [18], however, that the TTSs observed by Hipparcos are in fact at distances compatible with their location in associated molecular clouds, and that the variability and faintness of some CTTSs conspire in some cases to induce large errors in the Hipparcos parallax determination. Table 1 gives an example of Hipparcos results relevant to the Taurus-Auriga region. There, the first line indicates the weighted average parallax π_w (with its standard error σ_w) of the Taurus cloud distance indicators discussed by [132] and [51], the second line gives the weighted average parallax of the young stars observed by Hipparcos in the Taurus-Auriga region, and the third line indicates the average parallax obtained after solving the astrometric equations while constraining all Taurus YSOs to be located at the same distance. Obviously, all these parallaxes are fully compatible, indicating that young stars are indeed at the distance of their associated molecular cloud (cf. [18] for details).

1.2.2. *Location in the HRD and Lithium Abundance*

The last two criteria for youth are more indirect. First, the location of a given star in the HRD depends on the assumption made about the respective luminosities of photosphere and circumstellar matter in a given object [97], as well as on the detailed history of how matter was accreted onto the star. However, none of these effects is expected to change drastically the location of the TTSs above the main sequence, although these are critical issues when trying to derive precise ages and masses for these objects ([126],[139],[140]). Second, the lithium abundance may be dependent on other variables than age (e.g., [154]), and the depletion of lithium during the pre-main evolution of low-mass objects is not entirely understood yet. However, the ^7Li abundances derived from line profile analysis of CTTSs and WTTSs in their Hayashi evolutionary phase are generally consistent with the interstellar lithium abundance, indicating that the temperature has not yet reached $2\,10^6 K$ in the central regions of the stars (cf. [147],[11],[104],[107]).

After this brief introduction to TTSs and related objects, we turn to a more detailed analysis of their photometric variability, which is a topic of much current work.

2. The Nature of TTS Photometric Variability

2.1. HISTORICAL BACKGROUND

Stars associated with nebulosities were known to be variable before they were recognized as young stellar objects. They were classified according to the type of photometric variability they exhibited. With the introduction of independent classifications based on spectroscopic properties, interesting situations occurred; for example, the star T Tauri is a RW Aurigae variable, and RW Aurigae is a T Tauri star. In 1933, Paranego called RW Aurigae variables those irregular variable stars that belong to spectral class G and show rapid (up to one magnitude in several hours) and large (up to 3 or 4 magnitudes) variations of their brightness. Orion Nebula type variables denote those RW Aurigae variables that are associated with diffuse nebulae. In 1949, Hoffmeister proposed to extend the term RW Aurigae variable to all stars exhibiting rapid, non-periodic fluctuations, interrupted by intervals of rest, regardless of their spectral class (but excluding novae and U Geminorum stars). Commenting on the terminology of variable stars, Hoffmeister (1957)· wrote "The name T Tauri stars, which was introduced by Joy, points to certain properties of the spectra and therefore does not include all objects which might be assigned to the class on the basis of photometric behavior. Furthermore, T Tauri is photometrically not characteristic of the class". While Herbig was getting rid of references to photometry in his definition of the T Tauri class, Hoffmeister was thus trying to include all nebular variables in his definition of RW Aurigae variables. Later, Glasby [59] went the opposite way by dividing nebular variables in three distinct main categories: (i) RW Aurigae variables show rapid and totally irregular light variations, (ii) T Tauri variables display slow, small amplitude sinusoidal variations with a general tendency toward stronger variations when the star is bright, (iii) T Orionis variables show small fluctuations around maximum light with sudden, irregular fading by several magnitudes. The class of TTSs, as defined by Joy and Herbig, brings together nebular variables belonging to each of these categories. "Despite their ubiquity, the light variations of the T Tauri stars have shown no clearly distinctive group property, except that an occasional display of rapid activity has been reported for most well-studied objects." wrote Herbig in his 1962 review [84], and he dismissed Hoffmeister's classification of RW Aurigae variables on the grounds that "it has not been proven that the RW Aurigae variables as so defined are a physically significant group". This obvious conflict between photometrists

and spectroscopists is a thing of the past; every bit of information is useful
to help us unravel the nature of these stars.

2.2. LONG-TERM VARIABILITY

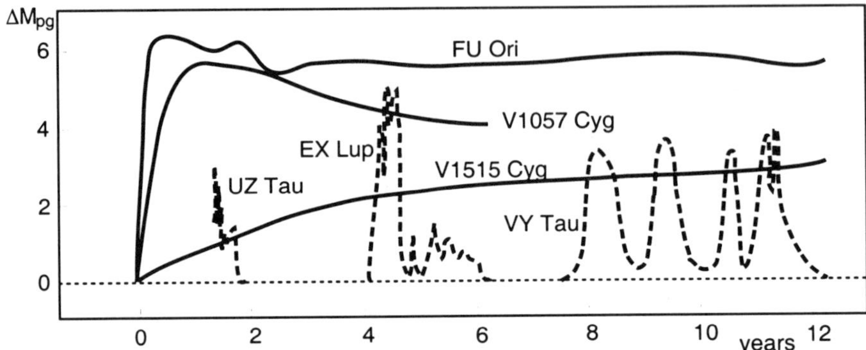

Figure 2.   Compared photographic light curves of FU Orionis objects with extreme TTSs
(UZ Tau, EX Lup, VY Tau), adapted from Herbig (1977).

The FU Orionis stars, briefly defined above, are characterized by a large
increase in brightness over several hundred days, followed by a long-term luminosity decline. This is illustrated in Fig. 2, where the light curves of three
FU Orionis objects are compared to light curves of a few TTSs with particularly large light variations such as EX Lup (from [85]). These large amplitude long-term variations are believed to be caused by disk instabilities
leading to non-steady accretion ([13],[14]), with the accretion rate changing
by several orders of magnitude (typically from 10^{-7} to 10^{-4} M$_\odot$/yr).

2.3. MID-TERM VARIABILITY

Our understanding of the mid-term (from a few days to about two weeks)
variability of TTSs has progressed largely over the last decade. This resulted
most notably from several multi-telescopes observing campaigns (e.g. [24],
[23]) as well as from an effort by Bill Herbst and collaborators [88] to gather
all available photometric data in an homogeneous database. These authors
distinguish between three types of mid-term variability. It is revealing of
the progress made over the last 30 years that the definitions of the first two
types refer to the physical model underlying the variability rather than to
the phenomenology of the light variations, in contrast to previous photometric classifications. The third variability type, on the other hand, is still
largely defined on a phenomenological basis; this reflects our current lack of
understanding of this particular pattern of variations. We summarize these

findings below, and discuss briefly some current work on these issues. The reader is referred to the excellent discussion in [88] for details.

2.3.1. *Type I: Rotational Modulation Caused by Cold Magnetic Spots*

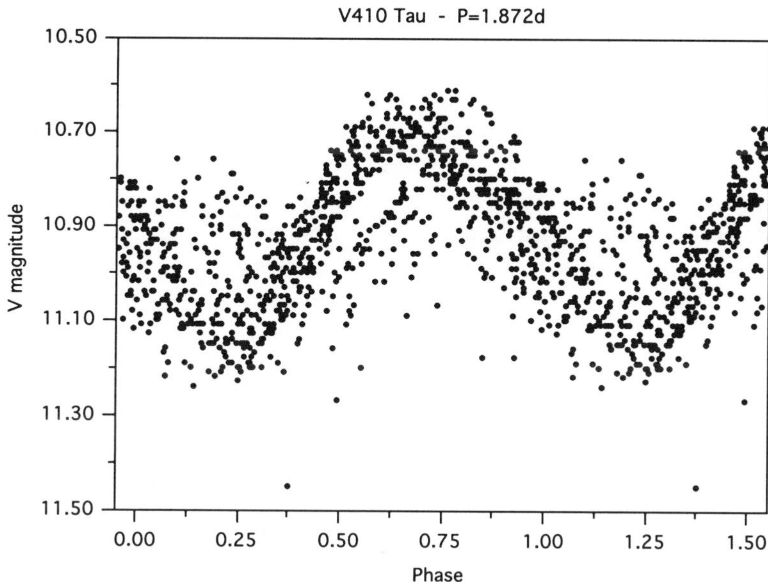

Figure 3. Phase-wrapped V-band photometric light curve for the WTTS V410 Tau, illustrating Type I periodic light variations due to dark magnetic spots on the stellar surface. The period used here is 1.872 days.

These periodic variations are seen in all young solar-type stars, but are most easily detectable in WTTSs, because it is the dominant mid-term variability cause in these stars. Figure 3, which displays all V data points available for the WTTS V410 Tau, illustrates this type of variations. The constancy of the 1.872 day period over the last 15 years is apparent on the figure, as well as some changes in the phase. While the overall spot structure remained surprisingly constant over the last 15 years, as witnessed by the nearly constant amplitude of the modulation, there are year to year changes in the shape of the light curve which reflect corresponding changes in the spot distribution and properties. The spots are often located at high-latitude in these stars, cover up to 40% of the projected surface, and are typically 1000K cooler than the remaining photosphere (see the instructive Doppler images of V410 Tau in [83] and [134]). In WTTSs, a single source of variability (the rotation of stars with cool spots) is sufficient to account for V, R, and I data. In U and sometimes B, short-term magnetic flaring is observed(e.g., [66]), especially when the stars are faint (this is similar to what is observed in dMe flare stars). The maximum amplitude that can be attributed to cold magnetic spots is about 0.8 mag in V and 0.5 mag in I.

2.3.2. *Types II and IIp: Variability Caused by Cold and Hot Spots*

Type II variations are either irregular or periodic, in which case they are called Type IIp. Both kinds of Type II variations, which occur only on CTTSs, can be understood in terms of a changing mix of cool and hot spots on their surfaces (e.g., [156],[155]). The localized hot zones on the stellar surface are attributed to accretion onto the star of circumstellar disk matter and also produce the spectroscopic veiling characteristic of the CTTS class. They are probably shorter-lived than the cold magnetic spots and can produce much larger amplitude variations (several magnitudes), particularly in the B and U filters, because of the temperatures of typically $7 \ 10^3$ - 10^4K reached in the accretion zones. However, the rise time for the amplitude variations is longer and smoother than for the flares described in the previous section.

The geometry and distribution of these zones remain unclear, although first tries at CTTS Doppler imaging appear promising (e.g.,[152]). Simple spot models indicate that the stellar fraction covered by hot spots is small, typically 1% or less. However, photometric observations of YY Ori [20] revealed a quasi-periodic modulation of its light curve indicating an accretion spot covering about 10% of the stellar disk surface, i.e., much larger than hot spots usually found in CTTSs. A comparison of YY Ori's characteristics to those of CTTSs shows that YY Ori has a much larger accretion rate than usually found in CTTSs [75], suggesting a link between the accretion spot size and the accretion rate. Further comparison of the spot luminosities and accretion luminosities of a small sample of CTTSs (including YY Ori) indicated that the spot's luminosity derived from the photometric variability is much smaller than the accretion luminosity of these stars. This and the fact that the accretion rates of CTTSs with no apparent accretion spots are similar to those of CTTSs with hot spots led [20] to suggest that a sizable fraction of the accretion power might be dissipated in axially or spherically symmetric regions that do not give rise to well-defined periodicities in the light-curve. This result also has important implications regarding the nature of the accretion process, as will be discussed in the next section.

Type IIp "periodic" variability remains somewhat mysterious, because the detected periods are apparently not stable: for instance, the period seen in DN Tau varied from 6.6 days in one observing run to 8.4 days in another [24]. Similarly for DF Tau, which was observed with periods ranging from 7.5 to 8.5 days. There have been several tentative explanations (e.g., [24], [151]) but none of them appears entirely convincing. Clearly, the mechanism which leads to these period changes is a key to understanding

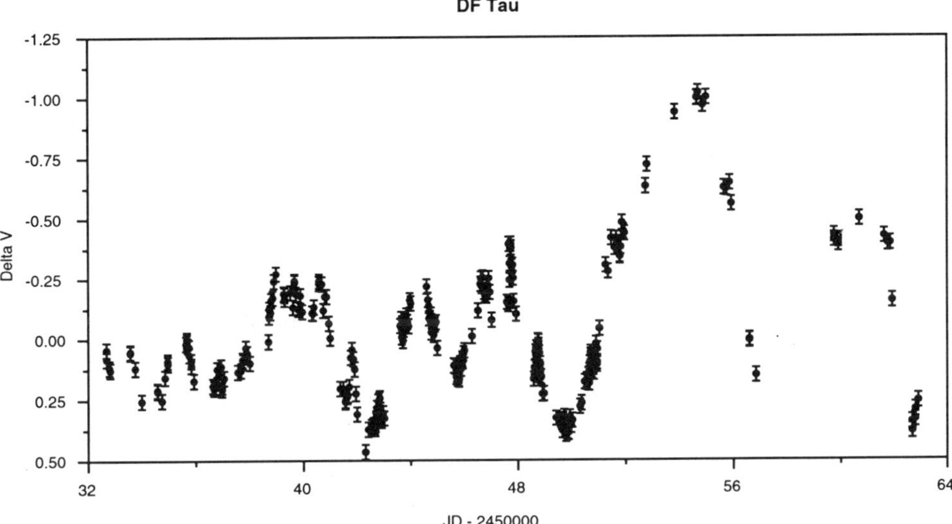

Figure 4. V-band photometric light curve of DF Tau in November 1995, illustrating Type IIp light variations due to a changing mix of cold and hot spots. See text for details.

the exact nature of the accretion process in CTTSs, and future work in this area should be rewarding. A fine example of Type IIp variability is shown in Figure 4, obtained by [33] and [26]. The periodogram of this time series shows two significant peaks at 7.45 and 4.0 days, and significant high-frequency variability is apparent in the data set. A possible model of DF Tau's light curve assumes that there are at least two accretion spots on the star and that variable accretion leads to the varying amplitudes seen in the Figure ([33] present a different interpretation).

2.3.3. *Type III: Variable Obscuration by Circumstellar Dust?*
This is the least understood variability type. It occurs mainly in Herbig Ae stars, the so-called UX Ori stars after the prototype, but also in stars with early K spectral types such as RY Tau and RY Lupi. It is characterized by large amplitude variations on stars with no evident veiling or effective temperature variations. The fact that the degree of linear polarization increases as the star fades is a major argument supporting the suggestion that variable obscuration is responsible for the star's fading. The pros and cons of the different models leading to such variability are discussed in some details by [88] and won't be repeated here. One should note, however, that Type III variability might also be present in some CTTSs with late spectral-types, most notably AA Tau [26] and possibly UY Aur [109], both of K7 spectral type [71], in contrast to previous findings[1].

---

[1]Note also that T Tau itself displays all three types of variability, and thus appears as a true prototype of the class [88].

2.3.4. *Are There Other Mid-Term Variability Components?*

While it is often assumed that Type I and II variabilities are enough to explain the mid-term photometric behavior of WTTSs and most late-type CTTSs, very little is known about possible variations on time-scales intermediary between the stellar rotation period (typically one week) and the slow trends observed over several years, which are thought to relate to long-term fluctuations in the accretion process. The cycle of variability in Type III objects, while not stricly periodic, recurs nevertheless on a preferred time-scale of typically 10-40 days. Herbst et al. [88] noted that this is hardly compatible with the idea that an obscuring object located in the disk causes the variability, but the timescale they considered is that of a blob crossing the stellar disk, which is probably not relevant here. Motivated by the recurrence of the observed time-scales in Type III objects, one might hypothetize instead that the relevant timescale is that of keplerian rotation within the disk, and that the obscuring objects are in fact located at distances in the range 10-20 stellar radii. The argument that the disk must be seen almost edge-on in order to get obscuration of the star by disk matter is valid only is the disk is flat, thin, and homogeneous. Instead of this idealized case, we might well have disk warps due to tidal interactions with companions, local flaring due, e.g., to anisotropic heating by radiation from the accretion spots, and density waves triggered by non-axisymmetric instabilities. All of these phenomena might lead to variability, either through partial, time-dependent obscuration of the star, or through anisotropic effects in the reprocessing of stellar light by disk matter. Whatever the mechanism leading to light variations - there has been no investigation of these effects so far - there is little reason at this point to dismiss the possibility of mid-term light variations which might give us information on physical processes at play in the inner circumstellar disk.

The problem is to detect these variations in spite of the shorter-term light variability on various timescales, which introduces a large scatter in any longer term time-series. To illustrate this point, we take a closer look at RW Aur, one of the best-studied CTTSs and a typical Type II variable. The periodogram of the V data available for this object reveals its extreme variability on all timescales, and it appears hopeless to use a periodogram analysis - a very useful tool for studying the WTTS periods - to say something about preferred timescales in RW Aur. But luckily RW Aur was one of the program star of the Hipparcos satellite, which happened to observed the star at a time when it displayed a clear periodicity.

The top panel of Figure 5 shows this time-series (black dots), which indicates a period of about 3.1 days[2]. Other Hipparcos observations (open

---

[2]Herbig [84] noted a preferred photometric time-scale of 5.5 days for this star. Period variations by 20% or more are commonly observed in CTTSs.

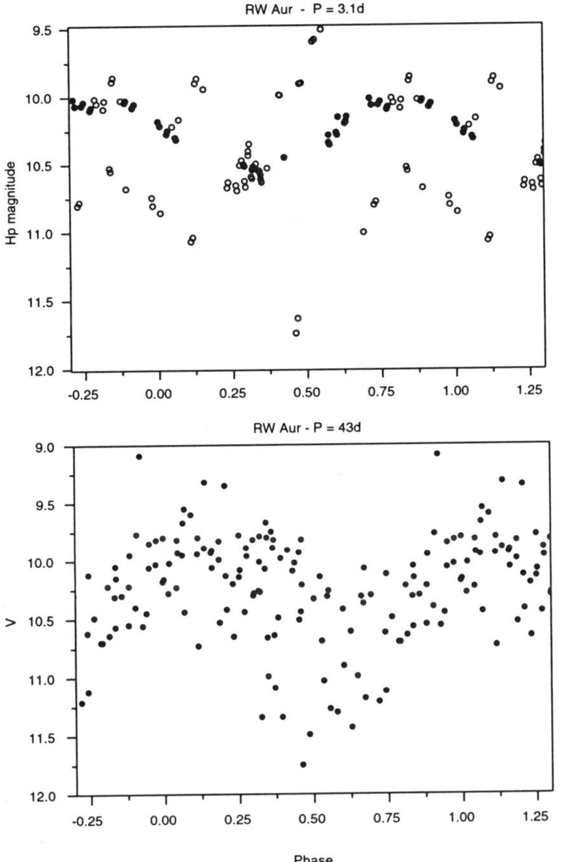

Figure 5. Variability of the CTTS RW Aur. **Top Panel:** Phase-folded Hipparcos satellite observations of RW Aur, using a period of 3.1d. The ordinate is the Hipparcos magnitude Hp. Black dots indicate a time-series of consecutive observations over more than 4 days. **Bottom Panel:** Phase-folded V data (compiled by W. Herbst) during the same time interval, showing a putative 43 day period uncovered by robust periodogram analysis. See text for details.

dots) were performed over time intervals too short to uncover a periodic behavior on the same time-scale but the phase-folded data (Fig.5) shows that the 3.1 day period was a preferred time-scale in the Hipparcos observing period, from about JD 2447900 to 2449100.

Both the amplitude of the 3.1d variations and the cycle's average magnitude vary in time, suggesting a modulation at longer time-scale. It was thus tempting to look at the large amount of UBV data obtained from the ground by various groups during the same time frame. Using a powerful period-search technique borrowed from geophysics, another preferred variation time-scale of about 43 days is uncovered in these data, as seen in the bottom panel of Fig.5, but no 3.1 day periodicity is apparent in the V data. The origin of this putative modulation is not known at present, but a full analysis of RW Aur's photometric data is currently under way [17]. In conclusion, one might say that the picture of CTTS variability is probably somewhat more complex than believed today, and is of potential interest to get information about physical conditions in the inner circumstellar disk.

2.4. SHORT-TERM VARIABILITY

Haro [73] discovered a number of flare stars in the Orion Nebula and in the Taurus-Auriga star-forming region. He called these objects flash stars in order to distinguish them from the UV Ceti stars. According to Haro [73], necessary criteria for classifying variables as flare stars are: (a) relative steadiness of the starlight during minima or "normal phase" and (b) a sudden, unpredictable rise to maximum and a slower, but still rapid, decline toward minimum. Haro classified as flash stars a number of TTSs by using his usual technique for discovering new flare stars, which consisted in repeatedly photographing the same sky region while shifting slightly the photographic plate between two exposures. A change in brightness on a time-scale of about 10 minutes is easily detected in this way. Well-known CTTSs which were classified as flash stars on the basis of one observed flare include YY Ori, NS Ori, SU Ori, as well as DF Tau, DS Tau and DN Tau. Haro in fact estimated that 30% of the Orion flare stars were also TTSs. As early as 1954, Ambartsumian speculated that TTSs were permanently flaring stars while Haro and Chavira proposed in 1955 that flash stars were post-TTSs (see [74]). None of these interpretations now appears likely.

Early UV/optical broad-band flaring observations often suffered from inadequate time coverage. It is currently believed ([58], [66], [65]) that strong ($\geq 0^m3$ in all wavebands) and rapid ($\leq 1h$ with sharp rise-time and slower decay time) variations are actually rare events on TTSs. Gahm et al. [58] emphasize that the normal short-term behavior of TTSs is to be completely constant in brightness or to vary slowly with small amplitudes over several hours. Rare but powerful rapid variations may be split in two categories: flare-like energetic bursts ($10^{33} - 10^{34}$ erg) in the Balmer continuum and line emission, and smoother, slower changes (with energies up to several 10^{35} erg) seen in all broad-band filters (such as those seen on a 2h time-scale in DF Tau - cf. Fig. 4). The latter short-term variations are likely caused by veiling variations in CTTSs, i.e., by an accretion related process (see [56]) and/or by type III variations (see [58] for details) while the Balmer continuum/line flares are likely to be genuine magnetic flares as seen in dMe stars. H_α-flares might also be associated with mass-loss events, further increasing the similarities between TTS and solar-like flares of magnetic origin. The X-ray and Radio variability also support this view.

Average X-ray luminosities for TTSs, both CTTSs and WTTSs, are typically a thousand times larger than the solar one. Repeated observations showed that flares are frequent. Monitoring of the ρ Ophiuchi region lead Montmerle et al. [117] to nickname the dark cloud an *"X-ray Christmas Tree"* due to the blinking of the low-mass PMS stars above and below Einstein's detection threshold.

During flares, the electron density ($n_e \sim 10^{10} cm^{-3}$), the plasma temperature ($T \sim 10^7$ K), and the magnetic field ($B_* \sim 10^3$G) are fairly similar to values found in solar flares. The probable causes for this large luminosity is therefore the availability of a large coronal volume, or of very large loop sizes (sometimes comparable to the stellar radius!). As for the Sun, radio emission is also expected from the large magnetic loops. And, similarly to X-rays, strong variability and large radio flares were detected on TTS using the VLA (e.g., [55],[143],[21]).

2.5. CONCLUSIONS

At least two distinct physical mechanisms are playing a role in YSO photometric variations. Magnetic fields are present on the surface of all late-type YSO. Their strengths appear compatible with expectations based on the rotation rates of other late-type dwarfs, and lead to the formation of large-sized magnetic spots and hot plages, leading to: (i) mid-term periodic light variations in the visual and near-infrared range; these are observed in both WTTSs and CTTSs; (ii) short-term UV/optical variability in WTTSs and CTTSs; and (iii) X-ray emission and variability in WTTSs and CTTSs.

The second well-documented physical process leading to variability is accretion of matter by the star from its circumstellar disk. This process is likely responsible for (i) the long-term eruptions in FU Ori objects, leading to an increased stellar effective temperature by a factor of almost 2 and a brightening of the star by several magnitudes, (ii) the mid-term periodic and more erratic optical light variations in CTTSs, due to the presence on the stellar surface of short-lived accretion spots that modulate the light curve, and (iii) the short-term veiling variability on CTTSs.

There is evidence that a third physical process is at work in some CTTSs and earlier-type objects, but its nature remains elusive today. While variable obscuration of the star by circumstellar matter is likely the source of this variability, many questions about the origin of the obscuring matter, its location in the star-disk-wind system, and the timescales involved remain unanswered today.

3. The Physics of CTTS Accretion Disks

The discussion of variability above provided us with some useful insight in the accretion process: its anisotropy, as witnessed by the hot accretion spots, and its non-stationarity, revealed by the short- and mid-term optical and UV variability. What is still lacking, however, is a physical picture connecting the large-scale processes in the accretion disk to the small-scale processes taking place on the stellar surface. In other words, how does the accretion of disk matter onto the star really proceed? This is the topic of this section, where we shall review the observational evidence and accretion models after giving some historical background information.

3.1. HISTORICAL BACKGROUND

Compelling observational evidence that rotationally supported disks exist around TTSs is only recent. In his pioneering observational study of YY Orionis stars, Walker [157] was first to propose disk accretion as a possibility to explain the inverse P Cygni profiles and strong UV excess displayed by these objects. When Lynden-Bell & Pringle (LBP; [103]) suggested, in their insightful theoretical paper of 1974, that viscous dissipation in a disk might account for the continuous excesses observed in TTSs, there was basically no data available to support, or refute, this hypothesis. Observations were restricted at that time to medium resolution optical spectroscopy and optical photometry, although the presence of a near-IR excess in some bright CTTSs had been discovered by Mendoza in 1968 [110] and was usually attributed to reprocessing of stellar photons by a dusty envelope [37]. A decade went by before two satellites provided observational means for testing LBP's model: IRAS (e.g.,[135]) in the infrared and IUE in the UV. Together with ground-based broad-band photometry, these facilities provided the data needed for an extensive study of the accretion disk model. But during the years 1974-1986, much work on TTSs focused on studies of atmospheres (e.g.,[38],[87]), winds (e.g.,[79]), as well as on studies of the infall hypothesis (e.g., [150],[161]). In a thoughtful paper, DeCampli [40] demonstrated in 1981 that the mechanism driving T Tauri winds must be extremely efficient, pointing to a magnetic origin. But his analysis also implied that a non-stellar energy source was likely needed to account for the overall energy expenditures of TTSs (wind, radiative excesses).

From 1983 on, the idea that CTTSs were surrounded by disks gained momentum steadily, starting with Cohen's paper on HL Tau [36] and the illuminating analysis of T Tauri forbidden line profiles by Appenzeller and collaborators [6], convincingly demonstrating the existence of an opaque layer (the disk) hiding the recessing part of the wind. Then came the era of SED interpretations using accretion disk models and finally the disk idea reached consensual status[3] in 1989. We should perhaps emphasize that consensus was found on the basis of converging indirect evidence rather than on firm observational facts: the first direct circumstellar T Tauri disk detections dating only a few years back.

---

[3]Worthy of mention is the alternative CTTS model proposed by Mitskevich & Grinin where the disk is replaced by an inhomogeneous, envelope-like medium ([116], [113],[114],[115])

Similar tests of the accretion disk hypothesis were made for all classes of solar-type YSOs at about the same time, and generally turned out to be successful. Hartmann & Kenyon ([80],[81]) convincingly argued that FU Orionis outburst were caused by massive accretion onto the central star and showed that infrared CO emission lines had the double-peaked profiles expected for a rotating flat disk. Lada & Wilking [101] surveyed the dense molecular cores of ρ Ophiuchi in the infrared and were able to classify all objects into three categories (later labelled as Class I, II, and III) according to the shape of their infrared spectral energy distribution. Adams, Lada, and Shu [2] proposed that this classification is in fact an evolutionary sequence, from embedded object to ZAMS star, an idea that still holds nowadays, although with a few modifications (see the review by Lada in this volume).

Since the Crete I meeting [100], our understanding of the accretion/ ejection process, and especially the pivotal role played by the disk, has evolved considerably. Most aspects of the classical LBP disk model are being challenged on both observational and theoretical grounds. The idea that the disk is flat and homogeneous has given way to the more complex view of a warped, inhomogeneous disk in which tidal interactions (induced by close companions or by non-axisymmetric instabilities) play a major role. Similarly, the classical view that disk material is accreted onto the star via an equatorial, steady boundary layer is not necessarily accurate in the complex environment of YSOs, although high-accretion rate FU Orionis objects do show evidence for boundary layer emission [95]. One should note here that the boundary layer hypothesis has been studied in some depth in recent years, and that boundary layer properties are now much better understood than they were at the Crete I meeting. The interested reader is referred to [133], [130] and references therein. In CTTSs it turns out that the stellar magnetic field is likely sufficient to disrupt the inner accretion disk, leading to magnetically controlled accretion along the stellar field lines of at least a fraction of the accreting matter. Although this accretion scenario was proposed as early as 1988 for DF Tau [19], it has become evident in recent years that it is valid for the entire CTTS class.

In the following paragraphs we shall briefly review the disk models and discuss the pros and cons of current ideas about the accretion process, using the newest observational support. We shall not discuss in detail the role played by the disk and magnetic field in the ejection process. The reader is referred, for example, to the review by Shu in this volume.

3.2. GLOBAL DISK OBSERVABLES

There are two kinds of observable quantities in accretion disk physics: global properties of the star/disk system that are accessible through unresolved photometric and spectroscopic observations, and observables which result from direct imaging and resolved polarimetry, giving access to the disk's detailed physical structure.

The global observables of disk accretion are (i) the accretion luminosity $L_{acc} = GM_*\dot{M}_{acc}/R_*$, which can – in principle – be determined from the IR and UV excesses (see below), (ii) the maximum disk temperature $T^D_{\max} = \zeta \left[L_{acc}/\sigma R_*^2\right]^{0.25}$ which is given roughly by the wavelength of the SED maximum, and (iii) the maximum disk velocity $v_K = \sqrt{GM_*/R_*}$, which measures the profile widths for spectral lines formed in the inner disk. These three quantities depend only on the stellar radius R_*, the stellar mass M_* and the mass accretion rate $\dot{M}_{acc}$ and are thus easily determined, at least to zeroth-order. The factor ζ depends on the disk model used; it is 0.287 for the standard LBP model.

From these three quantities, we can get a first reasonable guess of the physical conditions in the disk. Let's take the example of an "average" CTTS, with $R_* = 2R_\odot$, $M_* = 0.5M_\odot$, and spectral type K7. We can rewrite the equation for L_{acc} as

$$\frac{L_{acc}}{L_*} = 0.86 \left[\frac{\dot{M}_{acc}}{10^{-7}M_\odot\mathrm{yr}^{-1}}\right] \left[\frac{M_*}{0.5M_\odot}\right] \left[\frac{R_*}{2R_\odot}\right]^{-3} \left[\frac{T_{eff}}{4000\mathrm{K}}\right]^{-4} \tag{1}$$

and assuming that $\dot{M}_{acc} = 10^{-7}M_\odot\mathrm{yr}^{-1}$ we get $T^D_{\max} \approx 2000\mathrm{K}$ and $v_K \approx 220$ km/s. We thus expect mass accretion rates $\sim 10^{-7}M_\odot\mathrm{yr}^{-1}$ to be easily detected in spectroscopic and photometric work since the accretion luminosity is then comparable to the photospheric luminosity. In practice, one can detect CTTS mass accretion rates down to 10^{-9} $M_\odot\mathrm{yr}^{-1}$ by careful spectroscopic work ([9],[71]).

We also find that the dereddened spectral energy distribution (SED) should peak around 1 μm, which is observed in many CTTSs, that the disk should be mostly molecular and will contain dust, and that the near-IR lines formed in the disk should be easily resolved, which is also verified.

While these global observables thus allow for useful first-order estimates of disk properties, one needs to go into much more detail for a precise determination of the mass-accretion rate. In particular, one needs to make assumptions about the disk structure and IR emission properties.

The IR SED is made of a stellar photospheric component and an "excess emission" component, i.e., emission above what is expected from the photosphere alone (cf. Figure 1). How closely can this be related to the accretion luminosity? In general terms, the IR excess is composed of several components, depending on the disk physical model that one has in mind. When the standard LBP disk model is adapted to the case of CTTSs, the IR luminosity emitted by the disk can be written (e.g., see [1])

$$L_{IR}^{excess} = \frac{1}{2} L_{acc} + L_{rep} \tag{2}$$

where L_{acc} is the accretion luminosity and L_{rep} is the stellar luminosity reprocessed by the disk. The reprocessed luminosity depends on the amount of starlight absorbed by the disk which, in turns, depends on the dust scale height and/or the exact geometry of the disk. In the flat, geometrically thin but optically thick, viscous accretion disks envisioned by LBP, both L_{acc} and L_{rep} yield the same observational diagnostics: the disk temperature distribution and IR SED are power laws: $T(r) \propto r^{-3/4}$, and $\lambda F_\lambda \propto \lambda^{-4/3}$.

Observations suggest that the SEDs of many CTTSs depart significantly from that predicted by standard LBP flat disks. Rydgren & Zak [136] measured an average $\lambda F_\lambda \propto \lambda^{-3/4}$ over the range 3.5–20μm for 7 CTTSs, while Rucinski [135] found $\lambda F_\lambda \propto \lambda^{-1}$ in the range 12–100 μm. As noted by several authors, however, L_{rep} is highly dependent on the assumed disk geometry, and can be substantially increased if the disk intercepts more stellar photons than a flat disk, i.e., if its flaring vertical structure (in vertical hydrostatic equilibrium, the local scale height is $\propto r^{9/8}$) is taken into account (e.g., [96], [105], [39]), or if it is warped by tidal interactions ([148],[149],[102]). Realistic flaring disks do not differ much from flat disks in their innermost parts and effects of flaring is most apparent longward of about 5μm. On the other hand, a wide range of SEDs can be fitted by warped disks, depending on the disk mass and distance of the stellar companion that causes the tidal warp.

Several other causes can modify the IR SED. If the disk is disrupted and truncated by the stellar magnetic field, or if the presence of planets in the disk induces gaps, then IR emission deficit[4] occurs ([111], [108]). Nonstandard disk models such as the self-gravitating, instability-ridden disk envisioned by [141], MHD accretion/ejection structures [57] have flatter IR SEDs than LBP disks. Also, thermalization by the ambient medium [15], backscattering from a dusty envelope [121], and undetected infrared companions [16] are expected to contribute to the IR flux. So, using the IR SED alone to evaluate the mass accretion rate *in the disk* is usually impossible, except when $\dot{M}_{acc}$ is so large that L_{acc} dominates over all other IR luminosity components, so that $L_{IR}^{excess} \simeq \frac{1}{2} L_{acc}$. This is the case for FU Orionis objects.

However, the IR SED can be used in conjunction with *simultaneous* UV and optical data to constrain the mass accretion rate (e.g., [19],[9],[76]) or,

---

[4]However, see Boss & Yorke [22] for an alternate interpretation of "dips" in the SEDs.

alternatively, the UV/optical spectrum can be used alone to determine the mass accretion rate (e.g., [153],[75],[71]). These procedures involve additional assumptions about the physics of the interaction region between star and disk. As it turns out, separating the UV-optical excess from the true photospheric emission is not an easy task, especially if the excess is small with respect to the photospheric luminosity. Further complications come from the fact that the colors of a PMS late-type photosphere and *a fortiori* the colors of the excess are not well known, so the (large) extinction corrections to apply lead to large errors in the respective luminosities. Additional uncertainties come from the accretion geometry assumed. All these have resulted in estimation of the accretion rate that can vary by as much as a factor 10 for a given star. We shall summarize the current understanding of the complex star/disk interaction region in §3.4. We now turn to those properties of disks which are unveiled by direct imaging.

3.3. DIRECT OBSERVATIONS OF ACCRETION DISKS

In a broad sense, the observers benefit from two sources of information when trying to probe the disk structure directly. The first one is the *scattered* starlight the disk is sending back, the second one is the disk's own *thermal* emission. The scattering process occurs at the surface of an optically thick disk. At millimeter wavelengths the dust thermal emission is usually assumed optically thin, except at the very center. The CO gas may be optically thick however, depending on the line and isotope. Furthermore, at optical and near-infrared wavelengths, the grains interacting with the light are small, micron-sized at most, and likely smaller. The grains responsible for the mm continuum emission are bigger. Both approaches therefore complement each other nicely, as we will see below.

There is little doubt now that disks exist around TTSs. The images of the edge-on systems HH 30 [28] and HK Tau/c [144], the rotating gas rings around the binaries GG Tau [44] and UY Aur [45], the silhouette disks in Orion [124], are as many proofs. But these images contain much more: they contain the traces of the disk structure itself.

Models usually are simplified reproduction of nature's work. In what follows we will assume that the (parametrized) disks have axial symmetry and that quantities describing them are power-laws. For exemple, the surface density distribution, the vertical scale height, the radial temperature distribution, and the density profile can be respectively expressed by the following expressions:

$$\Sigma(r) = \Sigma_0(r/r_o)^{-p}, \quad H(r) = H_0(r/r_o)^{\beta}, \quad T(r) = T_0(r/r_0)^{-q}, \qquad (3)$$

$$\text{and} \quad \rho(r,z) = \rho_0(r/r_o)^{-\alpha} \exp[-|z|^2/2H^2(r)]. \qquad (4)$$

Figure 6. HST/WFPC2 image of the pre-main sequence binary star HK Tau. The F606W filter was used. It includes most of the Johnson V and R passbands. The separation between the two stars is $2''.4$. The companion (to the South) is entirely nebulous and the central star remains undetected at $\lambda \leq 2.2\mu$m. The dark lane is interpreted as a disk seen edge-on.

In equation (4), $\alpha \equiv p + \beta$ and we have assumed a Gaussian vertical density profile, valid for a vertically isothermal, hydrostatic, non-self gravitating disk. A summary of measured disk parameters for CTTSs is presented in Table 2 for comparison. Many of the parameters derived from the observations are best fits from multi-parameter χ^2-minimizations. Good descriptions of the dependences between the various parameters are given by, e.g., [28] for light scattering models, and by [68] for CO molecular emission models.

Outer Radius. Because the instruments have finite sensitivities, the outer radii measured are lower limits. Furthermore, a severe flaring of the disk could also cast shadow on the outer parts, reducing the amount of scattered light to zero. Nevertheless, the disks that have been imaged so far are in general *larger* than predicted by SED fitting, e.g., [12].

Whether the gaseous disk is actually larger than the dust disk remains an open issue. From existing mm-interferometric observations, one finds that either the dust disks are smaller or their emission drops below the detection limits faster (i.e., at 200-300AU) than the gas disks do (up to 850AU). However, in the case of GM Aur (a single star), UY Aur and

GG Tau (two binaries), the disks seen in scattered light extend over roughly the same distance as the CO disks. This suggests that both techniques are probing the same physical structure and the grains (at least those responsible for scattering) are mixed with the gas over large distances.

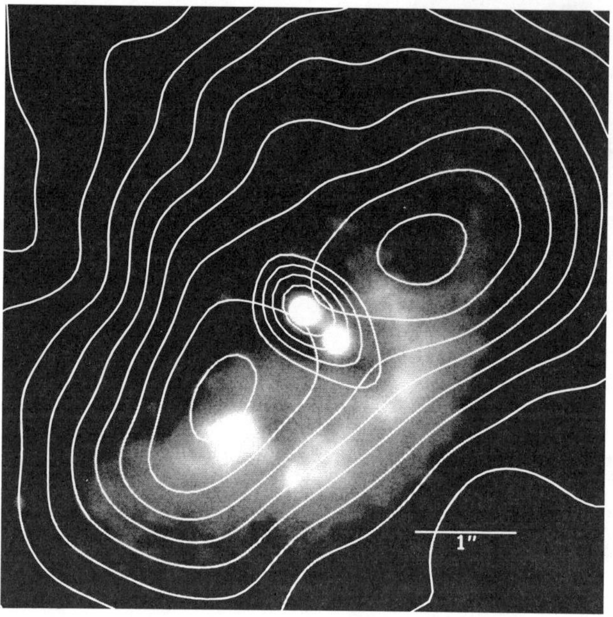

Figure 7.   Overlay of the ^{13}CO (2 → 1) integrated area (large contours) and 1.3mm continuum emission (smaller contour, close to the center) over an Adaptive Optics J-band image (grey shade) obtained at CFHT. As expected if the dust and gas are well mixed, the ^{13}CO countours trace roughly the scattered light. See Close et al. (1998) for the AO image, and Duvert et al.(1998) for the radio data.

Inclination.   The inclination of the disk is usually derived from the axis ratio of either the scattered light, the dust thermal continuum, or the CO gas total emission. The kinematic pattern can also be used to estimate the inclination in mm-line work. When disks are seen close to edge-on, model of the scattered light are used to estimate the inclination relying on the assumption that the nebula is uniform and both lobes should have equal brightness for $i = 90^o$. This technique has been applied for HH 30 and HK Tau/c for example.

Density Profiles and the Scale Height.   For edge-on systems, H_0 can be measured directly, in principle, from optical and infrared images. However, a large flaring or a thick envelope may forbid direct view of the central parts, reducing the radial coverage and forbidding an assessment of β. A

TABLE 2. Disk parameters for CTTSs

	R_{out}^{gas} (AU)	R_{out}^{dust} (AU)	i	p	β	q	M_{disk} ($M_\odot$)	reference
HH 30	—	250	> 80	0.75	1.45	—	0.006	[28]
HK Tau/c	—	105	85	0.3–1.5	1.2	—	6×10^{-4}	[144]
GG Tau	525	> 220	43	1.5	1.25	0.5	~ 0.17	[44]
DM Tau	850	—	33	~ 1.5	1.26	0.63	≤ 0.025	[68]
UY Aur	~ 1000	> 650	42	—	—	—	≤ 0.01	[45][34]
GM Aur	525	>200	56	1.57	1.18	0.64	0.025	[43]

The values in this table are difficult to estimate. They come from the best models and observations we found. But the reader should be aware they are subject to change in the future.

wavelength coverage as wide as possible is necessary to extract a more reliable estimate. For an arbitrary inclination, estimating $\Sigma(r)$ from scattered light is difficult because the disk is optically thick. For edge-on systems, $\Sigma(r)$ and $H(r)$ are coupled (degenerate) and isolating each one is next to impossible.

However, the size of millimeter continuum sources detected in the course of a survey was modelled by Dutrey et al. [42] who concluded that the radial dependence of the surface density is probably shallower than was imagined by previous models using single dish data. This shallowness is supported by yet better models for GM Aur and DM Tau for example, see Table 2.

Temperature Profile. The results presented in Table 2 suggest that the temperature profile is flatter than 0.75, the value expected for passive or viscous but otherwise geometrically thin disks. A discussion of various possibilities leading to flatter temperature distributions has been presented in §3.2. We note however, that it is compatible with a flaring disk, as suggested by the radial dependencies of the scale height, also quoted in Table 2.

Rotation Curve. From molecular line observations, solid evidence for rotating disks have been found in a number of CTTSs now, see Table 2 but also [46] for LkCa 15, [106] for MWC 480, an Herbig Ae stars, [91] for UZ Tau E. In many cases, models showed this rotation to be consistent with Keplerian rotation, with $V(r) \propto r^{-0.5}$ within the error bars.

Disk Mass. Disk mass estimates from scattered light models are severely affected by the large optical thickness of the disk and are mere lower limits, usually. An exception to that is found, once again, in the edge-on case. Then, the optically thick disk draws a dark lane in the equatorial plane, the *thickness* of which is directly related to the disk mass. For HH 30, the mass derived from such a model agrees reasonably with the mass estimated from mm continuum flux.

Estimating disk masses from mm observations is not any easier (see, e.g., [12],[4]). One usually assumes that the dust is optically thin throughout the disk and the continuum flux therefore a good tracer of the mass. However, the grain size distribution, the dust emission properties and the dust-to-gas ratio are not well known quantities. Similarly, mass estimates based on gas phase tracers can suffer from possible optical thickness and depletion of the molecular species used. Usually, dust masses are considered more reliable because the dust opacity at 1mm is considered accurate to within a factor of a few [128] for a wide-range of grain sizes and shapes.

Assuming that a representative CTTS is 10^6 years old and that its accretion rate is $10^{-7} M_\odot$ yr^{-1}, then the accreted mass is $0.1 M_\odot$. This is more than most of the disk masses listed in Table 2. However, the accretion rate may be lower, $10^{-8} M_\odot$ yr^{-1}, and it has been shown to decline with time [78]. The mass accreted over the pre-main sequence lifetime then becomes comparable to the disk masses deduced from mm continuum emission (typically a few $0.01 M_\odot$), and there is no need to "feed" the original disk to sustain the accretion process. For binaries however, the situation may be different. Their disks may be smaller, with the inner disks intact, but outer disks truncated, missing. This is based on their identical NIR excess compared to single TTSs [142], but mm continuum flux lower on average ([125], [42]). This may be supported by Fig.6, where the disk of the companion in this 2.''4 binary is the smallest CTTS disk imaged so far.

3.3.1. *Concluding Remarks*

Despite the large efforts deployed by many groups, the number of resolved disks remains decivingly small, much below the fraction of 50% inferred in Taurus [146], and $\sim 70\%$ in Orion [89]. If we assume that our vision of a disk around a TTS is right (but we can also question this affirmation!) various reasons may be invoked to explain the low detection rate. Apart from the obvious limits in resolution and contrast of the instruments, grain growth and molecular gas depletion may play a role. The presence of unknown companions may also lead to truncation and destruction of the outer disks.

Asymmetries are seen in the disk of HH 30 [145], and maybe also in HK Tau/c [144]. Although their origin is not clear at present, they may indicate that the disks are rapidly evolving. Hartmann et al. [78] have also suggested that accreting disk should evolve by increasing their outer radii, hence further lowering their surface density profile. This topic of disk evolution was, however, voluntarily ignored in the present section. The reader is referred to, e.g., [146] and [142] for an overview of our knowledge of disk dissipation and survival times in single and binary CTTSs. There is no need to emphasize that a better knowledge of the disk evolution is a necessary step to future studies of planet formation.

After this review of observed disk properties, we discuss in some details the (unresolved) inner part of the disk and disk/star interaction region, where accretion takes place. At the time of the Crete I conference, hints that the accretion might be channeled by the stellar magnetic field were sparse. In the next section we discuss the changes that took place since then.

3.4. STRUCTURE OF THE STAR-DISK INTERACTION ZONE

The interface zone between the accretion disk and the central star is small enough to prevent direct imaging. Yet, it is the site of energetic and unsteady phenomena that are as many finger-prints of low-mass stellar formation. It is likely instrumental in the accretion and ejection processes, and possibly in many of the variability events discussed before. Efforts to outline its morphology and isolate the relevant mechanisms at play have been numerous. A picture of this interaction zone is slowly emerging from indirect evidence. We will describe the current supportive observational evidence in the next paragraphs, keeping in mind that this picture is likely to evolve rapidly in the years to come as the physical processes at play are complex and not completely understood.

3.4.1. *Spectro-Photometric Evidence for Magnetospheric Accretion*

The observation of inverse P Cygni profiles in hydrogen Balmer lines and permitted metallic lines with pronounced redshifted absorption is frequent in CTTSs (see §1.1.1). These features are not easily explained by a boundary layer (where accretion proceeds at subsonic speeds) but are produced naturally in magnetospheric accretion funnels [30]. Radiative transfer models adapted for this geometry can reproduce many of the observed features of various Hydrogen lines ([82],[119]), in agreement with the idea that they are formed in the accretion flow rather than in a wind[5]. In that sense, the hot spots discussed in §2.3.2 are also suggestive of accretion columns as they may trace the shocks close to the stellar surface.

The red-displaced absorption features have velocities of a few hundred km/s, suggesting infall of material along the line-of-sight at velocities close to free-fall velocity. This is compatible with a magnetosphere that connects to (and possibly truncates) the disk at a few stellar radii, close to the

---

[5]One should probably add a word of caution here, as line profile shapes are usually non-unique, i.e., they can be modelled using different physical models. Basic features of CTTS Balmer line profiles can also be reproduced by wind models ([99],[116]) or by other accretion scenarios ([150]).

corotation radius, and channels material along its field lines. Independent support for that view comes from the near infrared colors of CTTSs that are best explained if the inner disks are absent (truncated) over a significant fraction of the corotation radius [111]. However, see Armitage & Clarke for a caution [7].

3.4.2. *Is There an Appropriate Stellar Magnetic Field?*

The magnetospheric accretion model assumes a dipolar field, tilted or not from the stellar rotational poles, that is steady or survives long enough for accretion to get established [98]. Estimations of the magnetic field are available for a few T Tauri stars only, many of them upper limits or non-detections ([27],[60],[10],[64],[41],[67],[92]). Magnetic fields of order kG were detected in T Tau and LkCa 15([64],[67]), with large filling factors, suggesting that field-free regions are not common at the stellar surface. Using Zeeman-Doppler imaging, Donati et al. [41] showed that V410 Tau and possibly HDE 283572, two bright WTTSs, had multi-component fields, ruling out a pure dipole.

On the other hand, Johns-Krull et al. [92] did not detect a net polarization in the absorption lines of the CTTS BP Tau but they found a strong circular polarization in the HeI $\lambda5876$ *emission* line, which is presumably formed in the accretion shock. They deduced a 2.4kG field in the line formation region. This result may well be the best evidence we have so far for magnetically mediated accretion proceeding along large-scale magnetic loops, the base of which covers only a small fraction of the stellar surface.

To summarize, magnetic fields of order of 1-2 kG have been measured on a few TTSs. There are signs that the field is not steady and its geometry may not be dominantly dipolar. As such, these measurements may still be compatible with but are not proofs that magnetospheric accretion occurs.

3.4.3. *Stellar Angular Momentum Evolution*

TTSs evolving along their Hayashi tracks are spinning up because of gravitational contraction and should reach break-up velocities on a time-scale comparable with the Hayashi evolutionary phase ($\sim 10^6$yr). Additionnally, an initially slow-rotating CTTS surrounded by an accretion disk and accreting material at a rate of $\dot{M} \sim 10^{-7}M_\odot yr^{-1}$ should reach break-up velocity on a time-scale of a few 10^4 years (e.g., [162]). In other words, if angular momentum was conserved during the Hayashi phase, all TTSs should be fast rotators, close to break-up speed. That velocities generally observed in both WTTSs and CTTSs are in the range $v \sin i \sim$10-25 km s^{-1} (about 0.1 of the break-up velocity) clearly indicates that a braking (or regulating) mechanism is efficient at confining the rotation to slow rates during the pre-main sequence phase ([98],[77]). MHD winds flowing along stellar or disk open field lines are often invoked as efficient carriers of angular momentum away from the stars (e.g., [127],[48]). However, indications that mass-loss is intimately linked to the accretion process are now available (e.g.,[29],[75]) and the disk itself could also play a determinant role in the momentum transport and spin regulation. Roughly speaking, one can imagine a scenario in which stellar field lines are anchored in the disk at or outside the corotation radius, effectively locking together the stellar and disk rotations. This idea follows from a model proposed by Gosh & Lamb, notably to explain the spin rates of pulsars ([61],[62]). Examination of the rotation periods of TTSs ([23],[49],[35]) confirmed that CTTSs are indeed slower rotators than WTTSs, and theoretical models showed that the stellar rotation could be regulated by transport of stellar angular momentum to the disk in the case of magnetic coupling ([31],[162],[32],[63],[7]).

Bouvier et al. [25] successfully modeled the evolution of angular momentum of low-mass stars with moderate and fast rotation rates from the pre-main sequence phase up to the main sequence phase. In their model, short-lived accretion disks result in main sequence stars with fast rotation because the stars are allowed to spin-up for a longer time during their contraction toward the main sequence. To the opposite, long-lived disks produce slow main sequence rotators.

From the spread of rotation periods of late-type main sequence stars, one can infer that disk lifetimes have to span a large range, from 0.1 Myr to a few 10^7yr, with a mean around 3 Myr. This is in rough agreement with disk lifetimes estimated by observations for TTSs (e.g., [146]), although 10^7yr or more may be too long (see [25] for a discussion of that issue).

3.4.4. *Problems and Prospective*

While the success of the magnetospheric accretion model to explain many observables of CTTSs is evident, Edwards [47] points out some of the weaknesses of the model. More discordantly, Safier [137] argues that there is little evidence for the long-lived, global dipole fields that are necessary in the models to insure disk truncation and steady accretion along field lines. He also shows [138] that some magnetic field detections are likely marred by systematic effects that tend to overestimate the field value. Popham [129] proposed that the angular momentum of CTTSs can be regulated by FU Orionis outbursts, and [131] discuss observational support to this hypothesis. While the angular momentum evolution models are impressive, one must note that the role of jets and winds in the angular momentum balance is not investigated in these models, in spite of its obvious relevance.

Recent 2D numerical simulations of the magnetosphere-disk interaction in fact hint at a much more complicated picture, in which various instabilities play a major role ([72], [112]). Studying several different magnetic field configurations, Miller & Stone [112] find notably that boundary layer-like, equatorial accretion occurs when the (highly time-dependent) ram pressure of the accreting material exceeds the magnetic pressure of the magnetosphere. The ram pressure depends on the angular momentum transport mechanism, which is either the Balbus-Hawley instability or magnetic braking in their simulations. They also find that the presence of a disk magnetic field comparable in strength to the stellar field itself can promote polar accretion – if the disk and stellar field are anti-aligned. They show that winds occur in all configurations they study, and that the evolution of angular momentum depends on the details of the star-disk interaction, making it difficult to predict. While this is an exploratory study that needs to be extended to a full 3D one, it is highly suggestive of the complexity of the accretion process, making it obvious that both the axisymmetric, steady boundary layer model and the dipolar magnetospheric accretion model are far too simplistic to give an accurate representation of reality.

Acknowledgements We gratefully acknowledge financial support from NATO and the Laboratoire d'Astrophysique de L'Obsevatoire de Grenoble. Thanks also to Jérôme Bouvier, Catherine Dougados, Gaspard Duchêne and Karl Stapelfeldt for their help in the preparation of this review.

References

1. Adams, F.C., Shu, F.H. 1986, ApJ, 308, 836
2. Adams, F.C., Lada, C.J., Shu, F.H. 1987, ApJ, 312, 788
3. Ambartsumian, J.R. 1947, Stellar Evolution and Astrophysics, Erevan, Acad. Sci. Armen. SSR
4. André, P., Montmerle, T. 1994, ApJ, 420, 837
5. Appenzeller, I. 1977, in *The Interaction of Variable Stars with their Environment*, eds. R. Kippenhahn, J. Rahe, and W. Strohmeier, Bamberg, Remeis-Sternwarte, p. 80
6. Appenzeller, I., Jankovics, I., Oestreicher, R. 1983, A&A, 141, 108
7. Armitage, P.J., Clarke, C.J. 1996, MNRAS, 280, 458
8. Armitage, P.J., Clarke, C.J. 1997, MNRAS, 285, 540
9. Basri, G., Bertout, C. 1989, ApJ, 341, 340
10. Basri, G., Marcy, G.W., Valenti, J.A. 1992, ApJ, 390, 622
11. Basri, G., Martin, E.L., Bertout, C. 1991, A&A, 252, 625
12. Beckwith, S.V.W., Sargent, A.I., Chini R.S., Güsten, R. 1990, AJ, 99, 924
13. Bell, K.R., Lin, D.N.C. 1994, ApJ, 427, 987
14. Bell, K.R., Lin, D.N.C., Hartmann, L.W., Kenyon, S.J. 1995, ApJ, 444, 376
15. Bell, K.R., Cassen, P.M., Klahr, H.H., Henning, Th. 1997, ApJ, 486, 372
16. Bertout, C. 1989, ARA&A, 27, 351
17. Bertout, C. 1999, in preparation
18. Bertout, C., Arenou, F., Robichon, N. 1999, A&A, submitted
19. Bertout, C., Basri, G., Bouvier, J. 1988, ApJ, 330, 350
20. Bertout, C., Harder, S., Malbet, F., Mennessier, C., Regev, O. 1996, AJ, 112, 2195
21. Bieging, J.H., Cohen, M. 1989, AJ, 98, 1686
22. Boss, A.P., Yorke, H.W. 1993, ApJ, 411, L99
23. Bouvier J., Cabrit S., Fernandez M., Martin E.L., Matthews J.M. 1993, A&AS, 101, 485

24. Bouvier, J., Cabrit, S., Fernandez, M., Martin, E.L., Matthews, J.M. 1993, A&A, 272, 176
25. Bouvier, J., Forestini, M., Allain, S. 1997, A&A, 326, 1023
26. Bouvier, J., Chelli, A., Allain, S., et al. 1999, A&A, submitted.
27. Brown, D.N., Landstreet, J.D. 1981, ApJ, 246, 899
28. Burrows, C.J., Stapelfeldt, K.S., Watson, A.M., et al. 1996, ApJ, 473, 437.
29. Cabrit, S., Edwards, S., Strom, S.E., Strom, K.M. 1990, ApJ, 354, 687
30. Calvet, N., Hartmann, L. 1992, ApJ, 386, 239
31. Cameron, A.C., Campbell, C.G. 1993, A&A, 274, 309
32. Cameron, A.C., Campbell, C.G., Quaintrell, H. 1995, A&A, 298, 133
33. Chelli, A., et al. 1999, A&A, in press
34. Close, L.M., Dutrey, A., Roddier, F. et al. 1998, ApJ, 499, 883
35. Choi, P.I., Herbst, W. 1996, AJ, 111, 283
36. Cohen, M. 1983, ApJ, 270, L69
37. Cohen, M., Kuhi, L.V. 1979, ApJ Supp, 41, 743
38. Cram, L.E. 1979, ApJ, 234, 949
39. D'Alessio, P., Canto, J., Calvet, N., Lizano, S. 1998, ApJ, 500, 411
40. De Campli, W.M. 1981, ApJ, 244, 124
41. Donati, J.F., Semel, M., Carter, B.D., Rees, D.E, Cameron, A.C. 1997, MNRAS, 291, 658
42. Dutrey, A., Guilloteau, S., Duvert, G., Prato, L., Simon, M., Schuster, K., Ménard, F. 1996, A&A, 309, 493
43. Dutrey, A., Guilloteau, S., Prato, L., Simon, M., Duvert, G., Schuster, K., M'enard, F. 1998, 338, L63
44. Dutrey, A., Guilloteau, S., Simon, M. 1994, A&A, 286, 149
45. Duvert, G., Dutrey, A., Guilloteau, S., Ménard, F., Schuster, K., Prato, L., Simon, M. 1998, A&A, 332, 867
46. Duvert, G., Guilloteau, S., Ménard, F., Simon, M., Dutrey, A. 1999, A&A, submitted
47. Edwards, S., 1997, in Proc. of IAU Symp. 182:*Herbig-Haro flows and the birth of low-mass stars*, eds. B. Reipurth and C. Bertout, Dordrecht, Kluwer, p. 433
48. Edwards, S. Mundt, R., Ray T. 1993, in *Protostars and Planets III*, eds. E.H. Levy and J.I. Lunine, Tucson, University of Arizona Press, p. 567
49. Edwards, S., Strom, S., Hartigan, P., Strom, K., Hillenbrand, L., Herbst, W., Attridge, J., Merrill, M., Probst, R., Gatley, I. 1993, AJ, 106, 372.
50. Edwards, S., Hartigan, P., Ghandour, L., Andrulis, C. 1994, AJ, 108, 1056
51. Elias, J.H. 1978, ApJ, 224, 857
52. Favata, F., Micela, G., Sciortino, S., D'Antona, F. 1998, A&A, 335, 218
53. Feigelson, E.D. 1996, ApJ, 468, 306
54. Feigelson, E.D., Jackson, J.M., Mathieu, R.D., Myers, P.C., Walter, F.D. 1987, AJ, 94, 1251
55. Feigelson, E.D., Montmerle, T. 1985, ApJ, 289, L19
56. Fernández, M., Eiroa, C. 1996, A&A, 310, 143
57. Ferreira, J., Pelletier, G. 1995, A&A, 295, 807
58. Gahm, G.F., Lodén, K., Gulbring, E., Hartstein, D. 1995, A&A, 301, 89
59. Glasby, J.S. 1974, *The Nebular Variables*, Oxford, Pergamon
60. Gnedin, Yu.N., Red'kina, N.P. 1984, Sov. Astron. Lett., 10(4), 255
61. Ghosh, P., Lamb, F.K. 1979, ApJ, 232, 259
62. Ghosh, P., Lamb, F.K. 1979, ApJ, 234, 296
63. Ghosh, P. 1995, MNRAS, 272, 763
64. Guenther, E.W. 1997, in *Herbig-Haro Flows and the Birth of Stars*, IAU Symp. No. 182, eds. B. Reipurth and C. Bertout (Dordrecht:Kluwer), p. 465
65. Guenther, E.W., Ball, M. 1999, A&A, submitted
66. Guenther, E.W., Emerson, J.P. 1997, A&A, 321, 803
67. Guenther, E.W., Lehmann, H., Emerson, J.P., Staude, J. 1999, A&A, 341, 768

68. Guilloteau, S., Dutrey, A. 1998, A&A, 339, 467
69. Guillout, P., Sterzik, M.F., Schmitt, J.H.M.M., Motch, C., Egret, D., Voges, W., Neuhäuser, R. 1998, A&A, 334, 540
70. Guillout, P., Sterzik, M.F., Schmitt, J.H.M.M., Motch, C., Neuhäuser, R. 1998, A&A, 337, 113
71. Gullbring, E., Hartmann, L., Briceño, C., Calvet, N. 1998, ApJ, 492, 323
72. Hayashi, M.R., Shibata, K., Matsumoto, R. 1996, ApJ 468, L37
73. Haro, G. 1968. in *Stars and Stellar Systems*, Vol. VII, eds. B.M. Middlehurst and L.H. Aller, Chicago, Univ. of Chicago Press, p. 141.
74. Haro, G. 1976, Bol. Inst. Tonantzintla, 2, 3
75. Hartigan, P., Edwards, S., Ghandour, L. 1995, ApJ, 452, 736
76. Hartigan, P., Hartmann, L., Kenyon, S.J., Hewett, R., Stauffer, J. 1991, ApJ Supp, 70 899
77. Hartmann, L. 1991, in *Angular momentum evolution of young stars*, eds. S. Catalano and J.R. Stauffer, NATO ASI series Dordrecht, Kluwer, p. 379
78. Hartmann, L., Calvet, N., Gullbring, E., D'Alessio, P. 1998, ApJ, 495, 385
79. Hartmann, L., Edwards, S., Avrett, A. 1982, ApJ 261, 279
80. Hartmann, L., Kenyon, S.J. 1985, ApJ, 299, 462
81. Hartmann, L., Kenyon, S.J. 1987, ApJ, 312, 243
82. Hartmann, L., Hewett, R., Calvet, N. 1994, ApJ, 426, 669
83. Hatzes, A.P. 1995, ApJ, 451, 784
84. Herbig, G.H. 1962, Advances in A&A, 1, 47
85. Herbig, G.H. 1977, ApJ, 217, 693
86. Herbig, G.H., Bell, K.R. 1988, Lick Obs. Bull. No.1111
87. Herbig, G.H., Goodrich, R.W. 1986, ApJ, 309, 294
88. Herbst, W., Herbst, D.K., Grossman, E.J., Weinstein, D. 1994, AJ, 108, 1906
89. Hillenbrand, L.A., Strom, S.E., Calvet, N., Merrill, M.K., Gatley, I., Makidon, R.B., Meyer, M.R., Skrutskie, M.F. 1998, AJ, 116, 1816
90. *The Hipparcos and Tycho Catalogues*, 1997, ESA SP-1200
91. Jensen, E.A., Koerner, D.W., Mathieu, R.D. 1996, AJ, 111, 2431
92. Johns-Krull, C.M., Valenti, J.A., Hatzes, A.P., Kanaan, A. 1999, ApJ, 510, L41
93. Jones, B.F., Herbig, G.H. 1979, AJ, 84, 1872
94. Joy, A.H. 1945, ApJ, 102, 168
95. Kenyon, S.J. 1999, this volume
96. Kenyon, S.J., Hartmann, L. 1987, ApJ, 323, 714
97. Kenyon, S.J., Hartmann, L. 1990, ApJ, 349, 197
98. Königl, A. 1991, ApJ, 370, L39
99. Kuhi, L.V. 1964, ApJ, 140, 1409
100. Lada, C.J., Kylafis, N.D. 1991, Proc. of the NATO ASI: *The Physics of Star Formation and Early Stellar Evolution*, Dordrecht, Kluwer, 746 pp.
101. Lada, C.J. 1984, Wilking, B.A. 1984, ApJ, 371, 171
102. Larwood, J.D., Nelson, R.P., Papaloizou, J.C.B., Terquem, C. 1996, MNRAS, 282, 597
103. Lynden-Bell, D. , Pringle, J.E. 1974, MNRAS, 168, 603
104. Magazzu, A., Rebolo, R., Pavlenko, I.V. 1992, ApJ, 392, 159
105. Malbet, F., Bertout, C. 1991, ApJ, 383, 814
106. Mannings, V., Koerner, D.W., Sargent, A.I. 1997, Nature, 388, 555
107. Martin, E.L., Rebolo, R., Magazzu, A., Pavlenko, Y. 1994, A&A, 282, 503
108. Mathieu, R.D., Adams, F.C., Latham, D.W. 1991, AJ, 101, 2184
109. Ménard, F., Bastien, P. 1987, in *Circumstellar Matter*, IAU Symp. 122, eds. I. Appenzeller and C. Jordan, (Dordrecht: Reidel), p. 133
110. Mendoza, V.E.E. 1968, ApJ, 151, 977
111. Meyer, M.R., Calvet, N., Hillenbrand, L.A. 1997, AJ, 114, 288
112. Miller, K.A., Stone, J.M. 1997, ApJ, 489, 890
113. Mitskevich, A.S. 1994, A&A, 281, 471

114. Mitskevich, A.S. 1995, A&A, 298, 219
115. Mitskevich, A.S. 1995b, A&A, 298, 231
116. Mitskevich, A.S., Natta, A., Grinin, V.P. 1993, ApJ, 404, 751
117. Montmerle, T., Koch-Miramond, L., Falgarone, E., Grindlay, J.E. 1983, ApJ, 269, 182
118. Muzerolle, J., Hartmann, L., Calvet, N. 1998, AJ, 116, 455
119. Muzerolle, J., Calvet, N., Hartmann, L. 1998, ApJ, 492, 743
120. Najita, J., Carr, J.S., Tokunaga, A.T. 1996, ApJ, 456, 292
121. Natta, A. 1993, ApJ, 412, 761
122. Neuhäuser, R., Sterzik, M.F., Schmitt, J.H.M.M., Wichmann, R., Krautter, J. 1995, A&A, 297, 391
123. Neuhäuser, R., Sterzik, M.F., Torres, G., Martin, E.L. 1995, A&A, 299, L13
124. O'Dell, C.R., Wen, Z., Hu, X. 1993, ApJ, 410, 696
125. Osterloh, M., Beckwith, S.V.W. 1995, ApJ, 439, 288
126. Palla, F., Stahler, S.W. 1993, ApJ, 418, 314
127. Pelletier, G., Pudritz, R.E. 1992, ApJ, 394, 117
128. Pollack, J.B., Hollenback, D., Beckwith, S., Simonelli, D., Roush, T., Fong, W. 1994, ApJ, 421, 615
129. Popham, R. 1996, ApJ, 467, 749
130. Popham, R. 1997, ApJ, 478, 734
131. Popham, R., Kenyon, S., Hartmann, L., Narayan, R. 1996, ApJ, 473, 422
132. Racine, R. 1968, AJ, 73, 233
133. Rekowski, M.v., Froehlich, H.-E., 1997, A&A, 319, 225
134. Rice, J.B., Strassmeier, K.G. 1996, A&A, 316, 164
135. Rucinski, S.M. 1985, AJ, 90, 2321
136. Rydgren, A.E., Zak, D.S. 1987, PASP, 99, 141
137. Safier, P.N. 1998, ApJ, 494, 336
138. Safier, P.N. 1999, ApJ, 510, 127
139. Siess, L., Forestini, M., Bertout, C. 1997, A&A, 326, 1001
140. Siess, L., Forestini, M., Bertout, C. 1999, A&A, 342, 480
141. Shu, F.H., Tremaine, S., Adams, F.C., Ruden, S.P. 1990, ApJ, 358, 495
142. Simon, M., Prato, L. 1995, ApJ, 450, 824
143. Stine, P.C., Feigelson, E.D., André, P., Montmerle, T. 1988, AJ, 96, 1394
144. Stapelfeldt, K.R., Krist, J.E., Ménard, F., Bouvier, J., Padgett, D.L., Burrows, C.J. 1998, ApJ, 502, L65
145. Stapelfeldt, K.R., Watson, A.M., Burrows, C.J., Krist, J.E. 1998, BAAS, 193, 1706
146. Strom, S.E. 1995, RevMexAA (serie de Conferencias), 1, 317
147. Strom, K.M., Wilkin, F.P., Strom, S.E., Seaman, R.L. 1989, AJ, 98, 1444
148. Terquem, C., Bertout, C. 1993, A&A, 274, 291
149. Terquem, C., Bertout, C. 1996, MNRAS, 279, 415
150. Ulrich, R.K. 1976, ApJ, 210, 377
151. Ultchin, Y., Regev, O., Bertout, C. 1997, ApJ, 486, 397
152. Unruh, Y.C., Cameron, A.C., Guenther, E. 1998, MNRAS, 295, 781
153. Valenti, J.A., Basri, G., Johns, C.M. 1993, AJ, 106, 2024
154. Ventura, P., Zeppieri, A., Mazzitelli, I., D'Antona, F. 1998, A&A, 331, 1011
155. Vrba, F.J., Chugainov, P.F., Weaver, W.B., Stauffer, J.S. 1993, AJ, 106, 1608
156. Vrba, F.J., Rydgren, A.E., Chugainov, P.F., Shakovskaya, N.I., Weaver, W.B. 1989, AJ, 97, 483
157. Walker, M.F. 1972, ApJ, 175, 8
158. Walker, M.F. 1983, ApJ, 271, 642
159. Walter, M.F. 1986, ApJ, 306, 573
160. Walter, F.M., Brown, A., Mathieu, R.D. Myers, P.C., Vrba, F.J. 1988, AJ, 96, 297
161. Wolf, B., Appenzeller, I., Bertout, C. 1977, A&A, 58, 163
162. Yi, I., 1994, ApJ, 428, 760

ANNEX 1: Properties of T Tauri Stars in Taurus-Auriga

HBC	name	RA (2000)	Dec (2000)	V	K	$v \sin i$ (km/s)	P_{rot} (d)	Sp.Ty.	W(Ha) (Å)	Type	sep. ('')	p.a. (o)	ratio @K	1.3mm (mJy)	M_{disk} (M$_\odot$)	log $\dot{M}_{acc}$ (M$_\odot$/yr)
351	034903+2431	3 52 02.3	+24 39 51	12.26	9.2	29	<3.5	K5	1.6	wt	0.61	317	0.22	<14		
352/353	035120+3154	3 54 29.4	+32 03 02	11.8/12.4	9.6/9.9			G8/K0	3/2	-/wt	8.7	71	0.77	<12/<14		
354	035135+2528NW	3 54 35.6	+25 37 12	13.79		<10		K3	hk					<16		
	SAO 76411	4 02 53.5	+22 08 12	8.85				G0		wt				<14		
355/354	035135+2528	3 54 35.9	+25 37 09	12.7/13.9	10.3/11.1			K0/K2		wt	6.3	298	0.46			
356/357	V1067 Tau NS	4 03 14.1	+25 53 00	12.91	10.2	>75		K2	0.9/1.8	?/wt	2			<16		
358	040047+2603W	4 03 49.2	+26 10 54	15.2	9.4			M3	13	wt	1.58	226	0.60			
359	040047+2603E	4 03 50.9	+26 10 55	14.17		<10		M2	2.4	wt				<13		
	SAO 76428	4 04 28.5	+21 56 04	9.3				F8		wt				<14		
360/361	040142+2150	4 04 39.3	+21 58 20	14.9/15.1	10.0/10.1			M3/M3	10/8	wt	7.2	65	0.89	<9/<26		
362	040234+2143	4 05 30.9	+21 51 11	14.67		14		M2	4.0	wt				<14		
365	V1095 Tau	4 13 14.2	+28 19 11	13.72		23		M4	4	wt				<14		
366n	V1096 Tau	4 13 27.2	+28 15 57	13.52:				M0	3	wt				<14		
367	V773 Tau	4 14 12.9	+28 12 12	10.62	6.5	55	3.43	K3	4	wt	0.112	295	0.47	42v	-	
23n	FM Tau	4 14 13.5	+28 12 50	14.22				M0	71	tt				31	<50	-8.45
24	FN Tau	4 11 08.6	+28 20 27	14.95				M5	25	tt				31	40	
25n	CW Tau	4 14 17.0	+28 10 59	12.36		36		K3	135	tt				96	<150	
368	V1098 Tau	4 14 48.0	+27 52 35	12.08	7.5	22	7.2	M1	2.5		0.491	77	0.96	<14		
26	FP Tau	4 11 43.5	+26 38 56	13.90		27		M4	38	tt				<50	<250	
369	FO Tau	4 14 49.6	+28 12 27	15.0:	8.1			M2	117	tt	0.166	182	0.65	<14	<17	-7.50
27	CX Tau	4 11 44.0	+26 40 42	13.72		20		M2.5	20	tt				<40	<110	-8.97
370	V1068 Tau	4 16 28.1	+28 07 36	12.54		26	3.37	K7:	5	wt				<14		
28	CY Tau	4 14 27.7	+28 13 29	13.33		10	7.5:	M1	70	tt				133	7000	-8.12
371	LkCa 5	4 17 39.0	+28 33 02	13.56		38:		M2	4	wt				<14		
29	V410 Tau	4 18 31.1	+28 27 16	10.82	7.5	71	1.87	K7	3v	wt	0.123	218	0.16	<30	09	-8.39
30n	DD Tau	4 18 30.6	+28 16 40	14.11	7.9			M1	182	tt	0.56	186	0.52	17	<30	-9.35
31n	CZ Tau	4 15 25.6	+28 09 44	15.33	9.3			M1.5	4	tt	0.33	84	0.46	<30		
372	041529+1652	4 18 21.4	+16 58 46	13.26		<10		K5	hk					<14		
374n	V1023 Tau	4 18 47.0	+28 20 08	12.73:		13		K7	3	wt				<25		
375	CoKu Tau/1	4 18 51.5	+28 20 28	19.3:				?	70					<12		
376	041559+1716	4 18 51.6	+17 23 15	12.28		68		K7	0.7	wt				<14		
377	FQ Tau	4 16 06.4	+28 22 21	15.6:	9.3			M2	114	tt	0.79	69	0.90	<40	<260	-6.45
32	BP Tau	4 19 15.8	+29 06 27	12.09		7.8	7.6	K7	40	tt				37	<110	-7.54
378	V819 Tau	4 19 26.4	+28 26 14	13.24		<15	5.6	K7	1.7	wt				34	320	
	FR Tau	4 19 35.5	+28 27 23	16.3										<15		
379	V1070 Tau	4 19 41.3	+27 49 39	12.6	8.3	12.9	5.64	K7	4	wt	1.05	25	0.56	<24		
	J2 157	4 20 52.6	+17 46 41							tt				<14		
33	DE Tau	4 21 54.2	+27 55 03	13.00		10	7.6	M2:	54	tt				36	70	-7.58
34n	RY Tau	4 21 57.4	+28 26 36	10.01		52		K1	20v	tt	0.02	317		229	390	
380	HD 283572	4 21 58.9	+28 18 06	9.04		>75	1.55	G5	abs	su				<15		
381n	FS Tau B	4 22 00.9	+26 57 38					?	pr	tt				134	>1700	
382	V1071 Tau	4 22 03.1	+28 25 39	13.47		60:		M3,4	6	tt				<12	-	
383n	FS Tau	4 22 02.1	+26 57 33	15.71	7.3			M1	57	tt	0.265	60	0.12	<35	<30	-8.09
35n	T Tau	4 21 59.4	+19 32 06	9.90	5.4	20	2.8	K1	38	tt	0.71	176	0.09	280	160	
	FU Tau	4 20 33.3	+24 56 08	15.1										<35		
	GT Tau	4 20 34.2	+24 48 22	14.4										<35		
384	FT Tau	4 23 39.2	+24 56 15	14.1				cont	pr	tt				130	340	
	J4423	4 21 40.3	+26 54 54							wt				<11		
385	IP Tau	4 24 57.1	+27 11 58	12.99		<11	3.25	M0:	11					16	48	-9.10
	J4872	4 22 13.8	+26 11 02	13.1				K7?		wt				<14		
386	FV Tau	4 26 53.6	+26 06 55	15.4:	7.5			K5	23	tt	0.73	92	0.64	15	6	-6.23
387n	FV Tau/c	4 26 54.4	+26 06 52	17.2:	8.5			M3	29	tt	0.74	293	0.17	<16		
36	DF Tau	4 27 02.8	+25 42 22	11.49	6.8	16	8.5	M0.5	60	tt	0.088	329	0.47	<25	<100	-6.91
37n	DG Tau	4 27 04.7	+26 06 17	12.01		22	6.3	cont.	113	tt				443	410	
388	V1072 Tau	4 27 10.6	+17 50 44	10.34		16	2.74	K1	hk	wt				<16		
	J 507	4 26 16.3	+26 27 09							wt				<14		

HBC	name	RA (2000)	Dec (2000)	V	K	v sin(i) (km/s)	P_{rot} (d)	Sp.Ty.	W(Ha) (Å)	Type	sep. ('')	p.a. (o)	ratio @K	1.3mm (mJy)	M_{disk} ($M_\odot$)	log M_{acc} ($M_\odot/yr$)
389n	GV Tau	4 26 21.9	+24 26 30	17.2:	7.0			K3	pr		1.21	355	0.13	116	17	
	FW Tau	4 29 29.6	+26 16 54	17.1(B)	9.2			M3-5	17		0.16	160	1.0	<15	<460	
38	DH Tau	4 26 37.1	+26 26 29	13.92		10.5	7.2	M1	53	tt				53	<170	-8.30
39	DI Tau	4 26 38.0	+26 26 20	12.86	8.4	11	7.5	M0	2	wt	0.12	294	0.13	<35	<110	-8.75
41	IQ Tau	4 26 47.7	+26 00 16	13.30		11.5	6.25	M0.5	7.8	tt				87	400	-7.55
42/43	UX Tau BA	4 27 09.6	+18 07 21	13.7/11.4	8.9/7.7			M2/K5	8/20		5.9	269	0.29	63	65	
43	UX Tau AC	4 27 10.0	+18 07 21	11.4/15.1(R)	7.7/10.5	27	2.7	K5	20	wt	2.7	181	0.07	63	65	
	ZZ Tau B	4 27 14.6	+24 33 02							wt				<20		
44	FX Tau	4 30 26.9	+24 26 35	13.90	8.1	10		M1	10	tt	0.91	292	0.55	<30	<50	-8.65
45	DK Tau	4 30 44.2	+26 01 24	12.40	7.0	11	8.4	K7/K7	19/34	tt	2.53	115	0.30	35	50	-7.42
46	ZZ Tau	4 27 49.3	+24 35 57	14.34	8.5			M3:	15	tt	0.029[l]	-	0.44	<15	<140	
	JH 56	4 31 14.4	+27 10 19	12.63						wt				<19		
47	V927 Tau	4 28 22.4	+24 04 30	14.63	8.7	19		M5.5	4.7	tt	0.30	290	0.73	<20	<120	
392	V1074 Tau	4 31 27.2	+17 06 25	12.53		17	3.38	K5	0.2	wt				<18		
393n	L1551/IRS 5	4 31 33.6	+18 08 15					K2		?				1276	370	
394	LkHa 358	4 28 42.2	+18 07 21	19.0:				M5.5	47					32	>31	
49n	HL Tau	4 31 38.4	+18 13 59	14.57				K7,M2?	55:	tt				879	1000	
50	XZ Tau	4 31 40.0	+18 13 58	13.30	7.1		2.6:	M3	274	tt	0.311	153	0.51	14	03	
48	HK Tau	4 28 48.9	+24 17 56	15.75	8.7/11.8	22		M0.5	29	tt/tt	2.4	175	0.06	41	50	
	J 665	4 28 55	+25 37 08							wt				<28		
51/395	V710 Tau	4 31 57.8	+18 21 37	14.2/14.4	8.6/8.6	17		M0.5/M2	89/11	tt/wt	3.24	357	0.83	71	540	
396	V806 Tau	4 32 15.7	+24 29 02	17.7:				cont	88	tt				124	140	
397	V1075 Tau	4 32 09.3	+17 57 23	12.06		27	2.43:	K7	0.5	wt						
398	V928 Tau	4 32 18.8	+24 22 28	13.7:	8.0	25		M0.5	1.2	wt	0.165	125	0.60	<11		
399	V827 Tau	4 32 14.6	+18 20 15	12.18		18.5	3.75	K7,M0	1.8	wt				<19	<120	
400	V826 Tau	4 29 22.0	+17 55 19	12.11		4.2	3.7	K7,M0	1.6	wt				<15	<450	
401	FY Tau	4 32 30.5	+24 19 58	15.4:				K7	59	tt				16	30	-7.41
402	FZ Tau	4 29 30.1	+24 13 44	15.0:				cont	204	tt				23	40	
54	GG Tau Aa	4 29 37.1	+17 25 22	12.24	7.5	10	10.3	M0	54	tt	0.288	3	0.25	593	290	-7.76
	GG Tau Bb	4 29 *	+17 25	17.3	10.0			M5	83	tt	1.4	135	0.19			
52/53	UZ Tau ew	4 32 43.1	+25 52 31	13.1/13.7	8.1/7.5	14/16		M2/M3	41/79	tt	3.78	273	0.49	172	540	
53	UZ Tau w	4 32 42.8	+25 52 32	13.7	7.5	16		M3	79	tt	0.360	359	0.46	172		
	JH 112	4 32 49.1	+22 53 03	15.54						tt				<18		
403	V1076 Tau	4 32 43.8	+18 02 58	13.22		11.5	6.20	K7	0.7	wt				<23		
55	GH Tau	4 33 06.1	+24 09 45	12.95	7.8	28:		M1.5	15	tt	0.314	299	0.56	<30	24	-7.92
404	V807 Tau S	4 33 06.6	+24 09 56	11.3	8.3			M0	13	tt	0.023[l]	-	1.0	17	24	
404	V807 Tau NS	4 33 06.6	+24 09 56	11.3	7.0			M0	13	tt	0.375	330	0.37	17		
405	V830 Tau	4 30 08.3	+24 27 27	12.21		29	2.75	K7,M0	3	wt				<9	<27	
56n	GI Tau	4 33 34.1	+24 21 18	13.01		11	7.2	K7	19	tt				<20	<30	-8.02
59	IS Tau	4 33 36.8	+26 09 50	14.29	8.0			K2	12	tt	0.221	92	0.16	<20	<230	
57n	GK Tau	4 33 34.5	+24 21 07	13.5/17.0	7.6	19	4.65	K7	16	tt	2.4	66	0.03	<21	<30	-8.19
58	DL Tau	4 33 39.1	+25 20 39	13.05		16		cont.	105	tt				230	870	
60/406	HN Tau	4 33 39.3	+17 51 53	14.2/18.1	8.2	53		K5/M4	138/108	tt	3.1	215	0.042	<15	05	-8.89
	IT Tau	4 30 50.5	+26 07 14	14.3	8.2			K0	4.3		2.48	225	0.23	<33	<60	
61	CI Tau	4 30 52.2	+22 44 17	13.09		11		K7	102	tt				190	630	-7.19
62	DM Tau	4 33 48.7	+18 10 12	13.99		<10		M1:	139	tt				109	340	-7.95
	J2-2041	4 33 55.4	+18 38 39							wt				<19		
	JH 108	4 34 10.9	+22 51 45	15.2						wt				<18		
407	V1077 Tau	4 34 18.1	+18 30 07	12.67		12.2	3.32	G8	0.3	wt				<64	-	
408	V1110 Tau	4 34 39.4	+25 01 01	10.37				K0	0.5	?				<19		
63	AA Tau	4 34 55.5	+24 28 54	12.37		11	8.2	K7	37	tt				88	210	-8.48
	V1078 Tau	4 35 14.1	+18 21 37	10.95		<5	3.2	F8		wt				<15		
64	HO Tau	4 32 20.8	+22 26 07	14.4/16:	9.6			M0.5	115	tt	6.9	109		<30	<190	-8.86
409	FF Tau	4 32 20.9	+22 48 17	13.67	8.7			K7	1.4	wt	0.037[l]	-	0.40	<27	<240	
65	DN Tau	4 32 25.7	+24 08 52	12.36		8	6.0	M0	12	tt				84	270	-8.46

HBC	name	RA (2000)	Dec (2000)	V	K	v sin(i) (km/s)	P_{rot} (d)	Sp.Ty.	W(Ha) (Å)	Type	sep. ('')	p.a. (o)	ratio @K	1.3mm (mJy)	M_{disk} (M$_\odot$)	log M_{acc} (M$_\odot$/yr)
411	CoKu Tau/3	4 35 41.2	+24 11 08	17.3:	8.2			M1	4.6		2.04	177	0.29	<16	<44	
412	043230+1746	4 35 34.1	+17 51 41	14.12	9.1	<30		M2	9	wt	0.70	68	1.0	<16		
	HQ Tau	4 32 47.4	+22 44 16	12.6	7.4						0.009[l]	-	0.14	<45	<110	
66	HP Tau	4 35 53.1	+22 54 24	13.16	7.3	15	5.9:	K3	35	tt	0.017[l]	-	0.26	62	50	
414	HP Tau/G3	4 35 53.5	+22 54 10	14.6	8.7			K7	2.0	wt	0.022[l]	-	0.40	<17	>37	
415/414	HP Tau G2/G3	4 35 54.1	+22 54 14	11.0/14.6	7.2/8.7	100:/	1.2/	G2/K5	2/2	su	10.0	243	0.24	<14/<17		
416	Haro 6-28	4 35 55.9	+22 54 36	17.3:	9.3			M5	92	tt	0.66	246	0.63	<14	<110	
	FG Tau	4 36 10.1	+24 02 55	16.4(B)				M5						<45	<210	
417	V1115 Tau	4 36 19.1	+25 42 59	13:		22	3.35	M0:	1	wt				<19		
	HD 283759	4 33 47.3	+24 06 58	10.36				F2						<35	<60	
	FI Tau	4 34 11.3	+23 30 16	16.3(B)				M5?		?				<22	<230	
	HT Tau	4 34 46.8	+26 18 01											<50	<460	
67n	DO Tau	4 38 28.6	+26 10 55	14.01				K7	109	tt				136	180	-6.84
418n	HV Tau	4 35 30.9	+26 04 45	14.0:	7.8	14		M1	4.5	wt	0.035[l]	-	0.58	40		
418n	HV Tau AB	4 35 30.9	+26 04 45	14.0:	7.8	14		M1	4.5	wt	4.00	45	0.03	40		
	GN Tau	4 39 20.8	+25 45 03	15.7	8.4			cont.	62	tt	0.041[l]	-	0.63	<50	<160	
68	VY Tau	4 36 17.4	+22 42 02	13.71	8.8	10	5.37	M0	4.9	tt	0.66	317	0.26	<17	<19	
419	V1079 Tau	4 39 17.8	+22 21 03	12.09		12.5	5.85	K5	13	wt				167	240	-8.87
	JH 223	4 40 49.5	+25 51 20	15.50						wt				<19		
	Haro 6-32	4 41 04.2	+25 57 57											<50		
420	IW Tau	4 41 04.7	+24 51 07	12.5	8.3	7	5.6	K7	4	wt	0.27	177	0.91	<19	<690	
421	CoKu Tau/4	4 41 16.8	+28 40 01	14.9:				M1.5	1.8					<15	<8	
422	LkHa 332/G2	4 42 05.5	+25 22 57	14.9	8.2	20		M0	3.1	wt	0.30	243	0.60	<15	-	
423	LkHa 332/G1	4 42 06.9	+25 23 04	15.4	8.1			M1	3.7		0.215	77	0.56	<14	-	-6.60
69	V955 Tau	4 42 07.7	+25 23 13	14.74	8.0	10		K7,M0	13	tt	0.33	204	0.22	<19	<11	-7.02
70	DP Tau	4 42 37.2	+25 15 31	14.22				M0.5	85	tt				<27	30	-7.88
71	GO Tau	4 43 03.1	+25 20 19	14.89		21		M0	81	tt				83	570	-7.93
72	DQ Tau	4 44 00.0	+16 54 40	13.16				M0,1	113	tt				91	250	-9.40
73/424	V1001 Tau	4 46 59.0	+17 02 39	13.5/14.7	7.5/8.9	10		K7/M1	21/166	tt	2.7	37	0.45	60	170	-7.00
74	DR Tau	4 47 06.3	+16 58 41	11.17v		<10	9.0	cont.	87	tt				159	280	
75	DS Tau	4 44 39.1	+29 19 56	12.4/13.6	8.0/11.3	10		K2/F8	70/0	tt	7.1	294	0.045	25	67	-7.89
76	UY Aur	4 48 35.7	+30 42 14	12.37	7.0			K7	73	tt	0.89	225	0.28	29	9	-7.18
	Haro 6-39	4 52 09.7	+30 37 46											24	50	
425	St 34	4 51 30.3	+17 05 07	13.5:				M	pr					<15		
77	GM Aur	4 51 59.8	+30 17 15	12.03		12	12.0	K3	96	tt				253	600	-8.02
426	V396 Aur	4 55 37.0	+30 17 55	10.85		19	2.24	K0	1:	wt						
79n	SU Aur	4 55 59.4	+30 34 02	8.93		67		G2	4v	su				<30	<30	
427	V397 Aur	4 56 02.2	+30 21 03	11.60		10.4	9.3:	K7	0.7	wt						
429	V836 Tau	5 03 06.6	+25 23 20	13.13		<15	6.78	K7,M0	9v	wt				37	400	
81	RW Aur BC	5 04 37.6	+30 20 14	12.7:	8.7						0.120	111	-	421		
80/81	RW Aur AB	5 04 37.7	+30 20 14	11.1/12.7	6.9	20:/	5.39/	cont.	84v	tt	1.50	258	0.23	421	3	

Columns 1,2 list the star's Herbig and Bell catalogue number and name respectively. **Col. 3,4** are the J2000 coordinates from SIMBAD. **Col. 5,6** are the V and K magnitudes. **Col. 7,8** are vsini and photometric rotation period. **Col. 9-11** give the spectral type, the Hα equivalent width and the type of the star: tt means CTTS, wt means WTTS, su means SU Aur object. **Col. 12-14** contain information on binarity: the projected separation, the apparent position angle, in degrees East of North, and the flux ratio (primary/secondary) in the K-band. A superscript l means lunar occultation. (mostly from Mathieu 1994, ARAA) **Col. 15-16** list the 1.3mm continuum flux and the correspondent disk mass ([12],125]). The last column, 17, is the mass accretion rate estimated by [71].

THE EVOLUTION OF PRE–MAIN-SEQUENCE STARS

FRANCESCO PALLA

Osservatorio Astrofisico di Arcetri

Largo E. Fermi, 5 - 50125 Firenze, Italy

1. The classical theory of pre–main-sequence evolution

The location of optically visible Pre–Main-Sequence (PMS) stars in the H-R diagram combined with the use of theoretical evolutionary tracks have played a central role in stellar evolution studies since the initial work of Hayashi, Cameron and Iben in the early '60s. Hayashi and his contemporaries lacked a theory of the protostar phase, and therefore could not specify with any degree of confidence the initial conditions for PMS contraction. A simple argument indicated that if the interstellar cloud radiated away little energy during its collapse, the starting values of the stellar radius would be some two orders of magnitude greater than the corresponding main-sequence values (Hayashi 1966). The contraction time, given by the Kelvin-Helmoltz time $t_{\mathrm{KH}} = GM_*^2/R_*L_{\mathrm{surf}}$, is so short for these large radii and high surface luminosities that the poorly understood early collapse would be much too brief to affect the subsequent evolution. Adopting the large initial radii, Iben (1965) and Ezer & Cameron (1965) followed numerically the hydrostatic contraction of stars with masses from less than 1 $M_\odot$ to over 100 $M_\odot$. The evolution of all stars was thought to be qualitatively the same, with quantitative differences scaling in a simple manner with the stellar mass M_*.

1.1. BASIC RESULTS

Let us first follow the arguments used by Hayashi and others to derive the initial stellar radii. A protostar of mass M_* and radius R_* forms out of cold, nearly static cloud material whose dimensions are enormous compared to R_*. Thus, we may effectively set the initial energy, both mechanical and thermal, to zero. The star itself, however, is a gravitationally bound entity with a *negative* total energy. Some of the energy difference is radiated into

375

C.J. Lada and N.D. Kylafis (eds.), The Origin of Stars and Planetary Systems, 375–408.
© 1999 *Kluwer Academic Publishers. Printed in the Netherlands.*

space during collapse, while most of the rest goes into dissociating and ionizing hydrogen and helium. We denote this latter, internal component as ΔE_{int}, where

$$\Delta E_{\text{int}} \equiv \frac{X\,M_*}{m_H}\left[\frac{\Delta E_{\text{diss}}(H)}{2} + \Delta E_{\text{ion}}(H)\right] + \frac{Y\,M_*\,\Delta E_{\text{ion}}(He)}{4\,m_H} . \tag{1}$$

Here, $\Delta E_{\text{diss}}(H) = 4.48$ eV is the binding energy of H_2, $\Delta E_{\text{ion}}(H) = 13.6$ eV is the ionization potential of HI, and $\Delta E_{\text{ion}}(He) = 75.0$ eV is the energy required to fully ionize helium. The hydrogen and helium mass fractions are taken to be $X=0.70$ and $Y=0.28$, respectively. Since the protostar's thermal energy is equal to half the gravitational potential energy by the virial theorem, we may write

$$0 = -\frac{1}{2}\frac{G\,M_*^2}{R_*} + \Delta E_{\text{int}} + L_{\text{rad}}\,t , \tag{2}$$

where L_{rad} is the average luminosity escaping over the formation time t. Suppose that we first take the extreme step of ignoring L_{rad} entirely. Then equation (2) yields the *maximum* radius R_{max} which the protostar could have at any mass M_*. We readily find

$$R_{\text{max}} = \frac{G\,M_*^2}{2\,\Delta E_{\text{int}}} = 61\,R_\odot\left(\frac{M_*}{M_\odot}\right) . \tag{3}$$

This numerical estimate would change slightly with a more careful treatment of the gravitational energy. No direct observations are available of the true radii of solar-type protostars, but even the youngest and largest T Tauri stars, the immediate descendants of these objects, have radii smaller by an order of magnitude than that predicted by equation (3).

Having established the initial conditions, the classical theory of PMS evolution then predicts that any star contracts due to heat loss at a rate given by

$$L_{\text{surf}} = 4\pi R_*^2 \sigma T_{\text{eff}}^4 , \tag{4}$$

where L_{surf} is the surface luminosity and T_{eff} the effective temperature of the star. The star is also fully convective because of the large radius. For a one solar mass star with an effective temperature of about 4000 K, $L_{\text{surf}} \sim 600\,L_\odot$ which is much higher than the luminosity that can be carried out radiatively

$$L_{\text{rad}} = L_0\left(\frac{M_*}{M_\odot}\right)^{11/2}\left(\frac{R_*}{R_\odot}\right)^{-1/2} , \tag{5}$$

where the precise value of L_0, a luminosity of order 1 $L_\odot$, depends on the star's detailed structure and where it has been assumed that the interior opacity obeys Kramer's Law (Cox & Giuli 1968). Thus, convection spreads inward from the surface (Von Sengbusch 1968).

The path followed by any star in the H-R diagram is the same. Stars contract *homologously* descending along nearly vertical tracks, commonly called *Hayashi tracks*), in a Kelvin-Helmoltz time scale which is much longer than the sound crossing time. Therefore, the star can effectively be considered to be in hydrostatic equilibrium at each instant, despite its slow, or *quasi-static*, contraction. At a certain point, a radiative core develops and the star moves away from the Hayashi track along a radiative track (discovered by Henyey, LeLevier & Levee in 1955), at almost constant luminosity. Eventually, hydrogen ignites in the center and contraction is halted: the PMS star has reached the *Zero-Age-Main-Sequence* (ZAMS). These results were obtained in the pioneering numerical models of Iben (1965) and Ezer & Cameron (1965) which established the *standard* set of PMS tracks in the H-R diagram and which can be found in any textbook on stellar evolution.

Before concluding this introductory section, let remind two fundamental properties discovered in the original derivation by Hayashi. First, stars of a given mass and radius have a minimum surface temperature, the *Hayashi temperature*; second, the Hayashi temperature is insensitive to the stellar radius, so that the convective portions of the tracks are nearly vertical. Both properties reflect the behavior of the opacity of the gas in the outer layers of the star. The minimum surface temperature is due to the steep decline of opacity with falling temperature below the hydrogen ionization zone near 10^4 K. For too low a surface temperature, the outer layers become optically thin, violating the condition that the photosphere be located at an optical depth of order unity from the surface. The extreme temperature sensitivity of the opacity is also responsible for the vertical nature of the Hayashi tracks. At temperatures below 10^4 K, the opacity varies as T^α with α between 7 and 13. Therefore, only a minor change of $T_{\rm eff}$ is needed for a substantial change in the photospheric pressure and entropy (which determine the boundary conditions). For solar-type stars, the effective temperature is about 4000 K. An excellent explanation of the role of the photosphere and entropy structure in determining the basic properties of PMS stars, together with the history of the early developments of PMS theory can be found in the review by Stahler (1988b).

1.2. THE INFLUENCE OF PROTOSTELLAR EVOLUTION

The development of protostar theory has altered the conception of PMS evolution, in particular for stars more massive than solar. Many such stars undergo a transient period of thermal relaxation prior to settling down to the type of evolution envisioned earlier. Furthermore, the union of protostar

and PMS theory implies that stars of 100 $M_\odot$ or even 20 $M_\odot$ should have
no PMS phase at all. To follow these changes in detail one must know how
stars of various masses are built up from collapsing clouds. An exhaustive
account of this phase can be found in the reviews by Stahler & Walter
(1993) and Palla (1994). Here, it suffices to point out some of the main
results, relevant for PMS evolution.

The formation of a protostar from the quiescient conditions typical of
dense molecular clouds occurs through the gradual accumulation of the interstellar gas onto an accreting core. This process requires a large decrease
in gravitational potential energy. Unlike the basic assumption that leads to
the expression for the maximum radius of the protostars (see eq. 3), most
of the energy during collapse is radiated away and only a small fraction
goes into the thermal energy of the protostar. As suggested early on by
Gaustad (1963) and others, the high efficiency of radiative losses keeps the
interstellar gas almost isothermal during the dynamical collapse phase. A
large fraction of the gravitational energy is radiated in radial or disk accretion shocks that form as a result of the abrupt change in the velocity of the
freely-falling gas. If no net energy is absorbed by the circumstellar material,
the resulting protostellar luminosity is given to a good approximation by

$$L_{\mathrm{proto}} = \frac{G\, M_* \, \dot{M}_{\mathrm{acc}}}{R_*} \, , \tag{6}$$

where M_* and R_* are the instantaneous mass and radius of the protostellar core and $\dot{M}_{\mathrm{acc}} \equiv dM_*/dt$ is the mass accretion rate. Thus, estimates
of the luminosity emitted during this phase rely on the knowledge of two
fundamental quantities: the accretion rate and the *mass-radius relation*.
The former is determined by the dynamics of the gravitational collapse,
while the relation between mass and radius are established by processes
occuring in the protostellar interior. The radiation produced at the shock
is absorbed, reradiated and thermalized in the optically thick, dusty infalling envelope. Most of the observable radiation is emitted at mid- and
far-infrared wavelengths. The exact shape of the emergent spectrum depends on the density and temperature distribution of the dust component,
which are set by the dynamics of collapse.

As shown by the first numerical models (e.g. Larson 1969), gravitational
collapse occurs in a highly *non-homologous* way. The central regions of
the cloud collapse faster than the outer parts, pressure gradients develop
fast, and the nonhomology sets in already during the initial isothermal
phase. Independently of the initial conditions, whether uniform or centrally
condensed, the matter develops a density distribution of the type $\rho \propto r^{-2}$,
where r is the radial distance from the cloud center. In a time slightly
longer than the free-fall time, the structure of the cloud is characterized

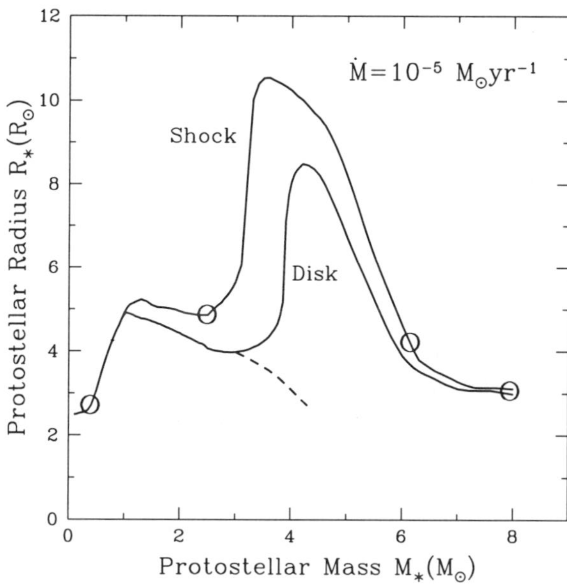

Figure 1. The evolution of the radius *vs.* mass in accreting protostars. The two curves refer to shock and disk accretion at a rate $\dot{M} = 10^{-5}\ M_\odot\ yr^{-1}$. The open circles represent, from left to right, the ignition of central deuterium; the start of D-shell burning; the ignition of central hydrogen burning via the CN-cycle; the final arrival on the ZAMS. To gauge the importance of D-shell burning, the dashed curve shows the mass-radius relation obtained ignoring it in the case of disk accretion (adapted from Palla & Stahler 1991).

by the presence at the center of a stable hydrostatic core, which contains only few percent of the total cloud mass, accreting mass at the typical rate $\dot{M}_{acc}$ (Stahler et al. 1980; Winkler & Newman 1980). At some point early in its growth, the hydrostatic core achieves central temperatures sufficient to ignite *deuterium burning* ($T_c \sim 10^6$ K). The exact value of the core mass when this occurs depends on the magnitude of the accretion rate. For a value $\dot{M} = 10^{-5}\ M_\odot\ yr^{-1}$, the ignition of deuterium starts at $M_* \sim 0.3\ M_\odot$. The resulting mass-radius relations for protostars accreting at such a rate is shown in Figure 1 (Stahler 1988; Palla & Stahler 1990). The two curves refer to the evolution computed assuming that matter is accreted directly onto the protostar through a strong shock and to the case where the core surface is treated as a normal photosphere, as in the case of accretion through a circumstellar disk. Apart from minor quantitative differences, the *qualitative* behavior is similar, suggesting that it does not depend on the details of the accretion process.

The large radii predicted by Hayashi (see eq. 3) are never achieved. On the contrary, the protostellar radius remains limited to a few solar radii by the efficient radiation losses at the accretion shock. Note also that any value of $\dot{M}_{\mathrm{acc}}$ below that used in Fig. 1 would yield even smaller radii. In order to reproduce the Hayashi initial conditions, $\dot{M}_{\mathrm{acc}}$ should increase by more than two orders of magnitude. The evolution of the radius shown in Fig. 1 is the result of the nuclear burning of interstellar deuterium (D) in the accreting protostellar core. The general evolution can be separated into three main stages.

- $0.1 \lesssim M_*/M_\odot \lesssim 1$: for low-mass protostars, deuterium burning occurs near the center for temperatures exceeding $\sim 10^6$ K. The energy input is sufficient to turn and maintain the star convectively unstable. As more mass is gathered, the newly accreted deuterium is instantly transported by convective eddies to the center, and a situation of steady-state burning is created. In this condition, the deuterium generated luminosity, L_D, is given by

$$L_D = \dot{M}_{\mathrm{acc}}\delta = 12L_\odot \left(\frac{\dot{M}_{\mathrm{acc}}}{10^{-5}M_\odot yr^{-1}} \right) , \tag{7}$$

where δ is the nuclear energy per unit mass available in interstellar matter (assuming a [D/H] value of 2×10^{-5} by number). The extreme temperature sensitivity of the energy generation rate is such that deuterium acts as an effective thermostat, preventing the central temperature from rising above $\sim 10^6$ K as the star gains mass. Since the amount of energy available in deuterium, δ, is comparable to the gravitational binding energy of the star, GM_*/R_*, the thermostatic effect results in a core radius that increases almost linearly with mass during the phase of active burning, as shown in Fig. 1 for $0.3 \lesssim M_*/M_\odot \lesssim 1$.

- $1 \lesssim M_*/M_\odot \lesssim 4$: the evolution of protostars more massive than solar is marked by the return of radiative stability and the onset of gravitational contraction. These events occur at about the same time, when the the deuterium generated luminosity, L_D, is equal to the radiative luminosity, L_{rad}. This condition is satisfied when

$$t_{\mathrm{KH}} \equiv \frac{GM_*^2}{R_*L_{\mathrm{rad}}} \approx \frac{GM_*^2}{R_*L_D} \approx \frac{GM_*^2}{R_*\dot{M}\delta} \approx \frac{M_*}{\dot{M}} \equiv t_{\mathrm{acc}} \tag{8}$$

Eq. (8) establishes that the star is losing heat by radiation at the same time as it accretes new matter, and therefore it undergoes a phase of rapid gravitational contraction. Note that for low-mass stars, the accretion time is always much shorter than t_{KH}. Because of the sensitivity of L_{rad} on M_* (see eq. 5), the equality of the two time scales is confined to a rather narrow range of masses, between 2 and 3 $M_\odot$. If nothing else happened in the interior, the rapid contraction would lead to the conditions appropriate for hydrogen burning in the center, and the star would then join the main-sequence while still accreting (as indicated by the dashed line in Fig. 1).

As a consequence, there should be no stars of mass greater than $2 - 3\ M_\odot$ in the PMS phase (e.g. Larson 1972). This result is in clear contradiction with the observational evidence of the Herbig Ae/Be stars, whose location in the H-R diagram is well above the ZAMS. The reason why this expected behavior does not occur is subtle, and is again related to the effects of deuterium burning.

When the thermodynamical conditions are appropriate, deuterium once more ignites, this time in a convective shell located at the transition between the inner radiative core and the external convective regions. This shell slowly shrinks toward the surface, due to the overall tendency of the star to become more and more radiatively stable. The release of nuclear energy in the lower density subsurface regions results in a dramatic swelling of the star, that doubles its radius. This is shown in the steep, almost vertical rise of the curve in Fig. 1. The expansion of the radius counteracts the gravitational pull, and retards the beginning of the phase of gravitational contraction.

- $4 \lesssim M_*/M_\odot \lesssim 8$: for more massive cores, the strength of the gravitational pull prevents a further increase of the radius, and the rapid expansion is replaced by a slow gravitational settling to the conditions of relaxed cores. Because of the swelling of the core, the interior temperature remains quite low, so that the approach to the ZAMS is naturally postponed to more massive cores. As shown in Fig. 1, central hydrogen burning is delayed until $M_* \sim 6\ M_\odot$, but the ZAMS is actually reached only at $M_* \sim 8\ M_\odot$. This finding reconciles the theoretical predictions of the protostellar models with observations of intermediate-mass PMS stars.

1.3. THE STELLAR BIRTHLINE

Under the assumption that protostars end their accretion phase in a time much shorter than t_{KH}, the mass-radius relation shown in Fig. 1 can be used to construct a theoretical *birthline* for stars of low- and intermediate-mass in the H-R diagram. The birthline is the locus along which young objects first appear as optically visible stars (Stahler 1983; Palla & Stahler 1990). Accordingly, there should be no observed PMS star whose position in the H-R diagram is *above* the birthline, since the largest radius a PMS star of a given mass can achieve is the protostellar radius at that mass. The resulting birthline is shown in Figure 2, together with the location of known T Tauri stars and Herbig Ae/Be stars. It is evident that the upper envelope of the stellar distribution is well matched by the theoretical birthline. The agreement also suggests that the specialized assumptions of the calculations (geometry, constant mass accretion rate) are not essential to the outcome of the mass-radius relation. Variations of the mass accretion rate induce a

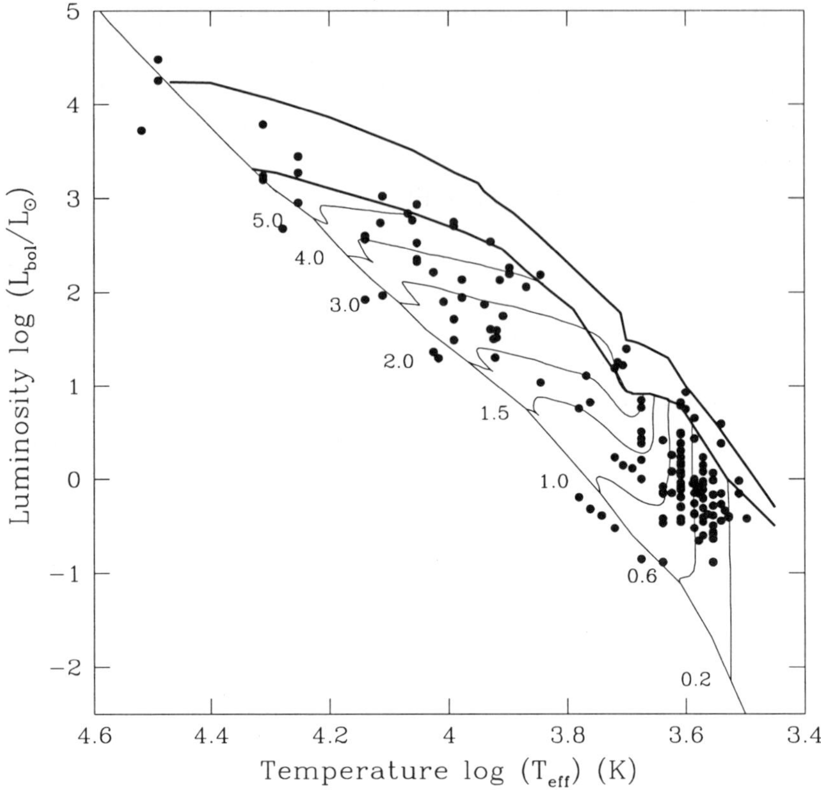

Figure 2.   Location of the stellar birthline in the H-R diagram for accretion rates of 10^{-5} (*lower curve*) and 10^{-4} $M_\odot$ yr^{-1} (*upper curve*). Symbols give the position of young PMS stars. Labeled curves are PMS evolutionary tracks for the indicated stellar mass. The location of the ZAMS is also indicated (adapted from Palla & Stahler 1993).

change in the mass-radius relation: for higher $\dot{M}_{\rm acc}$, the various evolutionary phases described above remain unaltered, even though each event is shifted to higher masses. The resulting birthline shown in Fig. 2 by the upper curve for $\dot{M}_{\rm acc} = 10^{-4}$ $M_\odot$ yr^{-1} does not improve significantly the agreement with the observational data. These considerations lead to the conclusion that an accretion rate $\dot{M}_{\rm acc} \sim 10^{-5}$ $M_\odot$ yr^{-1} can represent the typical value for the formation of stars in a mass range from few tenths of a solar mass to about 10 $M_\odot$.

2. Pre–main-sequence evolution

Once the main phase of accretion is completed, the stellar core emerges as an optically visible star along the birthline. The physical process by which infall stops is still not known, although stellar winds and bipolar outflows must play a fundamental role. The circumstellar matter surrounding these young stars, partly distributed in a disk and the rest in an extended envelope, still emits copiously at infrared wavelengths: the emergent spectral energy distribution departs substantially from that of a normal stellar photosphere, showing excess emission extending from the near- to the far-infrared. The origin of this excess has been commonly attributed to the presence of accretion disks. Depending on the mass of the star, young visible stars are classified as T Tauri (low-mass, typically ~ 1 $M_\odot$), or Herbig Ae/Be stars (intermediate-mass, $\lesssim 10$ $M_\odot$). Their early evolution can be followed in detail, once the initial conditions for the gravitational contraction phase toward the Main Sequence are specified. The birthline joins the ZAMS at $M_* \sim 8$ $M_\odot$. Thus, for stars with masses above this value there is no optical PMS phase: the rapid gravitational contraction in these objects has led to hydrogen ignition already during the accretion phase. Massive stars are born *on* the ZAMS. In this section, the basic features of theoretical models of PMS evolution will be outlined.

2.1. PROTOSTELLAR INITIAL CONDITIONS

The results of protostellar evolution described in Section 2 show that the classical assumptions are no longer appropriate. First, protostellar cores never reach large radii: the radius remains typically a factor of ten smaller than that predicted by eq. (3). It is reasonable to assume that the protostellar radii represent the *maximum* values that a star of a given mass can achieve prior to contraction. Second, the internal structure departs significantly, at least for intermediate-mass objects, from the assumption of thermal convection. In low-mass stars, convection is due to central nuclear burning. Protostars more massive about 2 $M_\odot$ are radiatively stable in the inner regions, and possess a thick, subsurface mantle of deuterium. This deuterium must ignite in a shell to fuse to helium during the approach to the main sequence. The disparate state of the two regions implies that the stars are *thermally unrelaxed* at the beginning of the PMS evolution. Thus, in contrast to the situation described by the classical theory, these stars must undergo *non-homologous* quasi-static contraction (Palla & Stahler 1993).

Another important aspect is that for a PMS of fixed mass, the amount of deuterium available is that which a protostar of the same mass did acquire in the course of its accretion history. The variation of the fractional deuterium concentration, f_D, for a protostar accreting at different rates is shown in Figure 3. Here, f_D is calculated relative to the interstellar concentration. We see that for the lowest values of the accretion rate f_D declines

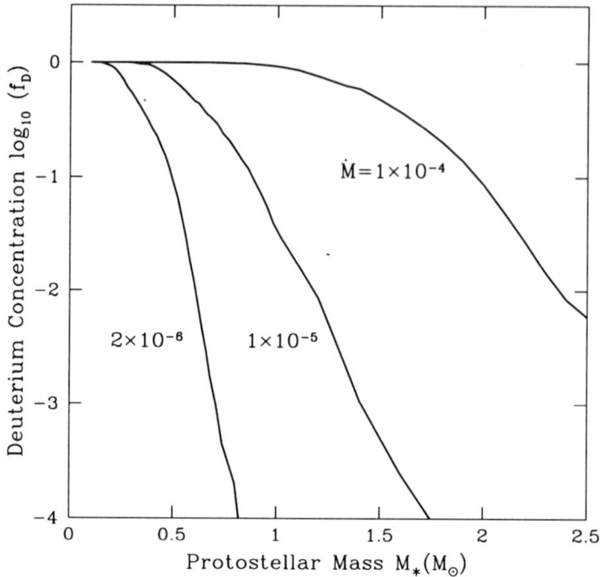

Figure 3. Evolution of the fractional deuterium concentration, f_D, in protostars accreting at different values of the accretion rate.

quite rapidly for $M_* \lesssim 1$ $M_\odot$, even thoguh the star is fully convective, and therefore accreting deuterium as fast as it is being consumed. Actually, consumption slightly exceeds accretion, and L_D differs correspondingly from the value in equation (7). The evolution of f_D in a fully convective protostar is given by

$$M_* \frac{df_D}{dM_*} = 1 - f_D - \frac{L_D}{\dot{M}\delta} \, , \tag{9}$$

from which it follows that

$$L_D = \dot{M}\delta \left[1 - \frac{d(f_D M_*)}{dM_*} \right] \, . \tag{10}$$

Since the product $f_D M_*$ is proportional to the total deuterium content, equation (10) shows that any net loss of deuterium implies that L_D exceeds $\dot{M}_{\rm acc}\delta$. Such loss is inevitable, because of the slowly rising interior temperature of the protostar. The fall of f_D shown in Fig. 3 continues until the radiative barries appears, and thereafer deuterium exists only in a small, subsurface mantle. For fast accretion rates, the drop of deuterium in the protostellar phase is postponed at masses $M_* \sim 1.5$ $M_\odot$.

2.2. STELLAR STRUCTURE EQUATIONS

Unlike the previous hydrodynamical protostellar phase, the PMS evolution of a star can be followed by models in hydrostatic and thermal equilibrium. These models are much easier to construct and less subject to numerical inaccuracies. The subject of stellar structure equations is well established and fully described in numerous textbooks. Here, I will present a brief description of the main equations needed for the calculation of PMS stellar models.

The *mechanical* stellar structure equations are simply a restatement of hydrostatic balance. If we neglect internal rotation, then a convenient spatial variable in a spherical star is the mass coordinate M_r, defined by

$$M_r \equiv \int_0^t 4\pi r^2 \rho dr \ , \tag{11}$$

where ρ is the mass density of the gas. The function M_r represents the mass enclosed within a sphere of radius r. From equation (11) we have immediately $\frac{dM_r}{dr} = 4\pi r^2 \rho$. The radius r then acts as a dependent variable, whose variation with M_r is governed by the inversion of this expression

$$\frac{dr}{dM_r} = \frac{1}{4\pi r^2 \rho} \ . \tag{12}$$

The internal pressure is related to the total weight by

$$\frac{dP}{dr} = -\frac{G\rho M_r}{r^2} \ . \tag{13}$$

Dividing (12) by (13) we derive the alternate form

$$\frac{dP}{dM_r} = -\frac{G M_r}{4\pi r^4} \ , \tag{14}$$

which again utilizes M_r as the independent variable. The pressure itself obeys the equation of state for an ideal gas:

$$P = \frac{\rho}{\mu} \mathcal{R} T \ . \tag{15}$$

Here the mean molecular weight μ depends on the state of ionization and dissociation of the gas, and is therefore a function of ρ and T. One may either calculate this function *ab initio*, by invoking statistical equilibrium, or obtain it from prior tabulations.

We turn next to the *thermal* stellar structure equations. Since the protostar's interior is highly opaque, the transport of radiation is governed by the diffusion equation

$$T^3 \frac{dT}{dM_r} = -\frac{3\kappa L_{\text{int}}}{256\pi^2 \sigma_B r^4} \ . \tag{16}$$

The Rosseland mean opacity κ is again a function of ρ and T. Like μ, it is available in numerical form.

Finally, we consider the spatial variation of L_{int}. Recall that the latter is the surface integral over a spherical shell of $|F_{\mathrm{rad}}|$, where the flux points radially outward. In general, a fluid element gains heat either by being irradiated externally or from internal nuclear reactions. Let $\epsilon(\rho, T)$ represent the rate of nuclear energy release per unit mass. Both this rate and F_{rad} enter as source terms in the *heat equation*:

$$\rho T \frac{ds}{dt} = \rho \epsilon - \nabla \cdot F_{\mathrm{rad}} \, , \tag{17}$$

where s is the entropy per unit mass of the fluid. For a spherical star, it is convenient to recast this relation as

$$\frac{dL_{\mathrm{int}}}{dM_r} = \epsilon - T \frac{ds}{dt} \, . \tag{18}$$

Equations (12), (14), (16), and (18) are the desired stellar structure equations for the dependent variables r, P, T, and L_{int}. They must be supplemented by the equation of state (15) and by knowledge of μ, κ, ϵ, and s as functions of ρ and T. In practice, tabulations of the entropy utilize the Second Law of thermodynamics:

$$T \Delta s = c_v \Delta T - P \Delta \rho / \rho^2 \, , \tag{19}$$

which governs small changes of the state variables. Here, c_v is the specific heat at constant volume. Knowing this latter quantity (or, equivalently, the specific internal energy) as a function of ρ and T allows one to integrate equation (19) in the $\rho - T$ plane and numerically obtain the function $s(\rho, T)$. Note that a purely monatomic gas would have $c_v = (3/2)\mathcal{R}/\mu$, where μ is now a constant. In this case, we may integrate equation (19) analytically to obtain

$$s = \frac{\mathcal{R}}{\mu} \ln\left(\frac{T^{3/2}}{\rho}\right) + s_\circ \, , \tag{20}$$

where $s_\circ$ is an arbitrary constant. Equation (20) is often a useful approximation in stellar interiors, where the gas is fully ionized and $\mu = 0.61$.

Solution of the stellar structure equations requires specification of four boundary conditions. Two of these are the statements that both $r(M_r)$ and $L_{\mathrm{int}}(M_r)$ vanish at the center of the configuration:

$$M = 0: \quad r(0) = 0 \quad L_{\mathrm{int}}(0) = 0 \, . \tag{21}$$

A third condition is that $P(M_r)$ must equal the appropriate photospheric value when $M_r = M_*$:

$$M = M_* : \quad P_e = \frac{2}{3}\frac{GM_*}{R_*^2 \kappa} . \tag{22}$$

The fourth boundary condition concerns the surface value of the temperature and its relation to the luminosity. For a PMS star, this relation is the standard photospheric one given by equation (4):

$$M = M_* : \quad L = 4\pi R_*^2 \sigma T_{\text{eff}}^4 . \tag{23}$$

The full four stellar structure equations can now be solved as a two-point boundary value problem, in which the values of the pressure and entropy are guessed at the center, while the stellar radius and luminosity are guessed at the surface. One can integrate the equations from both directions to an interior fitting point, where the four guesses must be changed until all variables match.

2.3. INPUT PHYSICS

The four basic ingredientes required by the stellar models are the equation of state, opacity, treatment of convection and the nuclear generation rate. Each of these physical processes would require a lengthy discussion, which is beyond the scope of this chapter. I will confine myself to a brief summary of some basic aspects and refer the reader to general reviews for more information.

2.3.1. *Equation of state*
The importance of using a good equation of state (EOS) for stellar models has been discussed by Däppen (1993) and by D'Antona (1993) for the case of PMS stars. The task is to develop a formalism for the computation of the EOS valid from the conditions of the photosphere, where the temperature is typically a few 10^3 K and the density some 10^{-7} g cm^{-3}, to the center of a star where $T \sim 10^7$ K and $\rho \sim 10^2$ g cm^{-3}. In the first approximation, PMS stars, because of contraction and of the relatively low density, should behave according to the ideal gas law. However, the Saha formula is just an approximation that no longer holds as soon as a small degree of degeneracy or other non-ideal gas effects are present. Mazzitelli (1989) has discussed in detail the implications of including non-ideal gas effects on the hydrogen and helium ionization. An important result is that because of the decrease of the value of the adiabatic gradient in regions of partial ionization, stellar models are charactized by *lower* effective temperatures The most recent efforts to derive reliable EOS for astrophysical use include

Mihalas, Hummer & Däppen (1988), as part of the international "Opacity Project" (OP), and Iglesias & Rogers (1991), as part of the opacity project at Livermore (OPAL). A clear explanation of the physical approach behind the two computations and comparisons between the results are presented by Däppen (1993).

2.3.2. *Opacity*

There has been a dramatic improvement in the calculation of the Rosseland mean opacity of gas mixtures over the last ten years. Prior to 1990, the only existing set of opacities were provided by the Los Alamos group, both at high ($T>10^4$ K) and low-temperatures ($T<10^4$ K) (Huebner et al. 1977; Cox & Tabor 1976). In the latter regime, an important contribution has been done by Alexander and collaborators who considered the presence of molecules and grains in the computations (Alexander et al. 1989; Alexander & Ferguson 1994). The physical processes considered in the calculations include continuum sources of opacity due to 14 elements (H, H^-, H_2^+, H_2^-, He, plus heavy elements); atomic and molecular lines (about 60 million transitions); electronic and Rayleigh scattering due to H, H_2 and He; absorption and scattering of grains with radius between 0.005 and 0.25 μm.

At temperatures above 8000 K, both the OPAL and OP projects provide extended tabulations of the opacity, including free-free, bound-bound and bound-free transitions for many elements from hydrogen to iron. The motivation for these calculations is interesting. In 1982, Simon made a plea for the re-examination of the heavy element contribution to opacities. He showed that an increase in the contributions from elements heavier than helium by factors of two or three in envelope regions, would resolve a number of outstanding problems in the theory of pulsating stars. Two groups, OPAL and OP, responded to the plea and performed calculations that showed that indeed a better treatment of atomic physics led to significant opacity enhancements that helped to solve the problem of Cepheid pulsation (Iglesias & Rogers 1991; see also Seaton 1993 for a review). The opacity tables are constantly updated, as improvements in atomic physics are obtained (Rogers & Iglesias 1992; Iglesias & Rogers 1996; Seaton et al. 1994). As an illustration of the properties of the Rosseland mean opacity at selected densities, Figure 4 shows the behavior of the opacity at low- and high-temperatures for different sets of tables. In the right panel, the peak at $T{\sim}10^5$ K, called the *Z-bump*, visualizes the impact of the OPAL calculations on the behavior of the opacity in the inner regions. This feature stems from bound-bound transitions involving the inner electrons of heavier metals, principally iron. The main peak at $T{\sim}10^4$ K is due to the photoionization of hydrogen. At lower temperatures, the opacity falls

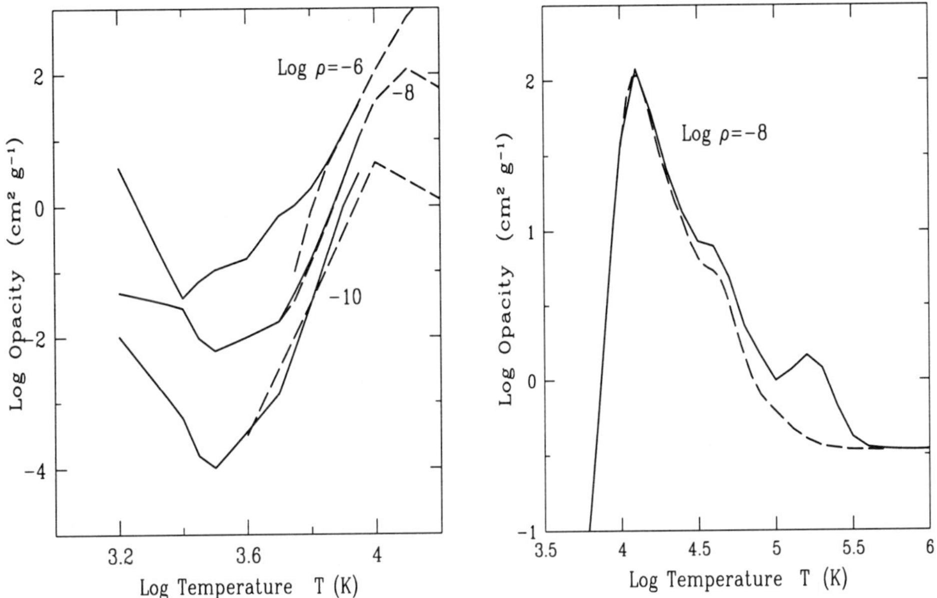

Figure 4. Rosseland mean opacities for a gas of solar composition. *Left panel*: comparison of the low-temperature opacities for different densities. The solid lines are from Alexander et al. (1994) and the dashed lines from Rogers & Iglesias (1992). Note that the latter do *not* include the contributions from molecules. *Right panel*: high-temperature opacities for a density of 10^{-8} g cm^{-3}. Solid line is from Rogers & Iglesias (1992) and the dashed line is from Cox & Tabor (1976). Note the enhancement of the opacity at T$\sim$10^5 K (*Z-bump*).

steeply because too few photons in the Planckian distribution have energies exceeding the ionization threshold of 13.6 eV. The decline in opacity with falling temperatures halts temporarily at near 2000 K, where the dominant process is now bound-bound transitions in molecules (left panel).

Molecules play a dominant role in the atmospheres of cool dwarfs, from mid K to late M stars and a mass from 0.6 M$_\odot$, to the hydrogen burning limit (0.075-0.085 M$_\odot$, depending on metallicity), and in the brown dwarf regime. In such stars, most of the hydrogen is locked in H$_2$ and most of the carbon in CO, with excess oxygen bound in molecules such as TiO, VO, and H$_2$O (e.g. Allard et al. 1997). These molecules can serve as useful diagnostics of the nature of the very low-mass stars. Opacities due to grains are also becoming available (e.g. Alexander et al. 1998). Grains can condensate in stellar atmospheres at temperatures below $\sim$2000 K, and their effect is to decrease the gas phase abundance of some important absorbers, such as TiO. The impact of grain opacity on the spectral distribution and atmosphere of cool stars is quite important, but a full understanding of the physics of grain condensation and absorption is still lacking.

2.3.3. *Treatment of convection*

The standard treatment of turbulent convection is a phenomenological one, based on Prandtl's theory of turbulence with the introduction of a quantity called the *mixing length* (Böhm-Vitense 1958). Despite its limitations, the mixing length theory (MLT) gives a satisfactory description of the convective zone since this region is to a large extent very close to being adiabatic. In the MLT theory, the length l over which a convective bubble merges with its surroundings is assumed to be proportional to the pressure scale height H_P, $l = \alpha H_P$, where α is a free parameter and $H_P = -dr/dlnP$. The actual value of α is chosen as to fit the solar radius and effective temperature. In general, α=1.5 yields a good fit. However, the surface properties of stellar models are quite sensitive to the value of α adopted in the computations. Lowering (rising) α to 1 (2) instead of 1.5 reduces (increases) the values of the effective temperature by about 10% (i.e. several hundred degrees), especially along the Hayashi vertical track (e.g. Forestini 1994). Thus, the poor knowledge of stellar convection introduces considerbale uncertainties in the location of the tracks in the H-R diagram. Also, the temperature at the base of the convection zone is very sensitive to the value of α. Therefore, important differences in the level of light elements depletion in the convective envelopes result from different choices of the mixing length parameter.

An alternative theory to the MLT has been presented by Canuto & Mazzitelli (1991, CM) and Canuto, Goldman & Mazzitelli (1996). Unlike MLT which considers a single element of turbulence ("one-eddy" description), the CM convection model takes into account the full spectrum of the eddy distribution. Also, the mixing length is chosen to be the harmonic mean between the distance z of the layer from the top and the distance from the bottom of the convective region. The properties of the Sun can then be fitted to high precision (less than 1%, as demanded by heliosismological constraints) by introducing a small amount of overshooting from the surface convective layer and the prescription for the scale lenght becomes $l = z + \beta H_P$. A full description of the CM convection model, together with a discussion on the reasons to adopt such a model instead of the MLT, can be found in the review by D'Antona & Mazzitelli (1998).

2.3.4. *Nuclear reactions*

The nucleosynthesis during the PMS phase is limited to the light elements. In addition to the destruction of deuterium, the nuclear burning of ^{6}Li, ^{7}Li, ^{9}Be, and ^{3}He takes place when the central temperature in solar-type stars reaches values of 2×10^6 K, 3×10^6 K, 3.5×10^6 K, and 8×10^6 K. However, the relatively tiny abundances of these species render the effect energetically and structurally unimportant. For example, the interstellar lithium abundance is [Li/H]$\sim 10^{-9}$: in order to halt PMS contraction, the abundance should be enhanced by a factor 10^5 or more over the observed ones. On the other hand, the depletion history of lithium in low-mass stars can be followed in detail to compare with observations and to test the internal structure of stars.

The nuclear network consists of the complete pp chain and the CNO cycle. Deuterium and ^{7}Li burning are considered separately, since they occur before the pp chain takes over. The computation of the energy generation requires the knowledge of the nuclear cross sections. These are tabulated by Caughlan & Fowler (1988), while weak and intermediate screening factors can be computed using the formalism of Graboske et al. (1973). Screening is also important for D and ^{7}Li burning reactions, especially for very low-mass stars and brown dwarfs. The relative concentration of the nuclear species follows the solar composition (e.g. Anders & Grevesse 1989) with a metal abundance $Z=0.019$. The location of PMS tracks is not sensitive to the exact value of the metallicity; however, the amount of ^{7}Li depletion is strongly affected by Z, increasing for larger values of Z (D'Antona & Mazzitelli 1984).

3. A new H-R diagram

The results of the calculations of PMS evolution incorporating the protostellar initial conditions and the input physics discussed in Section 3 are shown in the H-R diagram of Figure 5 for stars from 0.1 to 6.0 M$_\odot$ (Palla & Stahler 1999). The evolutionary tracks have been computed using the opacities of Alexander et al. (1994) at T<8000 K and of Rogers & Iglesias (1992) at higher temperatures. Convection has been treated using MLT with $\alpha=1.5$. From Fig. 5, we can see that while low-mass stars descend from the birthline on convective paths, stars more massive than $\sim$4 M$_\odot$ move to the left on radiative ones. Stars between 2 and 4 M$_\odot$ undergo thermal relaxation, looping back slightly behind the birthline and recrossing it before pursuing horizontal tracks. Compared to the classical set of PMS tracks, those shown in Fig. 5 occupy a much more limited portion of the diagram. The reason is that the older tracks correspond to stellar radii too large to have attained during protostellar accretion. Thus, the starting luminosities are much lower than previously envisioned. In addition, the surface temperature of these stars begin *higher*. For example, a 5 M$_\odot$ star starts with T$_{\rm eff}$=12000 K in contrast to the value of 4400 K obtained by Iben (1965). Notice also that the birthline intersects the main sequence at a mass M$_*$ $\sim$8 M$_\odot$: this point represents the critical stellar configuration in which hydrogen burning has stopped gravitational contraction while the star is still growing in mass (see Fig. 1). *More massive stars, therefore, have no PMS phase at all, but appear directly on the main sequence once they are optically revealed.*

An important result of the new calculations is that observed PMS stars should not be located to the right of the birthline. For both T Tauri and Herbig Ae/Be stars the agreement is excellent, despite the rather strong assumption of a constant mass accretion rate of 10^{-5} M$_\odot$ yr^{-1} used in the models (see Fig. 2). This fact indicates that a limited dispersion in the properties of molecular cloud cores (which determine the magnitude of $\dot{M}_{acc}$) somehow produces an impressive range of stellar masses. The case for even more massive stars is complicated by the difficulty in determining the appropriate stellar parameters to place them in the H-R diagram

Finally, another important result is the drastic reduction in age obtained for Herbig Ae/Be stars. The same is true, of course, for T Tauri stars, but the quantitative impact is minor. Even a large drop in the initial radius of a typical T Tauri star hardly changes its total contraction age, since t_{KH} scales as R_*^{-3}. For the Herbig stars, however, equation (8) implies that t_{KH} only increases as $R_*^{-1/2}$. The age reduction is far more significant for masses well below the upper cutoff at 8 M$_\odot$. Such a reduction is consistent with the observations of surface activity in these stars.

3.1. LOW-MASS STARS: $0.1 \lesssim M_*/M_\odot \lesssim 1$

The largest departure from the standard theory of PMS stars concerns the history of deuterium burning. Salpeter (1954) was the first to recognize that deuterium ignition via the reaction ^{2}H(p,γ)^{3}He at a temperature of about 1×10^6 K could slow down contraction or, for the least massive stars, stops it entirely for a relatively brief time. This effect was confirmed by the detailed calculations of Grossman & Graboske (1971) and Mazzitelli & Moretti (1980) who introduced the concept of a "deuterium main sequence" to denote the curve in the H-R diagram crossed by the PMS tracks at the point of ignition. However, Stahler (1988a) has shown that this concept is obsolete, since deuterium burning occurs already during the accretion phase for stars with masses above a few tenths of a solar mass (see Fig. 3). Thus, the locus of the deuterium main-sequence is nothing else than the birthline shown in Fig. 5. T Tauri stars, having consummed most of the available deuterium during the accretion phase, begin the contraction phase by simply descending along the vertical portion of the Hayashi track. The same is true for a Herbig Ae/Be star, which begins with a severely depleted deuterium supply. While this supply lasts, the outermost layers remain convectively unstable. But the luminosity provided by deuterium burning

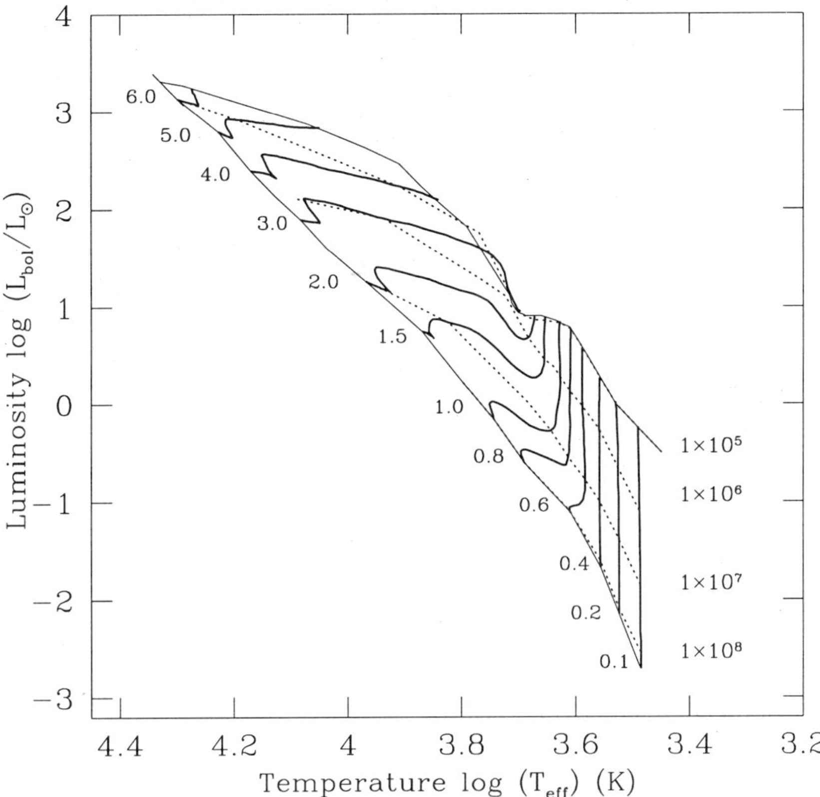

Figure 5. Evolutionary tracks in the H-R diagram for low- and intermediate-mass stars. Each track is labeled by the corresponding mass, in solar units. Selected isochrones are shown by the dotted lines. For each track, the evolution starts at the birthline (dotted curve), and ends at the ZAMS, also shown (from Palla & Stahler 1998).

L_D is now so minor compared to L_{rad} that even this limited convection vanishes by the time the star has joined the classical radiative portion of its evolutionary track. Finally, for stars of mass smaller than $\sim$0.3 M$_{\odot}$, protostar theory still implies deuterium burning during the PMS phase, while for masses as low as M$_*$=0.018 M$_{\odot}$ degeneracy effects set in and prevent nuclear burning (D'Antona & Mazzitelli 1994).

3.2. INTERMEDIATE-MASS STARS: $1 \lesssim$ M$_*$/M$_{\odot} \lesssim 8$

For stars more massive than solar, the imprint of the prior accretion history persists much longer than in the low-mass, convective case. The reason is that the stellar interior is not thermally relaxed, and the stars must undergo a phase of global readjustment. The main results can be schematically summarized as follows.

- *Fully convective stars*: Stars in the range $1 - 2.5\ M_\odot$ are fully convective due to surface cooling as soon as they appear as optically visible objects. The evolution is undisturbed by the fusion of residual deuterium: its concentration is so low, that it is quickly consumed (on a time scale of few years). Stars contract along the Hayashi track, but their path is much reduced (see the case for $M_* = 2\ M_\odot$ in Fig. 5).

- *Partially convective stars*: Stars in the range $2.5 - 4\ M_\odot$ undergo thermal relaxation and nonhomologous contraction. When they are first optically visible, they are *underluminous*: i.e., the evolutionary track begins at low luminosity and then moves up to join the radiative portion. The Hayashi phase is skipped entirely. In the H-R diagram, shown in Figure 6 for the case of a $3.5\ M_\odot$ star, the unrelaxed star first appears below the classical evolutionary track for that mass. At this early epoch, the star has a relatively shallow convection zone. During subsequent relaxation, the stellar radius, surface luminosity and temperature all increase, while the outer convection disappears. The variation of the distribution of the internal luminosity at different times is also shown in Fig. 6. Initially, the function $L_{\rm r}(M_{\rm r})$, the luminosity crossing any interior shell enclosing a mass $M_{\rm r}$, rises from zero at the center, peaks somewhere in the interior, then falls again to the relatively low surface values $L_{\rm surf}$. Such odd luminosity profile cannot be maintained indefinetely. According to the heat equation (18), matter between the star's center and the shell where $L_{\rm r}$ peaks is losing heat $(ds/dt < 0)$, while the gas between this shell and the surface is gaining heat. As shown in Fig. 6, this internal heat transfer, achieved entirely through radiative diffusion, causes the peak of $L_{\rm r}$ to move outward with an attendant rise in the surface luminosity. The approximate duration of this *thermal relaxation* is given by the Kelvin-Helmoltz time with $L_{\rm surf}$ replaced by the internal luminosity $L_{\rm rad}$ in equation (5).

The luminosity in stars of mass between $2.5 - 4\ M_\odot$ stems almost entirely from gravitational contraction. This contraction, however, is initially quite different from that envisioned in the classical PMS theory. That interior portion which is losing heat indeed shrinks, but the transfer of this energy actually expands the outer region and creates a net increase of the stellar radius. This *nonhomologous contraction* continues until $L_{\rm r}(M_{\rm r})$ is everywhere increasing, by which point the star as a whole is shrinking, as in the classical model.

- *Fully radiative stars*: Stars more massive than $4\ M_\odot$ appear immediately on the radiative portion of the classical track, and begin to contract homologously under their own gravity. As the star contracts, the average interior luminosity must rise as $R_*^{-1/2}$, according to equation (4). Such rise is evident in the evolutionary tracks of Fig. 5. The specific entropy decreases

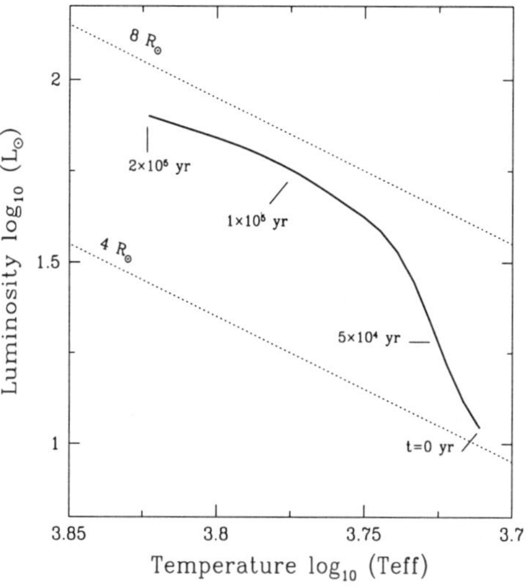
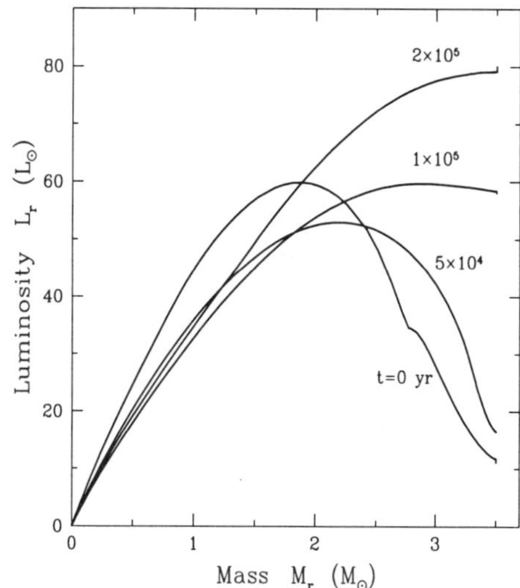

Figure 6. Left panel: Early evolution in the H-R diagram of a 3.5 M☉ star, covering the phase of thermal relaxation. The thin dashed lines are cruves of constant radius. Right panel: The evolution of the internal luminosity during the thermal relaxation process. Note the shift of the peak in luminosity toward the stellar surface as a result of nonhomologous contraction. (Adapted from Palla & Stahler 1993)

with time in all interior mass shells, a sign of homologous contraction. At any one time, the interior is radiative, i.e. the entropy is increasing outward, except for a very thin surface layer.

In comparison with standard Iben's calculations (1965), the new tracks shown in Fig. 5 occupy a much reduced portion of the H-R diagram. In addition, the surface temperatures are *higher*, since the outermost layers are fully ionized, and the dominant opacity source in this region is no longer the H^- ion. In conclusion, the combination of protostellar and PMS theory does seem to account for at least one basic property of T Tauri and Herbig Ae/Be stars: their distribution in the H-R diagram. As shown in Fig. 2, the upper envelope of the stellar distribution is well matched by the theoretical birthline obtained with the fiducial value of 10^{-5} $M_☉ yr^{-1}$.

3.3. MASSIVE STARS: $M_* \gtrsim 8$ M☉

The process of gravitational contraction during the accretion phase ends when the nuclear luminosity due to the CNO cylce becomes dominant. There is no precise way to specify the arrival of a protostar at the main sequence, since the models are never identical to ZAMS stars of the same mass. For example, at $M_*=8$ $M_\odot$, the nuclear luminosity still accounts for 75% of the total luminosity, with the remainder provided by gravitational settling in the outer layers. However, we know that the stellar radius must reach a minimum if accretion were to increase indefinetely. From Fig. 1, the value of the minimum R_* is 3.1 $R_\odot$. This radius is reached on the theoretical ZAMS at $M_* \approx 8$ $M_\odot$ (Iben 1965). Thus, we may take this mass as the value where the protostar has joined the main sequence. The consequence is obvious: there is no longer any PMS phase for stars more massive than this value. They can only be observed as infrared accreting protostars or as main-sequence or post-main sequence objects.

An implication of the evolution described so far is that most stars should form with nearly the same accretion rate, as supported by the distribution of known PMS star in the H-R diagram. One possible objection to this hypothesis is that stars forming at modest rates might complete their entire main-sequence evolution *before* reaching a sufficiently high mass. However, with an $\dot{M}_{\rm acc}$ of 10^{-5} $M_\odot$ yr^{-1}, the accretion and nuclear lifetimes become equal only at a stellar mass of 45 $M_\odot$, while with $\dot{M}_{\rm acc}=10^{-4}$ $M_\odot$ yr^{-1} the critical mass is higher than 100 $M_\odot$ (Schaller et al. 1992). These figures suggest that even high-mass stars, i.e. those with no visible PMS phase, form with accretion rates in the range considered above and do not require extreme conditions. This possibility has been considered by Beech & Mitalas (1995) and Bernasconi & Maeder (1996) who studied the formation and evolution of stars up to $\sim$100 $M_\odot$ under the accretion scenario. One of the interesting findings is the definition of an upper stellar birthline as an extension above $\sim$10 $M_\odot$ of that shown in Figs. 2 and 5. A massive star then begins its main-sequence evolution from these initial conditions that differ substantially from those at the classical ZAMS. In particular, massive stars begin their evolution with lower effective temperatures and with a reduced extent of the convective core.

However, accretion cannot proceed indefinetely because of the mechanical effect of the protostar's radiation on the infalling envelope. The accretion shock at the stellar and disk surfaces produces photons at optical and ultraviolet wavelengths. These are absorbed by the envelope's dust grains and reradiated into the infrared. Trapping of this infrared radiation creates thermal pressure that retards infall. The stellar luminosity also raises the envelope's temperature until the grains sublimate at the dust destruction front. The direct impact of short-wavelength photons on grains at this front provides an additional retarding force. Radiation pressure begins to become significant in protostars of intermediate mass. Infall cannot be stopped, however, as long as the accretion component dominates the total luminosity. Any retardation of the flow diminishes $\dot{M}_{\rm acc}$, which in turn lowers L_*. Once the luminosity falls, the flow picks up again. Numerical simulations of collapsing clouds with central stars display this effect. For stars in the proper mass range, $\dot{M}_{\mathrm{acc}}$ oscillates strongly without actually reversing (Yorke and Krügel 1977). The feedback between radiation pressure and infall is cut once L_{int} dominates L_*. Again, the transition occurs just where t_{KH} falls below t_{acc}, *i.e.*, for protostars that have joined the main sequence. The internal luminosity is now supplied by hydrogen fusion rather than contraction, but still scales as $M_*^{11/2} R_*^{-1/2}$. Hence the retarding force on the envelope increases rapidly with mass. Within the context of spherical collapse, the result is inevitable and the condition of balance between the forces yields

$$\frac{L_*}{M_*} = \frac{4\pi c G}{\kappa} = 1 \times 10^3 \frac{L_\odot}{M_\odot} \, , \tag{24}$$

where it has been used a value of $\kappa=10$ cm^2 g^{-1} for the opacity. Plausible variations in grain composition could alter this value by as much as a factor of 3 in either direction (Draine & Lee 1984; Pollack et al. 1994). Thus, the critical ratio occurs at a mass near 10 $M_\odot$, implying that accretion is significantly affected for central objects of greater mass. Wolfire and Cassinelli (1987) analyzed the dynamics of envelopes falling onto stars with M_* exceeding 60 $M_\odot$. They found that infall was reversed unless $\dot{M}_{\mathrm{acc}}$ were arbitrarily increased to *at least* 10^{-3} $M_\odot$ yr^{-1} *and* grains were depleted in abundance by an order of magnitude. Since these extreme conditions are unlikely to occur, we are faced with a basic dilemma. (For earlier discussion of this issue, see Larson and Starrfield 1971 and Kahn 1974.)

There is little doubt that a massive star produces more than enough energy to expel its envelope. In fact, the inequality $t_{\mathrm{KH}} < t_{\mathrm{acc}}$ implies that the radiated luminosity can easily unbind the parent cloud before the star has gained appreciable mass. The real issue is the *efficiency* of energy transfer. Here, any departure of the infall from spherical symmetry is likely to play a major role (Nakano 1989). Unfortunately, a consistent theory which includes rotation and the simulataneous presence of infall and outflow is still unavailable, so that the formation of massive stars via accretion remains a mistery. Perhaps, one should invoke a completely different scenario (e.g. collision/coalascence models in dense clusters; Bonnell, Bate & Zinnecker 1998; Stahler, Palla & Ho 1999).

4. Tests to PMS evolutionary diagrams

The location of young stars in the H-R diagram is commonly used to estimate their masses and approximate ages. When young clusters or associations are studied, the information on the mass of individual members can yield direct information on the rate and efficiency of star formation and, eventually, on the initial mass function to compare to that observed locally.

Also, the amount of depletion of the light elements due to nuclear activity in these early stages sheds light on the internal structure and on the mechanisms of angular momentum transport. It is important at this point to discuss the uncertainties that affect the computation of modern evolutionary tracks, such as those described earlier. The sources of uncertainty can be due to the uncertain knowledge of certain aspects of the input physics used in the theoretical models, or to the complex phenomena that accompany the evolution of young stars, such as the interaction with circumstellar disks and the accretion of residual matter, or the presence of companions and so on. In the following, I will first overview the uncertainties *intrinsic* to the models, and then discuss the effects of the environment.

4.1. COMPARISON OF TRACKS

Several recent studies have reconsidered the PMS evolution of stars with population I compositions over a wide mass range (e.g. D'Antona & Mazzitelli 1994, 1997; Forestini 1994; Swenson et al. 1996; Baraffe et al. 1997; Palla & Stahler 1999). The main motivation for these studies is the progress in the fields of the equation of state, radiative opacities and the treatment of convection, each of which greatly influences the surface properties, i.e. the observables, of PMS stars. Before summarizing the main properties of the new models, let first stress the following point: the average error in the determination of the observable quantities, L_{bol} and T_{eff}, is of the same order of magnitude than the theoretical ones. Typically, $\Delta T_{eff} = \pm 300$ K and $\Delta \log (L_*/L_\odot) = \pm 0.25$ dex, for stars observed in the nearby star forming regions (e.g. Strom & Strom 1994). On the other hand, the mean spacing between Hayashi tracks is of the order of ~ 130 K for $M_* = 0.5$ $M_\odot$, and of only ~ 60 K for a 1 $M_\odot$ star (Forestini 1994). Thus, for very young stars, still on the vertical portion of the tracks, the mass assignment based on their position can lead to errors of a factor of 2.

The largest sources of uncertainty in the position of the tracks in the H-R diagram are due to the poor description of turbulent convection and the limited knowledge of low-temperature opacities. In addition, to construct an H-R diagram, one should derive T_{eff} and L_{bol} from observable quantities. The derivation of T_{eff} relies on temperatures scales which are fairly well known for dwarfs between A and K types. For M dwarfs ($T_{eff} \lesssim 4000$ K), differences between temperature scales are much larger (Allard et al. 1997). The problem is that there are only two eclipsing binary systems, YY Gem (M0.5, 3770K) and CM Dra (M4.5, 3150 K), with accurately measured temperatures. Thus, it is difficult to plot a reliable H-R diagram for the low- and very low-mass stars and obtain good mass and age estimates. This warning is particularly important when attempting to derive the initial mass function in nearby star forming regions. An example is provided by the recent study of Luhman & Rieke (1998) of the the low-mass end of the IMF in L1495E in Taurus. They find *very* different results depending on the choice of the set of tracks: an object with a mass of 0.25 $M_\odot$ using D'Antona & Mazzitelli (1994) tracks corresponds to a 0.5 M_* using Baraffe et al. (1997) tracks. Thus, the shape of the IMF, as well as the star formation history, cannot be univoquely determined. The lesson is clear: it is still premature to draw firm conclusions from observations of the low mass end of the stellar distribution with the current uncertainties in both evolutionary tracks and in the conversion from observational to theoretical quantities (see also the discussion in D'Antona & Mazzitelli 1998).

4.2. LITHIUM DEPLETION

The depletion of lithium in both young and evolved stars is considered one of the strongest constraints imposed on stellar evolution theory. The surface. abundances in fact provide a critical test for the physical processes (convection, opacity, rotation, mass loss) which ought to be included in the theoretical models in order to reproduce the observed level of depletion. Due to its small binding energy, ^{7}Li (the most abundant of the two stable isotopes of lithium) is easily destroyed in stellar interiors via the reaction ^{7}Li$(p,\alpha)^4$He at temperatures around 2.5×10^6 K. According to standard models, depletion occurs only when the outer convection extends down to that temperature. Thus, it appears quite unavoidable that some ^{7}Li depletion due to nuclear burning takes place already during PMS contraction. Other processes, associated with the mixing due to transport of angular momentum through the star, are effective on much longer time scales and dominate the depletion history during the main sequence phase (e.g. Forestini 1994; Martin & Claret 1996; Venturi et al. 1998).

The theoretical predictions are rather well established, independently of the adopted models. Stars more massive than 1.7 $M_\odot$ should *not* show any ^{7}Li depletion at all; solar-type stars are significantly depleted, while for sub-solar stars there should be no ^{7}Li left over (e.g. Forestini 1994). For stars with $M_*\gtrsim0.5\ M_\odot$, most lithium burns after convection has halted in the stellar core. Predictions of lithium depletion then depend on the temperature at the bottom of the convection zone, whose exact location is sensitive to opacity, convection, and rotation. Stars with $M_*\lesssim0.5\ M_\odot$ are fully convective during lithium burning. Since the rate of contraction of such stars is controlled by the effective temperature, the amount of lithium depletion is also sensitive to $T_{\rm eff}$ and hence to the surface opacities. Several groups have followed numerically the phase of lithium burning, and the models make specific predictions about the time evolution of the lithium depletion boundary. At ages 30, 70 and 140 Myr, the lithium depletion edge should occur at $0.17\,M_\odot$, $0.09\,M_\odot$ and $0.07\,M_\odot$ (Baraffe et al. 1997; Burrows et al. 1997; D'Antona & Mazzitelli 1997). To bypass the uncertainties of numerical models, Bildsten et al. (1997) have developed a clever analytic approach based on the fact that T_{eff} remains approximately constant during the fully convective phase (see the low-mass tracks of Fig. 5). Rather than calculating T_{eff}, Bildsten et al. calculate the dependence of lithium depletion on T_{eff}, obtaining the radius, age and luminosity of a PMS star as a function of lithium depletion. This method can be used to derive the age of open clusters with higher accuracy than any other method.

One of the most successful applications of the theory of Li-burning has been in the field of brown dwrafs. The limiting mass for ^{7}Li-burning is $M_* \geq 0.06\,M_\odot$, since degeneracy effects in lower mass stars prevent the onset of nuclear reactions (D'Antona & Mazzitelli 1994). This value is below the substellar mass limit, usually taken between 0.08 and $0.07\,M_\odot$. Thus, it has been suggested that the presence of lithium lines in the observed stellar spectra can be used as a powerful probe to search for brown dwarf candidates (Rebolo, Martin & Magazzù 1992). The Li-depletion history for different masses is shown in Figure 7. Using this diagnostic, the first brown dwarfs have been discovered in the Pleiades (Rebolo et al. 1996; Basri, Marcy & Graham 1996) and about a dozen candidates have been identified to date (Martin et al. 1998; Stauffer et al. 1998).

4.3. EFFECTS OF MASS ACCRETION

The evolution described so far, although improved by the account of more realistic initial conditions and input physics, is still rather conventional. It assumes that each star evolves in isolation at constant mass from the birthline to the arrival on the main sequence. But we know that in reality the situation is more complex, owing to the interaction of the star with the environment. An important effect to consider is the occurrence of residual mass accretion from a circumstellar disk during the PMS phase. Accretion alters the total (star+disk) luminosity, the evolutionary path, and hence the age estimate of a star. Hartmann, Cassen & Kenyon (1997) have presented a detailed analysis of the effects of disk accretion on the initial evolution of fully convective low-mass PMS stars.

Current estimates of mass accretion rates derive from the analysis of the excess optical continuum radiation in T Tauri and Herbig stars that also show an infrared excess. The latter is interpreted as being due to the presence of an active circumstellar disk. In this case, the disk luminosity is the sum of two contributions,

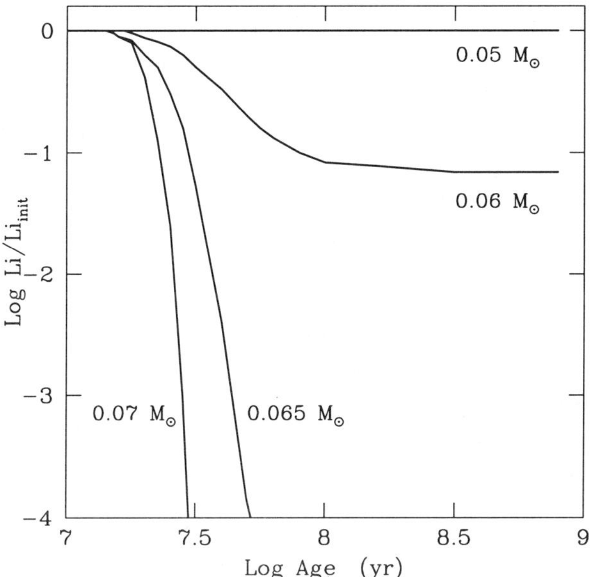

Figure 7. ^{7}Li-depletion as function of time for stars in the mass range 0.07 to 0.05 M$_\odot$. (From Rebolo et al. 1995)

$$L_{\mathrm{disk}} = \frac{GM_*\dot{M}_{\mathrm{acc}}}{2R_*} + L_{\mathrm{rep}}, \tag{25}$$

where the first term represents the accretion luminosity, while the second term is due to the absorption and reradiation of light from the central star. Typically, $L_{\mathrm{rep}} \sim 0.25$-$0.5$ L$_*$. Thus, unless $L_{\mathrm{disk}} \geq L_*$, the identification of the excess luminosity as *accretion* is not secure. On the other hand, observations indicate that at most $L_{\mathrm{disk}} \sim 0.5$ L$_*$.

If the optical excess is assumed to arise in a boundary layer (BL) between the star and a disk, then the standard theory of viscous accretion disks predicts that up to half of the accretion luminosity is emitted as radiation,

$$L_{\mathrm{BL}} \leq \frac{GM_*\dot{M}_{\mathrm{acc}}}{2R_*}. \tag{26}$$

Thus, measurements of L_{BL} can provide a direct estimate of the amount of mass actually falling onto a star. Measurements of the veiling in a large sample of T Tauri stars yield values in the range $\dot{M}_{\mathrm{acc}} = 10^{-8}$-$10^{-7}$ M$_\odot$ yr^{-1} (Hartmann & Kenyon 1990; Bertout & Regev 1992). Higher values of $\dot{M}_{\mathrm{acc}}$, up to 100 times, have been derived for the more massive Herbig

Ae/Be stars (e.g. Hillenbrand et al. 1992). Note, however, that in the absence of angular momentum loss, accretion at these rates would spin up the central star to half the break-up velocity in a few million years. Since young stars are known to rotate at typically 1/10–1/5 of the break-up speed (e.g. Bouvier, Forestini & Allain 1998), it is likely that strong winds carry away the angular momentum excess. Therefore, it is still not clear what fraction of the accreted mass is actually incorporated into the star.

The question, then, is whether the inferred rates are high enough to modify the evolutionary tracks. Hartmann & Kenyon (1990) argue that indeed the path is greatly affected, and estimate from qualititative arguments that about 10% of the final mass of a T Tauri is accreted during the PMS phase. Simple energy considerations allow to derive an estimate of the effect of mass accretion on fully convective stars (e.g. Palla 1994; Hartmann, Cassen & Kenyon 1997). The evolution of the stellar luminosity is described by the equation

$$\frac{d}{dt}\left(-\frac{3GM_*^2}{7R_*}\right) = -L_* - \frac{GM_*\dot{M}_{\mathrm{acc}}}{R_*} + \dot{M}_{\mathrm{acc}}\delta \,, \qquad (27)$$

where the term on the lhs represents the temporal change of the gravitational potential energy; the second and third terms on the rhs express the gain of (negative) energy by mass accretion and the released luminosity of freshly accreted deuterium. This equation can be recast in the form of a nondimensional differential equation, whose solution is straightforward. To test the effect of mass accretion, the evolution of stars with initial mass of 0.6 and 1.0 $M_\odot$ has been followed allowing for steady mass accretion at different rates for the first 3 million years (Parigi 1992). The result is that, as long as the value of $\dot{M}_{\mathrm{acc}}$ remains below $\sim 10^{-7}$ $M_\odot$ yr^{-1}, the new tracks do not differ substantially from those computed at constant mass. Moreover, if the *global* effect of mass accretion is limited to about 10% of the final mass, it would take only 1 million years to a 1 $M_\odot$ star to accrete such an amount of mass at $\dot{M}_{\mathrm{acc}}=10^{-7}$ $M_\odot$ yr^{-1}. Small departures can be appreciated only after at least 2 million years for $\dot{M}_{\mathrm{acc}}=10^{-7}$ $M_\odot$ yr^{-1}, implying that, unless accretion continues for a substantial part of the PMS evolution at high rates, the uncertainty in the location of a star in the H-R diagram due to this effect is small. Following a different approach, Siess & Forestini (1996) and Siess, Forestini & Bertout (1997) have computed the evolution of stars that accrete a substantial part of their final mass during the PMS phase. Clearly, in such a case the evolutionary tracks are greatly affected: compared to a non-accreting star, the evolution is accelerated, the central temperature remains larger, and the star appears systematically younger (with age estimates affected by up to factors 2 to 3). Although the physical effects are well understood, it remains to be seen if the assumption that most of the mass is accreted during PMS evolution is confirmed by observations.

4.4. EFFECTS OF BINARIES

The study of PMS binaries offers the best way to test the theoretical evolutionary tracks. The measurement of the dynamical mass of the components of any binary system would yield in fact an absolute mass calibration of the tracks, and thus provide a direct answer for the exact location of the tracks in the H-R diagram. The progress in the field of PMS binaries has been tremendous (e.g. Mathieu 1994). The population of binary systems in the nearest star forming regions has now been studied and a binary frequency of at least 50% has been derived for systems with angular separations in the range 3 to 1400 AU. It appears that there is even an excess of such systems in comparison to what found among nearby solar-type stars, possibly implying that binary formation is the main mode of star formation (Simon et al. 1995).

Unfortunately, there are very few cases where it has been possible to derive the dynamical mass of the components of an eclipsing binary system. One such case is TY CrA, a binary with a period of 2.89 days in the Corona Australis complex (Beust et al. 1997; Casey et al. 1998). Accurate spectroscopic measurements allow to derive the masses of the individual components, $M_1=3.16$ $M_\odot$ and $M_2=1.64$ $M_\odot$. The position of TY CrA in the H-R diagram is shown in Figure 8. The primary appears slightly below the ZAMS, near a mass of 3.0 $M_\odot$. This value compares well with the more precise spectroscopic determination of 3.16 $M_\odot$. The secondary indeed falls just above the 1.5 $M_\odot$ evolutionary track. Thus, the two methods (spectroscopic and photometric) for obtaining the component masses are in essential agreement. What about the stellar ages? Unfortunately, the location of the primary so close to the ZAMS makes it impossible to assign a reliable contraction age. However, the star illuminates a bright reflection nebula in the Corona Australis cloud, and the system's spectral energy distribution rises in the mid-infrared regime toward longer wavelengths. Thus, the primary is undeniably young, a typical Herbig Be star situated within both interstellar and circumstellar gas and dust. Suppose we tentatively assign it the pre-main-sequence age for a 3 $M_\odot$ star just arriving on the ZAMS, or 2×10^6 yr. Then, the corresponding isochrone (dashed curve in Fig. 8) passes close to the secondary's position. The components of TY CrA are therefore plausibly coeval, although the data do not permit a more quantitative assessment. Age assignments are more trustworthy when both stars lie well above the ZAMS. No eclipsing binaries of this kind are known, but there are a number of non-eclipsing, double-lined systems. In the absence

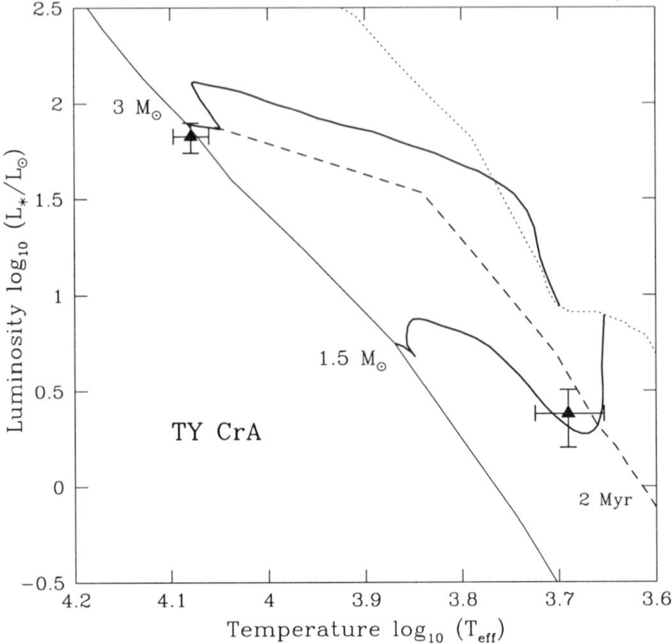

Figure 8.　Location of TY CrA in the H-R diagram. Stellar parameters are from Casey et al. (1998). Evolutionary tracks, isochrone and birthline are from Palla & Stahler (1999).

of eclipses, one must obtain $T_{\rm eff}$ and $L_{\rm bol}$ for each star by careful analysis of both the composite SED and the shifting, narrow-band spectrum.

Discerning the relative luminosities from an unresolved pair of stars is no simple matter, and the foregoing results are subject to some uncertainty. The situation further improves when we consider wider, spatially resolved binaries. In this case, however, we have no spectroscopic mass ratios for comparison. Studies conducted in several nearby star forming regions have shown that most of the pairs when placed on the H-R diagram apper to be coeval, within the observational errors, (Hartigan, Strom & Strom 1994; Brandner & Zinnecker 1997). This implies that binary systems consist of stars that were born nearly simultaneously. Thus, the typical binary must arise *in situ*, rather than through capture.

4.5.　THE INSTABILITY STRIP FOR PMS STARS

I will conclude this chapter with a discussion of a potentially important, but not yet fully explored aspect of PMS evolution: the pulsational properties of PMS stars. The fact that young stars during their evolution to the MS move across the instability region of post–MS stars raises the possibility that even a PMS could indeed pulsate (e.g. Baade & Stahl 1989). It is known that stars over essentially the whole mass spectrum can become pulsationally unstable during various stages of their evolution (Gautschy & Saio 1996). In particular, δ Scuti stars are intermediate-mass variables of spectral type A to F with pulsation periods less than $0\overset{d}{.}3$ and light amplitudes ranging from thousandths of magnitude to some tenths (Breger 1979). In spite of the traditional association of δ Scuti pulsation with stars in their core hydrogen-burning phase or evolving towards the base of the giant branch and burning hydrogen in a shell, some δ Scuti have been identified to be PMS objects. The best examples are two Herbig Ae stars, HR 5999 (Kurtz & Marang 1995) and HD 104237 (Donati et al. 1997). For HR 5999, the peak-to-peak pulsation amplitude is of about $0\overset{m}{.}013$ in Johnson V-band with a period of 4.99 hr, superimposed on a nonperiodic photometric variation of $0\overset{m}{.}35$. In HD 104237, the period of pulsation is approximately 40 minutes and the velocity amplitude about 1 km sec^{-1}.

Figure 9 shows the location of the boundaries of the instability strip for PMS computed by Marconi & Palla (1998). The red edge varies between $\sim$6500 K and $\sim$7100 K, whereas the blue edge between $\sim$7100 K and $\sim$7500 K. The width of the strip is nearly constant and equal to about 650 K. All the models on the red edge are pulsating in the fundamental mode, while those on the blued edge in either of the overtone modes. The time spent by each star within the strip is also a constant fraction of the total PMS contraction time, typically between 5 and 10%, although in absolute terms it decreases from $\sim$10^6 yr for M$_*$=1.5 M$_\odot$ to 8$\times$10^4 yr for M$_*$=4 M$_\odot$. Despite the relatively short times, Fig. 9 shows that about a dozen known Herbig Ae/Be stars have the right combination of luminosity and temperature to qualify as candidate pulsators. Searches for δ Scuti type variability in these stars are already under way. It is expected that the additional information provided by the detection of a larger number of pulsating stars will help constraining the internal structure of young stars and testing evolutionary models.

Acknowledgements

I am grateful to Steve Stahler who has worked with me on many aspects of the work presented in this review. I wish to acknowledge the frutiful collaboration with Marcella Marconi on the pulsational instability of PMS stars. It is a pleasure to thank Nick Kylafis and Charlie Lada for their kind invitation to attend Crete II. Partial support was provided by CNR grant 97.00018.CT02 to the Osservatorio di Arcetri.

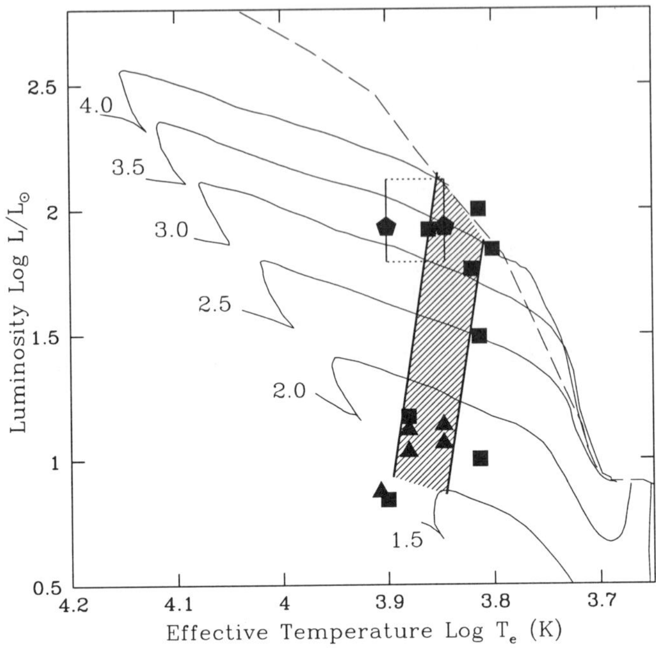

Figure 9. The shaded shows the location of the instability strip of PMS stars in the HR diagram. The various symbols indicate the distribution within or near the boundaries of the instability strip of known PMS stars that can be considered candidates for pulsational instability.

References

Alexander, D.R., Ferguson, J.W. 1994, *ApJ*, **437**, 879

Alexander, D.R., Augason, G.C., Johnson, H.R. 1989, *ApJ*, **345**, 1014

Alexander, D.R., Tamanai, A.I., Allard, F., Ferguson, J.W. 1998 *BAAS*, **67**, 16

Allard, F., Hauschildt, P.H., Alexander, D.R., Starrfield, S. 1997, *ARAA*, **35**, 137

Anders, E., Grevesse, N. 1989, *Geochim. Cosmochim. Acta*, **53**, 197

Baade, D., Stahl, O. 1989, *A&A*, **209**, 268

Baraffe, I., Babrier, G., Alla, F., Hauschildt, P.H. 1997, *A&A*, **327**, 1054

Basri, G., Marcy, G.W., Graham, J.R. 1996, *ApJ*, **458**, 600

Beech, M., Mitalas, R. 1995, *ApJS*, **95**, 517

Bernasconi, P., Maeder, A. 1996, *AA*, **307**, 829

Bertout, C., Regev, O. 1992, *ApJ*, **399**, L163

Beust, H., Corporon, P., Siess, L., Forestini, M., Lagrange, A-M. 1997, *A&A*, **320**, 478

Bildsten, L. Brown, E.F., Matzner, C.D., Ushomirsky, G. 1997, *ApJ*, **482**, 442

Böhm-Vitense, E. 1958, *Z. f. Ap.*, **46**, 108

Bonnell, I., Bate, M.R., Zinnecker, H. 1998, *MNRAS*, **298**, 93

Bouvier, J., Forestini, M., Allain, S. 1997, *A&A*, **326**, 1023

Brandner, W., Zinnecker, H. 1997, *A&A*, **321**, 220

Burrows, A., Marley, M., Hubbard, W., Lunine, J., Guillot, T., Saumon, D., Freedman, R., Sudarsky, D., Sharp, C. 1997, *ApJ*, **491**, 856

Canuto, V.M., Mazzitelli, I. 1991, *ApJ*, **370**, 295

Canuto, V.M., Goldman, I., Mazzitelli, I. 1996, *ApJ*, **473**, 550

Casey, B.W., Mathieu, R.D., Vaz, L.P.R., Andersen, J., Suntzeff, N.B. 1998, *AJ*, **115**, 1617

Caughlan, G., Fowler, W. 1988, *Atomic Data Nucl. Data Tables*, **40**, 283

Cox, J.P., Giuli, R.T. 1968, *Principles of Stellar Structure*, **II**, (New York: Gordon and Breach)

Cox, A.N., Tabor J.E. 1976, *ApJS*, **31**, 271

D'Antona, F. 1993, *Inside the Stars*, eds. W.W. Weiss & A. Baglin, A.S.P. Conf. Ser. 40 (San Francisco: Astr. Soc. Pac.), p.395

D'Antona, F., Mazzitelli, I. 1984, *A&A*, **138**, 431

D'Antona, F., Mazzitelli, I. 1997, priv. comm.

D'Antona, F., Mazzitelli, I. 1998 *Cool Stars in Clusters and Associations*, eds. G. Micela, R. Pallavicini and S. Sciortino , *Mem.S.A.It.*, **68**, 807

Däppen, W. 1993, *Inside the Stars*, eds. W.W. Weiss & A. Baglin, A.S.P. Conf. Ser. 40 (San Francisco: Astr. Soc. Pac.), p.208

Donati, J.-F., Semel, M., Carter, B.D., Rees, D.E., Collier Cameron, A. 1997, *MNRAS*, **291**, 658

Draine, B.T., Lee, H.M. 1984, *ApJ*, **285**, 89

Ezer, D., Cameron, A.G.W. 1965, *Canadian J. Phys.*, **43**, 1497

Forestini, M. 1994, *A&A*, **285**, 473

Gaustad, J.E. 1963, *ApJ*, **138**, 1050

Gautschy, A., Saio, H. 1996, *ARAA*, **34**, 551

Graboske, H., DeWitt, H.E., Grossman, H.S., Cooper, M.S. 1973, *ApJ*, **181**, 457

Grossman, A.N., Graboske, H.C. 1971, *ApJ*, **164**, 475

Hartigan, P., Strom, K.M., Strom, S.E. 1994, *ApJ*, **427**, 961

Hartmann, L., Kenyon, S.J. 1990, *ApJ*, **349**, 190

Hartmann, L., Cassen, P., Kenyon, S.J. 1997, *ApJ*, **475**, 770

Hayashi, C. 1966, *ARAA*, **4**, 171

Henyey, L.G., LeLevier, R., Levee, R.D. 1955, *Pub. A.S.P.*, **67**, 154

Hillenbrand, L.A., Strom, S.E., Vrba, F.J., Keene, J. 1992, *ApJ*, **397**, 613

Huebner, W.F., Merts, A., Magee, N.H., Argo, M.F. 1977, *Los Alamos Sci. Lab. Rep.*, LA-6760-M

Iben, I. 1965, *ApJ*, **141**, 993

Iglesias, C.A., Rogers, F.J. 1991, *ApJ*, **371**, L73

Iglesias, C.A., Rogers, F.J. 1996, *ApJ*, **464**, 943

Kahn, F.D. 1974, *A&A*, **37**, 149

Kurtz, D.W., Marang, F. 1995, *MNRAS*, **276**, 191

Larson, R.B. 1969, *MNRAS*, **145**, 121

Larson, R.B. 1972, *MNRAS*, **157**, 121

Larson, R.B., Starrfield, S. 1971, *A&A*, **13**, 190

Luhman, K.L., Rieke, G.H. 1998, *ApJ*, **497**, 354

Marconi, M., Palla, F. 1998, *ApJ*, **507**, L141

Martin, E.L., Claret, A. 1996, *A&A*, **306**, 408

Martin, E.L., Basri, G., Zapatero Osorio, M.R., Rebolo, R., López, R.J., Garcia, R.J. 1998, *ApJ*, **507**, L41

Mathieu, R.B. 1994, *ARAA*, **32**, 465

Mazzitelli, I. 1989, *Low-Mass Star Formation and Pre–Main-Sequence Objects*, ed. Bo Reipurth, ESO Proc. 33, p.407

Mazzitelli, I., Moretti, M. 1980, *ApJ*, **235**, 955

Mihalas, D., Däppen, W., Hummer, D.G. 1988, *ApJ*, **331**, 815

Nakano, T. 1989, *ApJ*, **345**, 464

Palla, F. 1994, *Infrared Astronomy*, eds. A. Mampaso & F. Sanchez (Cambridge: Cambridge Univ. Press), p.1

Palla, F., Stahler, S.W. 1990, *ApJ*, **360**, L47

Palla, F., Stahler, S.W. 1991, *ApJ*, **375**, 288

Palla, F., Stahler, S.W. 1993, *ApJ*, **418**, 414
Palla, F., Stahler, S.W. 1999, *ApJ*, in press
Parigi, G. 1992, Thesis, Univ. of Florence, unpublished
Pollack, J.B., Hollenbach, D., Beckwith, S., Simonelli, D.P., Roush, T., Fong, W. 1994,
    ApJ, **421**, 615
Rebolo, R., Magazzù, A., Martin, E.L. 1995, *Stellar and interstellar lithium*, eds. F. Spite
    and R. Pallavicini, *Mem.S.A.It.*, **66**, 375
Rebolo, R., Martin, E.L., Magazzù, A. 1992, *ApJ*, **389**, L83
Rebolo, R., Martin, E.L., Basri, G., Marcy, G.W., Zapatero Osorio, M.R. 1996, *ApJ*,
    469, L53
Rogers, F.J., Iglesias, C.A. 1992, *ApJS*, **79**, 507
Salpeter, E.E. 1954, *Mem. Roy. Soc. Liège*, **14**, 116
Schaller, G., Schaerer, D., Meynet, G., Maeder, A. 1992, *A&AS*, **96**, 269
Seaton, M.J. 1993, *Inside the Stars*, eds. W.W. Weiss & A. Baglin, A.S.P. Conf. Ser. 40
    (San Francisco: Astr. Soc. Pac.), p.222
Seaton, M.J., Yan, Y., Mihalas, D., Pradhan, A.K. 1994, *MNRAS*, **266**, 805
Siess, L., Forestini, M. 1996, *A&A*, **308**, 472
Siess, L., Forestini, M., Bertout, C. 1997, *A&A*, **308**, 472
Simon, M., Ghez, A.M., Leinert, Ch. et al. 1995, *ApJ*, **443**, 625
Simon, N. 1982, *ApJ*, **260**, L87
Stahler, S.W. 1983, *ApJ*, **274**, 822
Stahler, S.W. 1988a, *ApJ*, **332**, 804
Stahler, S.W. 1988b, *Pub. A.S.P.*, **100**, 1474
Stahler, S.W., Walter, F.M. 1993, *Protostars and Planets III*, eds. E.H. Levy & J.I.
    Lunine (Tucson: The Univ. of Arizona Press), p.405
Stahler, S.W., Shu, F.H., Taam, R.E. 1980, *ApJ*, **302**, 590
Stahler, S.W., Palla, F., Ho, P.T.P. 1999, *Protostars and Planets IV*, eds. V. Mannings,
    A.P. Boss, and S.S. Russell (Tucson: U. of Arizona Press), in press
Stauffer, J.J.R., Schultz, G., Kirkpatrick, J.D. 1998, *ApJ*, **499**, L199
Strom, K.M., Strom, S.E. 1994, *ApJ*, **424**, 237
Swenson, J.H., Faulkner, J., Rogers, F.J., Iglesias, C.A. 1996, priv. comm.
Venturi, P., Zeppieri, A., Mazzitelli, I., D'Antona, F. 1997, *A&A*, **331**, 1011
Von Sengbusch, K. 1968, *Zs. f. Ap.*, **69**, 79
Winkler, K.H., Newman, M.J. 1980, *ApJ*, **236**, 201

Francesco Palla and Ed Churchwell on the way to Santorini.

III – THE ROLE OF CLUSTERING IN STAR FORMATION, THE ORIGIN OF MASSIVE STARS

Students take an Uzzo Break!

OB ASSOCIATIONS

A.G.A. BROWN
Instituto de Astronomía UNAM,
Apartado Postal 877, Ensenada, 22800 Baja California, Mexico

A. BLAAUW
Kapteyn Instituut, Postbus 800, 9700 AV Groningen, The Netherlands
Sterrewacht Leiden, Postbus 9513, 2300 RA Leiden, The Netherlands

AND

R. HOOGERWERF, J.H.J. DE BRUIJNE AND P.T. DE ZEEUW
Sterrewacht Leiden, Postbus 9513, 2300 RA Leiden, The Netherlands

1. Introduction

Starting with the earliest studies of the distribution of the bright stars by Kapteyn, Rasmuson, and Pannekoek ([80], [81], [110], [118], and [119]), it was evident that O and B stars are not distributed randomly on the sky, but instead are concentrated in loose groups, which were subsequently called 'OB associations'. This inspired research on their individual properties, and on their motions and space distribution. From dynamical considerations it followed that OB associations must be young [4], a conclusion supported later by the ages derived from color-magnitude diagrams. Blaauw's 1964 review [9] already discussed the relation between OB associations and interstellar matter. Subsequent observations of molecular clouds (e.g., [151], [16]) indicated that these groups are usually located in or near star-forming regions, and hence are prime sites for the study of star formation processes and of the interaction of early-type stars with the interstellar medium.

An important aspect of the study of OB associations is the identification of their 'members'. Associations have small internal velocity dispersions (e.g., [99], [130]), so that the streaming motion of the association as a whole, as well as the Solar motion, is reflected as a motion of the members towards a convergent point on the sky (e.g., [6]). This can be used to establish membership of these 'moving groups' based on measurements of proper motions. At the time of the previous NATO-ASI on star formation [85] few astrometric membership studies existed for nearby OB associations because these generally cover tens to hun-

C.J. Lada and N.D. Kylafis (eds.), The Origin of Stars and Planetary Systems, 411–440.
© 1999 *Kluwer Academic Publishers. Printed in the Netherlands.*

dreds of square degrees on the sky. Ground-based proper motion studies therefore almost invariably had been confined to modest samples of bright stars ($V \lesssim 6$) in fundamental or meridian circle catalogues, or to small areas covered by a single photographic plate. Photometric studies can extend membership to later spectral types (e.g., [143]–[145]), but are less reliable due to, e.g., undetected duplicity, or the distance spread within an association. As a result, membership for many associations had been determined unambiguously only for spectral types earlier than B5 (e.g., [9], [13]).

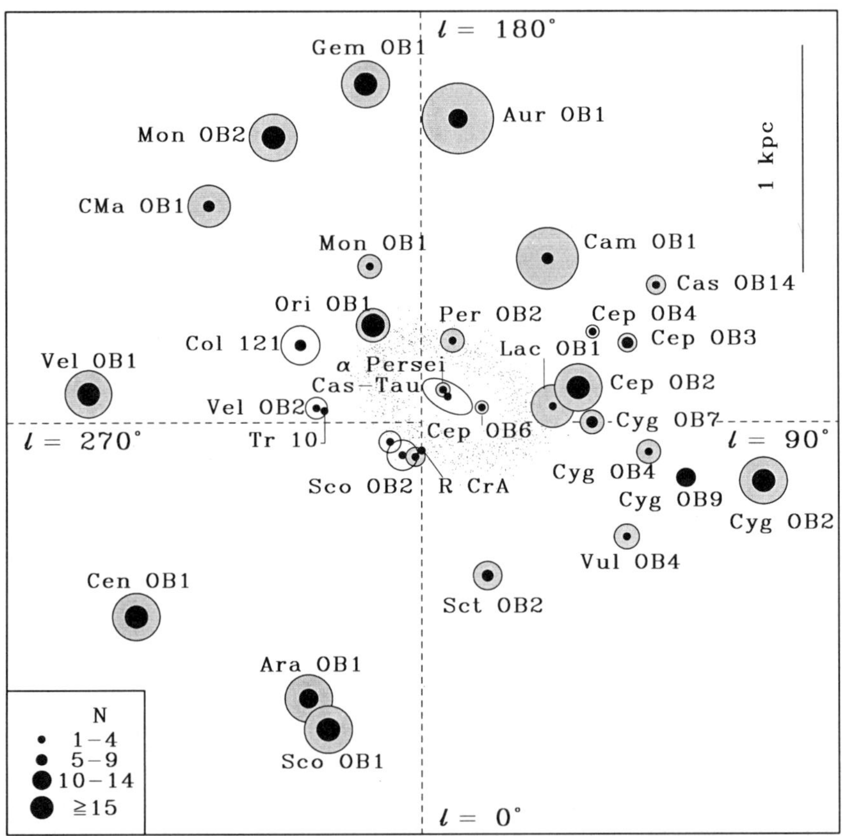

Figure 1. Locations of the OB associations within ∼1.5 kpc, projected onto the Galactic plane. This figure is an update of Figure 8 in [13] and is based on a list by Ruprecht [122] (see de Zeeuw et al. [150] for more details). The Sun is at the center of the dashed lines which give the principal directions in Galactic longitude, ℓ. The sizes of the circles represent the projected dimensions of the associations, enlarged by a factor 2 with respect to the distance scale. The sizes of the central dots indicate the degree of current or recent star formation activity, as given by the number N of stars more luminous than absolute magnitude $M_V \sim -5$ [76]. Associations which are absent from Ruprecht's list are represented as open circles. The distribution of small dots indicates the Gould Belt [115]. Figure 9 presents the post-Hipparcos map of the nearby OB associations.

This review is concentrated on OB associations located within ~ 1.5 kpc from the Sun, and updates and extends the chapter by Blaauw in the previous edition of this NATO-ASI [13]. The material discussed in Blaauw's chapter is still relevant but we chose to emphasize the developments in the field since 1991. The X-ray surveys by the EINSTEIN and ROSAT satellites enabled the identification of large numbers of low-mass stars in OB associations. Major progress in the identification of high-mass members of OB associations has been made possible through the use of precise astrometric data from the Hipparcos Catalogue [50]. Also, Blaauw's 1964 review [9] still is essential reading. Reviews concerning OB associations throughout the Local Group include [60], [97], and [103]. In addition there are conference proceedings that contain material on OB associations: [20], [33], [79], [100], and [132]. For more background on the historical development of OB association research we refer to [9], [30], [49], and [150].

This chapter proceeds as follows. §2 discusses the reasons for studying OB associations. §3 describes the problem of accurately defining what is meant by an OB association. §4 contains an overview of recent ground-based studies of OB associations, and §5 describes the Hipparcos census of the nearby OB associations [150]. §6 contains a discussion of Gould's Belt and the origins of OB associations. §7 points out directions for future research.

Most of the OB associations discussed here appear in a list compiled by Ruprecht [122], which contains field boundaries, some bright members, distances, and uses a consistent nomenclature, which was subsequently approved by the IAU. Figure 1 shows the distribution of the nearby OB associations, based on Ruprecht's list and including additional associations discussed by De Zeeuw et al. [150], projected onto the Galactic plane.

2. The Importance of Studying OB Associations

In the context of star formation a detailed examination of the stellar content, structure, and kinematics of OB associations allows us to address the following fundamental questions:

- What is the initial mass function? Young stellar groups are prime sites for the study of the initial mass function (IMF) because the corrections for stellar evolution and star formation history are minimal or at least straightforward [125]. Moreover, OB associations contain the entire range of stellar masses (cf. §§4.1 and 4.2).
- What are the characteristics of the binary population? A detailed answer to this question will furnish a better understanding of the processes through which binaries form, but also of the formation of massive stars and star clusters. Recent theories ([21], [22], and the chapter by Bonnell in this volume) predict the occurrence of few widely separated binaries among systems with

massive primaries. A characterization of the binary population in OB associations is a direct test of these theories.

- OB associations have traditionally been characterized as unbound groups of stars [9]. However, the youngest subgroup of Ori OB1, the Orion Nebula Cluster, may eventually evolve into a bound cluster [71]. What causes the distinction between the formation of bound open clusters and unbound associations? Various studies show that the fate of a particular young group of stars depends on the gas fraction, the time scale on which this gas is removed and the stellar velocity dispersion (e.g., [84], [135]). However, the processes that set the values of the above parameters during the formation of a star cluster are not well understood.

- Observations show that molecular clouds and cloud cores contain much more angular momentum than the stars that form from them (e.g., [18], [105]). How is angular momentum redistributed during star formation? What is the resulting distribution of rotational velocities of stars? Important clues to the answers of these questions may be obtained from studies of the distribution of rotational velocities of the members of OB associations in conjunction with studies of their binary population.

- O- and B-type runaway stars are a subset of the O and B stars characterized by their high space velocities, up to 200 km s^{-1}, and almost complete absence of multiplicity. For a number of these stars, the motion through space, when traced backwards in time, leads to the identification of a parent OB association. Possible mechanisms for the origin of the OB runaways are: release of the secondary following the supernova explosion of the primary member of a binary [8], and purely dynamical interactions amongst members of a protocluster [58]. The precise characterization of OB runaways as well as more identifications of parent associations can provide more insight into massive binary evolution [120] and massive star formation [34]. OB runaways were most recently reviewed by Blaauw [14].

Further questions related to star formation that can be addressed through studies of OB associations include the star formation rate and efficiency, the issue of the star formation history in clusters and associations (i.e., do all stars in a group form at the same time?), and the propagation of star formation throughout molecular cloud complexes (i.e., sequential star formation [45]). In a broader context the investigation of OB associations is important for numerous areas of Galactic and extragalactic research. A few examples follow:

- The calibration of the absolute luminosity of the upper main sequence is based largely on the bright members of the nearest OB associations (e.g., [7]).
- Massive stars influence the interstellar medium through their ionizing radiation, stellar winds, and supernovae. They may be responsible for determining the pressure of the gas and the velocity dispersion of atomic clouds [101],

and for the destruction of the molecular clouds in which they are born. Acting collectively, massive stars in OB associations have an even larger impact on the evolution of the interstellar medium and create H I supershells and superbubbles (e.g., [131]).

- The recent discovery of unexpectedly large amounts of gamma-ray emission in the direction of the Orion molecular cloud complex [15] have been linked to the presence of the nearby O and B stars in Ori OB1 (e.g., [111]).

The nearby associations contain few massive stars: even in Ori OB1 the upper mass limit is only about 50 $M_\odot$. More distant associations enable the determination of the IMF for the most massive stars. A well-known example of an OB association beyond 1.5 kpc is NGC 3603, which contains 50 O stars and ionizes the second most luminous H II region in the Galaxy [44]. The Carina H II region/ molecular cloud complex contains numerous star clusters, the most famous of which are Trumpler 14 and 16 [24]. The latter are among the most massive clusters in the Galaxy and contain many O stars, including some of type O3 ($\gtrsim 80\ M_\odot$).

A better characterization of the IMF at the high-mass end, the luminosities and spectral characteristics of massive stars, and the fraction of Wolf–Rayet stars in OB associations are essential items to consider when interpreting observations of extragalactic starforming regions and starburst galaxies.

3. Defining Characteristics of OB Associations

In 1947 Ambartsumian [3] introduced the term 'association' for groups of OB stars; he pointed out that their stellar mass density is usually less than $0.1\ M_\odot\ \mathrm{pc}^{-3}$. Bok [19] had already shown that such low-density stellar groups are unstable against Galactic tidal forces, which led Ambartsumian to conclude that OB associations must be young (~ 10 Myr) [4]. Because of these low stellar densities associations will quickly disperse. This makes them hard to recognize once they are older than about 25 Myr due to their large spatial extent (~ 10–50 pc). These properties were used in the earliest definitions of OB associations given by Ambartsumian and Blaauw [3], [9]. However, such definitions do not necessarily encompass all associations. For example, the Orion Nebula Cluster is considered a subgroup of the Ori OB1 association, but is likely to remain bound and thus compact [71]. If the cluster in reality is unbound a velocity dispersion of a few km s^{-1} will be enough to disperse it from its present 1 pc size to a size of a typical OB association within 5–10 Myr.

Another definition of an OB association was proposed by Lada & Lada [86]; a group of at least 10 members whose stellar density is less than 1 $M_\odot\ \mathrm{pc}^{-3}$. However, as the authors themselves pointed out, this would mean that the Hyades would be classified as an OB association and the association NGC 2264 as an open cluster. This led Brown et al. [27] to define OB associations as those stellar groups that are left unbound after the process of gas removal from the protocluster has been concluded. This is not a very practical definition as it is not easy to determine whether a particular stellar group is bound or not.

Recently, the problem of accurately defining the term 'OB association' has been discussed by Elmegreen & Efremov [49]. Numerous observational selection effects that occur when looking for OB associations are considered. For instance, how stellar groups are selected depends on distance, because a particular group of luminous stars can resemble a dense cluster if it is at a large distance, or a rarefied association if it is nearby. Observations of patches of Cepheid variables in the LMC and in local star formation regions [48] reveal a correlation between the duration of star formation in a region and its size. This means that if an OB association is defined as a grouping of OB and other young stars of, for example, less than 10 Myr, the association will be observed to have a maximum size of about 50 pc because larger regions will generally contain older stars. Elmegreen & Efremov argue that these difficulties arise because there is no physical scale associated with star formation. Rather, there is a whole hierarchy of scales varying from single/binary stars to star clusters, star complexes, and even up to small pieces of a spiral arm. In combination with the size-age correlation mentioned above this will lead to stellar groupings with particular dimensions being selected according to the age limits imposed. Hence, if one allows for enough range in age it could be argued that all of Gould's Belt, consisting of several associations and a wider distribution of early-type stars in the Solar neighbourhood, is actually one association. It is further argued [49] that the hierarchical and scale-free nature of star formation results from the fractal structure of interstellar gas. However, McKee & Williams ([102], [147]) showed that there is a physical upper limit to the size of OB associations and giant molecular clouds. This finding supports the idea that the characteristic sizes observed for OB associations ($\sim$10–50 pc) are a reflection of the sizes of giant molecular clouds from which they form. Moreover, it was recently inferred that the nearest molecular cloud complex, in Taurus, has a characteristic size scale and hence does not have a fractal structure [17].

Based on studies of Hertzsprung–Russell diagrams for OB associations it was realized early on [9] that several associations could be divided further into subgroups. These can primarily be distinguished on the basis of the ages of their members and their degree of association with interstellar matter ([9], [61], [143], [145]). Defining the boundaries of these subgroups is a thorny question. To date, one cannot distinguish subgroups kinematically, not even with proper motions of Hipparcos accuracy (see §5). A good example is Ori OB1. Originally this association was divided into four subgroups 1a–1d [9]. Subgroup 1b consists of the stars around Orion's Belt and was further subdivided in some studies, but was recently treated as a whole [25]. The latest finding on the Belt stars is that the star σ Ori is surrounded by a cluster of young stars, which may represent an older analogue of the Trapezium cluster [140], [141]. This suggests that more such clusters of pre-main sequence stars surrounding bright OB stars may be found and a further

subdivision of subgroup 1b may thus be warranted. This serves as a warning that
established subgroup boundaries in associations are not 'cast in stone' and that
future work may lead to different subdivisions.

In summary, it is not entirely trivial to accurately define what is meant by
the term 'OB association'. Nevertheless, for the purpose of this review we define
OB associations to be young ($\lesssim 50$ Myr) stellar groupings of low density—such
that they are likely to be unbound—containing a significant population of B stars.
Their projected dimensions range from ~ 10 to ~ 100 pc.

4. Recent Research

We now turn to work on OB associations carried out since 1991 but before the
availability of Hipparcos data, and grouped by subject. We first concentrate on the
high-mass stars and then discuss low-mass stars. This overview is not intended to
be complete but serves as an introduction to recent literature on OB associations.

4.1. HIGH-MASS STARS

Stellar Content. Extensive studies of the stellar content of OB associations were
carried out using photometric data. As explained in §1 this allowed an extension of
membership lists to later spectral types, for which no accurate large-scale proper
motion surveys were available yet. The work on Sco OB2 was summarized in [13]
and the details can be found in [61].

The Ori OB1 association consists of four subgroups: 1a, located to the northwest of Orion's Belt, 1b, located around the Belt, 1c, located around the Sword
of Orion, and 1d, which is the Trapezium cluster (see figure 7 in [13]). Brown
et al. [25] analyzed available Walraven photometry [94] for Ori OB1. Model atmospheres and empirical calibrations were used to determine effective temperatures, surface gravities, and absolute bolometric magnitudes of the stars, and to
derive ages for the association subgroups by isochrone fitting in the Hertzsprung–
Russell diagram. It was found that the distances to subgroups 1a–1c are 380,
360, and 400 pc, respectively. The distance to 1d could not be determined reliably due to the nebulosity in that region and insufficient stars. These distances
were smaller than the distances derived previously, a result which also followed
from a reanalysis of the data of Warren & Hesser [143] using a revised calibration of the $uvby\beta$ system [5]. The ages of the Ori OB1 subgroups were found to
be 11.4 ± 1.9 Myr for 1a, 1.7 ± 1.1 Myr for 1b, 4.6 ± 2 Myr for 1c, and less
than 1 Myr for subgroup 1d. The IMF was found to be a single power law with
$\mathrm{d}\log(\xi(\log m))/\mathrm{d}\log m = -1.7 \pm 0.2$ for all three subgroups (see also [30]).

Finally, we mention Cep OB3, which is considered to be an example of sequential star formation [45]. The most recent studies of its stellar content are by
Jordi et al. [77], [78]. Strömgren photometry was used to refine and extend the

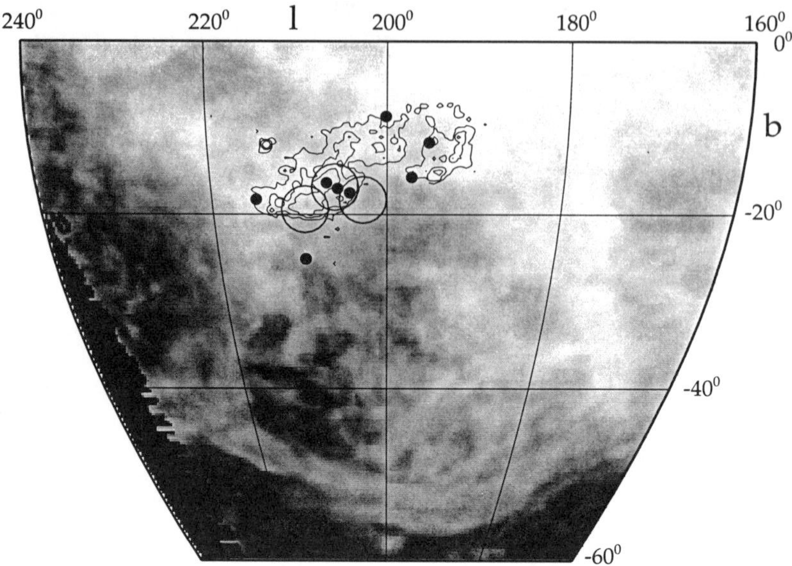

Figure 2. The Orion–Eridanus bubble. The grey-scale image is a logarithmically scaled representation of integrated H I emission from the Leiden–Dwingeloo survey in the velocity interval -1 km s$^{-1} \leq v_{\mathrm{LSR}} \leq +8$ km s^{-1}. The contours outline the 100 μm (IRAS) emission from the Orion A and B molecular clouds (the ring around $(\ell, b) = (195°, -12°)$ is the λ-Orionis ring). The dots show the brightest stars in the Orion constellation. The circles show the positions of the three main subgroups of Ori OB1. From right to left are shown 1a, 1b and 1c.

membership lists towards fainter stars. The existence of two subgroups in this association was confirmed and the ages were found to be 5.5 and 7.5 Myr.

Interstellar Medium.   The nearby OB associations offer a uniquely detailed view of the relation between early-type stars and the interstellar medium (cf. §2) which helps us understand the impact of associations on the interstellar matter throughout the Galaxy. Much work has been done since 1991, concentrating especially on the interstellar medium around Ori OB1. The characteristics of the interstellar medium related to Sco OB2 were summarized in [13] and [62].

For summaries of studies of the interstellar gas in the vicinity of Ori OB1 we refer to [100]. The impact of the early-type stars in this association has been addressed recently in a series of studies, [26], [32], [67], and [129], taking advantage of the availability of surveys such as IRAS, the ROSAT all sky survey, and the Leiden–Dwingeloo H I survey [69]. Figure 2 shows the distribution of H I around Orion over the velocity range -1 to $+8$ km s^{-1}. One can clearly distinguish a cavity surrounded by a shell of H I, the Orion–Eridanus bubble. The same features can be seen in an image at 100 μm from IRAS [26], and the whole of the cavity is filled with very hot gas, $\sim 10^6$ K, emitting in X-rays [32], [26], [129].

The shell has a measured expansion velocity of about 40 km s^{-1} and a mass of $2.3 \pm 0.7 \times 10^5$ M$_\odot$. Taking into account the IMF and the ages of the subgroups, the mechanical energy output in the form of stellar winds and supernovae over the lifetime of the association was estimated to be $\sim 10^{52}$ ergs [25]. Using semi-analytic models of wind-blown bubbles that take the density stratification of the Galactic H I layer into account [83], it was shown that this energy is indeed enough to account for the size as well as for the expansion velocity of the H I shell [26].

Early-type stars also have a large effect on the interstellar medium through their ionizing radiation, producing both localized H II regions and diffuse ionized gas. Based on the distribution of OB associations in the Galaxy, it was shown that their luminosity function can be fit with a truncated power law, and that there probably is a physical limit to the maximum size of H II regions in the Galaxy [102]. A comparison with the distribution of giant molecular clouds [147] showed that a 10^6 M$_\odot$ cloud is expected to survive about 30 Myr, and that on average 10 per cent of its mass is converted into stars by the time it is destroyed (see also the chapters by Blitz and by McKee). The overall distribution of associations is also important for understanding the hot-gas filling factor of the interstellar medium [55]–[57].

Formation of Stellar Clusters. Detailed studies of young clusters and OB associations lead to more insight into the formation of these systems (cf. §2). In an effort to address this question the Orion Nebula Cluster (ONC), with the Trapezium cluster at its core, was studied extensively by Hillenbrand [70]. An optical sample of ~ 1600 stars within 2.5 pc from the Trapezium stars was studied with photometry and spectroscopy. The overall IMF of this core region of the ONC was found not to be grossly inconsistent with 'standard' stellar mass spectra. The observed IMF appears to peak at ~ 0.2 M$_\odot$ and to fall off rapidly towards lower masses. Several substellar objects were identified. The total mass of the stars was found to be ~ 1800 M$_\odot$, and their mean age is less than 1 Myr, with the younger stars being concentrated towards the centre. Mass segregation was shown to be present in the ONC and it is probably not due to dynamical relaxation of the cluster, implying that the massive stars formed near the centre of the cluster. This confirms one of the predictions of cluster formation theories as discussed in the chapter by Bonnell. A study of the wider distribution of stars around the Trapezium shows that the ONC has an elongated structure similar to that of the molecular gas distribution in the region, suggesting that the cluster may still retain a memory of the geometry of the protocluster cloud [71].

Chemical Evolution. The interstellar medium in the Galaxy is continually enriched in chemical elements by stellar winds and by ejecta from supernovae. OB associations are generally still located near molecular clouds and thus are ideal sites for studying ongoing chemical evolution processes in the Milky Way. Abundance patterns of stars in Ori OB1 have been studied by Cunha et al. [38]–[41]. The abundance analysis shows that the stars in Ori OB1, in common with the Orion Nebula H II region, are underabundant in Oxygen with respect to the Sun. The lowest abundances are found in subgroups 1a and 1b. The Trapezium stars and some stars of subgroup 1c seem to have O abundances that are up to 40 per cent higher than those in subgroups 1a and 1b (although still subsolar). It is suggested that this is due to enrichment of the interstellar gas by mixing of supernovae ejecta from subgroup 1c with the gas that subsequently collapsed to form the Trapezium Cluster [38]. This enrichment scenario is confirmed by the fact that Cunha et al. [39] observe no abundance variations for C, N, and Fe, but do observe the same variations for Si as for O, as one would predict for cloud material enriched by Type II supernova ejecta. Supernovae must have occurred in Ori OB1 in the past given the presence of the Orion–Eridanus bubble. It is estimated that 1 to 2 supernovae have occurred in subgroup 1c [25].

Kinematic Ages. Because OB associations are unbound they will expand [4]. In principle one can use this expansion to trace back the motions of the stars in an OB association until some minimum configuration is reached (see e.g., figure 5 in [13]). The time at which this happens would then correspond to the kinematic age of the association, which can be compared to the age derived from the Hertzsprung–Russell diagram. Also, an estimate is obtained of the initial configuration of the association just after it was formed. Kinematic ages have in fact been determined for a number of associations [11], [12], [59], [87], and [89]. However, it was demonstrated that tracing back proper motions in OB associations always leads to underestimated ages and overestimated initial sizes [27]. The main reason is that the space motions of the stars are not rectilinear but are influenced by the N-body interactions in the initially more compact association, by the effects of the remnant molecular cloud, and by the Galactic tidal field.

Binaries. Among the most important clues to understanding the process of star formation are the characteristics of the binary population. Recently, a number of searches for spectroscopic binaries in Sco OB2 and Ori OB1 were carried out [90], [104], and [136], and the statistics of close binaries among early-type stars were summarized by Blaauw [13]. The problem in studying the binary population among early-type stars in OB associations is to obtain precise radial velocities for these stars, which is very difficult due to stellar rotation and the small number of spectral lines [137]. Nevertheless, we can expect progress to be made in the near future when Hipparcos data on binaries is analyzed.

With the advent of speckle and adaptive optics techniques, both in the optical and infrared, much attention has been devoted recently to studying the population of binaries among the stars in the Trapezium cluster and the pre-main sequence stars in Sco OB2 [113], [23]. Both studies conclude that the characteristics of the binary population depend on the star forming environment. In particular, Brandner & Köhler [23] suggest that the conditions which favor the formation of high-mass stars apparently lead to the formation of close binaries among low-mass stars, whereas conditions unfavorable to the formation of high-mass stars lead to the formation of wider binaries among low-mass stars. This bears directly on the issues discussed in the chapter by Bonnell in this volume.

4.2. LOW-MASS STARS

Extrapolating the mass function for OB stars, such as derived in Sco OB2 and Ori OB1 [62], [25], reveals that the bulk of the stars should be of low mass ($\lesssim 2$ $M_\odot$). Indeed, evidence was found early on for the presence of low-mass stars in the vicinity of the Orion Nebula [68], [138]. A search of the IRAS Point Source Catalogue for young stellar objects [117] shows how the distribution of these stars clearly outlines some of the OB associations, such as Ori OB1 and the Upper Scorpius (US) subgroup of Sco OB2. Large-scale proper motion searches for these fainter members of OB associations suffer from a much larger contamination by field stars, making it hard to identify the members without additional information. Hence, other techniques are employed in the search for low-mass members of OB associations. The two most widely used are objective prism $H\alpha$ surveys, which are sensitive to classical T-Tauri stars, and X-ray surveys, also sensitive to weak-line T-Tauri stars (see the chapter by Ménard & Bertout).

An EINSTEIN observatory X-ray search was used to look for the low-mass population of US [139]. After correcting for the incompleteness of the X-ray sampling, it was concluded that the association has a field star mass function between about 0.2 and 10 $M_\odot$. The total number of low-mass stars (< 2 $M_\odot$) is about 2000, which is in good agreement with the number of low-mass stars inferred by de Geus [62]. Recent X-ray studies, based on the ROSAT all-sky survey [116], [126] reveal the presence of many more X-ray selected pre-main sequence (PMS) candidates throughout Sco OB2. Preibisch et al. [116] performed spectroscopic follow-up observations to confirm the PMS character of the X-ray selected sources and also looked for PMS stars among objects without X-ray detections but with proper motions suggesting that they might be members of Sco OB2. No PMS stars were found among these proper motion selected candidates and these authors conclude that the X-ray selected sample of PMS stars is at least 75 per cent complete. However, Sciortino et al. [126], using ROSAT HRI, PSPC and all-sky survey observations, showed that the observations by Walter et al. [139] were much more incomplete than reported. They concluded that EINSTEIN and ROSAT data are far from giving a complete characterization of the X-ray population of US, implying the presence of even more PMS stars in Sco OB2.

To confirm the PMS nature of the objects detected in X-ray surveys the so-called Lithium test is used. Essentially one measures the strength of the Li line which is correlated with stellar age for PMS stars. Main sequence stars should have depleted all their Li in nuclear burning processes, and thus the presence of Li indicates the PMS character of the object under study. However, it has recently been pointed out that the Li test is not completely reliable [51]–[53]. Especially, when using low-resolution spectra to study the Li line one may confuse young and active main sequence stars with bona fide PMS stars, thus further complicating studies of the IMF down to the lowest masses.

In Ori OB1 the Kiso Hα survey [106], and the EINSTEIN [140], [141] and ROSAT [1] X-ray surveys have uncovered hundreds of emission-line and X-ray sources of which many are likely to be PMS or T-Tauri stars. Spectroscopic follow-up observations indeed confirmed the PMS nature of many of these stars [1], [82], [107]. However, it was shown [2] that the population of X-ray sources towards Orion consists of a mixture of true Orion PMS stars and young foreground stars. The latter may be related to Gould's Belt or may be $\sim 10^8$ yr old stars.

Recent studies have revealed a significant population of low-mass PMS stars in the region of subgroup 1b of Ori OB1 [140], [141]. The stars appear to cluster spatially around σ Ori and the narrowness of their PMS locus suggests coevality, at the 2 Myr age of Ori OB1b. The total inferred mass of this group of stars is comparable to that of the Orion Nebula Cluster. In fact the σ Orionis cluster may be an older analogue of the Trapezium cluster. This result illustrates the difficulty in defining subgroup boundaries in associations (cf. §3). It may be that subgroup 1b actually consists of a 'merger' of several Trapezium-like clusters. Most recently, the detection of brown dwarf candidates around σ Ori has been reported. We refer to [142] for the details and more material on low-mass stars in OB associations.

Enormous progress has been made towards characterizing the low-mass population in OB associations. The challenge now is to bring the studies of the high- and low-mass stars together into a comprehensive picture.

5. Results from Hipparcos

We now review the results on the stellar content of the nearby OB associations derived from analysis of Hipparcos data. The material in this section is essentially a summary of [150], [31], and [75] and we refer to those papers for all the details and further references.

The most important data contained in the Hipparcos Catalogue [50] are positions, proper motions, and trigonometric parallaxes for $\sim 120\,000$ stars. The median precision of the astrometric data for stars brighter than $V \sim 9$ is 0.88 and 0.74 milli-arcsecond yr^{-1} for the proper motions in right ascension and declination, respectively, and 0.97 milli-arcsecond (mas) for the parallaxes. Note that at the distance of the nearest associations (~ 150 pc) these numbers translate into individual stellar distances accurate to 15 per cent and transverse velocities accurate to 0.6 km s^{-1}. The astrometric data are supplemented with magnitudes and colors, variability information obtained from the photometry, and detailed data on the properties of binaries observed or discovered by Hipparcos. Simultaneously with the main mission the Tycho experiment was carried out. This resulted in a catalogue of over 1 million stars to $V \approx 11.5$ containing astrometric data with less precision (median $\sim$7 mas for $V \lesssim 9$) as well as accurate photometry.

The Hipparcos Catalogue has a limiting magnitude of $V \sim 12$, and is complete to $V \sim 7.3$ in the Galactic plane, and to $V \sim 9$ in the polar regions. Furthermore, for reasons having to do with the way the observations were carried out, the Catalogue suffers from severe selection effects in some regions of the sky. For OB associations the relevant selection biases are described in §3 of de Zeeuw et al. [150]. For more details on the Hipparcos Catalogue please refer to the extensive documentation provided with the Catalogue itself [50].

5.1. ASTROMETRIC MEMBERSHIP SELECTION

Two methods, based on the assumption of common space motions for stars, were used to select astrometric members of the nearby OB associations [150]. The first is a modification of the classical convergent point method [31], and the second is a new selection method which makes use of the parallaxes as well as the proper motions, and searches for members in velocity space [75]. Both methods were combined to find the members of the nearby associations. This leads to a very powerful membership selection tool in which many spurious interlopers that occur when either method is used alone are automatically removed. Even so, one should still expect that a fraction of the selected stars actually belong to the field but by coincidence have the right motion to be included as members. The fraction of expected interlopers was estimated for each association through extensive Monte Carlo simulations.

To illustrate the dramatic advances provided by the Hipparcos data, Figure 3 shows in detail the membership selection results for the Upper Centaurus Lupus (UCL) subgroup of Sco OB2. The first row displays the Hipparcos measurements for all stars in the UCL field. The panels show no clear sign of a physical group, except for the vector point diagram (panel two), which contains a concentration around $(\mu_\ell \cos b, \mu_b) \sim (-25, -10)$ mas yr^{-1} superimposed upon the broader Galactic disk distribution. The second row of Figure 3 shows 136 stars that were proposed as members of UCL, based on pre-Hipparcos kinematic and photometric studies. They are mostly B- and A-type stars, with fairly little concentration in the vector point diagram. Their parallax distribution is narrower than the one in the first row, and peaks around 9 mas. The characteristics of the set of astrometrically selected members are presented in the third row of Figure 3. The vector point diagram of these secure members is much more concentrated than that of the pre-Hipparcos members, and the parallax distribution is narrower. This is due to a reduced contamination by field stars. The spread and the elongated shape in the

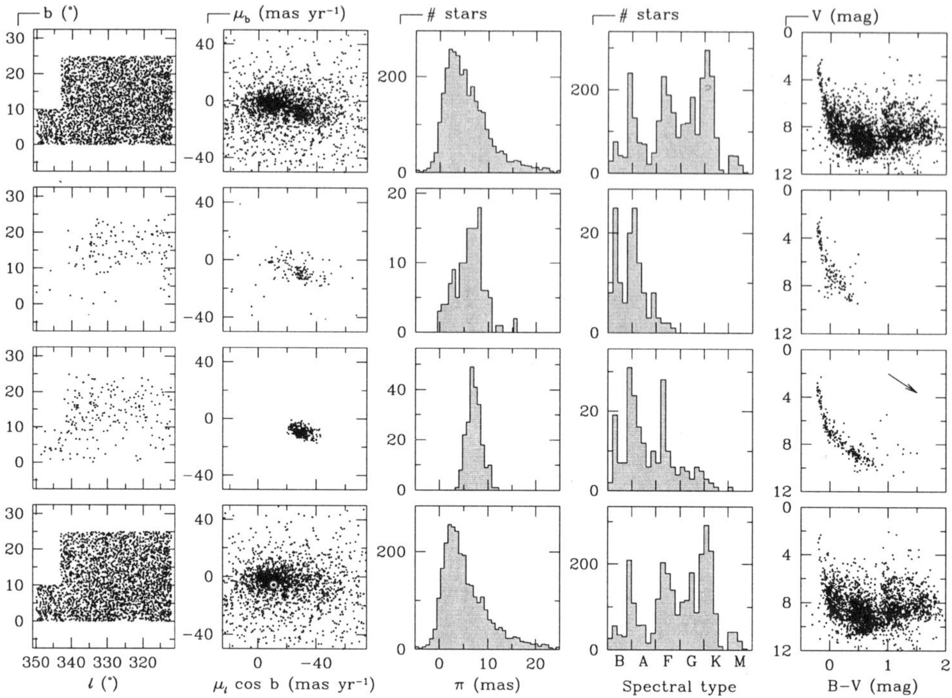

Figure 3. Hipparcos measurements for the subgroup Upper Centaurus Lupus of Sco OB2 (from the top row down): (1) all 3132 Hipparcos stars in the region; (2) the 136 pre-Hipparcos members; (3) the 221 Hipparcos members; (4) the remaining stars after member selection. The columns show (from left to right): (1) positions in Galactic coordinates; (2) Galactic vector point diagram; (3) trigonometric parallax distribution; (4) spectral type distribution; (5) color-magnitude diagram, not corrected for reddening. The arrow indicates the direction of reddening for the standard value $R = 3.2$ of the ratio of total to selective extinction.

vector point diagram are consistent with the combined effects of observational errors, the estimated internal velocity dispersion, and projection on the sky [150]. Note how much more extended the HR-diagram, not corrected for reddening, of the association is after Hipparcos membership selection. The data now extend to beyond spectral type F and in fact some PMS objects are included (see §7.1). Finally, the panels in the bottom row of Figure 3 show the not-selected stars, and demonstrate that the membership selection procedure does not leave 'holes' in the distributions of positions, proper motions, and parallaxes. This indicates that UCL was separated cleanly from the field stars.

5.2. SELECTED RESULTS

The outcome of the Hipparcos membership selection is described below with a couple of specifically chosen examples from de Zeeuw et al. [150], meant to illus-

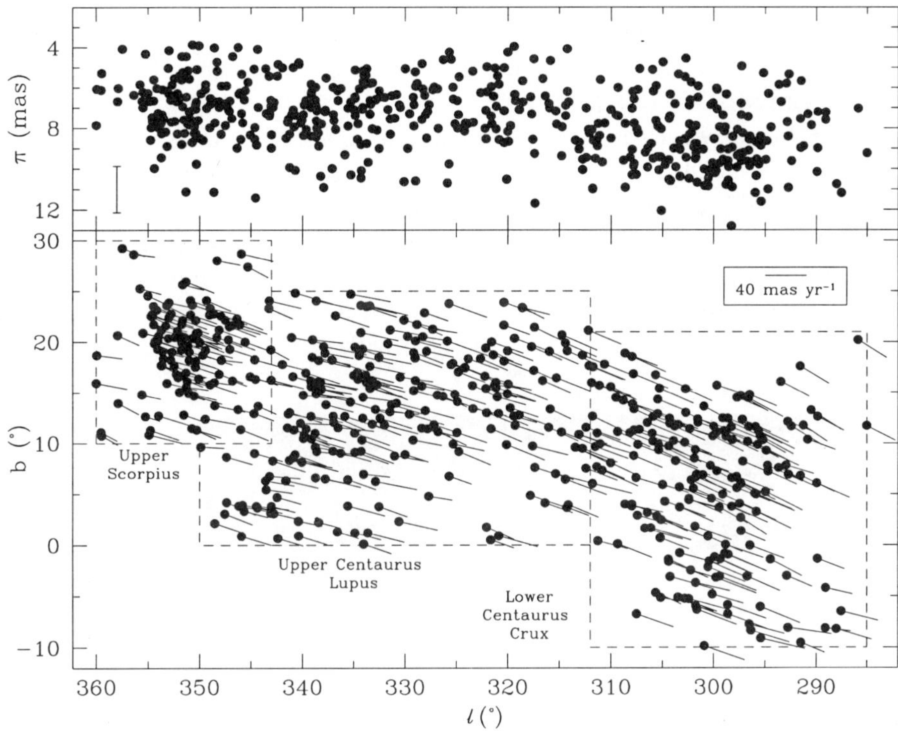

Figure 4. Positions and proper motions (bottom), and parallaxes (top), for 521 members of Sco OB2 selected from 7974 stars in the Hipparcos Catalogue in the area bounded by the dashed lines. The vertical bar in the top panel corresponds to the average $\pm 1\sigma$ parallax range for the stars shown. US is identified as a subgroup based on the concentration of members on the sky, and LCC can be distinguished from US and UCL based on its significantly smaller distance.

trate the various ways in which high-precision proper motion surveys can impact the field of OB associations. The case of Sco OB2 was chosen to show that the membership lists can now be extended into the PMS regime. The findings on Vel OB2 and Tr 10 show the dramatic improvements in the ability to identify OB associations (Vela), and how in some cases what was thought to be an open cluster actually turns out to be an association (Trumpler 10). The results on Col 121 show that we can now detect analogues of Sco OB2 at much larger distances (a factor four in this case). Cep OB6 is an example of a newly discovered association. Finally, we discuss Ori OB1 as an example of some of the difficulties encountered in identifying the members of associations.

Scorpius OB2. Figure 4 illustrates the proper motions of all the Sco OB2 members that have been identified, and also gives the subgroup boundaries. In US a total of 120 members were found located in a volume of $\sim$30 pc diameter at a mean distance of 145 $\pm$ 2 pc (see §5.3 for more information on the determination of mean distance). In UCL 221 members at 140 $\pm$ 2 pc were found, and in Lower Centaurus Crux (LCC) 180 at 118 $\pm$ 2 pc.

Sco OB2 clearly forms one coherent structure, although US stands out in the distribution of Sco OB2 members on the sky, and the parallax distribution clearly distinguishes UCL and LCC. The differences between the Hertzsprung–Russell diagrams of the groups [61] also indicate that a division of Sco OB2 into three separate subgroups is warranted. The field boundary separating UCL from US has, somewhat arbitrarily, been chosen in such a way that US comprises the stellar concentration centered on $(\ell, b) \sim (352°, 20°)$ with radius $\sim 5°$.

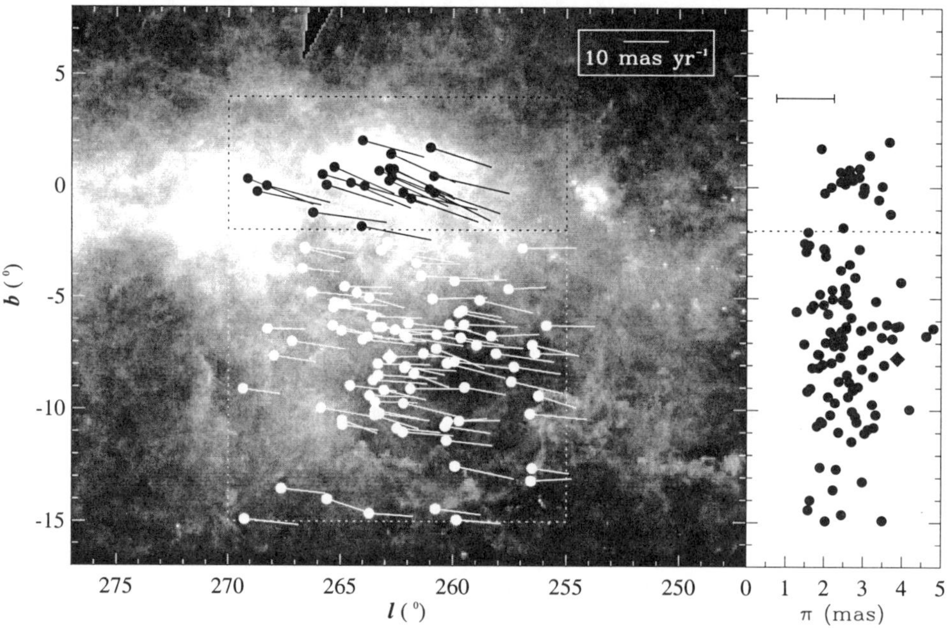

Figure 5.   Left: positions and proper motions for the Hipparcos members of Vela OB2 (white) and Trumpler 10 (black). The diamond denotes the Wolf–Rayet star γ^2 Velorum (WR11). The dotted lines indicate the field boundaries. The grey scale represents the IRAS 100 μm skyflux. The IRAS Vela shell is the ring-like structure centered on $(\ell, b) \sim (263°, -7°)$ with a radius of $\sim 6°$ surrounding Vel OB2. The intense emission in the area $260° \lesssim \ell \lesssim 273°$, $-2° \lesssim b \lesssim 2°$ corresponds to the Vela molecular ridge. Right: parallax distribution for Vel OB2 and Tr 10.

Vela OB2 and Trumpler 10.   In his 1914 paper, Kapteyn not only identified Sco OB2, but also discussed a group of bright stars in Vela and, based on proper motions, listed 15 probable members for this so-called Vela Group [80]. Ever since a clear identification of the members of this group has remained difficult

[150]. However, the new Hipparcos results, illustrated in Figure 5, show very clearly the presence of an OB association in the Vela region. In fact 93 members have been identified of which 89 are new! The color-magnitude diagram of this association shows that the earliest spectral type on the main sequence is B1, suggesting an age of $\lesssim 10$ Myr. The association is located at a distance of 410 ± 12 pc, and the new members are concentrated on the sky around $(\ell, b) \sim (263°, -7°)$ within a radius of $\sim 5°$. Sahu [123] reported the detection of the so-called IRAS Vela shell in the IRAS Sky Survey Atlas maps (cf. [124]). This is an expanding shell, centered on Vel OB2, with a projected radius of $\sim 6°$ (Figure 5). Sahu showed that the observed kinetic energy of the IRAS Vela shell is of the same order of magnitude as the total amount of energy injected into the interstellar medium through the combined effects of stellar winds and supernovae if the shell were to contain a 'standard' OB association [123]. The subsequent astrometric identification of Vel OB2 as a rich OB association, and the improved distance determination, confirm the relation of the association with the Vela shell.

Based on relative proper motion data for 29 stars, Lyngå [95], [96] identified 19 probable members of the sparse open cluster Trumpler 10. Analysis of the Hipparcos data showed that Tr 10 is actually a moving group. The Catalogue contains 23 members: 22 B-type stars (earliest spectral type B3V) and 1 A0V star. Figure 5 shows the members of Tr 10 seen in projection in front of the Vela molecular ridge. Tr 10 is located at a distance of 366 ± 23 pc. Based on its color-magnitude diagram this group is clearly older than Vel OB2, and a provisional age estimate based on the earliest spectral type is ~ 15 Myr.

Collinder 121. Collinder [35] studied the structural properties and spatial distribution of Galactic open clusters, and discovered a cluster of 20 stars at ~ 590 pc in an area of $1° \times 1°$ on the sky: Col 121. Schmidt–Kaler [127] noted a large number of evolved early-type stars in a field of $10° \times 10°$ centered on the bright supergiant o CMa located in the central part of Col 121. Subsequent studies revealed the possible existence of an OB association in this field [150].

This is indeed confirmed by the Hipparcos results. 103 stars were selected in the Col 121 field with a mean distance of 592 ± 28 pc. Figure 6 shows both the stars previously considered members of this group as well as the newly selected Hipparcos members. The figure also presents the Hipparcos color-magnitude diagram, not corrected for reddening. It shows that Col 121 contains a number of evolved stars. The abrupt cutoff of the main sequence near $V = 10$ is caused by the completeness limit of the Hipparcos Catalogue. Some of the late-type stars may well be interlopers. The presence of an O star and early-type B stars indicates this is a young group, of age ~ 5 Myr.

Col 121 has completely changed its appearance compared to the classical membership lists. The middle panel of Figure 6 suggests two subgroups, $(\ell, b) \sim (233°, -9°)$ and $(238°, -9°)$. A possible third subgroup lies at $(\ell, b) \sim (243°, -9°)$.

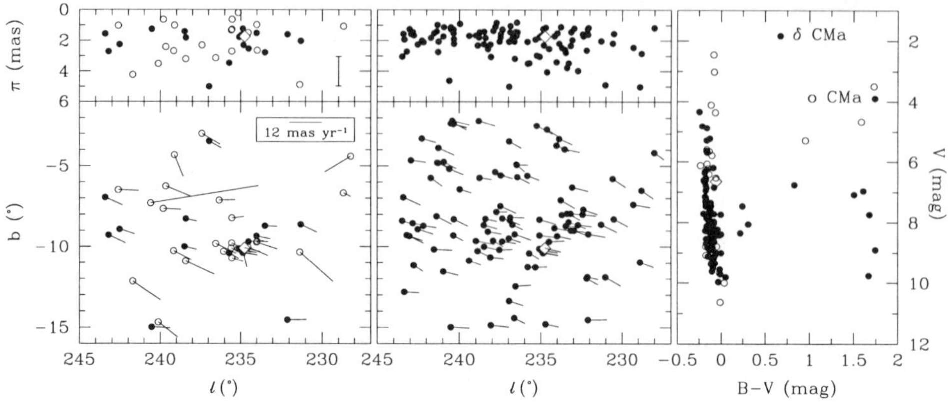

Figure 6. Left: positions and proper motions (bottom) and parallaxes (top) of the pre-Hipparcos members of Collinder 121. Filled circles are confirmed members. Open circles indicate stars not considered members after analysis of Hipparcos data. The confirmed member EZ CMa (WR6, see [150]) is indicated by an open diamond. Middle: same diagram for all stars selected as member of Col 121, illustrating the dramatic change from pre- to post-Hipparcos. Right: color-magnitude diagram, not corrected for reddening, for the Col 121 members (filled circles), and rejected classical members (open circles). The unusual position of EZ CMa (open diamond) may be caused by systematic effects in the Hipparcos photometry of this star [150].

The linear dimensions of this complex are 100×30 pc, similar to that of the Sco OB2 association. The color-magnitude diagram also resembles that of Sco OB2, in the sense that the earliest spectral type and amount of extinction are similar.

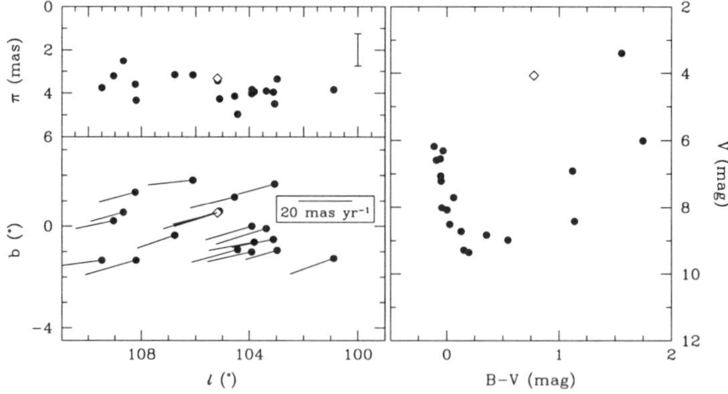

Figure 7. Left: positions and proper motions (bottom) and parallaxes (top) for the new moving group in Cepheus, designated Cepheus OB6. Right: color-magnitude diagram, not corrected for reddening. The open diamond denotes δ Cep.

Cepheus OB6.   Following a discovery by Hoogerwerf et al. [74], de Zeeuw et al. [150] examined the field $100° < \ell < 110°$ and $-4° < b < 3°$ in the Cepheus region and found a previously unknown moving group. This group, designated as Cepheus OB6, consists of 20 members. Their positions, proper motions and parallaxes are shown in Figure 7. The color-magnitude diagram is very narrow and strengthens the evidence that these stars form a moving group. It is probably an old OB association: the earliest spectral type is B5III, suggesting an age of $\sim$50 Myr. A noteworthy member of this group is δ Cep, the prototype classical Cepheid.

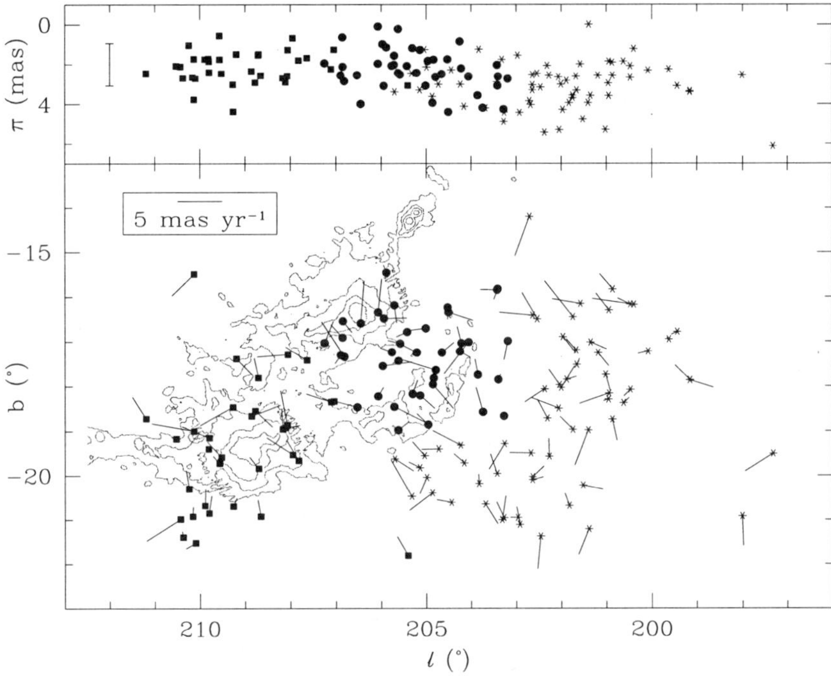

Figure 8.   Positions and proper motions (bottom) and parallaxes (top) for the stars of Ori OB1 selected by Brown et al. [30]. The proper motions are small because Ori OB1 lies near the direction of the Solar Antapex. The parallaxes in subgroup 1a (asterisks) are generally larger than those in 1b (filled circles) and 1c (filled squares). The contours indicate the 100 μm IRAS flux map.

Other Associations.   Astrometric membership detection for OB associations with Hipparcos data was successful out to $\sim$ 650 pc. The following associations were also detected: Per OB2, α Persei (Per OB3), Cas–Tau, Lac OB1, and Cep OB2. Astrometric evidence for moving groups in the fields of R CrA, CMa OB1, Mon OB1, Ori OB1, Cam OB1, Cep OB3, Cep OB4, Cyg OB4, Cyg OB7, and Sct OB2, is inconclusive. OB associations do exist in many of these regions, but they are either at distances where the Hipparcos parallaxes are of limited use, or they have unfavorable kinematics, so that the group proper motion does not distinguish them from the field stars in the Galactic disk.

Among the latter group we note especially Ori OB1. It is the best-studied nearby OB association but unfortunately it is located close to the Solar Antapex, and its space motion is almost purely radial with respect to the Sun. This means that it is very difficult to select members in Ori OB1 based on proper motions. A discussion on Ori OB1 is given in [150] based on the preliminary study described in detail in [30]. In that study a very crude selection of Ori OB1 members was done, based on Hipparcos proper motions, and the mean distances to the subgroups were calculated. These distances are: 336 ± 16 pc for 1a, 473 ± 33 pc for 1b and 506 ± 37 pc for 1c, where the quoted errors are the formal errors on the mean distances. The actual uncertainty is larger due to the simplified member selection. All of this is illustrated in Figure 8 which shows the proper motions and parallaxes of the selected members. Note that from the parallaxes alone it is clear that subgroup 1a is located much closer to the Sun than 1b and 1c [30].

5.3.　MEAN DISTANCES AND MOTIONS

Use of the Hipparcos trigonometric parallaxes of the secure members to determine mean distances to the associations or their subgroups requires some caution, as the inverse of the parallax is a biased distance indicator [128], [28], and the conversion of mean parallax to mean distance for a group of stars depends on the distribution of stars within the group. It can be shown [150] that for all spherical groups the expectation value of the mean of the measured parallaxes is equal to the true mean parallax, and corresponds to the true distance of the centre of the group. For elongated associations the bias in the mean parallax is small, typically less than 1 per cent. Hipparcos parallaxes measured in regions of high stellar density (in the Catalogue) have to be interpreted with care [93], [114], [121]. This is not a problem for the low-density associations.

Furthermore, the observed distribution of parallaxes may not be representative of the true underlying parallax distribution of an association. Magnitude limits and the characteristics of the Hipparcos Input Catalogue bias the selected members (see [150] for more details), and some stars in the Hipparcos Catalogue have a negative measured parallax. The member selection method described in §5.1 rejects these, which introduces a bias towards a smaller mean distance. De Zeeuw et al. [150] were able to correct the mean distances to the associations for this bias, by carrying out Monte Carlo simulations to estimate its magnitude.

The resulting distances and projected sizes of the kinematically detected nearby OB associations are listed in Table 1. This table also lists the mean proper motions. The means are based on the individual measurements for the association members. The Hipparcos Input Catalogue lists radial velocities from a variety of sources. They are not available for all secure association members, and therefore Table 1 lists the median radial velocity, and no error estimates are given.

The mean distances derived in Table 1 have very small formal errors due to the averaging over large numbers of stars. However, associations have large physical sizes, and the mean distances listed here should not be taken as 'the distance' to all association members, but rather as an indication of the location of their centroid.

The comparison between the new Hipparcos distances and the distances in the literature for OB associations reveals a tight correlation, although the Hipparcos distances are systematically smaller by about 0.2 magnitudes in the distance modulus. This issue is discussed in more detail by de Zeeuw et al. [150].

To conclude we return to the issues discussed in §3. The projected sizes of the associations correlate roughly with their distances. This is obviously not a real physical effect but merely a reflection of how the field boundaries were established prior to membership selection. The case of Col 121 is a good illustration of how an association like Sco OB2, which consists of three subgroups, may resemble a single entity at larger distances (see Figure 6). Moreover, any kinematic distinction between subgroups within an association is difficult to establish (cf. Figure 4), and extra information, such as photometry, is required. Thus, even though OB associations may possibly form with particular characteristic dimensions one should always bear in mind the observational biases that enter when addressing the question of initial configurations of young star clusters.

6. Gould's Belt and the Origin of the Nearby Associations

Mean space motions of the nearby OB associations, in km s^{-1}, were derived from the mean proper motions, mean distances, and the median radial velocities [150]. Figure 9 shows the result, after subtraction of Solar motion [42] (lower panel) and, additionally, differential Galactic rotation [54] (upper panel). When considering the motions with respect to the Local Standard of Rest, some of the associations in Figure 9 (lower panel) seem to fit a coherent pattern of expansion and rotation, which is very similar to that derived from Hipparcos measurements of OB stars with ages less than $\sim$30 Myr [92], [133]. This large-scale feature is known as Gould's Belt, which is the flat system of early-type stars within $\sim$500 pc [64], associated with a large structure of interstellar matter, including reflection nebulae, dark clouds, and H I. Its most striking feature is a tilt of $\sim$18° with respect to the Galactic plane. It has also been detected in the distribution of young stars observed in X-rays by ROSAT [66]. Pöppel [115] recently wrote a comprehensive review of this structure, with emphasis on the role and characteristics of the interstellar medium.

Following a suggestion by Blaauw [10], who studied the mean space motions of the nearby OB associations, Lindblad [91] interpreted the observations of the local H I gas associated with Gould's Belt in terms of a ballistically expanding

TABLE 1. Mean distances and mean motions of the nearby OB associations

Name	D	N	Size	$\langle \mu_\ell \cos b \rangle$	$\langle \mu_b \rangle$	$v_{\rm rad}$	Age	$N_{\rm br}$
US	145 ± 2	120	30	−24.5	−8.1	−4.6	5	5
UCL	140 ± 2	221	65	−30.1	−9.1	4.9	13	10
LCC	118 ± 2	180	45	−32.1	−13.1	12.0	10	4
Vel OB2	410 ± 12	93	70	−10.4	−1.3	18.0	10	12
Tr 10	366 ± 23	23	45	−14.3	−4.9	21.0	15	0
Col 121	592 ± 28	103	115	−5.1	−1.5	26.0	5	24
Per OB2	318 ± 27	41	45	8.4	−2.3	20.1	4–8	2
α Persei	177 ± 4	79	15	33.5	−8.7	−1.0	50	2
Lac OB1	368 ± 17	96	65	−2.3	−3.4	−13.3	16	7
Cep OB2	615 ± 35	71	105	−4.1	−0.5	−21.4	5	11
Cep OB6	270 ± 12	20	40	15.9	−4.4	−20.0	50	2
Ori OB1a*	336 ± 16	61	30	0.8	0.1	−23.0	11	6
Ori OB1b*	473 ± 33	42	30	0.8	0.1	−23.0	2	9
Ori OB1c*	506 ± 37	34	30	0.8	0.1	−23.0	5	12

All distances, in pc, include a correction for systematic effects as described in §5.3. The first column lists the association. The second column contains the mean distance and the error on the mean. The third column lists the number of Hipparcos members. Column 4 lists the projected size in pc of the associations derived from their extent on the sky and assuming spherical symmetry. Columns 5 and 6 list the average proper motions in the directions of Galactic longitude and latitude. The formal errors on the mean are typically 0.1–0.2 mas yr^{-1}. Column 7 contains the median radial velocity in km s^{-1} compiled from the Hipparcos Input Catalogue. Column 8 contains the (approximate) age in Myr; see [150] for references. Column 9 contains the number of stars with absolute magnitude brighter than −2. The absolute magnitudes were calculated from the mean distances and values for V as listed in the Hipparcos Catalogue. Due to its large physical size and unknown orientation, the Cas–Tau association cannot be represented adequately by a single mean distance, proper motion, and radial velocity, and has been excluded from this table.
* The distances and numbers of members for Ori OB1 are based on an *ad hoc* selection procedure and should be regarded as *provisional* numbers only.

ring. Subsequent kinematic models combined with new observations confirmed this picture (e.g., [46], [109]). However, these models seem inadequate for a description of the space motions of the clumpy distribution of early-type stars, as no unique expansion center and/or expansion age can be defined for them (e.g., [10], [36], [88], [146]). The lack of homogeneous and accurate radial velocities for all local early-type stars prevents an optimal exploitation of the Hipparcos data, and a full kinematic model is not yet available. Even so, we can use the mean motions of the OB associations to shed light on the systemic motions in Gould's Belt.

The associations Sco OB2, Ori OB1, Per OB2, and possibly Lac OB1, are thought to be components of the Gould Belt (e.g., [109]). The pattern displayed in the lower panel of Figure 9 seems to be shared by Tr 10, and perhaps also by Vel OB2 and Col 121. Moreover, all these associations are younger than ~

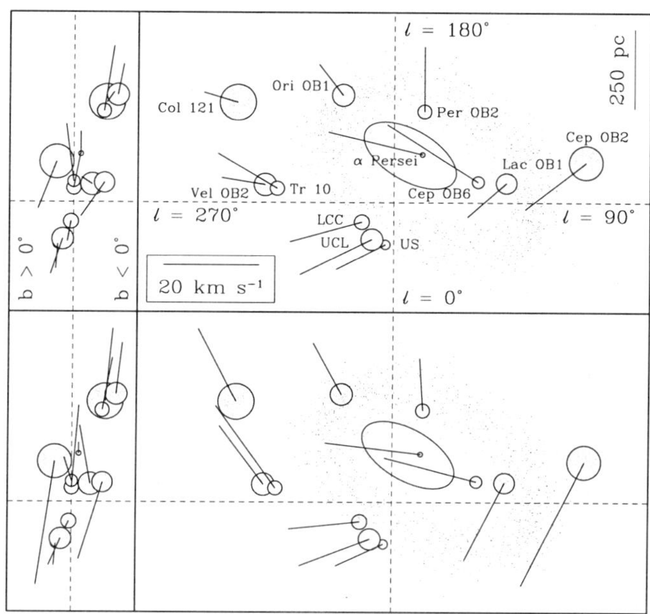

Figure 9.   Locations of the kinematically detected OB associations projected onto the Galactic plane (right, cf. Figure 1) and a corresponding cross section (left). The grey circles indicate the physical dimensions as obtained from the angular dimensions and mean distances, on the same scale. The lines represent the streaming motions, derived from the average proper motions, mean distances and median radial velocities of the secure members, corrected for 'standard' Solar motion and Galactic rotation (upper panel) or for Solar motion only (lower panel). The ellipse around the α Persei cluster indicates the Cas–Tau association. The small dots schematically represent a model of the Gould Belt [109].

20 Myr, suggesting that these may all belong to the same coherent structure. If it is part of Gould's Belt, the new distance of Lac OB1 reduces the extent of the Belt in the direction $\ell \sim 90°$. Cep OB2 is not located in the plane of the Belt, and its motion is parallel to the Galactic plane. It does not seem to belong to the Belt. The Cas–Tau complex which surrounds the α Persei cluster and shares its motion, as well as Cep OB6, are located inside the main ring of associations, have a different space motion, and are all significantly older at $\sim$50 Myr.

The following scenario was suggested for the formation of Gould's Belt and its constituent associations [47], [49]. The Carina spiral arm is presently $\sim$4 kpc from the Sun along the Solar circle at $\ell = 282°$ [65]. If the pattern speed of the associated spiral density wave is 13.5 km s^{-1} kpc^{-1} [149], then the physical speed of the local arm is 114 km s^{-1} in the tangential direction. This means that the time since the passage of the Carina spiral arm near the Sun would have been 35 Myr without streaming motions parallel to the arm. With streaming, the time may be twice this value. As the oldest components of Gould's Belt have a similar age, this suggests that the Belt began as a giant gaseous condensation in the Carina spiral arm when the location of the arm and the Sun last coincided. In this gas probably the Cas–Tau, α Persei, Cep OB6, and Pleiades clusters formed, as well as many dispersed B and later-type stars [88], [115]. The combined action of the stellar winds and supernovae from the massive stars in these older stellar groups may have led to the ring-like gaseous structure from which the younger associations formed [13]. An alternative scenario for the origin of Gould's Belt, invoking the oblique impact of a large high-velocity cloud onto the Galactic disk, was put forward to explain the tilt in Gould's Belt [36].

7. Future Perspectives

In his 1991 chapter on associations Blaauw concluded with the promise of major advances in the field through the availability of Hipparcos data as well as much improved radial velocities. Although precise radial velocities for the early-type stars are not yet available, the Hipparcos data have now been analyzed and have put the issue of association membership on much firmer ground. Below we will discuss how this provides an excellent start for future investigations and we briefly discuss what is in store for work on OB associations beyond the Galaxy.

7.1. THE NEARBY OB ASSOCIATIONS

Hipparcos has provided a much improved description of the ensemble of young stellar groups in the Solar neighbourhood, out to a distance of $\sim$650 pc. The list of members in the astrometrically detected associations has in all cases been refined and extended and a much better kinematical description of Gould's Belt is now available. With the number of candidate members greatly reduced, we are now in an excellent position to start gathering additional data such as photometry and radial velocities, which will help in further refining the membership lists and studying the physical properties of the association members.

The discovery of Cep OB6 in the Cep OB2 field [150] suggests that there might be other previously unidentified nearby associations. A systematic search in regions of the strip $-30° \leq b \leq 30°$ not covered by de Zeeuw et al. [150] might reveal such groups. It will also be interesting to investigate whether, as in the Cas–Tau/α Persei pair, more open clusters exist with an extended halo, which appears as an old association.

De Zeeuw et al. [150] presented color-magnitude diagrams based on the V and $B - V$ data in the Hipparcos Catalogue, but the physical parameters of the member stars were not discussed in detail, mostly because the required homogeneous multi-color photometry (and spectral classification) is incomplete. It is now relatively easy to acquire the necessary photometry. This is particularly interesting in Sco OB2, where the faintest astrometric members can be connected directly to the populations of PMS objects discovered through X-ray searches. This will provide accurate ages, initial mass functions [29], and the energy and momentum input into the interstellar medium, thus enabling a detailed investigation of the influence of young stellar groups on the surrounding distribution of gas and dust. The Leiden–Dwingeloo H I survey [69], which has recently been completed for the southern sky, can be used to undertake a comprehensive study of the interstellar medium surrounding the nearby associations. It will also be interesting for the nearest associations to combine the Hipparcos parallax measurements with absorption-line measurements of gas in order to gain insight into the 3D-structure of the interstellar medium.

An important next step is to complement the uniform astrometric measurements provided by Hipparcos with homogeneous radial velocities with accuracies of ~ 3 km s^{-1} or better, which is a considerable challenge for earlier spectral types (cf. §4.1). An unbiased member selection based only on radial velocities is still impractical, but measurement of the radial velocity of the proper motion members identified here is feasible. This will allow removal of a significant number of remaining interlopers [150] and will provide further information on the distribution of spectroscopic binaries in these groups. Combined with the many new astrometric binaries discovered by Hipparcos this will greatly improve our knowledge of the binary population in OB associations.

The Hipparcos parallaxes are not sufficiently accurate to resolve the internal structure of even the nearest associations. This would be interesting, as it might help to delineate substructure, and hence shed light on the details of the star formation process throughout an interstellar cloud. However, it is possible to improve the individual distance estimates by using the proper motions (and radial velocities) of established members to compute so-called secular parallaxes (e.g., [43]).

The lists of astrometrically selected members of OB associations are incomplete beyond $V \sim 7.3$, and a few genuine bright members (including some long-period binaries) may have been excluded (see [150] for details). The lists extend to $V \sim 10.5$, and include a few PMS objects in the nearest groups. Available ground-based studies which can now be put on the Hipparcos reference system can possibly complete the member list for the bright stars, and extend them to fainter magnitudes. Unfortunately, the space motions of the young groups are not large, and as a result the proper motions of the group members do not differ very much from those of the field stars. Reliable extension of the membership lists requires proper motions with accuracies of order 2 mas yr^{-1} or better. These are provided by the ACT Catalog [134] and the TRC Catalogue [73], through a combination of the positions in the Astrographic Catalog with the Tycho positions to $V \sim 11$. The individual stellar positions in these two catalogues have modest accuracy, but the ~ 80 yr epoch difference results in quite accurate proper motions. The first installment of the ACT and TRC catalogues both contain about 1 million measurements. The completion of the second edition of the Tycho Catalogue [72] will provide proper motions of similar quality for about 2.5 million objects to

$V \sim 12$. However, the 1 mas accuracy parallaxes obtained by Hipparcos, which are generally not available for fainter association members, play a crucial role in culling interlopers from the membership lists. A complete study of all OB associations in the Solar neighbourhood (extent, distance, structure, kinematics) has to await the future GAIA space astrometry mission [112], [63], which will gather astrometric data at the 10 micro-arcsecond precision level to 20th magnitude.

7.2. BEYOND THE GALAXY

Armed with a detailed understanding of associations in the Solar vicinity and throughout the Milky Way, we can now venture further and investigate the vast variety of extragalactic star forming regions, ranging from individual associations to complete star bursts.

The nearest extragalactic associations are to be found in the Large Magellanic Cloud. For example, at the heart of the 30 Doradus nebula one finds the spectacular R136 cluster, which has a stellar density ~ 200 times greater than that of a typical OB association and contains 39 identified O3 stars [98]. This cluster is even considered a possible analogue of a young globular cluster. Thus by studying R136 we may gain a much more detailed understanding of the 'super star clusters' observed with the Hubble Space Telescope in other galaxies (e.g., [37], [108], [148]), which are also thought to be progenitors of globular clusters.

One important lesson that has been learned from studying massive stars is that because of their very high effective temperatures ($> 30\,000$ K) it is vital to perform spectroscopy of these stars if one wants to obtain reliable physical parameters, such as masses [97]. High sensitivity combined with high spatial resolution are thus needed to perform detailed studies of the luminous young star clusters observed in galaxies in the Local Group and beyond. This will require the capabilities of the Hubble Space Telescope or the largest ground-based facilities. A good understanding of these extragalactic associations and the role they play will be very important if we want to make sense of the observations of the earliest epochs of star and galaxy formation that will eventually be carried out by future millimeter-arrays and by the Next Generation Space Telescope.

References

1. Alcalá, J.M., Terranegra, L., Wichmann, R., et al., 1996, A&AS, 119, 7
2. Alcalá, J.M., Chavarría–K., C., Terranegra, L., 1998, A&A, 330, 1017
3. Ambartsumian, V.A., 1947, in *Stellar Evolution and Astrophysics*, Armenian Acad. of Sci. (German translation, 1951, Abhandl. Sowjetischen Astron., 1, 33)
4. Ambartsumian, V.A., 1949, Dokl. Akad. Nauk SSR, 68, 22
5. Anthony–Twarog, B.J., 1982, AJ, 87, 1213
6. Bertiau, F.C., 1958, ApJ, 128, 533
7. Blaauw, A., 1956, ApJ, 123, 408
8. Blaauw, A., 1961, Bull. Astr. Inst. Neth., 15, 265
9. Blaauw, A., 1964, ARA&A, 2, 213

10. Blaauw, A., 1965, Proc. Royal Neth. Acad. of Sciences, 74, 54
11. Blaauw, A., 1978, in *Problems of Physics and Evolution of the Universe*, ed. L.V. Mirzoyan (Yerevan), 101
12. Blaauw, A., 1983, Irish Astron. J., 16, 141
13. Blaauw, A., 1991, in *The Physics of Star Formation and Early Stellar Evolution*, eds C.J. Lada & N.D. Kylafis, NATO ASI Ser. C, Vol. 342 (Dordrecht: Kluwer), 125
14. Blaauw, A., 1993, in *Massive Stars: Their Lives in the Interstellar Medium*, eds J.P. Cassinelli & E.B. Churchwell, ASP Conf. Ser., 35 (San Francisco), 207
15. Bloemen, H., Wijnands, R., Bennett, K., et al., 1994, A&A, 281, L5
16. Blitz, L., 1980, in *Giant Molecular Clouds in the Galaxy*, eds P.M. Solomon & M.G. Edmunds (Oxford: Pergamon Press), 1
17. Blitz, L., Williams, J.P., 1997, ApJ, 488, L145
18. Bodenheimer, P., Ruzmaikina, T., Mathieu, R.D., 1993, in *Protostars and Planets III*, eds E.H. Levy & J.I. Lunine (Tucson: University of Arizona Press), 367
19. Bok, B.J., 1934, Harvard College Obs. Circ., 384, 1
20. Boland, W., van Woerden, H., eds, 1985, *Birth and Evolution of Massive Stars and Stellar Groups*, Astroph. and Sp. Science Lib. 120 (Dordrecht: Reidel)
21. Bonnell, I.A., Bate, M.R., Clarke, C.J., Pringle, J.E., 1997, MNRAS, 285, 201
22. Bonnell, I.A., Bate, M.R., Zinnecker, H., 1998, MNRAS, 298, 93
23. Brandner, W., Köhler, R., 1998, ApJ, 499, L79
24. Brooks, K.J., Whiteoak, J.B., Storey, J.W.V., 1998, Publ. Astron. Soc. Aust., 15, 202
25. Brown, A.G.A., de Geus E.J., de Zeeuw P.T., 1994, A&A, 289, 101
26. Brown, A.G.A., Hartmann, D., Burton, W.B., 1995, A&A, 300, 903
27. Brown, A.G.A., Dekker, G., de Zeeuw, P.T., 1997, MNRAS, 285, 479
28. Brown, A.G.A., Arenou, F., van Leeuwen, F., Lindegren, L., Luri, X., 1997, ESA SP–402, 63
29. Brown, A.G.A., 1998, in *The Stellar Initial Mass Function*, eds G.F. Gilmore & D. Howell, ASP Conf. Ser., 142 (San Francisco), 45
30. Brown, A.G.A., Walter, F.M., Blaauw, A., 1998, in *The Orion Complex Revisited*, eds M.J. McCaughrean & A. Burkert, ASP Conf. Ser. (San Francisco), in press
31. de Bruijne, J.H.J., 1999, MNRAS, submitted
32. Burrows, D.N., Singh, K.P., Nousek, J.A., Garmire, G.P., Good, J., 1993, ApJ, 406, 97
33. Cassinelli, J.P., Churchwell, E.B., eds, 1993, *Massive Stars: Their Lives in the Interstellar Medium*, ASP Conf. Ser., 35 (San Francisco)
34. Clarke, C.J., Pringle, J.E., 1992, MNRAS, 255, 423
35. Collinder, P., 1931, Ann. Obs. Lund, 2, No. 1
36. Comerón, F., Torra, J., 1994, A&A, 281, 35
37. Conti, P.S., Vacca, W.D., 1994, ApJ, 423, L97
38. Cunha, K., Lambert, D.L., 1992, ApJ, 399, 586
39. Cunha, K., Lambert, D.L., 1994, ApJ, 426, 170
40. Cunha, K., Smith, V.V., Lambert, D.L., 1995, ApJ, 452, 634
41. Cunha, K., Smith, V.V., Lambert, D.L., 1998, ApJ, 493, 195
42. Dehnen, W., Binney, J.J., 1998, MNRAS, 298, 387
43. Dravins, D., Lindegren, L., Madsen, S., Holmberg, J., 1997, ESA SP–402, 733
44. Eisenhauer, F., Quirrenbach, A., Zinnecker, H., Genzel, R., 1998, ApJ, 498, 278
45. Elmegreen, B.G., Lada, C.J., 1977, ApJ, 214, 725
46. Elmegreen, B.G., 1982, in *Submillimeter wave astronomy*, eds J.E. Beckman & J.P. Phillips (Cambridge: Cambridge University Press), 3
47. Elmegreen, B.G., 1993, in *Protostars and Planets III*, eds E.H. Levy & J.I. Lunine (Tucson: University of Arizona Press), 97
48. Elmegreen, B.G., Efremov, Y.N., 1996, ApJ, 466, 802
49. Elmegreen, B.G., Efremov, Y.N., 1998, in *The Orion Complex Revisited*, eds M.J. McCaughrean & A. Burkert, ASP Conf. Ser. (San Francisco), in press
50. ESA, 1997, The Hipparcos and Tycho Catalogues, ESA SP–1200
51. Favata, F., Micela, G., Sciortino, S., 1996, A&A, 311, 951

52. Favata, F., Micela, G., Sciortino, S., 1997, A&A, 326, 647
53. Favata, F., Micela, G., Sciortino, S., D'Antona, F., 1998, A&A, 335, 218
54. Feast, M., Whitelock, P., 1997, MNRAS, 291, 683
55. Ferrière, K.M., 1995, ApJ, 441, 281
56. Ferrière, K.M., 1998, ApJ, 497, 759
57. Ferrière, K.M., 1998, ApJ, 503, 700
58. Gies, D.R., Bolton, C.T., 1986, ApJS, 61, 419
59. Garmany, C.D., 1973, AJ, 78, 185
60. Garmany, C.D., 1994, PASP, 106, 25
61. de Geus, E.J., de Zeeuw, P.T., Lub, J., 1989, A&A, 216, 44
62. de Geus, E.J., 1992, A&A, 262, 258
63. Gilmore, G.F., Perryman, M.A.C., Lindegren, L., et al., 1998, in *Astronomical Interferometry*, SPIE Proc. 3350, eds R.D. Reasenberg & M. Shao, in press
64. Gould, B.A., 1874, Proc. AAAS, 115
65. Graham, J.A., 1970, AJ, 75, 703
66. Guillout, P., Sterzik, M.F., Schmitt, J., Motch, C., Neuhäuser, R., 1998, A&A, 337, 113
67. Guo, Z., Burrows, D.N., Sanders, W.T., Snowden, S.L., Penprase, B.E., 1995, ApJ, 453, 256
68. Haro, G., 1953, ApJ, 117, 73
69. Hartmann, D., Burton, W.B., 1997, *Atlas of Galactic Neutral Hydrogen* (Cambridge: Cambridge University Press)
70. Hillenbrand, L.A., 1997, AJ, 113, 1733
71. Hillenbrand, L.A., Hartmann, L., 1998, ApJ, 492, 540
72. Høg, E., 1997, ESA SP–402, 25
73. Høg, E., Kuzmin, A., Bastian, U., Fabricius, C., Kuimov, K., Lindegren, L., Makarov, V.V., Röser, S., 1998, A&A, 335, L65
74. Hoogerwerf, R., de Bruijne, J.H.J., Brown, A.G.A., Lub, J., Blaauw, A., de Zeeuw, P.T., 1997, ESA SP–402, 571
75. Hoogerwerf, R., Aguilar, L.A., 1999, MNRAS, submitted
76. Humphreys, R.M., 1978, ApJS, 38, 309
77. Jordi, C., Trullols, E., Rosello, G., Lahulla, F., 1992, A&AS, 94, 519
78. Jordi, C., Trullols, E., Galadí–Enríquez, D., 1996, A&A, 312, 499
79. Janes, K.A., ed., 1991, *The Formation and Evolution of Star Clusters*, ASP Conf. Ser., 13 (San Francisco)
80. Kapteyn, J.C., 1914, ApJ, 40, 43
81. Kapteyn, J.C., 1918, ApJ, 47, pp 104, 146, 255
82. Kogure, T., Ogura, K., Nakano, M., Yoshida, S., 1992, PASJ, 44, 91
83. Koo, B.-C., McKee, C.F., 1990, ApJ, 354, 513
84. Lada, C.J., Margulis, M., Dearborn, D., 1984, ApJ, 285, 141
85. Lada, C.J., Kylafis, N.D., eds, 1991, *The Physics of Star Formation and Early Stellar Evolution*, NATO ASI Ser. C, Vol. 342 (Dordrecht: Kluwer)
86. Lada, C.J., Lada, E.A., 1991, in *The Formation and Evolution of Star Clusters*, ed. K. Janes, ASP Conf. Ser., 13 (San Francisco), 3
87. Lesh, J.R., 1968, ApJ, 152, 905
88. Lesh, J.R., 1968, ApJS, 17, 371
89. Lesh, J.R., 1969, AJ, 74, 891
90. Levato, H., Malaroda, S., Morrell, N., Solivella, G., 1987, ApJS, 64, 487
91. Lindblad, P.O., 1967, Bull. Astr. Inst. Neth., 19, 34
92. Lindblad, P.O., Palouš, J., Lodén, K., Lindegren, L., 1997, ESA SP–402, 507
93. Lindegren, L., 1989, ESA SP–1111, Vol. 3, 311
94. Lub, J., Pel, J.W., 1977, A&A, 54, 137
95. Lyngå, G., 1959, AfA, 2, 379
96. Lyngå, G., 1962, AfA, 3, 65
97. Massey, P., 1998, in *The Stellar Initial Mass Function*, eds G.F. Gilmore & D. Howell, ASP Conf. Ser., 142 (San Francisco), 17
98. Massey, P., Hunter, D.A., 1998, ApJ, 493, 180

99. Mathieu, R.D., 1986, in *Highlights of Astronomy 7*, 481
100. McCaughrean, M.J., Burkert, A., eds, 1998, *The Orion Complex Revisited*, ASP Conf. Ser. (San Francisco), in press
101. McKee, C.F., Ostriker, J.P., 1977, ApJ, 218, 148
102. McKee, C.F., Williams, J.P., 1997, ApJ 476, 144
103. Melnick, J., 1992, in *Star Formation in Stellar Systems*, eds G. Tenorio–Tagle, M. Prieto & F. Sánchez (Cambridge: Cambridge University Press)
104. Morrell, N., Levato, H., 1991, ApJS, 75, 965
105. Mouschovias, T.Ch., 1991, in *The Physics of Star Formation and Early Stellar Evolution*, eds C.J. Lada & N.D. Kylafis, NATO ASI Series C, Vol. 342 (Dordrecht: Kluwer), 61
106. Nakano, M., Wiramihardja, S.D., Kogure, T., 1995, PASJ, 47, 889
107. Nakano, M., McGregor, P.J., 1995, in *Future Utilisation of Schmidt Telescopes*, eds J Chapman, R. Cannon, S. Harrison & B. Hidayat, ASP Conf. Ser., 84 (San Francisco), 376
108. O'Connell, R.W., Gallagher, J.S., III, Hunter, D.A., Colley, W.N., 1995, ApJ, 446, L1
109. Olano, C.A., 1982, A&A, 112, 195
110. Pannekoek, A., 1929, Publ. Astron. Inst. Amsterdam, No. 2, 63
111. Parizot, E.M.G., 1998, A&A, 331, 726
112. Perryman, M.A.C., Lindegren, L., Turon, C., 1997, ESA SP–402, 743
113. Petr, M.G., Coudé du Foresto, V., Beckwith, S.V.W., Richichi, A., McCaughrean, M.J., 1998, ApJ, 500, 825
114. Pinsonneault, M.H., Stauffer, J., Soderblom, D., King, J., Hanson, R., 1998, ApJ, 504, 170
115. Pöppel, W.G.L., 1997, Fund. of Cosmic Phys., 18, 1
116. Preibisch, T., Günther, E., Zinnecker, H., Sterzik, M., Frink, S., Röser, S., 1998, A&A, 333, 619
117. Prusti, T., Adorf, H.–M., Meurs, E.J.A., 1992, A&A, 261, 685
118. Rasmuson, N.H., 1921, Lund Medd., Ser. II, 26, 1
119. Rasmuson, N.H., 1927, Lund Medd., Ser. II, 47b, 1
120. van Rensbergen, W., Vanbeveren, D., de Loore, C., 1996, A&A, 305, 825
121. Robichon, N., Arenou, F., Turon, C., Mermilliod, J.C., Lebreton, Y., 1997, ESA SP–402, 567
122. Ruprecht, J., 1966, IAU Trans., 12B, 348
123. Sahu, M.S., 1992, *PhD Thesis*, Groningen Univ.
124. Sahu, M.S., Blaauw, A., 1994, The Messenger, 76, 48
125. Scalo, J., 1998, in *The Stellar Initial Mass Function*, eds G.F. Gilmore & D. Howell, ASP Conf. Ser., 142 (San Francisco), 201
126. Sciortino, S., Damiani, F., Favata, F., Micela, G., 1998, A&A, 332, 825
127. Schmidt–Kaler, T., 1961, ZfA, 53, 28
128. Smith, H., Eichhorn, H., 1996, MNRAS, 281, 211
129. Snowden, S.L., Burrows, D.N., Sanders, W.T., Aschenbach, A., Pfeffermann, E., 1995, ApJ, 439, 399
130. Tian, K.P., van Leeuwen, F., Zhao, J.L., Su, C.G., 1996, A&AS, 118, 503
131. Tenorio–Tagle, G., Bodenheimer, P., 1988, ARA&A, 26, 145
132. Tenorio–Tagle, G., Prieto, M., Sánchez, F., eds, 1992, *Star Formation in Stellar Systems* (Cambridge: Cambridge University Press)
133. Torra, J., Gómez, A.E., Figueras, F., Comerón, F., Grenier, S., Mennessier, M.O., Mestres, M., Fernández, D., 1997, ESA SP–402, 513
134. Urban, S.E., Corbin, T.E., Wycoff, G.L., 1998, AJ, 115, 2161
135. Verschueren, W., David, M., 1989, A&A, 219, 105
136. Verschueren, W., David, M., Brown, A.G.A., 1996, in *The Origins, Evolution, and Destinies of Binary Stars in Clusters*, eds E.F. Milone & J.–C. Mermilliod, ASP Conf. Ser., 90 (San Francisco), 131
137. Verschueren, W., David, M., Vrancken, M., 1999, in *IAU Colloquium 170, Precise Stellar Radial Velocities*, eds J.B. Hearnshaw & C.D. Scarfe, ASP Conf. Ser. (San Francisco), in press
138. Walker, M.F., 1969, ApJ, 155, 447
139. Walter, F.M., Vrba, F.J., Mathieu, R.D., Brown, A., Myers, P.C., 1994, AJ, 107, 692

140. Walter, F.M., Wolk, S.J., Freyberg, M., Schmitt, J.H.M.M., 1997, in *Cool Stars in Clusters and Associations: Magnetic Activity and Age Indicators*, eds G. Micela, R. Pallavicini & S. Sciortino, Mem. Soc. Astr. It., Vol. 68, no. 4, 1081

141. Walter, F.M., Wolk, S.J., Sherry, W., 1998, in *Cool Stars, Stellar Systems, and the Sun*, eds R. Donahue & J. Bookbinder, in press

142. Walter, F.M., Alcalá, J.M., Neuhäuser, R., Sterzik, M., Wolk, S.J., 1999, in *Protostars and Planets IV*, eds V. Mannings, A.P. Boss, S. Russell (Tucson: Univ. of Arizona Press), in press

143. Warren, W.H., Hesser, J.E., 1977, ApJS, 34, 115

144. Warren, W.H., Hesser, J.E., 1977, ApJS, 34, 207

145. Warren, W.H., Hesser, J.E., 1978, ApJS, 36, 497

146. Westin, T.N.G., 1985, A&AS, 60, 99

147. Williams, J.P., McKee, C.F., 1997, ApJ, 476, 166

148. Whitmore, B.C., Schweizer, F., 1995, AJ, 109, 960

149. Yuan, C., 1969, ApJ, 158, 889

150. de Zeeuw, P.T., Hoogerwerf, R., de Bruijne, J.H.J., Brown, A.G.A., Blaauw, A., 1999, AJ, January issue

151. Zuckerman, B., Palmer, P., 1974, ARA&A, 12, 279

THE ROLE OF EMBEDDED CLUSTERS IN STAR FORMATION

ELIZABETH A. LADA

Department of Astronomy
University of Florida
211 Bryant Space Science Building
Gainesville, FL 32608 USA

1. Introduction

During the past century, astronomers have discovered that star formation is a continuous, ongoing process, occurring over the lifetime of our Galaxy and the universe. However, how stars form, still remains a mystery. Unraveling the process of stellar birth is of fundamental importance for understanding not only the origin of stars but also the origin and evolution of planets, life and even galaxies. For example, since stars are the basic constituents of galaxies, stellar formation and evolution are the engines that drive galaxy evolution. Once stars form, astronomers can use the elegant theory of stellar structure and evolution to predict the life histories of galaxies. However a comparable theory of star formation does not currently exist and therefore limits any future progress toward understanding galaxy evolution.

The reason that the process of stellar birth has remained a mystery for much of this century has to do with where stars form. Stars form deep within molecular clouds. The prenatal gas in these clouds is very cold and therefore emits only at millimeter and radio wavelengths. The clouds also contain large amounts of dust which surround forming stars, rendering them invisible at optical wavelengths. Consequently, the star forming process can not be observed via traditional methods of optical astronomy. Luckily, advances in observational technology over the last two decades, especially at millimeter and infrared wavelengths have permitted the direct observation of star forming regions. As a result many new and surprising discoveries about the star forming process have been uncovered. One such discovery is that rich embedded clusters, containing hundreds of newly formed stars, are more numerous than previously suspected. In fact, they may account for most of the star formation, independent of mass, occurring in our Galaxy.

C.J. Lada and N.D. Kylafis (eds.), The Origin of Stars and Planetary Systems, 441–478.

As we will discuss, the dense environments forming embedded clusters are quite different from regions forming only one or two isolated stars. Therefore one might expect to find fundamental differences in the physics of not only the star forming process but also the early evolution of young stars and their planets. In the following sections, I will discuss the role that young embedded cluster have in our attempts to understand the origins of stars and planets.

2. The Distribution of Star Formation in Giant Molecular Clouds

"The current views of stellar evolution allow us to date galactic clusters on the basis of their color-magnitude diagrams. The youngest known clusters, those containing O-type stars, are at most a few million years old. Obviously such clusters must be continually forming unless we are at the unlikely epoch when star formation has ceased. It would be important to know if these clusters and associations represent the only regions of star formation or whether individual stars may form in the general field. " - Roberts 1957

. Over the past decade, observations of nearby star forming regions have led to the suggestion that two distinct modes of star formation may exist, (1) a distributed, loosely aggregated or isolated mode and (2) a clustered, tightly packed or grouped mode. Distributed or isolated star formation can be characterized by low stellar density and low overall star formation efficiency. In this mode, one to a few stars form in single, low mass, dense cores distributed throughout a molecular cloud. An example of this mode is seen in the Taurus molecular cloud, where ~ 100 stars are forming over a 300 pc^2 area from $\sim 10^4$ M$_\odot$ of gas. In contrast, the clustered mode of star formation is characterized by high stellar densities and relatively high star formation efficiency. In this mode, groups of stars form from a massive concentration of gas. An example of clustered star formation is seen in the Rho Ophiuchus cloud core where 100 stars form over a 2 pc^2 region from ~ 600 M$_\odot$ of gas.

To determine the relative contributions of each of these modes of star formation to the current star forming activity in our Galaxy and thereby answer the question posed by Roberts, requires a thorough and systematic census of the youngest stellar populations in giant molecular clouds. Such surveys are best done at infrared wavelengths. Remember, molecular clouds are full of dust. This dust absorbs and scatters radiation from x-rays to infrared wavelengths (Mathis 1990) and hence renders the clouds and their new born stars nearly invisible at optical wavelengths. Luckily, the nature of the dust (e.g. size and type of grain) which is responsible for the observed extinction of light through the clouds, is such that it produces a continuous decrease in this extinction with increasing wavelength. Figure 1 shows the

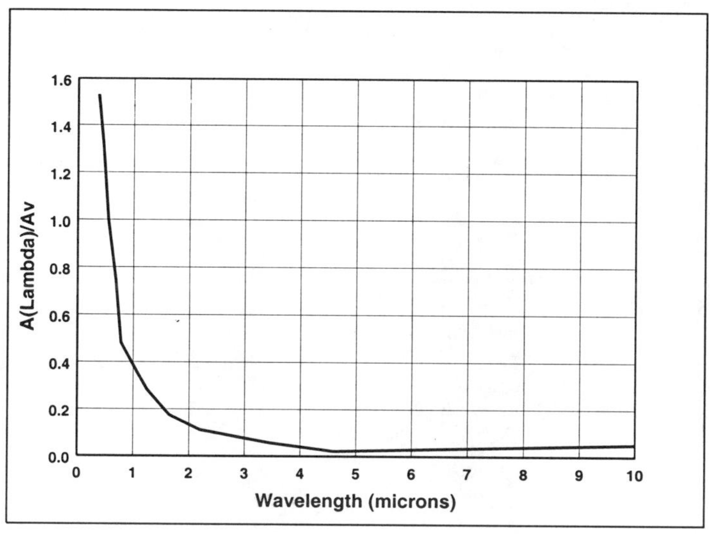

Figure 1.   The infrared extinction law. This figure was constructed using the data from the paper of Rieke & Lebofsky (1985).

extinction law measured by Rieke and Lebofsky (1985) for wavelengths ranging from 0.2 to 10 microns. From this figure, one can see a marked decrease in extinction from optical to infrared wavelengths. For example, the extinction at 2.2 μm is roughly a factor of 10 less than the extinction measured at visual wavelengths.

Advances in infrared detector technology have enabled us to probe the dusty interiors of molecular clouds and directly study their stellar progeny. For example, the IRAS satellite provided a flux-limited census of far-infrared sources in giant molecular clouds (GMCs). This data base was used for some of the first comprehensive studies of the distribution of young stellar objects in nearby molecular clouds such as Taurus (Kenyon et al. 1990). However, since the IRAS survey had limited sensitivity and young stars are typically weak far-infrared emitters, the majority of young stellar objects in all but the nearest molecular clouds ($\sim$ 100 pc), could not be detected by IRAS. In addition, the angular resolution of IRAS is coarse and consequently many star forming regions suffer from confusion due to source crowding. The recent development of sensitive, large format imaging arrays at near-infrared wavelengths has alleviated such problems and made it possible, for the first time, to search large areas of molecular clouds for their embedded stellar populations at unprecedented sensitivi-

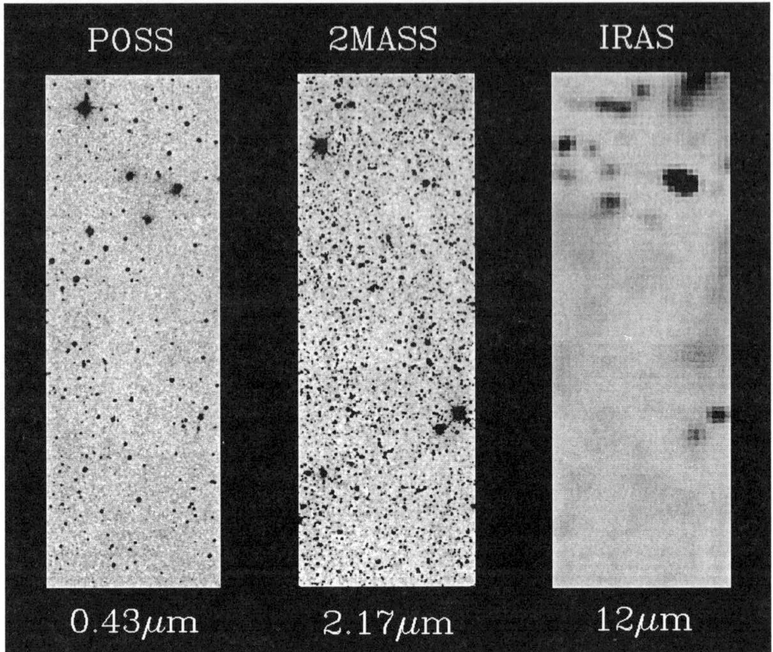

Figure 2. Comparison of an image of the sky taken at optical (POSS), near-infrared (2MASS) and mid-infrared (IRAS) wavelengths. This figure illustrates the advantage of near-infrared imaging observations for obtaining a thorough census of young objects.

ties and resolutions. Figure 2 compares an optical POSS image with a 2 micron image of the same field from the 2MASS survey and a 12 micron image from IRAS. The near-infrared observations clearly detect many more sources than either the optical or IRAS observations.

While, near-infrared imaging observations are the most effective means of obtaining a census of the youngest stellar populations in molecular clouds, they do have their own limitations. Namely, near-infrared surveys can be contaminated by field stars, in front of and behind the molecular clouds and can also suffer from background source confusion at very faint magnitudes. The general field star population consists mainly of late type dwarf and giants (Figure 3). These late type dwarfs and giants are bright at near-infrared wavelengths (Figure 4). Given the reduced extinction at these wavelengths such stars can contribute significantly to star counts obtained at 2 microns, even when they are observed through an intervening molecular cloud. It is therefore essential to account for the distribution of background field stars before assessing the young stellar content of the clouds. The most common and straight-forward way to determine the con-

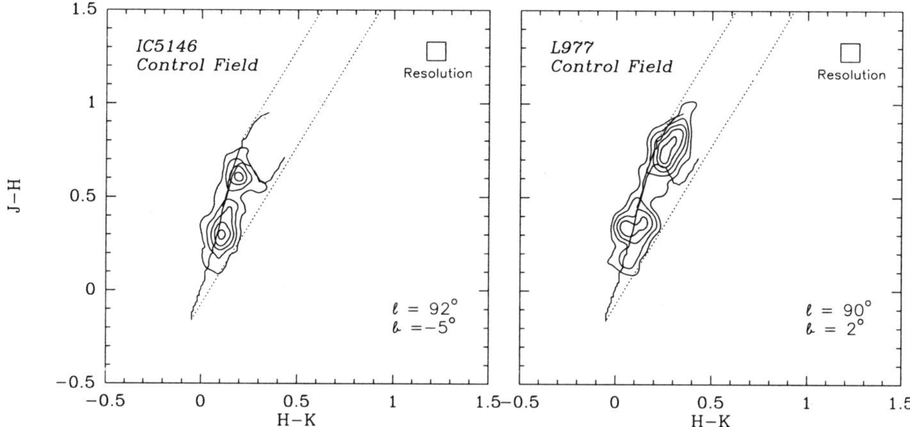

Figure 3.   Near-infrared (JHK) color-color diagrams for control (off) fields for the IC 5146 and L977 molecular clouds. The contours represent the surface density of stars in the JHK color-color plane. The dwarf and giant sequence (solid line) and the reddening vector (dashed line) are shown. This figure is from Alves et al., (1998)

tribution of field stars to a near-IR survey is to obtain star counts of control fields, large areas located nearby but off the cloud. By appropriately scaling the area of these control fields to match the area surveyed on the cloud, one can determine the membership of young stars in the cloud on a statistical basis. However, determining whether or not a specific star in the region is associated with the molecular cloud or is a field star is not possible by this method. In order to determine the membership of individual sources requires additional information, such as stellar proper motions, spectra, multi-color near-IR photometry or 2.3 μm CO narrow band photometry.

The first cloud to be systematically surveyed at near-infrared wavelengths was the L1630 molecular cloud (Lada et al. 1991). Approximately 0.8 square degrees of the cloud was observed at 2.2 μm to a completeness limit of 13th magnitude (Figure 5). The region surveyed was contained within the molecular cloud boundaries and roughly half of the area surveyed contained CS emission from dense gas and the remaining half did not. The survey was sensitive enough to sample both high and low mass stars and in particular to detect stars with masses comparable or less than the sun. The survey identified four embedded clusters, NGC 2024, NGC 2071, NGC 2068 and NGC 2023 within the cloud. These clusters are associated with previously known regions of star formation. The spatial distribution of the stars in L1630, shown in Figure 5, is highly concentrated. More than half the near-infrared sources detected were contained in the three most populous embedded clusters, NGC 2071, NGC 2068, NGC 2024. The sizes of these 3 clusters cover less than 18% of the cloud area surveyed, indicating that

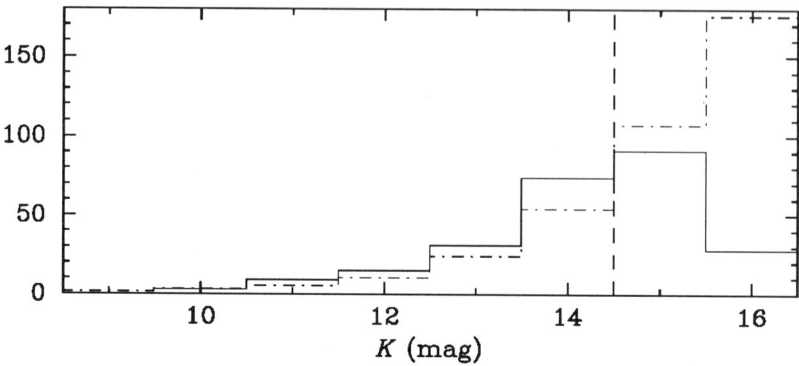

Figure 4. The observed K magnitude distribution of background sources for control fields near the L1630 molecular is shown as a histogram (solid line). Also shown are model estimates of the the field star distribution. This figure was taken from Li et al. (1999).

star formation occurs in very localized regions. In addition, the number of stars found outside the clusters is consistent with the expected number of background stars not associated with the cloud. After correcting for the presence of field stars, it was estimated that the vast majority ($\sim$ 96%) of the stars in the L1630 molecular cloud were formed within these three rich clusters!

Could there be a significant distributed population in L1630? Recently, Li, Evans & Lada (1997, 1999) addressed this question by searching for a distributed population of pre-main sequence stars in the cloud. They surveyed a 1320 arcmin^2 region of L1630, at J, H, and K. The region surveyed concentrated on an area that was away from the dense molecular cores and that exhibited modest visual extinctions ($\leq$ 10 mag) (Figure 6). Two different approaches, a near-infrared color analysis and a star counting analysis, were used to look for a distributed population. Since young stars are surrounded by large amounts of gas and dust, they radiate a significant fraction of their energy at infrared wavelengths and consequently many young PMS stars have near-infrared emission in excess of their photospheric emission. Therefore one can search for evidence of recent star formation by looking for sources exhibiting near-infrared excess. Using this method, Li, Evans & Lada (1997) estimated that the fraction of sources, located outside the cluster, with near-infrared excess is 3% - 8%. In addition, the surface density of near-infrared excess sources is 1/7 of that found in the distributed population of L1641 and 1/20 of that found in the low density, young clus-

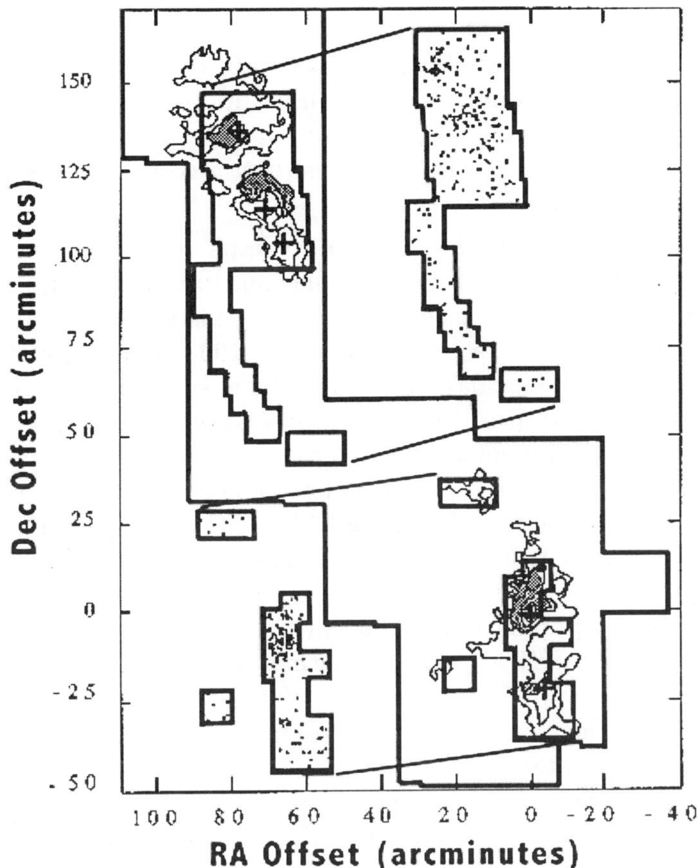

Figure 5. K-band survey of L1630 adapted from Lada et al. (1991) and Lada (1992). The survey boundaries are overlaid on a contour map of the CS(J=2-1) survey (outer boundary). The shaded areas indicate the cluster boundaries. The distribution of K-band sources with $m_K < 13$ is also shown in the in-set boxes. The (0,0) position corresponds to $\alpha = 5^h\ 39^m\ 12^s$ and $\delta = -01^\circ\ 55'\ 42''$.

ter NGC 2023. The low fraction and low surface density of excess sources indicates that recent star formation activity has been very low in the outlying regions of L1630. The second approach used to search for a distributed population, employed near-infrared star counts (Li, Evans & Lada 1999). Extensive and carefully studied control fields, off the molecular cloud, were used to establish the stellar surface densities of field stars and a method

was developed using the survey data and a ^{13}CO map to extract field stars
not associated with the cloud. Comparison of the star counts of the control
fields with the on-cloud fields reveals no evidence for a distributed population (Figure 6). Results from both the near-infrared color analysis and star
counting analysis both confirm and strongly support the earlier conclusions
that *most stars in L1630, independent of mass, formed in a clustered mode
of star formation, simi ʌr to the star formation process in the Rho Ophiuchi
cloud core* (Lada 1992).

If L1630 is a typical GMC, then most stars in the Galaxy may have
formed in embedded clusters, not in isolation. Indeed near infrared imaging observations of GMCs have revealed the existence of a large number
of young embedded clusters (e.g. Lada et al. 1993; Zinnecker et al. 1993).
However, only a few molecular clouds have been systematically surveyed to
the appropriate sensitivities to empirically test whether or not most stars
form in cluster environments. One cloud that has been extensively studied
for its young stellar content is the L1641 molecular cloud. L1641 is also
part of the Orion complex and is only located several degrees south of
L1630. Strom, Strom & Merrill (1993) found a population of isolated pre-
main-sequence (PMS) stars distributed throughout the cloud. The origin
of this distributed population is unclear at this time. Based on spectroscopic studies, Allen (1996) suggested that as much as half of the stars
in L1641 (excluding OMC1-2) may have formed in relative isolation, distributed through the cloud. This population may therefore have formed in
a similar manner to stars in the Taurus dark cloud. Is a distributed mode
of star formation more prevalent in L1641? In addition to the distributed
population of young stars, a number of rich clusters also have been identified in L1641 and several examples of these clusters are shown in Figure 7.
These include the famous Trapezium cluster, and a number of embedded
clusters found in the near-IR survey of Strom, Strom and Merrill (1993).
The clusters found in L1641 have properties very similar to the clusters in
L1630 (see Section 3, Table 1). Therefore the clustered mode of star formation also appears to be significant in this cloud. Meyer and Lada (1999)
recently examined the published studies of the young stellar population in
L1641 to assess the relative contributions of the clustered and distributed
modes of star formation and to compare the star forming activity in L1641
with that in L1630. When they considered the entire stellar population, including the OMC region, they estimated that $> 2/3$ of the total number of
stars < 10 Myr old are part of rich clusters (also see Lada, Strom & Myers
1993). They concluded that the star forming activity in L1641 is similar to
that in L1630 and that in both clouds, *most stars form as part of young
clusters compared to a minority that form in relative isolation.*

In addition to the studies in Orion, the cluster mode of star formation

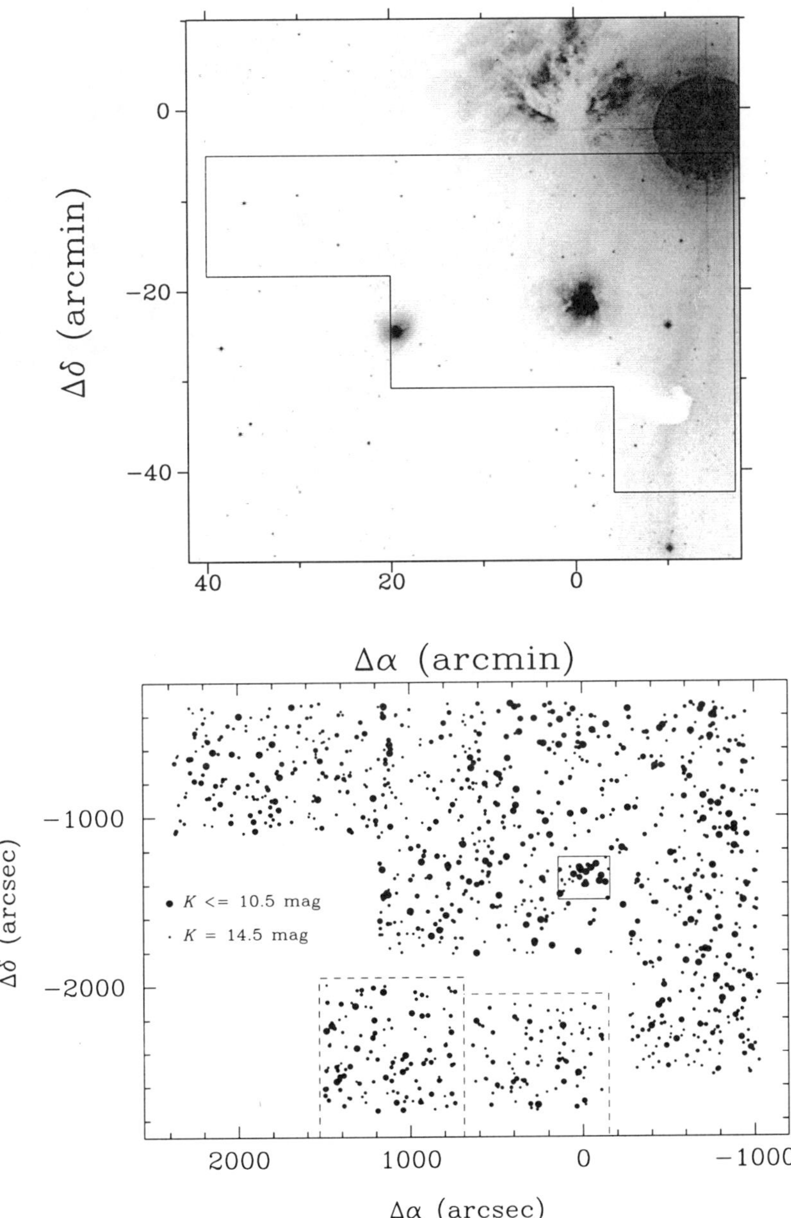

Figure 6.   Region of the L1630 molecular cloud surveyed at J, H and K by Li, Evans & Lada (1997, 1999) (Top Panel). The distribution of stars with K< 14.5 is shown in the bottom panel. For comparison, the distribution of field stars from control fields off the cloud are shown in the 2 dotted-lined inserts

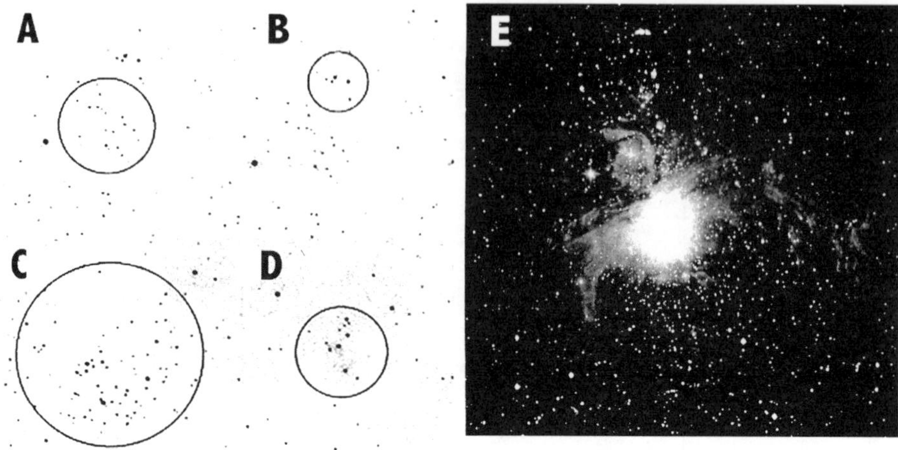

Figure 7. Clusters in the L1641 molecular clouds. On the left, 4 young clusters, L1641 C (A), KMS 36 (B) L1641 S (C) and CK Group (D) are shown. These clusters were identified from the near-infrared survey by Strom, Strom & Merrill (1993). On the right, a JHK composite image of the Trapezium cluster in L1641, courtesy of the 2MASS Survey is presented.

appears to contribute significantly to the star forming process in other molecular clouds. A recent J, H, K survey of roughly a square degree of the Rosette molecular cloud revealed the existence of at least seven embedded clusters in the RMC, five of which were previously unknown (Phelps & Lada, 1997). The locations of these clusters within the molecular cloud is shown in Figure 8. Sensitive follow-up observations of several of the Rosette clusters have shown them to be similar in richness to the L1630 clusters (Lada, Muench, Phelps, 1999 in prep.) Embedded cluster formation has also been found to be significant in the Gemini OB 1 cloud (Carpenter, Snell & Schloerb, 1995), NGC 2264 (Lada, et al. 1990), IC 5146 (Lada, et al. 1993). However, while clusters have been shown to be numerous in these other molecular clouds, most current observations are not sensitive or extensive enough to determine whether cluster formation is responsible for producing most of the stars in the clouds. Finally, even in regions, where the distributed mode of star formation was thought to dominate, such as the Taurus molecular clouds, clustering is also found to be important. Gomez et al. 1993 investigated the distribution of pre-main sequence stars in Taurus via the two-point correlation function and the nearest-neighbor analysis and found that young stars in Taurus preferentially form in small groups. As we shall see in Section 3, the basic properties of these small Taurus clusterings differ from the properties of the embedded clusters typically found in GMCs.

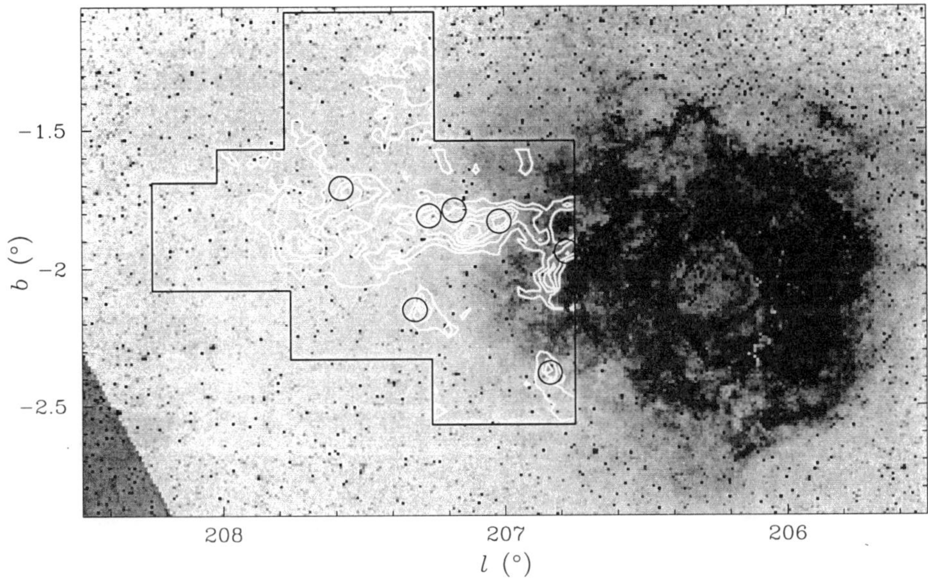

Figure 8.   The distribution of ^{13}CO gas (white contours) (Blitz & Stark 1986) and the location of embedded clusters (open circles) (Phelps & Lada 1997) in the Rosette Molecular cloud is overlaid on the Digital Sky Survey image of the Rosette region. The Digitized Sky Survey was produced at the Space Telescope Science Institute.

3. Properties of Embedded Clusters

3.1. DEFINITION & IDENTIFICATION OF EMBEDDED CLUSTERS

Before reviewing the properties of embedded clusters, it is useful to first define what an embedded cluster is. Empirically, a cluster or an association is defined as: *a group of stars of the same physical type whose surface density significantly exceeds that of the field for stars of the same physical type.* Clusters are distinguished from associations by their higher stellar space densities. To distinguish clusters from multiple star systems we arbitrarily require that a cluster consist of more than 10 members. Clusters can be characterized by their richness: clusters with 100 or more members are designated as rich and those with less than 100 members as poor. Further clusters as defined above can be classified into two environmental classes depending on their association with interstellar material. *Embedded clusters* (the subject of this review) are clusters which are fully or partially embedded in the interstellar gas and dust. Exposed clusters are clusters with little or no interstellar material within their boundaries. This basic definition of a cluster includes stellar systems of two dynamical states: bound and unbound. Bound clusters are systems whose total energy (kinetic + potential) is negative. Unbound clusters have total energies which are positive. Note, when determining the total energy, we include contributions from any interstellar material contained within the boundaries of the clusters. For a more in depth definition of clusters the reader is referred to the review by Lada & Lada (1991).

Infrared observations of molecular clouds are needed to reveal the presence of embedded clusters since many, if not all members will be heavily obscured. The initial identification of an embedded cluster is typically made by a survey at a single infrared wavelength or color, typically 2.2 μm or K-band. The existence of an embedded cluster is established by an excess density of stars of a given type (e.g. bright 2.2 μm sources) over the background. For example, the insets in Figure 5 show the location of the K-Band (K< 13) infrared sources detected in a survey of the L1630 GMC. About 1000 sources are present. Their distribution on the sky is not uniform and over half the sources are located in four well defined clusters as discussed in the previous section. These clusters are also shown in Figure 5 as shaded contours or regions overlaid on the CS(J=2-1) map. The boundaries of the shaded cluster regions correspond to a source surface density which is ten times that of the background. Here, star counts clearly reveal the presence of embedded clusters. In general, the ease of identifying an embedded cluster depends sensitively on the richness of the cluster, the apparent brightness of its members, its angular size or compactness, its location in the galactic plane, and the amount of obscuration in its direction. For example, it would be difficult to recognize a spatially extended, poor cluster of faint stars located in a direction where there is a high background of infrared sources.

3.2. STELLAR DENSITIES AND STRUCTURE

Once a cluster is identified, some basic properties can be determined. Table 1 summarizes the properties of several nearby embedded clusters. Embedded clusters exhibit high stellar surface densities. Typically these young clusters will have a few hundred to a few thousand stars per square parsec within their boundaries. Such densities are much higher than the surface densities for classical open clusters like the Pleiades or more distributed regions of star formation like Taurus (Gomez et al. 1993).

In general, the distribution of stars within a cluster is centrally condensed. The variation in surface density in young clusters such as the Trapezium (McCaughrean & Stauffer 1994), IC 348 (Lada & Lada 1995) and NGC 2282 (Horner, Lada & Lada 1997) falls off roughly as r^{-1}. Figure 9 shows the radial density profile of the K band sources for the NGC 2282 cluster. The best fit to the data is a r^{-1} power-law, represented in Figure

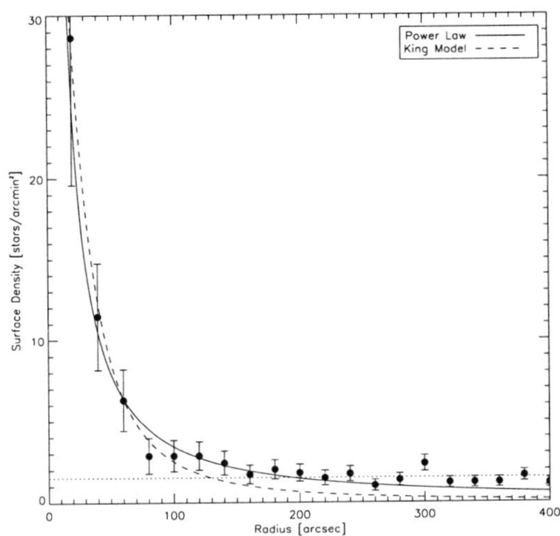

Figure 9.   Radial surface density profile of the NGC 2282 cluster. The dotted horizontal line corresponds to the density of field stars in this region. The power-law fit (solid line) gives a r^{-1} dependence while the King model (dashed line) gives a core radius of 0.19 pc. (From Horner, Lada & Lada, 1997)

9 as a solid line. For comparison a King model of the form $f(r) = f_0[1 + (r/r_c)^2]^{-1}$ is shown. It also fits the data fairly well. In addition to being centrally condensed, young embedded clusters also show evidence of significant structure and sub-clustering. Figure 10 presents the K-band image and corresponding contour map of the stellar surface density distribution for the IC 348 cluster located in the Perseus molecular cloud (Lada & Lada 1995). The lowest contour level in this figure corresponds to the estimated surface density of field stars in this region and subsequent levels are multiples of this value. The lowest solid contour level represents a surface density of three times the estimated field star density. Most of the stars in IC 348 are found in a large central concentration. However, outside this main cluster, eight smaller, but significant sub-clusterings can be readily identified. The stellar surface densities of these smaller sub-clusters are high, ranging from 70 - 270 stars pc^{-2}, typical of the overall densities of embedded clusters and more than an order of magnitude larger than the small clusters in the Taurus cloud (Gomez et al. 1993). In addition, the densities of the sub-clusters are consistent with those expected from a r^{-1} falloff in the overall surface density of the cluster and may be a result of expected statistical fluctuations in the outwardly decreasing surface density distribution of the cluster.

Table 1: Embedded Cluster Properties

Cluster	Average Properties			Central 0.1 pc Radius	
	Number of Stars	Radius (pc)	Stellar Density pc^{-2}	Number of Stars	Stellar Density pc^{-2}
NGC 2023	21	0.3	100		
NGC 2071	105	0.6	132		
NGC 2068	192	0.9	112		
NGC 2024	300	0.9	172	50	1600
L1641 S	134	1.3	100		
CK Group	25	0.7	72		
Trapezium	700	0.5	1000	98	3100
	1600	2.5	89		
IC 348	345	1.0	105	30	955
NGC 1333	275	0.5	140		
Taurus	9−15	0.5−1.1	5−14	1−few	30-90

It is interesting to consider the stellar densities of embedded clusters on the 0.1 pc scale of their central regions. A 0.1 pc radii corresponds to the size scale expected for the protostellar envelope of a single star (Adams et al. 1987) and also the characteristic size of a dense core in the Taurus molecular cloud which forms only one, or at most a handful of stars. The clusters IC 348, NGC 2024, and the Trapezium contain approximately 30, 50 and 100 stars in their central 0.1 pc regions. The implied stellar mass and equivalent molecular cloud densities of these clusters, correcting for projection effects (McCaughrean & Stauffer 1994), range from 2.5 - 8.2 x 10^3 $M_\odot$ pc^{-3} and 5-16 x 10^4 cm^{-3} respectively for the three clusters (Lada & Lada 1995). The latter range of densities is consistent with the range of molecular hydrogen densities measured on similar spatial scales in the densest regions of massive cluster forming cores (Lada, Evans & Falgarone 1997). If such high stellar densities were characteristic of those at formation, then this would have important consequences for star formation theory and our understanding of protostars. However, it is possible that the observed stellar densities do not reflect the initial cluster densities given that the dynamical time scales of such clusters are much shorter than their ages. Therefore it is possible

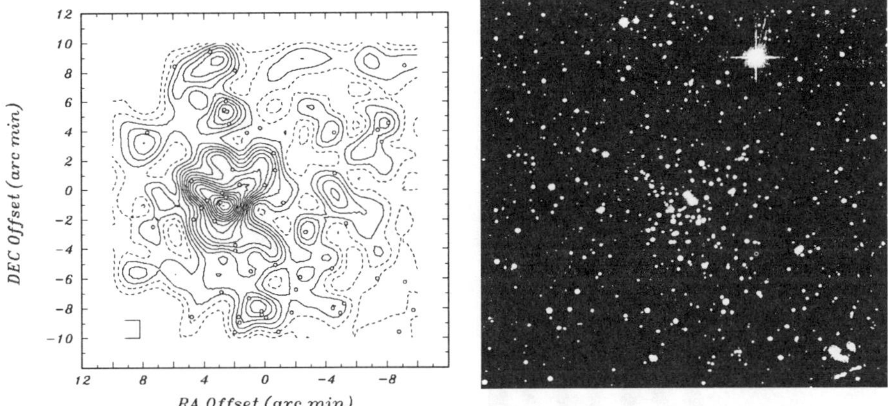

Figure 10.   A K-Band image of the IC 348 cluster, located in the Perseus molecular
cloud is shown on the left. A contour map of the stellar surface density as measured at
K is shown on the right. (Lada & Lada 1995). Open circles represent the distribution of
near-IR excess sources. See text for more details.

that the stars have evolved to denser configurations than they started out
with (Lada Margulis & Dearborn 1984; Patel & Pudritz 1994).

Finally, the inferred volume mass densities of young embedded cluster
are at least two orders of magnitude greater than that of optically visible
open clusters such as the Pleiades (Lada & Lada 1995). This large disparity
is consistent with the predictions of models for the dynamical evolution
of young clusters that form in molecular clouds with low star formation
efficiencies (e.g. Lada et al. 1984).

3.3. THE BINARY FRACTION IN YOUNG CLUSTERS

It is well known that a large fraction ($\sim 50\%$) of stars in the Galaxy are
members of binary, rather than single star systems (Abt 1983; Duquennoy
& Mayor 1991). Therefore, it is not too surprising that surveys of pre-main
sequence (PMS) stars in isolated nearby star forming regions, such as Taurus, have revealed that even stars in the earliest stages of stellar evolution
are also found in binaries. In fact, binaries are found even among the very
youngest PMS stars located near the stellar birthline in the HR diagram.
This indicates that binary formation occurs as a direct consequence of the
star formation process, e.g. binaries form in the protostellar phase, rather
than the PMS stage of stellar evolution at least in low density regions. At
the same time, however, such studies have suggested that PMS stars have
a significantly higher companion frequency than main sequence stars in the
separation ranges studies (i.e. 2-3 times higher than the field) (Ghez et al.

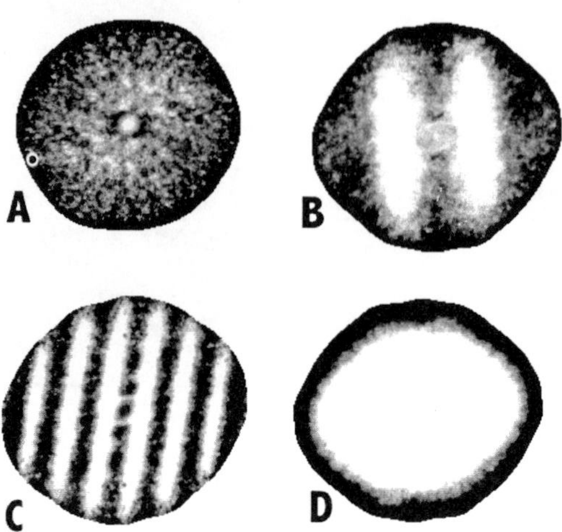

Figure 11.   Calibrated power spectra for stars in the NGC 2024 young embedded star cluster obtained with NIRC on Keck I. Panel A shows an essentially unresolved object. Panel B hows a binary with a separation of $0.09''$. Panel C shows a binary with a $0.35''$ separation. Finally, Panel D shows a single star with a luminous resolved envelope or circumstellar disk.

1993; Leinert et al. 1993; Reipurth & Zinnecker 1993; Simon et al. 1995). One implication of these findings is that stellar binarity may be more common in the earlier PMS stages of evolution than during the main sequence phase. If this is true then many binary systems must somehow be disrupted during the course of their PMS evolution.

Searches for binary stars in clusters such as the Pleiades (Bovier et al. 1997), IC 348 (Duchene et al. 1998) and the dense Trapezium cluster (Petr et al. 1998; Prosser et al. 1994) have revealed binary fractions in these clusters similar to that of the field. These findings reinforce the idea that most stars in our Galaxy form in clusters and not isolation. Recently, Lada and Elston (1999) obtained near-IR speckle observations of 43 stars in the NGC 2024 cluster using Keck I. Figure 11 shows the calibrated power spectra for a sample of stars observed. We found 12 of the 43 stars observed to be binaries in the separation range from 0.05 arc seconds to 1 arc second. This leads to a binary fraction in NGC 2024 of $28\pm8\%$, but the true fraction is slightly higher since the data have not yet been corrected for missed binaries with small separations and large magnitude differences. This binary frequency is very similar to the high binary frequency found in Taurus over the same physical separation range (32 ± 3%). However, given the large statistical uncertainty, it is also consistent with the expected binary fraction for G-dwarf field stars in the same physical separation range (19 ± 3%) and a better statistical sample is needed before we can differentiate between the Taurus fraction and the field. These new observations of the very young NGC 2024 cluster, while not definitive, leave open the possibility that binaries form with the same frequency whether they form in clusters or in isolation. If so, then binaries formed in clusters are quickly disrupted whereas binaries formed in isolation survive for longer periods of time. Again this picture would be consistent with observation that most stars form in embedded clusters, which later disperse to populate the field.

3.4. NEAR-INFRARED LUMINOSITY FUNCTIONS

Young embedded clusters offer unique opportunities to study the stellar initial mass functions (IMF) and its possible variation in space and time. Such clusters are not old enough to have lost significant numbers of members due to stellar evolution or dynamical effects. In addition, mass segregation is not as big a problem as in classical open clusters. The biggest advantage of using very young clusters for studies of the initial luminosity function is that the low mass stars in these clusters are brighter than at any other time in their pre-main sequence evolution. Consequently modern near-IR detectors even on modest telescopes can completely sample the entire IMF of nearby (D < 1Kpc) embedded clusters down to the hydrogen burning limit. The high stellar surface densities in young clusters combined with the high luminosities of the low mass stars also makes issues of membership determination much less of a concern than in older open clusters. Embedded clusters are rich (N>100 stars) and provide statistically significant samplings of the IMF at virtually all masses. However there are also disadvantages to studying young embedded clusters. First the youngest stars in clusters are still heavily embedded in their parental gas and dust. Consequently, the extinction toward cluster members can be highly non-uniform even at infrared wavelengths. In addition, due do the youth of the clusters, most members are still in the pre-main sequence phase of their evolution. Since there is no unique mass to luminosity relation for pre-main sequence stars, one must rely on using evolutionary models. Finally, because the clusters are young, many of the stars exhibit excess near-IR emission above their photospheric emission due to the presence of circumstellar dust or protostellar envelopes.

Comparison of the K-band luminosity functions or KLFs for such nearby young clusters has shown that these KLFs often exhibit two general characteristics. First, the shape of a cluster KLF often approximates a power-law for the brightest stars. Second, the KLF departs from this power-law, flattening out and turning over at the faintest magnitudes. An example of such a KLF is shown in Figure 12 for the IC 348 cluster. IC 348 is a relatively rich and young cluster located in the Perseus molecular cloud at a distance of 320 pc. In the top panel of Figure 12, the KLF for all stars observed toward the cluster is plotted as a histogram, as well as the KLF for stars in a nearby, equal sized, off or control field. The control field stars were artificially extincted by 0.5 magnitudes at K to approximately account for the extinction due to the associated molecular cloud. The shape of the KLF for the stars observed toward the cluster is somewhat different than that off the cluster. In the off field KLF, the number of sources increases with increasing K magnitude up to the completeness limits of our observations ($m_K = 14$). In contrast, the number of sources in the cluster KLF increases with increasing K magnitude until $m_K \sim 11$ after which the number of sources levels off and and then turns over near the completeness limit. The bottom panel shows the IC 348 KLF resulting from the subtraction of the KLFs of the cluster and extincted control field. This differential KLF represents the luminosity function of the cluster members. The slope of the power-law portion of the KLF is equal to 0.40 (± 0.03), similar to the slopes of the power-law portions of the KLFs of other young clusters. The IC 348 KLF departs from the power-law for fainter stars, flattening out at around $m_K \sim 11$ magnitude.

While the KLFs of most nearby clusters share these 2 general features, direct comparison of individual clusters shows that their KLFs can also differ significantly. In particular, both the widths of the KLFs and the points of departure from a power-law can differ from one cluster to another. For example, the KLFs for the Trapezium and NGC 1333 clusters are significantly narrower than that of IC 348 (see Figure 13) (Lada, Alves & Lada 1996).

In principle, the luminosity functions of young embedded clusters should provide fundamental constraints on issues concerning the IMF. Various groups have modeled the luminosity functions of young clusters using realistic stellar mass functions and appropriate mass-luminosity relationships (Zinnecker et al. 1993; Fletcher & Stahler 1994a, b; Megeath 1996; Lada & Lada 1995; Muench, Lada & Lada 1999). From a comparative analysis of published observations of the Trapezium cluster with such KLF models and observations of IC 348, Lada & Lada (1995) concluded that the underlying mass function of both clusters is similar to the IMF for field stars down to the hydrogen burning limit. In addition they found little evidence for a significant population of single, lower mass objects (brown dwarfs). Further comparisons of the KLFs of the young NGC 1333 cluster with IC 348 and the Trapezium show that the shape of the KLF of NGC 1333 is similar to that of the Trapezium but that both differ from the KLF of the older IC

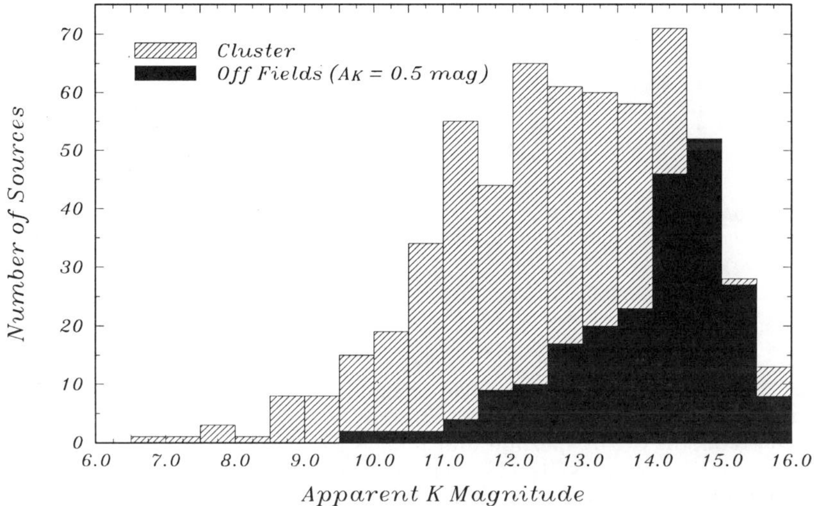

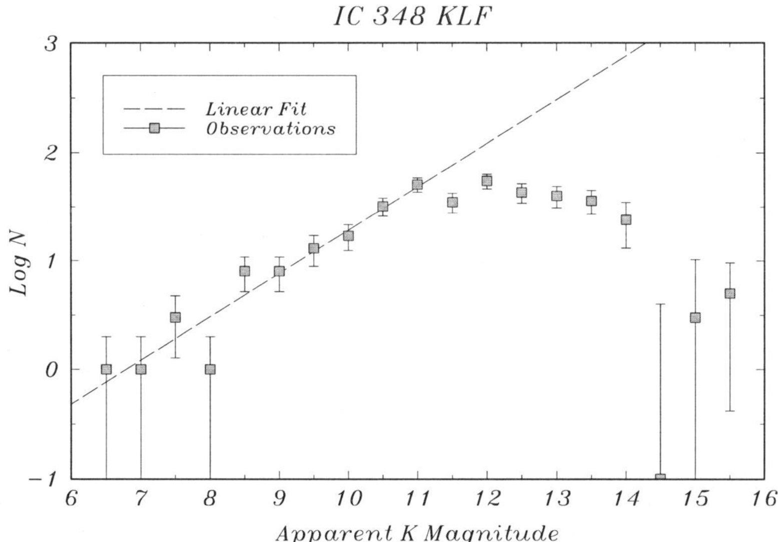

Figure 12. Top - The KLF of sources observed toward IC 348 along with the corresponding KLF for the control fields observed off the cluster. Bottom - The difference between the cluster and control field KLFs, representing the KLF of the cluster members. The source counts are complete to $m_K = 14$. Taken from Lada & Lada (1995).

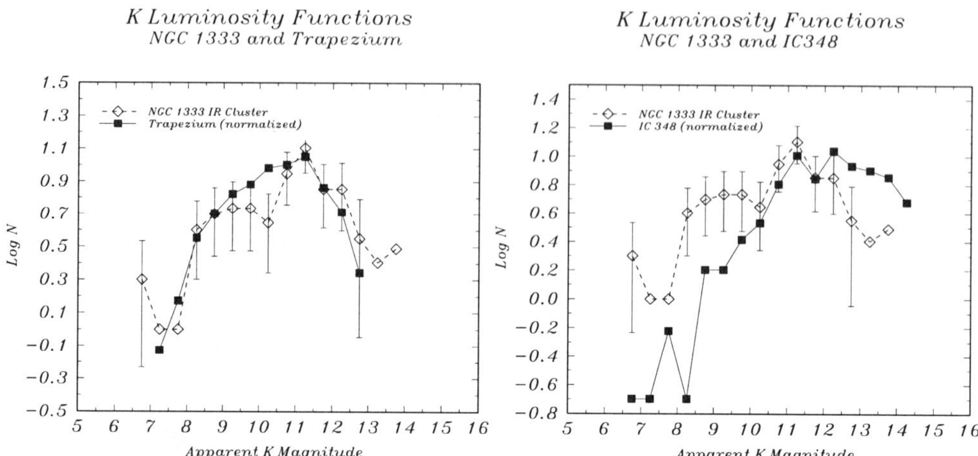

Figure 13. This figure from Lada, Alves & Lada (1996) compares the observed KLFs for three nearby young clusters: the Trapezium in Orion, NGC 1333 and IC 348 both in Perseus. The KLFs of the Trapezium and IC 348 clusters have been normalized to the size of the population of the NGC 1333 cluster. Additionally, the KLF of the Trapezium has been corrected to the distance modulus of NGC 1333, which we have taken to be the same as for IC 348 (m-M=7.5).

348 (Lada, Alves & Lada 1996). This can be attributed to the affects of luminosity evolution and suggests that the age of NGC 1333 is similar to that of the Trapezium but is significantly less than that of IC 348. For a more comprehensive review of young cluster luminosity functions and their interpretation with regards to issues concerning the IMF, the reader is referred to a review by Lada et al. (1998) and the chapter by C. Lada, this volume.

Despite the similarities in their mass functions, embedded clusters appear to be characterized by significantly different rates of star formation over their lifetimes. For example, the rate of star formation in the Trapezium has been more than 10 times greater than that in NGC 1333 and IC 348 (Lada & Lada 1995; Lada, Alves & Lada 1996). The difference in the star formation rates in these clusters suggests that the star formation mechanism can be very sensitive to initial conditions. It is possible that differences in environment (O stars in the Trapezium versus no O stars in either IC 348 or NGC 1333), could be responsible for the differing rates. However, evolutionary modeling of more clusters is needed to be able to test this or any hypothesis concerning the origin of the differing star formation rates observed in clusters.

4. The Formation of Embedded Clusters

Comparisons of the location of embedded clusters with the distribution of dense gas in L1630 (Lada 1992) (Figure 14) indicate that the young clusters are associated with and most likely formed from the most massive dense cores in the cloud. Similar results have been found in the Gem OB1 molecular cloud (Carpenter et al. 1995) and also the Rosette molecular cloud (Phelps & Lada 1997). Therefore, in order to understand how an embedded cluster forms, we must understand two basic physical processes: (1.) the formation of a massive, dense molecular core in a molecular cloud and (2.) the subsequent development of stars from the dense gas in the core (see chapters by Bonnell, McKee and Myers). Studies of molecular cloud structure provide insights into some important aspects of embedded cluster formation. For example, the slope of the mass spectrum of dense cores derived from the CS survey of the L1630 cloud indicates that a substantial fraction of the mass of dense gas is contained within the most massive dense cores (Lada, Bally, & Stark 1991; see Chapter by C. Lada, Figure 6). Consequently the observation, that the majority of the star formation in L1630 has occurred in clusters, may be naturally explained to a large extent by the fact that most of the dense gas in the cloud is contained within a few localized and massive, dense cores. Furthermore, investigations of the structure of other nearby clouds have strongly suggested that the mass spectra for clumps and cores in GMCs is universal (Blitz, 1993; see Chapter by Blitz). For example, Kramer et al. (1998) recently derived the mass spectra for seven molecular clouds (Figure 15) and found that mass spectra were consistent with a powerlaw, $dN/dM \propto M^{-\alpha}$, where $\alpha \sim 1.7$, similar to the mass spectra measured in other clouds and for the dense gas in L1630. This suggests that GMCs, in general, should produce most of their stars in embedded clusters.

Comparison of surveys for dense molecular gas and for young stars in GMCs can provide us with many insights into the star forming process in molecular clouds. Such surveys in the L1630 cloud have demonstrated that (1.) star formation within this cloud occurs almost exclusively within the dense ($> 10^4$ cm^{-3}) gas, and that (2.) star formation does not occur uniformly throughout the dense gas but is strongly favored in a few very massive (M$>$200M$_\odot$) dense cores, where the efficient conversion of molecular gas into stars has resulted in the production of rich stellar clusters (Figure 14). The three massive cores, NGC 2071, NGC 2068 and NGC 2024, associated with the rich embedded clusters in L1630 account for $\sim$ 30% of the total mass of dense gas in the cloud but only 8% of the total molecular cloud mass, suggesting that only a small fraction of a GMC has the conditions needed for efficient star formation.

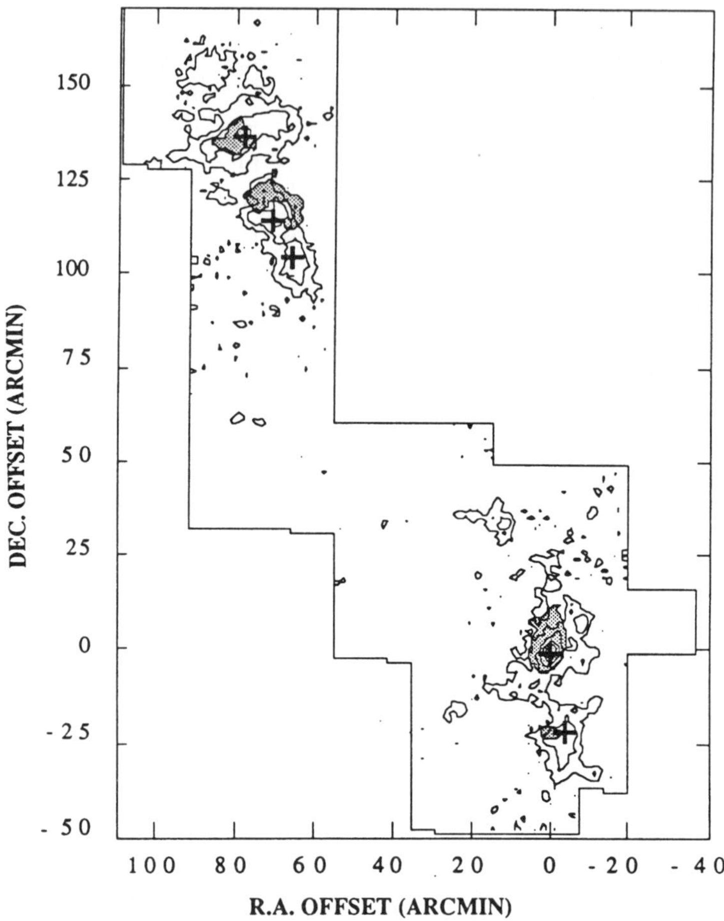

Figure 14. Comparison of embedded cluster locations and the distribution of dense gas in the L1630 molecular cloud. The shaded regions represent the location and extents of the embedded clusters. The contours show the map of CS(2-1) integrated intensity in the velocity range of 7-13 km s^{-1}. The (0,0) position is the same as in Figure 5. The lowest CS contour level is at 0.8 K km s^{-1}, which corresponds to a 3 sigma level detection. In addition the locations of the five most massive CS cores identified by Lada, Bally & Stark (1991) are represented by large crosses. The solid line represents the boundary of the CS survey.

While the observations of L1630 confirm the idea that clusters form from massive dense cores, they also suggest that high mass and high density may not be sufficient conditions for cluster formation. In addition to the massive cluster-forming cores, the L1630 cloud was found to contain a massive, dense core, LBS 23, that is not associated with a recognizable cluster. This core is however associated with current star formation (e.g. Bontemps et

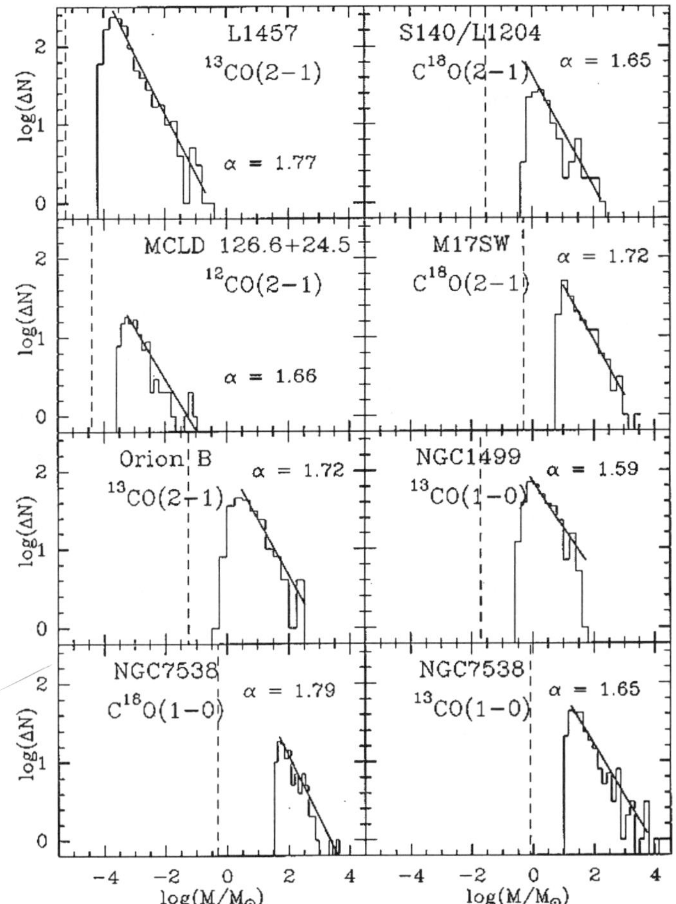

Figure 15.   The clump mass spectra for seven molecular clouds measured by Kramer et al. (1998). The slopes of these spectra are consistent with previous measurement of mass spectra of other nearby GMCs (e.g. Blitz, 1993) and the mass spectrum determined for the dense cores from CS observations of the L1630 molecular cloud (Lada, Bally & Stark, 1991).

al. 1995). In addition, another massive core, NGC 2023 was found to contain only a very poor cluster. In these two massive cores the level of star formation is much lower than in the three other cores of comparable mass that are producing rich clusters. This difference is reflected in estimates of the star formation efficiencies (SFEs) of these cores. Assuming an average or typical stellar mass of 0.6 $M_\odot$ for stars in the embedded clusters, Lada, Evans & Falgarone (1997) derived SFEs (SFE = $M_{stars}/M_{stars+gas}$) for the cores from the total number of stars in each cluster and the mass of the gas in each core (Lada 1992). As a result, the NGC 2071, NGC 2068 and NGC 2024 cores have star formation efficiencies on the order of 10% – 30%, whereas the LBS 23 and NGC 2023 cores exhibit SFEs $\sim$ 4%. The vast differences in the star formation activity in the five massive CS cores in L1630 could be due to either differences in the fundamental physical properties of the cores or differences in evolution.

In order to determine whether the physical properties of massive cores are related to their star formation efficiencies and, in particular their ability to form rich embedded clusters, the spatial and density structure of the massive cores and two low mass cores in L1630 were examined using multi-transition CS observations of the cores with spatial resolutions ranging from $11''$– $64''$(Lada, Evans, Falgarone 1997). The data were analyzed using an LVG model. Densities at least as high as 10^5 cm^{-3} were found to exist in all the massive cores. Comparison of the properties of massive cores with and without an associated embedded cluster reveal that cores having high SFE and consequently forming rich embedded clusters, tend to have larger regions of high density gas (i.e. have larger areas of detectable CS (J=5-4) emission) than those having low SFE and not associated with a populous cluster. In addition, the filling factor of gas at densities above 10^5 cm^{-3} is smaller for cores forming rich clusters than for the others. These differences suggest that regions forming rich embedded clusters require a large fraction of their gas to have densities exceeding 10^5 cm^{-3}, and that they are more fragmented. In contrast, regions with small amounts of dense gas and high volume filling factor (less fragmented) seem to produce only small numbers of stars.

5.　Fate of Embedded clusters

It is interesting to note that Roberts (1957) estimated that only 10% of all stars born in the galaxy formed in exposed clusters. This is roughly an order of magnitude lower than the estimate of the fraction of all stars formed in embedded clusters in the Orion clouds. Are the Orion clouds unusual in their cluster forming properties or do embedded clusters rarely form bound systems. The key to answering this question involves understanding the physical association of the clusters with the interstellar gas that they are embedded in.

The dynamical evolution of embedded clusters and their emergence from molecular clouds has been well studied both analytically (e.g. Hills 1980; Mathieu 1983; Elmegreen 1983; Elmegreen & Clemens 1985; Verschueren & David 1989; Verschueren 1990; Saiyadpour et al. 1997) and numerically (Lada, Margulis & Dearborn 1984; Kroupa 1995) and is also discussed in Lada & Lada 1991. As already discussed, embedded clusters form in dense, massive molecular cloud cores. These systems are strongly gravitationally bound systems, as seen from their high mass densities and moderate internal velocity dispersions. However, as we have seen, the star formation efficiency in these systems is relatively low (10% - 30%) and consequently a large fraction of gravitational binding energy for the system is provided by the gas. As long as the star formation efficiency is not very high, the evolution of the cluster depends sensitively on the evolution of the gas.

Molecular outflows, stellar winds and HII regions from the more massive stars in embedded clusters can violently disrupt the molecular gas in a dense core on time scales which are shorter than or comparable to the dynamical time of the system. When this occurs, the dynamical response of the stars which are left behind will depend primarily on the star formation efficiency achieved in the core at the moment of dispersal. In the face of rapid gas removal, a bound group will only emerge if the star formation efficiency is > 50% (i.e. Lada et al. 1984). Given the low star formation efficiencies of the known embedded clusters, such as the L1630 clusters, it is not too difficult to account for the low fraction of stars born in bound visible clusters. It is likely that most embedded clusters emerge from molecular clouds as unbound systems. Thus, although most stars may form in rich embedded clusters, these stellar groups evolve to become the cores of unbound associations, not bound open clusters.

6. Circumstellar Disks in Cluster Environments

The recent detections of extra-solar planets are among the most exciting and significant discoveries of the 20th century. Their discover has sparked intense interest in two of the most fundamental unsolved mysteries in modern astrophysics: (1.) how common are extra-solar planets in the Galaxy? (2.) what are the origins of planetary systems? Theoretical considerations suggest that a natural consequence of the collapse of a rotating gas cloud is the formation of a protostar with a surrounding accretion disk (Shu et al. 1993; see chapter by Ruden). It is within such disks that planets are thought to form. Over the past ten years, observers have assembled compelling evidence supporting this scenario: namely that circumstellar disks frequently exist around young stellar systems. This evidence has been based mainly on the unique emission characteristics of the disks at infrared and millimeter wavelengths (Shu et al. 1993; Welch & Sargent 1993; Lada, 1991). The evidence for proto-planetary systems rests largely upon assessments of the masses in disks (e.g. Beckwith & Sargent 1991, 1993) and the rate at which these masses evolve (Skrutski et al. 1990).

Most of our knowledge of young circumstellar disks has been derived mainly from considerations of isolated star systems in regions where stellar densities are low (e.g. Taurus). However, as we have already discussed, most stars are probably born in dense embedded clusters and not in isolated protostellar fragments. The natal environment of a massive core, which is forming a cluster, should be quite different than that of a low mass core producing only one or two isolated new stars and one might expect to find fundamental differences in the physics of the star forming process between such regions. For example, the dense stellar environments of the young clusters may influence the formation and lifetimes of disks around the newly formed cluster members and consequently influence planet formation. Indeed numerical simulations suggest that the disk lifetimes may be significantly shortened in dense clusters (Clark & Pringle 1991; Larson 1990). Given that a significant fraction of stars in the Galaxy form in clusters, establishing the properties of disks in crowded environments is essential to our understanding of the overall likelihood of planetary systems in the Galaxy.

In regions of star formation, one expects to find young stellar objects in differing stages of evolution, from protostars to pre-main sequence stars. In order to study the natures of circumstellar disks in clusters, we must be able to distinguish between the different evolutionary phases of young stellar objects in the clusters and identify those sources with disks. Since young stellar objects are associated with large amounts of gas and dust, they radiate a significant fraction of their energy at infrared wavelengths and consequently their infrared spectral energy distributions should reflect the nature and distribution of the material surrounding them and their evolutionary state (Lada 1987; Adams, Lada and Shu 1987). It has been demonstrated that YSOs can be meaningfully classified into at least three distinct classes based on the shapes of their broad band (1-100 μm) energy distributions (Lada and Wilking 1984; Adams, Lada and Shu 1987; Myers et al. 1987). These classes represent an evolutionary sequence of YSOs from protostar to young main sequence star (Lada 1987; Adams, Lada and Shu 1987). Class I sources have strong 'infrared excesses' (i.e. infrared energy distributions that are broader than a blackbody distribution) and their spectral energy distributions are characterized by positive slopes at wavelengths longer than 2 μm. Class I sources are likely protostellar in nature (Adams and Shu 1986; Adams, Lada and Shu 1987). Class II sources also have infrared excesses, but their energy distributions are characterized by flat or decreasing slopes at wavelengths longer than 2 μm. They are surrounded by considerably less circumstellar material than protostellar objects and their infrared excess is thought to arise in circumstellar disks (Rucinski 1985; Adams, Lada and Shu 1987; Kenyon and Hartmann 1987; Rydgren and Zak 1987; Strom et al. 1988). Class III objects have little or no infrared excess emission and their energy distributions are characterized by reddened blackbody functions. These sources are also optically visible and include young main sequence stars and also younger pre-main sequence stars that no longer have optically thick circumstellar disks such as the weak-lined or naked T Tauri stars (e.g. Walter 1987).

6.1. NEAR-INFRARED STUDIES OF CIRCUMSTELLAR DISKS

Ideally, one would like to obtain broad band energy distributions of significant numbers of embedded cluster members in order to study the frequency and lifetimes of the protostellar and disk phase of evolution. This is a difficult proposition given the current detector technology at longer wavelengths. Fortunately, Lada & Adams (1992) have shown that protostellar sources, AeBe stars and the majority of T Tauri stars are characterized by JHK infrared colors which place them in the infrared excess region of the JHK color-color diagram. Figure 16 shows a JHK color-color diagram for YSOs in Taurus. In this figure the locus of the main sequence is represented as a solid line and the region encompassing reddened main sequence stars is enclosed by the dotted lines. Stars that fall to the right of these reddening lines have an infrared excess. While not completely unambiguous, protostars (Class I), classical T-Tauri stars (ClassII and Weak-lined T-Tauri stars (class III) do separate, occupying differing regions of the color-color diagram. Further a large fraction ($\sim$ 50%) of Class II sources lie in the near-IR excess region.

An analysis of the infrared colors for several young clusters has been made and the results are presented in Table 2. For many clusters, a large fraction of the cluster sources exhibit significant near infrared excess indicating the possible presence of a circumstellar, proto-planetary disk. Typically, only half the stars with such disks exhibit measurable near-IR excess emission. The high fraction of excess sources in clusters such as NGC 2024, NGC 1333, Rho Oph, etc., suggests that there is a high fraction (100%) of protostellar sources or sources with disks in these young clusters. This is quite remarkable, given the high stellar densities in the cluster cores (i.e. Table 1). In contrast, a number of clusters, such as IC 348 and NGC 2362, have a small detectable near infrared excess fraction. This is consistent with these clusters being older. Figure 17 plots the near-IR excess fraction of young clusters versus their estimated age. The fraction of near-IR excess sources clearly decreases with increasing age and from this we estimate the lifetime of disks in such environments to be only a few million years. The comparison of the fraction of near-infrared excess sources in clusters of differing age, illustrates the potential power of using the infrared energy distributions of YSOs in clusters to constrain the timescale for the disk phase of stellar evolution.

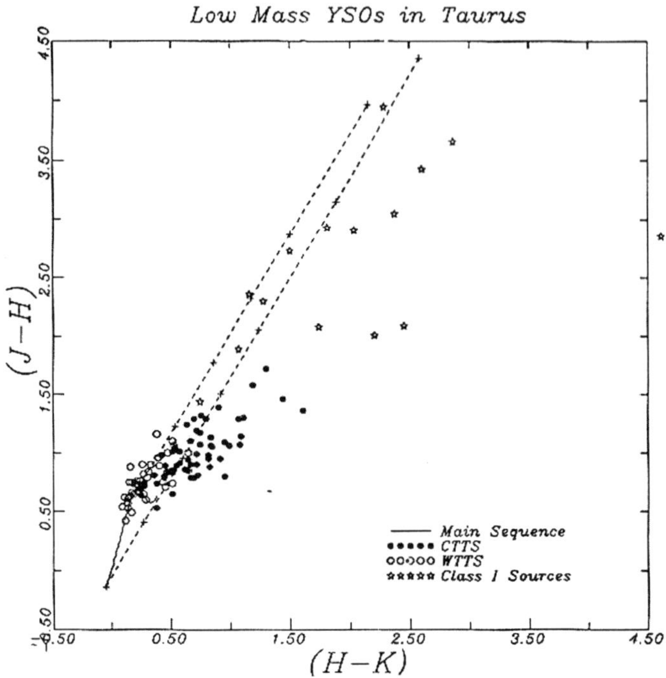

Figure 16. The JHK color-color diagram for low-mass YSOs in the Taurus-Auriga dark cloud complex. The figure is taken from Adams & Lada (1992) and the data were taken from Kenyon & Hartmann (1995). See text for a description of the figure.

Near-IR Excess Fraction in Clusters

Cluster	Excess Fraction	Reference
IC348	20%	Lada & Lada 1995
NGC 1333	60%	Lada, Alves & Lada 1996
NGC 2024	>60%	Comeron et al. 1996
		Lada et al. 1999
NGC 2068	38%	Lada,& Dahari 1999
NGC 2071	34%	Lada, & Dahari 1999
Trapezium	>50%	McCaughren & Stauffer 1996
		Hillenbrand et al. 1998
ρ Oph	50-70%	Green & Young 1992
		Strom, Kepner & Strom 1995
R Mon	62%	Carpenter et al. 1998
NGC 2362	0%	Alves et al. 1999

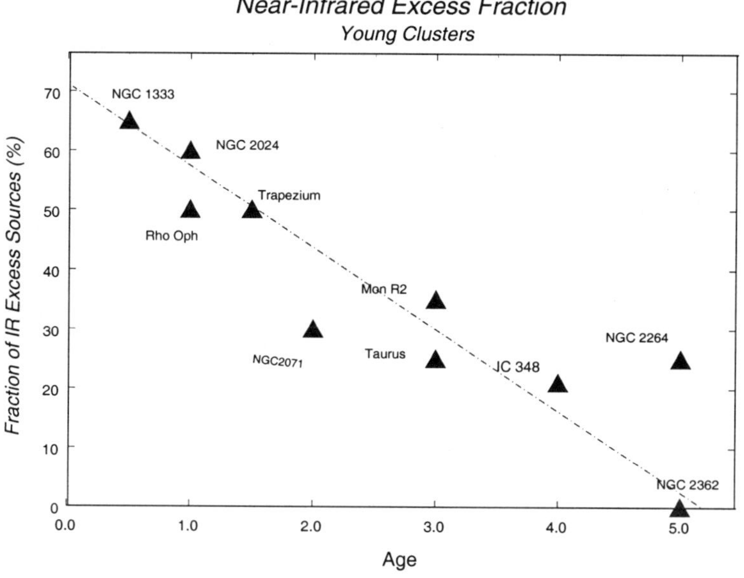

Figure 17.   The fraction of near-infrared excess sources in embedded clusters as a function of cluster age. The near-IR excess fraction decreases with increasing cluster age, suggesting estimates of the lifetime of circumstellar disks on the order of a few million years.

Near infrared observations by themselves, however, are of limited use in investigating the evolutionary status of young stellar objects in clusters because of the complications introduced by extinction and scattering. In addition these wavelengths are still not long enough to avoid significant contamination from normal stellar photospheric emission. Consequently, the near-infrared observations are capable of identifying only sources with relatively strong infrared excess and this accounts for only about half of the sources which actually possess circumstellar disks. For example, in the NGC 2024 cluster (Figure 18), it is not clear whether the large number of near-IR excess sources is due to the presence of disks, protostellar envelopes, or caused by emission nebulosity due to the HII region. Therefore such observations cannot distinguish between Class I, Class II and Class III sources especially in regions of high extinction. Combining near-infrared observations with longer wavelength (e.g. 3-20 μm) observations can help alleviate some of this ambiguity (Figure 19). Since contamination from photospheric emission is minimal at mid-infrared wavelengths, the presence of circumstellar, proto-planetary disks can be ascertained from observations

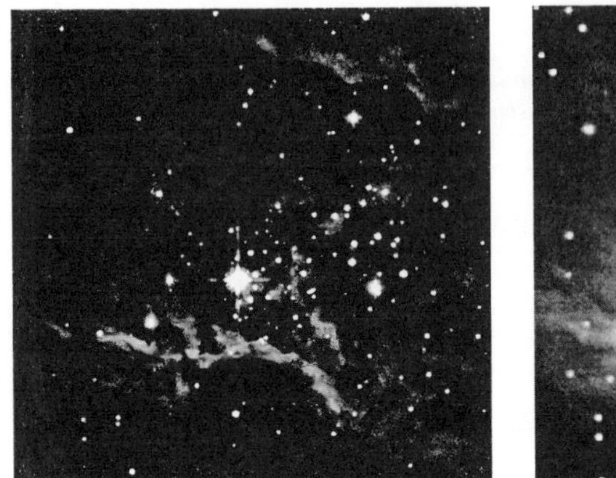

Figure 18. Near-Infrared images of the NGC 2024 embedded cluster. On the left is an L band image taken with the SAOIRCAM on the 48 inch telescope at Mt Hopkins. On the right is a K-Band image taken with SQIID on the 50 inch telescope at Kitt Peak. The images cover $\sim 20 \times 20$ arcmin region on the sky.

of 10 μm with little ambiguity if near-infrared data is also available. For his thesis, Karl Haisch is conducting an L-Band and mid-IR survey of a sample of young clusters to determine a more robust estimate of the disk fraction as a function of age in cluster environments. Recently obtained L-band and mid-IR images of the NGC 2024 region (Figure 18) confirm a large fraction of disk sources within this dense young cluster (Haisch, Lada, & Lada, 1999; Haisch, Lada & Pina 1999). An analysis of the JHK and L colors for the brightest young sources in NGC 2024 indicate a disk fraction of $\sim 60\%$. Furthermore, out of the 60 near-IR sources surveyed, approximately 60% exhibit 10 micron emission. This fraction is consistent with the fraction of near-IR (JHK) excess sources seen in the cluster (Comeron, Rieke & Rieke 1996; Lada & Dahari 1999).

6.2. MILLIMETER CONTINUUM OBSERVATIONS OF CIRCUMSTELLAR DISKS AND DISK MASSES

While near-infrared and mid-infrared data are essential for identifying sources having circumstellar disks, they are inadequate for determining the masses of the disks. Knowledge of disk masses is key for determining whether or not circumstellar disks can form planets. In almost all cases, disk masses can be estimated only from the dust continuum emission at millimeter wavelengths, where the dust is optically thin (e.g. chapter by Beckwith). An excellent place to investigate the formation and survival of circumstel-

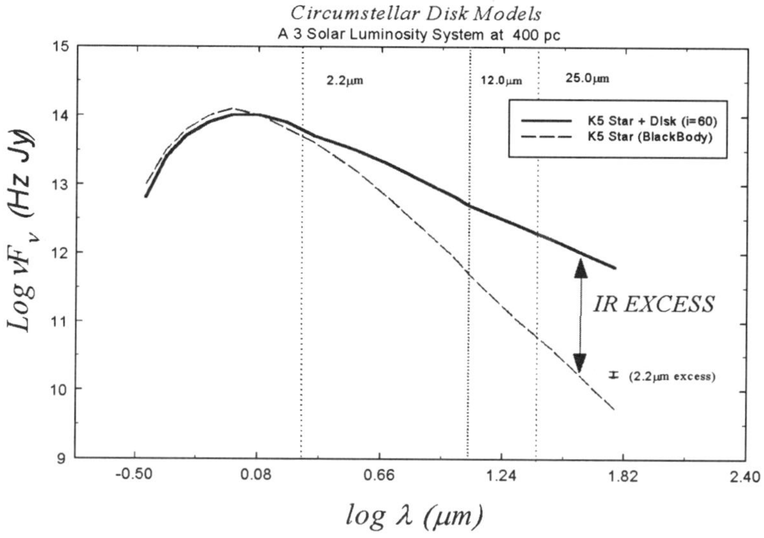

Figure 19.   A model of the spectral energy distribution for a 3 solar luminosity disk plus star system at 400 pc (solid line). The stellar black body distribution is plotted for comparison (dashed-line). This figure illustrates the advantage of observations longward of 2 microns for surveys for emission from circumstellar disks.

lar disks in cluster environments is the Trapezium cluster. Located in the Orion molecular cloud complex, it is one of the nearest (D $\sim$ 440 pc), rich, young clusters. The Trapezium has an age of $\sim$ 1Myr and exhibits stellar densities of 5×10^4 stars per pc^{-3} in its core (see Table 1). Strong evidence for the existence of circumstellar disks in the Trapezium come from centimeter emission (Churchwell et al. 1987; Garay, Moran & Reid 1987; Felli et al. 1993), near-IR excess (McCaughrean & Stauffer 1994) and HST observations of disks in emission (proplyds) and absorption (O'Dell, Wen & Hu 1993; O'Dell & Wen 1994). First attempts to detect millimeter continuum emission from circumstellar disks in the Trapezium and estimate their masses were carried out using the BIMA interferometer (Mundy, Looney & Lada 1995). While these observations sampled a large area of the cluster (4-8 arcmin^2), they were not sensitive enough to place strong constraints on disk masses. Only upper limits of < 0.15 M$_\odot$could be placed on the masses of individual disk systems identified by O'Dell and Wen (1994) in the vicinity of Theta 1 C. This limit is larger than the typical disk masses measured in the Taurus region. In order to obtain a meaningful census of disk masses in the Trapezium cluster, more sensitive observations are required.

Using the IRAM interferometer, Lada et al. (1999) have obtained sensitive 1.3 mm and 2.7 mm continuum observations of 2 fields in the Trapezium cluster containing 34 near-IR sources and 20 proplyds. The sensitivity of these observations permits the detection (at 4 sigma) of disks with masses 0.005 $M_\odot$, more than an order of magnitude smaller than previous interferometer observations of this cluster and sufficiently sensitive to detect the minimum mass solar nebula. These observations resulted in the *first* 1.3 mm continuum detections of circumstellar disk material associated with the near-IR sources and proplyds in the extreme environment of the Trapezium cluster and provide the first estimates of disk masses in this cluster. Three sources, HST 170-337, 171-340 and 158-327 were detected at approximately the 10 sigma level (Figure 20). The sources have 1.3 mm fluxes ranging from 9.9 to 11.3 mJy. The three detections are coincident with known proplyds and near-infrared sources. Disk masses were estimated in two ways: (1.) by comparing the measured fluxes for the Trapezium sources to the mm flux measured for HL Tau, which has a well established disk mass of 0.1 $M_\odot$ (Beckwith and Sargent 1991) and (2.) by assuming the dust is optically thin and adopting standard assumptions about the dust mass opacity, temperature and mass distributions (see chapter by Beckwith for a detailed discussion of how to calculate disk masses). Using either method, the disk masses for the three Trapezium detections were estimated to be on the order of 0.01 $M_\odot$. This is more than an order of magnitude smaller than HL Tau, the most massive disk in Taurus and comparable in mass to the minimum mass solar nebula.

Surprisingly, *only* 3 sources, coincident with previously identified proplyd and near-IR sources, were detected in this study. The area of the Trapezium cluster surveyed with the IRAM interferometer contained ~ 20 identified proplyds and ~ 34 identified near-infrared sources. This leads to disk detection rates of 15% for the proplyd sources and only 10% for the near-infrared sources. These disk detection rates are much lower than the detection rates measured by near-infrared excess ($> 50\%$), for the inner portions of the disks around young stars in this cluster (i.e. Table 2).

As part of a complementary survey for 1.3 mm emission from circumstellar disks in clusters, Lada & Beckwith (1999) surveyed approximately 48 near-infrared sources in the NGC 2071 cluster in L1630 for 1.3 mm continuum emission from disks, using the 19 element bolometer array on the IRAM 30 m telescope. As a result, 22 out of the 48 sources were detected at a > 3 sigma level. Most of the sources detected were compact and detected only in 1 beam of the array. Follow up observations of several detections have recently been carried out with the IRAM interferometer to confirm that the millimeter emission is coming from a disk rather than

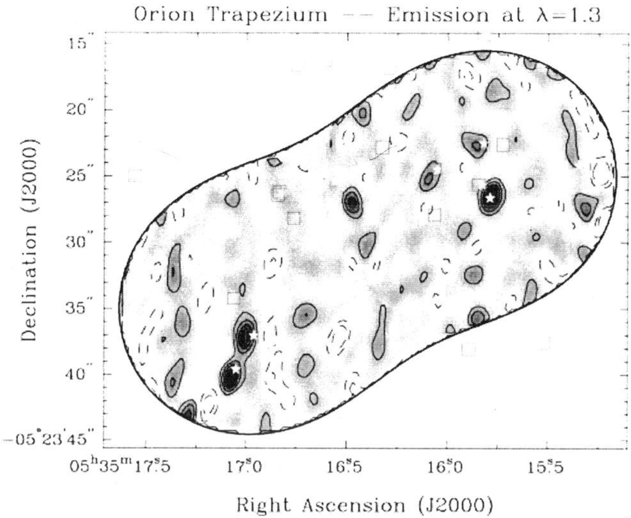

Figure 20. The first detection of dust continuum emission from the Orion proplyds. A contour map of 1.3 mm continuum emission of a region in the Trapezium cluster is presented. The data were taken with the IRAM interferometer (Lada et al. 1999)

from extended emission around the source (Haisch & Lada 1999). It is notable that the detection rate of 1.3 mm continuum sources in NGC 2071 (46%) is much higher than the rate found in the Trapezium cluster (10%), even though the two clusters have comparable ages and NGC 2071 has a lower fraction of near-infrared excess sources than the Trapezium. Since NGC 2071, unlike the Trapezium, does not contain any O stars this suggests that the presence of O stars in a cluster may significantly affect the lifetimes of disks in clusters.

The distribution of 1.3 mm flux for the young NGC 2071 cluster, is similar in range, down to the detection limits, to the distribution in Taurus (Beckwith et al. 1990; Osterloh & Beckwith 1995), corrected for distance (Figure 21). However, at least two sources in NGC 2071 have higher 1.3 mm fluxes than in Taurus. These results suggest that the range of disk masses in the NGC 2071 cluster may be similar to the more isolated Taurus environment.

Figure 21 also compares the distribution of 1.3 mm continuum fluxes for the detected Trapezium sources with the distributions for NGC 2071 and

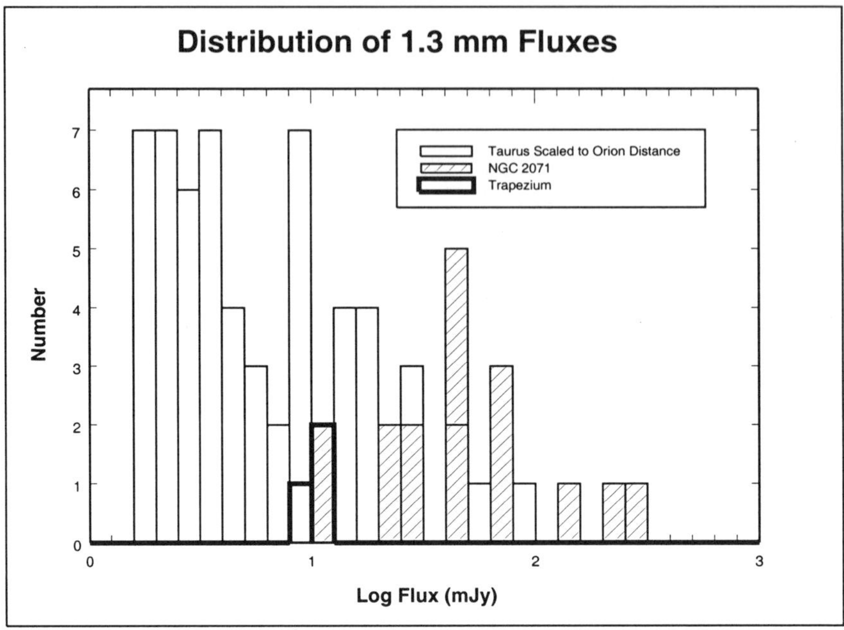

Figure 21. A comparison of the distribution of 1.3 mm continuum fluxes for sources in Taurus, the NGC 2071 cluster and the Trapezium cluster is presented. The fluxes for the Taurus sources were taken from Beckwith et al. (1990) and have been scale to the distance of Orion.

Taurus (corrected for distance to Orion). The largest fluxes measured for the Trapezium sources are an order of magnitude smaller than the largest fluxes measured for sources in NGC 2071 or Taurus. Remember that the majority of near-infrared sources in the Trapezium were not even detected at 5 mJy (5 σ level)). This result coupled with the HST optical observations of the proplyds suggests that circumstellar disks in the Trapezium OB association are being evaporated or destroyed. If we assume that the most massive disks in the Trapezium started out with masses as high as the most massive disks in NGC 2071 or Taurus (0.1 $M_\odot$), then we can estimate the mass loss rate and survival time for disks in the Trapezium environment.

$$mass\,loss\,rate\ =\ mass\,loss/cluster\,age\ =\ 0.1 M_\odot\ /10^6 yr \sim 10^{-7} M_\odot yr^{-1}$$

This mass loss rate implies disk lifetimes in the Trapezium of only 10^6 years, significantly shorter than lifetimes for disks in isolated environments such as Taurus (Strom, Edwards & Skrutskie 1993) and too short to allow for the formation of planets. This further implies that environment and stellar content play important roles in determining the properties and survival of protoplanetary disks.

These results have important ramifications for the likelihood of planet formation in the Galaxy. If a significant number of stars are born in clusters containing O stars, forming planetary systems may be a difficult proposition. What factors impact the formation and survival of proto-planetary disks in clusters, the presence of O stars or high stellar density? Are disks more massive and do they have longer lifetimes in clusters without O star members? It is clearly important to investigate disk properties in clusters with differing age and stellar content in order to access the ability to form planets in clustered environments.

Acknowledgements

I thank the organizers of this NATO ASI, Charles Lada and Nick Kylafis for inviting me to participate in this very lively and enjoyable conference and thank the participants of the conference for many stimulating discussions. I am grateful to Joao Alves, Karl Haisch and Gus Muench for interesting discussions and for providing data in advance of publication. I would also like to thank Richard Elston for help preparing the figures.

References

Abt, H. 1983 *Ann. Rev. Astron. & Astrophys.*, 21, 343.
Adams, F. C. and Shu, F. H. 1986 *Astrophys. J.*, 308, 836.
Adams, F. C., Lada, C. J., and Shu, F. H. 1987 *Astrophys. J.*, 312, 788.
Allen, L. E. 1996, Ph.D. thesis, Univ. of Massachusetts at Amherst.
Alves, J. Lada, C. J., Lada, E. A., Kenyon, S. J., & Phelps, R. 1998 *Astrophys. J.*, 506, 292.
Alves, J. Lada, C. J. and Lada, E. A. 1999 in prep.
Beckwith, S.V.W., Sargent, A.I., Chini, R.S., & Guesten, R. 1990 *Astron. J.*99, 924.
Beckwith, S.V.W., and Sargent , A.I. 1991 *Astron. J.*, 381, 250.
Beckwith, S.V.W., and Sargent, A. I. 1993 in *Protostars and Planets III*, ed. E.H. Levy and J.I. Lunine, Univ. of Arizona Press. p. 521.
Blitz, L., & Stark, A. A. 1986 *Astrophys. J. Letters*, 300, L89.
Blitz, L. 1993 in *Protostars and Planets III*, ed. E.H. Levy and J.I. Lunine, University of Arizona Press, p.125 .
Bontemps, S. Andre, P. & Ward-Thompson, D. 1995, *Astron. & Astrophys.*, 297, 98.
Bovier, J., Rigaut, F., & Nadeau, D. 1997 *Astron. & Astrophys.*323, 139.
Carpenter, J. M., Snell, R. L., & Schloerb, F. P. 1995 *Astrophys. J.*, 450, 201.
Carpenter, J. M., Meyer, M., Dougados, C., Strom, S.E., Hillenbrand, L.A. 1997 *Astrophys. J.*, 114, 127.
Churchwell, E., Felli, M., Wood, D.O.S., & Massi, M. 1987 *Astrophys. J.*, 321, 516.
Clarke, C., and Pringle, J. 1991 *Monthly Notices Roy. Astron. Soc.*, 249, 588.
Comeron, F., Rieke, G.H., & Rieke, M.J 1996 *Astrophys. J.*, 473, 294.
Duchene, G., Bouvier, J., Simon, T. 1999 *Astron. & Astrophys.*, 343, 831.
Duquennoy, A. & Mayor, M. 1991 *Astron. & Astrophys.*248, 48.
Elmegreen, B., 1983 *Monthly Notices Roy. Astron. Soc.*, 203, 1011.
Elmegreen, B., & Clemens, C. 1985 *Astrophys. J.*, 294, 523.

Felli, M., Taylor, G.B., Catarzi, M., Churchwell, E., & Kurtz, S. 1993 *Astron. & Astrophys. Suppl.*101, 127.

Fletcher, A. B., & Stahler, S. W. 1994a *Astrophys. J.*, 435, 313.

Fletcher, A. B., & Stahler, S. W. 1994b *Astrophys. J.*, 435, 329.

Garay, G., Moran, J.M., & Reid, M.J. 1987 *Astrophys. J.*314, 535.

Ghez, A., Neugebauer, G., Matthews, K. 1993 *Astron. J.*106, 2005.

Gomez, M., Hartmann, L., Kenyon, S. L. & Hewitt, R. 1993 *Astron. J.*, 105, 1927.

Greene, T.P. & Young, E. T. 1992 *Astrophys. J.*395, 516.

Haisch, K., Lada, E. A. & Pina, R., 1999 to be submitted.

Haisch, K., Lada, E. A. & Lada, C. J., 1999 to be submitted.

Haisch, K., & Lada, E. A. 1999 in prep.

Hillenbrand, L,. Strom, S., Calvet, N., Merrill., K.M., Gately., I. Russell, B., Meyer, M. Skrutski, M. 1998 *Astron. J.*, 116, 1816.

Hills, 1980 *Astrophys. J.*, 235, 986.

Horner, D.J., Lada, E.A. & Lada, C. J., 1997 *Astron. J.*, 113, 1788.

Kenyon, S. J. & Hartmann, L. W. 1987 *Astrophys. J.*, 323, 714.

Kenyon, S. J. & Hartmann, L. W. 1995 *Astrophys. J. Suppl. Ser.*, 101, 117.

Kenyon, S. J., Hartmann, L.W., Strom, K.M., & Strom, S.E. 1990 *Astron. J.*, 99, 869.

Kramer, C., Stutzki, J., Rohrig, R., Corneliussen, U. 1998 *Astron. & Astrophys.*, 329 249.

Kroupa, P. 1995 *Monthly Notices Roy. Astron. Soc.*, 277, 1522.

Lada, C. J. and Wilking, B. A. 1984 *Astrophys. J.*, 287, 610.

Lada, C. J., Margulis, M. & Dearborn, D. 1984 *Astrophys. J.*, 285, 610.

Lada, C. J. 1987, IAU Symposium No. 115: Star Forming Regions, ed. M. Peimbert and J. Jugaku, (Reidel Press), pg. 1.

Lada, C. J. 1991 in *The Physics of Star Formation and Early Stellar Evolution* ed. C.J. Lada & N. D. Kylafis p.329.

Lada, C.J. & Lada, E. A. 1991 in *The Formation and Evolution of Star Clusters*, ed. K. Janes (ASP, San Francisco) p. 3.

Lada, C.J. & Adams, F.C. 1992 *Astrophys. J.*, 387, 572.

Lada, C. J., Lada, E. A., Clemens, D. P. & Bally, J. 1995 *Astrophys. J.*429, 694.

Lada, C. J., Alves, J. and Lada, E. A. 1996 *Astron. J.*, 111, 1964.

Lada, E. A., DePoy, D., Evans, N. J. and Gatley I. 1991 *Astrophys. J.*, 371, 171.

Lada, E. A., Bally, J. & Stark, A. 1991 *Astrophys. J.*, 368, 42.

Lada, E. A. 1992 *Astrophys. J. Letters*, 393, L25.

Lada, E. A., Strom, K. M. & Myers, P.C. 1993 in *Protostars and Planets III*, ed. E.H. Levy and J.I. Lunine, University of Arizona Press, p. 245.

Lada, E. A. & Lada, C. J. 1995 *Astron. J.*, 109, 1682.

Lada, E. A., Evans, N.J., & Falgarone, E. 1997, *Astrophys. J.*, 488, 286.

Lada, E. A. & Beckwith, 1999 *Astrophys. J.*in preparation.

Lada, E. A., Dutrey, A., Guilloteau, S., & Mundy, L. 1999 to be submitted.

Lada, E. A., & Dahari, D. 1999 in prep.

Larson, R. 1990, in *Physical Processes in Fragmentation and Star Formation*, ed. R. Capuzzo- Dolcetta, C. Chiosio, and A. Fazio, Kluwer, pp 389-399.

Leinert, C., Zinnecker, H., Weitzel, N., Cristou, J., Ridgeway, S.T., Jameson, R., Haas, M., & Lenzen, R. 1993 *Astron. & Astrophys.*, 278, 129.

Li, W., Evans, N.J., & Lada, E. A. 1997 *Astrophys. J.*, 488, 277.

Li, W., Evans, N.J., & Lada, E. A. 1998 *Astrophys. J.*to be submitted.

Mathieu, R., D. 1983 *Astrophys. J. Letters*, 267, 97.

Meyer, M. & Lada, E. A. 1999 in Orion Resisted ed. M McCaughrean (ASP, San Francisco) in press.

Mathis, J. 1990 *Ann. Rev. Astron. & Astrophys.*28, 37.

McCaughrean M.J., & Stauffer, J.R. 1994 *Astron. J.*, 108, 1382.

Megeath, S.T. 1996 *Astron. & Astrophys.*, 311, 135.

Muench, A., Lada, E.A. & Lada, C.J. 1999 *Astrophys. J.*submitted.

Mundy, L. G., Looney, L.W., & Lada. E.A., *Astrophys. J. Letters*, 452, 137.

Myers, P. C., Fuller, G. A., Mathieu, R. D., Beichman, C. A., Benson, P. J., Schild, R. E., & Emerson, J. P. 1987 *Astrophys. J.*, 319, 340.
O'Dell, C.R., Wen, Z. & Hu, X. 1993 *Astrophys. J.*410, 696.
O'Dell, C.R. & Wen, Z. 1994 /apj 436, 194.
Osterloh, M. & Beckwith, S.V.W. 1995 *Astrophys. J.*429, 288.
Patel, K., & Pudritz, R.E. 1994 *Astrophys. J.*424, 688.
Petr, M., Codude DU Foresto, V., Beckwith, S.V.W., Richichi, A., & McCaughrean, M. 1998 *Astrophys. J.*500, 825.
Phelps, R. L. and Lada, E. A. 1997 *Astrophys. J. Letters*, 477, 176.
Prosser, C., Stauffer, J.R., Hartmann, L., Soderblom, D., Jones, B.F., Werner, M.W. & McCaughrean, M.J. 1994 *Astrophys. J.*, 421, 517.
Reipurth, B. & Zinnecker, H. 1993 *Astron. & Astrophys.*, 278, 81.
Rieke, G. & Lebofsky, M. 1985 *Astrophys. J.*, 288, 618.
Roberts, 1957 *P.A.S.P.*, 69, 406.
Rydgren, A. E., and Zak, D. 1987 *P.A.S.P.*, 99, 141.
Rucinski, S. M., 1985 *Astron. J.*, 90, 2321.
Saiyadopour, A., Deiss, B. M., Kegel, W. 1997 *Astron. & Astrophys.*, 322, 756.
Shu, F., Najita, J., Galli, D., Ostriker, E., & Lizano, S. 1993 in *Protostars and Planets III*, ed. E.H. Levy and J.I. Lunine, Univ. of Arizona Press, p.3.
Simon, M., Ghez, A.M., Leinert, CH., Cassar, L., Chen.W.P., Howell, R.R., Jameson, R.F., Matthews, K., Neugebauer, G., Richichi, A. 1995 *Astrophys. J.*443, 625.
Skrutskie, M. F., Dutkevitch, D., Strom, S. E., Edwards, S., and Strom, K. M. 1990 *Astron. J.*, 99, 1187.
Strom, K. M., Strom, S. E., Kenyon, S. J., and Hartmann, L. 1988 *Astron. J.*, 95, 534.
Strom, K. M., Strom, S. E. and Merrill, K. M. 1993 *Astrophys. J.*, 412, 233.
Strom, S. E., Edwards, S. & Skrutskie, M. F. 1993 in *Protostars and Planets III*, ed. E.H. Levy and J.I. Lunine, Univ. of Arizona Press, p. 837.
Strom, S.E., Kepner, J., & Strom, K. M. 1995 *Astrophys. J.*, 438, 813.
Walter, F. M. 1987 *P.A.S.P.*, 99, 31.
Welch, J. & Sargent, A. I. 1993 *Ann. Rev. Astron. & Astrophys.*, 31, 297.
Verschueren, W. & David, M. 1989 *Astron. & Astrophys.*, 219, 105.
Verschueren, W. 1990 *Astron. & Astrophys.*, 234, 156.
Zinnecker, H., McCaughrean, M.J. & Wilking, B. A. 1993 in *Protostars and Planets III*, ed. E.H. Levy and J.I. Lunine, University of Arizona Press, p.429.

Joao Alves receives a first place poster prize.

Lunch by the Sea

MULTIPLE STELLAR SYSTEMS: FROM BINARIES TO CLUSTERS

IAN A. BONNELL

Institute of Astronomy
Madingley Road, Cambridge, CB3 0HA, UK

1. Introduction

The study of star formation often concentrates on the easily detectable product of this process: the young stars. These stars are typically not in isolation but are found in groups ranging in sizes from binary systems to clusters containing thousands of stars. In this chapter we discuss the mechanisms for, and implications of, star formation in groups.

Although much theoretical work has gone into understanding the formation of a single star like our sun (e.g. Shu et al. 1987), most solar-type stars are found in binary systems (Duquennoy & Mayor 1992). Furthermore, surveys of pre-main sequence stars have shown that their binary frequency is at least as high as on the main sequence (e.g. Ghez 1995; Mathieu et al.1999). Indeed, there is strong evidence that in some nearby star forming regions such as Taurus and Chameleon, the binary frequency is significantly higher than among the main sequence populations.

A further complication is that star formation in giant molecular clouds (GMCs) such as Orion is predominantly forming clusters of young stars (Lada et al. 1991; Meyer & Lada 1998). Clustering appears to be ubiquitous around young stars more massive than $\approx 5M_\odot$ (Hillenbrand 1995; Testi et al. 1997) with the youngest O-type stars found in rich clusters of a thousand or more stars such as the Orion Nebula Cluster (e.g. Hillenbrand 1997).

The reason why star formation in groups is important is that the dynamical timescales of these systems is of the same order as the stellar formation timescale, the pre-collapse free-fall time. Thus, the stars interact as they are forming, and these interactions necessitate that we study their formation in the group as a whole and not as separate entities. In this chapter, we discuss the formation and early evolution of groups of stars, ranging from binaries to stellar clusters. In Sections 2 through 5 we discuss the formation

C.J. Lada and N.D. Kylafis (eds.), The Origin of Stars and Planetary Systems, 479–514.
© 1999 *Kluwer Academic Publishers. Printed in the Netherlands.*

of binary and multiple systems while in Sections 6 through 8 we discuss the formation of stellar clusters.

2. Observations of Binary Systems

As noted above, binary systems are common both amongst the main sequence (Duquennoy & Mayor 1991; Fisher & Marcy 1992; Mayor et al.; 1992) and amongst the pre-main sequence (PMS) (Simon et al. 1992; Ghez et al. 1993; Reipurth & Zinnecker 1993; Leinert et al. 1993; Mathieu 1994) populations. The fact that the PMS systems have all their discernible properties within 10^6 years implies that the binary pairing has to occur early, on the same timescale that the stars form.

2.1. MAIN SEQUENCE BINARIES

The best statistics on stellar multiplicity and binary properties come from recent volume limited samples. These surveys largely avoid the selection effects which have plagued earlier samples. They found that ≈ 60 % of G-dwarfs (Duquennoy & Mayor 1991), ≈ 45 % of K-dwarfs (Mayor et al. 1992), and ≈ 42 % of M-dwarfs are in binary systems (Fisher & Marcy 1992).

In addition to reproducing the frequency of binary systems, any theory for the formation of binary stars has to take into account the many observed properties of these systems, including their separations, eccentricities and mass ratios. Duquennoy & Mayor (1991), found that binary stars can have separations from a few $R_\odot$ to 10^4 AU. The distribution of separations appears to be unimodal, with a broad peak at ≈ 30 AU. In terms of the eccentricity, only the closest of binary stars have circular orbits, due to tidal circularization. Longer period binaries all have non-zero eccentricities, while the maximum eccentricity at a given period increases with the period. Considering the distribution of mass ratios, Duquennoy & Mayor (1991) show that most binaries do not have equal masses, but that the distribution is slowly increasing for smaller mass ratios. In contrast with the overall mass ratio distribution, the closest systems appear, although with only small number statistics, to have a nearly flat distribution with a possible rise towards equal mass systems (Mazeh et al. 1992).

2.2. PRE-MAIN SEQUENCE BINARIES

Even though a significant fraction of the earliest classified T Tauri stars were known to be binaries, it is only recently that extensive surveys have attempted to quantity the binary frequency of pre-main sequence stars in nearby star forming regions (Mathieu et al. 1989; Simon et al.1992; Ghez et al. 1993; Leinert et al.1993; Reipurth & Zinnecker 1993; Richichi et al. 1994;

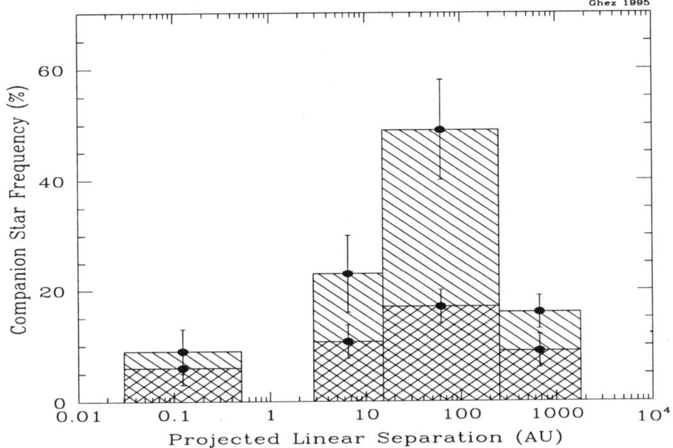

Figure 1. A comparison of companion star frequency for main sequence (cross hatched) and pre-main sequence (hatched) stars (Ghez 1995).

Ghez et al. 1997). The main result of these surveys is that there is an excess (*approx* twice) of pre-main sequence binaries in these regions, compared to the main sequence (Fig. 1; Ghez [1995]; Mathieu et al. [1999]). This unexpected excess is hard to explain if these regions are typical of most youngs stars. Interestingly, surveys of the binary frequency in young stellar clusters yield numbers more similar to the main sequence (e.g. Prosser et al. 1994; Petr et al. 1998) although that may depend on the cluster, and the location in the cluster, considered (Padgett et al. 1997). We will return to this in §8.

3. Binary Formation Theories: Fission and Capture

There have been many theories advanced to explain the origin of binary stars. These theories have generally involved either a dissipative mechanism to join to unbound stars together (capture mechanisms) or a mechanism to separate one body into two. Current thinking is that most binary systems form through some type of a fragmentation scenario, with a possible subsequent reorganisation of the system involving star-disc interactions (or star-disc capture). For the sake of completeness, we begin by discussing the currently discarded theories of fission and traditional capture mechanisms.

3.1. FISSION

Fission, first suggested by Kelvin and Tait in 1883, involves the rotational instability of a contracting, rapidly rotating protostar. It is hypothesised to occur after the initial formation phase has resulted in a quasi-equilibrium pressure-supported protostar. As the contraction proceeds, conservation of angular momentum increases the protostar's spin. The protostar then becomes unstable to non-axisymmetric perturbations once the ratio of rotational to the absolute value of the gravitational energies is greater than 0.27. This instability has the protostar deform into a bar-shape and subsequently separate into two distinct bodies. This process could only explain close binaries, with separations of the order of the protostellar core, so would be unable to form the majority of systems.

Although long an attractive binary formation mechanism, fission has recently been abandoned. This is because numerical simulations have shown that fission does not occur for compressible fluids like stars (e.g. Durisen et al. 1986). Instead, spiral arms form at the ends of the bar which transport angular momentum away from the central object by gravitational torques (e.g. Bonnell 1994). This results in the formation of a disk around a now stable protostar. This disk contains only a small fraction of the total mass but contains the majority of the system's angular momentum. Thus, fission is not a viable mechanism for the formation of any binary stars.

3.2. CAPTURE

Capture, first suggested as a binary formation mechanism by Stoney in 1867 (Tassoul 1978), involves some sort of process that can dissipate excess kinetic energy so that two unbound stars become bound during a close passage. Several variations of this mechanism have been proposed. These include *Tidal Capture*, in which two stars pass close enough together to excite tidal modes in each other which absorb much of their relative kinetic energy (e.g. Fabian, Pringle & Rees 1975; Mardling 1995); *Dynamic Capture*, where three stars pass close together and exchange energy such that one of them is ejected with most of the kinetic energy, leaving the other two bound; and *Star-Disk Capture*, where a circumstellar disk provides the dissipative medium and the relative kinetic energy goes into unbinding some of the disk material. The first two variants require high stellar density and are therefore not likely to contribute a significant fraction of the overall binary population (Hills & Day 1976), although tidal capture is still a leading mechanism for forming X-ray binaries in dense clusters. The last mechanism uses the fact that protostars are observed to have relatively large and massive circumstellar disks which can then act as a dissipative medium, significantly increasing the cross section (e.g. Larson 1990).

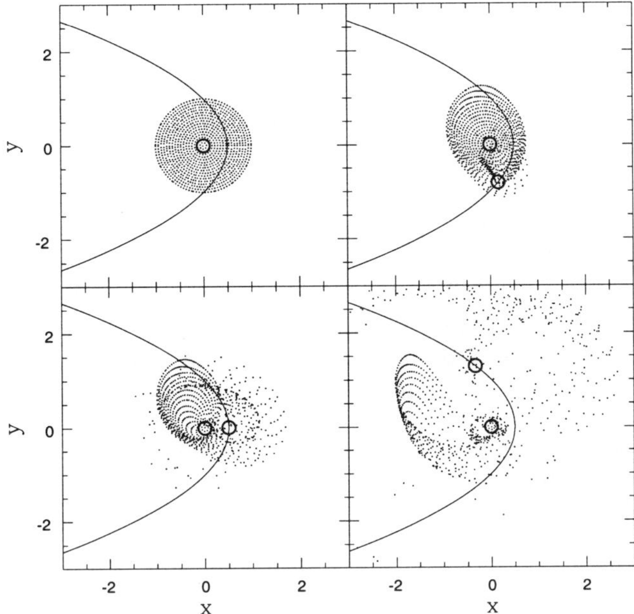

Figure 2. A star-disk interaction where the disk matter exterior to periastron is unbound in tidal tails (Hall, Clarke, & Pringle 1996).

3.3. STAR-DISC CAPTURE

The dynamics of star-disc captures have been investigated by Clarke & Pringle (1991; 1993), Heller (1993) and Hall, Clarke & Pringle (1996). During an encounter, where the minimum separation of the two stars is less than the disk radius, the disk matter exterior to this distance is unbound, removing the necessary energy from the two stars (see Fig. 2). The energy exchange is from perturbing star to disk as long as the periastron separation is $R_{\mathrm{peri}} \leq 1.5 R_{\mathrm{disk}}$. The actual energy liberated by this encounter is estimated to be between 2 and 5 times the binding energy of the material exterior to periastron, depending on the exact geometry of the encounter (Hall, Clarke & Pringle 1996).

Although this process will not form many binaries in the field stars where interactions are too rare, It can be a significant process in small stellar clusters (Clarke & Pringle 1991). These clusters are more appropriate than larger clusters (> 100 stars) as the higher velocity dispersion in dense stellar clusters tend to destroy the disk at the first encounter. The dynamics of these small clusters has been investigated for stars without (McDonald & Clarke 1993) and with star-disk interactions (McDonald & Clarke 1995). Without disks, interactions in a small cluster results in the formation of a single binary containing the two most massive stars. This dynamical biasing results in mass ratios consistent with that observed if the cluster contains

$\lesssim 4$ stars. Otherwise, the mass ratios peak towards equal masses. When star-disk interactions are included, the binary frequency is increased such that it is consistent with those observed for main sequence solar type binaries. Furthermore, the star-disk interactions remove the dynamical biasing so that the secondary mass function is compatible with random picking from the IMF. This happens as the star-disk interactions cause the binaries to be harder and thus are less likely to be disrupted by subsequent encounters, independent of the secondary mass (McDonald & Clarke 1995). In terms of the separation, the interactions are most likely to occur with periastron of order the disk radius. Thus, it is unclear at present if you can form very close systems from this mechanism.

An important point in this argument is that small clusters are necessary for this mechanism to work. Their formation would require some sort of "Prompt Fragmentation" as set out by Pringle (1989). An example of this is the fragmentation of a cooled shock layer in cloud-cloud collisions (Chapman et al. 1992). Star-disk capture has been found to occur among fragments formed in this layer (Turner et al. 1995).

4. Binary Formation through Fragmentation

Fragmentation as a mechanism for the formation of binary systems was first proposed by Hoyle (1953). He noted that if the collapse is isothermal, then as the density of a cloud is increased, the Jeans mass, the minimum mass for an object to be gravitationally bound decreases and smaller parts of the cloud become gravitationally unstable. He envisioned a repetitive process known as "hierarchical fragmentation". Although, in general, this does not occur due to the non-homologous nature of gravitational collapse with pressure (see below), the idea of fragmentation due to a decreasing Jeans mass does occur and can be thought of as "Thermally Driven Fragmentation". This generally occurs during the collapse process. Alternatively, fragmentation can occur with a near constant Jeans mass when the total mass available is increased. An example of this type of process is when rotation halts collapse at a certain stage (e.g a circumstellar disc) and mass is added through the continuing infall.

4.1. GRAVITATIONAL COLLAPSE AND FRAGMENTATION

In order to see when fragmentation may occur in the star formation process, it is useful to consider the Jeans mass, M_J and the Jeans radius, R_J the minimum mass and radius, respectively, for a cloud of given density, ρ and

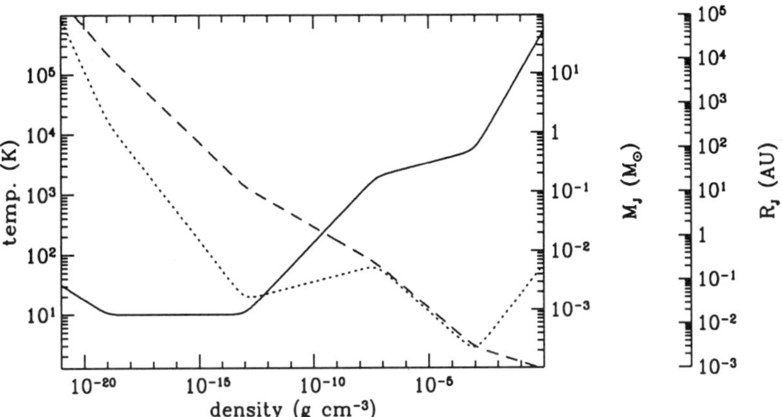

Figure 3. The temperature (solid line), Jeans mass (dotted line), and Jeans radius (dashed line) of interstellar gas as a function of density, showing the different phases the gas passes through during collapse (adapted from Tohline 1982).

temperature, T to be gravitationally bound:

$$R_J = \left(\frac{5R_gT}{2G\mu}\right)^{1/2}\left(\frac{4}{3}\pi\rho\right)^{-1/2}, \tag{1}$$

and

$$M_J = \left(\frac{5R_gT}{2G\mu}\right)^{3/2}\left(\frac{4}{3}\pi\rho\right)^{-1/2}, \tag{2}$$

where R_g is the gas constant, G is the gravitational constant, and μ is the mean molecular weight.

Once collapse is initiated, there are four main phases that the collapsing cloud passes through. Figure 3 shows a schematic of the temperature as a function of the density for a protostellar collapse (adapted from Tohline 1982). The collapse is parameterised by a polytropic equation of state $P = K\rho^\gamma$ ($T \propto \rho^{\gamma-1}$) where γ depends on the collapse phase. In the first phase, the collapse proceeds isothermally ($\gamma = 1$) until densities of the order of 10^{-14} to 10^{-13} g cm^{-3} are reached. At this point, the gas is no longer optically thin, and the trapped radiation increases the temperature (phase 2; $\gamma = 7/5$). This forms a first protostellar core which contracts quasistatically with increasing mass. Once the temperature has reached 2000 K, the molecular hydrogen starts to dissociate, absorbing some of the thermal energy and allowing for a second collapse stage (phase 3; $\gamma \approx 1.1$). Finally, once all the gas is in atomic form, a second protostellar core is formed which continues to contract quasistatically as it grows in mass (phase 4; $\gamma = 5/3$).

In order for fragmentation to occur, the total mass available for fragmentation has to be $M_{\text{tot}} \gtrsim 2M_J$, otherwise each object cannot be independently gravitationally bound. This occurs most readily when the Jeans mass decreases as the density increases during the collapse. Figure 3 show the evolution of the Jeans mass during the four phases of collapse. It decreases throughout the isothermal collapse phase where $\gamma = 1$, increases when the fist core is formed ($\gamma = 7/5$), decreasing again during the second collapse phase where $\gamma \approx 1.1$, and increasing once more when all the hydrogen has dissociated ($\gamma = 5/3$). Thus fragmentation is most likely to occur during or slightly after the two collapse phases.

The scale at which fragmentation is likely to occur can be understood in terms of the Jeans radius, the minimum radius for an object to be bound, and its evolution throughout the collapse (Fig. 3). The Jeans radius gives an indication of the minimal initial separation as objects with smaller separations would not be separately bound. Thus, in order for fragmentation to produce binaries with separations less than 1 AU, densities greater than 10^{-10} g cm^{-3} are required. This in turn implies fragment masses on the order of the Jeans mass being $M_{\mathrm{frag}} \lesssim 0.01 M_\odot$. Any close binary formed from a straight fragmentation will thus have to accrete most of its final mass in order to achieve stellar status.

A limitation on thermally driven fragmentation is that collapse proceeds non-homologously, limiting the mass available for fragmentation. From near-uniform density initial conditions, pressure gradients develop near the outer edge of the cloud and propagate inwards at the speed of sound. These pressure gradients act to slow down the collapse, allowing the inner regions to fall away from the outer part of the cloud. The central region of the cloud collapses with nearly uniform density while the outer part of the cloud has a steep density gradient (e.g. Larson 1969). Only in the central regions, where the dynamical timescale, the free-fall time,

$$t_{ff} = \left(\frac{3\pi}{32G\rho} \right)^{1/2}, \tag{3}$$

is approximately constant, can fragmentation occur. In the outer regions, the dynamical timescale increases with radius, so that mass at a given radius can never catch up with the mass interior to it, and can never be caught by matter exterior to it. Thus, due to tidal shearing, fragmentation can never occur in a region of a dynamically collapsing cloud with a steep density gradient. In an isothermal collapse, for example, the density profile approaches $\rho \propto R^{-2}$ (Larson 1969), so that the central mass decreases as $M_{\mathrm{cent}} \propto \rho^{-1/2}$, exactly as does the Jeans mass. This limitation is increased if the cloud is initially dominated by a steep density profile. Fragmentation from such initial conditions cannot occur during the dynamical collapse phase due to the tidal shearing from inner regions.

Fragmentation of a spherically-symmetric collapsing cloud is inherently difficult due to the arguments stated above. Furthermore, unless there ex-

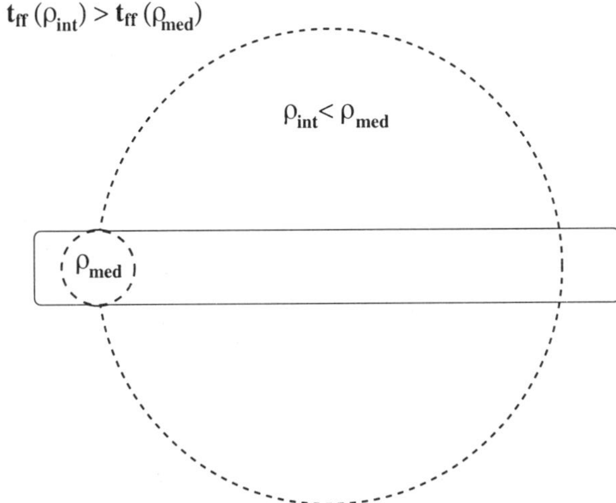

Figure 4.   Non-spherical structures are easier to fragment. The local density at a point inside a slab of gas is higher than the mean density interior to it. Thus, the local free-fall time is shorter than the free-fall time for the matter at this point to fall to the centre of the slab.

ists a significant off-centre density enhancement such as a shell, the uniform free-fall time implies that any small perturbation cannot grow before merging with the rest of the collapsing cloud. Fortunately, few non-stellar structures are perfectly spherical, including molecular clouds and their cores which exhibit predominantly elongated geometries (eg Myers et al. 1991). Such clouds, being either prolate or oblate are more susceptible to fragmentation due to their configurations. This occurs as subregions of the cloud are then more likely to collapse on themselves before merging with the rest of the cloud. This can be seen by considering the free-fall time for these regions and for the cloud as a whole (Fig. 4). A subregion in the slab has a density that is higher than the mean density of the material interior to it, as much of the interior region is empty. Thus, it's free-fall time is less than the mean free-fall time of the cloud at that radius. The subregion can thus collapse on itself, if it contains more than a Jeans mass, faster than the cloud as a whole collapses. This amplifies the asphericity of the cloud (e.g. Lin Mestel & Shu 1965), making it even more susceptible to fragmentation. Fragmentation is thus much easier in clouds which have significant non-spherical geometries, be it either initial (prolate or oblate) or generated during the collapse from smaller perturbations or rotation.

The last point we need to consider before discussing the fragmentation scenarios is the effect of rotation. Angular momentum perturbs the collapse, introducing a flattening into the cloud and eventually halting the collapse. The end-points for a collapse including rotation can be estimated as (e.g.

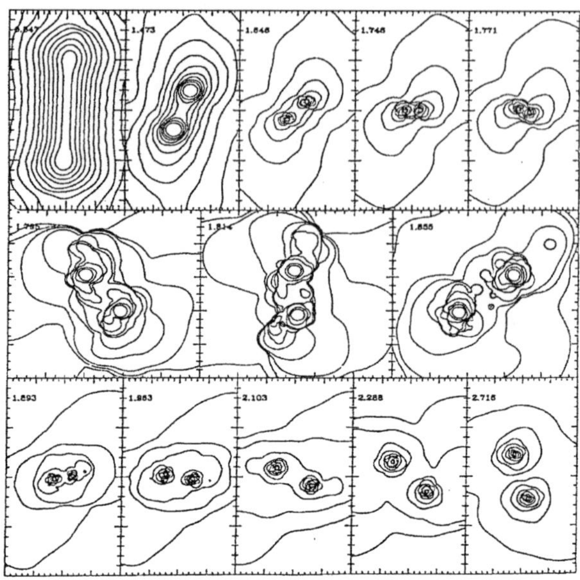

Figure 5. The fragmentation of a cylindrical prolate cloud and the formation of a binary system (Bonnell et al. 1991).

Tohline 1981)

$$R_{\text{final}} \approx \beta R_{\text{initial}}, \tag{4}$$

where $\beta = E_{\text{rot}}/|E_{\text{grav}}|$. Determinations of β in molecular cloud cores have yielded values of $\beta \lesssim 0.1$, with a mean value of $\beta \approx 0.02$ (Goodman et al. 1993). Thus for initial cloud radii of $R_{\text{initial}} \approx 0.05$ pc, rotationally supported objects should have sizes on the order of 200 AU.

4.2. FRAGMENTATION DURING COLLAPSE

The earliest three-dimensional simulations of gravitational collapse and fragmentation were of initially spherical clouds with several Jeans masses, and significant rotation. (e.g. Boss & Bodenheimer 1979; Bodenheimer et al 1980; Miyama et al. 1984; Monaghan & Lattanzio 1986; Boss 1986). In these simulations, fragmentation was due to large initial non-axisymmetric perturbations which grew during the collapse (e.g. Boss & Bodenheimer 1979), or due to significant initial rotation that halted the collapse allowing smaller perturbations to grow (e.g. Miyama et al. 1984).

Since then, observations have shown that molecular cloud cores deviate significantly from spherically symmetry, often having elongated (and probably prolate) shapes (Myers et al. 1991). Furthermore, their typically small rotation rates (Goodman et al. 1993), imply that rotation is generally not

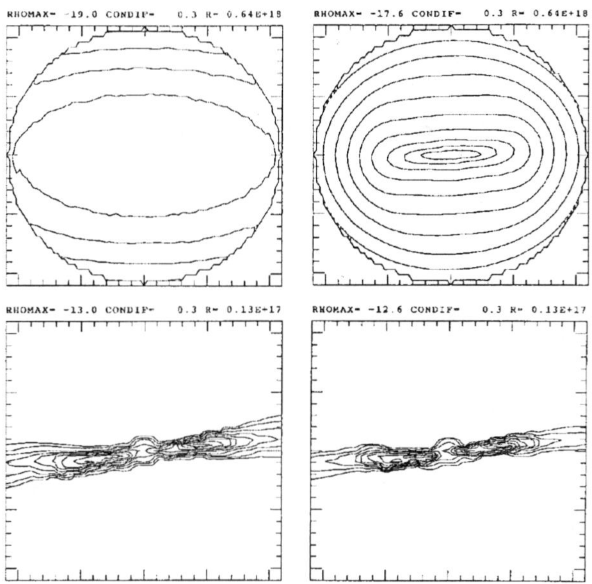

Figure 6. The fragmentation of a Gaussian density prolate cloud and the formation of a binary system (Boss 1993).

an important factor in driving fragmentation on scales greater than 200 AU.

This has driven the simulations towards the collapse of prolate clouds with little rotation. Simulations have been performed for uniform-density cylindrical clouds (Fig. 5 Bonnell et al. 1991, 1992), uniform-density prolate spheroidals (Nelson & Papaloizou 1993; Monaghan 1994; Nelson 1998) and Gaussian-density-profile prolate clouds (Fig. 6 Boss 1993, 1998; Sigalotti & Klapp 1997). All of these can result in the formation of binary systems. The fragmentation arises when collapse occurs preferentially towards the major axis such that the two sides of the cloud collapse on themselves before collapsing together. This occurs if the Jeans radius (see above) is less than the minor axis, or if the cloud has $M \gtrsim 2M_J$. The binary systems that result are highly eccentric and can have a variety of mass ratios (Bonnell & Bastien 1992a).

For prolate clouds that have many Jeans masses such that the Jeans radius is much less than the cloud's minor axis, then the collapse forms a narrow bar (Bastien 1983). This bar is unstable to fragmentation and can thus form many self-gravitating clumps (Bastien et al. 1991; Bonnell et al. 1991; Burkert & Bodenheimer 1993; Bate & Burkert 1997). A cautionary note on this bar-mode fragmentation is that it requires either significant initial perturbations (e.g. Bastien et al. 1991) which is not unlikely in molecular clouds, or else a mechanism that halts the bar's collapse (e.g. heating in the optically thick phase), at which the fragmentation can then proceed (e.g. Bate & Burkert 1997; Nelson 1998). The collapse of oblate clouds with many Jeans masses is also susceptible to fragmentation in an analogous manner (Boss 1996; Klapp & Sigalotti 1998). In this case, collapse occurs initially towards a sheet which can then fragment due to perturbations in the cloud.

According to the criteria of a decreasing Jeans mass, thermally driven fragmentation should, in principle, be possible during the second collapse phase (Fig. 3). Complications in this case are that before the second collapse phase commences, the matter is part of a pressure-supported outer protostellar core which can only contract quasistatically. Thus it should be near equilibrium and any perturbations should be removed before the collapse is initiated. Simulations of this collapse phase have not found any thermally driven fragmentation (Boss 1989).

Fragmentation during the collapse phase can also be driven by cooling (i.e. $\gamma_{eff} < 1$). The removal of thermal support is an efficient way of destabilising a cloud that would otherwise not fragment in the isothermal regime. Monaghan & Lattanzio (1991) have found that for spherical clouds that would normally collapse without fragmenting, the inclusion of molecular cooling can drive the fragmentation process. Chapman et al. (1992) have also invoked cooling to aid fragmentation. They modelled the collision of two stable clouds with cooling occurring in the shocked region. This resulted in a rapid decrease in the Jeans mass with resulting fragmentation into many protostellar fragments. In both cases, the fragmentation does not occur in a spherical cloud but rather in oblate slabs caused either by rotation (Monaghan & Lattanzio 1991) or the shock in the cloud-cloud collision (Chapman et al. 1992).

4.3. DISC FRAGMENTATION

Although early fragmentation simulations were performed with large amounts of rotational energy ($\beta \approx 0.2$) in which rotation soon impeded the collapse allowing for quick fragmentation, it is now clear that any rotationally driven fragmentation will generally only occur after the cloud has collapsed to a fraction of it's initial size (see above). Fragmentation at this stage requires either pre-existing non-axisymmetric perturbations or the generation of new non-axisymmetric structures. Any initial perturbation present in this object needs to have survived the collapse phase without growing non-linear and thus fragmenting at an earlier stage, and without being washed away in the earlier stages of the collapse where they could not have contained a Jeans mass.

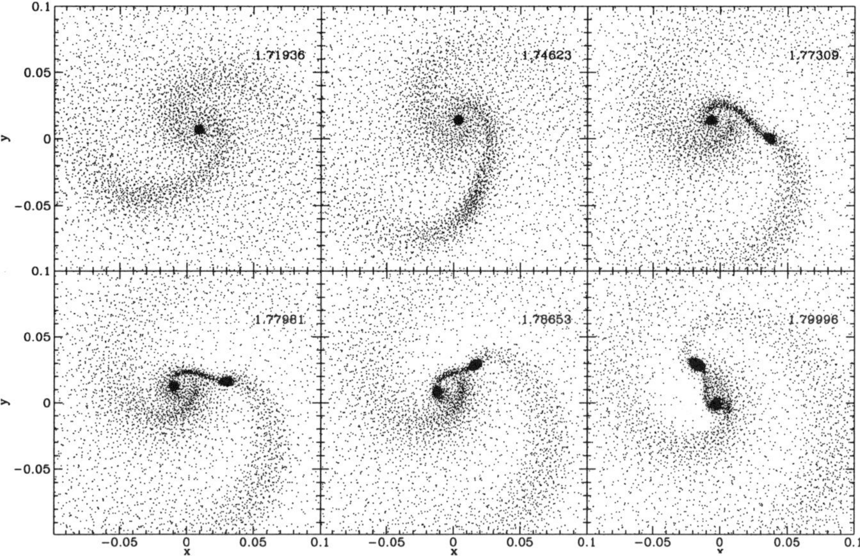

Figure 7.   Protostellar disk fragmentation due to a rotational instability (Bonnell &
Bate 1994a).

Alternatively, non-axisymmetric structure can be driven into the disk
due to a dynamical rotational instability (Bonnell 1994). In the absence of
pre-existing perturbations, the rotationally supported post-collapse object
will evolve towards a single central fragment and a surrounding disk. In
order to do this, it undergoes a rotational instability which generates an
$m = 2$ bar-mode and accompanying spiral arms which ultimately drives
the disk fragmentation (Bonnell 1994). This instability occurs due to the
necessary redistribution of angular momentum to form a stable pressure-
supported protostellar core and surrounding disk. This rotationally-driven
bar-mode instability is equivalent to that invoked in the fission mechanism.
The primary difference is that in this case the bar is surrounded by a
disk and interacts with the continuing infall. As this process occurs at the
end of the collapse phase, material is continually being added to the disk
on a dynamical timescale (equivalent to that of the disk). This material is
necessary to continually drive the instability and provide the mass-reservoir
for the subsequent fragmentation.

The outward transport of angular momentum from the bar-mode and
spiral arms drive an $m = 1$ mode into the system (Bonnell 1994; Bonnell &
Bate 1994a). Small asymmetries in the rate of angular momentum transport between the two ends of the bar, coupled with the dynamic nature of
the instability, then propel the central object from the centre of mass. The
combination of the two ($m = 1$ and $m = 2$) modes allow the spiral arms to

gather matter together, and even collide, such that a Jeans mass condensation is formed which can then collapse to form a secondary (Fig. 7). The
angular momentum transport by way of the spiral arms removes the rotational support inherent in the disk, thus allowing the fragment to collapse.
It should be noted that the continued infall onto the system is a necessary
condition, without which no fragmentation occurs (Bonnell 1994). A similar
mode of fragmentation has been found to occur where the disk is fed by an
accretion flow (Whitworth et al. 1995). In this case, subsequent accretion
from a shearing sheet formed in a cloud-cloud collision continually spins up
the disk until the bar-mode instability develops.

Disk fragmentation can also occur at the end of the second collapse
phase and in so doing, form a close binary system with separation of order
10 $R_\odot$ (Bonnell & Bate 1994a). A related study by Boss (1989) found
that an initial $m = 2$ perturbation grew during the second collapse phase
but that was forced to merge due to the gravitational torques. It is these
torques that cause the disk fragmentation in the Bonnell & Bate study,
which was able to resolve the fragmentation due to the Lagrangian nature
of the code and the longer periods over which the evolution was followed.
Due to the small value of the Jeans mass, the binary has a very small mass
($M_{\mathrm{binary}} < 0.01$ $M_\odot$), and thus will have to accrete the majority of its final
mass. A limitation on fragmenting a disk around the second core is that if
significant non-axisymmetric structure develops around the first core due
to a similar instability, then sufficient angular momentum is transferred
outwards ensuring that the second core and disk are rotationally stable
(Bate 1998a).

Massive circumbinary disks have also been shown to fragment due to
the interaction with the inner binary (Bonnell & Bate 1994; Burkert &
Bodenheimer 1996). In this case, the binary takes the role of the $m = 2$
bar-mode. The fragmentation follows the same evolution as above with the
binary driving an $m = 1$ mode into the system, forcing the binary to spiral
away from the system's centre of mass (see Fig. 8). The interaction of the
spiral arms then forms the additional fragment.

In all of these calculations, the role of the continued infall is of prime
importance as it adds mass to the disk helping to drive the instabilities.
This implies that this disk fragmentation process occurs only at the end of
a collapse phase as it is then that infall will be at its strongest.

An alternative method to fragment a disk can occur where the spiral
arms are triggered by a passing star (Watkins et al. 1998; see also Bonnell et al. 1992). In clusters or binary systems, the passage of a star near
the disk can induce spiral arms which removes the rotational support and
increases the local density, resulting in the disk fragmentation. Studies of
isolated, nearly stable circumstellar disks without infall have hypothesised

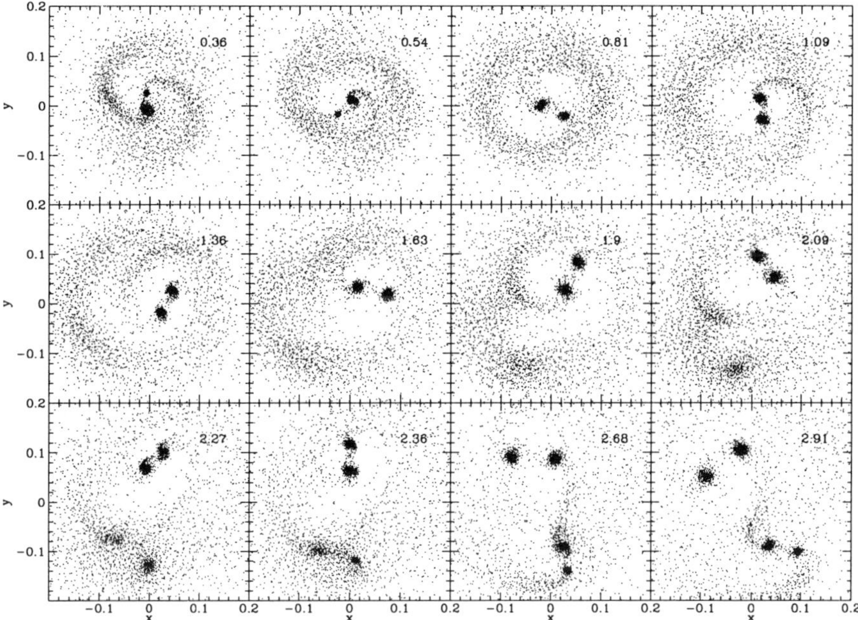

Figure 8. The fragmentation of a massive circumbinary disk due to the interaction with the central binary (Bonnell & Bate 1994b).

that fragmentation could result from the complicated interplay of waves in the disk that drive the $m = 1$ motion (Adams et al. 1989; Shu et al. 1990). To date this mechanism has not been found in simulations (eg Laughlin & Bodenheimer 1994) which may be due to questions as to the original boundary conditions (Heemskerk et al. 1992).

5. Early Evolution of Binary Systems

Once a young binary system has formed through one of the above processes, it can still be expected to undergo significant evolution before either component reaches the main sequence. This is because there will almost certainly be residual matter which will interact with the system. The fragmentation process is essentially inefficient in that most of the cloud mass has not ended up in one of the protostars by the time the system has formed. Instead, this matter infalls over longer periods, to be accreted or form surrounding disks. The disks themselves can further complicate the picture by removing angular momentum and energy from the system as in the star-disk interactions. Furthermore, gravitational interactions between the fragments in a multiple system will occur, forcing it to evolve into a more hierarchical and thus stable configuration (e.g. Burkert & Bodenheimer 1996).

5.1. ACCRETION IN BINARY SYSTEMS

Accretion on to a young binary system is likely to be one of the most important contributors to the final parameters of the system. This is especially true when the system is formed via fragmentation as it then contains but a fraction of the total mass available. In general, the mass of a protobinary soon after fragmentation is dependent on its separation such that closer systems have smaller initial masses and will subsequently have to accrete a larger percentage of their final masses (e.g. Boss 1986). In particular, systems that originate through fragmentation with separations of a few $R_\odot$ need to accrete $> 99\%$ of their final mass, without merging, to form a main sequence close binary (Bonnell & Bate 1994a).

To determine the final state of a binary formed through fragmentation, it is necessary to be able to follow the evolution until all of the infalling material has been accreted (Bate, Bonnell & Price 1995). Numerical simulations of accretion onto a binary system have found that the main dependencies are on the binary's mass ratio and the relative specific angular momentum, j, of the infalling material (Bate 1997; Bate & Bonnell 1997). For unequal masses, low j material is accreted by the primary or its disk. For higher j, the infall can also be accreted by the secondary or its disk. When $j >> j_{\rm binary}$ the infall is then accreted by neither component but forms a circumbinary disk (Fig. 9). Accretion can significantly alter the mass ratio and separation of the binary and is probably a main factor in determining the final binary properties. Probable final binary properties can be evaluated using these results under the assumption of various initial conditions (Bate 1998b). Thus, close binaries which formed from a disk fragmentation are likely to have near-equal masses if they accrete predominantly high j matter while if they accrete mostly low j matter, they would have been forced to merge.

5.2. STAR-DISC INTERACTIONS IN BINARY SYSTEMS

Star-disk interactions can be important even after a binary system has formed. In addition to the the penetrating circumstellar disk encounters which dissipate orbital motion and possibly binding initially unbound stars (e.g. Clarke & Pringle 1991, 1993 see §3.3 above), there are less dynamic interactions that occur over longer timescales. These include gap-clearing between circumstellar and circumbinary disks (e.g. Artymowicz & Lubow 1994,1996), the removal of angular momentum by a circumbinary disk (Pringle 1991; Artymowicz et al.1991) and the generation of disk-warps in non-aligned binary systems (e.g. Terquem & Bertout 1993, Larwood et

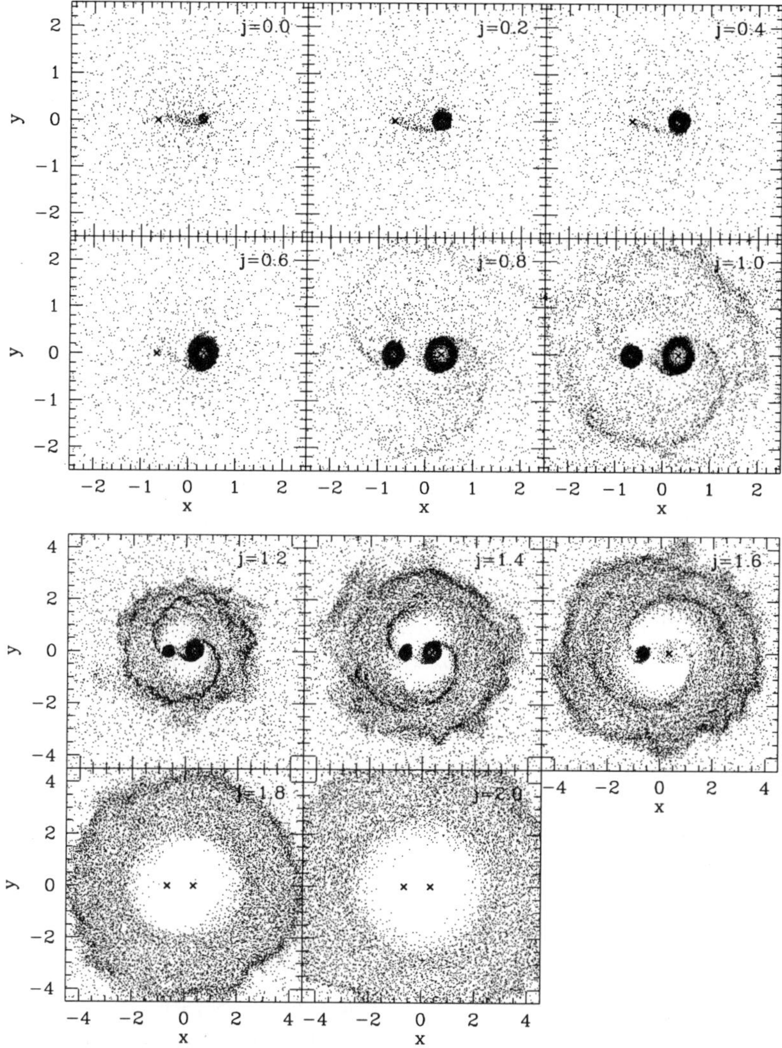

Figure 9. Accretion onto, and disk formation in, a binary system of mass ratio $m_2/m_1 = 0.5$. Each panel is for a different value of the specific angular momentum of the infalling gas in units of the binaries specific angular momentum. (Bate & Bonnell 1997).

al. 1996). Of particular interest in terms of the eventual binary properties is the possibility of the circumbinary disk driving the binary's orbital evolution towards lower separations and higher eccentricities (e.g. Artymowicz & Lubow 1994) and continued accretion through the gaps in the disks (Artymowicz & Lubow 1996).

6. Implications of Binary Star Formation

We now have many pieces to the puzzle of binary star formation. We can use them to draw some implications as to how star the formation process works. First of all, the existence of binary stars has significant implications as to the initial conditions when gravitational collapse is initiated. The fragmentation mechanisms discussed above all require a relatively flat density profile in the region where fragmentation occurs. Thus uniform initial conditions are the easiest (eg Bonnell et al. 1991) to fragment with Gaussian density profiles also being susceptible to fragmentation (Boss 1993; Burkert & Bodenheimer 1996), although only in the flat inner regions. Exceptions to this include having differential rotation in the initial conditions which helps drive a rotationally supported object to fragment (Myhill & Kaula 1992; Boss & Myhill 1995). Another possible except is the disk fragmentation of a $1/r$ density profile cloud (Burkert, Bate & Bodenheimer 1997) although such initial conditions are not in equilibrium such that the central regions expand while the rest of the cloud collapses.

All of the above exclude strongly centrally (e.g. $\rho \propto r^{-\alpha}, \alpha > 1$) condensed profiles with solid-body rotation such as is expected if the pre-collapse core forms through ambipolar diffusion. Furthermore, the requisite two Jeans masses in a near uniform density region mean that the core could not have formed through any quasistatic process. Instead, the core needs to have formed through a dynamical event such as originally discussed by Pringle (1989). Such initial conditions are unlikely to be smooth and will have much density structure that can aid the fragmentation process. Furthermore, the vast diversity in binary properties including non-aligned disks (Stapelfeldt et al. 1998) and non-coplanar multiple systems (Fekel 1981) imply that any angular momentum is independent of the structure of the core (e.g. Bonnell et al. 1992), again implying non-relaxed initial conditions.

There are many observational properties of PMS (and main sequence) binaries which should be used to constrain their formation. These include IR companions (e.g. Koresko, Herbst & Leinert 1997), circumbinary disks (e.g. Dutrey et al. 1994), spin-orbit alignment (Hale 1994), and relative disk evolution and accretion (e.g. Simon & Prato 1995).

7. Stellar Clusters

Star formation is a considerably more complex process than envisioned when considering the formation of single isolated objects. Not only are most stars in binary and multiple systems, it is also becoming apparent that most stars form in clusters containing from ten to thousands of stars (see chapter by E. Lada). Such clusters are best observed in the IR allowing us to peer through the surrounding gas (e.g. Fig. 10).

Figure 10. The Orion Nebula Cluster seen in the IR H-band (McCaughrean et al. 1996).

The question as to how clusters form is still very open. In this section we address some of the key processes that most likely play significant roles in the formation and early evolution of stellar clusters. Observations of young clusters are now beginning to yield the information necessary to constrain any cluster formation model. The knowledge of the gas content in clusters (e.g. chapter by E. Lada), the ability to peer through this gas using IR observations (e.g. McCaughrean & Stauffer 1994) in order to characterise the stellar content (e.g. Hillenbrand 1997) are crucial ingredients in constructing any cluster formation model.

7.1. FORMATION MECHANISM

The formation of stellar clusters is a significant unsolved problem in the field of star formation and for astronomy in general. There are currently no models that can adequately explain how hundreds to thousands of stars can form in a mutually bound environment. The problem involves both our lack of understanding of the initial conditions for cluster formation and the inherent difficulties in the fragmentation process to form the required number of fragments.

In order for a cluster of stars to form, the initial conditions must contain many (as many as stars) Jeans masses, yet are most likely supported against gravity for significant time periods before collapse is initiated. This is a requirement as otherwise the individual stars would form before the cluster is assembled. The initial support is unlikely to be thermal as the thermal energy content in molecular clouds is negligible on such length scales. Instead, molecular clouds are most likely supported by a combination of turbulent and possibly magnetic energies. Their removal is a key ingredient in the initial conditions for fragmentation and thus cluster formation. Whatever process removes the supporting energy needs to be dynamical 1) to avoid a relaxed pre-collapse cloud which will not fragment and 2) so that the stars thus formed are approximately coeval. Candidates for this process involve an external triggering such as a collision or shock wave (e.g. Chapman et al. 1992) or an internal dissipation of the supporting energy on a dynamical timescale. What is clear is that processes such as ambipolar diffusion cannot be the dominant removal mechanism.

Another key ingredient to the cluster formation puzzle is that the pre-collapse configuration must contain significant structure to allow for the fragmentation. This structure could either be on small scales such as kernels in a MHD turbulent medium (Myers 1998; see chapter by P. Myers) or on large scales (e.g. Klessen, Burkert & Bate 1998). The difficulty with the small-scale structure is that a triggering mechanism is required to initiate collapse in all the kernels near-synchronously.

The collapse and fragmentation of a cloud with large-scale structure has recently been investigated by Klessen et al. (1998). Simulating a piece of a molecular cloud, containing 200 Jeans masses and significant structure on larger scales, Klessen et al. (1998) showed that the collapse formed filamentary structures which fragmented into a significant number of stars. These stars then fall together to form a cluster (see Fig. 11).

In order to develop a model for the cluster fragmentation, we need to use the results of such simulations and determine some observational consequences that can be used to descriminate between the different scenarios. For example, the gas fraction in clusters can tell us about the relative strengths of large and small-scale structures, or stellar ages relative to the cluster's dynamical timescale can tell us about how coeval the population is, and thus from how far away the individual stars could have formed.

7.2. CLUSTER DYNAMICS

As soon as the stars form in a cluster, they start to interact gravitationally with each other. These interactions can remove or relax the initial conditions so have to be considered if we want to use the present cluster conditions to constrain the formation process. There are two types of in-

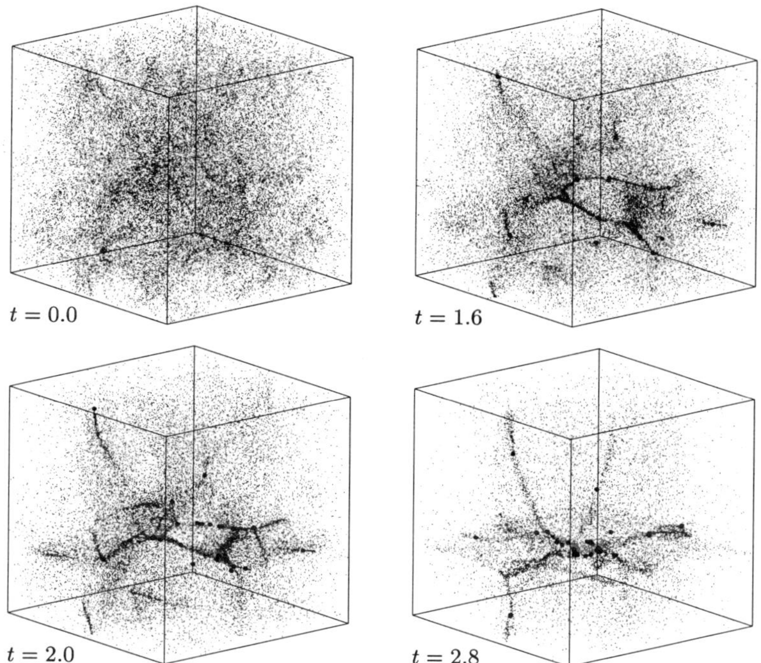

Figure 11. The collapse and fragmentation of a piece of a molecular cloud, leading to the formation of a stellar cluster (Kelssen et al. 1998).

teractions that we consider here. The first, *violent relaxation*, involves the global change of the cluster's potential due to a cold collapse. It occurs on the dynamical timescale of the cluster, the crossing time:

$$t_{\text{cross}} = \frac{2 \times R_{1/2}}{v_{\text{disp}}}, \qquad (5)$$

where $R_{1/2}$ is the cluster half-mass radius and v_{disp} is the cluster's one-dimensional velocity dispersion, which, in a virialised cluster is

$$v_{\text{disp}} = \sqrt{\frac{GM_{\text{clust}}}{R_{1/2}}} \qquad (6)$$

This is the shortest timescale for global cluster evolution. The second process, *two-body relaxation*, involves the gravitational interaction between individual stars. Two-body relaxation transfers kinetic energy to the lower-mass stars, driving the system towards equipartition and mass segregation. This process requires many interactions and thus occurs on longer

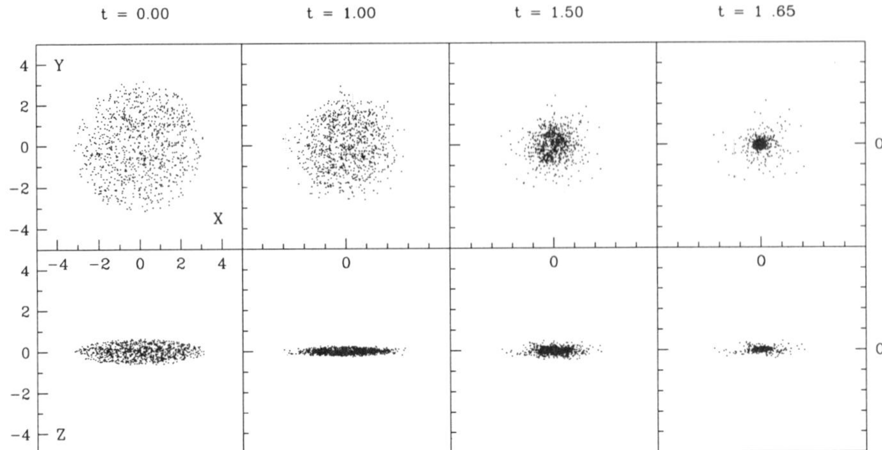

Figure 12. The cold collapse and violent relaxation of an initially flattened cluster (Boily et al. 1998). The post violent-relaxation cluster remains flattened, although less so than the original cluster.

timescales, called the relaxation time (Binney & Tremaine 1987):

$$t_{\text{relax}} \approx \frac{N}{8\ln N} t_{\text{cross}},\tag{7}$$

where N is the niumber of stars in the cluster. The relaxation time, t_{relax}, is a measure of the time it takes for the kinetic energy, E_{kin}, of a star to change by an amount similar to E_{kin}, and thus measures the time it takes for a stellar cluster to lose memory of its initial dynamical configuration.

It should be noted that sub-regions of the cluster, such as the central core, can evolve faster and thus independently of the rest of the cluster. The timescale for this is the crossing time of the sub-region or core, e.g., $t_{\text{core}} = 2 \times R_{\text{core}}/v_{\text{disp}}$. The same follows for the relaxation timescale.

7.2.1. *Violent Relaxation*

Violent relaxation involves the global collapse of a stellar system when it forms with sub-virial kinetic energy (Lynden-Bell 1967). This is likely to occur in the initial stages of a cluster as the the pre-cluster entity must have lost its support (thermal, kinetic or magnetic) in order to fragment. Unless it only loses this support on small, stellar scales, then it follows that the cluster should initially be "cold" and collapse.

The changing gravitational potential during the collapse introduces a large scatter into the distribution of stellar energies. Basically, the potential of the cluster during collapse and reexpansion are not the same so that individual stars "see" different potentials and thus react differently. Stars that fall in to the centre early see a low density core when they fall in but a high density core when they emerge on the other side, thus decreasing their energy. In contrast, stars that fall in later see a high density core and get accelerated towards the centre. By the time they pass through the centre the core has reexpanded and they thus suffer less deceleration and emerge with a higher energy.

Violent relaxation is mass-independent, such that the resultant stellar orbits depend solely on the stars' position in the cluster during the event. As it is mass independent, it results in a near uniform and mass-independent velocity dispersion, and a centrally condensed configuration (see Binney & Tremaine 1987).

Although violent relaxation tends to remove the initial conditions of the cluster, it does not completely erase large-scale non-spherical morphologies. This is important as most triggering mechanisms (shocks, cloud-cloud collisions) result in flattened systems and thus cluster morphology can be used as a constraint on the cluster formation mechanism. Boily, Clarke & Murray (1998; see also Goodwin 1997) have explored the relationship between the initial cluster morphology and the post violent relaxation configuration. They find that although the flattening decreases during the collapse and violent relaxation, it remains significant in the cluster morphology (Fig. 12). Thus, the apparent 2:1 elongation in the Orion Nebula Cluster (Hillenbrand & Hartmann 1998) would imply a pre-violent relaxation morphology that was significantly more aspherical with a 5:1 axis ratio (Boily et al. 1998). These calculations do not include any gas which could be a major contributor to the cluster potential. Still, the gas should collapse as do the stars, and even though it will not reexpand, the gas should not negate the violent relaxation process or the result that the stellar kinematics reflect to some degree the "cold" initial conditions.

7.2.2. *Mass Segregation*

Stellar clusters are commonly found to have their most massive stars in or near the centre (e.g. Raboud & Mermilliod 1998). This mass segregation is also found in the youngest clusters (e.g. Carpenter et al. 1997; Hillenbrand & Hartmann 1998). In the ONC, mass segregation appears to extend down to stars of a few solar masses (Hillenbrand & Hartmann 1998). This preference to have the massive stars near the centre of clusters is either due to the initial conditions of stellar clusters, in which case it is telling us something about the formation process and how massive stars form, or it could be due to dynamical mass segregation in the cluster.

We know that two-body relaxation drives a stellar system towards equipartition of kinetic energy and thus towards mass segregation. In gravitational interactions, the massive stars tend to lose some of their kinetic energies to lower-mass stars and thus sink to the centre of the cluster (see Fig. 13).

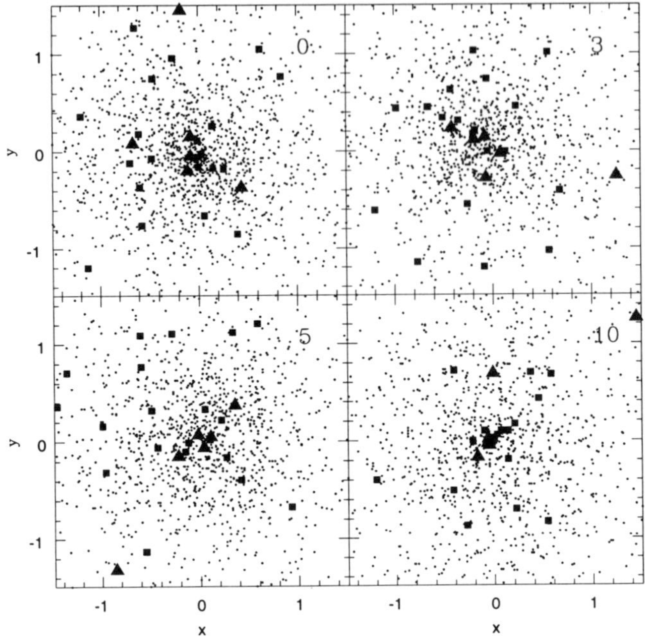

Figure 13. The dynamical mass segregation of the massive stars in a cluster containing 1500 stars (Bonnell & Davies 1998). The six most massive stars are indicated by triangles while the next 24 most massive stars are indicated by squares. The time is given in units of the crossing time.

The question is whether this process can occur quickly enough to explain the observations of mass segregation in young clusters. In order to test this possibility, we need to characterise the contents (masses, positions, ages) of young clusters which has only been accomplished for a few systems (e.g the ONC; Hillenbrand 1997; Hillenbrand & Hartmann 1998).

Two-body relaxation occurs on the relaxation timescale which for the ONC is about 25 t_{cross} whereas it appears to be only 3 t_{cross} old. Although this implies that the ONC cannot be fully relaxed, the relaxation timescale depends inversely on mass such that the most massive stars sink more quickly to the centre of the cluster. For this reason, Bonnell & Davies (1998) performed numerical N-body simulations of the cluster considering possible initial locations of the massive stars. Using initial conditions with uniform specific kinetic energies as is expected after cluster formation and violent relaxation, Bonnell & Davies showed that initially unsegregated clusters are not able to relax quickly enough to reproduce the observations. The only possibility for uniformly spatially distributed stellar masses is if the massive stars are initially in equipartition such that they are already en-ergy segregated and only have to sink to the centre of the cluster (on the

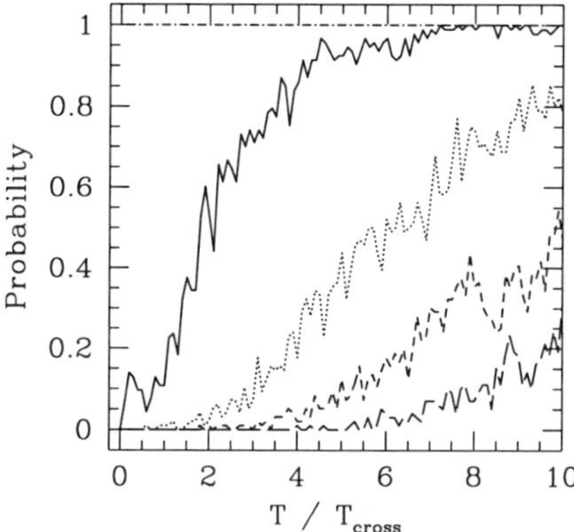

Figure 14.   The probability of observing a trapezium-like system at the centre of the cluster as a function of the time in units of the crossing time. The ONC is probably $\lesssim 3 t_{\mathrm{cross}}$ old. The five lines represent the cases when the 30 most massive stars are placed (from top to bottom): at Lagrangian radii containing 2 per cent (the 30 massive stars), 10 per cent, 20 per cent, 30 per cent, and 50 per cent of the stars in the cluster (from Bonnell & Davies 1998).

timescale of $\approx t_{\mathrm{cross}}$) in order to be spatially segregated. As stated above, such initial conditions are not expected for stellar clusters.

The best indication that the massive stars formed near the centre of the cluster comes from the probability to form a Trapezium-like system in the centre of the ONC. This dense knot consisting of a sizable fraction of the most massive stars requires that the massive stars are initially within the central 10 % (for a 50 % probability of having a Trapezium) to 20 % (≈ 10 % probability) of all stars (see Fig. 14).

7.3. ACCRETION AND STELLAR MASSES

Young stellar clusters are commonly found to be gas-rich with typically 50 to 90 % of their total mass in the form of gas (e.g. Lada 1991; see also chapter by E. Lada). This gas can interact with, and be accreted by, the stars as they both move in the cluster. The timescale for accretion can be estimated as the gas free-fall time which, for systems where the gas is a significant contributor to the cluster potential, is of the same order as the cluster's crossing time:

$$t_{\mathrm{ff}} \propto \rho_{\mathrm{gas}}^{-1/2},$$

$$t_{\mathrm{cross}} \propto \frac{R_{\mathrm{clust}}}{v_{\mathrm{disp}}} \propto \sqrt{\frac{R_{\mathrm{clust}}^3}{(M_{\mathrm{stars}} + M_{\mathrm{gas}})}} \tag{8}$$

$$\propto (\rho_{\mathrm{gas}} + \rho_{\mathrm{stars}})^{-1/2}.$$

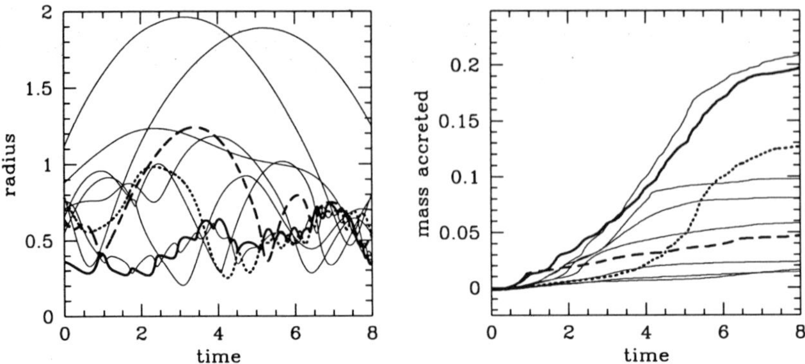

Figure 15.   The evolution of a cluster of 10 stars undergoing gas accretion is shown. The distance of each star from the centre of the cluster (left) and the gas mass fraction accreted by each star (right) are given as functions of time, in units where $t_{cross}=2.8$. The gas initially comprises 10 % of the total mass of the system. Three individual stars are highlighted (see text).

Thus, we expect that significant accretion can occur on the dynamical timescale of the cluster. If significant accretion occurs, it can affect both the dynamics and the masses of the individual stars (e.g. Larson 1992).

Simulations of accretion in clusters using a combination SPH and N-body code have found that accretion is a highly non-uniform process where a few stars accrete significantly more than the rest (Bonnell et al. 1997). Individual stars' accretion rates depend largely on their position in the cluster (see Fig. 15) with those in the centre accreting more gas than those near the outside. This process is termed "competitive accretion" (Zinnecker 1982) as the accretion depends on being in a favourable location and in having significant gravitational attraction to overcome that of the nearby stars.

The stars closest to the centre of the cluster (heavy-solid line in Fig. 15) accrete more than do the other stars because of the increased gas density near the centre. The density is greater because the gas preferentially falls into the deepest part of the cluster's gravitational potential. Furthermore, gas is efficiently funneled down to the cluster centre, thereby replenishing the gas as it is accreted. Stars that spend most of their time in the outer regions of the cluster do not have this benefit so they do not accrete nearly as much gas. Stars eject from the centre of the cluster also suffer a marked decrease in their accretion rates.

This competitive accretion process naturally results in a spectrum of stellar masses even if they are initially all equal (Bonnell et al. 1997). A large range of stellar masses is produced (factor 10) even though the median stellar mass increases by less than a factor 2. This occurs as the competitive

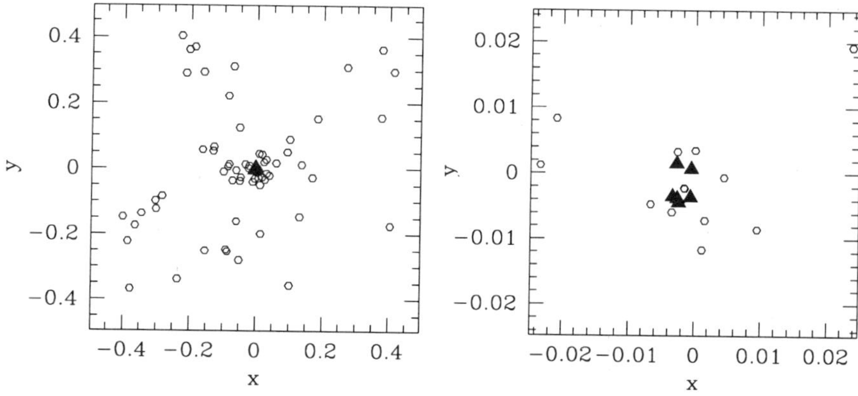

Figure 16. The position of the 6 most massive stars (with $M_\star \gtrsim 4M_{\mathrm{med}}$, filled triangles) in a cluster of 100 stars. The right-hand panel is a blow-up of the cluster core. The masses of the most massive stars are due to the competitive accretion process.

accretion is a runaway process where the few stars near the centre accrete more and are thus able to attract even more gas. Thus, if the initial gas mass-fraction in clusters is generally equal, then larger clusters will produce higher-mass stars and a larger range of stellar masses as the competitive accretion process will have more gas to feed the few stars that accrete the most gas. Finally, accretion in clusters also naturally results in a mass segregated cluster (see Fig. 16) where the most massive stars are formed near the centre of the cluster due to the accretion.

7.4. FORMATION OF MASSIVE STARS

The formation of massive stars is one of the most important unsolved problems in star formation. We know that they cannot form in an analogous manner to low-mass star-formation as their radiation pressure will reverse the infall process once they attain ≈ 10 M$_\odot$ (Yorke & Krugel 1977; Yorke 1993). Disk accretion is a possibility but the disks would have to 1) be thin enough to survive the radiation field and 2) extend far enough to be replenished without the infall being affected by the radiation pressure (Jijina & Adams 1996). The fact that young massive stars are generally found in the centre clusters suggests that their formation is linked to their special environment. They are unlikely to have formed elsewhere and evolved to the centre by the time we observe them (Bonnell & Davies 1998; and see above). Furthermore, a simple Jeans mass argument would imply that due to the crowded conditions in the centre of clusters, the gas density would have had to have been very high which implies a low Jeans (or fragment)

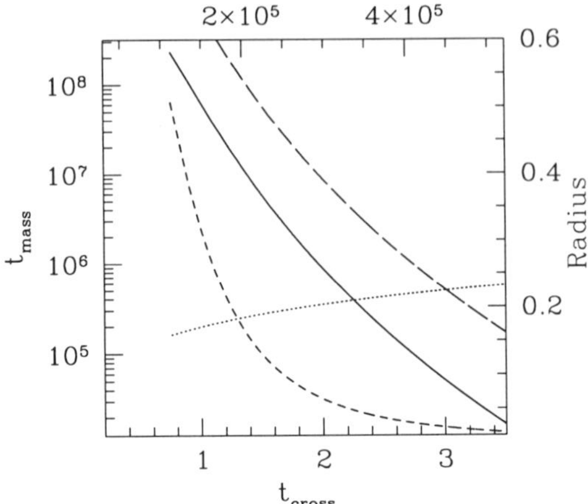

Figure 17. The timescale to double a star's mass either by collisions (solid and long-dashed lines) or through accretion (dotted line) is plotted versus time in units of t_{cross} (and in years) for a 0.1 pc cluster of 100 stars (Bonnell et al. 1998). The accretion rate is 2×10^{-6} $M_\odot\mathrm{yr}^{-1}$ and the merger radius is either 0.1 (solid line) or 0.01 au (long-dashed line). The half-mass radius is also plotted (short-dashed line).

mass at the time of formation (c.f. Zinnecker et al. 1993; Bonnell, Bate & Zinnecker 1998). Subsequent accretion (see above) can increase their mass to near $10 M_\odot$ but beyond that something further is required.

A secondary affect of accretion in clusters is that it can force it to contract significantly. If the gas has little net rotation, then its motions will generally be disconnected from the motions of individual stars, and have zero net momentum. The added mass increases the binding energy of the cluster while accretion of basically zero momentum matter will remove kinetic energy. If the core is sufficiently small that its crossing time is relatively short compared to the accretion timescale, then it will contract with added mass as (Bonnell, Bate & Zinnecker 1998)

$$R_{\mathrm{core}} \propto M_{\mathrm{core}}^{-3}. \tag{9}$$

This increases the stellar density dramatically, to the point where collisions become important (see Fig. 17). Collisions between intermediate mass stars ($2 M_\odot \lesssim m \lesssim 10 M_\odot$), whose mass has been accumulated through accretion in the cluster core, can then result in the formation of massive ($m \gtrsim 50 M_\odot$) stars (Bonnell et al. 1998). In this model, the massive stars that are found in the centre of rich young clusters are significantly younger than the mean stellar age. If this is how the massive stars in the Trapezium formed, then we would predict that $\theta^1 C$ is significantly younger than 10^6 years, possible as young as $\lesssim 10^5$ years.

7.5. CLUSTER DISSOLUTION

Although most stars appear to form in clusters, the vast majority of stars in the Galaxy are not in clusters. This implies that at some point the young clusters must dissolve to populate the galactic field population. Clusters can dissolve in one of two ways, they can be unbound either through stellar interactions with a central binary or if a significant percentage of their mass (residual gas) is removed.

In small clusters ($N \lesssim 100$), a central binary forms that can interact and eject other cluster members. Essentially, the binary absorbs the total potential energy of the cluster, unbinding all the other members. Estimates of the timescale for this process tell us that it is a strong function of the number of cluster members (Heggie 1974):

$$t_{\mathrm{diss}} \approx \frac{N^2}{100} t_{\mathrm{cross}}. \tag{10}$$

Thus clusters with $N \approx 100$ will dissolve on time scales of order 100 t_{cross}. This corresponds to lifetimes of several $\times 10^7$ years. Smaller clusters will dissolve on shorter timescales such that small clusters of order 10 stars will only last a few crossing times at most. These dissolved clusters then correspond to the Gomez groups in taurus (Gomez et al. 1993). The probability of seeing a cluster increases with the number of stars that it contains, and thus (for random picking from a universal IMF) with the expectation value of the mass of the most massive star. If we assume an age of a few million years as a median age of the young stellar population in the nearby star formation regions, then for an average crossing time of $t_{\mathrm{cross}} \approx 3 \times 10^5$ years, we should only see clusters with at least 25 stars. This also implies that we should only see clustering around stars of at least $m \gtrsim 3 M_\odot$. Hillenbrand (1995) and Testi et al. (1997) both found that clustering around Herbig AeBe stars is evident only for stars with masses greater than 3 to 5 $M_\odot$.

Larger clusters will not dissolve through interactions with a central binary as the timescale becomes too long for stars with greater than a few hundred members. Instead, these clusters are large enough that they are likely to contain a massive star ($m \gtrsim 8 M_\odot$) that can eject any residual gas from the cluster. Young clusters contain significant amounts of gas, typically comprising a majority of the total cluster mass (Lada 1991). This gas is therefore a major contributor in making the cluster gravitationally bound and its removal can unbind the cluster, dispersing the stars into the galaxy.

Gas removal in clusters occurs due to the energetics of the component massive stars (Whitworth 1979; Tenorio-Tagle et al. 1986; Franco, Shore & Tenorio-Tagle 1994). The photo-ionization and winds from these stars is capable of removing any residual gas from the cluster. The fate of a particular cluster depends on the gas fraction, the removal timescale and stellar velocity dispersion when the gas is dispersed (Lada, Margulis, & Dearborne 1984; Pinto 1987; Verschueren & David 1989; Goodwin 1997b). If the gas comprises a significant fraction of the total mass ($\gtrsim 50\%$) and is removed quickly compared to the cluster crossing time, t_{cross}, then the dramatic reduction in the binding energy, without affecting the stellar kinetic energy, results in an unbound cluster. Alternatively, if the gas is removed over several crossing times, then the cluster can adapt to the new potential and can survive with a significant fraction of its initial stars. For example, clusters with gas fractions as high as 80 per cent can survive with approximately half of the stars if the gas removal occurs over 4 or more crossing times (Lada et al. 1984).

8. Binary Stars in Clusters

In this last section, we address the overlap between binary systems and stellar clusters, namely the question of binaries in clusters. The finding that the PMS binary frequency in nearby clouds is higher than that found on the main sequence (Ghez 1995, and see above) is, at first, disturbing as it implies a decrease in binarity with age, Recently, binary surveys in clusters such as the ONC have found that the binary frequency in these dense regions is generally more typical of the field stars (Prosser et al. 1994; Petr et al. 1998; Padgett et al. 1997; Patience et al. 1998). We can thus explain the difference between the binary frequency in young and old systems by the hypothesis that most stars form in clusters such as Orion where the binary frequency is lower than in nearby dark clouds. The question remains why there is a difference between these two regions.

One possibility is that binary systems are intrinsically harder to form in rich dense clusters. For example, the fragmentation of a collapsing cloud could be inhibited by its environment. This hypothesis has not as yet been fully explored but the only relevant collapse calculations done to date indicate that fragmentation is easier in the presence of tidal fields (Boss 1981; Felice & Sigalotti 1991)).

Another possibility is that clusters have an initially high binary frequency but that a significant fraction of these binaries are destroyed through binary-binary interactions. Interactions involving binaries tend to disrupt wide (soft) ones and harden closer (hard) ones. The border between the two possibilities, the hard/soft limit is given by binary systems with orbital velocities similar to the velocity dispersion in the cluster. Thus, denser clusters will be able to disrupt closer binaries than can less dense clusters. Kroupa (1995) has investigated this scenario using N-body simulations and

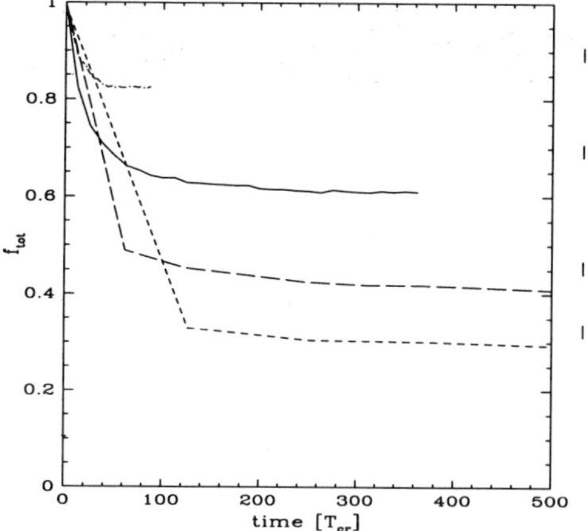

Figure 18. The binary fraction in clusters containing initially 100 % binaries is plotted as a function of time in units of the initial crossing time. The clusters considered have $R_{1/2} = 0.08$ pc (short-dashed line), $R_{1/2} = 0.25$ pc (long-dashed line), $R_{1/2} = 0.8$ pc (solid line), and $R_{1/2} = 2.5$ pc (dot-dashed line). Adapted from Kroupa 1995.

assuming a 100 % binary frequency. He found that the resultant binary frequency depends primarily on the cluster density (see Fig. 18), with preferentially the wider systems being disrupted. Figure 18 also shows that it takes several tens of crossing times before the soft binaries are all disrupted. Presumably, the wider systems are disrupted more quickly as their collisional cross section is larger, while closer systems have longer collisional timescales.

The binary fraction detected in the core of the ONC (Petr et al. 1998) is consistent with the above picture of binary disruptions as the core is $\approx 30 t_{\mathrm{cross}}$ old. It does imply that the binary fraction in the outer parts of the cluster should reflect the initial population as the cluster as a whole is only $\approx 3 t_{\mathrm{cross}}$ old. This may be a problem as there appears to be a lack of wide binaries even in the outer parts of the cluster (Scally, Clarke & McCaughrean 1998) where the collisional timescale is significantly longer than the local t_{cross}.

9. Summary

The formation of multiple systems and stellar clusters is a complex dynamical process where individual stars interact on the same timescale as they form. There are two main phases that have to be considered, 1) the gravitational collapse and fragmentation where the number of objects is decided and the subsequent 2) accretion of residual gas and interactions between individual stars where the eventual stellar and system parameters are fixed.

Fragmentation requires initial conditions for collapse that are far from equilibrium in that they must contain several Jeans masses, with generally elongated (prolate or oblate) structure, and not be very centrally condensed. Such initial conditions cannot have been formed from a quasistatic evolution but require a dynamic formation mechanism (e.g. Pringle 1989).

Binary and multiple systems can form through the fragmentation of a collapsing cloud or through the fragmentation of a circumstellar disk at the end of the collapse phase. Subsequent star-disk interactions can also form bound systems in small clusters. Binary systems formed through fragmentation can have a wide range of mass ratios and eccentricities, with separations from 10^4 AU down to a few $R_\odot$. Closer systems need to accrete the majority of their eventual mass and this accretion will be the determining factor of their system properties. Observations of the diverse properties ofyoung binary systems need to be used to constrain their formation and early evolution.

The formation of stellar clusters is a more difficult problem in that the cloud has to fragment into hundreds to thousands of self-gravitating objects. The first attempts at modelling this fragmentation are being made, but significant observational constraints are necessary in order to progress. Subsequent dynamics, including violent relaxation and mass segregation need to be considered in order to extrapolate backwards from the present-state of young stellar clusters to their initial post-fragmentation configuration. Using N-body simulations, it appears likely that the initial state of clusters is somewhat flattened and that mass segregation is initial and not due to two-body relaxation. These properties can then be used to constrain the cluster formation process.

Accretion in clusters can generate both a mass spectrum and mass segregation in a young cluster. It is a competitive process where a few stars, located near the cluster centre, accrete the majority of the gas. These stars benefit from the overall cluster potential that funnels gas down to the core. Accretion in clusters also tends to shrink the cluster. If sufficient accretion occurs in a dense cluster, the core may shrink sufficiently to allow stellar collisions to occur. Massive stars may then form by this process, alleviating the problem of accreting onto stars with $M_\star \gtrsim 10 M_\odot$.

Lastly, the binary frequency in different regions may reflect the stellar density when they were formed. Assuming a 100 % initial binary frequency, binary interactions in clusters can destroy binaries down to the hard/soft limit of the cluster such that denser clusters should have fewer binaries.

References

1. Adams F. C., Ruden S. P., Shu F. H., 1989, ApJ, 347, 959
2. Artymowicz P., Clarke C. J., Lubow S. H., Pringle J. E., 1991, ApJ, 370, L35

3. Artymowicz P., Lubow S. H., 1994, ApJ, 421, 651
4. Artymowicz P., Lubow S. H., 1996, ApJL, 467, L77
5. Bastien P., 1983, A&A, 119, 109
6. Bastien P., Arcoragi J.-P., Benz W., Bonnell I., Martel H., 1991, ApJ, 378, 255
7. Bate M.R., 1997, MNRAS, 285, 16
8. Bate M.R., 1998a, ApJ, 508, Nov 20 issue of ApJ Letters
9. Bate M.R., 1998b, in 'Brown Dwarfs and Extrasolar Planets', ASP Conference Series Vol. 134, eds. R. Rebolo, E.L. Martin, M.R. Zapatero Osorio (Brigham Young University; Provo, UT) p. 273-276.
10. Bate M.R., Bonnell I.A., Price N.M, 1995, MNRAS, 277, 362
11. Bate M.R., Bonnell I.A., 1997, MNRAS, 285, 33
12. Bate M.R., Burkert A., 1997, MNRAS, 288, 1060
13. Bate M.R., Clarke C.J., McCaughrean M.J., 1997, MNRAS, 297, 1163
14. Binney J., Tremaine S., 1987, Galactic Dynamics, Princeton Univ. Press, Princeton
15. Bodenheimer, P., Tohline, J.E., Black, D.C. 1980, ApJ 242, 209
16. Boily, C.M., Clarke, C.J. & Murray, S.D., 1998. MNRAS submitted.
17. Bonnell I.A., 1994, MNRAS, 269, 837
18. Bonnell I.A., Arcoragi J.-P., Martel H., Bastien P., 1992, ApJ, 400, 579
19. Bonnell I.A., Bastien P., 1992a, ApJ, 401, 654
20. Bonnell I.A., Bastien P., 1992b, ApJL, 401, L37
21. Bonnell I.A., Bate M.R., 1994a, MNRAS, 269, L45
22. Bonnell I.A., Bate M.R., 1994b, MNRAS, 271, 999
23. Bonnell I.A., Bate M.R., 1999, in preparation
24. Bonnell I.A., Bate M.R., Clarke C.J., Pringle J.E., 1997, MNRAS, 285, 201
25. Bonnell I.A., Bate M.R., Zinnecker H., 1998, MNRAS, 298, 93
26. Bonnell I.A., Davies M.B., 1998, MNRAS, 295, 691
27. Bonnell I.A., Martel H., Bastien P., Arcoragi J.-P., Benz W., 1991, ApJ, 377, 553
28. Boss A.P., 1981, ApJ, 246, 866
29. Boss A.P., 1986, ApJS, 62, 519
30. Boss A.P., 1989, ApJ, 346, 336
31. Boss A.P., 1993, ApJ, 410, 157
32. Boss A.P., 1996, ApJ, 468, 231
33. Boss A.P., 1998, ApJ, 501, L77
34. Boss A.P., Bodenheimer P., 1979, ApJ, 234, 289
35. Boss A.P., Myhill E. A., 1995, ApJ, 451, 218
36. Burkert A., Bate M. R., Bodenheimer P., 1997, MNRAS, 289, 497
37. Burkert A., Bodenheimer P., 1993, MNRAS, 264, 798
38. Burkert A., Bodenheimer P., 1996, MNRAS,
39. Carpenter, J.M., Meyer M.R., Dougados C., Strom S.E., Hillenbrand L.A., 199 7, AJ, 114, 198
40. Chapman, S., et al 1992, Nature, 359, 207
41. Clarke C.J., Pringle J.E., 1991, MNRAS, 249, 584
42. Clarke C.J., Pringle J.E., 1993, MNRAS, 261, 190
43. Duquennoy A., Mayor M., 1991, A&A, 248, 485
44. Durisen R.H., Gingold R.A., Tohline J.E., Boss A.P. 1986, ApJ, 305, 281
45. Dutrey, A., Guilloteau, S., Simon, M., 1994, A&A, 286, 149
46. Fabian A.C., Pringle J.E., Rees M.J., 1975, MNRAS 172, 15P.
47. Fekel, F. 1981, ApJ 246, 879
48. Felice F. de, Sigalotti , L. Di,1991, MNRAS, 249, 248
49. Fischer D. A., Marcy G. W., 1992, ApJ, 396, 178
50. Franco J., Shore S., Tenorio-Tagle G., 1994, ApJ, 436, 795
51. Ghez A.M., 1995, in Evolutionary Processes in Binary Stars, eds Wijers R., Davies M., Tout C., Kluwer Academic, p. 1.
52. Ghez A.M., Emerson J.P., Graham J.R., Meixner M., Skinner C., 1994, ApJ, 434, 707

53. Ghez A.M., McCarthy D.W., Patience J.L., Beck T.L., 1997,ApJ, 481, 378
54. Ghez A.M., Neugebauer G., Matthews K., 1993, AJ, 106, 2005
55. Gomez, M., Hartmann, L., Kenyon, S.J., Hewett, R., 1993, AJ 105, 1927.
56. Goodman, A., Benson, P., Fuller, G., & Myers, P. 1993, ApJ, 406 528
57. Goodwin S.P., 1997a, MNRAS, 286, 669
58. Goodwin S.P., 1997b, MNRAS, 284, 785
59. Hale A., 1994, AJ, 107, 306
60. Hall S.M., Clarke C.J., Pringle J.E., 1996, MNRAS, 278, 303
61. Heemskerk, H.M., Papaloizou, J.C., & Savonije, G.J. 1992, A&A, 260, 161
62. Heller C., 1993, ApJ, 408, 337
63. Heggie D.C., 1974, in *IAU Symp 62, The Stability of the Solar Sy stem and Small Stellar Systems*, ed Y. Kozai, Dordrecht, p. 225
64. Hillenbrand, L.A., 1995, PhD Thesis, University of Massachusetts.
65. Hillenbrand L.A., 1997, AJ, 113, 1733
66. Hillenbrand L.A., Hartmann L.W., 1998, ApJ, 492, 540
67. Hills J.G., Day C.A., 1976, Ap. Lett, 17, 87
68. Hoyle F., 1953, ApJ, 118, 513
69. Jijina J., Adams F.C., 1996, ApJ, 462, 874
70. Klapp J., Sigalotti L. Di., 1998, ApJ 504, 158
71. Klessen R., Burkert A., Bate M.R., 1998, ApJL, . 501, L205
72. Koresko C.D., Herbst T.M., Leinert Ch. 1997, ApJ, 480, 741
73. Kroupa P., 1995, MNRAS, 277, 1491
74. Lada C.J., 1991, in *The Physics of Star Formation and Early Stellar Evolution*, eds C. J. Lada, N. D. Kyfalis, Kluwer, p. 329
75. Lada C.J., Margulis M., Dearborn D., 1984, ApJ, 285, 141
76. Lada E.A., Depoy D.L., Evans N.J. Gatley I. 1991, ApJ, 371 171
77. Larson R.B. 1969, MNRAS, 145, 271
78. Larson R.B., 1990, in Capuzzo-Dolcetta R., Chiosi C., DiFazio A., ed., Physical Processes in Fragmentation and Star Formation. Kluwer, Dordrecht, p. 389
79. Larson, R.B., 1992, MNRAS, 256, 641
80. Larwood J.D., Nelson R.P., Terquem C., Papaloizou J.C.B., MNRAS, 282, 597
81. Laughlin G., Bodenheimer P., 1994, ApJ, 436, 335
82. Leinert Ch., Weitzel N., Zinnecker H., Christou J., Ridgeway S., Jameson R., Haas M., Lenzen R., 1993, A&A, 278, 129
83. Lin C.C., Mestel L., Shu F.F., 1965, ApJ, 142, 1431
84. Lynden-Bell D., 1967, MNRAS, 136, 101
85. Mardling R., 1995, in *Evolutionary Processes in Binary Stars*, eds Wijers R., Davies M., Tout C., Kluwer Academic, p. 81
86. Mathieu R.D., 1994, ARA&A, 32, 465
87. Mathieu R.D., Ghez A.M., Jensen E., Simon M., 1999, in *Protostars and Planets IV*, eds (V. Mannings A. Boss and S. Russell), in press
88. Mathieu R.D., Walter F.M., Myers P.C., 1989, AJ, 98, 987
89. Mayor M., Duquennoy A., Halbwachs J.-L., Mermilliod J.-C., 1992, in McAlister H.A., Hartkopf W.I., ed., Complementary Approaches to Double and Multiple Star Research (IAU Colloquium 135). ASP, San Francisco, p. 73
90. Mazeh T., Goldberg D., Duquennoy A., Mayor M., 1992, ApJ, 401, 265
91. McCaughrean M.J. , Rayner J.T., Zinnecker H., &. Stauffer J.R, 1996, in *Disks and Outflows around Young Stars*, S.V.W. Beckwith, J. Staude, A. Quetz, & A. Natta, eds., Lecture Notes in Physics 465, (Heidelberg: Springer), p33.
92. McCaughrean M.J., Stauffer J.R., 1994, AJ, 108, 1382
93. McDonald J.M., Clarke C.J., 1993, MNRAS, 262, 800
94. McDonald J.M., Clarke C.J., 1995, MNRAS, 275, 671
95. Meyer, M.R. & Lada, E.A. 1999 in The Orion Complex Revisited, eds M. McCaughrean, A. Burkert, in press
96. Miyama S., Hayashi C., Narita S., 1984, ApJ, 279, 621

97. Monaghan, J.J. 1994, ApJ, 420, 692

98. Monaghan, J.J., Lattanzio, J.C. 1986, A&A, 158, 207

99. Monaghan J.J., Lattanzio J.C., 1991, ApJ, 375, 177

100. Myers P.C., 1998, ApJL, 496, 109

101. Myers P.C., Fuller G.A., Goodman A.A., Benson P.J., 1991, ApJ, 376, 561

102. Myhill E.A., Kaula W.M., 1992, ApJ, 386, 578

103. Nelson R.P., Papaloizou, J.C. 1993, MNRAS, 265, 905

104. Nelson R.P., 1998, MNRAS, 298, 657

105. Padgett D.L., Strom S.E., Ghez A., 1997, ApJ, 477, 705

106. Patience J., Ghez A.M., Reid I.N., Weinberger A.J., Matthews K., 1998, AJ 115, 1972

107. Petr M.G., Coudé du Foresto V., Beckwith S.V.W., Richichi A., McCaughrean M.J., 1998, ApJ, 500, 825

108. Pringle, J.E. 1989, MNRAS, 239, 361

109. Pringle, J.E. 1991, MNRAS, 248, 759

110. Prosser C.F., Stauffer J.R., Hartmann L.W., Soderblom D.R., Jones B.F., Werner M.W., McCaughrean M.J., 1994, ApJ, 421, 517

111. Raboud D., Mermilliod J.C., 1998, A&A, 333, 897

112. Reipurth B., Zinnecker H., 1993, A&A, 278, 81

113. Richichi A., Leinert Ch., Jameson R., Zinnecker H., 1994, A&A, 287, 145

114. Scally A., Clarke C.J., McCaughrean M.J., 1998

115. Shu F. H., Adams F. C., Lizano S., 1987, ARA&A, 25, 23

116. Shu F.H., Tremaine S., Adams F.C., Ruden S.P., 1990, ApJ, 358. 495

117. Sigalotti L. Di., Klapp J., 1997, ApJ, 474, 710

118. Simon M., Chen W. P., Howell R. R., Slovik D., 1992, ApJ, 384, 212

119. Simon M., Prato L., 1995, ApJ, 450, 824

120. Stapelfeldt K.R., Krist J.E., Menard F., Bouvier J., Padgett D., Burrows C.J., 1998, ApJL, 502, 65

121. Tassoul J.L., 1978, Theory of Rotating Stars. Princeton University Press, Princeton.

122. Tenorio-Tagle G., Bodenheimer P., Lin D., Noriega-Crespo A., 1986, MNRAS, 221, 635

123. Terquem C., Bertout C., 1993, A&A, 274, 291

124. Testi, L., Palla, F., Prusti, T., Natta, A., & Maltagliati, S., 1997. AA 320,159

125. Tohline J. E., 1982, Fund. of Cos. Phys., 8, 1

126. Turner J., Chapman S.J., Bahtal A.S, Disney M.J., Pongracic H., Whitworth A., 1995, MNRAS, 277, 705

127. Verschueren W., David M., 1989, A&A, 219, 105

128. Watkins S., Boffin H.M., Francis N., Whitworth A.P., 1998, MNRAS, in press

129. Whitworth A.P., 1979, MNRAS, 186, 59

130. Whitworth A.P., Bahtal A.S, Chapman S.J., Disney M.J., Turner J.A., 1994, A&A, 290, 421

131. Whitworth A., Chapman S.J., Bahtal A.S, Disney M.J., Pongracic H., Turner J., 1995, MNRAS, 277, 727

132. Yorke H.W. 1993, in *Massive Stars: Their Lives in the interstellar Medium*, eds. J. Cassinelli, E. Churchwell, ASP Conf. Ser. 35, p. 45

133. Yorke, H.W., Krugel E., 1977, A&A, 54, 183

134. Zinnecker H., 1982, in *Symposium on the Orion Nebula to Honour Henry Draper*, eds A. E. Glassgold et al., New York Academy of Sciences, p. 226

135. Zinnecker H., McCaughrean M.J., Wilking B.A., 1993, in *Protostars and Planets III*, eds. E.H. Levy, J.I. Lunine, Univ. of Arizona Press, p. 429

Jos de Bruijne (top) and Jon Williams (bottom) receive their second place poster prizes.

MASSIVE STAR FORMATION

ED CHURCHWELL
University of Wisconsin
475 N. Charter St.
Madison, WI 53706
USA

1. Introduction

One of the most durable problems in all of astrophysics is how stars are formed. Since the time of LaPlace astromomers have postulated numerous scenerios, but to the present this problem has resisted solution. This is especially true for massive stars. Massive star formation (MSF) has received much less observational and theoretical attention than low mass star formation and as a consequence much less is known about its physics, indeed it could be argued that we know too little to ask the right questions. Some, however, are obvious. For example, the mechanism(s) of cluster formation is still a mystery. In particular, what determines the IMF? What are the roles of fragmentation, differential galactic rotation, and turbulence in regulating the rate, and possibly the IMF, of star formation in molecular clouds? Are bipolar outflows the mechanism for sheding angular momentum from accretion disks? How does matter actually get from an accretion disk onto a protostar? Are massive stars formed entirely by accretion processes or a combination of accretion and stellar mergers? When are stellar winds established, what is their relationship to bipolar outflows, and what role do they play in the formation process?

A few things are known. For example, massive stars typically, perhaps always, form in conjunction with an associated cluster of lower mass stars. They form in massive molecular cloud cores which are generally heavily obscured at optical and UV wavelengths. Consequently, they generally can only be observed at infrared and radio wavelengths. The velocity, density, and tempurature structure of the circumstellar matter around a forming massive star is expected to change by orders of magnitude over distances of <1000 AU. Although, little is known about the detailed distribution and

515

C.J. Lada and N.D. Kylafis (eds.), The Origin of Stars and Planetary Systems, 515–552.
© 1999 *Kluwer Academic Publishers. Printed in the Netherlands.*

kinematics of gas and dust at distances within several hundred AU of forming massive stars nor the physical process by which they form, there are
some critical size scales that we can approximately identify from present
observations and theoretical considerations. Stellar winds are expected to
accompany MSF with wind velocities up to a few thousand km/s, the effects of which may be detectable out to 1000 AU from the protostar in a
hot wind-shocked cavity which supports observed shell-like and cometary
ultracompact (UC HII) regions. Dense and massive molecular outflows extending out to several pcs are presumably driven by a central massive protostar. Although the mechanism for initiating and driving these outflows is
still unknown, ample observational evidence confirms their existence. Ultracompact HII regions surround young massive stars out to radii of about
10^{17} cm typically. See Churchwell (1990, 1991, 1995, 1997a) for reviews of
the properties of UC HII regions. During formation, it is expected that a
massive accretion disk forms which feeds the central protostar from infalling
ambient cloud core material; these disks, if they exist, may have radii of
a hundred AU or more and masses containing a significant fraction of the
central protostar mass. The protostar itself may have a diameter of 10^{12}
cm or more depending on the stage of evolution.

I will review some of the established properties of MSF regions in §2.
The morphologies, ages, and dynamics of UCHII regions will be discussed
in §3. §4 summarizes a recent determination of the angular expansion rate
of G5.89 and in §5 dust emission from UC HII regions is discussed. The
properties of hot, molecular, cloud cores where massive stars are formed
are addressed in §6. A summary is given of the recently discovered hard X-
ray emission from W3 in §7. The properties and implications of molecular
outflows associated with MSF are examined in §8 and a brief summary is
given in §9.

2. TYPICAL PROPERTIES OF MSF REGIONS

MSF occurs in the cores of massive molecular clouds. These cloud cores
generally have very large optical extinctions (>50 mag), kinetic temperatures 50-200 K, and densities 10^5 - 10^7 cm^{-3}. These molecular cores are rich
chemical laboratories in which many molecular species have been detected
and from which much information on the physical conditions have been
inferred (see §6). Radio continuum images often show multiple components
indicating the presence of several hot, young stars. In fact, massive stars
always seem to be accompanied by a cluster of lower mass stars. Within a
given region of space, star formation is often (perhaps always) not coeval; a
very good example of this is the G9.62+0.19 complex (Hofner *et al.* 1995)
in which four separate epochs of massive star formation are apparent. Sign

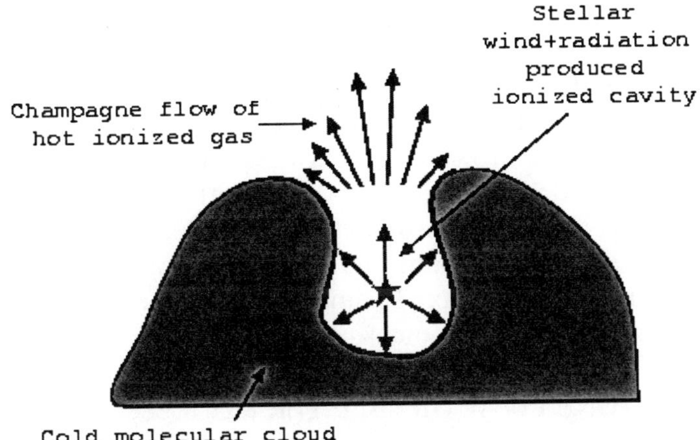

Figure 1. A schematic illustrating a champagne flow produced by a newly formed hot star whose wind and radiation evacuates a hot, ionized cavity at the edge of a large molecular cloud. At the cloud boundary, the ionized cavity "breaks out" and one sees a flow of ionized gas in directions where the gas is unimpeded by the molecular cloud. The flow is produced by a pressure gradient because of the over-pressure in the HII cavity relative to the ambient ISM pressure.

posts of massive star formation are UC HII regions, water masers, methanol masers, and luminous far infrared emission with colors in the range initially defined by Wood & Churchwell (1989b) and subsequently refined by others. The infrared flux distributions typically peak at about 100 μm and the typical dust temperatures are about 30 K, but of course the dust temperatures are a function of distance from the central YSO. The UC HII regions formed by massive YSOs have electron densities of $\sim 10^5$ cm^{-3}, emission measures of $\sim 10^8$ pc cm^{-6}, and diameters of a few times 10^{16} cm.

Massive stars are tightly confined to the Galactic plane (Wood & Churchwell 1989b; White, Becker, & Helfand 1991; Helfand, Zoonematkermani, Becker, & White 1992; Becker, White, Helfand, & Zoonematkermani 1994) There is some controversy over the exact value of the Z-scale height, but everyone agrees that it is ≤ 0.8 degrees.

3. MORPHOLOGIES, AGES, AND DYNAMICS OF UC HII REGIONS

UC HII regions have a surprisingly small range of morphologies which Wood & Churchwell (1989a; hereafter WC89a) classified as: cometary; core-halo; shell-like; irregular or multiply peaked; and spherical or unresolved. WC89a also pointed out that there are far too many UC HII regions to be consistent with classical Strömgren theory and concluded that UC HII regions must remain in a compact state $\sim 10^5$ yr, a factor of a hundred or so longer than the nebular sound crossing time. In fact, the morphology of a UC HII region is a complicated function of its age, the dynamics of both H^+ and H_2 gas, the density structure of the local ambient ISM, and the motion of the HII region relative to the ambient medium. Because the morphology of a UC HII region depends on so many factors, it has not been as easy as one might have supposed to isolate the primary mechanism that determines the morphology and maintains UC HII regions in a compact state. In this section, I will review the mechanisms proposed to explain the morphologies and lifetimes of UC HII regions.

3.1. CHAMPAGNE FLOW OR "BLISTER MODELS"

These models have their origin in the 1970s when it was believed that massive stars form preferentially near the boundaries of molecular clouds and as the stars become self-luminous they ionize a cavity at the surface of the cloud. The ionized cavity is confined in directions where the stellar photons are intercepted by the molecular cloud but ionized gas freely flows away from the cavity in directions away from the molecular cloud. This class of models produce cometary morphologies and are illustrated schematically in Figure 1. Further references to and variants of this class of models can be found in York, Tenorio-Tagle, & Bodenheimer (1983) and references therein. Garay, Lizano, & Gomez (1994) incorporated the effects of stellar winds and Forster *et al.* (1990) and Fey *et al.* (1995) and others have shown that star formation in a medium with a steep density gradient will also produce cometary morphologies. This class of models, however, has the basic problem that they evolve on a dynamical time scale and are therefore unlikely to be able to account for ages a factor of 10-100 times greater, as required by the large number of observed UC HII regions. There are several cases where this model seems applicable (e.g. Mon R2, Massi, Felli, & Simon 1985). Certainly, in situations where an embedded O-star is emerging from its natal molecular cloud, the champagne flow will apply.

3.2. IN-FALL MODELS

Reid *et al.* (1981), Zheng *et al.* (1985), Ho & Haschick (1986), Keto, Ho, & Reid (1987); and Keto, Ho, & Haschick (1987; 1988) proposed that UC HII regions can be maintained in a compact state well in excess of their sound crossing times (i.e. dynamical ages) due to gravitational in-fall of molecular gas. This is illustrated schematically in Figure 2. Although this class of models might possibly account for the ages of UC HII regions, it is not clear how they can account for UC HII morphologies. Also, the required

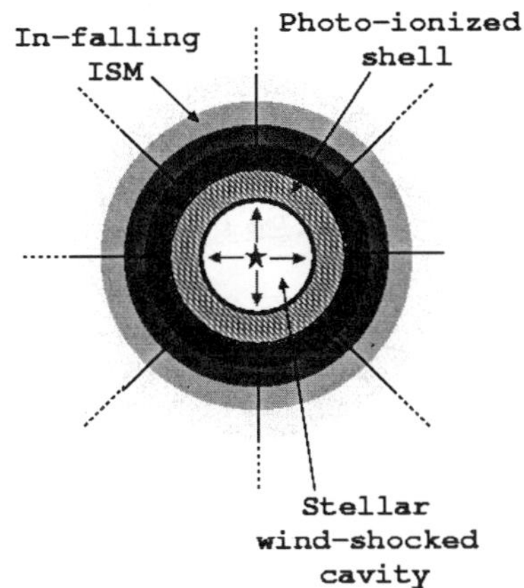

Figure 2.   A schematic illustrating confinement of a UC HII region by in-falling ambient molecular gas. The in-falling molecular gas provides both an increase in density and pressure at the I-front of the UC HII region.

in-fall rates have not been worked out so it is not clear if typical cloud cores can actually supply them over the required $\sim 10^5$ yr timescale.

3.3. PHOTO-EVAPORATING DISK MODEL

This model, developed by Hollenbach *et al.* (1993, 1994) uses the expected presence of a massive accretion disk and the strong UV radiation field of a massive YSO to explain the longevity of UC HII regions. The idea is basically that the UV radiation from the central star ionizes the outer envelope of the accretion disk. Inside the radius r_g, the photoevaporated gas flows away from the disk and forms a compact HII region which is continuously replenished as long as the accretion disk persists. Outside the radius r_g, the gas flows away from the disk at speeds greater than the escape velocity and produces a wind from the disk. In cases where the star has a strong stellar wind, the radius at which the disk wind occurs is greater than r_g due to stellar wind pressure. This is shown schematically in Figure 3 (Fig. 1 of Hollenbach *et al.* 1993). Although this picture gives a natural explanation for the longevity of UC HII regions in terms of properties that we believe are integral to such regions, it does not address the observed variety of morphologies. It may very well be applicable to those compact nebulae that are unresolved by the VLA as suggested by Hollenbach *et al.* (1993, 1994).

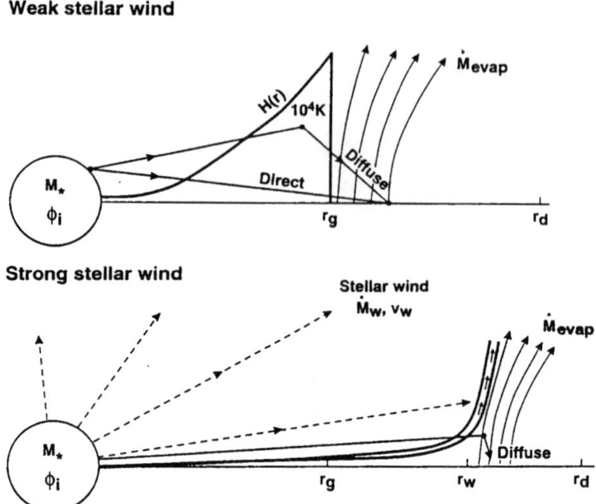

Figure 3.   A schematic illustrating the photoevaporation of a disk around a massive star. The upper panel illustrates the weak stellar wind case where both direct stellar and diffuse photons ionize the disk. Beyond the radius r_g the ionized photons escape to form a compact HII region around the star. The lower panel illustrates the strong wind case where the region of escaping ionized gas is pushed further from the central star by the stronger stellar wind. These illustrations are from Hollenbach, Johnstone, & Shu (1993).

3.4. PRESSURE CONFINED UC HII REGIONS

In recent years it has been discovered that the molecular density in the immediate vicinity of UC HII regions is much higher than originally thought, in several cases as high as 10^7 cm^{-3}. DePree, Rodriguez, & Goss (1995) and DePree, Goss, & Gaum (1998) have postulated that at such densities, the ambient medium around UC HII regions can provide enough pressure to confine an HII region for periods up to 10^5 years as required. This idea is shown schematically in Figure 4. Xie *et al.* (1996) have suggested an alternative version of this idea. They suggest that turbulent pressure could provide the high ambient pressures needed for confinement. In this case, because the pressure does not necessarily arise from particularly high densities, it avoids the high emission measures (EM$\geq 10^{10}$ pc cm^{-6}) predicted by the DePree *et al.* picture. Although confinement by external pressure (gas or turbulent) may be able to explain the longevities of UC HII regions, it does not address their morphologies. Also, it is unclear if the required high densities are extensive enough to provide confinement for the required time.

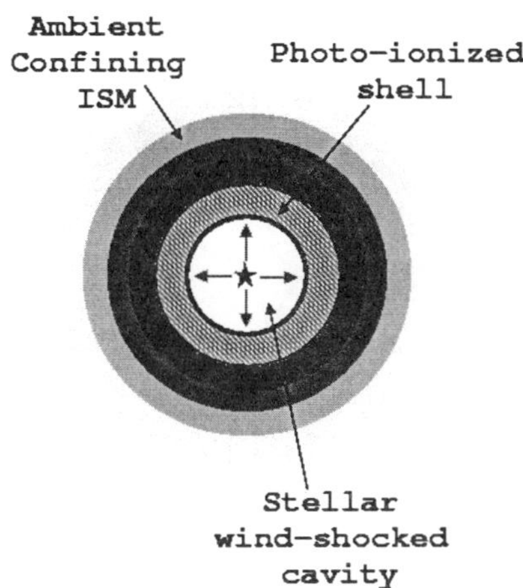

Figure 4.   A schematic illustrating the confinement of a UC HII region by dense, ambient, molecular gas which provides the thermal pressure required for confinement. The sketch implies that the density rises toward the I-front of the HII region, as suggested by observations.

3.5. STELLAR WIND SUPPORTED BOW SHOCK MODEL

Wood & Churchwell (1989a) suggested that the cometary UC HII regions could be produced by a stellar wind supported bow shock. MacLow *et al.* (1991), and VanBuren *et al.* (1990) have calculated bow shock models which satisfactorily reproduce both cometary and core-halo morphologies. The idea here is that the YSO has a small velocity relative to the natal molecular cloud core (simply due to the velocity dispersion of the OB association to which the YSO belongs). This relative motion causes a bow shock due to ram pressure of the ambient medium in the direction of motion. This is opposed by the pressure produced by the stellar wind which shocks a low density, but very hot cavity inside the observed HII shell. This hypothesis provides a static configuration for the UC HII region which is neither expanding nor contracting and will remain in this state as long as the YSO remains in the molecular cloud. In this picture, a cometary morphology is observed when viewed "from the side" and a core-halo morphology is observed when viewed "head-on" or "tail-on". It also follows that spherical and shell-like morphologies can prevail when the YSO has little or no motion relative to the ambient medium, and in this case, the nebula would probably be expanding rather than in a static state. This scenerio is shown schematically in Figure 5. It has been questioned by several groups either because the velocity structure does not seem to follow model predictions (Lumsden & Hoare 1996) or because the model calculated by Van Buren *et al.* (1990) required a relatively high velocity so the dwell time in the natal

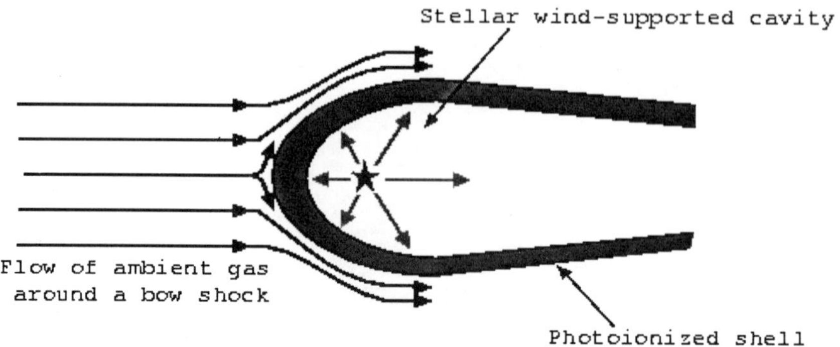

Figure 5.   A schematic illustrating the confinement of a UC HII region by a bow shock produced by ambient molecular gas flowing past a newly formed massive star. The ionized shell is supported from inside by the stellar wind-shocked cavity. The assumption here is that the star is moving supersonically relative to the ambient natal molecular cloud (> a few tenths of a km/s).

molecular core would be too short. Neither of these may be fatal as there are counter arguments to both criticisms.

3.6. MASS-LOADED STELLAR WINDS

Dyson, Williams, & Redman (1995), Williams, Dyson, & Redman (1996), Dyson (1994), Redman, Williams, & Dyson (1996), and Lizano *et al.* (1996) proposed that UC HII regions can be maintained in a compact state via mass loaded stellar winds. The idea is that when the medium surrounding a massive young star contains many small, dense, and neutral condensations, the stellar wind plus UV radiation will ablate and photoionize the envelopes of the condensations, thereby continuously replenishing the nebular gas and producing small cometary structures like those observed in the Orion Nebula by O'Dell and coworkers. The continuous release of newly ionized gas into the nebula via photoionization of the embedded condensations traps the I-front and maintains the nebula in a compact state. This is illustrated in Figure 6 which is similar to a figure in Lizano *et al.* (1996). This model addresses the longevity of UC HII regions, but does not explain their morphologies.

Which model is correct? Possibly all the above may be valid in certain nebulae at particular evolutionary stages. The bow shock model is compelling because it can explain three of the observed morphological types and the lifetime problem, but it may not be the only explanation. For example, when a star exits its natal cloud, it should produce a champagne flow. It is possible to test the various hypotheses by obtaining high reso-

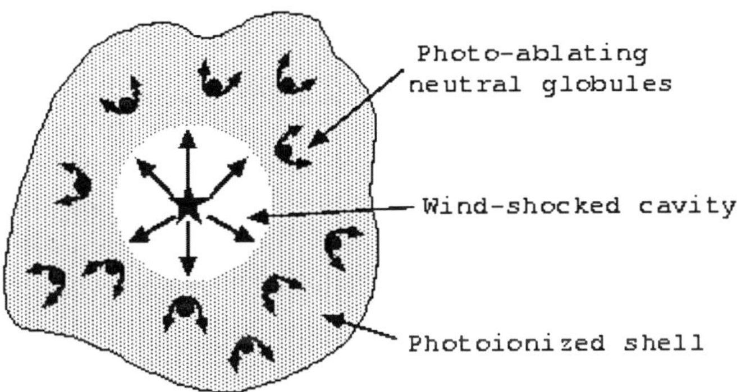

Figure 6. A schematic illustrating confinement of a UC HII region by mass loading the nebula from embedded molecular globules which are slowly photoevaporated and thereby continuously replenish the ionized gas maintaining the HII region in a dense, compact state.

lution velocity distributions of the ionized gas (via recombination lines or infrared fine structure lines) since each of the models predict rather different kinematics of the ionized gas. At present, such information does not exist for most nebulae.

4. THE ANGULAR EXPANSION RATE OF G5.89-0.39

In a effort to test some of the ideas proposed in §3, Acord, Churchwell, & Wood (1998; hereafter ACW98) used the VLA at three epochs over a period of 5.3 years to determine the angular expansion rates of G5.89-0.39 (shell-like) and G34.26+0.15 (cometary). Both of these UC HII regions are good candidates for measuring their angular expansion rates because they are among the brightest UC HII regions known at radio wavelengths and have approximately symmetric morphologies. G5.89-0.39 has an approximately spherical geometry which simplifies interpretation of its angular motions. Two independent techniques were used: 1) by scaling an earlier epoch image until it coincides with that of a later epoch image; and, 2) by differencing images obtained at two different epochs using the "cross-calibration" technique developed by Masson (1986, 1989). ACW98 found an expansion rate of 4 ± 1 mas yr^{-1}. From an analysis of the H42α line, ACW98 further inferred a radial expansion velocity of $v_{exp} \sim 35$ km s^{-1} under the assumption that random turbulent motions are no more than twice the thermal motions (i.e. turbulent FWHM≤ 30 km s^{-1}). With an estimate of the expansion velocity of the nebula, they derived the distance to the nebula from the relation $D = v_{exp}/\mu$ where μ is the angular expansion rate. ACW98 find $D = 2.0^{+0.7}_{-0.4}$ kpc. The kinematic distance of this source is in the range 2.5 to 3.0 kpc, but at galactic longitudes $< 6^0$, kinematic distances are quite uncertain. For an expansion velocity of $\sim$35 km s^{-1} and an angular radius of the bright shell of 0.010$\pm$0.002 pc, ACW98 infer a dynamical age of only 600^{+250}_{-125} yr, making this one of the youngest nebulae known. The fact that this nebula is expanding does not eliminate the bow shock model as this source has a spherical shell morphology which implies that it has little or no motion relative to the local ISM and therefore it is not confined by a bow shock. The program to obtain angular expansion rates is continuing.

5. DUST ASSOCIATED WITH UC HII REGIONS

MSF regions are among the brightest objects in the Galaxy in the far infrared. The reason for this is that massive YSOs are very luminous and deeply embedded in natal molecular gas and dust clouds. The dust ultimately absorbs all the radiation from all the stars in the cluster and reradiates it as infrared radiation at wavelengths determined by the temperature of the dust. At increasing distances from a UC HII region the dust temperatures decrease. The dust at greater radii radiate at ever longer IR wavelengths. At wavelengths shorter than about 3mm thermal dust emission dominates free-free emission from the gas. Because the dust has a temperature gradient with radius, the SED is broader than a single temperature Planck function. The SED of a typical MSF region is shown schematically in Figure 7 from radio wavelengths to the near infrared. The radiation mechanisms and properties of each region of the SED are indicated in Figure 7. The peak IR flux is a factor of 10^3 to 10^6 greater than that of the optically thin radio free-free emission. The radio flux density is a measure of the stellar ionizing flux and the FIR emission is a measure of the total radiation over all wavelengths from all stars in the cluster. Kurtz, Churchwell, & Wood (1994) showed that the value of the FIR to radio flux ratio is a function of the spectral type of the star responsible for ionization of the nebula; in the log the ratio is $\sim$6 for a B3 star and $\sim$3 for an O4 star. They also showed that MSF regions with a given 100μ flux to 2cm flux density essentially always have a larger total luminosity than can be produced by a single hot star of that luminosity. From this, they argued that MSF is always accompanied by a cluster of stars. This seems to be born out by observations as all MSF regions which have been imaged in the near infrared show an associated cluster of stars.

The IR SEDs of MSF regions have been modeled by several groups including Krügel&Mezger (1975); Leung (1976); Tielens&de Jong (1979); Tutukov&Shustov (1981); Lee&Draine (1985); Chini, Krügel, &Kreysa (1986); Hoare *et al.* (1988); Churchwell, Wolfire, & Wood (1990; hereafter CWW);

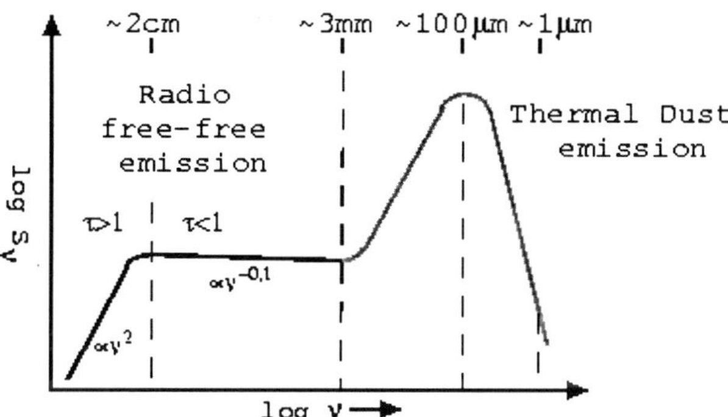

Figure 7. A schematic of the spectral energy distribution of a UC HII region from radio to the near infrared. S_ν is the flux density and ν is the frequency. Several key wavelengths are indicated along the top of the figure.

Wolfire & Churchwell (1994); and Faison *et al.* (1998). The latter three papers included a range of dust sizes at each position. In all models, spherical symmetry and a Planck function for the input stellar spectrum were assumed. The multi-source models of Faison *et al.* (1998) give an overview of the predicted range of dust properties associated with MSF regions heated by stars of type O5 to B0. In all cases, a central dust free cavity is required to fit the observed SEDs. No significant amounts of dust can exist close to the star(s) because they would produce too much NIR emission. The dust free cavities are typically about a factor of ten larger than the dust sublimation radius, thus the inner radius of the dust shell is determined not by photodestruction but probably by radiation pressure and/or stellar winds. The outer radii of the dust shell, defined by the radius at which the dust can no longer be distinguished from the ambient dust, are very large ranging from $\sim$1 to $\sim$4 pc. The ratio of outer to inner dust shell radii are in the range $\sim$50 to $\sim$150. The large outer radii are required in order to have dust cool enough that they emit most of their radiation at $\sim$100 μm. The dust temperatures drop very sharply in the inner 20% of the dust shell radius, then decrease more slowly (as expected for an optically thin spherically symmetric cloud) in the outer 80% of the shell (see CWW90). The average dust temperatures at the inner dust shell boundary are 340 K to 200 K, depending on the region, and $\sim$20 K at the outer boundary. Constant density models fit the data better than power-law density gradients.

The 9.7 μm silicate feature is often strong and always in absorption with spatial resolutions used so far (see Faison *et al.* 1998; Roelfsema *et al.* 1996). Polycyclic aromatic hydrocarbon features in the range 3 to 12 μm are often detected toward MSF regions and in some cases are the dominant features in this wavelength range (Roelfsema *et al.* 1996).

6. HOT MOLECULAR CLOUD CORES

What about the nature of the natal molecular cloud cores in which massive star formation occurs? Considerable work has been done on these regions by many groups in the past few years. I will try to summarize some of the main results of the past few years. It will be impossible to reference all the work done in this area and I apologize to those whose work I do not quote. I will not discuss molecular abundances, chemistry, grain mantle destruction and its impact on gas phase abundances, nor freezing out of gas phase molecules onto grains as these topics will be covered by other speakers. Molecular cloud cores where massive stars form are hot, dense, compact, and dynamic. Their temperature and density decrease with distance from the center and the enclosed mass increases. Cesaroni *et al.* (1991, 1992), Churchwell, Walmsley, & Wood (1992), Hofner *et al.* (1995) and others have measured hot core properties toward a number of MSF regions. Using $C^{34}S$ (2-1; 3-2; and 5-4) and NH_3 (1,1; 2,2; 4,4; and 5,5) Cesaroni *et al.* (1991, 1992) showed that typical hot cores have densities $\sim 10^6$ cm^{-3}; temperatures $\sim$100-200 K; mass $\sim$1000 M$_\odot$, and diameters $\sim$0.5 pc. A particular example is the MSF core in which G9.62 is embedded. Hofner *et al.* (1995) has observed this region in several molecular line probes as well as radio continuum emission. From a $C^{18}O$ (1-0) image obtained at OVRO, they find a core of extent 0.6x0.3 pc, mass $\sim$1000 M$_\odot$, average hydrogen density $2x10^5$ cm^{-3}, and average temperature over the entire core $\geq$30 K. Embedded in this core are two very young B stars (components D and E) which have formed very dense and compact UC HII regions and a currently forming massive protostar in a rapid stage of accretion (component F) all aligned along a SE-NW ridge. In addition, a string of H_2O, OH, CH_3OH, and NH_3 (5,5) masers are found along the central ridge between the D and E UC HII regions. This is illustrated in Figure 8 which shows the molecular core by contours, the radio continuum emission in grey scale, and the various masers by different symbols. This is the only object in the sky where the NH_3 (5,5) maser has been detected. It is located close to component E. This region also illustrates another morphological feature that seems fairly common in MSF regions, namely, the tendency for masers and star formation to occur along linear features. In a high resolution study of H_2O masers, Hofner & Churchwell (1996) noted that water masers often occur in linear features possibly indicating the tendency of cloud cores to form disk-like morphologies (Cesaroni *et al.* 1998), as discussed below. Finally, the G9.62 region appears to show at least four different epochs of massive star formation implied by the diameters and mean densities of HII regions

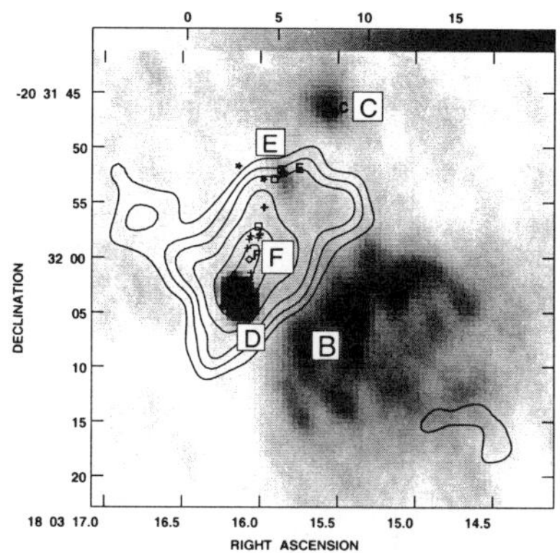

Figure 8. The $C^{18}O$ (1-0) core toward the massive star formation complex known as G9.62. The CO core is indicated by contours. The gray scale is the radio free- free emission produced by ionized Hydrogen. Discrete radio continuum sources are designated as components B, C, D, E, and F. The youngest and densest components D, E, and F are all embedded in the $C^{18}O$ core. There are also numerous water, ammonia, methanol, and hydroxyl masers located along a ridge running along the major axis of the CO core. See the caption for figure 3 in Hofner et al. (1995) for a definition of the various symbols. There are HII regions with a range of sizes and densities in this complex which Hofner et al. (1995) have argued indicate a range of ages. This figure is from Hofner et al. (1995).

A (oldest); B (intermediate age); C, D, and E (just emerging HII regions); and F (a protostar).

Cesaroni *et al.* (1994) pointed out an apparent luminosity paradox toward MSF cores. They found that the luminosities obtained by assuming that the cores are spherical, constant temperature blackbodies with temperatures and diameters derived from molecular line observations were about 5 times those obtained from integrated IRAS fluxes. The blackbody luminosities were calculated using the temperatures obtained from NH_3 line ratios (or excitation models) and diameters measured from the half-power distributions of the the NH_3 emission. This apparent paradox was resolved by Cesaroni *et al.* (1998) who measured the temperature profiles toward three MSF cores and modeled their temperature profiles using spherical and disk-like geometries. They found in all three cases that disk-like geometries reproduced the observed profiles quite accurately, but spherical models did not. This is illustrated in Figure 9 taken from Cesaroni *et al.* (1998). Further, cores with the observed temperature gradients and disk-like geometries have luminosities in agreement with those derived from the IRAS fluxes. The tendency of star-forming cloud cores to evolve to disk-like geometries may be the reason why masers and star formation often seem to occur in "elongated filaments".

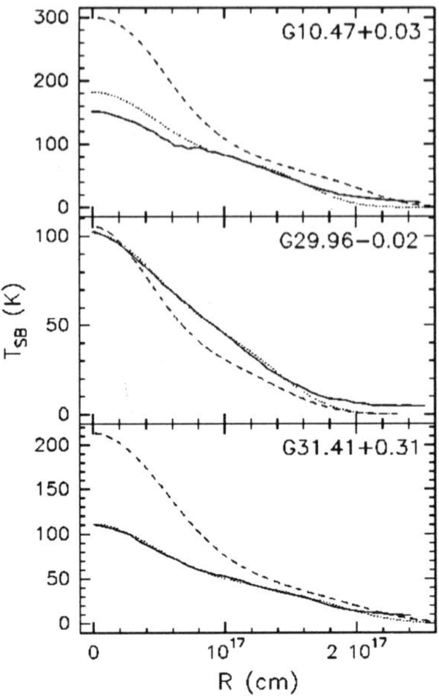

Figure 9. A comparison of model temperature profiles with observationally determined temperature distributions of three hot cores from Cesaroni *et al.* (1998). The dashed curves are for best fit spherical geometries, the solid curves are the observed temperatures, and the dotted curves are for disk-like geometries. See Cesaroni *et al.* (1998) for the input parameters and model results. The main point these models emphasize is that a disk-like geometry produces far better fits to the observed temperature profiles than a spherical geometry in all three sources.

7. HARD X-RAY EMISSION FROM MSF REGIONS

Main sequence OB stars and evolved stars are X-ray emitters. It has been postulated that this emission is produced either in a hot corona at the base of a stellar wind (Cassinelli&Olson 1979) and/or by shocks throughout the stellar wind (Lucy 1982). This emission generally occurs at energies ≤ 4 kev. X-rays have been observed in a small number of star forming regions such as ρOph and the Orion Nebula (Koyama *et al.* 1992; Yamauchi and Koyama 1993a). In both regions many of the X-ray sources have been identified with low-mass premain-sequence (PMS) stars. A further very intriguing result is the discovery of FeXXV line emission at 6.7 kev whose distribution is tightly confined to the disk of the Galaxy (Yamauchi and Koyama 1993b), similar to that of CO and O stars. Gas that can produce this line requires a temperature $\geq 70 \times 10^6$ K and it is very difficult to understand how it can be so tightly confined to the Galactic disk. Motivated by the detection of X-ray emission from low-mass PMS stars and the discovery of the disk component of FeXXV line emission, Hofner & Churchwell (1997; hereafter HC97) used

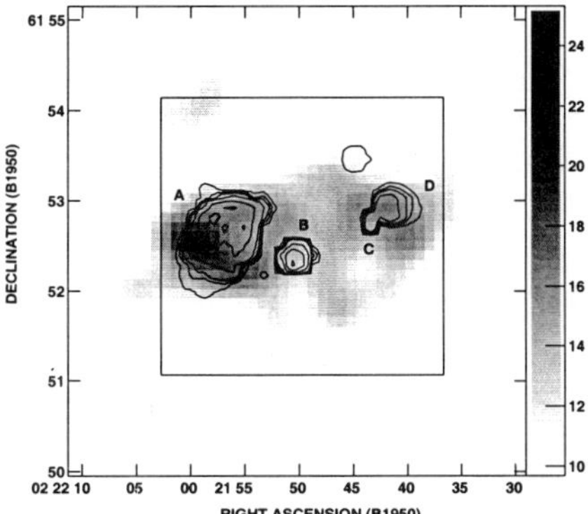

Figure 10.   A comparison of radio continuum emission at 8.3 GHz from Adler, Wood, & Goss (1996) (contours) with X-ray emission in the energy range 0.5-10 kev (gray scale) toward the W3 complex. This is Fig. 1 from HC97 and it shows that the most intense hard X-ray emission coincides well with the UC HII regions in this complex.

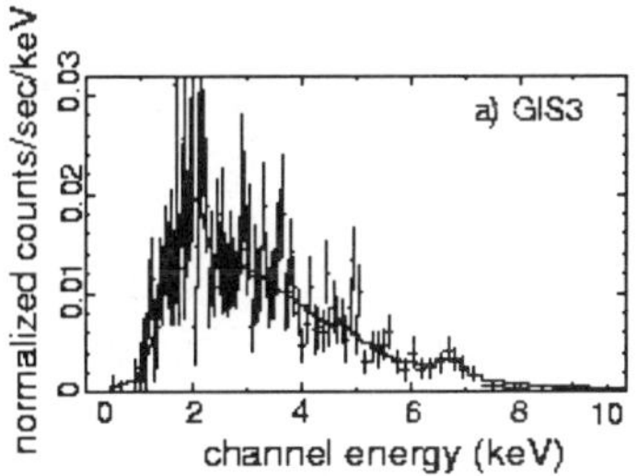

Figure 11.   The observed X-ray spectrum integrated over a 6' diameter toward the core of W3 obtained with the GIS3 spectrometer on the ASCA satellite. The FeXXV line at ~6.7 kev is apparent in this spectrum. From HC97.

the ASCA satellite to search for X-ray emission from the MSF region W3. Their rationale was that massive protostars have fast and massive winds which are thought to produce a very hot ($\sim 10^8$ K), shocked cavity that supports the observed UC HII shell. This cavity should emit rather hard X-rays and possibly provide an explanation for how the FeXXV line can be so tightly confined to the Galactic disk, as O stars are also tightly confined to the disk. HC97 detected X-ray emission coincident with the UC HII regions which define the core of the W3 MSF region. The distribution of X-ray emission in the energy range 0.5-10 kev toward W3 is shown in Figure 10.

The X-ray spectrum toward the W3 core is shown in Figure 11. The X-ray flux is maximum at $\sim$2 kev but emission is seen as high as $\sim$8 kev. The FeXXV line at 6.7 kev, shown in Fig. 11, has a flux of $\sim$1.2x10^{-5} photons cm^{-2} s^{-1}. A Raymond & Smith (1977) model fit to the observed spectrum, including extinction by interstellar hydrogen, implies the following properties of the emitting gas: T$\sim$7x10^7 K; a volume emission measure of 2.1x10^{56} cm^{-3}; an X-ray flux in the 2-10 kev range of 2.6x10^{-12} erg cm^{-2} s^{-1}; and a luminosity of $\sim$2x10^{33} ergs s^{-1} in the 2-10 kev energy range. An absorbing hydrogen column density of 2.1x10^{22} cm^{-2} is inferred by the model, which is consistent with that measured by Dickel, Goss, & Condon (1996).

What is the origin of the observed hard X-ray emission in W3? The stellar coronal emission and shocks within stellar winds do not appear to be energetic enough to produce the emission above a few kev and there are probably not enough low-mass PMS stars to produce the observed flux from W3. The X-ray emission from W3 is a factor of 30 to 3000 times higher than Orion and ρ Oph, respectively. Thus, if low-mass PMS stars provide the X-ray emission, W3 would have to contain a substantially larger number of such stars than either Orion or ρ Oph. PMS stars also do not produce FeXXV line emission, so another explanation for this would be required. HC97 suggested that the wind shocked cavity associated with UC HII regions would provide a natural explain for both the presence of the FeXXV line and its confinement to the Galactic plane. However, this does not explain the apparent extended X-ray component toward the W3 complex. If the stenght of the FeXXV line from W3 is typical of MSF regions, then it does not appear that UC HII regions could be the major contributor to the disk component of FeXXV emission. However, only one such region has been observed so far, and it could well be that W3 is under-luminous relative to other such regions. More MSF regions need to be observed in the 1-10 kev energy range to resolve this issue.

8. MOLECULAR OUTFLOWS ASSOCIATED WITH MSF

8.1. THE FREQUENCY OF OUTFLOWS FROM MSF REGIONS

By 1995 several MSF regions had been observed to have molecular outflows. Among these were the Orion IRc2 outflow (Kwan & Scoville 1976; Zuckerman, Kuiper, & Rodriguez Kuiper 1976; Downes et al. 1981; Erickson et al. 1982; Wright et al. 1983; Allen & Burton 1993, and others); G5.89 (Harvey & Forveille 1988; Cesaroni et al. 1991; Choi, Evans, & Jaffe 1993); DR21 (Garden et al. 1991a,b; Garden & Carlstrom 1992); NGC 6334I (Bachiller & Cernicharo 1990); G34.26 and G10.62 (Cesaroni et al. 1991); W75N (Hunter et al. 1994); W3IRS 5, GL490, GL2591, S140, and CephA (Choi, Evans, & Jaffe 1993); and several low-mass and high-mass outflows are reported by Cabrit & Bertout (1992) and Lada (1985). In an effort to better determine the frequency of molecular outflows in MSF regions, Shepherd & Churchwell (1995; hereafter SC95) observed and analyzed the CO(1-0) line profiles with high signal-to-noise toward 94 UC HII regions (HPBW~60"). They found that fully 90% of this sample had line wing emission in excess of a gaussian profile fit to the core of the line. These results indicate gas motions in excess of thermal plus turbulent motions, but they do not necessarily indicate bipolar outflows as rotation, or expansion, or contraction, or different velocity components along the ling of sight could also produce excess line wings. To determine if the excess line wings are due to outflows Shepherd & Churchwell (1996; hereafter SC96) mapped a selection of 10 of the sources observed to have excess line wings by SC95. They found five to have well defined outflows, two had multiple velocity components along the line of sight and were not outflows, and three had either S/N ratios that were too low to confidently map or weak CO emission was in the reference position such that the images were unusable. They conclude that at least 50%, and possibly as high as 80%, of the selected sample have bipolar outflows.

8.2. MOLECULAR BIPOLAR OUTFLOW PROPERTIES

In Table 1, I have tabulated the sources toward which massive outflows have been detected. These are listed in order of increasing RA. Along with the source names and their positions, I also give the distances from the Sun D, the HII luminosities L_{HII} as determined from radio free-free flux densities or as given by the reference in Table 2, and the total luminosity of the HII region and associated cluster L_{IR} as determined from the integrated infrared flux. The distances are taken from the references listed in Table 2 and the outflow properties were derived on the basis of this distance.

In Table 2, representative properties of bipolar outflows thought to be driven by massive protostars are tabulated. I have reported values as given by the authors without insisting that the data be analyzed in a uniform and systematic way. As a consequence, there are disagreements among different authors on some flow properties perhaps at the factor of two level. Many of the sources have been observed by numerous observers with different techniques. In these cases, either the most complete set of published values or the latest values are generally given. Several sources, known to have outflows, are included for which no outflow properties are measured (eg. G10.62 and G34.26). The data given in Table 2 are from single dish observations which do not suffer from the "missing flux" problem that plagues interferometers and therefore provides more reliable global properties of the outflows such as mass, extent, and energetics but gives little information

Table 1
Massive Outflows

Source	RA(1950)	DEC(1950)	D kpc	L_{HII} $L_\odot$	L_{IR} $L_\odot$	Refs	Notes
W3(IRS5)	02 21 53.1	61 52 22	2.3		2.00×10^5	1	1
GL 490	03 23 39.2	58 36 36	0.9		$1.7\text{-}2.4\times10^3$	1,2	
Orion IRC2	05 32 46	-05 24 00	0.5	$>1.00\times10^4$	2.10×10^5	2,16	5
G173.58+2.45	05 36 06.2	35 39 06	1.8	3.00×10^3	1.80×10^3	3,5,14	
G192.16-3.82	05 55 20.0	16 31 44	2.0	3.00×10^3	3.26×10^3	3,5,14	
MonR2(IRS1)	06 05 20.0	-06 22 39	1.0	4.00×10^4	5.00×10^4	4,5,6	
G240.31+0.07	07 42 45.1	-24 00 24	6.4	1.26×10^4	5.01×10^4	7	
NGC6334 I	17 17 32.3	-35 44 04	1.7	3.00×10^4	8.00×10^4	8	
G5.89-0.39	17 57 26.8	-24 03 55	3.0	1.20×10^5	3.00×10^5	4,5	
G9.62+0.19 F	18 03 16.09	-20 31 59.7	5.7			13	2
G10.62-0.38	18 07 30.7	-19 56 29	6.5	2.10×10^5	1.24×10^6	4,5	
G25.65+1.05	18 31 40.1	06 02 06	2.8	8.90×10^3	1.76×10^4	9,5	
G34.26+0.15	18 50 46.1	01 11 12	3.7	1.40×10^5	5.93×10^5	4,5	
G35.2N (IRS1)	18 59 14.1	01 09 03	2.5	9.10×10^4	1.23×10^5	4,5,14	
G45.07+0.13	19 11 00.4	10 45 43	8.3	1.40×10^4	1.42×10^6	4,5	
G45.12+0.13	19 11 06.2	10 48 25	8.3	5.80×10^5	1.66×10^6	4,5	
IRAS20126+4104	20 12 41.0	41 04 20.5	1.7	5.01×10^3	6.31×10^4	7,15	
G75.78+0.34NE	20 19 52.7	37 27 22	5.5	4.80×10^4		3,5	
GL 2591	20 27 35.8	40 10 14	2.0	2.00×10^4	9.00×10^4	1,5	
W75NBb	20 36 50.0	42 26 57	2.0	5.50×10^3	1.40×10^5	10,5	
DR21	20 37 14.4	42 08 55	3.0	4.00×10^5	$\sim6\times10^5$	11,5	
G98.04+1.45	21 41 21.3	54 42 32	7.4	2.30×10^5	1.39×10^5	3,5,14	3

Table 1 (continued).

Source	RA(1950)	DEC(1950)	D (kpc)	L_{HII} ($L_\odot$)	L_{IR} ($L_\odot$)	Refs	Notes
S140 (IRS1)	22 17 41.2	63 03 45	0.9		$0.5\text{-}9.0\text{x}10^4$	1,2	1
Ceph A	22 54 19.0	61 45 47	0.7	$3.98\text{x}10^3$	$1.58\text{x}10^4$	7	
NGC7538(IRS9)	23 11 52.8	61 10 59	2.7		$4.00\text{x}10^4$	12	1
G111.25-0.77	23 13 57.9	59 39 00	3.5	$2.00\text{x}10^4$	$9.80\text{x}10^3$	3,5,14	4
IRAS23385+6053	23 38 31.4	60 53 50	4.9		$1.60\text{x}10^4$	17	6

References

1. Choi (et al.)(1993)
2. Cabrit & Bertout (1992)
3. Shepherd & Churchwell (1996)
4. Wood & Churchwell (1989)
5. Panagia (1973)
6. Thronson et al. (1980)
7. MacLeod et al. (1998)
8. Bachiller & Cernicharo (1990)
9. Shepherd & Churchwell (1995)
10. Hunter et al. (1994)
11. Garden & Carlstrom (1992)
12. Mitchell & Nasegawa (1991)
13. Hofner et al. (1995)
14. Shepherd (private communication)
15. Cesaroni et al. (1997)
16. Menten & Reid (1995)
17. Molinari et al. (1998)

Notes

1. The IR source does not coincide with a detected HII region.
2. No cm emission detected, mm emission is from warm dust. FIR emission is confused.
3. Two UC HII regions are present, neither of which is likely powering the outflow.
4. There may be multiple outflows toward this UC HII region.
5. The outflow source is not the radio source, as its luminosity is much too small.
6. No cm emission detected, mm emission is probably due to dust.

Table 2

Typical Properties of Massive Bipolar Outflows

Source	R_f pc	τ_d 10^4 yr	M_f $M_\odot$	$\dot{M}_f$ $M_\odot$ yr^{-1}	P_f $M_\odot$ km s^{-1}	E_f 10^{45} ergs	L_f $L_\odot$	Refs.	Notes
W3 (IRS5)		0.3	~8	4.8×10^{-5}	110	16		1	1
GL 490	1.2	1.8	15.4	8.6×10^{-4}	493	158	19	2.1	2
Orion IRC2	0.1	0.075	8.2	1.1×10^{-2}	533	346	1000	2	3
G173.58+2.45	0.6	5.3	32	6.0×10^{-4}	170	9.4	1.4	3	4
G192.16-3.82	2	27	58	2.2×10^{-4}	210	9.4	0.2	3	4
MonR2 (IRS1)	$\geq$3.2	15	202	1.3×10^{-3}	~2000	290	16	4,5,6,7	
G240.31+0.07	0.76	2.3	8.3	3.6×10^{-4}	93	11	3.9	8	7
NGC6334F	0.36	0.23	>2.3	1.0×10^{-3}	92	43	170	9	
G5.89-0.39	0.1	0.3	77	2.6×10^{-2}	990	500	1300	10,11,1	5,2
G9.62-0.19F		0.8	~50					20	
G10.62-0.38								11	6
G25.65+1.05	1.1	1.8	7.6	4.2×10^{-4}	136	27	12.4	8	7
G34.26+0.15								11	6
G35.2N (IRS1)	1.4	4.6	73	1.6×10^{-3}	1100	164	51	2	
G45.07+0.13	0.3	3.2	50	1.4×10^{-3}	631	36	~11	12	
G45.12+0.13	1.8	20	~4800	$\geq1.3\times10^{-3}$	~31000	5600	10-100	12	8
IRAS20126+4104	0.3	0.58	>90-310	1.6×10^{-2}	540		80	13	
G75.78+0.34NE	0.8	3	72	2.4×10^{-3}	970	130	36	3	
GL 2591		0.2	33	6.8×10^{-5}	269	23		1	1
W75N	0.7	$\geq$4.4	48	1.2×10^{-3}	~700	110	22	14,15	
DR21	~2.2		>3000			>2000		16,17,18	9
G98.04+1.45	1.2	9.1	40	4.4×10^{-4}	240	15	1.4	3	
S140 (IRS1)	1.1	11	24.4	9.4×10^{-4}	488	98	8.4	2,1	
Ceph A		0.2	~8	2.8×10^{-5}	86	10		1	1

Table 2 (continued)

Source	R_f	τ_d	M_f	$\dot{M}_f$	P_f	E_f	L_f	Refs.	Notes
	pc	10^4 yr	$M_\odot$	$M_\odot$ yr^{-1}	$M_\odot$ km s^{-1}	10^{45} ergs	$L_\odot$		
NGC7538 (IRS9)	0.97	~1.6	51	$4.2\mathrm{x}10^{-3}$	490	100	790	19	
G111.25-0.77	0.6	3.7	16	$4.3\mathrm{x}10^{-4}$	110	8.4	1.8	3	
IRAS23385+6053	>0.01	<0.73	11-20	$>1.5\mathrm{x}10^{-3}$	$\geq$67	>4.7	>5.4	21	

References

1. Choi *et al.* (1993)
2. Cabrit & Bertout (1992)
3. SC96
4. Wolf, Lada, & Bally (1990)
5. Loren (1981)
6. Tafalla *et al.*(1997)
7. Meyers-Rice & Lada (1991)
8. SC95
9. Bachiller & Cernicharo (1990)
10. Acord *et al.* (1997)
11. Cesaroni *et al.* (1991)
12. Hunter *et al.* (1997)
13. Cesaroni *et al.* (1997)
14. Hunter *et al.* (1994)
15. Fischer *et al.* (1985)
16. Garden *et al.* (1991a)
17. Garden *et al.* (1991b)
18. Garden & Carstrom (1992)
19. Mitchell & Hasegawa (1991)
20. Hofner *et al.* (1995)
21. Molinari *et al.* (1998)

Notes

1. The mass and other reported flow properties assume the mass is swept-up ISM.
2. Ref. 2 only mapped the inner part of the outflows: they find lower values than other observers.
3. Many observers have studied this source; ref 2 values are reported here.
4. HPBW$\approx$60".
5. Tabulated values are from ref 10, see note 2 above.
6. Images are shown but no outflow properties were given.
7. Optical depth unity was assumed.
8. Values from ^{12}CO are reported rather than those from ^{13}CO.
9. At least 2 outflows and probably more are present.

on multiplicity, detailed morphologies, and velocity distributions. The headings in Table 2 are, from left to right: the HII region or IR source toward which the outflow is observed, the projected separation of the outflow lobes R_f, the dynamical time scale of the outflow $\tau_d = R_f/2v_f$, the total mass in the outflow M_f in solar masses, the rate of mass outflow $\dot{M}_f$, the outflow momentum P_f, the momentum supply rate or force of the outflow F_f, the total energy of the outflow E_f, and the mechanical luminosity of the outflow L_f in solar luminosities.

Aside from the usefulness of assembling the list of currently known massive outflows and their properties in one place, the main reason for assembling Table 2 was to provide an overview of massive outflow properties. These outflows are thought to be driven by massive protostars as implied by Fig. 13 below and the fact that massive outflows only occur in MSF regions (Table 2). Single dish spatial resolutions are not good enough to identify the source of the outflow with a particular star or protostar. Identification of the driving source is exacerabated by the fact that massive star formtion occurs in distant (typically $d > 1$ kpc), dense stellar clusters. This has been demonstrated by direct near infrared imaging of MSF regions (McCaughrean 1993; Mc Caughrean and Stauffer 1994; Aspin and Walther 1990; and others); by optical imaging of evolved HII regions which contain star clusters; and, indirectly by Kurtz, Churchwell, & Wood (1994) who showed that UC HII regions have 100 μm-to-2 cm flux density ratios well in excess of those produced by single stars of the same luminosity. In all cases where IR imaging has been carried out, a dense cluster of stars or protostars have been found to be associated with the UC HII region toward which outflows have been observed. Only radio interferometry in combination with high spatial resolution NIR imaging can isolate the sources of bipolar molecular outflows. In a few cases it appears that an outflow may be driven by the star responsible for ionizing a UC HII region, but outflows are often offset from the nearby UC HII region and probably driven by an even younger star than the ionizing star of the UC HII region. A consequence of large distances and crowded star fields is that the ionizing star of only one UC HII region (the relatively lightly extincted nebula G29.96) has been identified (Watson *et al.* 1998) to the present time and no star has been unambiguously identified as the driving source for a massive outflow.

Interferometric studies of G45.12, G75.78, and NGC2024 by Hunter, Phillips, & Menten (1997); Shepherd, Churchwell, & Willner (1997); and Chernin (1996) have shown that these regions harbor two or more outflows when observed with high spatial resolution. We also know that multiple outflows are present in DR21, even from single dish observations. The mass of each of the outflows (with the exception of the one toward NGC2024) is still quite large (10s to 100s $M_\odot$). The interferometric images have also shown that massive outflows do not appear to be well collimated. The Orion IRC2 outflow, traced by high resolution shocked H_2 emission (Allen & Burton 1993), is shown in Figure 12. This outflow is certainly not well collimated nor is it a smooth gaseous flow. It does not come to a point at its origin but rather appears to originate from a large surface area, most of which is well away from the central position of the outflow. Allen & Burton (1993) referred to it as an explosive event. The "fingers" of this outflow which reach out into the ambient interstellar medium defines a jagged interface not a smooth working surface. It has been suggested that the fingers might be jets from individual low-mass protostars at the center of this feature, although possible this explanation seems unlikely: one, because it would require many stars to have their jets aligned within about 45^0 of each other; two, NIR photometry does not implicate such stars along the waist of this bowtie feature (although this could be because of large 2μm optical depths); and three, the outflow looks like a coherent feature, not a montage of many separate features (althought this could be a resolution effect).

The outflows found in MSF regions have energetics, masses, and outflow rates much larger than those associated with low-mass star formation. This is illustrated quite clearly in Table 3 where the average outflow mass, force, and mechanical luminosity are tabulated from Lada (1985) for star formation regions with bolometric luminosity $L_{bol} > 10^3\ L_\odot$ versus those with $L_{bol} < 750\ L_\odot$. If the more recent data given in Table 2 had been used, the mean outflow mass and luminosity from MSF regions would be >130 $M_\odot$ and $>180\ L_\odot$, respectively. The mean properties of low-mass outflows do not seem to have changed significantly excluding the discovery of much

TABLE 3. Comparison of Mean Outflow Properties

Selected Outflow Parameter	$L_{bol} > 10^3\ L_\odot$ Average	$L_{bol} < 750\ L_\odot$ Average
$M_f\ (M_\odot)$	44	1.8
$F_f\ (10^{-3}\ M_\odot\ km\ s^{-1}\ yr^{-1})$	8.3	0.8
$L_f\ (L_\odot$	8	1.1

larger outflow radii. Thus, MSF regions have outflows at least a factor of ten more massive and more luminous than those from low-mass stars. SC96 have shown that the mass outflow rate $\dot{M}_f$ is a continuously increasing function of stellar bolometric luminosity over the range 1 to $10^6\ L_\odot$ as shown in Fig. 13. The solid curve is a quadratic least squares fit to the

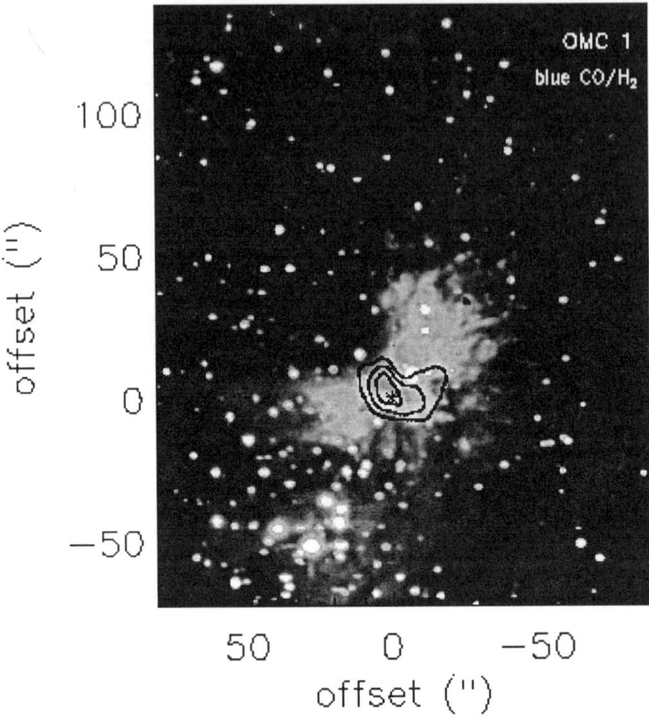

Figure 12. The Orion IRC2 outflow as delineated by shocked H_2 (in grey scale) from Allen & Burton (1993) with CO(1-0) emission contours superimposed from Chernin & Wright (1996).

data. A linear least squares fit to the data implies that

$$\dot{M}_f \approx L_{bol}^{0.7} \propto M_*^{2.1} \tag{1}$$

The dependence on stellar mass assumes a mass-luminosity relation (M-L) $L \propto M^3$ (appropriate for the upper end of the M-L relation). The importance of the tight relation between $\dot{M}_f$ and L_{bol} is that one can measure $\dot{M}_f$ relatively easily from which it is possible to infer the luminosity of the driving source to relatively high accuracy, even in the absence of radio continuum or IR fluxes. Further, the product of $\dot{M}_f$ with the stellar luminosity function $\Phi(M)$ (Allen 1972), when plotted against L_{bol} indicates that A-F protostars are the major contributors to mass outflow during star formation. In principle, it is only legitimate to make such a plot for outflow sources of the same age; however from Table 2, we see that most massive outflows have ages within a factor of 10 of each other.

SC96 also showed that within a given luminosity range, the outflow mass increases with age as one would expect and that flows driven by luminous

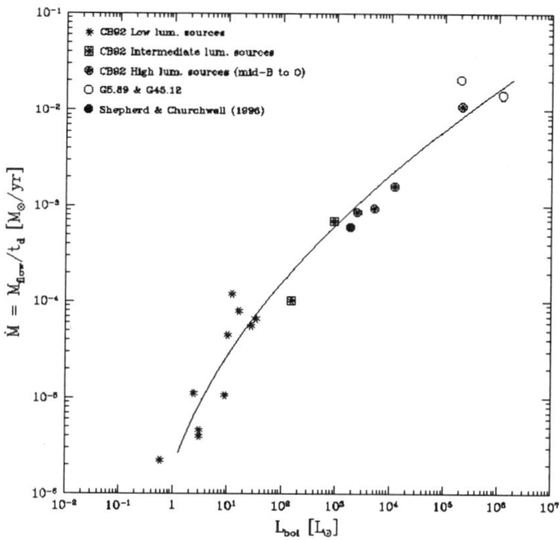

Figure 13. The mass outflow rate $\dot{M}_f$ versus the bolometric luminosity L_{bol} inferred from integrated far infrared flux densities. This figure is a compilation of data from several sources as described in SC96. A tight and continuous relationship between $\dot{M}_f$ and L_{bol} exists from 1 to 10^6 solar luminosities.

stars are substantially more massive than those driven by lower luminosity stars of the same age. For example, the Orion IRC2 outflow has almost 10^3 times more mass than that of L1448C even though the L1448C outflow is about 10 times older than the Orion IRC2 outflow.

8.3. ISSUES RAISED BY MASSIVE OUTFLOWS

The large masses and energetics associated with young molecular outflows from forming massive stars raise some very puzzeling questions. Among these are: How can several tens of solar masses of gas be accelerated to supersonic speeds on timescales of $\sim 10^4$ years and remain cold and neutral?; What supplies the momentum to the outflows?; What is the origin of the mass in the outflows?; What is the driving mechanism of the outflows?; and, Why is there no clear evidence of the outflow in the morphologies of UC HII regions? That is, why does one not see obvious disruptions in the shells of cometary or shell-like HII regions where the outflow has broken through the ionized shell. A good example of this is the shell-like UC HII region G5.89 with which a massive and very energetic molecular outflow has been identified (Harvey & Forveille 1988; Acord, Walmsley, & Churchwell 1997). This HII region is shown in Figure 14. Although there is a low intensity extension of the nebula in the SE-NW direction, there is no evidence that the dense, inner, bright shell has been disrupted by a bipolar outflow.

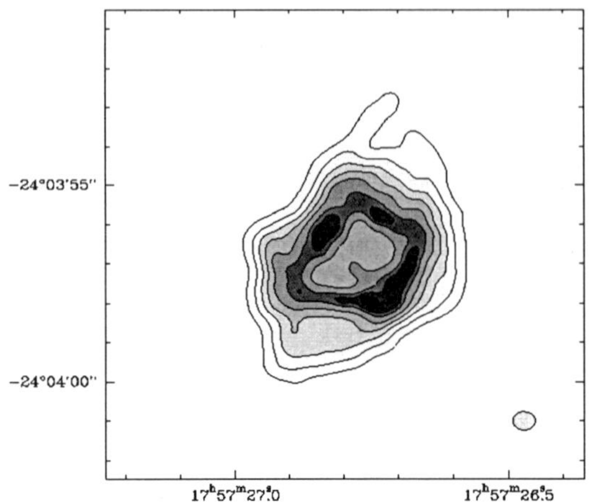

Figure 14. A 3.6 cm VLA image of the shell-like UC HII region G5.89-0.39 from ACW98. The synthesized beam (HPBW=0.57"x0.47") is shown in the lower right corner. The coordinates are RA and Dec (1950) and the coutours are similar to those given in Fig. 1 of ACW98.

A possibility is that the outflow does not originate from the ionizing star of the HII region but from another massive star at an earlier evolutionary age which has not been able to form a detectable HII region yet because of rapid accretion of matter onto the star. Another is orientation of the outflow relative to the line of sight. If the outflow were aligned close to the line of sight, one would not necessarily see evidence of this in the morphology of the HII shell. It is unlikely that this explanation holds for all shell-like or cometary nebulae, however. In the following section, I will consider possible origins of the large outflow masses and discuss their implications.

8.4. ORIGIN OF THE MASS IN MASSIVE OUTFLOWS

Churchwell (1997b) noticed that outflow masses are often more massive than the mass of the central protostar believed to be responsible for driving the outflow (inferred from Fig. 13 in conjunction with the mass-luminosity relation). Possible mechanisms to achieve such massive outflows were discussed by Churchwell (1997b) which I will review here with some updating and illustrations. Four possible mechanisms are: accumulated stellar winds; entrained interstellar medium in bipolar jets; swept-up interstellar medium; and, in-falling matter deflected into bipolar outflows.

8.4.1. *Accumulated Stellar Winds*

The idea is illustrated in Figure 15 where a high mass flux, bipolar stellar wind is blown away from the protostar in directions perpendicular to the

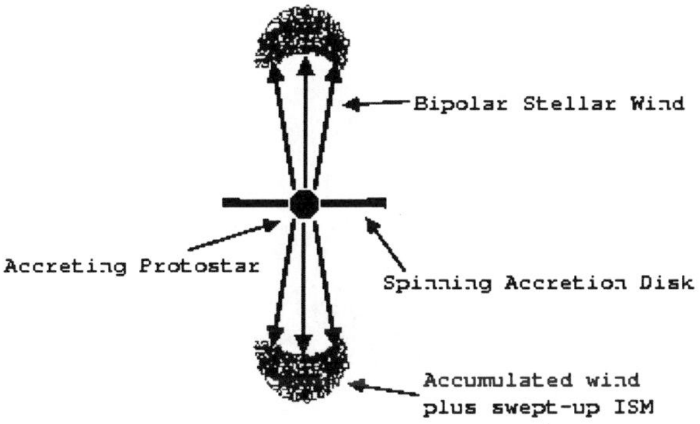

Figure 15.   A schematic illustrating accumulation of stellar wind material from a polar wind. A segment of the accretion disk is shown edge-on.

accretion disk. The accumulated wind material plus whatever ISM has been accelerated by the wind is indicated at the ends of the outflows.

The winds from O and B stars are radiation driven (Abbott 1985; Lamers & Groenewegen 1990; Pauldrach *et al.* 1990; Castor 1993). For stellar winds to supply an outflow of 50 $M_\odot$ over a typical period of 10^4 yr, the average mass loss rate would have to be $\sim 5x10^{-3} M_\odot$ yr^{-1}. This is about two orders of magnitude greater than the highest mass loss rate observed toward O-stars and WR stars. A mass loss rate of this magnitude, if it is radiation driven, cannot be greater than $(L_*/c)/v_\infty$ where L_* is the luminosity of the star, c is the speed of light, and v_∞ is the wind terminal velocity. This requires that the wind terminal velocities must be in the range 0.4 to 4 km s^{-1} for $L_* = 10^{5-6}$ $L_\odot$. Such velocities are several times lower than observed molecular outflow velocities.

Mass accumulations could be greater if the dynamical lifetimes (determined from the outflow lobe separation divided by the outflow speed) under-estimates the actual ages possibly because of confinement by ambient pressure. Inclination of the outflow axis to the line of sight foreshortens the lobe separation and the flow velocity in the same way so τ_d is independent of inclination effects. The typical pressure applied by massive outflows is 4-5 orders of magnitude greater than that of the ambient ISM. It therefore seems unlikely that the ambient pressure can significantly increase the lifetime beyond that implied by the dynamical age. The effective radiative momentum can be increased, and therefore mass-loss rates increased, if multiple scattering is important. Lucy & Abbott (1993) and Gayley, Owocki, & Cranmer (1995) have shown that multiple scattering can increase the mass-loss rate by as much as a factor of 10 in extreme cases (factors ≤ 3 are more typical), but only in a spherically symmetric geometry and only

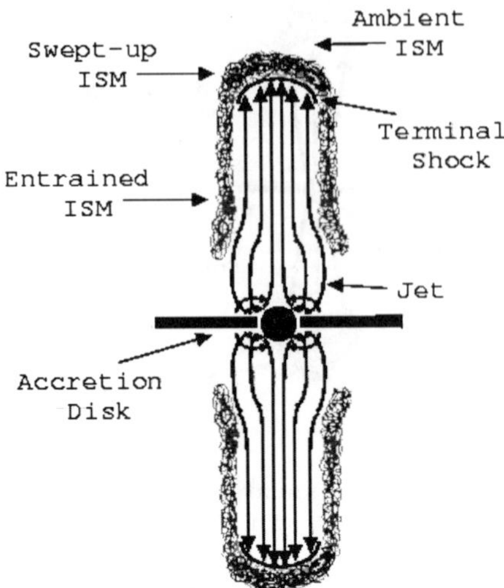

Figure 16. A schematic of entrained plus swept-up interstellar matter along the interface of a bipolar jet with the ambient interstellar medium. A segment of the accretion disk is shown edge-on with arrows showing matter going from the disk to the star and some arrows indicating a flow of matter into the jet from both the star and the inner accretion disk. It is not known where the jets actually originate (i.e. disk, star, or both).

when comoving opacities are large. In a bipolar outlfow, multiple scattering is unlikely to be important because of scattering losses perpendicular to the flow axis. Thus, neither increases in τ_d by ambient pressure nor multiple scattering appears to be capable of increasing the accumulated outflow masses by large enough factors to account for the observed values. The fact that $\dot{M}_f \propto L_{bol}^{0.7}$ (Fig. 13) also indicates that a wind cannot drive the outflows since radiatively driven winds have $\dot{M}_w \propto L_{bol}^{1.7}$. The outflow masses are unlikey to be the result of accumulated stellar winds plus swept-up ISM, because they do not have enough radiative momentum.

8.4.2. *Entrained ISM in Stellar Bipolar Jets*

We consider here a highly collimated, high speed, bipolar jet ejected along the spin axis of a rapidly accreting, massive protostar. It is not known whether the jets originate from the inner disk, the protostar, or both. In this scenerio, a mixing layer at the interface of the jet walls with the ambient ISM entrains matter and accelerates the entrained matter in the direction of the jet flow. Also, some ISM will be accelerated ("swept-up") at the leading shock fronts ("working surfaces") of the jets. This scenerio is sketched in Figure 16.

The issue here is how much ISM can be entrained in a bipolar stellar jet. The mass entrainment rate per unit area is

$$\dot{M}_e = \epsilon \rho_o c_o \tag{2}$$

(Cantó and Raga 1991) where ϵ is the entrainment efficiency, ρ_o is the ambient ISM density, and c_o is the speed of sound in the ambient ISM. For T = 20-50 K and $N_{H_2} \approx 10^5$ cm^{-3},

$$\dot{M}_e = \epsilon(0.9 - 1.4)x10^{-14}gm\ cm^{-2}\ s^{-1} \tag{3}$$

If we take the central flow in G75.78NE as an example (see Shepherd, Churchwell, & Willner 1997), we have a lobe separation of 0.69 pc and an age of $3.7x10^4$ yr. To estimate its area, an approximate cone shape is assumed. For a flow mach number in the range 10-20, the jet opening angle is expected to be $\theta = 2/\sigma \approx 0.04$ rad=2.3^0 from experimental results for a "mixing layer limited" entrainment regime (see fig. 2 in Cantó and Raga 1991). In the above, σ is the "spreading parameter" which is inversely proportionaly to the opening angle. This gives a total surface area (both lobes) of about $1.5x10^{35}$ cm^2 and an entrained mass

$$M_f = R_e A_{flow}\tau_d \approx \epsilon(0.8 - 1.2)M_\odot \tag{4}$$

where A_{flow} is the area of the turbulent mixing layer of the jet and τ_d is the dynamical lifetime of the outflow. In the case of stellar jets with temperature $T_j \approx 10^4$ K and an ambient temperature $T_o < 100$ K, Cantó and Raga (1991) find that the "mixing layer limited" regime is appropriate and that the entrainment efficiency in this regime is $\epsilon < c_o/2c_j$ where c_o and c_j are the sound speed in the ambient medium and the jet, respectively. For the above conditions, $\epsilon < 0.1$ and the total mass of entrained ISM would be $M_f < 1M_\odot$. One could increase the entrained mass by assuming that the stellar mass loss occurs not as a narrow jet but as a flow of matter with much larger opening angles. For example, if $\theta = 30^0$, $M_f \approx \epsilon(13-20)M_\odot \leq 2M_\odot$. Thus, even in the large opening angle case, one can only entrain a few solar masses in outflows; certainly nothing like the several tens of solar masses observed.

Three caveats are in order here. One, the entrained mass is proportional to the third power of the lobe separttion. An under-estimate of this parameter would increase the estimated entrained mass. On average this effect would under-estimate the lobe separation by a factor of $\sqrt{2}$, which results in an average under-estimate of M_f by a factor of $\sim 35\%$, not enough to account for the observed masses. Two, this whole analysis depends on the entrainment efficiency ϵ being small relative to unity. The conclusion of Cantó and Raga (1991) that ϵ is small is based on experimental data and theoretical arguments. If, however, the experimental data are not applicable

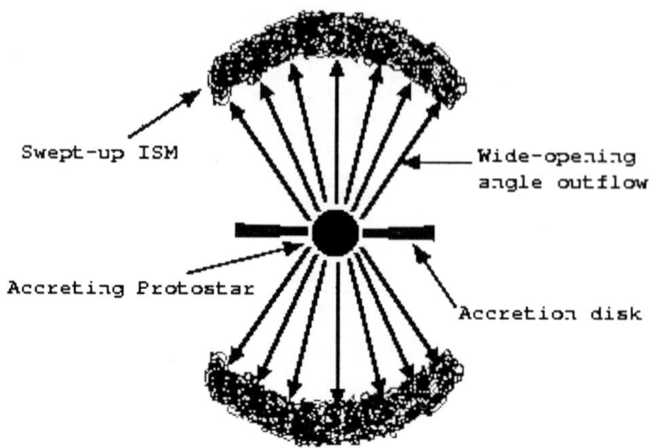

Figure 17.   A schematic of swept-up interstellar matter due to a wide-angle outflow with a farily continuous terminal shock surface.

to the conditions in the ISM, then the entrained mass could be substantially greater. Three, there is clearly evidence for entrainment of ISM in low-mass outflows and one will surely find similar evidence in massive outflows when the data become precise enough to measure it. However, the masses involved in low-mass outflows are typically a small fraction of a solar mass. Such masses are easy to account for by entrainment. It is much more problematic to account for 10s to 100s of solar masses by entrainment if Cantó and Raga (1991) are correct. However, caution is in order because it is not absolutely clear that the experimental data on which their conclusions rest are applicable to the ISM.

8.4.3. *Swept-Up ISM*

It is also possible to accelerate a large mass of ambient ISM into motion along an outflow axis if: 1) the ISM density surrounding a forming star is large; 2) the opening angle of the outflow is large (i.e. has a large working surface); and, 3) the interface of the outflow lobe (working surface) with the ISM is a reasonably continuous surface. Under these conditions, the outflow lobes could act like pistons or snow plows which sweep-up ISM along their leading shock fronts rather than sweeping aside the ISM as narrow jets tend to do. This scenerio is sketched in Figure 17.

ISM densities are large toward MSF regions (Cesaroni *et al.* 1991; Churchwell, Walmsley, & Wood 1992; Cesaroni, Walmsley, & Churchwell 1992; Hofner *et al.* 1995; and others). Thus the main issues are whether the opening angles are systematically large for MSF and if the interfaces of the outflows with the ISM indeed form reasonably impervious surfaces or are

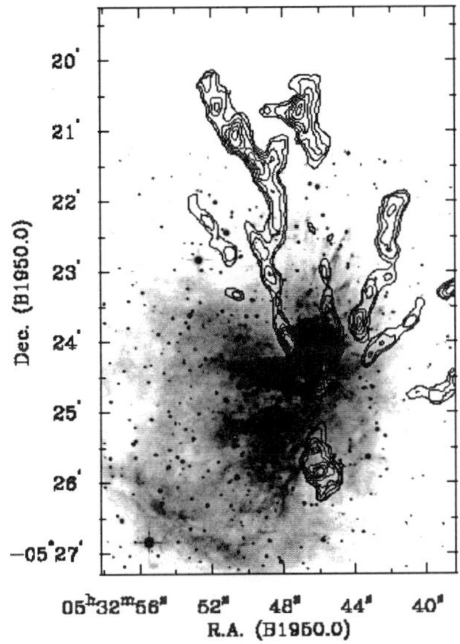

Figure 18. An overlay of NH_3 streamers (contours) from Wiseman and Ho (1998) onto the Allen & Burton (1993) shocked 2 μm H_2 outflow from Orion IRC2. This shows that all the streamers seem to be pointing toward the origin of the outflow and possibly shaped by the outflow flowing past the dense molecular clumps.

leaky pistons. None of the highest resolution data currently available (for G75.78, SCW97; G45.12, HPM97; and G192.16, SWSC98) indicate well collimated outflows. The opening angles of the outflows in all three of these regions seem to be several tens of degrees. Unfortunately, all these sources are $\geq$2 kpc away and the best millimeter interferometer resolutions do not match that of the best images toward Orion. Therefore, let us examine the Orion IRC2 shocked H_2 outflow (shown in Fig.13) as a possible prototype of an outflow from a massive protostar. First, it is clear that this outflow has a wide opening angle. Second, the interface with the ISM is not a smooth surface. In fact, Wiseman and Ho (1996;1998) have imaged NH_3(1,1 and 2,2) emission toward the outflow lobes of Orion IRC2 and find long filaments ($\geq$ 0.5 pc) of dense ammonia gas elongated parallel to the outflow axis (see Figure 18). These filaments are density enhancements as

confirmed by $^{12}C^{18}O$ observations. They find a systematic change in velocity and temperature of the filaments with distance from the origin. These results are strongly suggestive that the elongation of the filaments is a result of the outflow stretching the dense molecular core condensations out into long filaments as the flow passes around the dense condensations and strike them on both sides from acute angles because of the wide opening angle of the outflow. It appears that rather than sweeping-up dense ISM components, the flow simply sweeps around them while stretching them out into long filaments aligned with the outflow axis and possibly superimposing a weak velocity gradient on the filaments. Thus, while the data suggest that there is a strong interaction of the outflow with the ISM, it is not a simple "snow plow" type interaction, but rather a very leaky snow plow that is not particularly effective at sweeping-up large masses of ISM. Much more work needs to be done before one can make a definitive statement on this issue. From the currently available data one can only get an impression of what might be happening, but there is certainly not enough data to be more quantitative or more certain about how effective wide-angle, massive outflows might be in sweeping-up large amounts of ISM.

8.4.4. *Accretion Driven Outflows*

Protostars, or more precisely their accretion disks, must shed angular momentum in order to accrete matter. The discovery of optical jets and radio bipolar molecular outflows associated with YSOs suggests that this is how they shed angular momentum and permit stars to form. The process by which accreting matter is diverted into outflows is not understood, especially for massive stars. Several models have been proposed (Pudritz 1988; Shu 1991, 1995; Shu *et al.* 1991, 1994; Fiege & Henriksen 1996a,b; Henriksen & Valls-Gabaud 1994; and others). In particular, Fiege & Henriksen (1996a,b) have developed protostellar bipolar outflow models in which infalling material is diverted into bipolar outflows due to high central pressures produced by radiative and/or dissipative heating. They find that these models are consistent with radio observations of molecular outflows. Tomisaka (1998) has developed a collapse-driven outflow model which achieves a mass outflow rate of $\sim 1/3$ the accretion rate for a Class 0 low-mass protostar. Since entrainment and swept-up mass do not appear to be able to account for the very large observed outflow masses, it seems likely that deflection of infalling matter into bipolar outflows may be an important (perhaps primary) mechanism that feeds outflows of massive YSOs. This is illustrated in Figure 19.

Here, I will examine some of the consequences of this scenerio. First, it implies that massive YSOs must go through a very rapid accretion phase with a mass accretion rate $\dot{M}_{acc} > \dot{M}_f \sim 10^{-3} \ M_\odot \ \mathrm{yr}^{-1}$. $\dot{M}_{acc}$ must be larger than $\dot{M}_f$ since some of the accreting matter ultimately ends up on the central star. $\dot{M}_{acc}$ also places restrictions on the density structure of the accreting matter. Since the infall cannot exceed the free-fall rate, if gravity is the only inward force, we have:

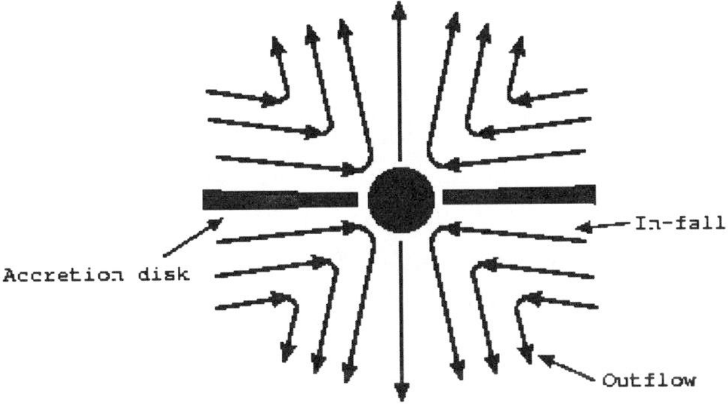

Figure 19.   A schematic of in-falling matter diverted into a broad bipolar outflow. The physical mechanism for achieving this is not understood, although several models have been proposed (see text).

$$\rho(r) \geq \frac{\dot{M}_{acc} r^{-3/2}}{4\pi (2GM_r)^{1/2}} \tag{5}$$

where r is the distance from the central protostar and M_r is the mass (protostar, disk and core material) interior to r. This implies a very rapid increase in density toward accreting protostars, consistent with observations (Hofner *et al.* 1995; Bronfman, Nyman, and May 1996).

Second, the outflow plus stellar mass gives the <u>minimum</u> mass that a molecular cloud core must have to form a star of a given mass. For example, the central outflow in the G75.78NE region is apparently driven by an early B-star (M_* about 16 $M_\odot$) and it has an outflow mass of about 58 $M_\odot$. Thus the fraction of the infalling matter actually incorporated into the star is only about 15% and about 85% is deflected back outward into the outflow. In this sense, the efficiency of getting matter onto the central star is only about 15% and one concludes that a cloud core with a mass at least 6-7 times that of the final star is required to form the star. But we know that massive stars only form in clusters, so the initial cloud mass is probably $\geq$100 times that of the most massive star. This may also be part of the reason why massive stars are so rare; typical cloud cores generally do not achieve such large masses.

Third, accretion rates of 10^{-2} to 10^{-3} $M_\odot$ yr^{-1} will result in most of the stellar UV radiation being absorbed close to the central star, thus delaying the formation of a detectable UC HII region until after this phase of star formation. For example, a density greater than 10^8 cm^{-3} of in-falling matter toward an O5 star would result in a Strömgren sphere of radius $\leq$ 100 AU. This implies that the massive outflows observed to date are likely to be associated with YSOs that have not yet formed a detectable UC HII region. Those that appear to be driven by the central star of a UC HII region, such as G5.89, may either be relics of an earlier rapid accretion phase and no longer being driven or are misidentified. This has the further implication that detection of massive YSOs in their rapid accretion phase or earlier can only be achieved either via their molecular outflows or their thermal dust emission at FIR to mm wavelengths or high excitation molecular line probes, but not from radio free-free emission.

9. SUMMARY

A very brief review of the established properties of MSF regions was given, followed by an abbreviated description of six models that have been proposed to explain the morphologies, and/or ages, and/or dynamics of UC HII regions. The proposed models are: champagne flow or "blister models", in-fall models, photo-evaporating disk models, pressure confined UC HII regions, stellar wind supported bow shock models, and mass-loaded stellar winds. Each of these models predict different kinematics of the ionized gas and are therefore distinguishable, but relevant data are available for a very small number of nebulae. The angular expansion rate for one UC HII region, G5.89-0.39, has been measured (ACW98) to be 4 ± 1 mas yr^{-1}. Combined with an expansion rate of ~35 km s^{-1}, estimated from radio recombination lines, a distance of $2^{+0.7}_{-0.4}$ kpc, independent of a galactic rotation model, was derived. The expansion age of this nebula is <1000 yr.

Because dust converts all radiation from a MSF region to far infrared (FIR) radiation, UC HII regions and their associated stellar cluster are among the most luminous objects in the Galaxy at FIR wavelenghts. Their emission generally peaks at ~100 μm. Models of the spectral energy distributions of UC HII regions indicate that the dust resides in a shell surrounding the central star. A dust evacuated cavity of radius $\sim10^{16}$ cm is required by the models to avoid too much near infrared emission. The average dust temperatures at the inner face of the dust shell are typically ~300 K and they drop very rapidly in the inner 20% of the shell radius and then decrease more slowly in the remaining 80% of the dust shell. The 9.7 μm silicate feature is observed to be in absorption and is often very strong and broad.

Hot molecular cloud cores are heated by the ionizing stars of UC HII regions and their associated cluster members. These cores are dense, hot, compact, and dynamic. Typical properties are: densities $\sim10^6$ cm^{-3}, temperatures ~100-200 K, masses ≥1000 $M_\odot$, and diameters ~0.5 pc. The densities and temperatures are not this high over the entire cloud core, just in the vicinity of the UC HII regions. Water and methanol masers are generally seen toward MSF regions. High resolution observations show that water maser spots are often found in linear structures. It is speculated that this may reflect the tendency of a collapsing cloud to form flattened disk-like structures which would produce linear structures when seen edge-on. This idea is also supported by the temperature profiles of hot cores which are well fit by disk-like models but poorly fit by spherical models.

Hard X-ray emission (up to 8 kev) has been detected toward one MFS region (W3). A model fit to the observed X-ray spectrum indicates a temperature $\geq 70x10^6$ K which is postulated to be associated with hot, stellar wind-shocked cavities immediately surrounding the ionizing stars of UC HII regions at the core of the W3 MSF complex. The 6.7 kev line of FeXXV has also been detected probably from the same gas responsible for the continuum X-ray emission. Presumably this is a general phenomenon associated with most if not all MSF regions, although only one has been observed to date.

During the past several years, massive molecular outflows have been detected toward MSF regions. The main conclusions regarding this phenomenon are as follows. Molecular outflows are found toward $\geq 50\%$ of MSF regions. Typical outflow properties are: Mass outflow rates are in the range 10^{-2} to 10^{-4} $M_\odot$ yr^{-1} ($\dot{M}_f \propto L_{bol}^{0.7}$); outflow masses are in the range of 10s to 100s of solar masses; outflow ages are typically $\sim 10^4$ yr; collimation seems to be poor; opening angles seem to be wide (several 10s of degrees); and the energetics of massive outflows are ≥ 10 times those of low-mass outflows Massive outflows raise the interesting question of how several 10s of solar masses of material can be accelerated to supersonic speeds (relative to the ambient medium) over time scales $\leq 10^4$ yr and remain neutral (molecular) and cool. Possible origins of the mass in massive outflows associated with MSF were considered. It was shown that accumulated stellar winds cannot provide the momentum nor the mass observed. It does not appear possible to entrain 10s to 100s of solar masses of material into outflows if the Cantó and Raga (1991) analysis is correct. Also, the Orion IRC2 outflow seems to indicate that rather than sweeping-up large masses of ISM, outflows simply sweep by dense condensations of ISM in their paths. These arguments seem to favor the possibility that massive outflows may be accretion driven, but do not by any means constitute proof of such.

Questions raised in this review regarding molecular outflows will require much more observational underpinning to achieve further progress in their resolution. In particular, details of the outflow morphologies, their velocity structures, and their relationship to accretion disks need to be studied in much more detail.

ACKNOWLEDGEMENTS I thank Tim Drexler who helped me with the sketches for this review and Dorothy Churchwell who proofed the text for spelling and grammatical errors. I acknowledge partial support from NSF grant AST-9617686.

References

Abbott, D. C. 1985, in *Radio Stars*, eds. R. M. Hjellming & D. M. Gibson, Reidel, Dordrecht, p. 61

Acord, J. M., Churchwell, E., Wood, D. O. S. 1998, ApJ Lett., 475, L107 (ACW98)

Acord, J. M., Walmsley, C. M., Churchwell, E. 1997, Ap. J., 475, 693

Allen, C. W. 1973, *Astrophysical Quantities*, 3rd ed., The Athlone Press, Univ. London

Allen, D. A., Burton, M. G. 1993, Nature, 363, 54

Aspin, C., Walther, D. M. 1990, A&A, 235, 387

Bachiller, R., Cernicharo, J. 1990, A&A, 239, 276

Becker, R. H., White, R. L., Helfand, D. J., Zoonematkermani, S. 1994, ApJS, 91, 347

Bronfman, I., Nyman, L.-., May, J. 1996, A&A Suppl., 115,81

Cabrit, S., Bertout, C. 1992, A&A, 261, 274

Cant, J., Raga, A. C. 1991, Ap. J., 372, 646

Castor, J. I. 1993, in ASP. Conf. Proc. 35, *Massive Stars: Their Lives in the Interstellar Medium*, eds. J. P. Cassinelli & E. B. Churchwell, San Francisco, PSP, p.297

Cassinelli, J. P., Olson, G. L. 1979, ApJ, 229,304

Cesaroni, R., Churchwell, E., Hofner, P., Walmsley, C. M., Kurtz, S. 1994, Ast. Ap., 288, 903

Cesaroni, R., Felli, M., Walmsley, C. M., Olmi, L. 1997, A&A, 325, 725

Cesaroni, R., Hofner, P., Walmsley, C.M., Churchwell, E. 1998, Ast. Ap., 331, 709

Cesaroni, R., Walmsley, C. M., Churchwell, E. 1992, A&A, 256, 618

Cesaroni, R., Walmsley, C.M., Kmpe, C., Churchwell, E. 1991, A&A, 252, 278

Chernin, L. M. 1996, ApJ, 460,711

Chernin, L. M., Wright, M. C. H. 1996, 467, 676

Choi, M., Evans, N. J. II, Jaffe, D. T. 1993, Ap. J., 417, 624

Chini, R., Krgel, E., Kreysa, E. 1986, A&A, 167, 315

Churchwell, E. 1990, A&A Rev., 2, 79

Churchwell, E. 1991, in *The Physics of Star Formation and Early Stellar Evolution*, eds. C. J. Lada and N. D. Kylafis,Kluwer Acad. Pub., pp. 221-268

Churchwell, E. 1995, Ast. Sp. Sci., 224, 157

Churchwell, E. 1997a, in *Herbig-Haro Flows and the Birth of Low Mass Stars*, IAU Symp. No. 182, eds. B. Reipurth and C. Bertout, Kluwer Acad. Pub., Dordrecht, pp. 525-536

Churchwell, E. 1997b, ApJ Lett., 479, L59

Churchwell, E., Walmsley, C.M., Wood, D.O.S. 1992, A&A, 253, 541

Churchwell, E., Wolfire, M., Wood, D. O. S. 1990, ApJ, 354, 247 (CWW)

De Pree, C. G., Rodriguez, L. F., Goss, W. M. 1995, Rev. Mex. de Ast. y Astrofis.,31, No. 1, 39

De Pree, C. G., Goss, W. M., Gaum, R. A. 1998, ApJ, 500, 847

Dickel, H. R., Goss, W. M., Condon, G. R. 1996, ApJ, 460, 716

Downes, D., Genzel, R., Becklin, E. E., Wynn-Williams, C. G. 1981, Ap. J., 244, 869

Dyson, J. E. 1994, in *Lecture Notes in Physics*, Vol 431, ed. T. P. Ray and S. V. W. Beckwith, Berlin: Springer-Verlag, p93

Dyson, J. E., Williams, R. J. R., Redman, M. P. 1995, MNRAS, 277, 700

Erickson, N. R., Goldsmith, P. F., Snell, R. L., Berson, R. L., Huguenin, G. R., Ulich, B. L., Lada, C. J. 1982, Ap. J. Lett., 261, L103

Faison, M., Churchwell, E., Hofner, P., Hackwell, J., Lynch, D. K., Russell, R. W. 1998, ApJ, 500, 280

Fey, A. L., Gaume, R., Claussen, M. J., Vrba, F. J. 1995, ApJ, 453,308

Fiege, J. D., Henriksen, R. N. 1996a, MNRAS, 281, 1038

Fiege, J. D., Henriksen, R. N. 1996b, MNRAS, 281, 1055

Fischer, J., Sanders, D. B., Simon, M., Solomon, P. M. 1985, Ap. J., 293, 508
Forster, J. R., Caswell, J. L, Okamura, S. K., Hasegawa, T., Ishiguro, M. 1990, A&A, 231, 473
Garay, G., Lizano, S. , Gomez, Y. 1994, ApJ, 429, 268
Garden, R. P., Geballe, T. R., Gatley, I., Nadeau, D. 1991a, Ap. J., 366, 474
Garden, R. P., Hayashi, M, Gatley, I., Hasegawa, T. Kaifu, N. 1991b, Ap. J., 374, 540
Garden, R. P., Carlstrom, J. E. 1992, Ap. J., 392, 602
Gayley, K. G., Owocki, S. P., Cranmer, S. R. 1995, ApJ, 447, 296
Harvey, P. M., Forveille, T. 1988, A&A, 197, L19
Helfand, D. J., Zoonematkermani, S., Becker, R. H., White, R. L. 1992, ApJS, 80, 211
Henriksen, R. N., Valls-Gabaud, D. 1994, MNRAS, 266, 681
Ho, P. T. P., Haschick, A. D. 1986, ApJ, 304, 501
Hoare, M. G., Glencross, W. M., Roche, P. F., Clegg, R. E. S. 1988, in *Dust in the Universe*, eds. M. E. Bailey and D. A. Williams, Cambridge Univ. Press.
Hofner, P., Churchwell, E. 1996, A&A Suppl., 120, 283
Hofner, P., Churchwell, E. 1997, ApJ Lett, 486, L39 (HC97)
Hofner, P., Kurtz, S., Churchwell, E., Walmsley, C. M., Cesaroni, R. 1995, ApJ, 460, 359
Hollenbach, D., Johnstone, D., Shu, F. 1993, in *Massive Stars: Their Lives in the Interstellar Medium*, eds. J. P. Cassinelli and E. Churchwell, A. S. P. Conf. Series, 35, p26
Hollenbach, D., Johnstone, D., Lizano, S., Shu, F. 1994, ApJ, 428, 654
Hunter, T. R., Phillips, T. G., Menten, K. M. 1997, Ap. J., 478, 283 (HPM97)
Hunter, T. R., Taylor, G. B., Felli, M., Tofani, G. 1994, A&A, 284, 215
Keto, E. R., Ho, P. T. P., Haschick, A. D. 1987, ApJ, 318, 712
Keto, E. R., Ho, P. T. P., Haschick, A. D. 1988, ApJ, 324, 920
Keto, E. R., Ho, P. T. P., Reid, M. J. 1987, ApJ Lett, 323, L117
Koyama, K., Asaoka, I., Kuriyama, T., Tawara, Y. 1992, PASJ, 44, L255
Krgel, E., Mezger, P. G. 1975, A&A, 42, 441
Kurtz, S., Churchwell, E., Wood, D. O. S. 1994, ApJ Suppl, 91, 659
Kwan, J., Scoville, N. Z. 1976, Ap. J., 459, 638
Lada, C. J. 1985, ARA&A, 23, 267
Lamers, H. G. J. L. M., Groenewegen, M. A. T. 1990, in ASP. Conf. Proc. 7, *Properties of Hot Luminous Stars*, ed. C. D. Garmany, San Francisco, ASP, p 189.
Lee, H. M., Draine, B. T. 1985, ApJ, 290, 211
Leung, C. M. 1976, ApJ, 209, 75
Lizano, S., Cant, J., Garay, G., Hollenbach, D. 1996, ApJ, 468, 739
Loren, R. B. 1981, Ap. J., 249, 550
Lucy, L. B. 1982, ApJ, 255, 286
Lucy, L. B., Abbott, D. C. 1993, ApJ, 405, 738
Lumsden, S. L., Hoare, M. G. 1996, ApJ, 464, 272
MacLeod, G. C., Scalise, E. Jr., Saedt, S., Galt, J. A., Gaylard, M. J., 1998, AJ, in press
MacLow, M.-M., Van Buren, D., Wood, D. O. S., Churchwell, E. 1991, ApJ, 369, 395
Massi, M., Felli, M., Simon, M. 1985, A&A, 152, 387
Masson, C. R. 1986, ApJ, 302, L27
Masson, C. R. 1989, ApJ, 336, 294
Mc Caughrean, M. 1993, in *Massive Stars: Their Lives in the Interstellar Medium*, eds. J. P. Cassinelli and E. Churchwell, ASP Conf. Series, 35, 80
Mc Caughrean, M., Stauffer, J. R. 1994, AJ, 108, 1382
Menten, K. M., Reid, M. J. 1995, ApJ Lett., 445, L157
Meyers-Rice, B., Lada, C. J. 1991, Ap. J., 368, 445
Mitchell, G. F., Hasegawa, T. I. 1991, Ap. J., 371, L33
Molinari, S., Testi, L., Brand, J., Cesaroni, R., Palla, F. 1998, ApJ Lett., in press
Panagia, N. 1973, AJ, 78, 929
Pauldrach, A. W. A., Puls, J., Gabler, R., Gabler, A. 1990, in ASP. Conf. Proc. 7, *Properties of Hot Luminous Stars*, ed. C. D. Garmany, San Francisco, ASP, p. 171

Pudritz, R. E. 1988, in *Galactic and Extragalactic Star Formation*, eds. R. E. Pudritz and M. Fich, Kluwer Acad. Pub. Dordrecht, Netherlands, p. 135

Raymond, J. C., Smith, B. W. 1977, ApJS, 35, 419

Redman, M. P., Williams, R. J. R., Dyson, J. E. 1996, MNRAS, 280, 661

Reid, M. J., Haschick, A. D., Burke, B. F., Moran, J. M., Johnston, K. J., Swenson, G. W. Jr. 1981, ApJ, 239, 89

Roelfsema, P. R. *et al.* 1996, A&A Lett., 315, L289

Shepherd, D. S., Churchwell, E. 1995, Ap. J., 457, 267 (SC95)

Shepherd, D. S., Churchwell, E. 1996, Ap. J., 472, 225 (SC96)

Shepherd, D. S., Churchwell, E., Willner, D. J. 1997, ApJ, 482,355 (SCW97)

Shepherd, D. S., Watson, A. M., Sargent, A. I., Churchwell, E. 1998, ApJ, in press (SWSC98)

Shu, F. H. 1991, in *The Physics of Star Formation and Early Evolution*, eds. C. J. Lada and N. D. Kylafis, Kluwer Acad. Pub., Dordrecht, Netherlands, p. 365

Shu, F. H. 1995, Rev. Mex. A. A. (Ser. de Conf.), 1, 375

Shu, F. H., Najita, J., Ostricker, E., Wilkin, F., Ruden, S., Lizano, S. 1994, ApJ, 429, 781

Shu, F. H., Ruden, S. P., Lada, C. J., Lizano, S. 1991, ApJ, 370, 31

Tafalla, M., Bachiller, R., Wright, M. C. H., Welch, W. J. 1997, Ap. J., 474, 329

Thronson, H. A. Jr., Gatley, P. M., Sellgren, K., Werner, M. W. 1980, ApJ, 237, 66

Tielens, A. G. G. M., de Jong, T. 1979, A&A, 75, 326

Tomisaka, K. 1998, ApJ Lett., submitted

Tutukov, A. V., Shustov, B. M. 1981, Soviet Ast., 58, 109

Van Buren, D., Mac Low, M.-M., Wood, D. O. S., Churchwell, E. 1990, ApJ, 353, 570

Watson, A. M., Coil, A. L., Shepherd, D. S., Hofner, P., Churchwell, E. 1998, ApJ, in press

White, R. L., Becker, R. H., Helfand, D. J. 1991, ApJ, 371, 148

Williams, R. J. R., Dyson, J. E., Redman, M. P. 1996, MNRAS, 280, 667

Wiseman, J. J., Ho, P. T. P. 1996, Nature, 382, 139

Wiseman, J. J., Ho, P. T. P. 1998, ApJ, in press

Wolfire, M. G., Churchwel, E. 1994, ApJ, 427,889

Wood, D.O. S., Churchwell, E. 1989a, ApJ S, 69, 831(WC89a)

Wood, D. O.S., Churchwell, E. 1989b, ApJ, 340, 265

Wolf, G. A., Lada, C. J., Bally, J. 1990, A.J., 100, 1892

Wright, M. C. H., Plambeck, R. L., Vogel, S. N., Ho, P. T. P., Welch, W. J. 1983, Ap. J. Lett., 267, L41

Xie, T., Mundy, L. G., Vogel, S. N., Hofner, P. 1996, ApJ, 473, L131

Yamauchi, S., Koyama, K. 1993a, ApJ, 405, 268

Yamauchi, S., Koyama, K. 1993b, ApJ, 404, 620

York, H. W., Tenorio-Taglio, G., Bodenheimer, P. 1983, A&A, 127, 313

Zheng, X. W., Ho, P. T. P., Reid, M. J., Schneps, M. H. 1985, ApJ, 293, 522

Zoonematkermani, S., Helfand, D. J., Becker, R. H., White, R. L. 1990, ApJS, 74, 181

Zuckerman, B., Kuiper, T. B. H., Rodriguez Kuiper, E. N. 1976, Ap. J. Lett., 209, L137

MASERS IN STAR-FORMING REGIONS

N. D. KYLAFIS
University of Crete
Physics Department
P.O. Box 2208
710 03 Heraklion, Crete
Greece

AND

K. G. PAVLAKIS
University of Leeds
Department of Physics and Astronomy
Woodhouse Lane
Leeds LS2 9JT
United Kingdom

1. Introduction

The subject of astronomical masers is now more than forty years old, and as a consequence, several good reviews have appeared in print. For the readers who want to enter the subject of astronomical masers in a serious way, we want to recommend two reviews (Elitzur 1982; Reid and Moran 1988) and a book (Elitzur 1992a). They contain reviews of the observations as well as the basic ideas and equations that will enable the readers to comprehend the research papers on the subject. The most recent review in the Annual Reviews of Astronomy and Astrophysics is that of Elitzur (1992b).

In the present review, we want to avoid the mathematical details but stress instead a qualitative understanding of astronomical masers. Needless to say, some overlap with the above reviews is unavoidable. Section 2 contains the basic concepts regarding masers. In §3 the differences between laboratory and astronomical masers are briefly discussed and in §4 a few examples which demonstrate the usefulness of astronomical masers are presented. Section 5 describes what it takes to build a maser model and how well we know the various input parameters. The observations and proposed

553

C.J. Lada and N.D. Kylafis (eds.), The Origin of Stars and Planetary Systems, 553–575.
© 1999 *Kluwer Academic Publishers. Printed in the Netherlands.*

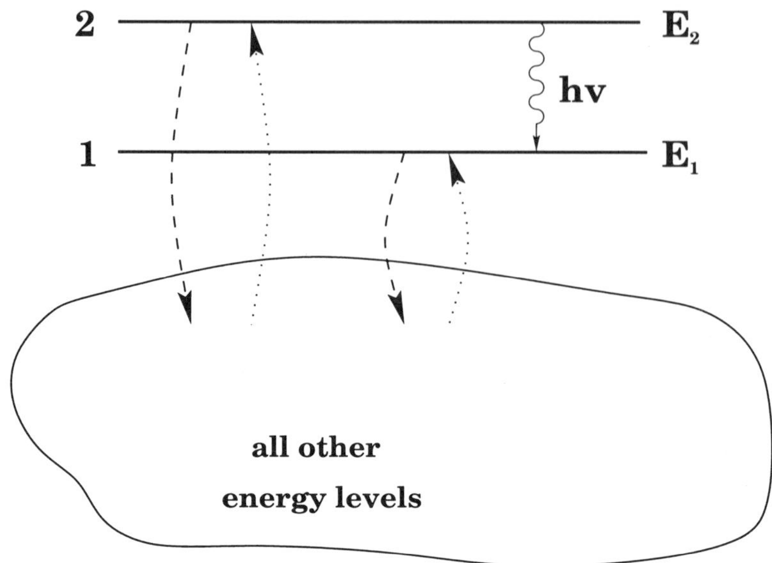

Figure 1.   Schematic energy level diagram of a molecule. Only two levels are explicitly shown. The rest are lumped together and are considered as a reservoir. The dotted and dashed arrows indicate respectively the transitions from the reservoir to the two levels and the opposite.

pumping mechanisms are given in §6 for OH and H_2O masers. The location of OH and H_2O masers in star-forming regions is discussed in §7 and the conclusions are given in §8.

2. Basic Concepts

The energy level diagram of a molecule consists of an infinite number of levels. Nevertheless, one need only consider a finite and relatively small number of energy levels in order to perform an accurate model calculation regarding a specific molecule. Even better, for the purposes of a qualitative discussion of masers, one can consider only two energy levels and treat the rest of them as a reservoir. Figure 1 shows such a simplified model of a molecule.

Level 1 is the lower maser level and level 2 is the upper one. The energy difference between the two levels is $\Delta E \equiv E_2 - E_1 = h\nu$, where ν is the frequency of the maser. Collisional and radiative transitions cause the molecules to circulate between the reservoir and the maser levels. Thus, the dotted lines in Figure 1 indicate the transitions of molecules from the reservoir to the maser levels. The opposite transitions are indicated by dashed lines. Let n_1 and n_2 be the populations *per magnetic sublevel* of the two maser levels. The maser phenomenon occurs if, in steady state, the populations satisfy the condition $n_2 > n_1$. This is because, as a photon of frequency ν propagates, it is more likely to cause stimulated emission than to be absorbed.

2.1. AMPLIFICATION

Consider radiation of frequency ν and intensity I propagating through molecules whose energy level diagram is that of Figure 1. The change dI in the intensity that occurs as the radiation propagates by a length dl through the medium is

$$dI = -\kappa I dl \ , \tag{1}$$

where $\kappa \propto (n_1 - n_2)$ is the absorption coefficient corrected for stimulated emission. Here it is assumed for simplicity that the molecules do not emit radiation by spontaneous emission. Defining the optical depth τ by $d\tau \equiv \kappa dl$, one finds that the intensity as a function of optical depth is given by

$$I = I_0 e^{-\tau} \ , \tag{2}$$

where I_0 is the intensity of a background source that is incoming to the medium. If the populations of levels 1 and 2 in Figure 1 are such that $n_1 > n_2$, then $\tau > 0$ and the radiation is attenuated as it propagates through the medium. This situation is typical in astronomical environments and it leads to the formation of absorption lines. On the other hand, if $n_1 < n_2$, the optical depth is negative and the radiation is amplified. This is because a propagating photon is more likely to cause a stimulated emission than it is to be absorbed. Thus, one photon leads to two propagating in the same direction, two photons to four and so on. In other words, the amplification of the radiation is *exponential*.

As mentioned above, it was only for simplicity that spontaneous emission in the medium was neglected. In reality, such an emission always occurs and the radiation thus produced is also amplified. If there is no background source, as it is the case in most astronomical masers, it is only the spontaneous emission that gets amplified.

2.2. SATURATION

The exponential growth of the radiation intensity in a medium that exhibits inverted populations is truly fascinating. It is therefore important to investigate whether this exponential growth has any limits, assuming that the maser region is long enough. Since a stimulated emission event causes the transition of a molecule from the upper maser level to the lower, it is clear that stimulated emission tends to *reduce* the population inversion. At low intensities, the stimulated emission rate is relatively small and the population difference $n_1 - n_2$ remains essentially unaffected. As the intensity grows, so does the stimulated rate. Thus, a point is reached where the stimulated emission rate becomes comparable to the rate that causes the population inversion. At this point, *saturation* has set in because the rate of extraction of photons has reached its limiting value, which is the pumping rate (the rate of population inversion). Increasing the volume of the emission region under saturation conditions will only increase the photon emission rate in proportion to the volume.

It is important to remark here that the astronomical definition of *saturation* given above is exactly opposite to that used by experimental physicists for laboratory masers and lasers. Experimental physicists call a maser *saturated* when its intesity grows exponentially.

2.3. THERMALIZATION

The establishment of inverted populations in a medium containing a certain kind of molecules is not as difficult as one might think. In fact, it is not an exaggeration to say that if the molecules are abused, they will most likely exhibit inverted populations in *some pair* of energy levels. One cannot easily predict which pair it will be, but inversion in some pair is almost certain. Common methods of abuse of molecules in astronomical environments are a) collisions with other molecules, b) irradiation of the molecules and c) a combination of the above, which is the usual case. As an illustration, let us consider collisions with other molecules (e.g., H_2) as the cause of population inversion. If the density of H_2 molecules is low, the collision rate is low, the rate of population inversion is low and therefore the maser luminosity is correspondingly low. For low rates of abuse, the maser luminosity increases in proportion to the rate of abuse of the molecules. However, this proportionality has an upper limit. When the collision rate becomes very large, the collisional transitions of the molecules dominate all other transitions (see § 5.3 below), the molecules thermalize at the temperature of the H_2 molecules and the population inversion is destroyed. At thermalization, the level populations obey the Boltzmann distribution, namely

$$\frac{n_2}{n_1} = \exp(-\Delta E/kT) < 1 \;, \tag{3}$$

where T is the temperature of the H_2 molecules.

2.4. BEAMING

The majority of the astronomical masers *do not* emit isotropically. This effect can be understood as follows: Consider a maser region which is not spherically symmetric but instead it is elongated in some direction, as shown

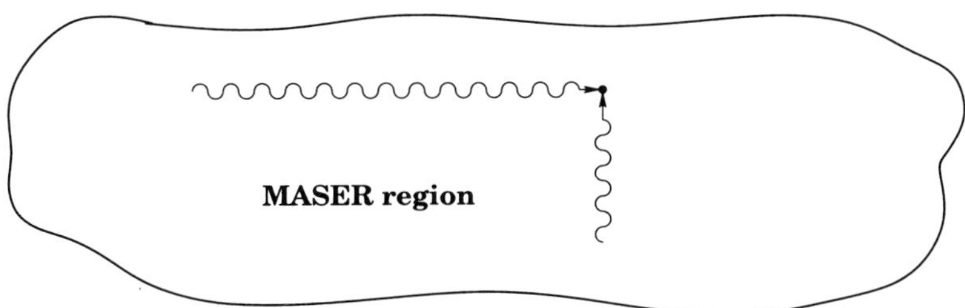

Figure 2. Schematic representation of a maser region. The two wiggly arrows indicate the intensity of the maser in two directions.

in Figure 2. The gain $|\tau|$ along the direction of elongation is larger than in directions perpendicular to it. Consequently, the intensity of the maser is largest along the direction of elongation. Thus, a departure from spherical symmetry introduces an anisotropy in the intensity.

Once an anisotropy in the intensity is established, it is further enhanced through the following process. Consider the maser photons that reach a certain point in the maser region from all possible directions (see Figure 2). These beams of photons compete for the available pump photons there. The rate of stimulated emission caused by a given beam is proportional to its intensity. Therefore, the stronger beam wins and becomes even stronger (Alcock and Ross 1985a,b; 1986). As a result, a non-spherically symmetric maser region results in maser emission in two opposite cones along the direction of elongation.

Radio observers measure the flux of photons that reach their telescope and, not knowing the degree of beaming, make the assumption that the emission is isotropic. Thus, from the observed flux, an equivalent *isotropic* luminosity is quoted for the source. This procedure certainly overestimates the luminosity of a *maser spot*, but it gives the right value when one considers a region that contains many maser spots. This is because the cones of emission of the various spots have random orientations, but we only see the ones that are directed towards the earth.

2.5. GEOMETRY AND APPARENT SIZE

The geometry of a maser region and its apparent size are intimately related. Furthermore, the apparent size of a maser region may be significantly smaller than its actual size. For the purposes of illustration we will consider two shapes of maser region, elongated and nearly spherical.

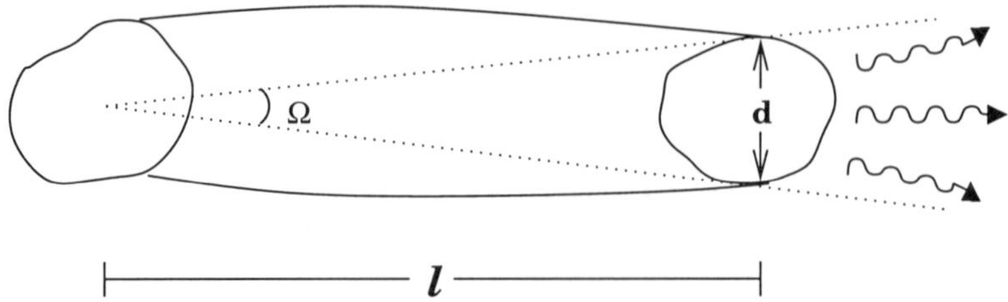

Figure 3.   Schematic of a maser region which is approximately cylindrical. The wiggly lines indicate the emergent maser intensity.

2.5.1. *Elongated Structures*

Consider a maser region which is more or less cylindrical as shown in Figure 3. Let the length of the region be l and its characteristic width be d, with $l \gg d$. As discussed in § 2.4, most of the maser photons escape through the bases of the cylinder and into solid angles $\Omega \sim (d/l)^2$.

If the observational direction falls in this beaming angle Ω, then the observer sees the maser spot and determines an apparent size $\sim d$. If, on the other hand, the observational direction falls significantly outside Ω, the observer *does not* see the maser spot, because the intensity of the maser in the direction of the observer is below the detection limit.

2.5.2. *Nearly Spherical Structures*

Consider now a maser region which is nearly spherical with diameter D, as shown in Figure 4. In this case, all observers see the maser spot as having the same apparent size. However, the apparent size is *less* than D because the rays that come out of the region have traversed different lengths and therefore the maser photons have experienced different gains. It is straightforward to estimate the apparent size of the maser spot once the detection limit of the telescope is given.

2.6. VARIABILITY

All masers, even the very strong ones, are variable (see, e.g., Clegg and Cordes 1991; Lekht, Mendoza-Torres and Sorochenko 1995; Claussen *et al.* 1996). This, of course, is not surprising because even a small change in the gain introduces a significant change in the intensity. Small changes in the gain can be caused either by changes in the aspect angle or by changes in the local physical parameters such as the density and the velocity field. For example, if the maser region is elongated, a small change in the aspect angle may result in a relatively large change in the intensity (see § 2.5.1).

Despite the short timescale of variability, some maser spots can be followed for years. The lifetime of individual spots can be as large as one to

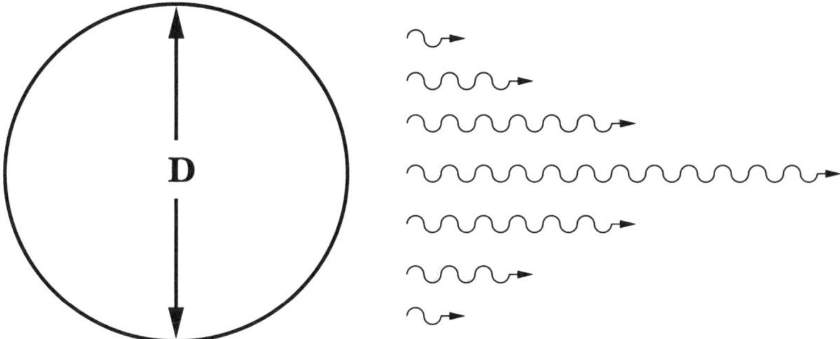

Figure 4. Schematic of a maser region which is approximately spherical. The wiggly
lines indicate the emergent maser intensity.

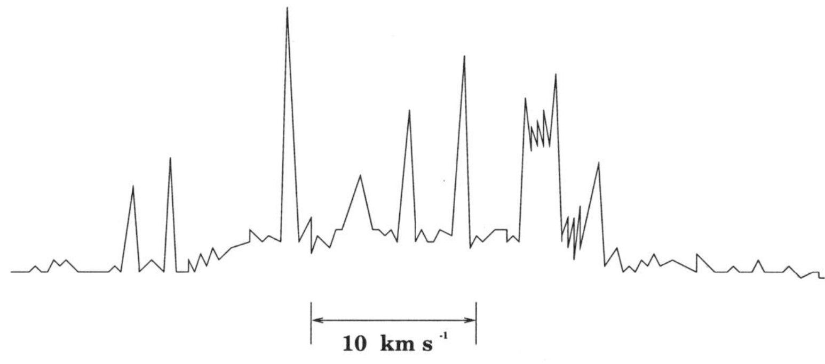

Figure 5. A sketch of a spectrum of H_2O emission in a region of star formation.

two years for H_2O and more than ten years for OH masers. This relatively
long lifetime has allowed studies of proper motion of the maser spots and
through this the determination of their distance (see § 4).

2.7. SPECTRA AND LINE WIDTHS

The maser spectra observed in star-forming regions are fairly complex (see
Genzel and Downes 1977 for a collection of such spectra). A sketch of a
typical spectrum of H_2O emission in a region of star formation is shown in
Figure 5.

H_2O maser spectra in star-forming regions extend over a frequency range
corresponding to a velocity difference of a few tens km s^{-1} (typically less
than 10 km s^{-1} for OH) and consist of many spikes. VLBI maps which resolve the maser region have shown that a spike arises from one or more distinct spots. The line width of a strong spike is typically a few times smaller
than the width expected from the thermal motions of the molecules. This
is expected because unlike a thermal line, a maser line grows exponentially

(in the unsaturated regime) and therefore the full width at half maximum is smaller than thermal. As the line center saturates, the line wings continue to grow exponentially and the line width approaches the thermal one.

The reader should not be left with the impression that unsaturated lines have narrow widths while saturated ones have thermal widths. It has been demonstrated by Goldreich and Kwan (1974) that the broadening of the line during saturation can be inhibited if trapping of the infrared radiation involved in the pumping occurs.

2.8. POLARIZATION

Polarization is fairly common in masers observed in star-forming regions. The OH masers are typically highly polarized while the H_2O ones show polarized features in about 25% of the sources. In order for a maser source to emit polarized radiation, the molecules must not only exhibit inverted populations but also unequal populations in the magnetic sublevels. It has been demonstrated (Goldreich, Keeley and Kwan 1973) that an ordered magnetic field in the maser region can establish unequal populations in the magnetic sublevels and therefore result in polarized emission. The magnetic field required is of order a few milligauss for OH and somewhat larger for H_2O.

The maser polarization would be an ideal way of measuring the magnetic field strength in maser regions if the magnetic field were the only cause of such polarization. As it has been demonstrated (Western and Watson 1983a,b,c, 1984; Deguchi and Watson 1986a,b; Deguchi, Watson and Western 1986; Deguchi and Watson 1990; Nĕdoluha and Watson 1990a,b,c; Elitzur 1991, 1993; Watson 1994; Nedoluha and Watson 1994; Elitzur 1995, 1996; Wallin and Watson 1997), anisotropic radiation fields in the maser region can also cause polarization. Thus, the origin of the observed polarization is not always clear.

Despite the above ambiguity, the unmistakable Zeeman pattern has been seen by Moran *et al.* (1978) for OH and Fiebig and Güsten (1989) for H_2O. These observations have determined the magnetic field strength to be 2 to 9 milligauss in W3(OH) and of order 50 milligauss in Orion KL and W49N.

The Zeeman splitting of OH maser lines can also yield the line-of-sight direction of the magnetic field. Reid and Silverstein (1990) have in this way been able to infer the direction of the magnetic fild in maser sources over large regions of the Galaxy.

3. Laboratory versus Astronomical Masers

Both laboratory and astronomical masers are based on the same principle, namely, some transition exhibits inverted populations. Nevertheless, there are significant differences in the size and the technical characteristics of the two.

In laboratory masers, the maser region is typically ~ 1 m long (to astronomical accuracy). In order to have significant gain, the radiation is bounced between two mirrors, one of which is semi-transparent to allow the amplified radiation to escape. On the other hand, the astronomical masers have significantly lower densities but gain lengths of order a few AU or larger. The absense of reflecting boundaries in astronomical masers makes them single pass as opposed to multi pass in the laboratory ones. Furthermore, the fixed size of the laboratory masers introduces a mode selection, i.e., not all wavelengths can fit in the length of the maser. On the other hand, the astronomical masers allow all possible wavelengths and they are broad band. Finally, laboratory masers emit coherent radiation, while the astronomical ones emit non-coherent. This is because in astronomical masers the beaming angle (see § 2.4) is fairly large and therefore a significant phase difference is introduced between two rays in the beam.

Because the radiative transfer in astronomical masers is relatively simple (see § 5), the main theoretical effort has concentrated on inventing efficient pumping mechanisms and understanding the origin of polarization.

4. Usefulness of Masers

The astronomical masers, because of their unique properties, provide very useful information. Below are listed some of the most important results that have been obtained from maser observations.

As mentioned in § 2.8, when a maser line shows the Zeeman pattern, the magnetic field in the maser region is determined unambiguously. Thus, we now have an accurate knowledge of the magnetic field strength in some OH and H_2O maser regions.

Observations of maser spots have provided detailed information on the dynamics of a region. As an interesting example, we mention the work of Chapman and Cohen (1986, see their Figures 5 and 6), who studied OH amd H_2O maser emission from the circumstellar envelope of the supergiant star VX Sagittarius and inferred the velocity field there.

VLBI proper motion studies of H_2O maser spots have provided a means to determine distances across the Milky way. Thus, the distances to Orion KL and W51(Main) have been determined to be respectively 480 ± 80 pc (Genzel *et al.* 1981a) and 7.0 ± 1.5 kpc (Genzel *et al.* 1981b). The distance to W51(North) is 7 ± 2 kpc (Schneps *et al.* 1981), the distance to Sgr-B2(North) is 7.1 ± 1.5 kpc (Reid *et al.* 1987) and the distance to W49N is 11.4 ± 1.2 kpc (Gwinn, Moran and Reid 1992).

The long-standing problem of locating the position of the compact non-thermal radio source Sgr A* on infrared images of the Galactic center region has been solved with the use of maser observations (Menten et al. 1997).

The astronomical masers can also be useful on earth. The brightest maser sourse in the sky (Orion KL) has been used in a holographic technique to measure the figure of large radio telescopes (Moran 1989).

With the development in maser modeling, we are now in a position to infer physical conditions in star-forming regions from observations of all the maser lines exhibited by a molecule (Kaufman and Neufeld 1996; Pavlakis and Kylafis 1996b,c; 1999).

Submilliarcsecond VLBI mapping of the H_2O maser emission in NGC 4258 has shown that the masers are located in a thin disk around the central black hole (Miyoshi *et al.* 1995). It is the first time that we see structure [probably a warped disk (Herrnstein, Greenhill and Moran 1996)] at a distance of 0.2 pc from the central black hole of an AGN.

5. Maser Models

In what follows, the requirements for the construction of a maser model are discussed and the basic equations that need to be solved are given. Also, a qualitative discussion of how collisional and radiative pumping works is presented.

5.1. REQUIREMENTS

In order to construct a maser model one needs the following:

a) The energy levels of the molecule that exhibits maser emission. These are generally well known.

b) The einstein coefficients A_{ij} for the transitions from any upper level i to any lower level j. These also are well known, at least for the most common molecules.

c) The collision rates C_{ij} for transitions from any level i to any level j of the maser molecule. The colliding particles are the other molecules in the region (e.g., H_2). These rates are only approximately known. Therefore, when maser models are constructed, it should always be kept in mind that the currently published collision rates may be revised.

d) The velocity field in the maser region. This is unknown and it is only guessed in maser models. Two types of velocity field have generally been assumed. In one, the medium is static and the molecules obey a distribution of velocities (typically Gaussian). In the other, the medium exhibits a large velocity gradient, i.e., the macroscopic velocity differences are significantly larger than the thermal velocities of the molecules.

e) The geometry of the maser region is also unknown and in maser models a simple (typically spherical or cylindrical) geometry is assumed.

5.2. BASIC EQUATIONS

The set of equations that need to be solved consists of the statistical equilibrium equations for the level populations and the radiative transfer equations for the various line frequencies. For simplicity, we will assume that the molecules under study have only three, non-degenerate energy levels with energies E_1, E_2 and E_3 and corresponding populations n_1, n_2 and n_3. If only collisional and radiative transitions are taken into account, the steady state equations describing the level populations are

$$\frac{dn_3}{dt} = -(A_{32} + A_{31})n_3 + R_{13}(n_1 - n_3) + R_{23}(n_2 - n_3)$$

$$+C_{13}n_1 + C_{23}n_2 - (C_{32} + C_{31})n_3 = 0 \;, \qquad (4a)$$

$$\frac{dn_2}{dt} = A_{32}n_3 - A_{21}n_2 - R_{23}(n_2 - n_3) + R_{12}(n_1 - n_2)$$

$$+C_{12}n_1 + C_{32}n_3 - (C_{23} + C_{21})n_2 = 0 \;, \qquad (4b)$$

$$n_1 + n_2 + n_3 = N \;, \qquad (4c)$$

where N is the total number of maser molecules and R_{ij} are the radiative absorption rates corrected for stimulated emission. Generalization of these quations to any number of levels is straightforward.

If ν is the frequency of the transition from upper level j to lower level i and I_ν is the corresponding intensity, then the radiative transfer equation is given by

$$\frac{dI_\nu}{dl} = -\kappa_\nu I_\nu + \eta_\nu \;, \qquad (5)$$

where κ_ν is the absorption coefficient (which is a function of ν and $n_i - n_j$) and η_ν is the emissivity (which is a function of ν and n_j).

The system of equations (4) and (5) must be solved simultaneously (Goldreich and Keeley 1972; Elitzur 1990a,b). In the case where there is a large velocity gradient in the medium, equations (4) and (5) are decoupled (Sobolev 1960; Castor 1970). For a complete formalism in this case see, for example, Kylafis and Norman (1991).

5.3. COLLISIONAL AND RADIATIVE PUMPING

Collisional pumping works as follows: The maser molecules are excited by collisions with the other particles (typically H_2) in the maser region. Their de-excitation is done both radiatively and collisionally. If the radiative de-excitations dominate the collisional ones, it is likely that some transition will exhibit inverted populations. However, when the collisions dominate both excitations and de-excitations, the molecules thermalize and the population inversion is destroyed. Equations (4), with only collision terms, yield the Boltzmann distribution for the level populations.

Radiative pumping works in a similar way. The excitation of the molecules is done with infrared radiation from an external source. The de-excitation is done via spontaneous and stimulated emission. In this "flow" of the molecules from level to level, it is likely that inverted populations will be established in some transition. If the external radiation dominates in the de-excitation, thermodynamic equilibrium is established at the temperature of the external radiation.

6. Masers in Star-Forming Regions

Maser emission is a common phenomenon in star-forming regions. The best studied masers in such regions are the OH and H_2O ones. In recent years, the methanol and the ammonia masers have received a lot of attention. The SiO masers are rare (they have been seen only in Orion, W51 and Sgr B2), while other molecules exhibit fairly weak maser emission.

Here we will concentrate on the OH and the H_2O masers in star-forming regions.

6.1. OH MASERS

The energy levels of OH are grouped into two ladders, the $^2\Pi_{3/2}$ and the $^2\Pi_{1/2}$ (see, e.g., Figure 1 of Pavlakis and Kylafis 1996b for an energy level diagram). In each ladder, the energy levels are characterized by the total angular momentum J and are split into Λ-doublets. Each component in a given Λ-doublet is further split into two hyperfine levels. Thus, there are four allowed ground state tansitions and many excited ones in the $^2\Pi_{3/2}$ and $^2\Pi_{1/2}$ ladders.

The OH molecules in star-forming regions are rich in maser emission. They exhibit four maser lines (at 18 cm) in the ground state $^2\Pi_{3/2}$, $J = 3/2$ (e.g., Gaume and Mutel 1987; Cohen, Baart and Jonas 1988), three maser lines (at 5 cm) in the first excited state $^2\Pi_{3/2}$, $J = 5/2$ (Knowles, Caswell and Goss 1976; Guilloteau et al. 1984; Smits 1994; Caswell and Vaile 1995; Baudry et al. 1997; Desmurs et al. 1998; Desmurs and Baudry 1998), three maser lines (at 6 cm) in the next level $^2\Pi_{1/2}$, $J = 1/2$ (Gardner and Martin Pintado 1983; Gardner and Whiteoak 1983; Palmer, Gardner and Whiteoak 1984; Gardner, Whiteoak and Palmer 1987; Baudry et al. 1988; Baudry and

Diamond 1991; Cohen, Masheder and Walker 1991) and one maser line (at 2 cm) in the level $^2\Pi_{3/2}$, $J = 7/2$ (Turner, Palmer and Zuckerman 1970; Baudry et al. 1981; Baudry and Diamond 1998).

The four transitions in the ground state are the best known ones. They have wavelengths approximately 18 cm and frequencies 1665, 1667 MHz (main lines) and 1720, 1612 MHz (satellite lines).

6.1.1. *Observations*

From the extensive observational study of star-forming regions by Gaume and Mutel (1987), the following observational characteristics have appeared.

a) All four ground-state transitions of OH have been seen as masers (not all four, though, from the same maser spots).

b) The main-line masers at 1665 and 1667 MHz are usually the most common masers with the intensity of the 1665 MHz line typically larger than the intesity of the 1667 MHz line.

c) Most (but not all) of the 18 cm masers are found at the projected edges of H II regions.

d) The observations *do not support* a simple shell model for the OH maser distribution.

e) VLBA observations of the excited-state 13.4 GHz maser in W3(OH) with spatial resolution 670 μarcsec (Baudry and Diamond 1998) reveal that no simple geometry exists for the individual maser spots. Perhaps they are related with local shocks.

f) No correlation has been found between the fluxes of individual features and their distances from the centers of the H II regions.

g) Proper motion studies (Bloemhof, Reid and Moran 1992; Migenes, Cohen and Brebner 1992; Bloemhof, Moran and Reid 1996) show motion *away* from the central H II regions.

For some time, the observations of OH masers were way ahead of the theoretical interpretation. For example, in W3(OH) *all* radio transitions between rotational states below 510 K have been observed (for a summary see Table 1 of Cesaroni and Walmsley 1991). They include nine maser lines and twelve emission or absorption ones. In the review of Reid and Moran (1981, see their Table 1), one can see that several OH maser lines have been seen in tens of sources. Also, correlations between ground and excited state masers have been observed (Gardner, Whiteoak and Palmer 1987; Baudry et al. 1988; Cohen, Masheder and Walker 1991). Fortunately, the situation has changed recently (see below) and now the theoretical calculations can be used as diagnostics of the physical conditions in star-forming regions.

6.1.2. *Theoretical calculations*

Before discussing the latest developments on our understanding of the OH masers, we will give a brief account of the various attempts that were made to explain the ground state OH masers.

An attractive suggestion by Andresen (1986) that the 18 cm OH masers are pumped by photodissociation of H_2O is also ruled out. This is because the H_2O has to be reformed at a rate which is larger than the collision rate. This is not possible.

The work of Kylafis and Norman (1990) showed that the ground state OH masers *are not* pumped by collisions alone.

Model applications to H II/OH were performed by Cesaroni and Walmsley (1991), Gray, Doel and Field (1991) and Gray, Field and Doel (1992) using both collisional and radiative excitation of OH, and taking into account radiative trapping and line overlap. Cesaroni and Walmsley (1991) succeeded, with some combination of the parameters of their model, to reproduce qualitatively the observed lines of OH for the best studied star forming region W3(OH). This was very encouraging for further detailed studies of OH masers in star-forming regions. The work of Gray, Doel and Field (1991) and Gray, Field and Doel (1992) is quite extensive, but it is a pity that the collision rates they used were outdated and the new ones are so different that qualitatively different conclusions are introduced in some cases. Also, their treatment of line overlap is not sound, because the formulae they have used are unproven.

In view of the wealth of available data and the recently calculated collision rates between OH and H_2, we undertook a systematic study of OH maser pumping in star-forming regions with the hope that we will be able to invert (at least partially) the maser problem. We added one pumping mechanism at a time (collisions, local line overlap, non-local line overlap, external infrared radiation field) and explored the various parameter regimes for each combination of pumping mechanisms.

Our work was an improvement over the work of previous researchers in two main aspects. First, we treated local and non-local line overlap with the use of the recently developed formalism (Pavlakis and Kylafis 1996a). Previous workers have either assumed complete (i.e., 100%) line overlap (Cesaroni and Walmsley 1991) or treated line overlap in a heuristic, unproven way (Doel, Gray and Field 1990; Gray, Doel and Field 1991; Gray, Field and Doel 1992).

Second, we used the recently calculated collision rates between OH and H_2 (Offer, van Hemert and van Dishoeck 1994) and for the first time we distinguished between ortho-H_2 and para-H_2. All previous calculations considered para-H_2. As we found out, it makes a big difference in which form H_2 is in the interstellar medium. It is thus hoped that our work will help distinguish between the two forms of H_2 in maser regions and help determine the abundance of ortho-H_2.

The results of our calculations appeared in Pavlakis and Kylafis (1996b,c) and our major conclusion was that with reasonable values of the parameters it was possible to explain the general characteristics of the 18 cm and the 6 cm OH masers. In addition, with our exploration of the parameter regimes, we were able to propose a number of diagnostics of the physical conditions in star-forming regions using maser observations. For example, if both the 1665 MHz and the 1667 MHz maser lines are seen *in the same spatial region*, then a far infrared radiation field must be present and the density must be less than or of order 10^6 cm^{-3} there.

In our study of the OH maser pumping we had also computed the maser emission of the 5 cm lines of OH, but we decided to not show or discuss the maser lines in the excited state $^2\Pi_{3/2}$, $J = 5/2$, because this state is directly connected with the state $^2\Pi_{1/2}$, $J = 7/2$, which was not included in our calculations. The omission of this state could, in principle, affect the results.

Soon after our calculations were published, Baudry et al. (1997) reported an extensive study of the 5 cm maser lines of OH in star-forming regions. To our surprise, *our already performed calculations explained the observed characteristics.* This probably means that, for temperatures between 100 and 200 K that are thought appropriate for OH maser regions, the missing collision rates are not important for the 5 cm masers. Thus, now we can claim that the essential observational characteristics of the 18 cm, the 6 cm and the 5 cm OH masers in star-forming regions can all be explained with the same set of reasonable values of the parameters. Of course, when the missing collision rates are computed, it will be reassuring to show that they are indeed not important for the 5 cm masers.

6.2. H_2O MASERS

The rotational energy levels of H_2O are characterized by three quantum numbers and are denoted by $J_{K_+ K_-}$, where J is the total angular momentum and K_+ and K_- are its projections on two molecular axes. For an energy level diagram of H_2O see, e.g., Neufeld and Melnick (1991).

The first H_2O maser transition was detected thirty years ago (Cheung et al. 1969). It is the transition from level 6_{16} to level 5_{23}, giving rise to radiation with frequency $\nu = 22$ GHz or wavelength $\lambda = 1.35$ cm. This transition lies 640 K above the ground state.

6.2.1. *Observations*

Up to 1990, the words H_2O masers were synonymous with the transition at 22 GHz. In 1990, three mm lines were reported as masers. These are: $10_{29} - 9_{36}$ at 321 GHz (Menten, Melnick and Phillips 1990), $3_{13} - 2_{20}$ at 183 GHz (Cernicharo *et al.* 1990) and $5_{15} - 4_{22}$ at 325 GHz (Menten *et al.* 1990). Of these, the 321 GHz transition lies 1860 K above the ground state. Therefore, it is likely that this transition is excited in a warm (say, 1000 K) environment. Three years later, three sub-mm maser lines were observed (Melnick *et al.* 1993). These are: $7_{53} - 6_{60}$ at 437 GHz, $6_{43} - 5_{50}$ at 439 GHz and $6_{42} - 5_{51}$ at 471 GHz.

Water maser emission is a common feature of deeply embedded yound stellar objects (YSOs). It must be stressed though, that H_2O maser emission *does not guarantee* that star formation is taking place there.

It is now widely accepted that the H_2O masers occur in shocks and that the pumping is collisonal.

6.2.2. *Theoretical Calculations*

As with the OH masers, we will give a brief account of the attempts to explain the H_2O masers in star-forming regions.

Radiative pumping of the H_2O masers can be ruled out because, even with 100% efficiency, one infrared pumping photon is needed for every maser photon emitted. However, the observed number of infrared pumping photons is significantly lower than the observed maser photons.

Collisions with H_2 molecules on the other hand *can* invert the 22 GHz transition (Elitzur, Holenbach and McKee 1989; Kylafis and Norman 1991) as well as the three mm transitions (Neufeld and Melnick 1990; 1991) and the three sub-mm ones (Kaufman and Neufeld 1996). The detected transitions, as well as a few more that will undoubtedly be detected in the near future, will constrain the models significantly.

Strelnitskij (1984) proposed the very interesting idea that the H_2O molecules collide with two kinds of particles (e.g., H_2 and electrons) at different kinetic temperatures, but with comparable collision rates. In other words, the H_2O molecules find themselves in contact with two heat baths, and not knowing which of the two Maxwellian distributions to obey, they obey none. Thus, even at very high densities the H_2O molecules are not thermalized. Kylafis and Norman (1986; 1987) showed that, with the then available collision rates, collisional pumping of H_2O in environments where the temperature of the neutral particles is significantly less than that of the electrons, *can* sustain inverted populations. Such environments exist in the magnetic precursors of MHD shocks (Draine, Roberge and Dalgarno 1983).

This idea is now temporarily abandoned because the recalculated collision rates (Palma *et al.* 1988) are significantly different than the older ones.

Anderson and Watson (1990) have shown that with the new rates inversion does not occur.

Elitzur, Hollenbach and McKee (1989) proposed that all H_2O masers in star-forming regions, even the very powerful ones, can be explained with conventional collisional pumping. They modeled the maser regions as static, in the shape of cylinders with aspect ratios (see Figure 3) $10 \lesssim l/d \lesssim 100$. This is fine, but it is a little difficult to imagine how such long and narrow structures can exist in star-forming regions. Furthermore, if there is a velocity gradient $\Delta v/l \gtrsim \Delta v_{thermal}/l$ along the axis of the cylinder (which is not unreasonable), then the gain is lower and one needs aspect ratios $l/d > 100$ to explain the powerful H_2O masers (Kylafis and Norman 1991).

An interesting idea regarding the powerful H_2O masers was proposed by Deguchi and Watson (1989). Consider two cylindrical masers, of length l and base diameter d each, separated by a distance D and having their axes aligned. The observer is located on their common axis. If there were only one maser, its radiation would be beamed into a solid angle $\Omega \sim (d/l)^2$ (see § 2.4). However, the radiation of the back maser causes by stimulated emission the radiation of the front maser to be emitted in a much smaller solid angle $\Omega \sim (d/D)^2$. Thus, the observer detects radiation that would not be detected if the front maser emitted its radiation into $\Omega \sim (d/l)^2$. The aligned masers do not emit more than their sum, but *they give you the impression of a powerful maser.*

All the previous model calculations addressed the 22 GHz H_2O maser line and included energy levels up to about $E/k = 1000$ K. For the 321 GHz line, energy levels up to about $E/k = 7000$ K had to be included (Neufeld and Melnick 1991; Kaufman and Neufeld 1996). These impressive calculations explain not only the pumping of all the observed H_2O maser lines, but they also predict that more maser transitions should be seen.

7. Location of Masers

Significant progress has been made in the last few years in pinpointing the positions of masers close to the YSO (e.g., Torrelles *et al.* 1996; Fiebig *et al.* 1996; Hofner and Churchwell 1996; Meehan *et al.* 1998; Torrelles *et al.* 1998) and in associating the positions of the masers with the evolutionary state of the YSO.

Through extended surveys, the group in Firenze and others showed that most of the H_2O masers (up to 80%) are not associated with diffuse H II regions (Palla *et al.* 1993). Instead, it was shown that the H_2O masers trace the earliest evolutionary phases of massive stars, much before the onset of an ultra compact H II region (Codella *et al.* 1994; Palumbo *et al.* 1994;

 N. D. KYLAFIS AND K. G. PAVLAKIS

Codella and Felli 1995; Codella *et al.* 1995; Codella, Felli and Natale 1996; Codella, Testi and Cesaroni 1997; Testi *et al.* 1997; Felli *et al.* 1997).

An excellent example that shows the connection of H_2O and OH masers with the earliest evolutionary phases of YSOs is the work of Torrelles *et al.* (1997) on W75N(B). They observed with the VLA the star-forming region W75N(B) and detected three continuum (at 1.3 cm) sources (VLA 1, VLA 2, and VLA 3) in a region of 1.5 arcseconds. VLA 1 is elongated approximately in the direction of the bipolar molecular outflow. VLA 2 appears unresolved, while VLA 3 shows a bright core plus extended emission. They observed simultaneously the H_2O maser emission toward these sources and detected a total of 29 H_2O maser spots. These masers are mainly distributed in two clusters, one associated with VLA 1 (11 maser spots) and the other one associated with VLA 2 (8 maser spots). Only one maser spot is associated with VLA 3. The masers associated with VLA 1 are distributed along the major axis of the radio jet and are delineating the outflow at scales of 1 arcsecond. The masers coincident with VLA 2 are distributed in a shell and are probably bound.

When the OH maser spots of Baart *et al.* (1986) are plotted in the same Figure as the H_2O masers (see Figure 7 of Torrelles *et al.* 1997), the following picture emerges:

a) The OH and the H_2O masers are close to each other *but never coincident.*

b) No OH maser spot is associated with VLA 3.

c) In VLA 2, there are three OH maser spots, but all of them are further away from the continuum source than the H_2O masers associated with it.

d) In VLA 1, the OH masers like the H_2O ones are excited in the jet, but in its outer parts.

The non-coincidence of the OH and H_2O masers is easy to understand theoretically. The H_2O masers thrive in regions where the density is high ($n_{H_2} \sim 10^9$ cm^{-3}). At such high densities the OH molecules are completely thermalized. Nevertheless, the above observations are consistent with the picture that both H_2O and OH masers are pumped in shocks, with the OH masers pumped in less dense regions than the H_2O ones.

The interpretation, in terms of evolution, that Torrelles *et al.* (1997) give is the following: VLA 3 is an ultracompact and less developed H II region. Somewhat more developed is VLA 2, with its associated masers probably tracing bound motions in the surrounding circumstellar gas (rotating disk?). The most developed of all is VLA 1, which exhibits a large-scale molecular outflow. Thus, H_2O masers appear first and then follow the OH masers. This is consistent with the proposal of Forster and Caswell (1989).

8. Conclusions

From the above presentation we hope it has become evident that the astronomical masers are very useful. They are particularly useful in probing the dense star-forming regions. The resolution (μarcsec) achieved with masers cannot be achieved with any other method. Now we have reached the point where several maser lines from the same molecule are observed. This places significant constraints on the theoretical models and for the first time model predictions have been verified observationally. It is also reasonable to expect that maser observations will reveal the physical conditions in the dense parts of star-forming regions.

9. References

Alcock, C., and Ross, R. R. 1985a, *Ap. J.*, **290**, 433.

————. 1985b, *Ap. J.*, **299**, 763.

————. 1986, *Ap. J.*, **306**, 649.

Anderson, N., and Watson, W. D. 1990, *Ap. J.*, **348**, L69.

Andresen, P. 1986, *Astr. Ap.*, **154**, 42.

Baart, E. E., Cohen, R. J., Davies, R. D., Norris, R. P., and Rowland, P. R. 1986, *M.N.R.A.S.*, **219**, 145.

Baudry, A., Desmurs, J. F., Wilson, T. L., and Cohen, R. J. 1997, *Astr. Ap.*, **325**, 255.

Baudry, A., and Diamond, P. J. 1991, *Astr. Ap.*, **247**, 551.

————. 1998, *Astr. Ap.*, **331**, 697.

Baudry, A., Diamond, P. J., Booth, R. S., Graham, D., and Walmsley, C. M. 1988, *Astr. Ap.*, **201**, 105.

Baudry, A., Walmsley, C. M., Winnberg, A., and Wilson, T. L. 1981, *Astr. Ap.*, **102**, 287.

Bloemhof, E. E., Moran, J. M., and Reid, M. J. 1996, *Ap. J.*, **467**, L117.

Bloemhof, E. E., Reid, M. J., and Moran, J. M. 1992, *Ap. J.*, **397**, 500.

Castor, J. I. 1970, *M.N.R.A.S.*, **149**, 111.

Caswell, J. L., and Vaile, R. A. 1995, *M.N.R.A.S.*, **273**, 328.

Cernicharo, J., Thum, C., Hein, H., John, D., Garcia, P., and Mattioco, F. 1990, *Astr. Ap.*, **231**, L15.

Cesaroni, R., and Walmsley, C. M. 1991, *Astr. Ap.*, **241**, 537.

Chapman, J. M., and Cohen, R. J. 1986, *M.N.R.A.S.*, **220**, 513.

Cheung, A. C., Rank, D. M., Townes, C. H., Thornton, D. D., and Welch, W. J. 1969, *Nature*, **221**, 626.

Claussen, M. J., Wilking, B. A., Benson, P. J., Wootten, A., Myers, P. C., and Terebey, S. 1996, *Ap. J. Suppl.*, **106**, 111.

Clegg, A. W., and Cordes, J. M. 1991, *Ap. J.*, **374**, 150.

Codella, C., and Felli, M. 1995, *Astr. Ap.*, **302**, 521.

Codella, C., Felli, M., and Natale, V. 1996, *Astr. Ap.*, **311**, 971.

Codella, C., Felli, M., Natale, V., Palagi, F., and Palla, F. 1994, *Astr. Ap.*, **291**, 261.

Codella, C., Palumbo, G. G. C., Pareschi, G., Scappini, F., Caselli, P., and Attolini, M. R. 1995, *M.N.R.A.S.*, **276**, 57.

Codella, C., Testi, L., and Cesaroni, R. 1997, *Astr. Ap.*, **325**, 282.

Cohen, R. J., Baart, E. E., and Jonas, J. L. 1988, *M.N.R.A.S.*, **231**, 205.

Cohen, R. J., Masheder, M., and Walker, R. N. F. 1991, *M.N.R.A.S.*, **250**, 611.

Deguchi, S., and Watson W. D. 1986a, *Ap. J.*, **300**, L15.

———. 1986b, *Ap. J.*, **302**, 750.

———. 1989, *Ap. J.*, **340**, L17.

———. 1990, *Ap. J.*, **354**, 649.

Deguchi, S., Watson, W. D., and Western, L. R. 1986, *Ap. J.*, **302**, 108.

Desmurs, J. F., and Baudry, A. 1998, *Astr. Ap.*, **340**, 521.

Desmurs, J. F., Baudry, A., Wilson, T. L., Cohen, R. J., and Tofani, G. 1998, *Astr. Ap.*, **334**, 1085.

Doel, R. C., Gray, M. D., and Field, D. 1990, *M.N.R.A.S.*, **244**, 504.

Draine, B. T., Roberge, W. G., and Dalgarno, A. 1983, *Ap. J.*, **264**, 485.

Elitzur, M. 1982, *Rev. Mod. Phys.*, **54**, 1225.

———. 1990a, *Ap. J.*, **363**, 628.

———. 1990b, *Ap. J.*, **363**, 638.

———. 1991, *Ap. J.*, **370**, 407.

———. 1992a, *Astronomical Masers*, (Dordrecht: Kluwer)

———. 1992b, *Ann. Rev. Astr. Ap.*, **30**, 75.

———. 1993, *Ap. J.*, **416**, 256.

———. 1995, *Ap. J.*, **440**, 345.

———. 1996, *Ap. J.*, **457**, 415.

Elitzur, M., Hollenbach, D. J., and McKee, C. F. 1989, *Ap. J.*, **346**, 983.

Felli, M., Testi, L., Valdettaro, R., and Wang, J.-J. 1997, *Astr. Ap.*, **320**, 594.

Fiebig, D., Duschl, W. J., Menten, K. M., and Tscharnuter, W. M. 1996, *Astr. Ap.*, **310**, 199.

Fiebig, D., and Güsten, R. 1989, *Astr. Ap.*, **214**, 333.

Forster, J. R., and Caswell, J. L. 1989, *Astr. Ap.*, **213**, 339.

Gardner, F. F., and Martin-Pintado, J. 1983, *Astr. Ap.*, **121**, 265.

Gardner, F. F., and Whiteoak, J. B. 1983, *M.N.R.A.S.*, **205**, 297.

Gardner, F. F., Whiteoak, J. B., and Palmer, P. 1987, *M.N.R.A.S.*, **225**, 469.

Gaume, R. A., and Mutel R. L. 1987, *Ap. J. Suppl.*, **65**, 193.

Genzel, R., and Downes, D. 1977, *Astr. Ap. Suppl.*, **30**, 145.

Genzel, R., Reid, M. J., Moran, J. M., and Downes, D. 1981a, *Ap. J.*, **244**, 884.

Genzel, R., Downes, D., Schneps, M. J., Reid, M. J., Moran, J. M., Kogan, L. R., Kostenko, V. I., Matveyenko, L. I., and Ronnang, B. 1981b, *Ap. J.*, **247**, 1039.

Goldreich, P., and Keeley, D. A. 1972, *Ap. J.*, **174**, 517.

Goldreich, P., Keeley, D. A., and Kwan, J. Y. 1973, *Ap. J.*, **179**, 111.

Goldreich, P., and Kwan, J. 1974, *Ap. J.*, **190**, 27.

Gray, M. D., Doel, R. C., and Field, D. 1991, *M.N.R.A.S.*, **252**, 30.

Gray, M. D., Field, D., and Doel, R. C. 1992, *Astr. Ap.*, **262**, 555.

Guilloteau, S., Baudry, A., Walmsley, C. M., Wilson, T. L., Winberg, A. 1984, *Astr. Ap.*, **131**, 45.

Gwinn, C. R., Moran, J. M., and Reid, M. J. 1992, *Ap. J.*, **393**, 149.

Herrnstein, J. R., Greenhill, L. J., and Moran J. M. 1996, *Ap. J.*, **468**, L17.

Hofner, P., and Churchwell, E. 1996, *Astr. Ap. Suppl.*, **120**, 283.

Kaufman, M. J., and Neufeld, D. A. 1996, *Ap. J.*, **456**, 250.

Knowles, S. H., Caswell, J. L., and Goss, W. M. 1976, *M.N.R.A.S.*, **175**, 537.

Kylafis, N. D., and Norman, C. 1986, *Ap. J.*, **300**, L73.

————. 1987, Ap. J., **323**, 346.

————. 1990, Ap. J., **350**, 209.

————. 1991, Ap. J., **373**, 525.

Lekht, E. E., Mendoza-Torres, E., and Sorochenko, R. L. 1995, *Ap. J.*, **443**, 222.

Meehan, L. S. G., Wilking, B. A., Claussen, M. J., Mundy, L. G., and Wooten, A. 1998, *A. J.*, **115**, 1599.

Melnick, G. J., Menten, K. M., Phillips, T. G., and Hunter, T. 1993, *Ap. J.*, **416**, L37.

Menten, K. M., Melnick, G. J., and Phillips, T. G. 1990, *Ap. J.*, **350**, L41.

Menten, K. M., Melnick, G. J., Philips, T. G., and Neufeld, D. A. 1990, *Ap. J.*, **363**, L27.

Menten, K. M., Reid, M. J., Eckart, A., and Genzel, R. 1997, *Ap. J.*, **475**, L111.

Migenes, V., Cohen, R. J., and Brebner, G. C. 1992, *M.N.R.A.S.*, **254**, 501.

Miyoshi, M., Moran, J., Herrnstein, J., Greenhill, L., Nakai, N, Diamond, P., and Inoue, M. 1995, *Nature*, **373**, 127.

Moran, J. M. 1989, in *Handbook of Laser Science and Technology*, ed. M. J. Weber.

Moran, J. M., Reid, M. J., Lada, C. J., Yen, J. L., Jonston, K. J., and Spencer, J. H. 1978, *Ap. J.*, **224**, L67.

Nedoluha, G. E., and Watson, W. D. 1990a, *Ap. J.*, **354**, 660.

————. 1990b, *Ap. J.*, **361**, 653.

————. 1990c, *Ap. J.*, **361**, L53.

————. 1994, *Ap. J.*, **423**, 394.

Neufeld, D. A., and Melnick, G. J. 1990, *Ap. J.*, **352**, L9.

————. 1991, *Ap. J.*,, **368**, 215.

Offer, A. R., van Hemert, M. C., and van Dishoeck, E. F. 1994, *J. Chem. Phys.*, **100**, 362.

Palla, F., Cesaroni, R., Brand, J., Caselli, P., Comoretto, G., and Felli, M. 1993, *Astr. Ap.*, **280**, 599.

Palma, A., Green, S., DeFrees, D. J., and McLean, A. D. 1988, *Ap. J. Suppl.*, **68**, 287.

Palmer, P., Gardner, F. F., and Whiteoak, J. B. 1984, *M.N.R.A.S.*, **211**, 41p.

Palumbo, G. G. C., Scappini, F., Pareschi, G., Codella, C., Caselli, P., and Attolini, M. R. 1994, *M.N.R.A.S.*, **266**, 123.

Pavlakis, K. G., and Kylafis, N. D. 1996a, *Ap. J.*, **467**, 292.

————. 1996b, *Ap. J.*, **467**, 300.

————. 1996c, *Ap. J.*, **467**, 309.

————. 1999, *Ap. J.*, submitted.

Reid, M. J., and Moran, J. M. 1981, *Ann. Rev. Astr. Ap.*, **19**, 231.

————. 1988, in *Galactic and Extragalactic Radio Astronomy*, eds. G. L. Verschuur and K. I. Kellerman, (New York: Springer-Verlag), p. 255.

Reid, M. J., Schneps, M. H., Moran, J. M., Gwinn, C. R., Genzel, R., Downes, D., and Ronnang, B. 1987, in *Star Formation*, eds. M. Peibert and J. Jugaku, IAU Symposium 115 (Dordrecht: Reidel).

Reid, M. J., and Silverstein, E. M. 1990, *Ap. J.*, **361**, 483.

Schneps, M., Lane, A. P., Downes, D., Moran, J. M., Genzel, R., and Reid, M. J. 1981, *Ap. J.*, **249**, 124.

Smits, D. P. 1994, *M.N.R.A.S.*, **269**, L11.

Sobolev, V. V. 1960, *Moving Envelopes of Stars*, (Cambridge: Harvard University Press).

Strelnitskij, V. S. 1984, *M.N.R.A.S.*, **207**, 339.

Testi, L., Felli, M., Omont, A., Perault, M., Seguin, P., Comoretto, G., and Gilmore, G. 1997, *Astr. Ap.*, **318**, L13.

Torrelles, J. M., Gomez, J. F., Rodriguez, L. F., Curiel, S., Anglada, G., and Ho, P. T. P. 1998, *Ap. J.*, **505**, 756.

Torrelles, J. M., Gomez, J. F., Rodriguez, L. F., Curiel, S., Ho, P. T. P., and Garay, G. 1996, *Ap. J.*, **457**, L107.

Torrelles, J. M., Gomez, J. F., Rodriguez, L. F., Ho, P. T. P., Curiel, S., and Vazquez, R. 1997, *Ap. J.*, **489**, 744.

Turner, B. E., Palmer, P., and Zuckerman, B. 1970, *Ap. J.*, **160**, L125.

Wallin, B. K., and Watson, W. D. 1997, *Ap. J.*, **481**, 832.

Watson, W. D. 1994, *Ap. J.*, **424**, L37.

Western, L. R., and Watson, W. D. 1983a, *Ap. J.*, **268**, 849.
————. 1983b, *Ap. J.*, **274**, 195.
————. 1983c, *Ap. J.*, **275**, 195.
————. 1984, *Ap. J.*, **285**, 158.

Nick Kylafis

Anneila Sargent really enjoys the first week's banquet.

IV – THE PHYSICS OF CIRCUMSTELLAR DISKS AND PLANET FORMATION

The Beckwith family with Anneila Sargent, Geoff Marcy, Ralf Klessen, Mchael Meyer, Joao Alves, and Paul and Aspa Kalas.

CIRCUMSTELLAR DISKS

STEVEN V. W. BECKWITH
Max-Planck-Institut für Astronomie
& Space Telescope Science Institute

1. Introduction

Circumstellar disks usually contain only a few percent of the total material going into a young star after the main collapse has stopped and the surrounding molecular cloud is cleared away. Yet the disks are of great interest to the study of star formation, perhaps as great as the stars themselves, because the disks may build planetary systems. The planet Earth contains less than one millionth of the mass of the Sun, but it is probably the most interesting body in the Solar System, certainly to us. Beckwith & Sargent (1996) argue that the currently known properties of disks are evidence that other planetary systems are common in the Galaxy and discuss the reasons for the interest in disk properties; that article provides a broad introduction to the subject.

The purpose of this chapter is to provide a tutorial on how observations of the radiation from disks may be used to elicit their physical characteristics. In keeping with the spirit of the Crete meeting, the treatment is not a comprehensive review nor will it give a complete analysis of each method used to tease the disk properties from faint light observed with telescopes. Rather, the idea is to show that basic intuition about disk physics is easily related to what is observed and to provide a general foundation for the understanding of more elaborate theoretical calculations.

Early disk models assumed that matter is confined to a very thin plane extending from the stellar surface to a sharp outer edge more than 100 AU from the star. The disk energy balance was attributed to accretion of matter through the disk. This oversimplified picture has been modified by a careful treatment of the underlying physics, and the more modern view is that the disk flares gently, often with an inner edge at some distance from the star, and is heated mainly by radiation as opposed to accretion. Some disks are surrounded by spheroidal "halos" that trap radiation and contain strong

C.J. Lada and N.D. Kylafis (eds.), The Origin of Stars and Planetary Systems, 579–612.

outflows. Most of the young disks are accompanied by mass loss in columns along the polar axes that contribute to the total energy budget. Although it is not always possible to derive disk characteristics unambiguously from observations, most of the intuitive interpretations have been supported by increasingly better data and improved angular resolution images allowing us to separate t! ! he different components of a star/disk system directly.

The article is organized along the following questions:

1. What are the expected disk properties based on the theory of Solar System formation?
2. How do we identify disks?
3. How do we determine physical properties of disks from radiation?
4. Do the observed properties show that disks are interesting?

Because this article is a tutorial, some of the material is adopted from articles that I co-authored for Nature (Beckwith & Sargent 1996) and Protostars and Planets IV (Beckwith, Henning, & Nakagawa 1999).

2. The early Solar System

It is generally agreed that a flat layer of gas and dust - a disk - orbited the early Sun and provided the material which later made up the Earth, Mars, Jupiter, and the other planets (Safronov 1969; Wood & Morfill 1988; Cameron 1988). The young Sun and the circumsolar disk were born from an extended cloud of gas and dust that was assembled from the detritus of dying stars and remnants of the early universe that collapsed under its own gravity. The material accumulated quickly onto the central proto-Sun but with enough residual angular momentum to prevent some from spiraling inwards - the exact proportion remaining in orbit is not known but should have been a considerable fraction of the total mass (Shu, Adams, & Lizano 1987; Bodenheimer 1995). The average angular momentum of the collapsing region defined a rotation axis around which the orbits quickly stabilized, creating a disk with a thickness much smaller than its radius, at least within the regions now containing the! giant planets. The formation of the stable disk probably occurred over about 10^5 years after the onset of free fall collapse (Shu et al. 1993), almost instantaneously in cosmic time.

As the central proto-Sun evolved, the solid particles in the orbits settled to a dense layer in the mid-plane of the disk and began to stick together as they collided (Safronov 1969; Weidenschilling 1987; Mizuno et al. 1988). During the next 10^4 to 10^5 years, large rocks and small asteroids grew gradually from the small dust particles (Weidenschilling & Cuzzi 1993). When the gravitational pull of the largest asteroids was sufficient to attract neighboring pebbles and rocks, they grew even more rapidly to the size of small planets (Wetherill & Stewart 1993). Gravity was important for bodies more than 10 km across. The terrestrial planets are large accumulations of solid particles that grew from the collisions between these smaller bodies. In the outer parts of the disk, a few such solid cores became large enough (10 Earth masses) to accrete gas (Mizuno 1980; Stevenson 1982), the dominant reservoir of mass, and gave rise to the giant gas planets (Wetherill 1990). Te! m! peratures close to the proto-Sun were presumably too high to allow gas accretion. The planet building phase is thought to have taken between $\sim 10^7$ and a few times 10^8 years, although the cores probably developed quickly, within the first 10^6 years or so. These timescales are not very well constrained by data, and it may well be observations of developing planetary systems around other stars that tell us how planets are really built.

By analogy, we expect circumstellar disks surrounding other stars to contain a few percent or more of the stellar mass, to extend at least 50 AU from the central star, to be relatively flat, and to be free from disruption for at least a few million years if they are to create the rocky cores needed to build large planets. These characteristics certainly represent only a subset of the disks that accompany star formation. Conceivably, a disk with substantially lower mass ($10^{-6} M_\odot$), size (a few AU), and lifetime ($\sim 10^6$ yr) could create terrestrial-like planets suitable for life without the presence of gas giants. In principle, even larger, more massive disks could accompany the birth of stars much more massive than the Sun. Although we must keep an open mind about the characteristics that constitute a disk, the early Solar disk provides a framework to identify those disks that may become interesting for planet formation.

3. How do we know that disks exist?

Soon after the discovery that T Tauri stars – very young stars of approximately solar mass – had more radiation at infrared wavelengths than the photospheres should emit, Lynden-Bell & Pringle (1974) suggested that most of their peculiar characteristics might be explained by circumstellar disks. Their suggestion was based on the unusual spectral energy distributions: the stars radiated too much ultraviolet light *and* too much infrared light at the same time. A disk could account for the ultraviolet light through emission from the *boundary layer* between the star and the inner edge of the disk, in which matter from the disk suddenly accreted onto the star, slowing down from Keplerian speeds to essentially zero speed so rapidly that the radiation temperatures are tens of thousands of Kelvins. The infrared light was radiation from the outer parts of the disk resulting from energy liberated as the matter slowly spiraled to smaller radii eventually to accrete throu! gh the boundary layer. One of their strong predictions was that the long wavelength infrared radiation would follow a power law, $F_\nu \propto \nu^{\frac{1}{3}}$, where F_ν is the flux density, and ν is the frequency of the radiation. This result for the release of accretion energy through a disk is quite general.

Even with no accretion, a disk will be heated by radiation from the star itself. The dust grains in the disk absorb stellar radiation and re-radiate in the infrared to maintain thermal balance. Remarkably, the spectral energy distribution also follows a power law with the same exponent as for accretion over a broad range of wavelengths: $F_\nu \propto \nu^{\frac{1}{3}}$ between about 5 and $100\,\mu$m depending on the luminosity of the star. At wavelengths shortward of about $3\,\mu$m, the flux density stops increasing due to the inner radius of the disk, and at long wavelengths the power law becomes steeper due to the outer edge of the disks. The most complete treatment for a flat disk is given by Adams, Lada, & Shu (1988).

When Lynden-Bell & Pringle made their suggestion, the long wavelength SEDs of disks could only be measured to about $10\,\mu$m. The SEDs could easily be explained over this limited spectral range by other distributions of dust near the stars. Although prescient, the predictions of the early disk theory went untested for nearly a decade.

A few years after the SED calculations, Elsässer & Staude (1978) discovered that several young stars had rather high degrees of linear polarization in their optical light. They explained the polarization as scattered light from dust grains arranged symmetrically above and below a star that was obscured by a planar or toroidal distribution of dust oriented perpendicular to the line of sight and parallel to the direction of the polarization. These observations suggested that the dust distribution around the stars was axisymmetric and flattened relative to a spherical halo.

The most striking demonstrations of axisymmetry in T Tauri stars are the well collimated jets seen in images of ionized lines. Mundt & Fried (1983) discovered the first of many young stellar jets in their image of HL Tau. The jets implied a strong axisymmetry near the stars, one component of which may be flattened disks perpendicular to the axis of the jets. In the same year, Cohen (1983) used the Kuiper Airborne observatory to measured the first far infrared SED of HL Tau, and he inferred the presence of a thin disk following the reasoning of Lynden-Bell & Pringle (1974). One year later, two groups observed HL Tau at angular resolutions several times better than previously possible and discovered elongated infrared emission reminiscent of disks (Beckwith et al. 1984; Grasdalen et al. 1984). Following the evidence accumulated for some time on this object, both groups suggested that the disk might be the progenitor of a planetary system similar to the early solar nebula! . Subsequently, Beckwith et al. (1986) and Sargent & Beckwith (1987) made inteferometric maps of the CO and ^{13}CO emission and showed that the gas, too, was elongated as if in a disk; the direction of elongation was perpendicular to the jet and aligned with the major axis of the near infrared emission. Sargent & Beckwith also observed the velocity field and interpreted it as gas orbiting the star. At this time, the evidence for a disk surrounding HL Tau appeared to be quite strong.

These observations implied the distribution of material was disk-like but did not determine how much material was in a disk. The interferometric maps showed that HL Tau had continuum emission at 2.7 mm that appeared to be thermal emission from dust particles. With conservative assumptions about the radiative properties of the dust, Sargent & Beckwith derived a mass for the HL Tau disk of $\sim 0.1\,M_\odot$ with a considerable uncertainty. This mass is well above the minimum mass for the disk around the primitive solar nebula, $0.01\,M_\odot$, thereby strengthening the case for a protoplanetary disk surrounding HL Tau.

At about the same time, astronomers analysing the far infrared radiation from disks from the IRAS survey recognized that many T Tauri stars had strong far infrared excesses indicative of disks along the lines first noted by Cohen (1983). Rucinski (1995), Beall (1987), and Adams, Lada, & Shu (1987) recognized that the IRAS observations were most easily interpreted as disk emission, although there was some controversy about the applicability of Lynden-Bell & Pringle's flat, black disk calculations. Adams, Lada, & Shu (1988) showed that a class of these sources, termed "flat-spectrum sources" (see the next section), required disks that were quite a bit warmer in their outer regions than predicted by the flat, black disk model.

Two major advances followed in 1989 and 1990. Working with data from the IRAS sky survey, Cohen, Emerson, & Beichman (1989) and Strom et al. (1989) discovered that almost half the T Tauri stars in the Taurus-Aurgia dark clouds had far infrared SEDs characteristic of heated dust, and the relatively low, visual extinctions implied that the dust was in flattened distributions – most likely disks. This result meant that disks were common around young stars; HL Tau was not an anamoly. In the following year, Beckwith et al. (1990) found that a similar fraction of stars had emission at 1.3 mm, from which they placed an even stronger limit on the flattening, essentially showing that disks were the only viable explanations for the data. For the first time, they derived the distribution of masses for the disks around stars. Not only were half the T Tauri stars surrounded by disks, the great majority of the disks had sufficient mass to build planetary systems like our own.

Most specialists in star formation had by then accepted disks as the best explanation for the accumulated data on T Tauri stars. But the evidence was largely indirect, and it was often possible to interpret the data sets with different theories for the distribution of dust near these stars. The

first pictures of disks taken with the Hubble Space Telescope demonstrated
clearly that the dust distributions followed the theoretical pattern of a disk.
The image of HH 30 by Burrows et al. (1996) shows a jet perpendicular to a
dark disk seen edge-on with a gently flared shape, as expected for thermally
supported disks (Kenyon & Hartmann 1987). Striking examples of disks are
seen in silhouette against the Orion Nebula by O'Dell & Wen (1994) and
McCaughrean & O'Dell (1996). Several more images of disks like HH 30
have appeared recently (Stapelfeldt 1998; Padgett et al. 1999). Figure 1
shows several images of disks taken with the Hubble Space Telescope.

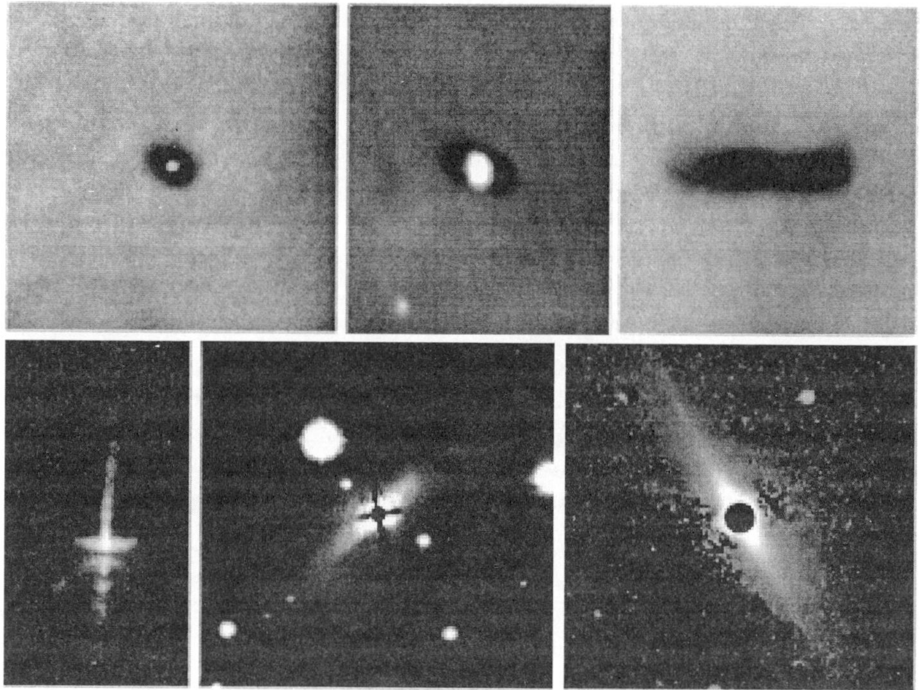

*Figure 1.   These images of disks were obtained with the Hubble Space Telescope. The
upper three images are disks seen in silhouette against the Orion nebula (McCaughrean &
O'Dell 1996). The bottom three images show disks close to edge on. The image of HH 30
(Burrows et al. 1996) n the lower left shows flaring and illumination of cone-shaped
cavities in the surrounding halo due to residual cloud material. The two images on the
lower right are of main sequence disks: BD+31 643 (Kalas & Jewitt 1997) and β Pictoris
(Kalas & Jewitt 1995, and references therein).*

The images leave no doubt about the distribution of matter: the particles surrounding T Tauri stars are confined to flattened disks. Many other
observations – very high resolution spectra (Hartmann et al. 1986), inteferometric maps of the gas (Dutrey, Guilloteau, & Simon 1994; Koerner &

Sargent 1995; Saito et al. 1995) and polarization maps (Bastien & Menard 1990; Whitney & Hartmann 1992; Piirola, Scaltriti, & Coyne 1992) – are easily interpretted if disks are present and difficult to explain with other distributions of the matter. Disks exist.

Compounded, the evidence for disks is overwhelming, but that does not mean that disks are present around every star or that individual stars may not have other distributions of matter. For those stars that do have disks, we turn our attention to their properties and how we observe them.

4. Spectral Energy Distributions

Disks contain a mixture of gas and dust. Early in a disk's evolution – the first few million years, say – there are enough small particles to absorb and reradiate light from the star. Dust warmed by the starlight radiates as a blackbody modified by the emissivity of the grains, which for small particles is approximately proportional to the frequency throughout the infrared region. The dust temperatures depend on the distance of a grain from the star. Because a continuous disk contains particles at a wide range of distances from close to the star's surface to several hundred AU or more, dust temperatures span several orders of magnitude. The result is a mixture of thermal emission across a broad spectrum from about 1 μm to more than 100 μm.

The dust may be heated radiatively – by the central stars, for example – or by the gas via the loss of gravitational energy as it spirals through the disk to the star, called accretion heating. Accretion heating might be important in the earliest stages of disk formation and could dominate the total energy budget for a disk within the innermost 10 stellar radii or so, but it is probably unimportant for the infrared spectral energy distributions discussed in this section (Chiang & Goldreich 1996). For this reason, we will examine only radiative heating mechanisms for disks in this article.

The study of accretion onto the star is a large topic in itself and well treated by other authors. The interested reader is referred to Bertout (1989). It has been suggested that disks may be heated by wave-driven accretion (Shu et al. 1990), where the waves are first moment modes ($m = 1$). This hypothesis has yet to find favor in the community, because it adds a complication to the physics for which there is no observational evidence, and radiative heating appears to be adequate to explain almost all observations. The subject is not closed, however, and it may be that dynamical heating will become important when the angular resolution of telescopes is good enough to map the density distributions of disks.

Young objects display a variety of different spectral energy distributions depending on their state of evolution. At the very earliest stages, the SEDs

are dominated not by the disks but by envelopes surrounding a central
luminosity source most likely in free fall onto the star/disk system. After
some time, these envelopes dissipate, either because all the matter is accreted onto the star or disk or because winds from the central objects blow
away the remaining material. Lada (1987) invented a classification scheme
to describe the sequence of SEDs that result from this process, a sequence
generally used today to classify ages on the basis of SEDs. In the language
of this scheme, we are concerned with Class II sources, sources in which
the disk dominates the SEDs, and the envelope emission is negligible. The
younger objects are treated by C. J. Lada in his chapter. However, Chiang
& Goldreich (1998) show that some Class I sources can be explained as
standard T Tauri star! ! disks viewed at high inclination, and Padgett et al.
(1999) recently published infrared images of Class I sources that show almost all of them to be edge-on disks, so we must reserve judgement about
the interpretation of individual objects based solely on their spectral class.

4.1. FLAT, BLACK DISKS

To understand the infrared spectral energy distribution of a disk, it is useful
to calculate the simplest case, a thin, black disk extending from the star
to a distant boundary, specified by the outer radius, R_{max}. Figure 2 is a
schematic diagram of the star/disk system viewed exactly in the plane of
the disk and showing the variables used for the following anaylsis.

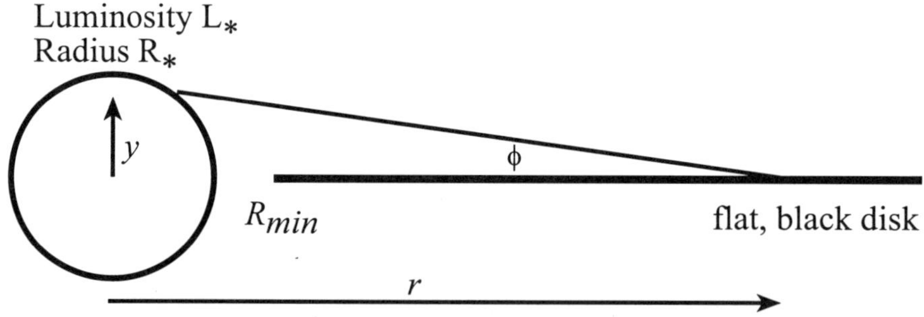

Figure 2. This sketch shows a star surrounded by a thin disk viewed in the plane of the
disk. It extends from an inner radius, R_{min}, to outer radius, R_{max} (not shown), illuminated by a star of luminosity, L_*. The disk is axisymmetric in all physical parameters.
A point at radius, r, in the disk is illuminated by a point on the star as shown with the
variables labelled. The variable, x, is into the plane of the paper.

The flux from the star that illuminates a point at radius, r, in the disk
is given by integrating over the stellar surface as seen from the disk. It

is sufficient to consider only one surface (side) of the disk. If r is sufficiently large, so that the stellar diameter subtends a small angle, the flux is approximately:

$$F_{illum} \approx \frac{1}{r^2} \int_0^{R_*} 2\sin\phi \int_0^{\sqrt{R_*^2 - y^2}} \sigma T_*^4 \, dx \, dy, \tag{1}$$

$$\approx \frac{2}{r^2} \sigma T_*^4 \int_0^{R_*} \left(\frac{y}{r}\right) \int_0^{\sqrt{R_*^2 - y^2}} dx \, dy, \tag{2}$$

$$\approx \frac{2}{3} \left(\frac{R_*}{r}\right)^3 \sigma T_*^4, \tag{3}$$

where T_* is the effective temperature of the star. The factor of $\sin\phi$ accounts for the illumination of the flat disk by a point on the star at distance, y, above the disk plane at distance, r, which is $\approx \left(\frac{y}{r}\right)$ for large enough r. This flux is totally absorbed by a black disk.

The disk radiates thermally as a blackbody, so:

$$F_{rad} = \sigma T^4(r), \tag{4}$$

from which we derive,

$$T(r) \approx \left(\frac{2}{3}\right)^{\frac{1}{4}} T_* \left(\frac{r}{R_*}\right)^{-\frac{3}{4}}. \tag{5}$$

This result is nice, because it is easy to calculate the spectral energy distribution with free parameters that are, in princple, observable. Assuming the star and disk are at distance, D, and oriented at angle, θ, to the line of sight ($\theta = 0$ corresponds to face-on viewing), the flux density is:

$$F_\nu = \frac{1}{D^2} \int_{R_{min}}^{R_{max}} \cos\theta \, B_\nu \left[T(r)\right] 2\pi r \, dr \tag{6}$$

$$= \frac{1}{D^2} \frac{4\pi h\nu^3}{c^2} \left(\frac{kT_0}{h\nu}\right)^{\frac{8}{3}} R_{min}^2 \cos\theta \int_{x_{min}}^{x_{max}} \frac{x\,dx}{exp(x^{\frac{3}{4}}) - 1}, \tag{7}$$

where we define $T_0 \equiv T(R_{min})$, the other physical constants are labelled conventionally, and we substituted variables to create the definite integral. It is easy to evaluate (7) for any choice of inner and outer radii, but the beauty of this development is that for a wide range of wavelengths, the limits are essentially 0 and ∞, and the integral is a number. To see this result, note that the limits are:

$$x_{min} \equiv \left(\frac{h\nu}{kT(R_{min})}\right)^{\frac{4}{3}} \tag{8}$$

$$x_{max} \equiv \left(\frac{h\nu}{kT(R_{max})}\right)^{\frac{4}{3}} \tag{9}$$

For most wavelengths, the emission from the inner and outer parts of the disk is negligible; $x_{min} \approx 0$ and $x_{max} \approx \infty$ for all practical purposes. The definite integral is then:

$$\int_0^\infty \frac{x\,dx}{exp(x^{\frac{3}{4}}) - 1} \approx 2.576 \tag{10}$$

and the spectral energy distribution becomes:

$$F_\nu \approx \frac{2.576}{D^2} \cos\theta \frac{4\pi h\nu^3}{c^2} \left(\frac{kT_0}{h\nu}\right)^{\frac{8}{3}} R^2_{min} \tag{11}$$

$$\approx C_1 D^{-2} \cos\theta\, T_0^{\frac{8}{3}} R^2_{min} \nu^{\frac{1}{3}} \tag{12}$$

Equation (5) shows how to determine $T(R_{min})$ for any value of R_{min}, if the stellar luminosity and source distance are known, and the disk orientation, θ, is observed with high enough angular resolution. The spectral energy distribution can be determined exactly. This predicted SED is the standard by which most SEDs have been measured to determine if a star is surrounded by a disk.

The approximations made to arrive at (12) break down for high frequencies – short wavelengths – where the emission comes from near the inner disk boundary, R_{min}, and at low frequencies where the emission comes from near the outer disk boundary, R_{max}. There is a further modification that becomes important when the inner disk boundary is close to the stellar radius, R_*. In this case, the star and disk mutually heat one another to increase the temperatures of both objects slightly – a kind of greenhouse effect – and the actual SED is more complicated: Adams, Lada, & Shu (1988) calculate this effect exactly for the interested reader. For our purposes, the approximations make very little error, since, as we will see, there are almost *no* examples of disks whose SEDs are fitted by this simple model.

We derived (12) assuming a specific temperature distribution for the disk given by radiative heating. We can generalize the result for arbitrary power law temperature distributions for disks that have other ways of establishing temperature distributions. Assume that $T(r) = T_0(r/R_{min})^{-q}$, then (7) becomes:

$$F_\nu \approx \frac{1}{D^2} \frac{4\pi h\nu^3}{c^2} \left(\frac{kT_0}{h\nu}\right)^{\frac{2}{q}} R^2_{min} \cos\theta \int_{x_{min}}^{x_{max}} \frac{x\,dx}{exp(x^q) - 1}, \tag{13}$$

$$\approx C_2 D^{-2} \cos\theta\, T_0^{\frac{2}{q}} R^2_{min} \nu^{3-\frac{2}{q}}, \tag{14}$$

where we used the approximation that $x_{min} \approx 0$ and $x_{max} \approx \infty$. The constant, C_2 includes the definite integral evaluated numerically for values of q that allow convergence.

Equation (14) is useful for SEDs that are power law in frequency. If the spectral index is defined as, $F_\nu \propto \nu^\alpha$, then (14) tells us that $\alpha = 3 - \frac{2}{q}$. The spectral index, α is observed directly. Thus, it is always possible to derive the disk temperature distribution from the observed SED. This result generally applies to black disks, and they need not be very flat to produce this result. Even the absolute temperature at any radius may be derived if D and θ are known. It is, therefore, possible to estimate disk temperatures with only modest uncertainties from the observed SED, assuming only the disk contributes (see Beckwith et al. 1990).

The flat disk model is quite useful for developing intuition to make more complicated models easier to understand. At a wavelength, λ, the thermal emission comes from a limited range of radii in the disk. The way to see this is to note that the integrand in (7) is maximum when $x \sim 0.5$, meaning $T(r) \approx 1.7hc/k\lambda$, or $r(\lambda) \approx (0.6kT(R_{min})\lambda/hc)^{\frac{4}{3}} R_{min} \approx (\lambda/\lambda(R_{min}))^{\frac{4}{3}} R_{min}$. For a young star with $L_* = L_\odot$, $T_* = 4000\,\mathrm{K}$, and adopting $R_{min} = R_* \approx 2R_\odot$, the wavelength reference is $\lambda(R_{min}) \approx 7\,\mu\mathrm{m}$. At the outer edge of the disk, we can define a similar "reference" wavelength, that, for $R_{max} = 100\,\mathrm{AU}$, say, is $\lambda(100\,\mathrm{AU}) = 7\,\mathrm{mm}$. Emission from the disk dominates the SED throughout the far infrared, roughly $10 < \lambda < 100\,\mu\mathrm{m}$.

It is also interesting to see what happens as the inner radius is increased. A disk with a hole in the center, such as seen around the main sequence stars Vega and β Pictoris, will not radiate at short wavelengths, because there is not enough material near the star. The effect is to lower the contribution from the disk in the near infrared without changing the far infrared radiation. SEDs characteristic of disks with inner holes are seen, and the changes brought about by a hole are pretty much the same for complicated disk geometries with flaring and radiative transfer as they are for the flat, black disk. Figure 3 shows a number of SEDs for different assumptions about the flat disks developed in this section.

If a gap in the disk develops, it will also change the SED. Gaps are thought to be created when the first planets are assembled in disks. Unfortunately, a gap will have only a modest effect on the SED, unless it is unusually large. The reason for this is that every wavelength gets contributions from the disk over a fairly large range of radii, because of the width of the integrand in (7) and the slow variation of temperature as a function of radius. Therefore, gaps will be difficult to detect from the SEDs alone, unless they are very large. Figure 4 gives several examples.

The flat, black disk model makes a number of predictions that can be compared with observations:

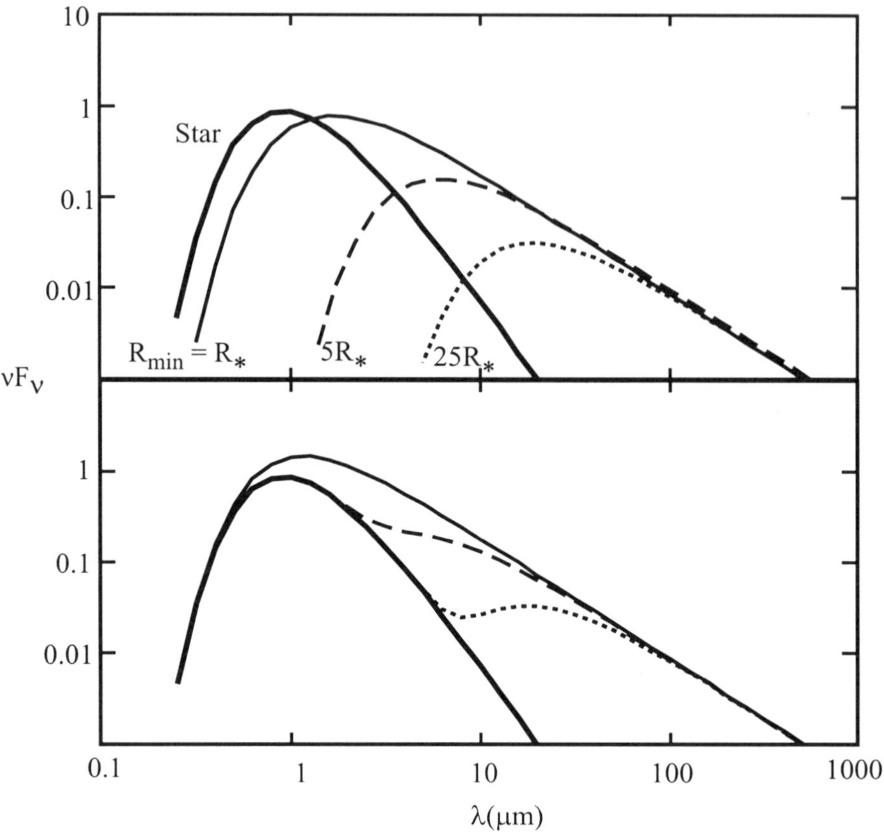

Figure 3. These spectral energy distributions were calculated from the equations for a flat disk with different assumptions about the inner disk radii as denoted in the figure. The top half shows the separate contributions to the SEDs from the star and the disks with various inner radii. The bottom half shows the total SED consisting of the star alone, then the star plus the three disks with differing inner radii.

1. The infrared spectral energy distribution is a power law: $F_\nu \propto \nu^{\frac{1}{3}}$.

2. The total disk luminosity, as measured by emission in excess of that from the stellar photosphere alone, is about $\frac{1}{4}$ that of the stellar luminosity. This fraction is exact for a disk extending from the stellar surface to infinity, and is in any case an excellent approximation for a disk extending a few AU from the star, since most of the luminosity comes from the inner radii.

3. The spectrum of the disk should be smooth, if it is truly black. If there is some radiative transfer – i.e. if the disk has finite thickness – spectral features should appear in emission, because the disk atmosphere is

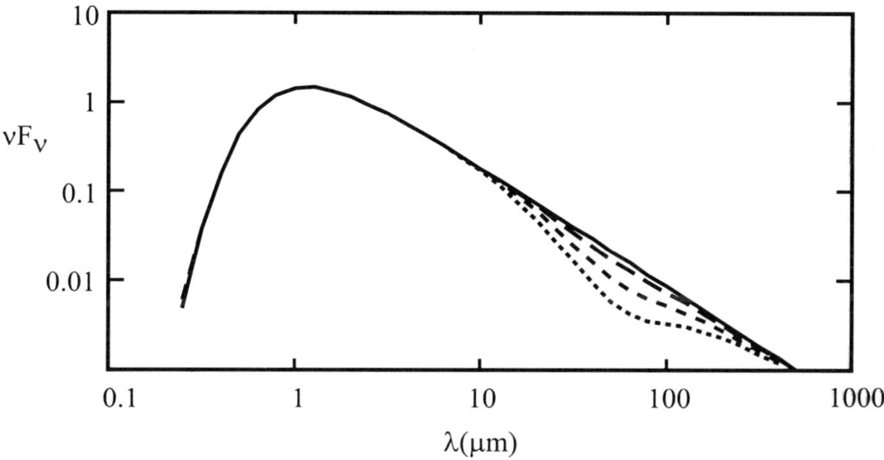

Figure 4. These spectral energy distributions were calculated assuming gaps of different sizes are created in an otherwise continuous disk. The solid line is a continuous disk; the long dash is a disk with a gap between 0.75 and 1.25 AU; the short dash is a disk with a gap between 0.5 to 2.0 AU; the dotted line is a disk with a gap between 0.3 and 3 AU. Notice that the effect of a gap is slight until the gap cuts out almost a decade of radii from the disk.

    heated from above.

4. If there is additional luminosity very close to the star, as may occur if radiation is released by accretion of matter onto the stellar surface, the disk emission will also increase in response to this additional heating.

It is easy to test these predictions. Figure 5 shows several observed spectral energy distributions of T Tauri stars with excess infrared emission compared with the predictions of the flat, black disk calculations. It is immediately obvious from the figure that the simple model does *not* fit the data. T Tauri SEDs generally have more emission at long wavelengths than the disk model permits. The infrared luminosities are, therefore, higher, and the slope of the long wavelength SED is flatter than the model. Modifications of the model by increasing the inner radius, R_{min}, or decreasing the outer boundary, R_{max} relative to the ideal model would exacerbate the differences with the data. The data show that if the T Tauri stars are surrounded by disks, the outer regions of the disks are *hotter* than they would be if heated by the central sources in a flat disk model.

This problem was recognized as soon as the first SEDs were observed (Rucinski 1985) and subsequently treated in an ad hoc manner by simply modifying the disk temperatures to empirically fit the SEDs (Adams, Lada, & Shu 1988). The ad hoc assumptions about disk temperature are unsat-

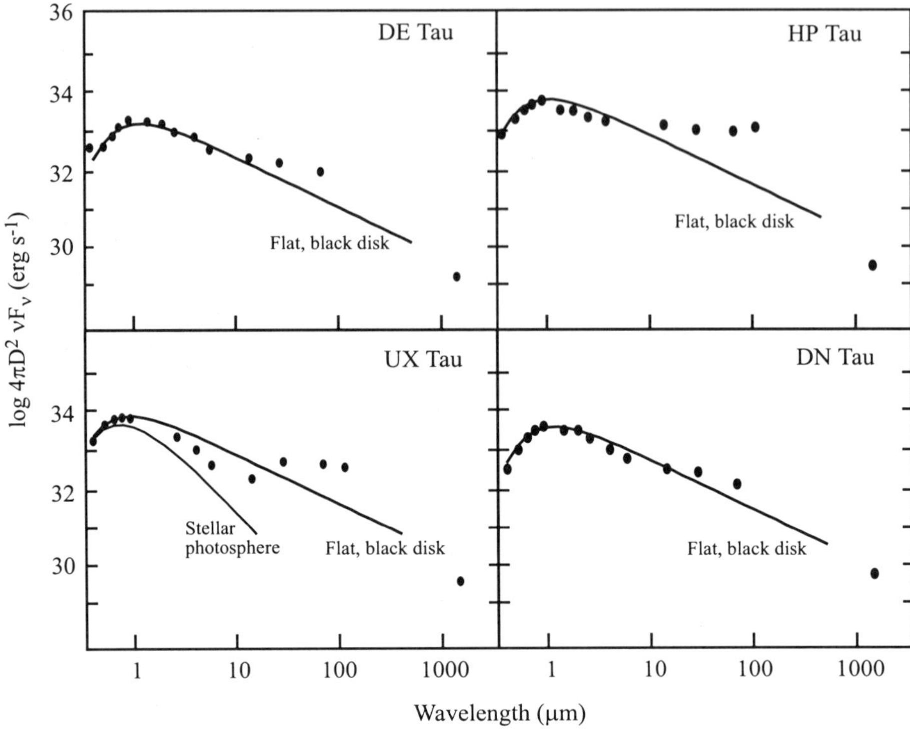

Figure 5. Observed spectral energy distributions are plotted on calculations of SEDs using the flat, black disk model. In general, the actual SEDs have more excess infrared radiation than predicted by the flat disk model.

isfying in absence of a physical theory. The flat, black disk model is not adequate to fit the great majority of data on circumstellar disks.

The observations indicate that there is more far infrared radiation than expected from flat disks. Either the outer radii are hotter than would be the case for flat disks, or there are additional components – other reservoirs of dust – that can absorb and radiate along with the disk. To heat the outer radii, one must either increase the radiation field at large distances from the star or modify the disk in some way to absorb more energy from the star. There are models in the literature for all three approaches, additional dust components, increased radiation field, and increased absorption by the disk, and there is not yet a consensus on whether any or all of the possibilities are important in different objects.

4.2. MODIFYING THE DISK SHAPE: FLARING

Disks should not be completely flat. Flaring will occur naturally in the outer parts of a disk due to the warming of the material. The disk should be in vertical hydrostatic equilibrium – i.e. normal to the surface – and the ratio of thermal to gravitational energy increases with disk radius. In the outer parts, thermal energy fattens the disk where gravity is too weak to confine the material to a thin plane. For example, if the temperature as a function of radius, $T(r) \sim r^{-3/4}$, as for a flat disk, and the vertical gravitational potential is dominated by the central star, $E_{vert} \sim -\frac{z}{r}\frac{GM_*}{r} \sim kT(r)$, then the scale height will vary with radius as, $h_{scale} \sim \frac{k}{GM_*}r^{5/4}$, and the disk will flare – we ignore factors of order unity for this example. Figure 6 shows schematically what this means.

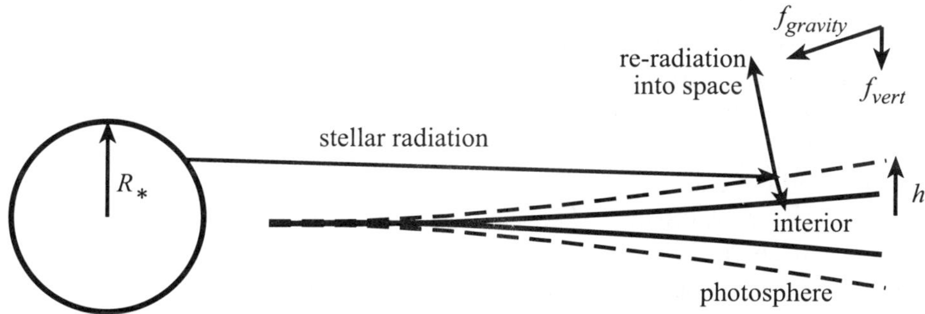

Figure 6. The flaring of a disk occurs naturally for a disk in hydrostatic equilibrium. The disk mass is assumed to be negligible; gravity from the star acts to keep the material in a plane. The scale height of the disk increases with radius, because the thermal energy decreases more slowly than the vertical gravitational energy as radius increases. The vertical gravitational force, f_{vert}, is shown as a component of the stellar gravitational force, $f_{gravity}$. The ray from the star shows the point at which short wavelength stellar radiation from the star is absorbed in the disk photosphere. The two other rays from this point show how the energy is reradiated into space and into the interior of the disk, thus heating the interior from the above.

Flaring decreases the angle between the star and the normal to the surface, thus increasing the amount of light absorbed - the angle, ϕ, in (1) is increased. Kenyon & Hartmann (1987) recognized the importance of this effect as it applies to SEDs of disks. They noted that many SEDs can be fitted well by adopting some modest flaring of the disk surface and modelled the flaring necessary to produce good fits to extant data with black disks. Calvet et al. (1992) added the important effect of radiative transfer in a flared disk. They realized that the disk atmosphere would have a temperature inversion if heated by the star, and they explicitly calculated models for flared disks with realistic atmospheres. Their numerical calculations reproduced the main features of disk SEDs for a large number of disks and specifically predicted the presence of emission features, both molecular lines and resonant dust features, if the disks are irradiated from the central ! source.

Chiang & Goldreich (1997) refined this idea by showing how the main physics, especially the flaring, could be understood analytically in a self-consistent calculation. They showed that realistic models of disk atmospheres do a good job of fitting data for most but not all disks. They also pointed out that because the dust particles that absorb the stellar radiation are in an optically thin part of the disk as far as re-emission is concerned and are small compared to the wavelengths of radiation, their equilibrium temperatures are warmer than they would be for purely black particles (cf Spitzer 1978). The upper disk atmosphere is warmer at larger distances from the star than predicted by the black disk calculations. Radiation at a specific wavelength must, therefore, have its origin over a larger area in the disk than expected from the black disk models. It will be easier to resolve these regions when very high angular resolution becomes available than predicted by some ! estimates (e.g. Bertout, Reipurth, & Malbet 1995).

The results of these calculations produce two main changes to our thinking about the physics of disks. First, the disk flares approximately as a power law with radius, where the height of the visible photosphere above the midplane, h_{phot}, as a function of distance from the star, r, is: $h_{phot} \sim r^{\frac{58}{45}}$ (Chiang & Goldreich 1997) or approximately $r^{\frac{4}{3}}$.

Second, in contrast to the black disk where radiation is effectively absorbed and emitted from a black surface, light from the star at short wavelengths – 0.5 to 2 μm, say – is absorbed in the visible photosphere at height, h_{phot}, where it heats the dust. The dust then reradiates at longer wavelengths, where it is optically thin above this layer and also downward into the disk until it is absorbed by dust in a lower layer that is opaque to the longer wavelength radiation. Because this longer wavelength radiation contains no more than half the energy originally absorbed in the photosphere, the inner layer is at a lower temperature than the photosphere. The result is a disk atmosphere in which the temperature increases vertically from the midplane outward, and most of the radiation seen by the observer is optically thin. Figure 6 shows the geometry of the radiative transfer and the different layers.

The combination of optically thin and optically thick emission from the photosphere and interior, respectively, creates an SED that flattens out or rises toward long wavelengths. Figure 7 reproduces the calculation by Chiang & Goldreich (1997) showing how these models fit the data from the T Tauri star, GM Aur (see also Calvet et al. 1992). The flared disk model can fit many of the disks in Figure 5 with modest adjustments to the free parameters. Since it is self-consistent and invokes no other components in the model except a star and a disk, it is a reasonable explanation for many observed SEDs. The superheated, optically thin layer obeys the scaling, $F_\nu \sim \nu^{-5/7}$ ($q = \frac{14}{33}$), and the interior, optically thick layer scales as, $F_\nu \sim \nu^{-2/3}$ ($q = \frac{6}{15}$).

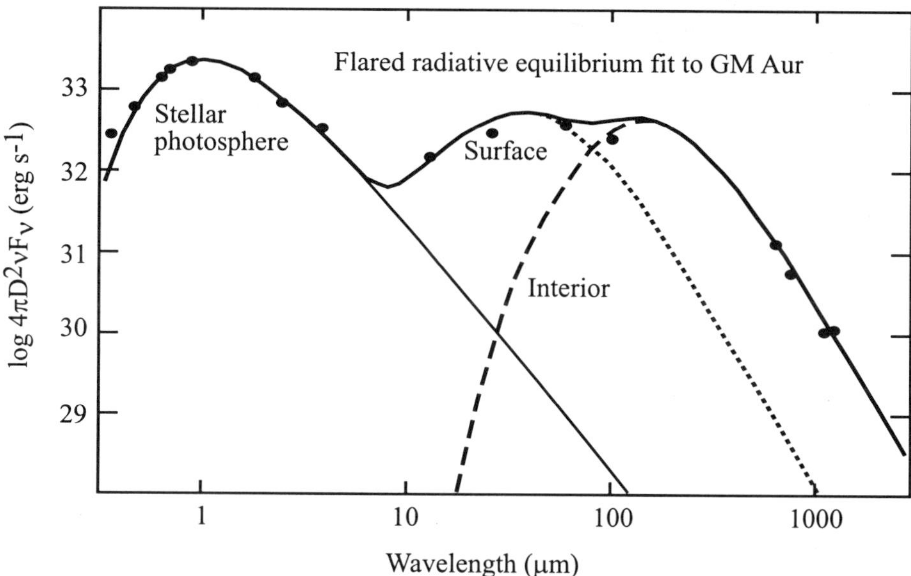

Figure 7.   Figure 8 from Chiang & Goldreich (1997) showing how a flared disk with a photosphere reproduces one SED that differs substantially from a flat, black disk. They need a hole in the inner disk to account for the lack of disk emission shortward of about $5\,\mu$m.

The prediction of Calvet et al. (1992) and Chiang & Goldreich (1997) that irradiated disks produce emission features can be tested. A strong emission feature near $10\,\mu$m usually attributed to silicate resonance in dust was first observed for T Tauri star disks by Cohen & Whiteborn (1985) and more recently by Robberto et al. (1999). Figure 8 reproduces Figure 5 from Robberto et al. (1999) showing spectra of 9 disks in the region of the silicate feature at $10\,\mu$m, that can be compared with Figure 10 from Chiang & Goldreich (1997) showing the expected strength of emission features for a radiatively heated disk. The most beautiful example of emission features is given by Waelkens et al. (1998) and Malfait et al. (1998) from their ISO spectra of the Herbig Ae/Be star HD 100546 compared with the spectrum of comet Hale-Bopp. Figure 9 reproduces these spectra and demonstrates not

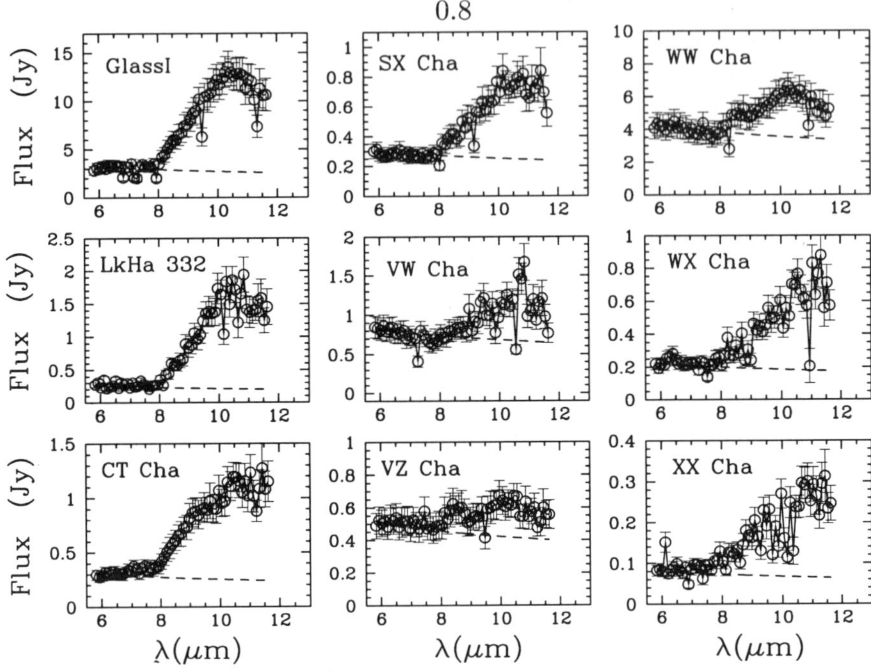

Figure 8.   Observed spectra for the 9 stars in the sample from Robberto et al. (1999). The dotted line shows the continuum fitted to the data. The emission features are due to a resonance feature in silicate dust around $10\,\mu m$ showing that these disks almost all have emission from optically thin dust, most likely in the superheated layers in the disk photospheres.

only that emission features come from optically thin dust in the d! isk, but that these features indicate the dust composition is almost identical with that in the early solar system based on the comet spectrum. The results support the conclusion that disks are radiatively heated from above, as in the flared disk model.

4.3. EXCEPTIONS: FLAT SPECTRUM SOURCES

The disk SEDs that are most difficult to fit with the models described above are the flat spectrum sources (Adams, Lada, & Shu 1988). The principal problem is that they emit more luminosity in the far infrared than at optical/near-infrared wavelengths, and so they defy explanations in which the disk temperatures are produced by radiation from the central star. Since a classical accretion disk produces an SED of identical shape to that of a flat, black disk, $T(r) \sim r^{-3/4}$, it does not help to increase the accretion rate. What is needed is to increase the temperatures in the outer disk, add another component for the far infrared radiation, or decrease the optical radiation in a very selective way; the optical/near-infrared colors are usually inconsistent with pure extinction by normal interstellar dust. All ideas to understand the flat spectrum sources include additional components to the star and disk.

A disk may be warmed by increasing the radiation flux on the outer parts if there is a way to scatter radiation onto the disk from the surrounding region. Natta (1993) suggested that an optically thin spheroidal cloud, a "halo", might scatter starlight in the outer regions of the disk to increase the energy absorbed and, therefore, the temperature. The spheroidal cloud may be a tenuous remnant of the molecular cloud from which the star and disk were created or it may be a transient part of the environment, such as residual infall or outflow in a wind. The halo acts like a greenhouse, trapping radiation at large distances from the star where the total scattering optical depth approaches unity.

Calvet et al. (1994) expanded upon this theme to compute models of multi-component systems with a star, disk, outflowing wind, and a halo out of which the wind carves a cavity. These calculations can successfully reproduce almost all observed SEDs in star/disk systems, but they include quite a few adjustable parameters, and they cannot be unique fits to the data. Calvet et al. (1994) argue that the flat spectrum sources are really closer to Class I sources in the current nomenclature than the Class II sources that concern us here, and so these models are perhaps less relevant to understanding disk physics than the ones described in the previous subsection. Chiang & Goldreich (1998) fit Class I sources using only a disk viewed edge-on. They probably need an extra scattering component to account for the visible and near infrared light from some of these stars, but the number of assumption is, nevertheless, smaller.

Calvet's calculations demonstrate that realistic models of young stellar environments can fit the data in even the most truculent of SEDs. The data do not require new physics or surprising physical conditions near the stars. Figure 10 is an example of Calvet's model applied to the flat spectrum source, HL Tau. Her models require many assumptions about the distribution of matter near the stars, including infalling envelopes with cavities, winds, and non-spherical density distributions. In principle, most of these assumptions can be checked by observation, and it should be possible to determine how well the models correspond to reality.

The introduction of additional parameters into these models does call into question our ability to derive physical parameters from the limited data sets provided by SEDs. Figure 5 shows that the number of independent points in a typical SED is limited. Often, there are at most five data

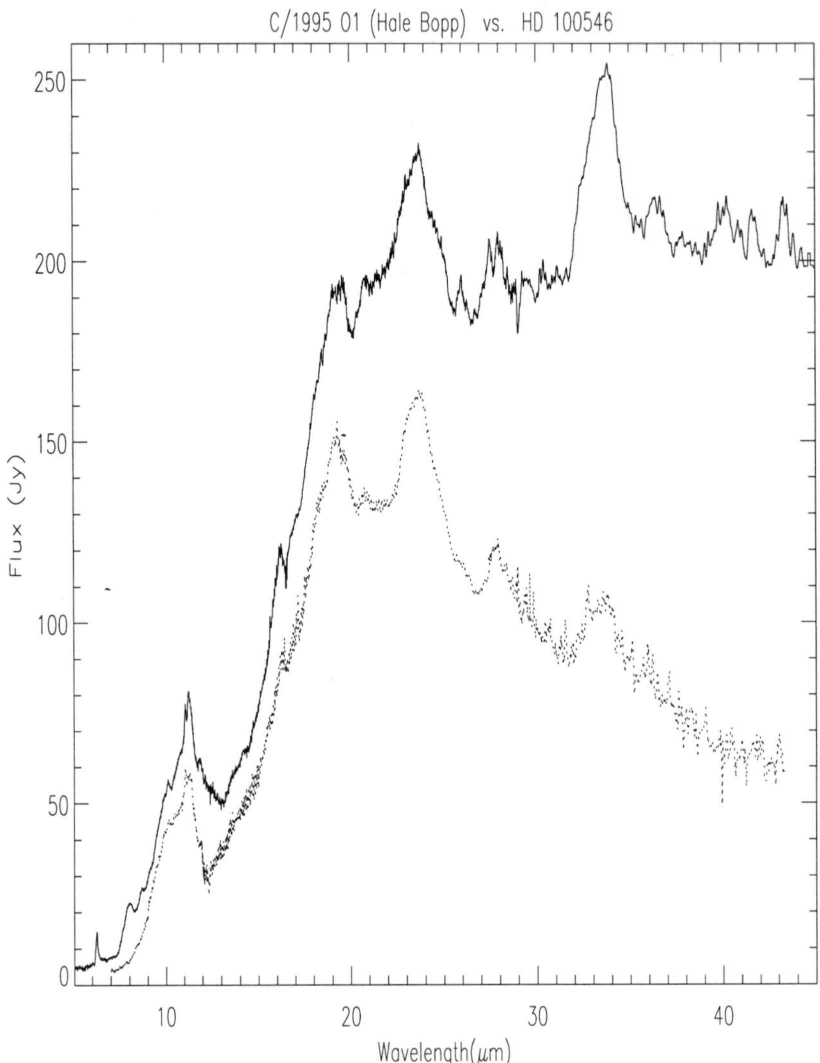

Figure 9. This is Figure 5 from Malfait et al. (1998) showing the detection of emission features in the disk around the Herbig Ae/Be star, HD 100546, (solid line at top) as predicted if the disk is heated by radiation from the central source. It is compared with the spectrum of comet Hale-Bopp (dotted line underneath). Notice the close correspondence between emission features in the comet and in the disk spectrum, indicating that the particles in the disk are made from similar material as particles in the early solar system that made up the comet.

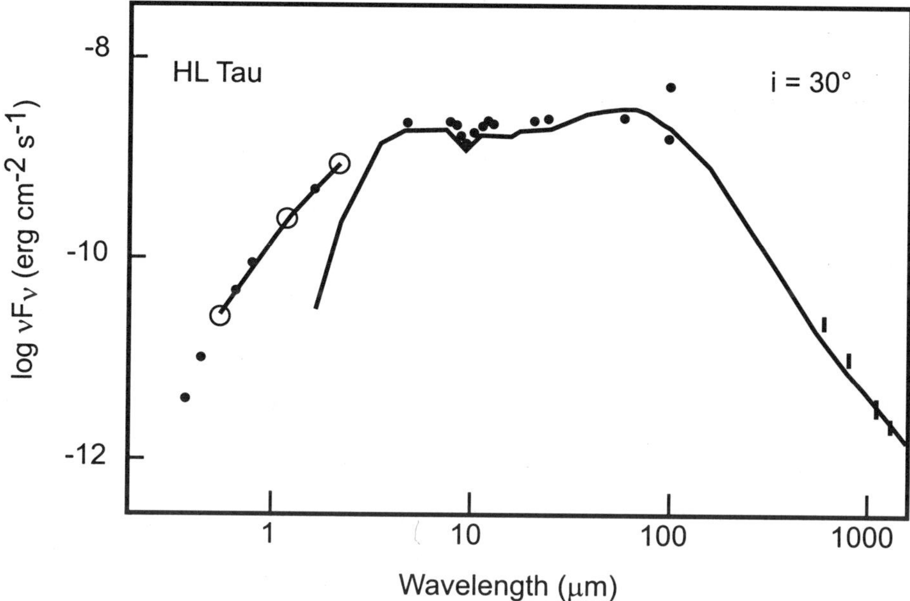

Figure 10. This is Figure 8 from Calvet et al. (1994) showing their fit to the flat spectrum source, HL Tau. Their model includes a disk, an infalling envelope with a cavity along the polar axes with an opening angle of 10°, and a scattering halo filling this cavity to produce the near infrared and optical light – shown as a separate line connecting open circles at the shortest wavelengths. This model successfully fits the SED at the expense of many adjustable parameters.

points between 10 μm and 2 mm, the four IRAS measurements and a single millimeter-wave observation. The theoretical treatment of flat disks shows that each point derives its emission from a wide range of radii in the disk, and that rather large changes in the disk – large gaps, for example – impress modest changes on the detailed SED, details that may often be confused with noise when only a few points are available. The poor angular resolution of the IRAS observations means that far infrared data are susceptible to extended emission from the surrounding molecular cloud or a very tenuous but large halo surrounding the star. It is challenging, perhaps impossible, to sor! t! out the various contributions to the SED in this case. Therefore, the analysis of SEDs can give us only limited information about the disk upon which we can rely. The most important way to disentangle the various components will need very high angular resolution in the infrared (e.g. Padgett et al. 1999) and submillimeter.

4.4. VERY LONG WAVELENGTH EMISSION: MEASURING DISK MASS

At wavelengths longer than about $300\,\mu m$, these disks become optically thin; you can see every particle in them. Therefore, the spectral energy distributions at these wavelengths are linearly proportional to the total amount of dust, and it is possible in principle to figure out the disk mass from the observations. The dust temperatures are almost always high enough to put this long wavelength emission on the Rayleigh-Jeans side of the Planck function, and the total emission becomes particularly simple to express.

The flux density, F_ν, from an optically thin disk at distance, D, is:

$$F_\nu = \frac{1}{D^2} \int_{R_{min}}^{R_{max}} B_\nu\left[T(r)\right] \tau_\nu(r) 2\pi r dr, \tag{15}$$

where $\tau_\nu(r)$ is the optical depth. The optical depth can be written in terms of the surface density, $\Sigma(r)$, and mass opacity coefficient, κ_ν as: $\tau_\nu(r) = \kappa_\nu \Sigma(r)$. If the emission is in the Rayleigh-Jeans regime, $B_\nu \approx 2kT\nu^2/c^2$. In this limit:

$$F_\nu \approx \kappa_\nu \frac{2k\nu^2}{c^2 D^2} \int_{R_{min}}^{R_{max}} T(r)\Sigma(r) 2\pi r dr \tag{16}$$

$$\approx \kappa_\nu \frac{2k\langle T\rangle\nu^2}{c^2 D^2} M_{dust}, \tag{17}$$

where $\langle T\rangle$, discussed explicitly in the next paragraph, is defined implicitly in (17). Regardless of the disk structure, the flux density is directly proportional to the particle mass opacity, κ_ν, and the total particle mass, M_{dust}, through the integral. The absolute value of κ_ν is difficult to determine accurately, although a good estimate may be made with reference to interstellar clouds. The frequency dependence of κ_ν can be observed directly.

If the temperature were constant, the integral is just the mass of the disk times the temperature:

$$\int_{R_{min}}^{R_{max}} \Sigma(r)\, 2\pi r\, dr = M_{dust}. \tag{18}$$

The variable temperature has the effect of weighting the hotter (inner) parts of the disk more than the cooler (outer) parts. If the weighting function is not too steep, we can replace the integral in (17) with $\langle T\rangle M_{dust}$, where $\langle T\rangle$ is some suitable average temperature that takes care of the weighting functions. We can calculate $\langle T\rangle$ explicitly for power law forms, $T(r) = T_0(r/R_{min})^{-q}$, and $\Sigma(r) = \Sigma_0(r/R_{min})^{-p}$:

$$I \equiv \int_{R_{min}}^{R_{max}} T_0\Sigma_0 \left(\frac{r}{R_{min}}\right)^{-q-p} 2\pi r dr \tag{19}$$

$$= 2\pi T_0 \Sigma_0 R_{min}^2 \frac{\left(\frac{R_{max}}{R_{min}}\right)^{2-q-p} - 1}{2 - q - p} \tag{20}$$

$$= T_0 M_{dust} \frac{\left(\frac{R_{max}}{R_{min}}\right)^{2-q-p} - 1}{\left(\frac{R_{max}}{R_{min}}\right)^{2-p} - 1} \frac{2 - p}{2 - q - p} \tag{21}$$

$$\approx T_0 M_{dust} \left(\frac{R_{max}}{R_{min}}\right)^{-q} \frac{2 - p}{2 - q - p}, \tag{22}$$

where we have assumed that $2 - q - p > 0$ for the last approximation and note that $R_{max}/R_{min} \gg 1$; typically, $R_{min} \sim 0.01\,\mathrm{AU}$ and $R_{max} \sim 100\,\mathrm{AU}$ making this an excellent approximation. If $2 - q - p = 0$ or $2 - p = 0$, one of the integrals is logarithmic. The approximation does not hold for $2 - q - p < 0$, but it is easy to work out the answers for those cases, too. For the case in (22):

$$\langle T \rangle \approx T_0 \left(\frac{R_{max}}{R_{min}}\right)^{-q} \frac{2 - p}{2 - q - p}. \tag{23}$$

Thus, for $q = 0.5$ and $p = 1$, $\langle T \rangle \approx 2T(R_{max})$, and $F_\nu \approx \kappa_\nu \frac{4kT(R_{max})\nu^2}{D^2} M_{dust}$.

Using the analytical results from Chiang & Goldreich's analysis for the interior disk layer that produces almost all of the long wavelength radiation, we can work out $\langle T \rangle$ for a centrally irradiated, flared disk of the sort in their model. Noting that $q = 6/14$ and $p = 3/2$ in their calculations:

$$\langle T \rangle \approx T_0 \frac{\left(\frac{R_{max}}{R_{min}}\right)^{\frac{1}{14}} - 1}{\left(\frac{R_{max}}{R_{min}}\right)^{\frac{1}{2}} - 1} \frac{\frac{1}{2}}{\frac{1}{14}} \tag{24}$$

$$\approx \frac{1}{2} T_0 \left(\frac{R_{max}}{R_{min}}\right)^{-\frac{1}{2}} \ln\left(\frac{R_{max}}{R_{min}}\right) \tag{25}$$

$$\approx \frac{1}{2} T(R_{max}) \left(\frac{R_{max}}{R_{min}}\right)^{\frac{1}{14}} \ln\left(\frac{R_{max}}{R_{min}}\right) \tag{26}$$

$$\approx 2.4\, T(R_{max}). \tag{27}$$

For the numerical answer in (27), I assumed that $(R_{max}/R_{min}) = 10^4$, but the very weak dependence of $\langle T \rangle$ on this ratio means that it is not a major uncertainty. This result is useful as a simple way to estimate disk mass from the temperature in the outer regions and the observed flux density. In most models including Chiang & Goldreich's, the temperature in the outer parts of the disk is around $20\,\mathrm{K}$. This means that $\langle T \rangle \approx 50\,\mathrm{K}$ for the purposes of estimating disk mass. In principle, it can be determined directly from millimeter-wave maps of molecules in the outer disk. In any case, it is not very sensitive to the conditions in the disk for these models.

The major uncertainty in mass estimates is in κ_ν, not in the model parameters. Theoretically, κ_ν should be well determined for particles that are much smaller than the wavelength of observation, but it can vary substantially with changes in particle composition – silicates vs. iron, for example – and to some extent with particle size and shape. Attempts to determine κ_ν from observations have given disparate results at millimeter wavelengths to date (Draine 1989; Beckwith & Sargent 1991), with variations of an order of magnitude published for $\sim 1\,\mathrm{mm}$ wavelength.

Theoretically, κ_ν is expected to vary as a power of the frequency, ν: $\kappa_\nu = \kappa_0(\nu/\nu_0)^\beta = \kappa_0(\lambda/\lambda_0)^{-\beta}$. A commonly adopted value at the time of this writing is $0.02(1.3\,\mathrm{mm}/\lambda)\,\mathrm{cm}^2\,\mathrm{g}^{-1}$ from Beckwith et al. (1990) and applies to total gas mass assuming $M_{gas}/M_{dust} = 100$, but this value is by no means certain and could be wrong by a factor of five in either direction. Adopting this value, the disk mass can be written as a function of the continuum flux density at millimeter wavelengths as:

$$M_{disk} = 0.03\,M_\odot\,\frac{F_\nu}{1\,\mathrm{Jy}}\left(\frac{D}{100\,\mathrm{pc}}\right)^2\left(\frac{\lambda}{1.3\,\mathrm{mm}}\right)^3\frac{50\,\mathrm{K}}{\langle T\rangle}\frac{0.02\,\mathrm{cm}^2\,\mathrm{g}^{-1}}{\kappa_{1.3\,\mathrm{mm}}}. \quad (28)$$

4.5. SUMMARY

The spectral energy distributions of disks can reveal quite a bit of information about their physical properties. Some of this information is robust, and some is model dependent and must be regarded with skepticism.

- Disks have broad thermal emission from the near infrared to the submillimeter, much broader than a Planck function (blackbody). The emission is often well fitted by a power law over an order of magnitude or more in wavelength.
- The radial temperature dependence of the photospheric layer has an almost one-to-one relationship with the SED. A good estimate of the temperature as a function of radius, $T(r)$, can be derived from the SED.
- Typical disk temperatures are warmer than predicted by pure accretion heating or pure irradiation of a flat, black disk by a central source. Flaring of the disks is an economical means of explaining the warmer temperatures for many but not all disks.
- Large gaps, including inner holes, will create "holes" in the SED, as well, although the effect of gaps is diminished because each wavelength derives contributions from a wide range of radii in the disk.

- Disks are expected to be optically thin (transparent) at millimeter wavelengths, and the total mass may be estimated from a measurement of the optical depth.
- Disks with residual material from their formation or from central outflows often display complex SEDs that require several components to reproduce. The SEDs alone are inadequate to demonstrate that the model distributions reflect the actual disk environment, but the model predictions can be checked from observations with high enough angular resolution.

5. Properties of Disks

It is not possible to image most disks in the manner of Figure 1. They are rarely situated in front of bright background sources, such as those in the top half of the figure, and they are normally not sufficiently edge-on to be seen in scattered light as they are in the bottom half of the figure. Detecting disks for the purposes of counting is normally done by looking for infrared radiation in excess of the stellar photosphere and assuming that any excess is due to disks.

5.1. HOW MANY YOUNG STARS HAVE DISKS?

As a general rule, it is easier to infer the presence of disks from radiation at wavelengths longward of $10\,\mu$m, say, because that radiation is generated far from the star and requires more than trace amounts of dust to be observable. The first attempts to count disks in samples of young stars used far infrared and millimeter wave observations. Strom et al. (1989) and Cohen, Emerson, & Beichman (1989) found that nearly half of the young stars in their samples were detected in one or more of the IRAS bands – 12, 25, 60, and $100\,\mu$m – and were likely to have disks. Beckwith et al. (1990) looked at similar samples at 1.3 mm detecting about 40% of them. These early studies asserted that disks were common among classical T Tauri stars.

A variety of different surveys since then have generally supported the finding that disks are common, although the exact frequency depends on the sample and the selection criteria for young stars. It is much easier to detect light in the near infrared, at $2\,\mu$m, for example, where large arrays of sensitive detectors are readily available, but the amount of radiation from a disk is a much smaller fraction of the total – the star is bright, too – and very little dust is needed to produce an excess. Surveys of dark clouds at $2\,\mu$m show disk fractions in anywhere from 10% to 90% of the sample. The high figure, 90%, is seen in the innermost regions of the Orion Trapezium

cluster (McCaughrean & Stauffer 1994), whereas the low figure includes
large numbers of stars selected on the basis of x-ray emission (Walter et
al. 1988). It is still controversial if x-ray emission is sufficient by itself to
reveal very young stars, so the 10% figure should be regarded with ca! ution.
Recent surveys at $15\,\mu$m with the ISO satellite give numbers in the 30% to
50% range (Nordh et al. 1996; Nordh et al. 1998; Bontemps et al. 1998). It
is reasonable to conclude that disks accompany young stars about half the
time, and it may well be that all stars are born with disks.

5.2. DISK LIFETIMES

The lifetime of the dust is determined by counting disk fractions in young
clusters with different ages. The first systematic attempts were made by
Beckwith et al. (1990) at 1.3 mm and Skrutskie et al. (1990) at $10\,\mu$m. The
latter paper found that the sample fraction with disks decreased markedly
between ages of a few million and ten million years. More recent studies
using near-infrared excesses shows a steady decrease to almost zero at 10
million years (E. Lada; this volume) indicating that the inner parts of the
disks disappear over this time. A similar study at 25 and $60\,\mu$m shows the
outer parts of the disks disappear just as rapidly (Robberto et al. 1999;
Meyer et al. 1999).

Main sequence stars can also have tenuous disks, although the disk
masses are many orders of magnitude less than their young counterparts
(Backman & Paresce 1993). These disks are presumably the remnants of
those in the young stars, and they may be generated by the grinding of
larger bodies – an asteroid belt, for example – orbiting the stars. Zuckerman & Becklin (1993) inferred a decrease of disk mass with time following
approximately: $M_{disk} \sim t^{-2}$ during the first 300 Myr. This result suggests
a steady loss of small particles over almost the entire pre-main sequence
lifetime of a star.

Small particles should coagulate into larger ones on even faster timescales.
The early solar nebula is thought to have created small planets within
10 Myr. The observed disk lifetimes are, therefore, consistent with expectations for disks making planets, although there is no strong evidence that
the particles are being accumulated into planet-sized bodies.

5.3. DISK MASSES

Most of the disk mass should be gaseous hydrogen and helium. The particles
probably contain only 1% of the mass when a disk is first assembled from
interstellar clouds. This ratio may change as the disk evolves.

The particles are easy to observe, and the total particle mass can be
estimated from the observed optical depth at millimeter wavelengths, as

we saw in a previous section. Millimeter wavelength observations of large samples show that typical disks are a few percent of a solar mass with a wide range. Figure 11 shows a histogram of disk masses for samples in the Taurus and Ophiuchus dark clouds (Beckwith et al. 1990; André et al. 1994; Osterloh & Beckwith 1995). These samples yield consistent results. New data (E. Lada 1999, this volume) indicate that the disks in the Orion star forming region may have lower masses on average, although the sensitivities of the observations are not yet good enough to make a robust conclusion.

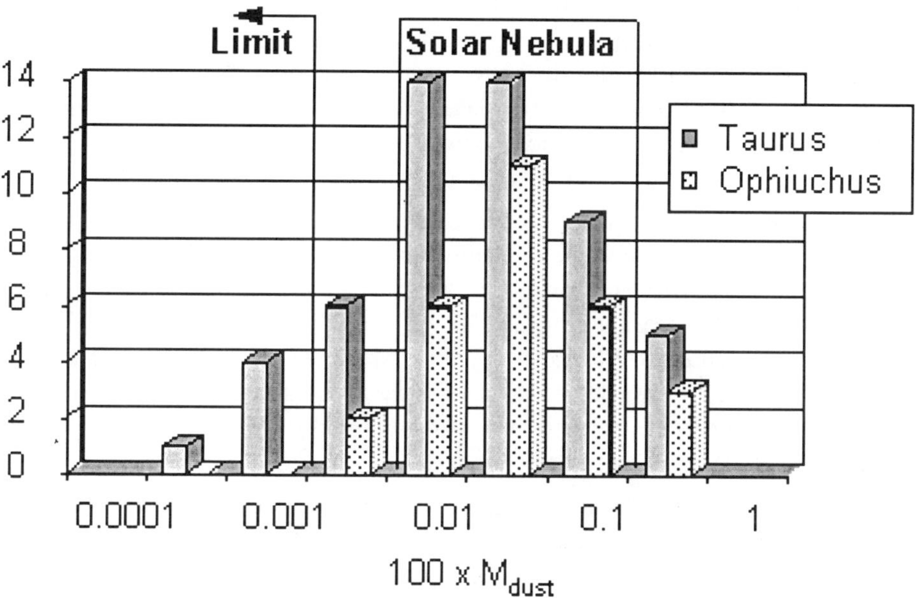

Figure 11. The histograms show the distribution of disk masses among stars in the Taurus and Ophiuchus star forming regions determined by Beckwith et al. (1990) and André et al. (1994).

Theories of the early solar system require between 0.01 and 0.1 $M_\odot$ in the solar disk to create our planetary system. Disk masses around other stars fall in this range. It is clear that there is enough mass in a typical disk to create a planetary system like our own, but there is no proof that it will do so.

If disks do make planets, there should be evidence of particle growth in the older disks. We know that particles are lost as the disks age. If the loss mechanism is accumulation into larger bodies, it seems likely that planetary systems are the final product of this evolution; if there is no particle growth, the disks may simply disperse into space or be accreted onto the central stars.

5.4. PARTICLE SIZES

At submillimeter wavelengths and longer, κ_ν is expected to vary as a power of the frequency or wavelength: $\kappa_\nu \propto \nu^\beta \propto \lambda^{-\beta}$. For compact spherical particles smaller than the observing wavelength, $\beta = 2$ for metals and insulators under a wide range of conditions (Bohren & Huffman 1983; Emerson 1988). The power law exponent, β, can be as small as 1 for certain types of materials - amorphous carbonaceous material, for example - and even smaller over limited ranges, but it must go to 2 at long enough wavelengths to fulfill causality relations. It is expected that $\beta \approx 2$ for interstellar dust particles at long wavelengths. Careful attempts to measure both κ_ν and β in interstellar clouds have, indeed, yielded $\beta \approx 2$ and values of $\kappa_\nu = (0.002 - -0.004)(\lambda/1.3\,\mathrm{mm})^{-2}\,\mathrm{cm^2\,g^{-1}}$ (Hildebrand 1983; Draine and Lee 1984). If a mixture of particle types and si! zes is present, the observed value is an average over the different constituents.

Rocks, asteroids and planets are opaque to radiation at $\lambda \sim 1\,\mathrm{mm}$, in which case $\beta = 0$. Pebbles with sizes of order $1\,\mathrm{mm}$ should have exponents with intermediate values: $0 < \beta < 2$. If dust grains grow large enough to put most of the mass in bodies larger than pebbles, they should have an observable signature at these wavelengths. Because the radiation does not penetrate far below the surface of a large body, meaning most of the material never interacts with it, the absolute value of the opacity per unit mass must decrease as the particles grow.

Changes in the spectral energy distributions (SEDs) of disks relative to those typical of interstellar clouds are observed (Beckwith and Sargent 1991; Mannings & Emerson 1994; Koerner et al. 1995). The flux densities, F_ν, of disks tend to fall more slowly at wavelengths longer than about $400\,\mu\mathrm{m}$ than those of interstellar clouds. The spectral index, α - where $F_\nu \propto \nu^\alpha$ - is typically between 2 and 3, whereas $\alpha \sim 4$ for the interstellar medium. Assuming optically thin emission in the Rayleigh-Jeans limit, the exponent, β, is between 0 and 1 for disks compared to the theoretical value of 2 for small particles in the long wavelength limit. The right hand part of Figure 12 shows the distribution of emissivity exponents, β, determined from the observed spectral slope of disk emission near $\lambda \sim 1\,\mathrm{mm}$. Many of the disks in this sample have $\beta \leq 1$. It is tempting to interpret this change ! as the result of particle growth.

Figure 12 includes the distribution of β's derived from model fits assuming the surface density is proportional to $r^{-3/2}$. Including the surface density distribution tends to increase the derived value of the emissivity exponent relative to the observed spectral slope. For example at $1.3\,\mathrm{mm}$, observations of a disk with $M_D = 0.03\,\mathrm{M_\odot}$, $R_D = 100\,\mathrm{AU}$, and $\kappa_\nu = 0.02\,\mathrm{cm^2\,g^{-1}}$, the average optical depth, $\bar{\tau}_\nu = 0.24$, and $\beta \approx 1.36(\alpha - 2)$. Any conclusions

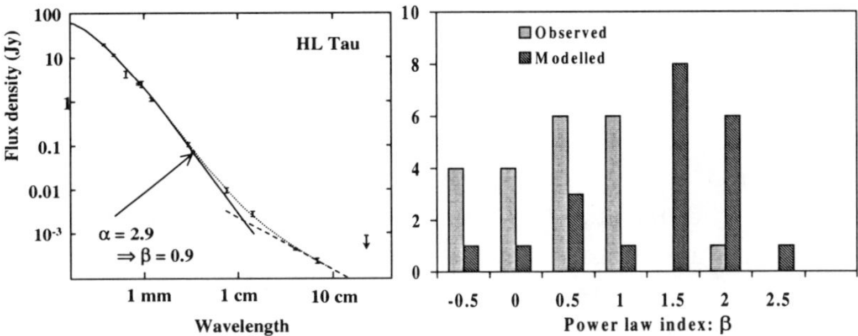

Figure 12. The left hand part shows the spectral energy distribution of HL Tau from about $350\,\mu$m to 6 cm wavelength. The solid line fits the short wavelengths – thermal emission by dust – and the dashed line fits the long wavelengths – free-free emission from ionized gas. The long wavelength spectral index of the dust is 2.9, meaning $\beta = 0.9$ with no corrections for optically thick emission (Wilner, Ho, & Rodriguez 1996). The right hand portion shows the distribution of β's derived in two ways for a larger sample of disks. The "observed" β's are derived directly from the spectral index as described in the text: $\beta_{obs} = \alpha - 2$. The "model" β's are derived by assuming a surface density distribution as discussed in the text (Beckwith & Sargent 1991).

about particle growth depend on knowledge of how much optically thick parts of the disks contribute to the spectral energy distribution at long wavelengths. It is probably for this reason that the samples of disks with measured α's at long wavelengths have not been enlarged much since the work of Beckwith & Sargent (1991) and Mannings & Emerson (1994).

To measure the optical depth, it is essential that the disk emission be resolved at wavelengths where it is relatively transparent so that the surface density distribution may be determined directly from the distribution of optical depth. Resolving the disks is time-consuming with the present suite of millimeter facilities. Several groups have now managed to resolve the emission from one star, HL Tau, by carrying out novel experiments with existing millimeter-wave telescopes. Lay et al. (1994) and Lay, Carlstrom & Hills (1997) combined the CSO and JCMT into a single-baseline interferometer to observe at 0.65 and 0.87 mm and combined it with data from the Owens Valley millimeter interferometer at 1.3 mm. Mundy et al. (1996) extended the BIMA array to measure a size of the disk at 2.7 mm wavelength, and Wilner, Ho & Rodriguez (1996) used the VLA to get sub-arcsecond resolution of the disk at 7 mm. Each group resolved the emission but with beam sizes of the same order! ! as the disk size, and the sub-millimeter observations had $u - v$ plane coverage too poor to map the brightness distribution.

All groups measured the average brightness temperature and overall disk orientation and constrained subsequent model fitting much more than is possible from the SEDs alone.

6. Are Disks Important?

Disks are common around young stars, and they contain enough mass to make planetary systems like our own.. At the time of this writing, sixteen planets orbiting stars other than the Sun have been inferred using radial velocity observations of the stars – although there is some controversy as to whether each of the objects is a planet according to usual definitions. Since we know some extrasolar planets exist and there are many disks, it is a small leap to suppose that disks make planets, as the theory asserts, and there are many planetary systems. This is an assumption, however, and proof is not likely to be forthcoming until we have actually observed some planets built up within the disks.

If planetary systems are common and disks are the necessary precursors, disks are important for the understanding of planetary systems. They are likely to be our only means of filling in many missing links in our understanding of early Solar System evolution. To understand our own origins and to assess the likelihood of other planetary systems, further study of disks around other stars presents one of the most fruitful avenues.

On the other hand, the disks do not appear to be of great importance to the early evolution of the star after the main collapse has occurred. From the T Tauri stage until the main sequence, disks probably contain only a few percent of a stellar mass, and the amount of material accreted onto the star during this period is not likely to be sufficient to add significantly to a star's mass. It is possible that magnetic coupling between the disk and the star governs stellar rotation and determines the star's final rotation rate. There are some hints of this process but no proof.

Many of us suspect that the very early stages of disk formation are crucial to the buildup of a star. During the initial collapse, there is almost certainly too much angular momentum in the molecular cloud to make a centrally condensed star unless a mechanism exists to shed angular momentum during the collapse. A disk is a natural repository of such angular momentum. In this way, disks may be essential to a star's birth. But the disk angular momentum appears to be insufficient to take up that from the molecular cloud by the T Tauri phase (Beckwith etal. 1990), suggesting a large redistribution of angular momentum within the disk itself subsequent to the collapse phase. It is still difficult to study disks in great detail during and immediately after the collapse phase. There are not many candidates, yet – the entire phase is very short-lived – and the extinction toward the very young disks is too high to allow even near infrared observations. Very high resolution mill! ! imeter wave images enabled by planned large interferometers (MMA/LSA) will likely change this situation and make observations of the physics of young disks routine.

ACKNOWLEDGMENTS

I am grateful to Charlie Lada and Nick Kylafis for their generosity in inviting me to the Crete conference and to the postdocs and students of the conference for their stimulating questions and comments on circumstellar disks. Portions of the text were taken from articles that I co-authored with Anneila Sargent, Thomas Henning, and Yoshitsugu Nakagawa; I am grateful for their permission to reprint those parts in this article.

References

Adams, F. C., Lada C.J. & Shu, F.H. 1987, ApJ, 312, 788
Adams, F. C., Lada C.J. & Shu, F.H. 1988, ApJ, 326, 865
André, P., & Montmerle, T. 1994. ApJ, 420, 837.
Backman, D. E., & Gillett, F. C., 1987, in *Protostars & Planets III*, eds. E. Levy & J. Lunine, (Univ. Arizona Press:Tucson) p. 1253
Bastien, P. & Menard, F. 1990 ApJ, 364, 232
Beall, J.H. 1987, ApJ, 316, 227
Beckwith, S. V. W., Henning, Th., & Nakagawa, Y. 1999, in Protostars and Planets IV, V. Mannings ed., Tucson: U. Arizona Press, in press
Beckwith, S. V. W., & Sargent, A. I. 1991, ApJ, 381, 250
Beckwith, S. V. W., & Sargent, A. I. 1996, Nature, 383, 139
Beckwith, S., Sargent, A. I., Scoville, N. Z., Masson, C. R., Zuckerman, B., & Phillips, T. G. 1986, ApJ, 309, 755
Beckwith, S. V. W., Sargent, A. I., Chini, R. S., & Gusten R. 1990, AJ, 99, 924
Beckwith, S., Zuckerman, B., Skrutskie, M. F., & Dyck, H. M. 1984, ApJ, 287, 793
Bertout, C. 1989, ARAA, 27, 351
Bertout, C., Reipurth, B., & Malbet, F. 1995, in *Science with the VLT*, eds. J. R. Walsh & I. J. Danziger (Springer-Verlag, Heidelberg), p. 41
Bodenheimer, P. 1995, ARAA, 33, 199
Bohren, C. F., & Huffman, D. R. 1983. *Absorption and Scattering of Light by Small Particles* (New York:Wiley)
Bontemps, S. et al. 1998, in *Star Formation with the Infrared Space Observatory*, 24-26 June 1997, Lisbon, Portugal, eds J. Yun & R. Liseau, ASP Conference Series, 132, 141
Burrows, C. J., Stapelfeldt, K. R., Watson, A. M., Krist, J. E., Ballester, G. E., Clarke, J. T., Crisp, D., Gallagher, J. S. III, Griffiths, R. E., Hester, J. J., Hoessel, J. G., Holtzman, J. A., Mould J. R., Scowen P. A., Trauger, J. T., & Westphal, J.A., 1996, ApJ, 473, 437
Calvet, N., Magris, G. C., Patino, A., & D'Alessio, P. 1992, RMxAA, 24, 27
Calvet, N., Hartmann, L., Kenyon, S. J., & Whitney, B. A. 1994, ApJ, 434, 330
Cameron, A. G. W. 1988, ARAA, 26, 441
Chiang, E.I., & Goldreich, P. 1997, ApJ, 490, 368
Chiang, E.I., & Goldreich, P. 1999, astro-ph/9812194
Cohen, M. 1983, ApJ, 270, L69
Cohen, M., & Witteborn, F.C. 1985, ApJ, 294, 345

Cohen, M., Emerson, J.P., & Beichman C.A. 1989, ApJ, 339, 455

Draine, B. T. 1989, Proceedings of the 22nd ESLAB Symposium, Infrared Spectroscopy in Astronomy, ed. B. H. Kaldeich (ESA Publications, Noordwijk)

Draine, B. T., & Lee, H. M. 1984, ApJ, 285, 89

Dutrey, A., Guilloteau, S. & Simon, M. 1994, AA, 286, 149

Elsässer, H. & Staude, J. 1978, AA, 70, L3

Emerson, J. P. 1988, In *Formation and Evolution of Low Mass Stars*, eds. A. K. Dupree & M. T. V. Lago (Dordrecht: Kluwer Acad. Publ.), p. 21

Grasdalen, G. L., Strom, S. E., Strom, K. M., Capps, R. W., Thompson, D., & Castelaz, M. 1984, ApJ, 283, 57

Hartmann, L., Hewett, R., Stahler, S., & Mathieu, R. 1986, ApJ, 309, 275

Hildebrand, R. H. 1983, *Q.J. RAS*, 24, 267

Kalas, P. & Jewitt, D. 1995, AJ, 110, 794

Kalas, P. & Jewitt, D. 1997, Nature, 386, 52

Kenyon, S. J., & Hartmann, L. 1987, ApJ, 323, 714

Kenyon, S. J., Yi, I., & Hartmann, L., 1996, ApJ, 462, 439

Koerner, D. W., Chandler, C. J., & Sargent, A. I. 1995, ApJ, 452, L69

Koerner, D. W., & Sargent, A. I. 1995, AJ, 109, 2138

Lada, C. J. 1987, in *Star Forming Clouds*, IAU Symp 115,eds. M. Peimbert & J. Jugaku, (Dodrecht, Reidel), p. 1

Lada, E. 1999, this volume.

Lay, O. P., Carlstrom, J. E., Hills, R. E., & Phillips, T. G. 1994, ApJ, 434, L75

Lay, O. P., Carlstrom, J. E., & Hills, R. E. 1997, ApJ, 489, 917

Lynden-Bell, D., & Pringle, J. E. 1974, MNRAS, 168, 603

Malfait, K., Waelkens, C., Waters, L. B. F. M., Van den Bussche, B., Huygen, E., & de Graauw, M. S. 1998, AA, 332, 25

Mannings, V., & Emerson, J. P. 1994, MNRAS, 267, 361

McCaughrean, M. J., & O'Dell, C. R. 1996, AJ, 111, 1977

McCaughrean, M. J. & Stauffer, J. R. 1994, AJ, 108, 1382

Meyer, M. R., Beckwith, S. V. W., Stauffer, J. R., & Schultz, B. 1999, in preparation

Miyake, K., & Nakagawa, Y., 1993, Icarus, 106, 20

Mizuno, H., Markiewicz, W., & Völk, H. 1988, AA, 195, 183

Mundt, R. & Fried, J. 1983, ApJ, 274, L83

Mundy, L. G. et al. 1996, ApJ, 464, L169

Natta, A., 1993, ApJ, 412, 761

Natta, A., Meyer, M. R., & Beckwith, S. V. W., 1999, in preparation

Nordh, L. et al. 1996, AA, 315, L185

Nordh, L. et al. 1998, in *Star Formation with the Infrared Space Observatory*, 24-26 June 1997, Lisbon, Portugal, eds J. Yun & R. Liseau, ASP Conference Series, 132, 127

O'Dell, C. R. & Wen, Z., 1994, ApJ, 436, 1940

Osterloh, M. & Beckwith S. V. W., 1995, ApJ, 439, 288

Padgett, D. L., Brandner, W., Stapelfeldt, K. R., Strom, S. E., Tereby, S., & Koerner, D. 1999, Astro-ph/9902101

Piirola, V., Scaltriti, F., & Coyne, G. V. 1992, Nature, 359, 399

Robberto, M., Beckwith S. V. W., & Meyer, M. R., 1999, Paris meeting on observations with the ISO satellite.

Rucinski, S. M., 1985, AJ, 90, 2321

Saito, M., Kawabe, R., Ishiguro, M., Miyama, S. M., & Hayashi, M. 1995, ApJ, 453, 384

Safronov, V. 1969, in *Evolution of the Protoplanetary Cloud and the Formation of the Earth & Planets*, NASA TTF-667, (translation from Russian)

Sargent, A. I. & Beckwith, S. V. W. 1987, ApJ, 323, 294

Shu, F. H., Adams, F. C., & Lizano, S. 1987, ARAA, 25, 23

Shu, F. H., Najita, J., Galli, D., Ostricker, E., & Lizano, S. 1993, in *Protostars & Planets III*, eds E. Levy & J. Lunine (Univ. Arizona Press:Tucson) p. 3

Shu, F. H., Tremaine, S., Adams, F. C., & Ruden, S. P. 1990, ApJ, 358, 495

Skrutskie, M. F., Dutkevitch, D., Strom, S. et al. 1990, ApJ, 99, 1187
Spitzer, L., Jr. 1978,*Physical Processes in the Interstellar Medium*, (Wiley, New York) p. 191-192.
Stapelfeldt, K. R. 1998, Exo-zodiacal Dust Workshop, p. 289
Stevenson, D. J. 1982, Planet. & Space Sci. 30, 755
Strom, K. M., Newton, G., & Strom, S. E. 1989, ApJSS, 71, 183S
Waelkens, C., Malfait, K., & Waters, L. B. F. M. 1998, Ap&SS, 255, 25
Walter, F. M., Brown, A., Mathieu, R.D. et al. 1988, AJ, 96, 297
Weidenschilling, S. J. 1987, J. Gerlands Beitr. Geophys, 96, 21
Weidenschilling, S. J., & Cuzzi, J. N., 1993, in Protostar & Planets III, p.1031
Wetherill, G. 1990, Ann. Rev. Earth Planet. Sci., 18, 205
Wetherill, G. W. & Stewart, G. R. 1993, Icarus, 106, 190
Whitney, B. A. & Hartmann, L. 1992, ApJ, 395, 529
Wilner, D. J., Ho, P. T. P., & Rodriguez, L. F. 1996, ApJ, 470, L117
Wood, J. A. & Morfill, G. E. 1988, in *Meteorites and the Early Solar System*, eds. J. F. Kerridge & M. S. Matthews, 329-347 (U. of Arizona Press, Tucson)
Zuckerman, B. & Becklin, E. E. 1993, ApJ, 414, 793

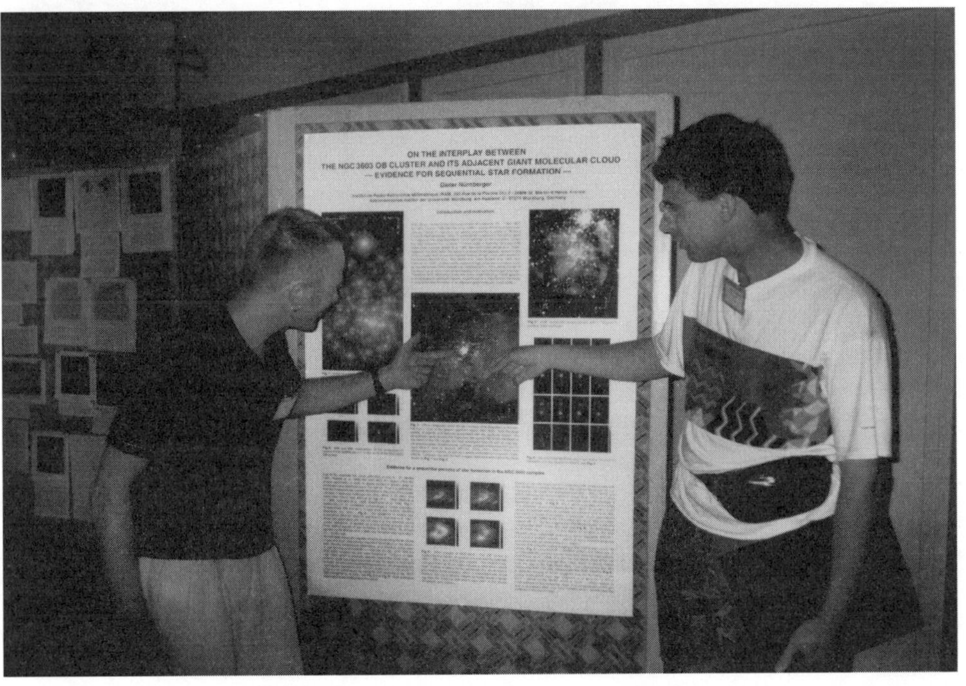

"The source is here,.....no, it's here!"

Scott Kenyon explains his poster.

ACCRETION DISKS AND ERUPTIVE PHENOMENA

SCOTT J. KENYON
Smithsonian Astrophysical Observatory
60 Garden Street, Cambridge, MA 02138 USA

1. Introduction

In the 1700's, Immanuel Kant and the Marquis de Laplace proposed that the solar system collapsed from a gaseous medium of roughly uniform density ([79], [105]). A flattened gaseous disk – the protosolar nebula – formed out of this cloud. The Sun contracted out of material at the center of the disk, while the planets condensed in the outer portions. Despite its simplicity, this model suffered from the *angular momentum problem* inherent to star formation: a cloud of gas and dust with the diameter of the solar system and the mass of the Sun has too much angular momentum to collapse to the Sun's present size. It took 150 years to solve this problem. C. von Weizsäcker worked out the basic physics of a viscous accretion disk, building on previous realizations that a turbulent viscosity could move material inwards and angular momentum outwards through the protosolar nebula ([178], [179]). With Lüst's steady-state solution ([119]), Kant's nebular hypothesis no longer suffered from the angular momentum problem and became the leading model for solar system formation.

Disk accretion is now known to power a wide range of astronomical sources. Starting with Kuiper's landmark study of β Lyr, viscous accretion disks have become central to our understanding of interacting binaries, where one star fills its inner Lagrangian surface and transfers matter into a disk surrounding a smaller companion ([103]). A disk surrounding a supermassive black hole is widely believed to drive the phenomena observed in active galactic nuclei (e.g., [175]). Accretion disks remain an important issue in star and planet formation, as described throughout this volume.

One striking feature of recent observations is the common phenomena observed in different accreting systems. All disks vary in brightness. These fluctuations range from rapid flickering on time scales as short as a few milliseconds up to long-lived eruptions with durations of decades or centuries.

C.J. Lada and N.D. Kylafis (eds.), The Origin of Stars and Planetary Systems, 613–642.
© 1999 *Kluwer Academic Publishers. Printed in the Netherlands.*

Most accreting systems lose mass; practically all disks that lose mass eject material in a well-collimated jet ([115]). These common phenomena occur in systems with luminosities ranging from much less than 1 $L_\odot$ in some interacting binaries up to 10^{11} $L_\odot$ in active galactic nuclei.

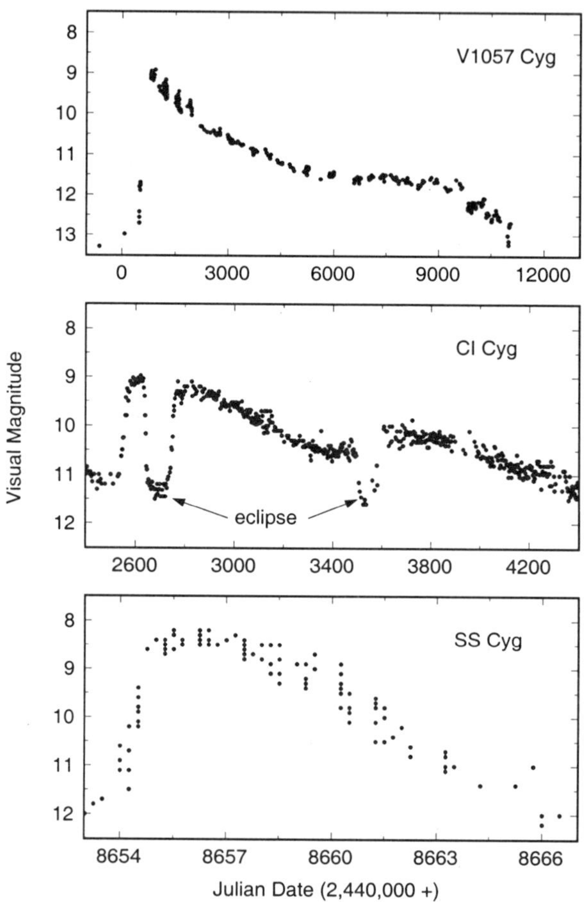

Figure 1 - Light curves for SS Cyg, a cataclysmic binary; CI Cyg, a symbiotic binary; and V1057 Cyg, an FU Ori variable star. Despite a different size scale in each system, the light curves have common features that indicate a similarity in their underlying physics.

As an example of these common phenomena, Figure 1 shows light curves for SS Cyg – a cataclysmic binary with an orbital period of $P_{orb} = 6.6$ hr ([30]); CI Cyg – a symbiotic binary with $P_{orb} = 855$ days ([81], [16]); and V1057 Cyg – an FU Ori variable that is apparently a single star ([84]). The 3–5 mag eruptions of these systems occur when the accretion rate through the disk increases. Aside from the deep eclipses in CI Cyg, the three light curves have common features, with a rapid rise, a brief interval at maximum, and a long decay. All three systems resemble A- or F-type stars at maximum light. This spectrum cools as the brightness fades. Substantial mass loss is associated with each system; the mass loss probably increases with the rise in brightness and decreases as the brightness declines. These similarities suggest that the same physical processes govern the evolution of disks ranging in size from 0.1 $R_\odot$ to several tens of AU.

TABLE 1. Types of disk systems

System	Distance (pc)	Angular size (arcsec)	N_{res}
BH + MS	1000	$\lesssim 10^{-5}$	100
NS + MS	1000	$\lesssim 10^{-5}$	100
WD + MS	100	$\lesssim 10^{-3}$	100+
MS + SG	100	$\lesssim 10^{-3}$	10+
MS + RG	1000	$\lesssim 10^{-3}$	10–100
β Pic	10+	1–10	10–100
Young star	100+	1–2	10–20
AGN	10^8	$\lesssim 10^{-3}$	1–10

The common features of accreting systems provide good tools to test physical models of disks. Table 1 summarizes the main types of disk-accreting systems known today and provides simple comparisons of their angular sizes and the number of resolution elements across the face of the disk, N_{res}, that can be derived from modern observations. The first set of systems includes (i) compact binaries containing a disk surrounding a black hole (BH), a neutron star (NS), or a white dwarf star (WD), and (ii) wider binaries with a disk surrounding a main sequence star. A low mass main sequence star usually feeds the disk in compact binaries; a more evolved subgiant or red giant star feeds the disk in the wider binaries. In both types of binaries, the angular size of the disk is small compared to the resolution possible with current ground-based or space-based observatories. However, eclipses of the disk by the secondary star can be used to divide the disk into many resolution elements and to map out the physical structure of the disk with a few simplifying assumptions ([70], [71]). The dusty disks in the second group of systems are large enough to image directly at optical and radio wavelengths, although few systems have been mapped in much detail ([164], [77]). The final group of active galactic nuclei have only recently been mapped with VLBI techniques; N_{res} will probably increase as these techniques become more sophisticated ([68]).

2. Steady Disks

Accretion disks work to transport mass radially inwards and angular momentum radially outwards. To understand how a disk manages this feat, consider a thin ring with two adjacent annuli at distances r_1 and r_2 from a central star (Figure 2). Material in these annuli orbits the central star with velocities, v_1 and v_2. The velocity difference between the two annuli, $v_1 - v_2 > 0$, produces a frictional force that attempts to equalize the two orbital velocities. The energy lost to friction heats the annuli; some disk material then moves inwards to conserve total energy. This inward mass motion increases the angular momentum of the ring; some disk material moves outwards to conserve angular momentum. Energy and angular momentum conservation thus lead to an expansion of the ring in response to frictional heating. The ring eventually expands into a disk, which generates heat (and radiation) at a level set by the accretion rate.

Figure 2 - Schematic view of two adjacent annuli in a disk surrounding a compact star. Annulus '1' lies inside annulus '2' and orbits the star (filled circle) at a higher velocity, $v_1 > v_2$.

2.1. ACCRETION LUMINOSITIES AND TEMPERATURES

Figure 3 shows the standard disk geometry. In a *steady* disk, material drifts radially inward at a constant rate, $\dot{M}$. For an infinite disk, the total luminosity generated by accretion is $L_{acc} = GM_\star \dot{M}/R_\star$, which is

$$L_{acc} = 314 \ \mathrm{L_\odot} \left(\frac{M_\star}{1 \ \mathrm{M_\odot}} \right) \left(\frac{\dot{M}}{10^{-5} \ \mathrm{M_\odot \, yr^{-1}}} \right) \left(\frac{1 \ \mathrm{R_\odot}}{R_\star} \right) \qquad (1)$$

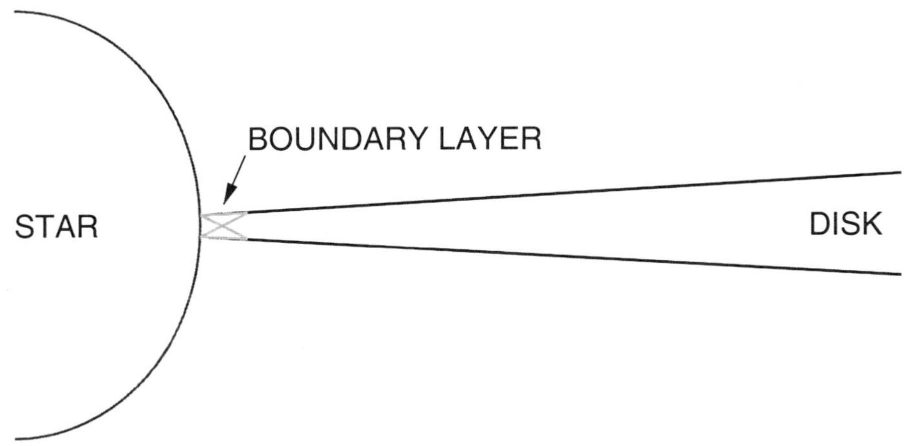

Figure 3 - Schematic view of a disk that extends to the stellar photosphere. Disk material drifts radially inward until it reaches the boundary layer, where the rotational velocity slows to match the stellar rotational velocity.

in familiar units. As material drifts inwards, half of this accretion energy is radiated by the disk,

$$L_{disk} = \frac{1}{2} L_{acc} \, . \tag{2}$$

The other half of the accretion energy becomes kinetic energy of orbital motion around the central star. In most cases, disk material drifts inwards until it reaches the stellar photosphere (Figure 3). At this point, disk material with an angular velocity of $\Omega = (GM_\star/R_\star^3)^{1/2}$ must slow down to match the stellar angular velocity, $\Omega_\star$. This transition occurs in the "boundary layer," a narrow ring with a radial size,

$$R_{bl,d} \sim h_{bl}^2/R_\star \ll R_\star, \tag{3}$$

where h_{bl} is the local scale height ($h_{bl}/R_\star \ll 1$; [138], [140], [141], [132], [96], [143], [95]). The energy lost in the boundary layer is the difference between the rotational energy per unit mass in the disk and the rotational energy per unit mass in the central star:

$$L_{BL} = \frac{1}{2} L_{acc} \left(\frac{\omega_{disk}^2 - \omega_\star^2}{\omega_{disk}^2} \right) \, . \tag{4}$$

In some accreting systems, the magnetic field of the central star is strong enough to truncate the disk at a radius, $R_m > R_\star$. A rough estimate of the truncation radius is the Alfvén radius,

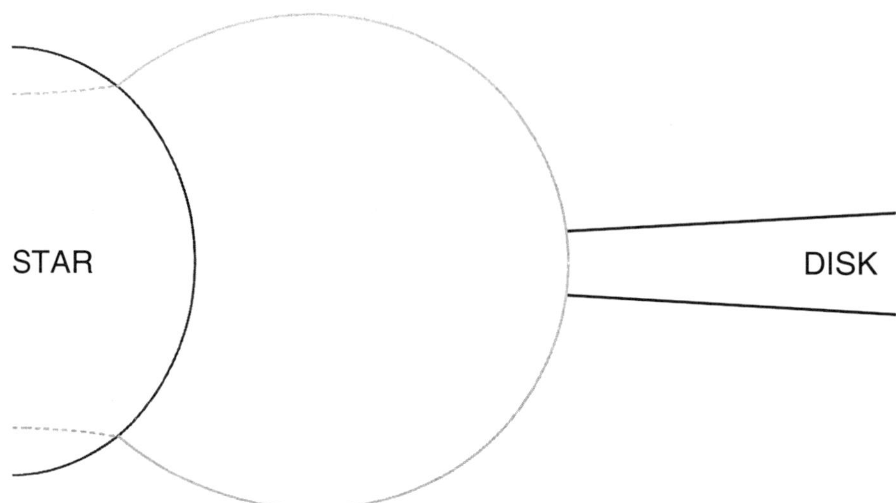

Figure 4 - Schematic view of a disk truncated by a strong dipolar magnetic field. The inner radius of the disk lies close to the corotation radius, where the disk and the stellar photosphere have the same angular velocity. Material falls from the inner disk onto the star along the magnetic field lines and produces two circular rings surrounding the magnetic axis.

$$R_m \approx R_A \approx \left(\frac{\mu_\star^4}{2GM_\star \dot{M}^2} \right)^{1/7} \tag{5}$$

where $\mu_\star$ is the magnetic dipole moment of the central star. More rigorous estimates of R_m depend on the magnetic field geometry ([113], [46], [114], [29], [101], [191], [4], [107], [182]). The disk then has a total luminosity

$$L_{disk} = \frac{1}{2} L_{acc} \left(\frac{R_\star}{R_m} \right) \tag{6}$$

In this geometry, disk material falls onto the star along magnetic field lines (Figure 4). The infalling gas shocks and produces two rings at stellar latitudes, $\pm b$, if the magnetic and rotational axes are parallel ([46], [101]). These rings are compressed into tilted ellipses if the magnetic axis is inclined with respect to the rotational axis, as expected in most systems ([88], [122]). In both cases, the luminosity of one ring is

$$L_{ring} = \frac{1}{2} L_{acc} \left(1 - \frac{1}{2} \frac{R_\star}{R_{mag}} \right) \tag{7}$$

To derive the temperature structure of the disk and the boundary layer or magnetic accretion ring, we again use conservation of energy. The disk temperature assumes an approximate balance between the energy lost from blackbody radiation and the torque due to frictional heating from mass motion through a distance ΔR in a gravitational potential:

$$GM_\star \dot{M} \left(\frac{1}{R} - \frac{1}{R + \Delta R} \right) \approx 2 \cdot 2\pi R \Delta R \ \sigma T_d^4 \tag{8}$$

For $\Delta R \ll R$, this expression is

$$GM_\star \dot{M} \frac{\Delta R}{R^2} \approx 4\pi R \Delta R \sigma T_d^4 \tag{9}$$

which results in

$$T_d \approx \left(\frac{GM_\star \dot{M}}{4\pi \sigma R^3} \right)^{1/4} \propto R^{-3/4} \tag{10}$$

The exact temperature of a steady disk depends on how the torque transports energy through the disk. Standard models assume that frictional heating vanishes at the stellar photosphere, which leads to an extra term in the energy equation (8) above ([120], [161]). The temperature is then ([13]):

$$T_d(R) = T_{acc} \left(\frac{R}{R_\star} \right)^{-3/4} \left(1 - \sqrt{R_\star/r} \right)^{1/4} \tag{11}$$

where

$$T_{acc} = 13000 \text{ K} \left(\frac{M_\star \dot{M}}{10^{-5} \text{ M}_\odot^2 \text{ yr}^{-1}} \right)^{1/4} \left(\frac{1 \text{ R}_\odot}{R_\star} \right)^{3/4}. \tag{12}$$

This expression for the temperature vanishes at the stellar surface, because the frictional heating vanishes. The temperature reaches a maximum at $R/R_\star = 49/36$ ([13]):

$$T_{max} = 6500 \text{ K} \left(\frac{M_\star \dot{M}}{10^{-5} \text{ M}_\odot^2 \text{ yr}^{-1}} \right)^{1/4} \left(\frac{1 \text{ R}_\odot}{R_\star} \right)^{3/4} \tag{13}$$

Different boundary conditions at the stellar surface yield slightly different, non-analytic, temperature laws ([134], [135]). These results are usually close to the standard temperature law, equation (11).

The temperature of the boundary layer depends on the mass accretion rate. At high $\dot{M}$, the region is optically thick ([120], [174]). Energy generated in the thin "dynamical boundary layer" (equation 3) should diffuse over a broader ring with a size comparable to the thermal scale height, h_{bl} ([132]). The scale height is smaller than the stellar radius in most cases. This "thermal boundary layer" is then much larger than the dynamical boundary layer and has a temperature 4–5 times larger than the maximum disk temperature ([91]). For low $\dot{M}$, the boundary layer is optically thin and cannot radiate energy efficiently ([138], [140]). The thermal scale height is then comparable to the stellar radius, and the boundary layer resembles a hot stellar corona ([138], [140]). The details of boundary layer structure are sensitive to properties of the stellar photosphere and the viscosity mechanism, but these general considerations hold for many physical settings ([18], [17], [136], [72], [73], [47], [48], [137], [76], [97], [98]).

The temperature of magnetic accretion rings also depends on geometry. If the magnetic axis is aligned with the rotational axis, the rings have equal luminosities and temperatures. The temperature is also constant with azimuth along each ring ([122]). The two rings have unequal luminosities when the magnetic axis is tilted with respect to the rotational axis, as needed to produce the observed light variations in magnetic cataclysmic variables and T Tauri stars ([36], [88]). The luminosity also varies with azimuth along each ring ([122]). In the aligned case, the infalling matter shocks above the stellar surface and forms identical optically thick accretion columns that radiate in the Balmer and Paschen continua ([104], [26]). The surface area of this emission is comparable to the surface area of the boundary layer, so the temperature is still 3–5 times larger than the maximum disk temperature ([26]). The accretion columns are probably quite different in non-aligned cases, but their structure has not been addressed.

TABLE 2. Accretion and nuclear energies

Object	ϵ_g (erg g^{-1})	ϵ_{nuc} (erg g^{-1})
T Tauri star	5×10^{14}	5×10^{13}
main sequence star	2×10^{15}	4×10^{18}
white dwarf star	1×10^{17}	4×10^{18}
neutron star	1×10^{20}	6×10^{18}

To judge the effectiveness of accretion as an energy source, Table 2 compares the gravitational potential energy, $\epsilon_g = GM_\star/R_\star$, of various 1 M$_\odot$ stars with the nuclear energy generation rate of the dominant fusion source, ϵ_{nuc}. If a star can accrete *and burn* material from a disk, then the ratio of the accretion luminosity to the nuclear luminosity is simply the ratio of these two quantities, $r \equiv L_{acc}/L_{nuc} = \epsilon_g/\epsilon_{nuc}$. This expression ignores the difference in accretion and nuclear time scales, but it provides a simple guide to the importance of accretion as a long-term energy source. Table 2 shows that accretion is rarely the major energy source for main sequence stars and white dwarfs; nuclear energy will always overwhelm accretion by factors of 40–2000 on long time scales. Accretion is an excellent energy source for pre-main sequence stars and neutron stars. It is roughly 10 times more effective than deuterium burning in pre-main sequence stars and nearly 20 times more productive than complete hydrogen burning in neutron stars.

2.2. TURBULENT VISCOSITY AND DISK TIMESCALES

The source of the frictional heating in accretion disks remains controversial. Ordinary molecular viscosity is too small to generate mass motion on a reasonable time scale, $\sim$ days for disks in interacting binary systems and $\sim$ years to decades for disks in pre-main sequence stars. The large shear between adjacent disk annuli suggests that disks might be unstable to turbulent motions, which has led to many turbulent viscosity mechanisms. Convective eddies, gravitational instabilities, internal shocks, magnetic stresses, sound waves, spiral density waves, and tidal forces have all been popular turbulence mechanisms in the past three decades (see, for example, [120], [161], [25], [112], [176], [7], [60], [172], [168], [24], [177], [157], [144], [5]).

Recent work has shown that magnetic stresses in a differentially rotating disk inevitably lead to turbulence ([9], [8], [110]). The growth time and effectiveness of these magnetohydrodynamic mechanisms make them the current leading candidate for viscosity in most applications. How this turbulence leads to significant mass motion in a real accretion disk remains an unsolved problem.

Shakura & Sunyaev side-stepped the basic uncertainties of viscosity mechanisms when they developed the popular "α-disk" model ([161]; see also [179]). In this approach, the frictional heating in the disk is due to a turbulent viscosity, $\nu = \alpha c_s h$, where c_s is the sound speed, h is the local scale height of the gas, and α is a dimensionless constant. This concept is similar to the mixing length theory of convection, with α serving the role of the mixing length. In most applications, $\alpha \lesssim 1\text{--}10$; α needs to exceed $\sim 10^{-4}$ to allow material to move inwards on a reasonable time scale.

This viscosity definition orders the important time scales for the disk ([120]). The shortest disk time scale is the dynamical (orbital) time scale, which increases radially outward:

$$\tau_D \approx 0.1 \text{ day} \left(\frac{R}{1 \text{ R}_\odot}\right)^{3/2} \left(\frac{1 \text{ M}_\odot}{M_\star}\right)^{1/2} \tag{14}$$

The thermal time scale measures the rate that energy diffuses through the disk,

$$\tau_T \approx 1 \text{ day} \left(\frac{R}{1 \text{ R}_\odot}\right)^{11/8}. \tag{15}$$

The thermal time scale is intermediate between the dynamical time scale and the viscous time scale, which measures the rate matter diffuses through the disk,

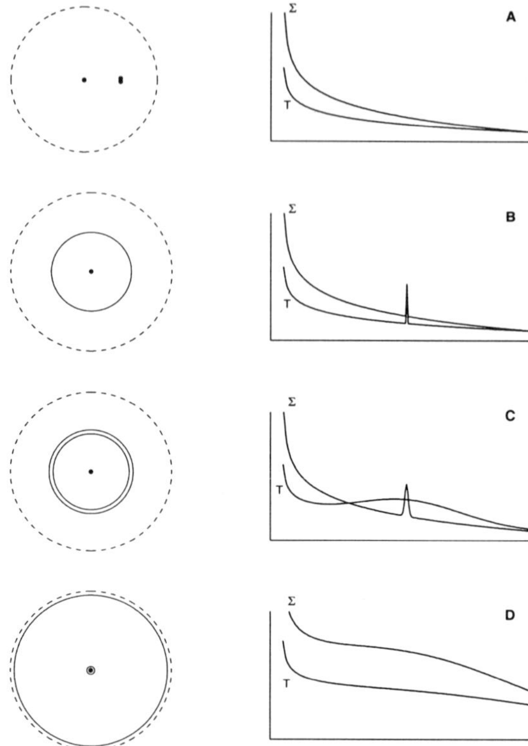

Figure 5 - Schematic evolution of a point-like mass enhancement, a blob, in a viscous accretion disk. A central star lies at the center of the disk with an outer edge indicated by the dashed circle. The disk initially has a smooth radial decrease in the effective temperature, T, and the surface density, Σ (panel A). Differential rotation between adjacent annuli smooths the blob into a ring on the orbital time scale, τ_D, which produces δ-function increases in T and Σ (panel B). The thermal energy in the blob moves inwards and outwards on the thermal time scale, τ_T, which produces a gaussian-like perturbation in the temperature (panel C). The mass motion on this time scale is small; Σ changes little on the thermal time scale. The mass perturbation spreads out through the disk on the viscous time scale, τ_V, as indicated in panel D.

$$\tau_V \approx \frac{40 \text{ day}}{\alpha} \left(\frac{R}{1 \text{ R}_\odot} \right)^{5/4} \tag{16}$$

Figure 5 illustrates the evolution of the disk on these time scales. The top panel shows a blob of material superimposed on the smooth density and temperature structure of a steady disk. In this example, the blob has a mass, δM, larger than the mass of an annulus, and a thermal energy content, δE, larger than the local thermal energy of an annulus. Disk rotation broadens the blob into a narrow ring in one or two rotation periods, which produces a narrow spike in the radial distributions of surface density and temperature. The thermal energy of this ring moves inwards and outwards on the thermal time scale. The temperature distribution broadens considerably, but the surface density hardly changes. The mass in the ring finally moves inwards and outwards on the viscous time scale. This evolution produces an extra increase in the temperature due to viscous dissipation.

Table 3 compares numerical values for the three disk time scales. These results assume $M_\star = 1\,M_\odot$ and $\alpha = 10^{-1}$. For most interacting binaries, the disk radius is ~ 0.1–$10\,R_\odot$; the time scales range from a fraction of a day for τ_D up to several tens of days for τ_V. The time scales increase dramatically for the larger disks in pre-main sequence stars and active galactic nuclei. The viscous time scale at the edge of a typical solar system is comparable to the disk lifetime.

TABLE 3. Disk time scales

Disk radius	h/r	T (K)	τ_D	τ_T	τ_V
$0.1\,R_\odot$	0.02	2×10^4	0.004 d	0.04 d	20 d
1.0	0.02	5×10^3	0.120 d	1.00 d	400 d
1 AU	0.04	250	1 yr	4 yr	1×10^4 yr
10	0.06	60	32 yr	105 yr	1×10^5 yr
100	0.08	25	1000 yr	2500 yr	2×10^6 yr

2.3. DISK ENERGY DISTRIBUTIONS

Figure 6 shows the broadband spectral energy distribution of a steady accretion disk with a boundary layer. The model assumes that the disk is optically thick and that each annulus radiates as a blackbody ([120], [13], [174], [158]). There are two main features in this spectrum: (i) radiation from the boundary layer at short wavelengths, and (ii) disk radiation at longer wavelengths. If the disk has a large ratio of outer radius to inner radius, $R_{out}/R_{in} \gg 1$, the disk spectrum follows

$$\lambda F_\lambda \propto \lambda^{-4/3} \qquad\qquad \lambda_{in} < \lambda < \lambda_{out} \qquad\qquad (17)$$

where $\lambda_{in} T_d(R_{in}) \approx \lambda_{out} T_d(R_{out}) \approx 0.36$ ([120]). The disk spectrum follows the Wien tail of a hot blackbody, $T_d(R_{in})$, at short wavelengths and the Rayleigh-Jeans tail of a cool blackbody, $T_d(R_{out})$, at long wavelengths.

Although Figure 6 assumes a blackbody disk, the general features of the model change little if the disk radiates as some type of stellar atmosphere. The most important modifications for pre-main sequence stars involve irradiation from the central star and the infalling envelope, as described by Beckwith in this volume.

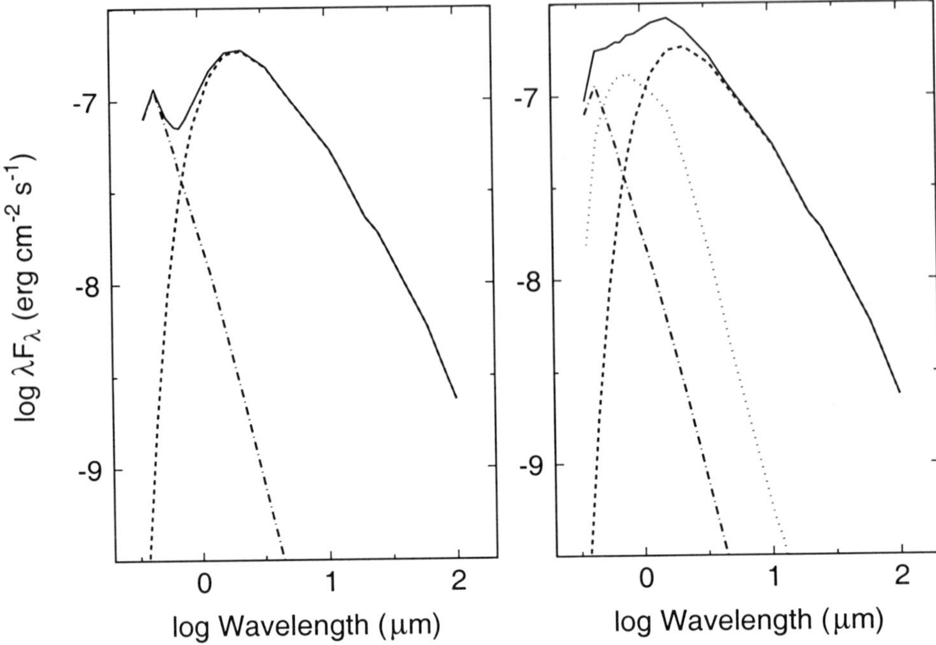

Figure 6 - Energy distributions for accretion disks. The left panel shows the energy distribution of a disk (dashed line), a boundary layer (dot-dashed line), and the total energy output (solid line) for $\dot{M} = 2 \times 10^{-7}$ $M_\odot$ yr^{-1} onto a solar-type star. The right panel adds the energy distribution (dotted line) for a star with $L_\star \approx L_d$.

Lynden-Bell & Pringle first applied these concepts to disks in interacting binaries and pre-main sequence stars. They added radiation from the central star to the spectrum, as indicated in the right panel of Figure 6. This model suggests that identifying disk radiation can be difficult when the stellar luminosity exceeds the accretion luminosity. Table 4 compares temperatures and luminosities of pre-main sequence stars and steady disks for two accretions rates, $\dot{M} = 10^{-6}$ and 10^{-4} $M_\odot$ yr^{-1}. At low masses, the contrast between the disk and star is considerable unless the accretion rate is $\lesssim 10^{-8}$ $M_\odot$ yr^{-1}. This contrast decreases for more massive stars. At 10 $M_\odot$, the star is hotter and as luminous as the disk, because massive young stars burn hydrogen before reaching the main sequence. Table 3 shows that accretion cannot be a dominant energy source for main sequence stars; the difficulty of seeing an accretion disk against the background of a young O or B star therefore is not surprising.

3. Unstable Accretion Disks

Despite the elegant simplicity of the steady-state model, all disks vary their energy output. These fluctuations range from small, 10%–20%, amplitude flickering on the local dynamical time scale ([23]) to large-scale eruptions,

TABLE 4. Accretion and stellar luminosities in pre-main sequence stars

$M_\star(M_\odot)$	$T_\star(K)$	$L_\star(L_\odot)$	$10^{-6}\ M_\odot\ yr^{-1}$		$10^{-4}\ M_\odot\ yr^{-1}$	
			T_{max} (K)	L_{acc} ($L_\odot$)	T_{max} (K)	L_{acc} ($L_\odot$)
0.1	3100	0.1	1850	3	5850	275
0.4	3600	0.5	1825	7	5800	675
1.0	4250	7.0	1800	12	5700	1200
3.0	8650	80	1700	25	5350	2350
6.0	17000	1000	2100	50	6650	5000
10.0	21000	5000	1850	60	5850	6000
20.0	29000	45000	1600	75	5000	7450

$\sim$ 3–5 mag or more, that can last for several times the local viscous time scale (Figure 1). In well-studied dwarf novae, the outbursts occur in cyclical patterns with quasi-periods of 10 or more days ([30], [130]).

Osaki and Pringle $et\ al.$ first noted the possibility for thermal instabilities in accretion disks ([139], [131]). In the standard picture, the structure of a steady disk is derived from balancing heat generated by viscosity with radiative cooling. However, the viscous energy generation and radiative losses are set by local disk parameters; the input $\dot{M}$ is an external parameter. Cooling can balance heating only if the local physical quantities can adjust to the input $\dot{M}$. If the local physical variables cannot adjust, cooling cannot balance heating and a thermal instability is likely.

Figure 7 displays a simple illustration of a thermal instability in a single disk annulus. The solid curves indicate loci of stable disks, where cooling precisely balances heating. To the right of these lines, heating exceeds cooling because viscous energy generation – set by the local surface density Σ through $\dot{M} = \nu\Sigma$ – exceeds the radiative losses set by the effective temperature, T_d. Cooling exceeds heating to the left of the stability lines.

To illustrate the evolution of the instability, assume the annulus has (Σ, T_d) at A$'$ and an input accretion rate, $\dot{M}_1$. This input accretion rate exceeds the stable accretion rate at A$'$, $\dot{M}_s = \dot{M}_{A'}$. The annulus evolves up the solid line on the viscous time scale to find a solution where $\dot{M}_s = \dot{M}_1$. Before reaching this point, the annulus arrives at $\dot{M}_B$, the last stable solution for this opacity source. At B, the annulus needs a larger surface density to accommodate the larger $\dot{M}$, and a larger T_d to radiate away the extra heat generated by the larger $\dot{M}$. To do so, the annulus evolves to C; it makes this transition at constant Σ because the thermal time scale is shorter than the viscous time scale. Once at C, the annulus radiates more energy than generated by viscous dissipation ($\dot{M}_s = \dot{M}_C > \dot{M}_1$) and evolves down the solid line towards a solution where $\dot{M}_s = \dot{M}_1$. The annulus moves towards lower Σ on the viscous time scale until it reaches point D, the last stable solution on this branch of the equilibrium curve. At D, the annulus drops to A to try to find the smaller T_d needed to accommodate a smaller $\dot{M}$. Once at A, the annulus retraces the steps around this 'limit-cycle' curve as long as $\dot{M} = \dot{M}_1$.

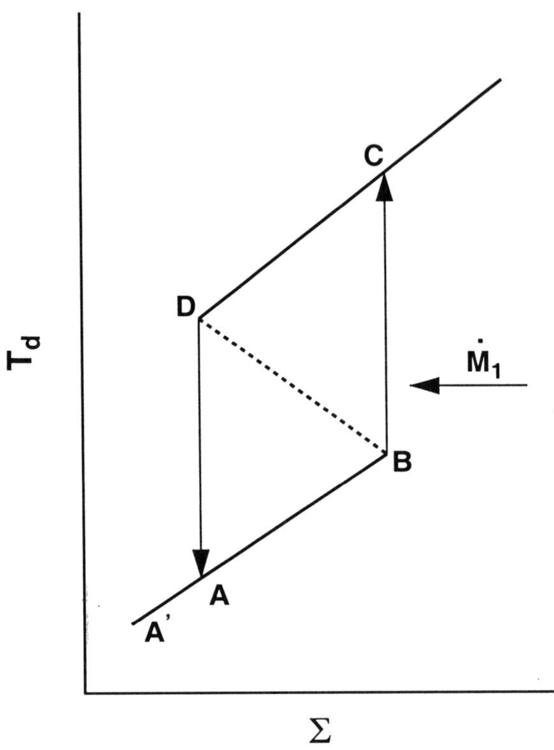

Figure 7 - Outline of a disk instability in a single annulus. The disk evolves along the solid lines ABCD for an input $\dot{M}$ (horizontal arrow). Heavy solid lines correspond to stable equilibrium solutions; a dashed line indicates the locus of unstable equilibria.

The evolution of an instability in a complete accretion disk follows the simple illustration. In a real disk, most annuli are close to B when a single annulus makes the transition from B to C. The increased temperature of a single annulus transfers heat to neighboring annuli on the thermal time scale; these annuli then jump to the 'high state' and propagate the eruption to their neighbors. Figure 8 shows this evolution for five snapshots in the evolution of a dwarf nova accretion disk. Time increases upwards in the Figure. The bottom panel shows the temperature distribution of the disk at the onset of the eruption. The inner edge of the disk jumps to the high state first; annuli at larger radii progressively follow until all of the disk resides in the high state in the top panel. The disk maintains this state for several days, and then retraces its path back to the low state.

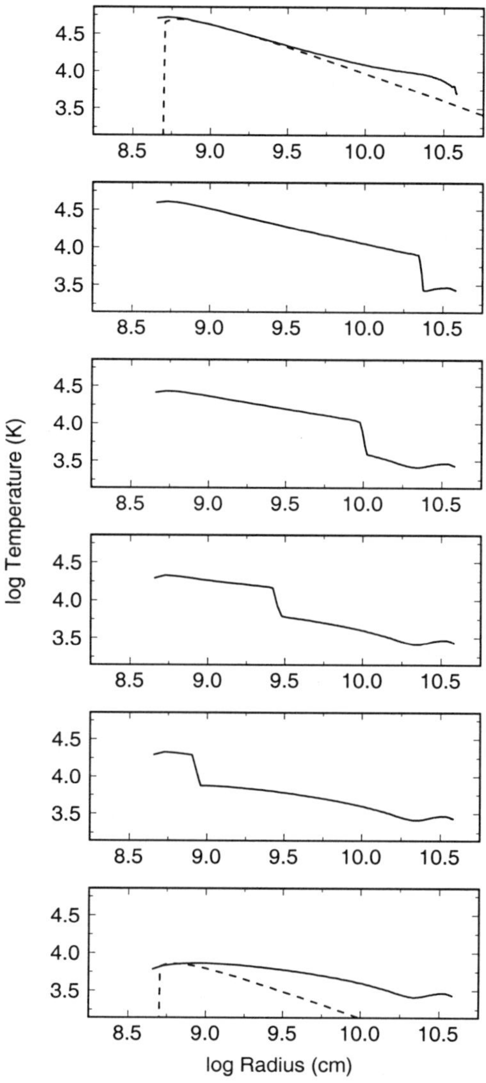

Figure 8 - Evolution of a disk instability in a dwarf nova disk. The bottom panel compares the radial temperature distribution of a time-dependent disk (solid line) with the temperature distribution of a steady disk (dashed line). The eruption begins in the next panel with a sharp increase in the temperature at the inner edge of the disk. This increase creates a 'wave' of increased temperature that propagates radially outwards in the disk as time moves forward (up in the figure). In the top panel, the entire disk has reached the hot state; the actual temperature distribution (solid line) is then close to the steady-state temperature distribution (dashed line)

One feature of the disk instability picture is that the disk is never in a steady-state. The dashed lines in Figure 8 plot steady-state temperature distributions normalized to the temperature at the inner edge of the disk.

The actual temperature gradient is much flatter than the steady-state gradient in the low state (bottom panel). The $\dot{M}$ through the disk increases radially outwards and is below the input $\dot{M}$ at each point in the disk. The disk approaches but never reaches the steady-state gradient in the high state (top panel).

Actual disk temperature distributions have been derived from eclipse light curves similar to those shown in Figure 1 for CI Cyg. K. Horne & collaborators ([70], [71], [125], [10]) have used maximum entropy techniques to recover $T_d(R)$ from multi-wavelength light curves and spectroscopic data assuming (i) the disk is azimuthally symmetric and (ii) the brightness temperature is close to the local blackbody temperature. Their results indicate that quiescent disks are rarely close to steady-state, with temperature distributions usually much flatter than $T_d(R) \propto R^{-3/4}$. Systems in eruption – dwarf novae and symbiotic stars at maximum light – more closely resemble, but never achieve, the steady-state temperature distribution with $q = 3/4$. These results generally agree with the model predictions in Figure 8.

4. Disk Eruptions in Pre-Main Sequence Stars

Most pre-main sequence stars vary in brightness. Irregular brightness variations of $\lesssim 1$ mag are a defining feature of T Tauri stars and many Herbig AeBe stars ([78], [61]). Recent studies show that many of these variations are due to dark spots rotating with the stellar photosphere ([66], [21], [22], [67]) or bright spots produced at the base of a magnetic accretion column ([21], [88]). Small eruptions of 1–3 mag lasting several years occur in the EXors, a poorly-studied class which includes EX Lup and DR Tau ([64]). More spectacular 3–6 mag eruptions occur in the FU Orionis variables, also known as FUors ([63], [59]).

G. Herbig first associated the eruptions of FUors with pre-main sequence stars ([62], [63]; see also [3], [94]). FU Ori, which lies at the apex of a fan-shaped nebula within the dark cloud B35, brightened by a factor of over one hundred in 100–200 days (Figure 9; [180], [181], [62], [63], [74], [162], [75]). Thirty years later, Welin discovered the 5 mag brightening of V1057 Cyg within an eccentric ring of reflection nebulosity (Figure 9; [185], [63], [84], [99]). Herbig later noted the similarity between the optical spectra of these two stars with spectra of V1515 Cyg, a faint variable star embedded in arc-shaped nebulosity ([61]). He collected archival photographic photometry and identified a slow rise from $m_{pg} \approx 15.5$ in the late 1940s to $m_{pg} \approx 13.5$ in the late 1970s ([63]). This brightness increase continued until 1980, when the star experienced a dramatic decline and slow recovery ([100], [90]).

Herbig's demonstration that FU Ori – and by analogy other FUors – is a pre-main sequence star is straightforward. FUors are clearly associated with dark clouds, having radial velocities indistinguishable from the cloud velocity. The optical spectrum, including the high lithium abundance, has similarities with spectra of T Tauri stars. The event statistics are plausible for pre-main sequence stars. Finally, the rise in brightness is a real luminosity increase that is *not* a nova outburst, the main alternative.

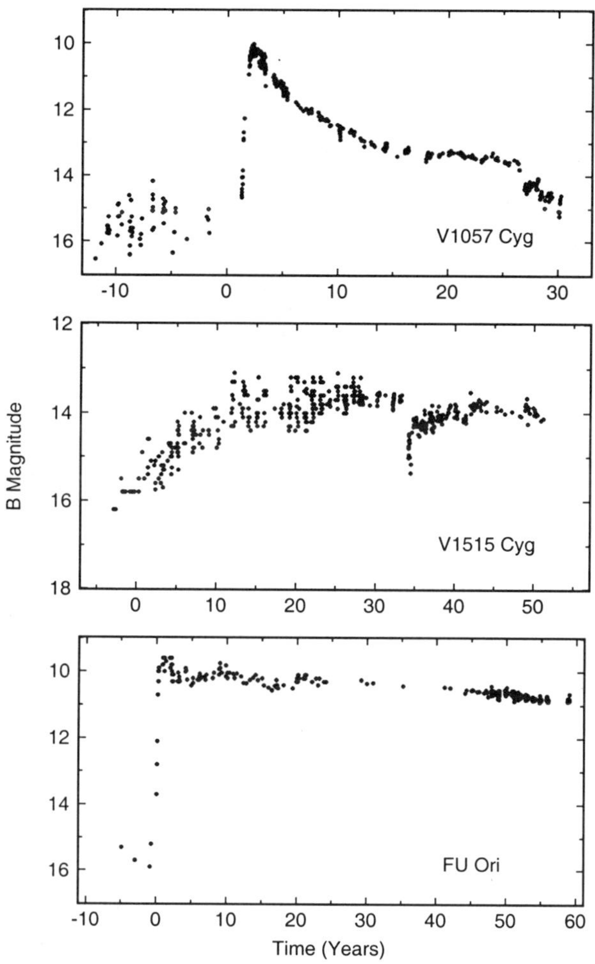

Figure 9 - Blue light curves for three FU Ori variables.

Many new FUors have been discovered since V1515 Cyg. The catalogued number now stands at 11 ([59], Table 5). Most have been observed to rise 3–5 mag in optical or near-IR brightness in less than one year, as indicated by the 'Y' in the outburst (OB) column of Table 5. V1515 Cyg is the only known example to require a decade to rise to visual maximum, but the historical light curves for some systems are poorly documented. A few objects have been called FUors based on common spectroscopic properties, which are described more completely below. Most recent FUors are more intimately associated with the densest dark clouds than the first members of the class, which suggests that many eruptions might have been missed.

There are also 3 candidate FUors listed as a second group of objects in Table 5. These objects have some properties in common with FUors, but more data are needed to see if they have the other characteristics as well ([169], [159]). A few other objects have one or two properties of known FUors, such as an outburst or unusual light curve (e.g., [41], [69], [2], ([163], [192]). These are more probably EXors or Ae/Be stars than FUors.

4.1. BASIC PROPERTIES OF FU ORIONIS OBJECTS

FUors share a distinctive set of morphological, photometric, and spectroscopic characteristics (Table 5). Three are known binary stars, L1551 IRS5, Z CMa, and RNO 1B/1C ([102], [86], [11], [171], [156]). Most have delicate fan-shaped or comma-shaped reflection nebulae (RN) on optical and near-IR images ([50], [148], [127], [118]). Optical jets and HH objects are common ([148], [38], [59]). Most appear associated with large-scale molecular outflows ([189], [43], [42], [121], [190], [44], [117], and references therein). These features – together with the broad, blue-shifted Na I and H I absorption features described below – demonstrate that FUors drive powerful winds that interact with the surrounding medium ([12], [37], [188]).

TABLE 5. Selected properties of FU Orionis objects

Object	OB	Opt ST	IR ST	RN	HH/Jet	Wind	Outflow	Radio
L1551 IRS 5	?	K	M	Y	Y	?	Y	Y
FU Ori	Y	F–G	M	Y	N	Y	N	N
Z CMa	?	F–G	M	Y	Y	Y	Y	Y
BBW 76	@	G	M	Y	?	Y	N	?
V346 Nor	Y	?	C	Y	Y	?	Y	?
Par 21	?	F	M	Y	Y	Y	?	?
V1515 Cyg	Y	G	M	Y	N	Y	Y	N
V1057 Cyg	Y	G	M	Y	N	Y	Y	Y
V1735 Cyg	Y	?	M	Y	N	?	Y	Y
RNO 1B/1C	Y	G	M	Y	N	?	Y	?
IC 430	?	G	?	Y	Y	Y	N	Y
PP 13S	Y	?	M	?	?	?	Y	?
Re 50	?	G	?	Y	Y	Y	Y	Y

FUors also have very unusual spectroscopic characteristics. The optically visible sources have F–G, and sometimes K-type, giant or supergiant spectra ([63]; Table 5, see also [58], [149], [153])). The optical reflection nebulae of several embedded FUors also show G-type absorption features ([167], [166]). All FUors but Z CMa and V346 Nor have very deep CO absorption bands on near-IR spectra at 1.6 μm and 2.3 μm ([126], [40], [31],

[160], [82], [19]). These features resemble the CO absorption bands observed
in red giants and are much stronger than those observed in any other pre-
main sequence star ([56], [57], [51]). CO absorption in Z CMa is weakened
by dust emission from an embedded companion; V346 Nor has weak CO
emission instead of CO absorption ([102], [170]; [153]). Many FUors also
display strong water absorption features, which strengthens the evidence
for a ~ 2000 K photosphere ([27], [160], [28]).

All FUors show large *excesses* of radiation over normal G supergiants
at both ultraviolet and infrared wavelengths. The near-IR excess is clearly
photospheric in origin, because the CO and H_2O absorption features are
so intense. The UV excesses in Z CMa and FU Ori appear associated with
an A- or F-type photosphere that is hotter than the G-type photosphere
observed at longer wavelengths ([89]). In addition to significant far-IR and
submm emission ([183], [184], [84]), many FUors are strong radio continuum
sources at cm wavelengths ([35], [154], [155], [116], [156]). This emission is
not photospheric in some FUors and may be produced in the outflow or
the jet.

Herbig first noted broad absorption lines on optical spectra of several
FUors ([63]). Hartmann & Kenyon later confirmed large rotational velocities of $v \sin i \approx 15$–60 km s^{-1} for V1057 Cyg, V1515 Cyg, and FU Ori
([55]). This property is now characteristic of the class; all FUors have broad
optical or infrared absorption lines or both ([58], [165], [166], [82]). The rotational velocity further depends on wavelength. The near-IR CO lines in FU
Ori and V1057 Cyg have significantly smaller rotational velocities than the
optical lines ([56], [57]). The optical rotational velocity smoothly increases
with decreasing wavelength in Z CMa and V1057 Cyg ([186], [187]). In FU
Ori itself, a powerful wind masks weak absorption lines and makes it difficult to detect any variation of rotational velocity with wavelength should
one exist.

Many FUors display *doubled* absorption lines on optical and near-IR
spectra ([55], [56], [57], [58], [165], [166]). The two absorption components
in V1057 Cyg are separated by 30–40 km s^{-1}; these features have much
larger separations in Z CMa and FU Ori. The long-term stability of the
doubled absorption lines indicates that the lines are not produced by two
stellar components in a binary system ([87]).

Most FUors also show strong evidence for mass loss, in addition to
the HH objects and CO outflows described above. Practically every FUor
displays deep, blueshifted absorption components on Hα and Na I D ([12],
[37], [39], [187], [54]); weak blueshifted CO absorption might be present in
FU Ori ([57]). The optical line profiles can also change on month to year
time scales ([12], [37], [187]). The near-IR line profiles may also change
([19]). Both V1057 Cyg and V1515 Cyg have very pronounced line profile
changes and dips in their optical light curves. These fluctuations suggest
dust formation in a variable wind, but the data are not very extensive ([90]).

Finally, FUor eruptions must be repetitive ([63], [55]). The event statistics of known FUors suggest that a young star must undergo 10–20 FUor

eruptions before reaching the main sequence. This estimate may be a lower limit, because some outbursts have certainly been missed. Reipurth's proposal that FUor eruptions produce HH objects suggests a recurrence time scale of ~ 1000 yr, based on the identification of multiple bowshocks (ejection events) with dynamical separations of 500–2000 yr ([145], [146], [53], [152], [6]). The discovery of giant HH flows with dynamical time scales of 10^4–10^5 yr ([151]) allows 10–100 eruptions if eruptions actually power the outflow (see below).

4.2. FU ORIONIS OBJECTS AS ACCRETION DISKS

The observations described above place severe constraints on possible FUor outburst mechanisms. With a luminosity of a few hundred $L_\odot$ (Table 6), a typical FUor emits 10^{45} to 10^{46} erg during the course of a 10–100 yr eruption. Outbursts also recur every 10^3–10^5 yr. This process produces an object resembling a rapidly rotating F-G supergiant in the optical and a more slowly rotating M giant in the infrared. It must be unique to young stars, because we have not observed a FUor event in an older star system.

Accretion is the most plausible energy source for FUor eruptions ([55], [109]). By analogy with dwarf novae, FUor eruptions are the high state of a thermal instability in a disk surrounding a pre-main sequence star. This instability occurs if the accretion rate from the molecular cloud core into the disk lies between the stable accretion rates in the low state and the high state (Figure 7). The system cycles between these two states as long as the surrounding cloud can replenish the disk between outbursts.

TABLE 6. Disk Properties of FU Orionis Objects

Object	L_{bol} ($L_\odot$)	$M_\star \dot{M}(10^{-5}\ M_\odot^2\ \mathrm{yr}^{-1})$	$R_\star$ ($R_\odot$)
L1551 IRS 5	30	0.1	0.4
FU Ori	220	1.7	1.2
Z CMa	400	4.0	1.6
BBW 76	200	1.4	1.1
V346 Nor	160	1.0	1.0
Par 21	200	1.4	1.1
V1515 Cyg	135	0.8	0.9
V1057 Cyg	400	4.0	1.6
V1735 Cyg	200	1.3	1.1
RNO 1B/1C	750	3.5	1.5

Table 6 lists the disk properties needed to power observed FUor luminosities from accretion. To make these estimates, I assume a steady disk with $L_{disk} = 157$ L$_\odot (M_\star \dot{M}/R_\star)$ and $T_{max} \approx 6500$ K $(M_\star \dot{M}/R_\star{}^3)^{1/4}$. Combining these and setting $T_{max} = 6500$ K for an F-G supergiant star yields

$$R_\star \approx \left(\frac{L_{disk}}{157 \text{ L}_\odot} \right)^{1/2} \qquad (18)$$

and

$$M_\star \dot{M} \approx \left(\frac{L_{disk}}{157 \text{ L}_\odot} \right)^{3/2} \qquad (19)$$

These expressions produce the results in Table 6 for known FUors, assuming a typical inclination angle, $\cos i = 1/2$. The median inner radius for the FUor sample is $R_\star \approx$ 1.1–1.2 R$_\odot$; the median mass accretion rate is $M_\star \dot{M} \sim$ 1–2 $\times 10^{-5}$ M$_\odot^2$ yr^{-1}. For a mass-radius relation for pre-main sequence stars with ages of 10^5 yr, $M_\star \approx$ 0.2–0.3 M$_\odot$ and $\dot{M} \approx$ 0.5–1 $\times 10^{-4}$ M$_\odot$ yr^{-1}.

This very simple analysis indicates that a young star must accrete material at $\sim 10^{-4}$ M$_\odot$ yr^{-1} during an eruption to power the observed luminosity. For an adopted outburst duration of 100 yr, this accretion rate results in a total accreted mass of $\sim$ 0.01 M$_\odot$ per eruption. The total mass accreted during a FUor eruption must be replenished during a low state lasting $\sim$ 10^3–10^5 yr, which results in an infall rate of $\sim 10^{-5}$–10^{-7} M$_\odot$ yr^{-1}. This range is close to the infall rates envisioned – and in some cases observed – for typical cloud cores ([1], [128], [124]).

4.3. OBSERVATIONAL TESTS OF DISK MODELS

Steady accretion disk models successfully explain many observations of FUors ([55], [87], [59]). Theoretical models of disk instabilities produce eruptions that generally resemble FUor events, although comparisons with observations are still in their early stages ([109], [32], [33], [80], [14], [15], [34], [129], [173]). The next few paragraphs summarize these results; Herbig & Petrov present a different interpretation ([65], [133]; see also [106]).

Figure 10 shows light curves for V1057 Cyg, one of the best-studied FUors. These light curves closely resemble those of other accreting systems and provide strong support for an accretion model ([87], [33], [84], [15], [34], [111], [173]). The amplitude of the decline is largest in the UV and decreases monotonically from 0.3 μm to 5 μm. This behavior is characteristic of accreting systems, which evolve towards cooler temperatures as the luminosity declines ([91], [15], [111], [30]). The spectral type variation also agrees with disk model predictions. If we assign stellar effective temperatures to the optical spectrum at maximum (A5 star) and at the current epoch (G5 star), the UBV decline indicates a source with a roughly constant radius. This requirement is easy to achieve with an accretion disk if the radius of the central star remains constant. Most stars cannot evolve at constant radius if their effective temperatures change.

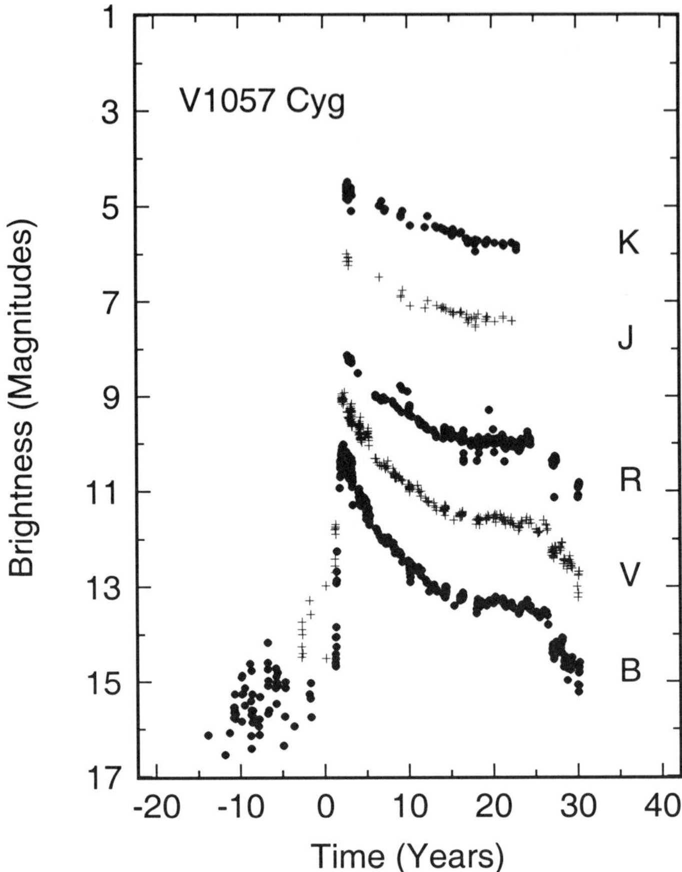

Figure 10 - Light curves for V1057 Cyg. The symbols alternate between filled circles (at B, R, and K) and plus signs (V and J). The system varied erratically at B = 14–17 prior to outburst and then rose 5–6 mag in less than one year. The visual observations indicate a similar rise time, although few pre-outburst observations are available. No pre-outburst near-IR data exist; K-band observations began shortly after the outburst was detected. The amplitude of the decline clearly depends on wavelength, with larger amplitudes at shorter wavelengths (see [84], [162], [99]).

The light curves of other FUors also generally agree with disk model predictions. The rise times of 200–400 days in FU Ori and V1057 Cyg are comparable to the thermal time scale, as expected for an eruption that begins at the inner edge of the disk (Table 3; [15]). The several decade rise in V1515 Cyg is too long for such an "inside-out" instability, but could be caused by an outburst that begins in the outer part of the disk . These "outside-in" eruptions can result from perturbations of the disk by a companion star or planet, in addition to the disk instability mechanism described above ([33], [20], [15], [34]). In all systems, the decay times are comparable to the viscous time scale (Table 3).

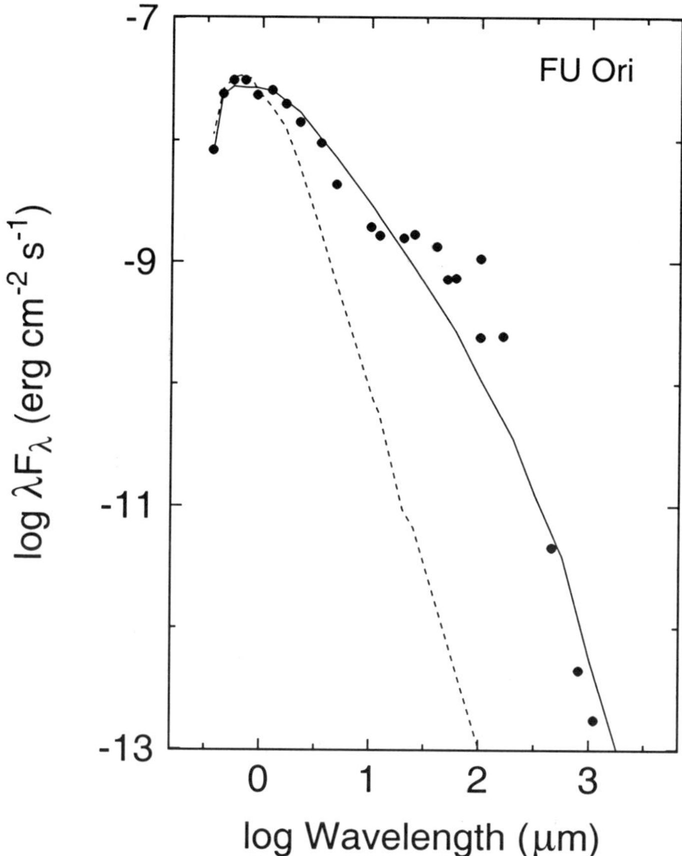

Figure 11 – Spectral energy distribution for FU Ori. The observations (filled circles) indicate a clear excess over a G-type star (dashed line). A disk model that produces a G-type optical spectrum accounts for most of the data (solid line). The modest far-IR excess over the disk spectrum at ~ 100 μm is probably produced by the surrounding nebula responsible for the fan-shaped reflection nebula.

Disk models also successfully explain FUor SEDs ([55], [87], [84]). Figure 11 shows a dereddened SED for FU Ori. The deep UV and near-IR absorption features indicate that the UV and IR excesses over a G-type supergiant spectrum are photospheric ([27]). If disks radiate like stars at the effective temperatures appropriate for the observed absorption features, the SEDs for FU Ori and other FUors require that the surface area of the emitting region increases with increasing wavelength. In particular, the surface area of the near-IR source must be 10–20 times larger than the surface area of the optical source; the 300 K material responsible for the 10 μm excess must have roughly 100 times the emitting area of the optical continuum region. A disk in which the temperature decreases radially outward (§2) naturally explains this observation. Disk models account for the 0.3–10 μm SEDs of V1057 Cyg, V1515 Cyg, and FU Ori quite well (Figure 11; see also

[55], [1], [87], [84], [15], [173]). Several other FUors with modest reddening have similar SEDs ([84]).

Recent interferometric observations have resolved FU Ori at near-IR wavelengths. The angular size at 2.2 μm agrees with predictions of the accretion disk model ([123]).

The disk model *fails* to explain 10–100 μm data of many FUors, although the model fits both FU Ori and BBW 76 from 1–100 μm ([84]). However, the decline of the 10–20 μm light of V1057 Cyg follows the optical light curve very closely ([84]). This behavior suggests that the mid-IR radiation is optical light absorbed and reradiated by a surrounding envelope. An infall rate of 1–5 $\times$ 10^{-6} M$_\odot$ yr^{-1} produces an optically thick envelope that can reprocess optical light from the inner disk and account for the 10–100 μm SED ([84], [15]). This rate is sufficient to replenish the disk in 1000 yr, which allows recurrent FUor eruptions in this system. The observational evidence for infall at comparable rates in two FUors, Z CMa ([108]) and L1551 IRS5 ([128], [124]), lends support to this interpretation.

Finally, an accretion disk also naturally explains the optical and IR line profiles observed in some FUors ([56], [57], [87], [58], [186], [187]). The gradual decrease in the rotational velocity with increasing wavelength occurs because the longer wavelength emission is produced in more slowly-orbiting material at larger disk radii than the more rapidly moving inner disk material responsible for the short wavelength emission. At a given disk temperature, a larger fraction of the disk surface rotates at high line-of-sight velocities and the lines appear doubled ([87], [28], [137]).

4.4. THE IMPORTANCE OF FU ORI ERUPTIONS

Despite their relative rarity, FUors are an important part of pre-main sequence stellar evolution. FUor eruptions add a significant fraction of a stellar mass to the central star. FUor eruptions also eject a large amount of mass into the surrounding cloud. FUor eruptions may be frequent enough to provide nearly all of the mass of a typical low mass star. If this hypothesis is correct, FUor eruptions can power most molecular outflows and may be a dominant source of cloud heating in molecular cloud cores. This view is controversial, but can be tested with observations as outlined below.

Although the statistics are still crude, young stars clearly accrete much of a stellar mass in FUor events. A typical eruption adds $\sim$ 0.01 M$_\odot$ to the central star; a rate of 10–20 eruptions per pre-main sequence star leads to a total accreted mass of $\sim$ 0.1–0.2 M$_\odot$. These eruptions eject considerable amounts of mass and momentum into the surrounding cloud. For example, FU Ori itself has lost material with a momentum of $\sim$ 0.3 M$_\odot$ km s^{-1} since its eruption began. With 10–20 such eruptions, FU Ori could supply a small fraction of the momentum of a typical molecular outflow, $(Mv)_o \sim$ 1–100 M$_\odot$ km s^{-1} ([45]).

Two conditions must be satisfied for a young star to accrete *all* of its mass in FUor events: 1. The disk must undergo regular thermal instabilities and cycle between the low and high states on a reasonable time scale. 2. FUors must be young, $\lesssim$ a few $\times$ 10^5 yr, to allow the surrounding cloud to fuel the recurring instabilities.

As a group, FUors clearly are young objects. FUors have many more properties in common with young class I protostars than older optically visible T Tauri stars. FUors have distinctive reflection nebulae and large far-IR excesses; most are associated with jets, HH objects, and molecular outflows. These properties are common in class I sources and rare in T Tauri stars ([145], [147], [49]). The typical age of a FUor is therefore $\sim$ a few $\times$ 10^5 yr ([85], [184], [84]).

The conditions needed for the disk to cycle between the high and low states are probably satisfied in many pre-main sequence stars. For a typical infall rate of 1–10 $\times$ 10^{-6} $M_\odot$ yr^{-1}, theoretical models indicate that the disk spends most of its time in a low state where the mass accretion rate through the disk is much lower than the infall rate ([14], [15], [111]). The disk requires only $\sim$ 10^3–10^4 yr to accumulate the $\sim$ 0.01 $M_\odot$ needed to power a FUor eruption. The disk will continue to cycle between the low and high states as long as the cloud supplies mass to the disk.

If young stars accrete most of their mass in FUor events, it follows that FUors can power molecular outflows. In most outflow models, the mass ejected by the central star is $\sim$ 10%–30% of the accreted mass. For a wind velocity of $\sim$ 200 km s^{-1}, the momentum in the wind is

$$(Mv)_w \ \sim \ 20 \ (M_\star/M_\odot) \ M_\odot \ \mathrm{km \ s^{-1}} \tag{20}$$

for a typical wind velocity of 200 km s^{-1}. The observed range of outflow momenta – $(Mv)_o = 1$–100 $M_\odot$ km s^{-1} ([45]) – requires stellar masses of $M_\star \sim 0.1$–5 $M_\odot$ if the flows conserve momentum. The winds of FUors can power this outflow **if** a young star accretes most of its mass in FUor events.

These simple estimates establish that FUors are important events if they recur on time scales of $\sim$ 10^3 yr during the protostellar phase of pre-main sequence stellar evolution. In this picture, the fraction of time in the FUor state is simply the ratio of the infall rate to the FUor accretion rate: $f \sim \dot{M}_i/\dot{M}_{FU}$. FUor disk models require $\dot{M}_{FU} \sim 10^{-4}$ $M_\odot$ yr^{-1}, so $f \sim 0.05$ for a typical infall rate of $\dot{M}_i \sim 5 \times 10^{-6}$ $M_\odot$ yr^{-1}. This prediction currently agrees with the statistics of FUors among protostars: 1 out of $\sim$ 20 protostars in the Taurus dark cloud contains a FUor (L1551 IRS5; [85]) and 5 out of 17 protostars with HH objects contains a central star that resembles known FUors spectroscopically ([150]; see also [52]). Sensitive photometric and spectroscopic surveys of nearby molecular clouds can improve these statistics considerably. If the FUor frequency among protostars turns out to be more than a few per cent, then FUors may represent the main accretion phase of early stellar evolution.

I thank Janet Mattei for the SS Cyg data and John Cannizzo for the disk instability models.

References

1. Adams, F. C., Lada, C. J., & Shu, F. H. 1987, ApJ, 308, 788
2. Alves, J., Hartmann, L., Briceño, C., & Lada, C. J. 1997, AJ, 113, 1395
3. Ambartsumian, V. A. 1954, Comm. Byurakan Obs, No 13
4. Armitage, P. J. 1995, MNRAS, 274, 1242
5. Armitage, P. J. 1998, ApJL, 501, L189.
6. Bachiller, R., Tafalla, M., & Cernicharo, J. 1994, ApJo, 425, L93
7. Balbus, S. A., & Hawley, J. F. 1991, ApJ, 376, 214
8. Balbus, S. A., & Hawley, J. F. 1998, Rev Mod Phys, 70, 1
9. Balbus, S. A., Hawley, J. F., & Stone, J. M. 1996, ApJ, 467, 76
10. Baptista, R., Horne, K., Wade, R. A., Hubeny, I., Long, K. S., & Rutten, R. G. M. 1998, MNRAS, 298, 1079
11. Barth, W., Weigelt, H., & Zinnecker, H. 1994, A&A, 291, 500
12. Bastian, U., & Mundt, R. 1985, A&A, 144, 57
13. Bath, G.T., Evans, W.D., Papaloizou, J., & Pringle, J.E. 1974, MNRAS, 169, 447
14. Bell, K. R., & Lin, D. N. C. 1994, ApJ, 427, 987
15. Bell, K. R., Lin, D. N. C., Hartmann, L., & Kenyon, S. J. 1995, ApJ, 444, 376
16. Belyakina, T. S. 1992, Izv. Krym. Ap. Obs., 84, 45
17. Bertout, C., Bouvier, J., Duschl, W.J., & Tscharnuter, W.M. 1993, A&A, 275, 236
18. Bertout, C., & Regev, O. 1992, ApJ, 399, L163
19. Biscaya, A. M., Rieke, G. H., Narayanan, G., Luhman, K. L., & Young, E. T. 1997, AJ, 491, 359
20. Bonnell, I., & Bastian, P. 1992, ApJ, 401, L31
21. Bouvier, J., & Bertout, 1989, A&A, 211, 99
22. Bouvier, J., Cabrit, S., Fernandez, M., Martin, E. L., & Matthews, J. M. 1993, A&A, 272, 176
23. Bruch, A. 1994, in *Flares and Flashes*, IAU Colloquium No. 151, edited by J. Greiner, H. W. Duerbeck, & R. E. Gershberg, Berlin, Springer, p. 288
24. Cabot, W. 1996, ApJ, 465, 874
25. Cabot, W., Canuto, V. M., Hubickyj, O., & Pollack, J. B. 1987, Icarus, 69, 387
26. Calvet, N. & Gullbring, E. 1999, ApJ, in press
27. Calvet, N., Hartmann, L., & Kenyon, S. J. 1991, ApJ, 383, 752
28. Calvet, N., Hartmann, L., & Kenyon, S. J. 1993, ApJ, 402, 623
29. Campbell, C. G. 1987, MNRAS, 229, 405
30. Cannizzo, J. K., & Mattei, J. A. 1998, ApJ, 505, 344
31. Carr, J. S., Harvey, P. M., & Lester, D. F. 1987, ApJL, 321, L71
32. Clarke, C. J., Lin, D. N. C., & Papaloizou, J. C. B. 1989, MNRAS, 236, 495
33. Clarke, C. J., Lin, D. N. C., & Pringle, J. E. 1990, MNRAS, 242, 439
34. Clarke, C. J., & Syer, D. 1996, MNRAS, 278, L23
35. Cohen, M., Beiging, J. H., & Schwartz, P. R. 1982, ApJL, 289, L5
36. Cropper, M. 1990, Sp. Sci. Rev., 54, 195
37. Croswell, K., Hartmann, L., & Avrett, E. 1987, ApJ, 312, 227
38. Davis, C. J., Mundt, R., Eislöffel, J., & Ray, T. P. 1994, AJ, 110, 766
39. Eislöffel, J., Hessman, F. V., & Mundt, R. 1990, A&A, 232, 70
40. Elias, J. H. 1978, ApJ, 223, 859
41. Eislöffel, J., Günther, E., Hessman, F. V., Mundt, R., Carr, J. S., Beckwith, S., Ray, T. P. 1991, ApJ, 383, L19
42. Evans, II, N. J., Balkum, S., Levreault, R. M., Hartmann, L., & Kenyon, S. J. 1994, ApJ, 424, 793
43. Fridlund, C. V. M., Knee, L. B. G. 1993, A&A, 268, 245
44. Fridlund, C. V. M., & Liseau 1998, ApJ, 499, L75
45. Fukui, Y. 1989, in *ESO Workshop on Low Mass Star Formation and Pre-Main Sequence Objects*, edited by B. Reipurth, Garching, ESO, p. 95
46. Ghosh, P., & Lamb, F. K. 1979, ApJ, 232, 259

47. Godon, P., Regev, O., & Shaviv, G. 1995, MNRAS, 275, 1093
48. Godon, P. 1996, MNRAS, 279, 1071
49. Gómez, M., Whitney, B. A., & Kenyon, S. J. 1997, AJ, 114, 1138
50. Goodrich, R. 1987, PASP, 99, 116
51. Greene, T. P., & Lada, C. J. 1997, AJ, 114, 2157
52. Hanson, M. M., & Conti, P. S. 1995, ApJ, 448, L45
53. Hartigan, P., Raymond, J., & Meaburn, J. 1990, ApJ, 362, 624
54. Hartmann, L., & Calvet, N. 1995, AJ, 109, 1846
55. Hartmann, L., & Kenyon, S.J. 1985, ApJ, 299, 462
56. Hartmann, L., & Kenyon, S.J. 1987a, ApJ, 312, 243.
57. Hartmann, L., & Kenyon, S.J. 1987b, ApJ, 322, 393
58. Hartmann, L., Kenyon, S. J., Hewett, R., Edwards, S., Strom, K. M., Strom, S. E., & Stauffer, J. R. 1988, ApJ, 338, 1001
59. Hartmann, L., & Kenyon, S. J. 1996, ARA&A,
60. Hawley, J. F., & Balbus, S. A. 1991, ApJ, 376, 223
61. Herbig, G. H. 1960, ApJS, 4, 33
62. Herbig, G. H. 1966, Vistas in Astr, 8, 109
63. Herbig, G. H. 1977, ApJ,
64. Herbig, G. H. 1989, in *ESO Workshop on Low-Mass Star Formation and Pre-Main Sequence Objects* ed. B. Reipurth, Garching, ESO, p. 233
65. Herbig, G. H., & Petrov, P. P. 1992, ApJ, 392, 209
66. Herbst, W., *et al.* 1987, AJ, 94, 137
67. Herbst, W., Herbst, D. K., & Grossman, E. J. 1994, AJ, 108, 1906
68. Herrnstein, J. R., Greenhill, L. J., Moran, J. M., Diamond, P. J., Inque, M., Nakai, N., Miyoshi, M. 1998, ApJ, 497, L69
69. Hodapp, K.-W., Hora, J. L., Rayner, J. T., Pickles, A. J., & Ladd, E. F. 1996, ApJ, 468, 861
70. Horne, K., & Cook, M. C. 1985, MNRAS, 214, 307
71. Horne, K., & Steining, R. F. 1985, MNRAS, 216, 933
72. Hujeirat, A. 1995a, A&A, 295, 249
73. Hujeirat, A. 1995b, A&A, 295, 268
74. Ibragimov, M. A. 1993, Astr. Zh., 70, 339
75. Ibragimov, M. A. 1997, Pis'ma Astr. Zh., 1, 125
76. Idan, I. & Shaviv, G. 1996, MNRAS, 281, 604
77. Jayawardhana, R., Fisher, S., Hartmann, L., Telesco, C., Pina, R., & Fazio, G. 1998, ApJ:, 503, L79
78. Joy, A. H. 1945, ApJ, 102, 168
79. Kant, I. 1755, *Universal Natural History and Theories of the Heavens*
80. Kawazoe, E., & Mineshige, S. 1993, PASJ, 45, 715
81. Kenyon, S. J. 1986, *The Symbiotic Stars,* Cambridge University Press
82. Kenyon, S. J., Calvet, N., & Hartmann, L. 1993a, ApJ, 414, 676
83. Kenyon, S. J., & Hartmann, L. 1987, ApJ, 323, 714
84. Kenyon, S. J., & Hartmann, L. 1991, ApJ, 383, 664
85. Kenyon, S. J., & Hartmann, L. 1995, ApJS, 101, 117
86. Kenyon, S. J., Hartmann, L., Gómez, M., Carr, J., & Tokunaga, A. 1993b, AJ, 105, 1505
87. Kenyon, S. J., Hartmann, L., & Hewett, R. 1988, ApJ, 325, 231
88. Kenyon, S. J., *et al.* 1994, AJ, 107, 2153
89. Kenyon, S.J., Hartmann, L.W., Imhoff, C.L., & Cassatella, A. 1989, ApJ, 344, 925.
90. Kenyon, S. J., Hartmann, L., & Kolotilov, E. A. 1991, PASP, 103, 1069
91. Kenyon, S. J., & Webbink, R. F. 1984, ApJ, 279, 252
92. Kenyon, S. J., Whitney, B., Gómez, M., & Hartmann, L. 1993c, ApJ, 414, 773
93. Kenyon, S. J., Yi, I. & Hartmann, L. 1996, ApJ, 462, 439
94. Kholopov, P. N. 1959, Sov. AJ, 3, 291
95. Kley, W. 1991, A&A, 247, 95

96. Kley, W., & Hensler, G. 1987, A&A, 172, 124
97. Kley, W., & Lin, D. N. C. 1996, ApJ, 461, 933
98. Kley, W., & Papaloizou, J. C. B. 1997, MNRAS, 285, 239
99. Kolotilov, E. A., & Kenyon, S. J. 1998, IBVS, No. 4494
100. Kolotilov, E. A., & Petrov, P. P. 1983, Pis'ma Astr. Zh., 9, 171
101. Königl, A. 1991, ApJL, 370, L39
102. Koresko, C. D., Beckwith, S. V. W., Ghez, A. M., Matthews, K., Neugebauer, G. 1991, AJ, 102, 2073
103. Kuiper, G. P. 1941, ApJ, 93, 133
104. Lamzin, S. A. 1998, Astr. Rept., in press
105. Laplace, P. S. 1796, *Mécanique Céleste*
106. Larson, R. B. 1980, MNRAS, 190, 321
107. Li, J. 1996, ApJ, 456, 696
108. Liljeström, T., & Olofsson, G. 1997, ApJ, 478, 381
109. Lin, D. N. C., & Papaloizou, J. 1985, in *Protostars and Planets II,* ed. D. C. Black and M. S. Matthews, Tucson, University of Arizona Press, p. 981
110. Lin, D. N. C., & Papaloizou, J. C. B., 1996, ARA&A, 33, 505
111. Lin, D. N. C., & Papaloizou, J. C. B., 1996, ARA&A, 34, 703
112. Lin, D. N. C., & Pringle, J. E., 1987, MNRAS, 225, 607
113. Lipunov, V. M. 1978, Astron. Zh., 55, 1233
114. Lipunov, V. M. 1980, Astron. Zh., 57, 1253
115. Livio, M. 1997, in ASP Conf. Ser 121, *Accretion Phenomena and Related Outflows,* edited by D. T. Wickramasinghe, G. V. Bicknell, & L. Ferrario, San Francisco, ASP, p. 845
116. Looney, L. W., Mundy, L. G., & Welch, W. J. 1997, ApJ, 484, L157
117. López, R., *et al.* 1998, AJ, 116, 845
118. Lucas, P. W., & Roche, P. F. 1996, MNRAS, 280, 1219
119. Lüst, R. 1952, Zs. f. Nat., 7a, 87
120. Lynden-Bell, D., & Pringle, J. E. 1974, MNRAS, 168, 603
121. McMuldroch, S., Blake, G. A., & Sargent, A. I. 1995, AJ, 110, 354
122. Mahdavi, A., & Kenyon, S. J. 1998, ApJ, 497, 342
123. Malbet, F., *et al.* 1998, ApJL, 507, 149
124. Mardones, D., Myers, P. C., Tafalla, M., Wilner, D. J., Bachiller, R., & Garay, G. 1997, ApJ, 489, 719
125. Marsh, T. R., & Horne, K. 1988, MNRAS, 235, 269
126. Mould, J. R., Hall, D. N. B., Ridgway, S. T., Hintzen, P., & Aaronson, M. 1978, ApJL, 222, L123
127. Nakajima, T., & Golimowski, D. A. 1995, AJ, 109, 1181
128. Ohashi, N., Hayashi, M., Ho, P.T.P., Momose, M., & Hirano, N. 1996, ApJ, 466, 957
129. Okuda, T., Fujita, M., & Sakashita, S. 1997, PASJ, 49, 679
130. Oppenheimer, B. D., Kenyon, S. J., & Mattei, J. A. 1998, AJ, 115, 1175
131. Osaki, Y. 1974, PASJ, 26, 429
132. Papaloizou, J. C. B., & Stanley, G. Q. G. 1986, MNRAS, 220, 593
133. Petrov, P., Duemmler, R., Ilyin, I., Tuominen, I. 1998, A&A, 331, L53
134. Popham, R., & Narayan, R. 1991, ApJ, 370, 604
135. Popham, R., & Narayan, R. 1991, ApJ, 394, 255
136. Popham, R., Narayan, R., Hartmann, L., & Kenyon, S. J. 1993, ApJL, 415, L127
137. Popham, R., Narayan, R., Kenyon, S. J., & Hartmann, L. 1996, ApJ, 473, 422
138. Pringle, J. E. 1977, MNRAS, 178, 95
139. Pringle, J. E., Rees, M. J., & Pacholczyk, A. G. 1973, A&A, 29, 179
140. Pringle, J. E., & Savonije, G. J. 1979, MNRAS, 187, 777
141. Regev, O. 1983, A&A, 126, 146.
142. Regev, O., & Bertout, C. 1995, MNRAS, 272, 71
143. Regev, O., & Hougerat, A. A. 1988, MNRAS, 232, 81

144. Regös, E. 1997, MNRAS, 286, 104
145. Reipurth, B. 1985, A&A, 143, 435
146. Reipurth, B. 1989, Nature, 340, 42
147. Reipurth, B. 1990, in *Flare Stars in Star Clusters, Associations, and the Solar Vicinity*, IAU Symposium No. 137, Dordrecht, Kluwer, p. 229
148. Reipurth, B. 1991, in *Physics of Star Formation and Early Stellar Evolution*, NATO Adv. Study Inst., edited by C. J. Lada and N. D. Kylafis, p. 497
149. Reipurth, B. 1997, in *Low Mass Star Formation – from Infall to Outflow*, Poster Proceedings of IAU Symposium No. 182 on Herbig-Haro Objects and the Birth of Low Mass Stars, edited by F. Malbet & A. Castets, p. 309
150. Reipurth, B., & Aspin, C. 1997, AJ, 114, 2700
151. Reipurth, B., Bally, J., & Devine, D. 1997, AJ, 114, 2708
152. Reipurth, B., & Heathcote, S. 1992, A&A, 257, 693
153. Reipurth, B., Olberg, M., Gredel, R., & Booth, R. S. 1997, A&A, 327, 1164
154. Rodriguez, L., Hartmann, L. W., & Chavira, E. 1990, PASP, 102, 1413
155. Rodriguez, L., & Hartmann, L. 1992, Rev. Mex. A&A, 24, 135
156. Rodriguez, L., D'Alessio, P., Wilner, D.J., Ho, P.T.P., Torrelles, J.M., Curiel, S., Gómez, Y., Lizano, S., Pedlar, A., Canto, J., & Raga, A.C. 1998, Nat, 395, 355
157. Rozyczka, M., Bodenheimer, P., & Lin, D. N. C. 1996, ApJ, 459, 371
158. Rucinski, S. M. 1985, AJ, 90, 2321
159. Sandell, G., & aspin, C. 1998, A&A, 333, 1016
160. Sato, S., Okita, K., Yayamshita, T., Mizutani, K., Shiba, H., Kobayashi, Y., & Takami, H. 1992, ApJ, 398, 273
161. Shakura, N. I., & Sunyaev, R. A. 1973, A&A, 24, 337
162. Shevchenko, V. S. 1995, rotor project
163. Shevchenko, V. S., Ezhkova, O., Tjin A Djie, H. R. E., van den Anckner, M. E., Blondel, P. F. C., & de Winter, D. 1997, A&AS, 124, 33
164. Smith, B. A., & Terrile, R. J. 1984, Science, 226, 1421
165. Staude, H. J., & Neckel, Th. 1991, A&A, 244, L13
166. Staude, H. J., & Neckel, Th. 1992, ApJ, 400, 556
167. Stocke, J. T., Hartigan, P. M., Strom, S. E., Strom, K. M., Anderson, E. R., Hartmann, L. W., & Kenyon, S. J. 1988, ApJS, 68, 229
168. Stone, J. M. & Balbus, S. A. 1996, ApJ, 464, 364
169. Strom, K. M., & Strom, S. E. 1993, ApJL, 421, L63
170. Teodorani, M., Errico, L., Vittone, A. A., Giovanelli, F., & Rossi, C. 1997, A&AS, 126, 91
171. Thiebaut, E., Bouvier, J., Blazit, A., Bonneau, D., Foy, F.-C. & Foy, R. 1995, A&A, 303, 795
172. Tout, C. A., & Pringle, J. E. 1992, MNRAS, 259, 604
173. Turner, N. J. J., Bodenheimer, P., & Bell, K. R. 1997, ApJ, 480, 754
174. Tylenda, R. 1981, Acta Astr., 31, 267
175. Ulrich, M.-H., Maraschi, L., & Urry, C. M. 1997, ARA&A, 35, 445
176. Vishniac, E. T., & Diamond, P. 1989, ApJ, 347, 435
177. Vishniac, E. T., & Zhang, C. 1996, ApJ, 461, 307
178. von Weiszäcker, C. F. 1943, Zs. f. Ap., 22, 319
179. von Weiszäcker, C. F. 1948, Zs. f. Nat., 3a, 524
180. Wachman, A. A. 1939, Beob. Zirk., 21, 60
181. Wachman, A. A. 1954, Zs. f. Ap., 35, 74
182. Wang, Y. M. 1997, ApJL, 475, 135
183. Weintraub, D. A., Sandell, G., & Duncan, W. D. 1989, ApJ, 340, 69
184. Weintraub, D. A., Sandell, G., & Duncan, W. D. 1991, ApJ, 382, 270
185. Welin, G. 1971, A&A, 12, 312
186. Welty, A. D., Strom, S. E., Strom, K. M., Hartmann, L., Kenyon, S. J., Grasdalen, G. L., & Stauffer, J. R. 1991, ApJ, 349, 328
187. Welty, A. D., Strom, S. E., Edwards, S., Kenyon, S. J., & Hartmann, L. W. 1992,

ApJ, 397, 260
188. Whitney, B. A., Clayton, G. C., Schulte-Ladbeck, R. E., Calvet, N., Hartmann, L. & Kenyon, S. J. 1993, ApJ, 417, 687
189. Yamashita, T., & Tamura, M. 1992, ApJL, 387, L93
190. Yang, J., Ohashi, N., & Fukui, Y. 1995, ApJ, 455, 175
191. Yi, I. 1994, ApJ, 428, 760
192. Yun, J. L., Moreira, M. C., Alves, J. F., & Storm, J. 1997, A&A, 320, 167

Neal and Leslie Evans, Doris and Ed Churchwell and Antonella Natta take a lunch break in the Sammaria Gorge.

THE FORMATION OF PLANETS

STEVEN P. RUDEN

University of California, Irvine
Department of Physics and Astronomy
Irvine, CA 92697
USA

1. Introduction

Mankind has attempted to understand the origin of the Earth and planets
for countless generations. In the past few decades, starting largely from the
pioneering work of Safronov in the 1960s, philosophical speculation about
their birth has been replaced with detailed analytical and numerical simulations whose goal is to reproduce the observed properties of our own solar
system. With the recent groundbreaking discoveries of planets around ordinary stars like our Sun (Mayor & Queloz 1995; Marcy & Butler 1996;
Butler & Marcy 1996; see also the review by Marcy in this volume), we
now have a reasonable sample of nearby planetary systems that we can use
to test and validate theoretical models and from which we can learn new
and unexpected features of planet formation. In the upcoming decade the
launches of NASA's Space Interferometry Mission (SIM) and ESA's Global
Astrometric Interferometer for Astrophysics (GAIA) (and their longer term
counterparts Terrestrial Planet Finder [TPF] and Infrared Space Interferometry Mission [IRSI]) will allow us to detect *directly* planets around nearby
stars, which will give us an even more complete census of the number and
content of planetary systems in the Solar neighborhood. To place this observational data in a proper astrophysical context, we need to understand
how planets are born and how they interact with their environment. In this
review, I will discuss the dynamics of the growth of solid particles from
initially micron-sized dust grains through kilometer-sized planetesimals to
their final planetary mass. I will also discuss the formation of the gas–rich
giant planets and their tidal interaction with the surrounding nebular disk.
The reader interested in further details of the planet formation process

643

C.J. Lada and N.D. Kylafis (eds.), The Origin of Stars and Planetary Systems, 643–680.

is encouraged to consult the excellent reviews by Lissauer (1993) and by Marcy and Butler (1998).

1.1. OVERVIEW: FROM DUST TO PLANETS

The low eccentricities and inclinations of the planets (see Table 1) have long been used as evidence that the Solar System formed from a flattened disk of gas and dust: the *primordial solar nebula*. The recognition that the collapse of magnetized, rotating molecular cloud cores naturally leads to protostellar disks with roughly the right mass and size ($\sim$ 100 AU) is one of the successes of modern star formation theory (for a thorough review see Shu, Adams, & Lizano 1987). The solid content in these disks is comprised of the interstellar grains that survive their passage into the disk (through a surface disk shock) plus nebular volatiles that condense as dust within the disk. In the current "standard model" of planet formation, the rocky cores of all planetary bodies grow from the accumulation of small dust grains; the increase in growth is tremendous – nearly a factor of 10^{13} in size and 10^{40} in mass. In our Solar System, this model is sufficient to explain the formation of the terrestrial planets (Mercury, Venus, Earth, Mars), which have densities comparable to rock (see Table 1) and low-mass gas atmospheres that were most likely outgassed from the rocky material from which they accreted (Brown 1949; Prinn 1982; Pepin 1989). The giant planets (Jupiter, Saturn, Uranus, Neptune), which have much lower mean densities (*cf.* Table 1) and massive gas atmospheres, are also thought to have formed from initially rocky cores[1]. However, once the core mass grows to be larger than a critical value ($\gtrsim 10 M_\oplus$), rapid accretion of nebular gas transforms them into gas giant planets (Mizuno 1980).

The growth of planet-sized solid bodies conventionally occurs in two loosely defined sequential phases. In the first phase (§3), dust grains, which are uniformly distributed in and strongly coupled to the nebular gas, grow through binary collisions. As they grow larger they settle gravitationally towards the nebular midplane forming a thin layer. Mutual collisions (and possibly gravitational instability of the dust layer) induce further growth in the midplane forming roughly kilometer-sized bodies called *planetesimals*. Planetesimals are, by definition, massive enough to have largely decoupled from the gas and move on nearly circular, Keplerian orbits about the central star. The first phase is believed to be relatively short, occurring on a timescale as little as a few thousand orbital periods (Weidenschilling & Cuzzi 1993). In the second phase, planetesimals continue to grow through inelastic collisions. The essential feature is that collision cross-sections are

---

[1]Models invoking dynamical gravitational instabilities to form the giant planets (Cameron 1978) have fallen out of favor (see Lin & Papaloizou 1985).

TABLE 1. Properties of Solar System Planets

Planet	Orbital Semimajor Axis (AU)	Mass ($M_\oplus$)	Mean Density (g/cm^3)	Eccentricity	Inclination (degrees)
Mercury	0.39	0.055	5.4	0.20	7
Venus	0.72	0.82	5.2	0.007	3
Earth ($M_\oplus$)	1.0	1.0	5.5	0.02	0
Mars	1.5	0.11	4.0	0.09	1
Jupiter	5.2	318	1.3	0.05	1
Saturn	9.6	95.1	0.70	0.06	2
Uranus	19	14.5	1.6	0.05	1
Neptune	30	17.2	1.6	0.009	2

Notes: Pluto is not included because it is more typical of the low–mass outer Solar System Kuiper belt objects.
$1M_\oplus = 6 \times 10^{27}\mathrm{g} = 3 \times 10^{-6} M_\odot$

increased above their geometrical value by the focusing nature of the gravitational interaction between the colliding planetesimals (Safronov 1972). If the relative velocity dispersion among the planetesimals remains sufficiently small, the increase in cross-section is greatest for the most massive planetesimal. This planetesimal will then collide more frequently and gain mass more rapidly than its neighbors, which causes its collision cross-section to increase even further. "Runaway" growth occurs in which the most massive planetesimal accretes all the available solid matter in its local "accretion zone" – the region of the disk over which its gravity can perturb nearby planetesimals into colliding orbits. The runaway process produces a series of rocky protoplanetary bodies that are dynamically and physically isolated from each other and occurs on timescales of $\sim 10^6$ years (Wetherill & Stewart 1989; Aarseth, Lin, & Palmer 1993; Weidenschilling, *et al.* 1997). After the runaway stalls because all the available nearby solid matter has been accreted, further collisional growth of the protoplanets can occur but only on much longer timescales ($\sim 10^8$ years) as mutual gravitational perturbations cause orbits to cross and highly inelastic collisions to occur (Wetherill 1980; Chambers, Wetherill, & Boss 1996).

As the planets grow, they find themselves in a nebular environment that changes with time as the gas in the disk is accreted by the central star (Ruden & Lin 1986). Eventually, the disk is cleared of gas via processes such as stellar winds or UV irradiation by the central star (Shu, Johnstone, & Hollenbach 1993). The growth of the terrestrial planets to their current masses could have continued long after the dispersal of the gas disk. However, because the more massive atmospheres of the giant planets were accreted from the nebular gas disk, the growth of their rocky cores to the critical mass had to be complete *before* disk dispersal. Observations indicate protostellar disk lifetimes are $\lesssim 10^7$ years (Strom, *et al.* 1989; Beckwith, *et al.* 1990; Zuckerman, Forveille, & Kastner 1995; see also the review by Beckwith in this volume), placing an important time constraint on the planet formation process. Current models have difficulty forming the giant planet cores on this timescale unless the mass of the solar nebula is significantly larger than the minimum-mass solar nebula.

The giant planets are massive enough that gravitational tidal interactions between the planet and the surrounding disk can create annular gaps largely devoid of gas at the orbit of the planet. This mechanism of *tidal truncation* (Lin & Papaloizou 1993) has been invoked to explain why the giant planets stop their rapid gas accretion and thus attain their final masses (§5.2). The tidal interaction also causes the orbital semimajor axes of the planets to decrease in time. This *orbital migration* of the giant planets is a serious problem that will be discussed in more detail in §5.3.

1.2. FUTURE RESEARCH QUESTIONS

Throughout this review I will list a selection of the major unsolved questions in planet formation theory. I hope these questions will provoke and encourage the reader to consider devoting some thought to their solution.

- When does the planet building epoch begin?

  - Is it a continuous process beginning from the initial formation of the disk?

  - Does planet formation require some critical conditions (such as a sufficiently cool or quiescent nebula) before it begins?

- How does planet formation modify the disk environment and its evolution?

  - Are their unambiguous observable signatures such as holes or gaps?

- What outcomes are likely?

  - A diverse planetary system like our own?

  - Giant planets close to the central star as in 51 Pegasus?

  - Rubble disks like β Pictoris?

2. Structure and Evolution of Disks

Protoplanetary disks are just one example of an accretion disk – a geometrically thin disk of gas (and dust) revolving around a central mass point (Pringle 1981; Lin & Papaloizou 1985; and the review chapter by Kenyon in this volume). We can understand their basic properties by examining the force balance equations assuming the gas is in orbit around a central protostar with mass M_*. Adopting cylindrical coordinates (r, ϕ, z) and assuming axisymmetry, the disk is in vertical hydrostatic equilibrium where the vertical pressure force balances the vertical component of the central star's gravity

$$\frac{1}{\rho} \frac{\partial P}{\partial z} = -\frac{GM_*}{r^2} \left(\frac{z}{r}\right) \equiv -\Omega^2 z, \tag{1}$$

where P and ρ are the gas pressure and density, and where we have defined the Keplerian rotation rate

$$\Omega = \sqrt{\frac{GM_*}{r^3}}. \tag{2}$$

We can scale equation (1) to find an expression for the vertical height of the nebula, H,

$$H = \frac{c}{\Omega}, \tag{3}$$

where the nebular sound speed c is defined by the relation $P = \rho c^2$. We say the disk is geometrically thin if $H \ll r$, which from equation (3) is equivalent to the disk being dynamically cold, $c \ll \Omega r$, i.e., the nebular sound speed is much less than the orbital Kepler speed. Because disks are thin, it is convenient to use vertically averaged quantities such as the surface density, Σ:

$$\Sigma = \int_{-\infty}^{+\infty} \rho \, dz \approx 2\rho H = 2\frac{\rho c}{\Omega}, \tag{4}$$

where we interpret ρ in the last two terms on the right side of the equation as the value of the density at the midplane of the nebula.

Force balance in the radial direction is given by

$$\frac{v^2}{r} = \frac{GM_*}{r^2} + \frac{1}{\rho} \frac{\partial P}{\partial r}, \tag{5}$$

where v is the orbital speed of the gas. For thin, cold disks, the radial pressure gradient is smaller than the stellar gravitational force by $\sim (H/r)^2 \ll 1$ (as may be verified by scaling eq. [5] and using eq. [3]) and is typically neglected in most accretion disk applications. In this case, equation (5) shows the gas rotates in centrifugal balance at the Keplerian orbital speed, $v = \Omega r$.

In most disks, the pressure is largest near the star and decreases outward so that the radial pressure term in equation (5) is negative. This means the pressure of the gas partially supports the disk against gravity, and the disk gas rotates at a slightly *sub-Keplerian* speed $v = \Omega r - \Delta v$ with $\Delta v > 0$. If we substitute this expression for v into equation (5), we find

$$\Delta v = \frac{1}{2} \left| \frac{\partial \ln P}{\partial \ln r} \right| \frac{c^2}{\Omega r} \approx \frac{c^2}{\Omega r} = \left(\frac{H}{r} \right) c \ll c, \tag{6}$$

where we have used equation (3) in the last terms on the right. The departure from Keplerian motion is indeed very small being much less than the sound speed which is itself much less than the Keplerian rotation speed. We shall see in §3.1 that this small departure from exact Keplerian rotation plays a critical role in the motion of solids in protoplanetary disks.

In two seminal papers, Lynden-Bell & Pringle (1974) and Shakura & Sunyaev (1973) demonstrated that the evolutionary behavior of accretion disks is governed by the *outward* transport of angular momentum, which allows mass to flow *inward* to be accreted by the central object. The precise mechanisms that cause the transport in *any* astrophysical disk are still far from certain. Numerous suggestions have been made in the case of protostellar disks (see the review by Adams & Lin 1993): gravitational instabilities (Lin & Pringle 1987; Shu, Tremaine, Adams, & Ruden 1990), magnetic instabilities (Stepinski & Levy 1990; Balbus & Hawley 1991), thermal convective instabilities (Lin & Papaloizou 1980; Ruden & Lin 1986), stellar or disk winds (Hartmann & MacGregor 1982; Wardle & Königl 1993) and fluid dynamical shear instabilities (Dubrulle 1993). In most applications, the details of the physical cause of the transport are ignored and angular momentum is assumed to be transported by some form of localized fluid dynamical turbulence that creates an eddy viscosity in the gas. The disk evolution is governed by the magnitude of the turbulent viscosity, ν, which is the product of the turbulent eddy size, ℓ_t, and velocity, v_t. The eddies are assumed to move subsonically and to have sizes smaller than the vertical height of the disk, which leads to the well-known parameterization of the viscosity as (Shakura & Sunyaev 1973)

$$\nu \sim v_t \cdot \ell_t \equiv \alpha c H, \tag{7}$$

where $\alpha \lesssim 1$ is a dimensionless parameter. Under the action of viscous stress, an initial surface density distribution evolves diffusively by spreading outward while transferring mass inward to be accreted by the central protostar. The viscous evolutionary timescale is

$$t_{\text{vis}} \approx \frac{r^2}{\nu} \approx \frac{1}{\Omega} \frac{1}{\alpha} \left(\frac{r}{H} \right)^2, \tag{8}$$

$$\approx\ 10^7 \left(\frac{10^{-2}}{\alpha}\right) \left(\frac{r/H}{25}\right)^2 \left(\frac{r}{100\mathrm{AU}}\right)^{3/2}\ \text{years.}$$

The solids in protoplanetary disks find themselves in a continuously changing nebular environment in which temperature, density, pressure, and turbulent velocity typically decrease with time (after transients from the initial distribution of matter decay; see Fig. 1). Detailed calculations using reasonable values for the eddy viscosity ($\alpha \approx 10^{-2}$) yield evolutionary times of order $10^6 - 10^7$ years (Ruden & Lin 1986; Ruden & Pollack 1991), which are in agreement with the observational disk lifetimes cited above.

2.1. THE MINIMUM-MASS SOLAR NEBULA (MMSN)

The Solar System was formed from the collapse of a molecular cloud having solar element abundances, yet the present day planets are clearly of non-solar composition. We can ask the important cosmogonical question, "How much *solar composition* gas is required to make the planets?" The mass and inferred spatial distribution of this gas are known as the *Minimum-Mass Solar Nebula* (hereafter, MMSN). The answer to this question has been given by several authors (Weidenschilling 1977a; Hayashi 1981) who take the present amount of condensed solid matter in the Solar System and augment it to solar composition to determine the minimum mass. The radial surface density distribution of this mass is deduced by smearing out the augmented mass of a planet over an area defined by its nearest neighbors.[2]

A key ingredient in the analysis is the dust-to-gas ratio, ζ, which measures the mass fraction of dust (*i.e.*, high-Z elements) in a solar composition gas. In cool regions of the nebula with temperatures below about 170 K, both rocky and icy material (such as water, methane and ammonia) can condense from the gas giving the dust-to-gas ratio a value $\zeta \approx 1/60$ (Hayashi 1981). In warmer regions (above 170 K but below about 1500 K where much of the dust evaporates) only the more refractory (high condensation temperature) solids survive and the dust-to-gas ratio drops by a factor of four to $\zeta \approx 1/240$ (Hayashi 1981). To find a rough value for the mass in the MMSN we must estimate the present amount of solids in the planets. This comes almost entirely from the high-Z material inside the giant planets, which is estimated to range from 40 - 80 $M_\oplus$ (Hubbard & Marley 1989; Chabrier, *et al.* 1992) compared to only 2 $M_\oplus$ in the terrestrial planets. Adopting a value of 60 $M_\oplus$ for the present mass of condensed solids and an average dust-gas-ratio of 1/100 gives a rough estimate of 6000 $M_\oplus \sim 0.02 M_\odot$ for the mass of solar composition gas needed to make the

---

[2]An important assumption is that the planets formed at their present locations.

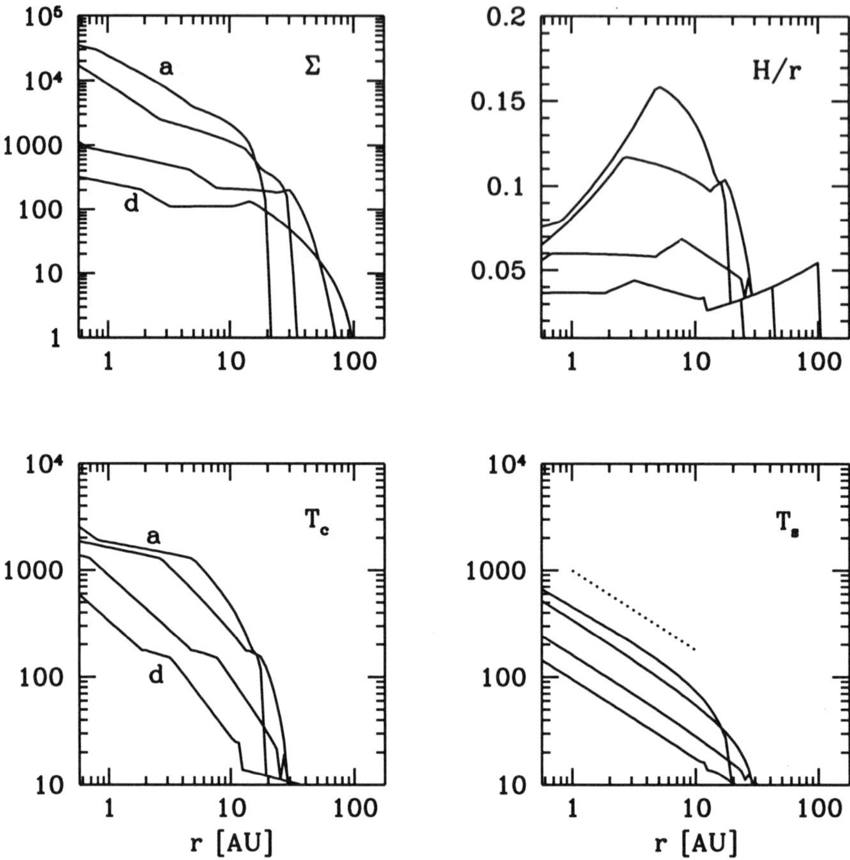

Figure 1. Models for the evolution of the primordial solar nebula. Surface density, Σ, aspect ratio, H/r, central (midplane) temperature, T_c, and surface (effective) temperature, T_s, are plotted versus distance from the sun (in AU) for the times a) 10^3, b) 10^4, c) 10^5, d) 10^6 years. Note the diffusive spreading of the nebula from less than 10 AU to 100 AU, and the cooling of the nebula with time. The disk always remains geometrically thin ($H/r \ll 1$), and at late times $H/r \approx 1/25$. The surface temperature falls off as $r^{-3/4}$ (marked as the dotted line) away from the outer disk edge. The viscous α parameter is 10^{-2} in these calculations adapted from Ruden & Lin (1986).

planets.[3] A more detailed calculation of the MMSN by Hayashi (1981) is illustrated in Figure 2. Disk masses derived from millimeter observations of nearby star forming regions are found to range from $0.005 - 0.2$ $M_\odot$ (Beck-

---

[3] Adding the mass of solar composition gas already present in all the planetary atmospheres ($\lesssim 400 M_\oplus$) is only a small correction.

with & Sargent 1996; also see the review by Beckwith in this volume), comparable to the MMSN.

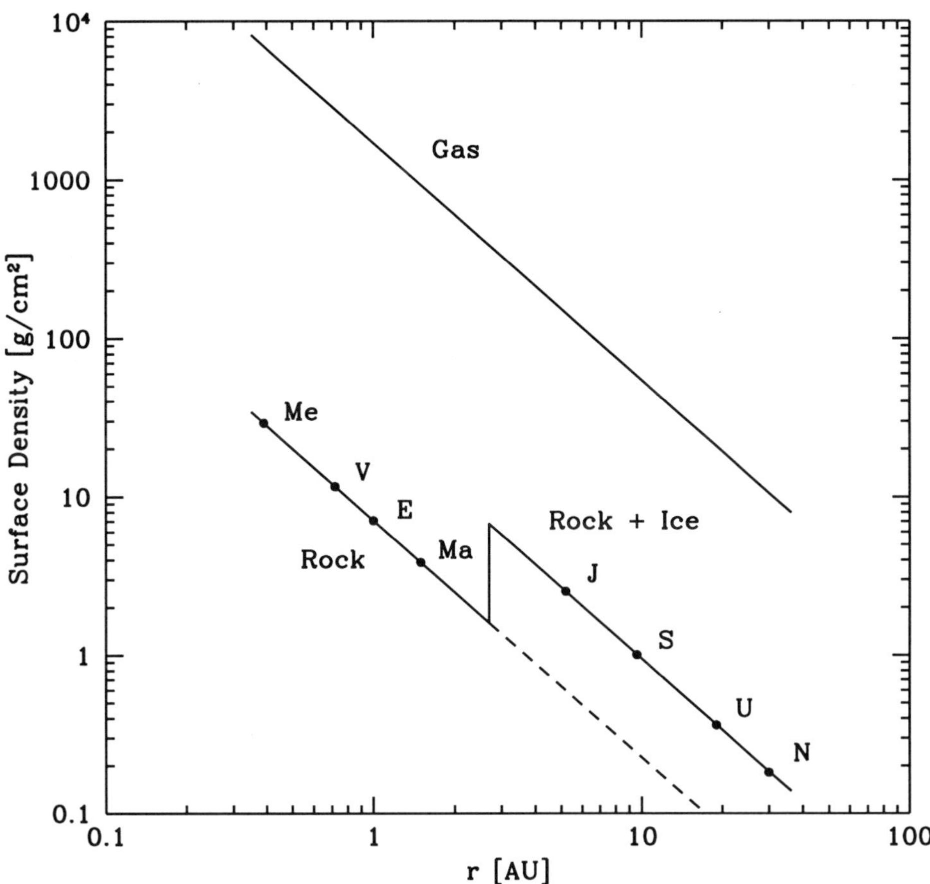

Figure 2. Surface density distribution in the Minimum-Mass Solar Nebula (MMSN) adapted from Hayashi (1981). The planets are denoted by Me, V, E, Ma, J, S, U, N. Outside 2.7 AU the temperature is deduced to be cool enough for ices to condense, which causes the surface density of solids to jump by a factor of four. The gas distribution shows no discontinuity, and the surface density of both gas and solids falls off as $r^{-3/2}$. The total mass between 0.35 AU and 36 AU is 0.013 $M_\odot$.

The MMSN, while playing a useful role as a conceptual guide, is perhaps becoming overused as a quantitative "reference" standard in model calculations of the planet formation epoch. The planets formed in an *evolving* disk whose properties (such as the position where ices condense) changed with time; the static $r^{-3/2}$ surface density distribution presented by the

MMSN may not only be quantitatively wrong, but misleading is well. If it does accurately describes any part of the evolution of the solar nebula, it is probably the *final* stages when planet formation is complete rather than the *initial* stages when planet formation began. Furthermore, as we will see in §5.3, it is now thought that orbital evolution of the giant planet semimajor axes may have occurred, which would modify any current reconstruction of the MMSN surface density distribution from that depicted in Figure 2.

2.2. NONLINEAR NATURE OF DISK PHYSICS

It is obvious there is a very complicated interrelationship among all the physical processes that can occur in protoplanetary disks. Figure 3 is an intentionally complicated schematic diagram indicating some of the predominant physical processes, their dependencies and their feedbacks. Although there are many models by many different research groups addressing the key underlying physics in any one of the boxes in Figure 3, successful integration of the different processes together into a coherent whole has been elusive. There are several reasons for this: the inherent complexity of the physics involved, the lack of experimental guidance as to the magnitude of important physical parameters, the inability of current observational techniques to probe the relevant scales (which occur at angular resolutions of $\sim 10^{-2}$ arcseconds in the nearest star forming complexes) and the nonlinear interplay among the processes themselves. For these reasons, we are far from having a theory that can predict the type of planetary system formed from a given nebular environment. Furthermore, researchers have typically focused their modeling on explaining the properties of our own Solar System, perhaps rejecting scenarios that, although inadequate for our own system, might be good representations of the physics in other protoplanetary disks. As is often the case, more complete observational data will enable modelers to explain the diversity of planetary systems found in nature.

3. From Grains to Planetesimals

3.1. PARTICLE–GAS DYNAMICS

The motion of solid particles in protoplanetary disks is affected by the drag they experience as they move through the gaseous background of the nebula (Whipple 1972; Adachi, Hayashi, & Nakagawa 1976; Weidenschilling 1977b). We will consider the solids as spheres of radius a having internal mass density δ (to be distinguished from the ambient gas density ρ) and mass $m = 4\pi\delta a^3/3$. The form of the gas drag force, $\mathbf{F}_D$, depends on the size of the particle, its speed through the gas, and local gas properties. The Epstein drag law, which is appropriate for particles with sizes smaller

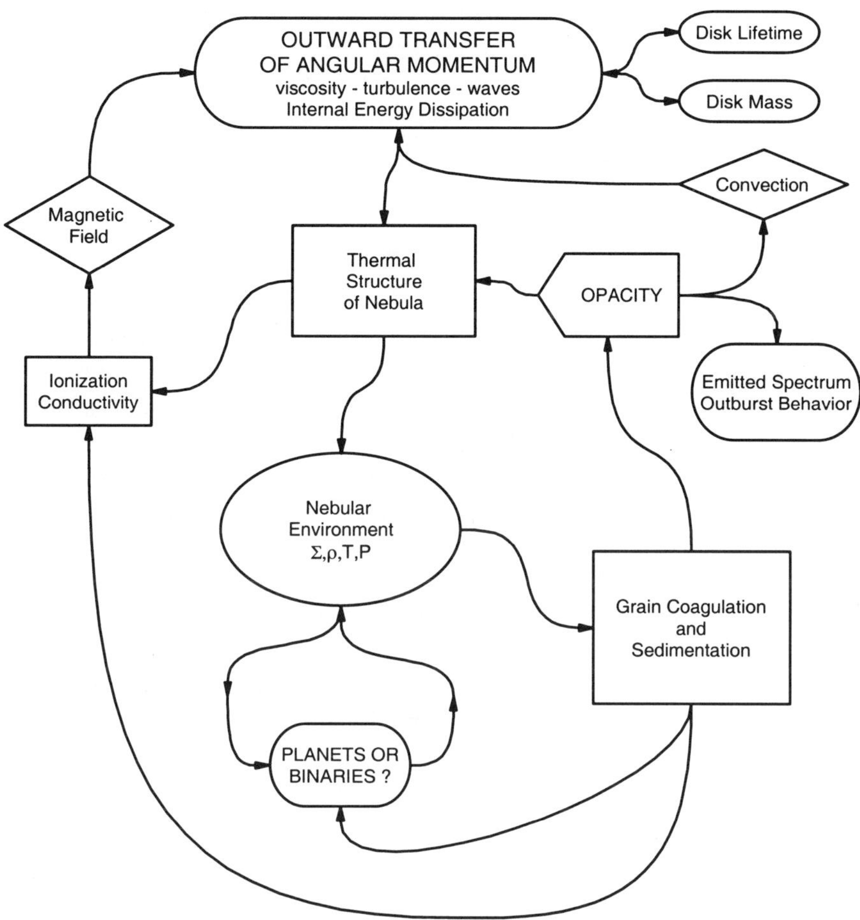

Figure 3. Schematic diagram showing the relationships and feedbacks among nebular processes, many of which are only poorly understood. Small modifications to the underlying physics for any one process may have profound effects on the ultimate evolutionary outcome of the star/disk system.

than the mean free path in the gas, is

$$\mathbf{F}_D = -\frac{4}{3}\pi a^2 \rho\, c\, \mathbf{v}, \qquad (9)$$

where c is the gas sound speed and $\mathbf{v}$ is the velocity of the particle relative to the gas. In most regions of the solar nebula, this law is valid for particles with sizes smaller than a few centimeters (see the references above for a complete list of the different drag law regimes); for simplicity of presentation in this review, I will adopt equation (9) as the principal drag formula. The *stopping time* for a particle is defined to be the time needed for drag to dissipate its momentum, $t_{\mathrm{s}} \equiv mv/F_{\mathrm{D}}$. Substituting equation (9) into this definition and eliminating the volume gas density in favor of the surface density from (4), we find the relation

$$\Omega t_{\mathrm{s}} = \frac{\delta\, a}{\Sigma}. \tag{10}$$

The dimensionless quantity Ωt_{s} is the ratio of the particle stopping time to the local orbital time in the disk and is a key determinant of the dynamical behavior of the particle. In the limit $\Omega t_{\mathrm{s}} \ll 1$, the particle, considered "small", is very strongly coupled to the gas. Drag will force it to move at the local gas velocity, which we saw in §2 to be sub-Keplerian. In the opposite limit, $\Omega t_{\mathrm{s}} \gg 1$, the particle, considered "large", is only weakly perturbed by gas drag and orbits at the Keplerian speed.

Let us consider the dynamics in more detail. At each radius, gas drag forces small particles to orbit at the local gas speed, which is smaller than Keplerian. The net effective gravity (stellar plus centrifugal) acting on the particle points toward the central star, and the particle drifts inward at the terminal speed. This radial velocity is easily shown to be

$$v_r = -2\,\Omega t_{\mathrm{s}}\, \Delta v, \qquad (\Omega t_{\mathrm{s}} \ll 1) \tag{11}$$

where Δv is the difference between the Kepler and gas speeds defined in equation (6). Large particles move at the Kepler velocity and experience a headwind as they move through the more slowly moving gas. The gas drag torque causes the orbit of the particle to decay with radial velocity

$$v_r = -\frac{2\Delta v}{\Omega t_{\mathrm{s}}}. \qquad (\Omega t_{\mathrm{s}} \gg 1) \tag{12}$$

Detailed calculations by Weidenschilling (1977b; see Fig. 4) verify these limiting behaviors and show that the maximum inward drift speed is $\approx \Delta v$ and is achieved by particles with sizes that satisfy $\Omega t_{\mathrm{s}} \approx 1$. There are several points to note about this inward drift. First, it is size dependent leading to differential speeds among particles with different radius, which can enhance the rate of collisions. Also, the relative speed between bodies entrained in the flow can easily exceed their escape velocities (see Fig. 4). Second, the affect is largest for particles having radii for which $\Omega t_{\mathrm{s}} \approx 1$. Using equation (10), we find that the size of these particles is $a \approx \Sigma/\delta$, which is of order 1 m at the position of the Earth. Third, although the magnitude of the drift speed is highly subsonic, particles satisfying $\Omega t_{\mathrm{s}} \sim 1$ decay into the central star in a time very much less than the evolutionary time of the disk. The drift time for these particles is $t_{\text{drift}} = r/v_r = \Omega^{-1}(r/H)^2 \approx \alpha t_{\text{vis}} \approx 100(r/\text{AU})^{3/2}$ years. Of course such particles are likely to collide with other solids on their way in and through shattering or growth move to size regimes with smaller drift rates. Nonetheless, gas drag-induced decay is a potentially large sink for the solid component of the disk.

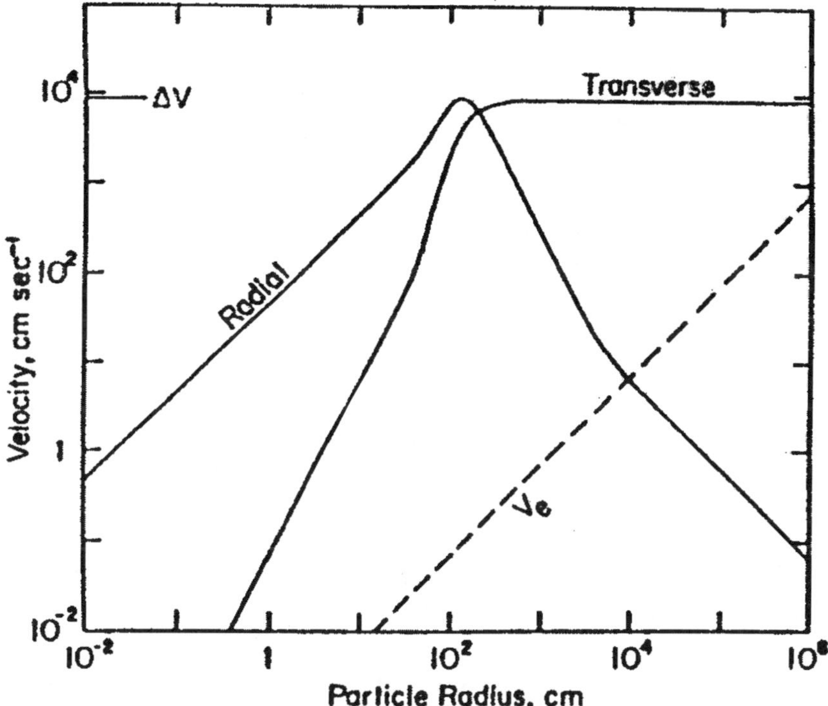

Figure 4. Radial and transverse velocity components, relative to the gas, as a function of particle radius. The difference between the Kepler speed and the gas speed is labeled ΔV. Small particles have transverse speeds equal to the gas speed, while large particles move at the Kepler speed, *i.e.*, have transverse speeds relative to the gas of ΔV. The maximum inward drift speed is $\approx \Delta V$ and is achieved by particles with $\Omega t_s \approx 1$. The changes in slope are transitions between drag regimes, and V_e is the escape speed from the particle. From Weidenschilling (1977b).

3.2. MIDPLANE SETTLING AND GROWTH

As discussed earlier, the formation of solid bodies in the midplane of the nebula begins with agglomeration and sedimentation of smaller grains initially suspended throughout the nebula. Small particles settle in the vertical gravitational field ($g_z = (z/r)GM_*/r^2 = \Omega^2 z$) at the terminal velocity. Using the Epstein law (9), the settling speed and the time to reach the midplane are ($\Omega t_s \ll 1$)

$$v_z = -(\Omega t_s)\,\Omega z, \tag{13}$$

$$t_{\text{settle}} = \frac{z}{|v_z|} = \frac{1}{\Omega}\left(\frac{1}{\Omega t_s}\right) = \frac{1}{\Omega}\left(\frac{\Sigma}{\delta a}\right). \tag{14}$$

The last equation shows that the settling time is always much longer than the rotation period for small grains ($\Omega t_s \ll 1$) and that small grains settle more slowly than larger grains ($t_{\text{settle}} \propto 1/a$). For typical solar nebula conditions the settling time for micron-sized grains is about 10^6 years, comparable to the lifetimes of protoplanetary disks. Without collisions to increase grain size (leading to more rapid settling), small grains, which are the dominant contributor to the opacity, will remain suspended in the disk, which will remain optically thick.

Safronov (1972) was one of the first to realize that collisions not only make particles grow but also rapidly decrease the settling time. Equations (13) and (10) show that the settling speed is proportional to the size of the particle. Larger grains will sink towards the midplane faster and will sweep up smaller grains in their path. This causes them to grow larger and to sink even more rapidly. In this manner, the dust component of the disk will therefore "rainout" and form a thin layer of much larger bodies in the midplane of the nebula. We can estimate the size of these bodies by using mass conservation. The total surface density of solids in the disk is $\zeta\Sigma$, where ζ is the dust-to-gas ratio introduced above. A body that reaches the midplane with final radius a will have swept up a column of dust having a mass $p_s \cdot \zeta\Sigma \cdot \pi(a/2)^2$, where we have taken its average radius on the descent to be $a/2$ and have introduced the probability of sticking in a collision, p_s. Equating this to its mass $m = 4\pi\delta a^3/3$ yields a size

$$a = \frac{3\,p_s\,\zeta\,\Sigma}{16\,\delta}, \tag{15}$$

which is about 1 cm at the Earth in the primordial solar nebula (for $p_s = 1$). The time for solids to sediment to the midplane can be estimated by inserting this value for a into equation (14); more accurate integrations yield (Nakagawa & Hayashi 1981)

$$t_{\text{sediment}} \approx \frac{100}{\Omega}\,\frac{1}{p_s\,\zeta}, \tag{16}$$

which is only a few thousand orbital periods at any point in the disk (again for $p_s = 1$).

There have been several detailed numerical simulations of the collisional growth and sedimentation of dust in the solar nebula (Weidenschilling 1984; Nakagawa, Nakazawa, & Hayashi 1981; Mizuno 1989; Weidenschilling & Cuzzi 1993). A major difficulty with simulating particle growth is our inadequate knowledge of the physical inputs. The sticking probability is a complicated and currently only poorly understood function that depends on the impact velocity and the size, shape and internal strength of the impactors, but it is usually simply taken to be unity for all collisions in the simulation. Realistic nebular particles may have low internal strengths and may shatter in high speed impacts. Some experimental work has been performed on this important issue. Blum & Münch (1993) did not find *any* sticking in collisions between millimeter-sized dust aggregates at speeds $\lesssim 1$ m/s and found fragmentation at speeds above this. More recently, this group (Wurm & Blum 1998) has investigated lower speed ($\lesssim 1$ cm/s) collisions between small, porous, fractal grains with sizes ~ 10 μm and found perfect sticking ($p_s = 1$). In the outer, cooler regions of the nebula volatile ices can form frosts on particle surfaces, which may enhance the sticking probability (Bridges, *et al.* 1996; Supulver, *et al.* 1997).

Turbulence in the gas can have profound affects on the evolutionary simulation. A given sized particle will respond to the fluctuating turbulent velocity field according to whether its stopping time is longer or shorter than the local eddy turnover time. Stirring by turbulence can inhibit sedimentation of particles that have sizes small enough that they are strongly coupled to and co-move with the turbulent eddies. The turbulent velocity field can also greatly increase the relative velocity between particles with different sizes (Markiewicz, Mizuno, & Völk 1991), which can increase the collision rate but also lead to collisions where the bodies shatter (Weidenschilling 1984). Given our poor understanding of turbulence in protoplanetary disks, its effects on grain growth (and vice versa) and grain sedimentation are presently far from clear. Even with these uncertainties and although the individual details vary, the different numerical simulations nonetheless yield midplane particle sizes and sedimentation timescales that are not too dissimilar from the simple estimates above (eqs. [15] & [16]).

3.2.1. *Radial Distribution of Solids in Protoplanetary Disks*

The sedimentation of dust grains produces a thin layer of solids in the nebular midplane. The surface density of this distribution is often taken to be just the gas-to-dust ratio times the local gas surface density, $\zeta\Sigma$. To test whether this is a valid approximation and to see whether the solid surface density distribution of the MMSN is reproduced, I have calculated simplified models of the midplane accumulation of solids using the nebular evolutionary models of Ruden & Pollack (1991). The models follow track the condensed and vapor phases of refractory rocky and volatile icy matter, with ices condensing only in nebular regions with temperatures below 170 K. Early in the evolution, the nebula is too hot for ices to condense anywhere, although rocky matter can condense everywhere outside about 0.1 AU. As the nebula evolves and cools, the radius where ices can condense, *i.e.*, the $T = 170$ K boundary, moves inward and icy solids sediment. The treatment of grain growth and sedimentation is phenomenologically modeled by allowing the condensed solids to rainout on a timescale given by $\tau_0/\zeta\Omega$ (*c.f.* eq. [16]), where τ_0 is an adjustable constant. This model is nonlinear because as the solids sediment to the midplane, the dust-to-gas ratio ζ decreases, and the sedimentation time lengthens. Figure 5 shows the results of two evolutionary scenarios: the first with rapid sedimentation, $\tau_0 = 600$, and the second with much slower sedimentation, $\tau_0 = 6 \times 10^4$. Both models started from the same initial conditions. Significantly more solids are deposited in the midplane in the rapid scenario, but at the time shown in the plot (10^5 years) the nebula is warm enough that ices only condense outside 6 AU. The results of the slow scenario give a surface density distribution comparable to the MMSN after $\sim 10^6$ years, but with a more shallow slope (and more solid matter) in the outer cool regions.

It seems clear from these idealized models that very different distributions of midplane solid matter are possible, from the same initial mass reservoir, if the sedimentation physics (primarily the average sticking probability) is different. The solid distributions shown in Figure 5 do *not* bear a simple constant multiplicative relationship ($\zeta\Sigma$) to the instantaneous gas surface density distribution. The reason for this is the solids at any given radius were deposited over time and reflect accumulation from different nebular conditions, with primarily rocky material sedimenting early on when the temperature was high followed by icy sedimentation when the nebula was cooler. Note also that the locus at which ice can condense is different in the two models; in general, the midplane distance at which $T = 170$ K (the ice condensation front) is a function of time.

3.3. PLANETESIMAL FORMATION

Planetesimals are bodies large enough to move on Keplerian orbits largely unaffected by gas drag forces over the age of the disk. For typical nebular conditions, bodies larger than about 1 km satisfy this constraint. A key unanswered question is how the smaller sized bodies (~ 1 cm; eq. [15]) produced in the sedimentation process assemble into planetesimals. The most elegant and efficient mechanism proposed is that the thin layer of solids produced by sedimentation becomes gravitationally unstable and rapidly forms fragments that collapse to produce planetesimals (Goldreich & Ward 1973).

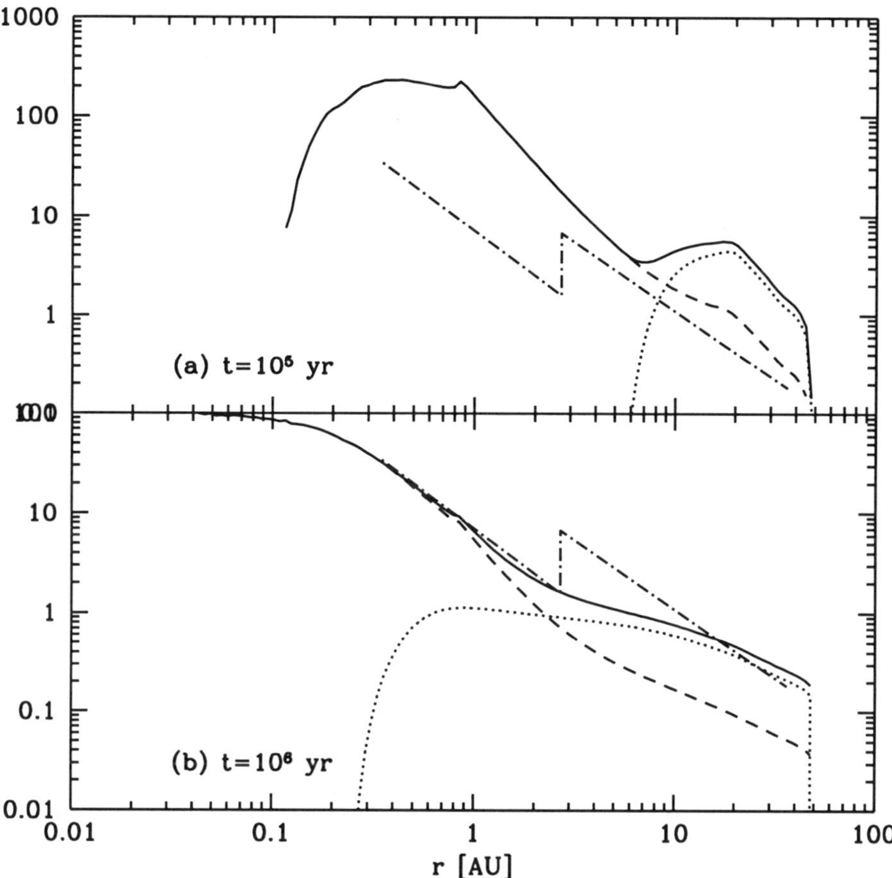

Figure 5.   Radial distribution of solids in a model of grain growth and sedimentation (Ruden 1998). Surface density (in g/cm^2) is plotted versus distance in AU. The surface densities of icy (dotted), rocky (dashed), and rock + ice (solid) matter are plotted. The MMSN surface density distribution is shown for comparison (dot-dashed). Panel (a) [(b)] is for the rapid [slow] accumulation scenario, with the evolution times indicated. Note the difference in vertical scale between the panels. See text for further details.

They showed the critical fragmentation lengthscale, λ_c, is

$$\lambda_c \approx 4\pi r \left(\frac{\pi \zeta \Sigma r^2}{M_*} \right), \tag{17}$$

and the typical fragment mass, M_{GW} is

$$M_{\mathrm{GW}} = \zeta \Sigma \lambda_c^2 = 16\pi M_* \left(\frac{\pi \zeta \Sigma r^2}{M_*} \right)^3. \tag{18}$$

In these equations, it is the surface density of *solids*, $\zeta\Sigma$, that is important, and the ratio in parenthesis is roughly the mass of the solid disk to the protostellar mass. Using nebular parameters from the MMSN, Goldreich & Ward (1973) showed that kilometer-sized bodies can be produced by this gravitational collapse process in a few thousand years.

The Goldreich-Ward instability occurs only in very quiescent dust disks. If the random velocities of the particles composing the dust disk are too large (or equivalently if the dust layer is too thick)[4] then the instability is quenched. The critical random speed is $\pi G\zeta\Sigma/\Omega$, which is only about 10 cm/s. In the thin dust disk the solids rotate at nearly the Keplerian velocity while the gas above the layer rotates at a sub-Keplerian rate (eq. [6]). It is thought that turbulence generated by this shear layer will stir up the random velocity in the dust disk and prevent the Goldreich-Ward instability (Weidenschilling & Cuzzi 1993). Detailed hydrodynamical models of the particle-gas interaction have been made by Cuzzi, Dobrovolskis, & Champney (1993) who find that it is unlikely the dust layer can achieve such low random speeds.

If the Goldreich-Ward mechanism does not proceed in protostellar disks, then it is likely that planetesimals are produced in the thin dust layer by inelastic binary collisions. Recall that gas drag causes radially inward flow of solid particles (eqs. [11] & [12]). In the dense particle layer, solids dominate the local density and the drift speed is reduced (Cuzzi, Dobrovolskis, & Champney 1993). Since larger particles have lower drift speeds (eq. [12]), Cuzzi, Dobrovolskis, & Champney (1993) have argued that planetesimals can grow in this layer as the smaller solids drift past them and are collisionally accreted. They estimate that growth to 100 km sizes can be achieved in $\lesssim 10^6$ years.

3.4. PROSPECTS FOR OBSERVING THE EARLY PLANET FORMATION EPOCH

It would be satisfying if theoretical models gave an unambiguous prediction for the observational signature of grain growth in young protoplanetary disks, for example, the existence of a "hole" in the infrared spectrum (such as the one inferred observationally in HR 4796 by Koerner, *et al.* 1998). Unfortunately, this is presently not the case. In order to "see" forming planets or other structures in the midplane of the disk, the optical depth to the nebular midplane must be small. The disk optical depth is equal to the gas surface mass density times the opacity. The midplane regions can remain cloaked to the eyes of observers if *either* is sufficiently large.

---

[4]The factor in parentheses in equations (17) & (18) is the maximum value of the aspect ratio, H/r, the layer can have to be gravitationally unstable.

The primary contributors to the opacity are small grains, but its value even for the standard interstellar grain-size distribution is subject to debate (Beckwith & Sargent 1993; see the review chapter by Beckwith); grain size evolution only complicates the picture (Miyake & Nakagawa 1993).

As noted in §3.2, micron-sized grains take times comparable to the age of disks to settle in quiescent nebulae. Any small grains left suspended in the disk after the bulk of the solid mass has sedimented to the midplane will remain suspended in the disk as long as it is quiescent. The effects of nebular turbulence on a suspended small grain population are harder to discern. By itself, turbulence causes small particles to remain stirred up and suspended in the nebula. However, turbulence also mixes these particles into the midplane region where they may collide and stick to larger bodies (Weidenschilling & Cuzzi 1993). The net result of these two competing processes on the small grain population is unknown. Also uncertain is how the turbulent properties change as the disk optical depth changes, *i.e.* whether the disk will remain turbulent as grain growth proceeds.

Current theoretical models cannot reliably predict how much of the initial solid mass will remain leftover as small particles, from processes such as particle shattering and erosion. The nebular models of Ruden & Lin (1986, Fig. 1) and Ruden & Pollack (1991) remain optically thick through mid-infrared wavelengths at all disk radii as long as $\gtrsim 1\,\%$ of the initial grain mass remains suspended. The MMSN (Fig. 2) will also remain optically thick out to $\lesssim 10$ AU under the same conditions. Finding observable signatures may require waiting until the disk gas is cleared (probably taking with it the small grain population) or searching for planet formation signatures in long wavelength regimes that remain optically thin.

3.5. FUTURE RESEARCH QUESTIONS

— What are the appropriate conditions in protostellar disks at the inception of the planet building epoch?

- Do they resemble the MMSN?
- Do different initial conditions lead to diverse planetary systems?

— What is the sticking probability in grain-grain collisions?

- Do simulations using $p_s = 1$ adequately model grain growth?

— Is planet formation inefficient?

- How much solid mass is lost via gas-drag induced radial flow?

— How are planetesimals formed?

- Via the Goldreich-Ward instability or by collisions?

— What are observable signatures of planet formation?

- Holes, or gaps, or warps?

4. From Planetesimal to Planet

Planets are formed from the midplane planetesimal distribution by inelastic collisions. The key feature of the collisional evolution is that the cross-section for pairwise collisions can be increased above the geometrical value by gravitational attraction between the colliders (Safronov 1972). This result can be shown quite easily by considering energy and angular momentum conservation in the collision between two particles with masses m_1 and m_2 and radii a_1 and a_2. In the center-of-mass system, the constant orbital energy and angular momentum are

$$E \;=\; \frac{1}{2}\mu v^2 - \frac{Gm_1 m_2}{r} = \frac{1}{2}\mu V^2, \tag{19}$$

$$J \;=\; \mu\,|\mathbf{r} \times \mathbf{v}| = \mu r v \sin\theta = \mu b V, \tag{20}$$

where $\mu = m_1 m_2/(m_1 + m_2)$ is the reduced mass. In the rightmost equalities I have evaluated the constants E and J when the particles are at large separation where b is the impact parameter and V is the relative collision velocity. Without gravity the maximum impact parameter that results in a collision is the sum of the radii $a_1 + a_2$. With gravity, a grazing collision results when the minimum separation is $r = a_1 + a_2$ and the velocity is purely tangential, $\theta = \pi/2$. Substituting these two conditions into equation (20) gives the speed at impact as $v = bV/(a_1 + a_2)$, which can be substituted into equation (19) to give the gravitationally enhanced cross-section for collision

$$\pi b^2 = \pi \left(a_1 + a_2\right)^2 \left[1 + \frac{v_e^2}{V^2}\right], \tag{21}$$

where the mutual escape speed, v_e, is

$$v_e = \sqrt{\frac{2G\left(m_1 + m_2\right)}{a_1 + a_2}}. \tag{22}$$

The term in brackets in equation (21) is the gravitational focusing factor and can greatly increase the cross-section above geometrical if the relative collision speed, V, is much smaller than the escape speed, v_e. Also note that if the focusing factor is large the cross-section is proportional to the fourth power of the particle radius (because $v_e^2 \propto a^2$) rather than the second. In honor of Safronov's pioneering contributions to the study of collisional planet growth, the gravitational focusing factor is often written as $1 + 2\theta$, where $\theta \equiv v_e^2/2V^2$, is called the Safronov number. Detailed three-body numerical orbit integrations confirm that the cross-section (21) is a good approximation (Ida & Nakazawa 1989; Greenzweig & Lissauer 1990, 1992).

The equation for the collisional growth of a planet with mass $m_p = 4\pi\delta a^3/3$ (recall δ is the internal density of the planet) accreting matter from a background "swarm" of planetesimals that move with respect to it at a relative speed V is (using the cross-section [21] and neglecting the size of the planetesimals)

$$\frac{dm_p}{dt} = \rho_s \cdot V \cdot \pi a^2 \left[1 + \frac{v_e^2}{V^2}\right], \tag{23}$$

where ρ_s is the volume density of the planetesimal swarm. We can rewrite this as an equation for the rate of change of the planetary radius

$$\frac{da}{dt} = \frac{1}{8} \left(\frac{\Omega\Sigma_s}{\delta}\right) \left[1 + \frac{v_e^2}{V^2}\right], \tag{24}$$

where $\Sigma_s = 2\rho_s V/\Omega$ is the surface density of the planetesimal swarm (c.f. eq. [4]).

4.1. ORDERLY GROWTH

The parameter that plays the most crucial role in determining the type of planetary growth is the Safronov number, i.e., the magnitude of the planetary escape speed relative to the planetesimal velocity dispersion. Using a variety of analytical approaches, Safronov (1972) argued that as planetesimals accumulated, gravitational scattering off the largest member of the swarm would cause the velocity dispersion to keep pace with the escape speed from that member, regulating θ to be near unity. In this scenario planets grow in an orderly fashion with most of the mass in the large bodies. From equation (24), the accumulation time is

$$t_{\text{orderly}} = \frac{a}{da/dt} \approx \frac{1}{\Omega} \left(\frac{\delta\, a}{\Sigma_s}\right), \qquad (\theta \sim 1) \tag{25}$$

$$\approx 10^8 \left(\frac{r}{\text{AU}}\right)^3 \text{ years.}$$

The growth time is determined by both the Kepler clock and the solid surface density. In the last line I have evaluated the growth time for Earth-sized bodies and taken Σ_s from the MMSN distribution (Fig. 2). These timescales are quite long, far greater than the lifetime of nebular disks, with the orderly growth process being unable to produce the cores of the outer giant planets in less than the age of the universe!

4.2. RUNAWAY GROWTH

The key to forming planets more rapidly is keeping the velocity dispersion small enough so that the gravitational focusing factor becomes very large ($\theta \gg 1$). In this case, the most massive planetesimal in the swarm will have the largest collisional cross section ($\propto a^4$), will grow more rapidly than any other body and will separate itself from the remainder of the swarm. Numerical models by Greenberg, *et al.* (1978) were the first to find this "runaway" path to planetary growth; subsequent calculations have confirmed these models (Wetherill & Stewart 1989, 1993; Ohtsuki & Ida 1990; Aarseth, Lin, & Palmer 1993). The basic mechanism behind runaway growth was elucidated by Wetherill & Stewart (1989) & (1993) Planetesimal velocities are determined by a balance among 1) stirring by gravitational scattering, 2) stirring by inelastic collisions, 3) damping due to energy dissipation in inelastic collisions, 4) damping due to gas drag, and 5) energy transfer from large to small bodies via dynamical friction (Lissauer & Stewart 1993). The most critical factor is dynamical friction, which tends to lead to an equipartition of kinetic energy among the bodies (Binney & Tremaine 1987), with the largest bodies having the smallest random velocities and thus the largest cross-sections. Figure 6 illustrates the runaway process in which the largest body grows the fastest as long as the velocity dispersion remains low.

Runaway growth can proceed until all the available material within the "accretion zone" (sometimes called the "feeding zone") of the planet is consumed (Wetherill 1980; Lissauer 1987, 1993). The accretion zone is an annulus over which the planet can exert its gravitational influence to perturb nearby planetesimals into crossing orbits. The most common measure of the gravitational range of a planet of mass m_p a distance r away from a primary body of mass M_* is the Hill sphere (or Roche lobe radius) defined to be

$$r_\mathrm{H} = r \left(\frac{m_\mathrm{p}}{3M_*}\right)^{1/3}. \tag{26}$$

The origin of the cube root mass dependence comes from finding the region over which the gravity of the planet, $Gm_\mathrm{p}/r_\mathrm{H}^2$, dominates the tidal force of the primary object, $(GM_*/r^2)(r_\mathrm{H}/r)$. The radial size of the accretion zone has been estimated to be a numerical factor $B \approx 4$ times larger than the Hill sphere of the planet (Lissauer 1993). If the planet accretes all the available solid mass (with surface density Σ_s) in an accretion zone having width $B\,r_\mathrm{H}$ on either side, its final "isolation" mass will be

$$m_\mathrm{p} \;=\; 2\pi r \cdot 2Br_\mathrm{H} \cdot \Sigma_\mathrm{s} = 4\pi B r^2 \Sigma_\mathrm{s} \left(\frac{m_\mathrm{p}}{3M_*}\right)^{1/3}$$

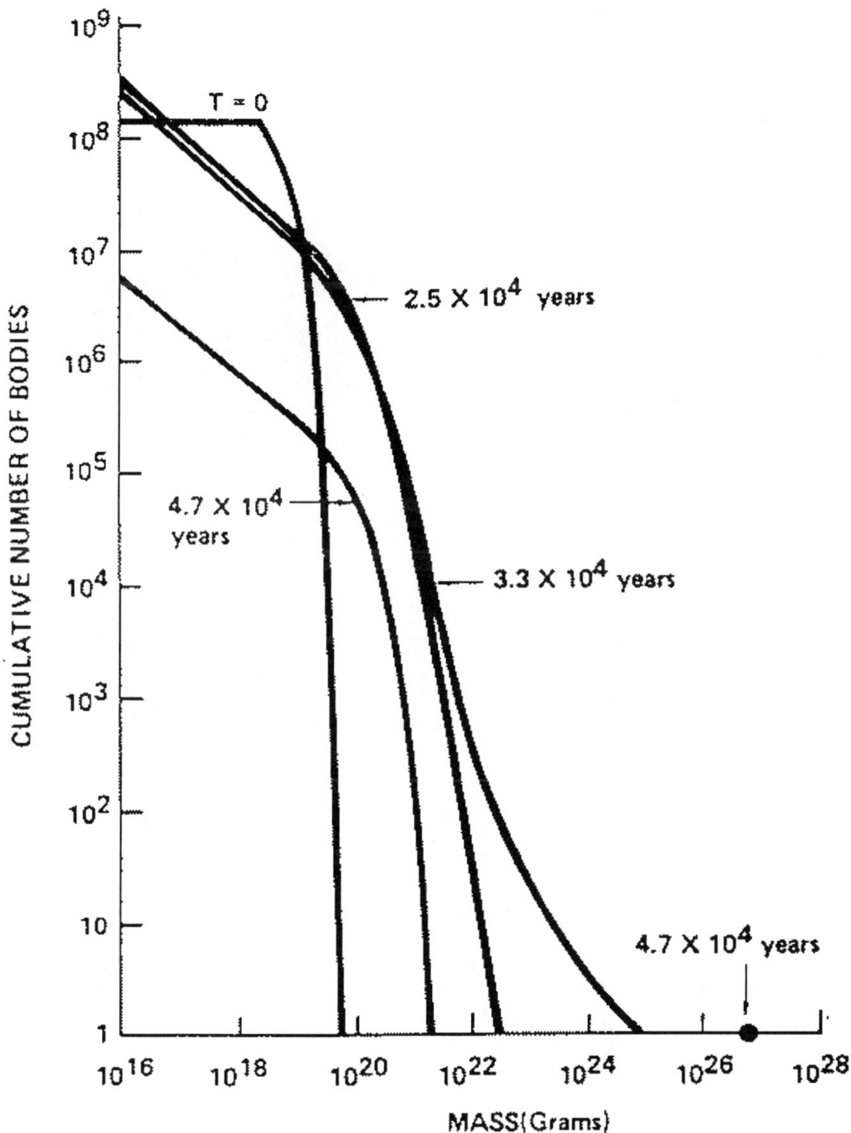

Figure 6. The evolution of the mass distribution of a planetesimal swarm between 0.99 and 1.01 AU illustrating the rapid runaway growth of the largest bodies. The most massive body becomes detached from the remainder of the swarm in less than 5×10^4 years. See Wetherill & Stewart (1989).

$$= \frac{\left(4\pi B r^2 \Sigma_{\mathrm{s}}\right)^{3/2}}{\left(3M_*\right)^{1/2}}. \tag{27}$$

The radial spacing of planets that have each become isolated is $\approx 2Br_{\mathrm{H}}$. For the MMSN, we find an isolation mass of $0.05M_{\oplus}$ at 1 AU and $1.4M_{\oplus}$ at Jupiter's distance.

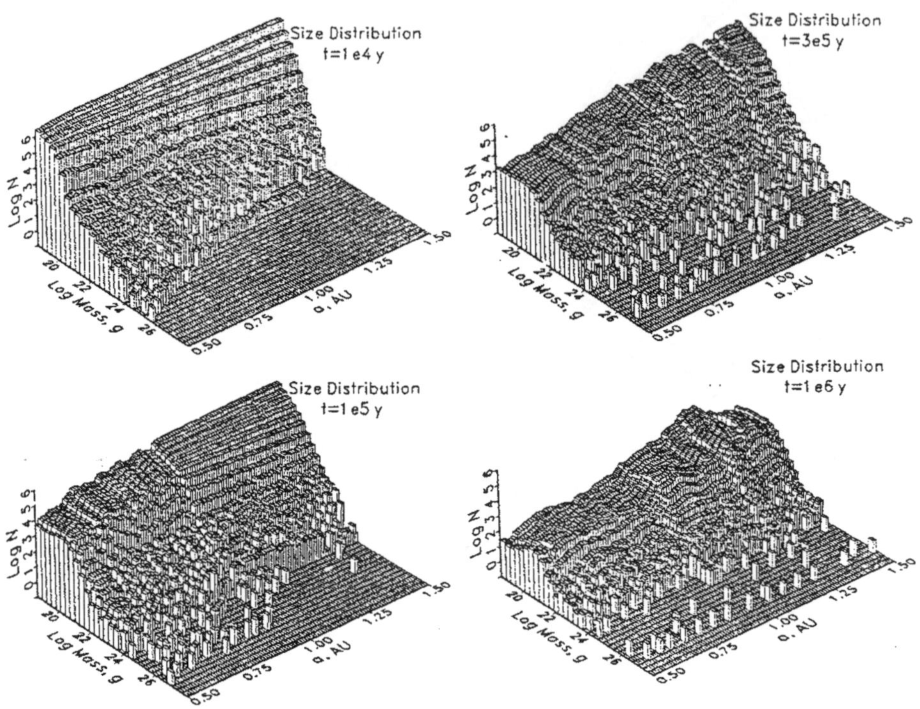

Figure 7.   Accretional evolution of a planetesimal swarm near 1 AU. The number of planetesimals are plotted per logarithmic interval in mass as a function of distance from the Sun. The evolution is clearly more rapid at 0.5 AU than at 1.5 AU, but the final mass of the protoplanets ($\sim 10^{27}$ g) is not dependent on distance from the Sun. Runaway growth is evident as the largest bodies separate themselves from the rest of the planetesimal population. After 10^6 years there are more bodies with lower mass than in the terrestrial region of the Solar System. From Weidenschilling, *et al.* (1997).

A recent detailed simulation of the accumulation of planetesimals in the terrestrial planet region has been performed by Weidenschilling, *et al.* (1997) that includes the processes mentioned above that control the velocity dispersion plus orbital perturbations caused by distant planetary bodies. The results of a simulation of the accumulation of 2 $M_{\oplus}$ of initially 4.8×10^{18} g (15 km) planetesimals spread between 0.5 and 1.5 AU is illustrated in Figure 7. These authors verified the critical role dynamical friction plays in producing rapid runaway growth. The first panel in Figure 7 ($t = 10^4$ years) clearly shows the evolution to larger masses occurs more rapidly at smaller radii where the Kepler frequency is larger. The other panels show that more massive bodies runaway and detach themselves from the distribution of lower mass planetesimals. Runaway growth produces bodies with masses $\sim 10^{26}$ g in $\lesssim 10^5$ years, but the growth rate slows as the velocity dispersion of the population of lower mass bodies increases, and the runaway stalls. By 10^6 years, about a dozen bodies with masses $\approx 10^{27}$ g have been produced, but they are lower in mass and more closely spaced than the terrestrial planets. The masses (and separations) found in the simulation are about a factor of three larger than the analytic isolation mass in equation (27).

4.3. FINAL ACCUMULATION STAGES

The outcome of runaway growth is self-limiting for two reasons. First, if the mass of the largest bodies becomes ~ 100 times the median mass of the continuum bodies, the velocity dispersion of the lower mass planetesimals is pumped up by gravitational scattering, and the runaway slows (Ida & Makino 1993; Weidenschilling, *et al.* 1997). Second, even if the velocity dispersion remains small, all the mass within the accretion zone will be consumed by the planet, and it will become dynamically isolated with a mass given by equation (27). The evolution following runaway is much longer term as gravitational encounters slowly perturb the bodies into crossing orbits and violent impacts occur. The models described in §4.2, which start from MMSN-type conditions, do not yield masses and spacings similar to those of the terrestrial planets, hence, longer term collisional evolution is necessary to produce a final model that is similar to the Solar System. Figure 8 shows the results of six accumulation calculations by Wetherill (1988) that started with 500 bodies each of mass 2×10^{25} g. The outcome of each simulation is a small number of planets after about 10^8 years. The differences in the calculations illustrate the stochastic nature of the collisional accumulation. The long accumulation timescales do not present any serious difficulties for the terrestrial planets, which as noted in §1.1 could have completed their evolution in a gas-free environment. However the giant planet cores had to form in less than the lifetime of the gas disk, which seems to require a rapid runaway growth process occurring in a disk with solid mass larger than the MMSN (Lissauer 1987, 1993; eq. [27]).

4.4. FUTURE RESEARCH QUESTIONS

- Can the Earth and the cores of giant planets be built directly from the runaway or is further long-term collisional growth needed?

  - How do you build the giant planet cores before the gas is dis-

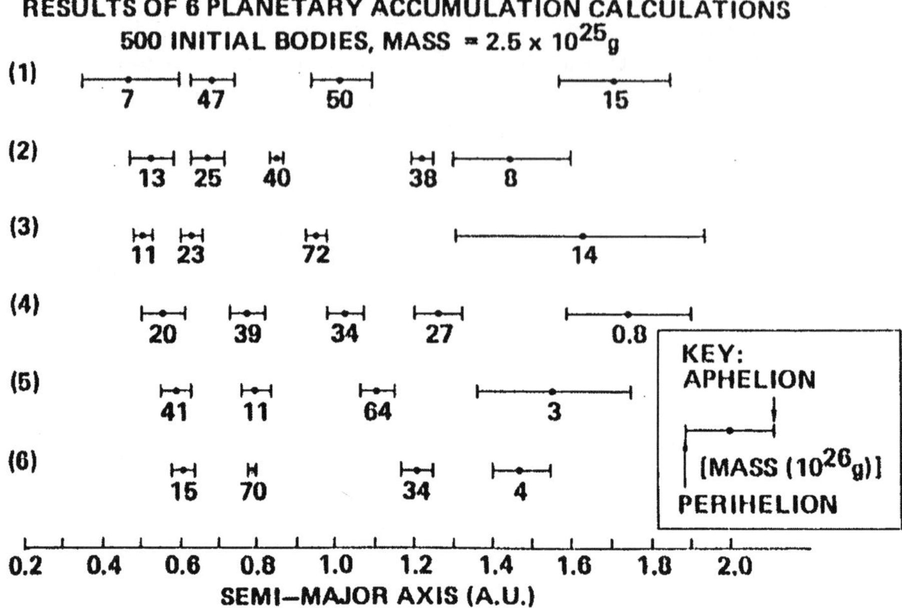

Figure 8. Six calculations of terrestrial planet accumulation by Wetherill (1988). The final mass of each planet in units of 10^{26} g is indicated ($1M_\oplus = 60$ in these units). The semimajor axis, with a bar extending from perihelion to aphelion, is also shown. From Lissauer (1993).

persed?

— How does the final planetary system depend on the initial disk mass distribution?

  • Do higher mass disks produce *more* planets or fewer, but higher-mass, planets?

— Can *rocky* planets with masses comparable to Jupiter be built in protoplanetary disks?

5. The Formation of Gas Giant Planets

5.1. THE CORE-INSTABILITY SCENARIO

The current model for the formation of gas giant planets is that they follow the path outlined above and begin as rocky cores. Mizuno (1980) showed that a gas envelope gravitationally bound to a rocky core cannot remain in hydrostatic equilibrium if the core mass exceeds a critical value $\sim 10 - 15\ M_\oplus$, comparable to the core masses inferred for the giant planets (Hubbard & Marley 1989; Chabrier, *et al.* 1992). Subsequent time-dependent calculations (Bodenheimer & Pollack 1986; Pollack, *et al.* 1996; see top panel in Fig. 9) have shown the nebular gas is rapidly accreted by the rocky core once its mass is above critical, leading to the conversion of a rocky planet into a gas giant planet. Before the critical core mass is reached, the radiated luminosity of the planet is supplied primarily by the gravitational energy of the accreted planetesimals. As the gravitationally bound gas envelope becomes more massive, planetesimal accretion is unable to supply the luminosity needed to maintain the extended envelope and rapid (but not hydrodynamic) collapse of the envelope takes place, with the gravitational contraction of the envelope supplying the luminosity. As the envelope contracts, surrounding nebular gas flows in, and the gas accretion rate rises to become orders of magnitude larger than the solid accretion rate. The evolutionary timescale shortens; the gas envelope of Jupiter is accreted in $\lesssim 10^5$ years (Pollack, *et al.* 1996). Detailed calculations show the value of the critical core mass ($\sim 10 - 30$ M$_\oplus$) is only moderately dependent on orbital radius and nebular properties and depends most strongly on the planetesimal accretion rate (see Stevenson [1982] for a simple analytical calculation of the critical core mass).

In the core-instability model, giant planets are composite bodies with the rocky cores forming first and the gas envelopes accreting afterwards. If this model is correct, there is a logical simplicity to the formation of planetary systems - all planets are born by the accumulation of solids; giant planets are higher mass cores that happen to accrete large gas envelopes. The core-instability model is not without problems, however. First, it is not obvious that $\gtrsim 10$M$_\oplus$ solid cores can form before the nebular gas is dispersed. Runaway growth to the critical mass requires solid densities about four times larger than in the MMSN (see eq. [27]). Forming the cores in $\sim 10^6$ years requires a large net gravitational focusing factor ($\theta \gtrsim 10^3$), which requires a very cold population of accreting planetesimals. Even if such cores can be formed at the orbit of Jupiter, it is not clear they can be formed so quickly at the orbits of Uranus or Neptune where the Kepler clock runs more slowly. Second, this model does not explain what *halts* the accretion of gas onto the planets. While it is possible the nebular gas was dispersed rapidly during the giant planet gas accretion phase, severe timing constraints (dispersal in $\lesssim 10^5$ years) make this unlikely. The most plausible explanation is that planetary gravitational tides become strong enough to "repel" nearby gas once the planet is sufficiently massive (Lin & Papaloizou 1993; see the next section). Finally, it is not clear from the internal structure of the giant planets in our Solar System that all four actually went critical. The large gas masses of Jupiter and Saturn seem to require their cores to have reached critical mass, but the smaller envelope

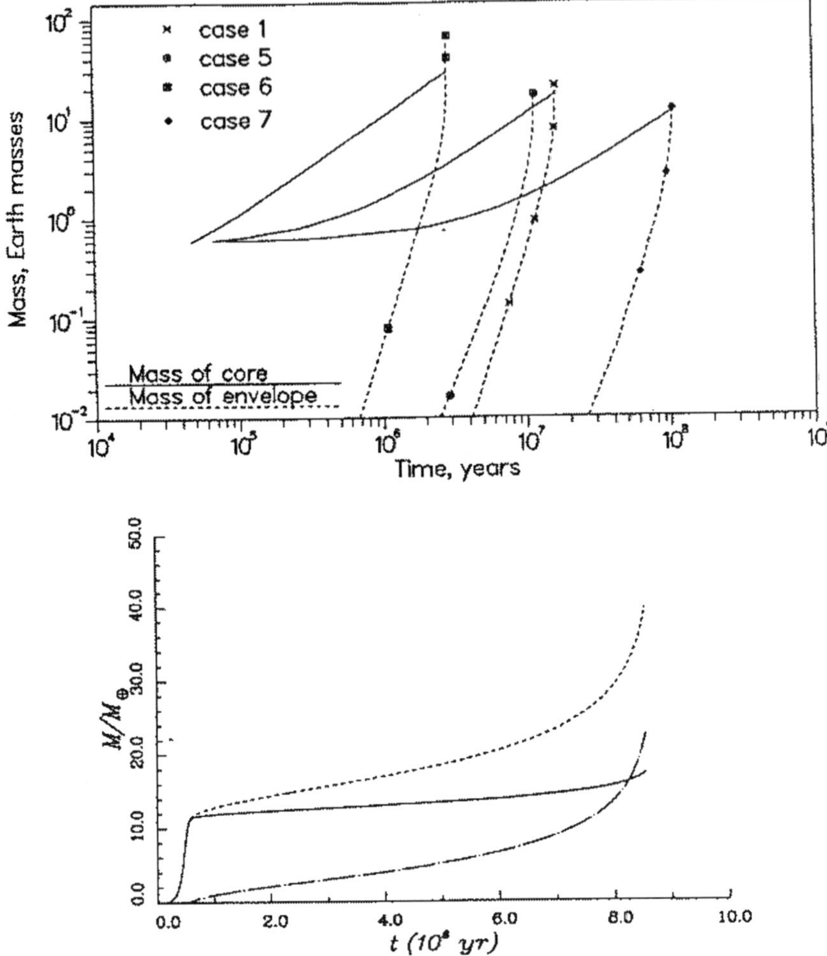

Figure 9. Core-Instability Model. **Top panel** from Bodenheimer & Pollack (1986) shows the core mass (solid) and envelope mass (dotted) as a function of time. Cases 6, 1 & 5, and 7 have constant planetesimal accretion rates of 10^{-5}, 10^{-6}, and 10^{-7} $M_\oplus$/yr, respectively. Case 5 has grain opacity reduced by 50 to reflect the decreased dust-to-gas ratio after solids have sedimented from the gas. The envelope mass is very small until the critical core mass is reached (defined to occur when $M_{\rm core} = M_{\rm env}$). At that point rapid gas accretion ensues. **Bottom panel** from Pollack, *et al.* (1996) shows the core mass (solid), envelope mass (dot-dashed) and total mass (dotted) as a function of time. Rather than assuming constant planetesimal accretion rates (as in the top panel), this calculation uses more detailed and self-consistent planetesimal accretion rates. The initial core forms by runaway in less than 10^6 years after which the growth rate slows. The planet becomes critical after $\approx 8 \times 10^6$ years, when rapid gas accretion begins.

masses of Uranus and Neptune ($M_{\rm env} \lesssim 0.5 M_{\rm core}$) imply these planets did not. Shu, Johnstone, & Hollenbach (1993) have discussed how photoevaporation of the disk by the ultraviolet flux from the central star preferentially removes disk gas beyond ~ 10 AU (the orbital distance of Saturn), which may explain the smaller gas mass reservoir available for the outer planets. Recent calculations of the core-instability model that adopt more realistic planetesimal accretion rates (Pollack, *et al.* 1996) indicate the planets experience a long phase ($\sim 7 \times 10^6$ year) after core runaway in which the envelope mass is smaller than the core mass (see bottom panel in Fig. 9). If the nebula were removed during this long phase of "stalled" gas accretion, the resulting planets would be similar to the outer giants.

5.2. TIDAL INTERACTION

Gravitational interactions between orbiting mass points and disks has been considered by many authors (Goldreich & Tremaine 1979; Lin & Papaloizou 1979; Ward & Hourigan 1989; Takeuchi, Miyama, & Lin 1996; see review by Lin & Papaloizou 1993). The essential physics can be elucidated by using the impulse approximation (Lin & Papaloizou 1979). As gas flows by the planet interior (exterior) to its orbital radius, the planet's gravitational field causes a slight deflection in the gas streamlines causing the angular momentum of the gas to decrease (increase). The gravitational tidal interaction allows angular momentum to be transferred from the gas on the inside to the planet and from the planet to the gas on the outside. Because orbital angular momentum increases outward in dynamically stable disks (the Rayleigh criterion; Drazin & Reid 1981), gas interior (exterior) to the planet that loses (gains) angular momentum in the tidal exchange must move farther inward (outward). In effect, the gravitational tides of the planet act to *repel* nearby gas. In a fluid dynamical context (Goldreich & Tremaine 1979), the tides of the planet excite spiral density waves that propagate away from the planet (in the absence of disk self-gravity). The waves are excited at Lindblad resonances, locations in the disk where the natural frequency of radial oscillation (the epicyclic frequency) is an integer multiple of the frequency at which the gas is forced by the planet.[5] Spiral density waves carry angular momentum, which is deposited in the gas as the waves are damped by viscous stresses. Figure 10 illustrates the situation schematically. The total torque on the gas is found by summing the contributions from all the Lindblad resonances; it is negative (positive) for the gas interior (exterior) to the planet. The magnitude of the torque at either the

---

[5]For Kepler disks, the Lindblad resonances are at radii satisfying $\pm m(\Omega(r) - \Omega_\mathrm{p}) = \Omega(r)$, where Ω_p is the orbital frequency of the planet and m is a positive integer. The positive (negative) sign is for inner (outer) resonances.

672　　　　　　　　　　STEVEN P. RUDEN

inner or outer resonances, T_{tidal}, is approximately (Lin & Papaloizou 1993)

$$T_{\text{tidal}} \approx f \, \Sigma \, \Omega_{\text{p}}^2 \, r_{\text{p}}^4 \left(\frac{r_p}{H}\right)^3 \left(\frac{m_{\text{p}}}{M_*}\right)^2, \tag{28}$$

where Ω_{p} is the rotation frequency evaluated at the planetary radius r_{p}, H is the vertical height of the disk and $f \approx 0.2$ is a numerical constant. This torque is deposited in the vicinity of the planet unless the viscosity is too small to damp the waves locally. The *net* tidal torque, which is the sum of the two nearly equal and opposite inner and and outer contributions, is difficult to calculate precisely but is roughly a factor H/r smaller than (28) (Ward 1986).

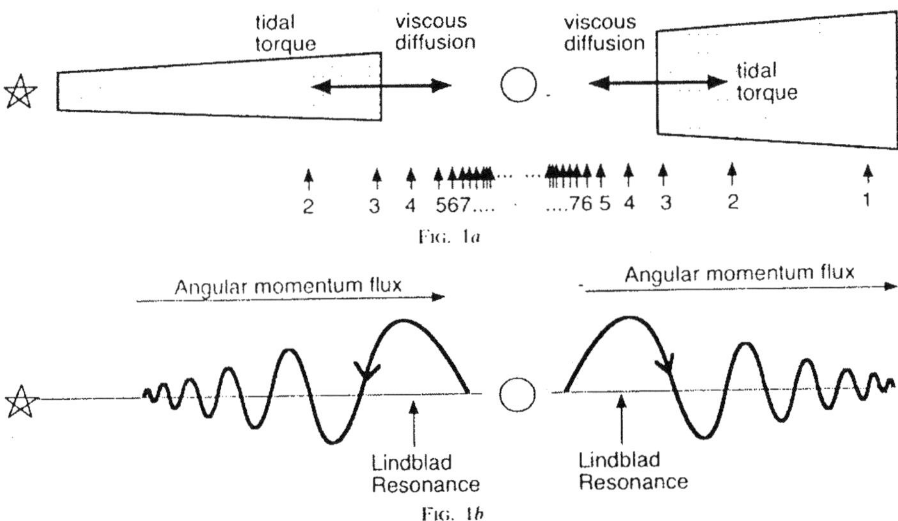

Figure 10. Formation of gaps in disks from Takeuchi, Miyama, & Lin (1996). **Top Panel.** Schematic illustration shows that gaps are formed from a balance between tidal torques that repel gas from the planet and viscous stresses that diffuse gas back into the gap. The tidal torque is actually deposited at the Lindblad resonances (indicated by arrows). **Bottom Panel.** Schematic illustration of wave propagation in disks. Waves are excited at the inner and outer Lindblad resonances and propagate away from the planet. As they damp, the angular momentum they carry is deposited into the gas as a tidal torque. A net outward flux of angular momentum results.

The action of the planetary tidal torques, which tend to push gas away from the planet, is opposed by the viscous stress of the gas, which tends to diffuse gas back toward the planet. The viscous torque on the disk gas, T_{vis}, is (Lynden-Bell & Pringle 1974)

$$T_{\text{vis}} = 3\pi \Sigma \nu \, \Omega r^2 = 3\pi \alpha \Sigma \Omega^2 r^4 \left(\frac{H}{r}\right)^2, \tag{29}$$

using equations (3) and (7) to get the rightmost form. The criterion for the planet to open an annular gap at its orbital radius is that the gravitational tidal torque exceed the viscous torque, $T_{\rm tidal} \gtrsim T_{\rm vis}$. Using equations (28) and (29) the minimum planet mass, $m_{\rm p}$, needed to open a gap is (Takeuchi, Miyama, & Lin 1996)

$$\frac{m_{\rm p}}{M_*} \gtrsim \sqrt{\frac{3\pi}{f}} \left(\frac{H}{r}\right)^{5/2} \alpha^{1/2}. \tag{30}$$

For typical solar nebular parameters this gives a mass of $\sim 75 M_\oplus$, slightly less than the mass of Saturn. It thus seems reasonable that Jupiter and Saturn (but not Uranus or Neptune unless the disk was extremely cold) were able to form gaps in the surface density distribution, reducing the local gas density to small values and possibly terminating the rapid gas accretion phase. I note however that some recent hydrodynamical simulations of the gap formation process though have questioned whether gap formation completely halts gas accretion onto the planet. The simulations indicate that even if a low density gap is formed, the planet remains connected to the gap walls by arc-like streams through which gas can flow (Artymowicz & Lubow 1996, personal communication 1998). Clearly far more work is needed to determine whether the formation of a gap completely halts gas accretion, and if not, what determines the final mass of the giant planets.

5.3. TIDAL MIGRATION

An essential consequence of the tidal exchange of angular momentum between the planet and disk is that a net torque can act on the planet, causing its orbital radius to change with time. This process is known as orbital or tidal *migration* (Lin & Papaloizou 1986; Ward 1997). The angular momentum of the planet (on a circular orbit) is $m_{\rm p}\Omega_{\rm p}r_{\rm p}^2 = m_{\rm p}(GM_*r_{\rm p})^{1/2}$ and writing the *net* tidal torque exerted by the planet on the gas disk as $T_{\rm net}$, the orbital evolution of the planet is governed by

$$\frac{d}{dt}\left(m_{\rm p}\Omega_p r_{\rm p}^2\right) = -T_{\rm net}, \tag{31}$$

or equivalently

$$\frac{dr_{\rm p}}{dt} = -\frac{2}{m_{\rm p}\Omega_{\rm p}r_{\rm p}} T_{\rm net}. \tag{32}$$

The orbital migration timescale is

$$t_{\rm mig} = \frac{r_{\rm p}}{|dr_{\rm p}/dt|} = \frac{\Omega_{\rm p}r_{\rm p}^2}{2T_{\rm net}},$$

$$\sim \frac{1}{\Omega_{\rm p}} \left(\frac{M_*}{m_{\rm p}}\right) \left(\frac{M_*}{\Sigma r_{\rm p}^2}\right) \left(\frac{H}{r}\right)^2, \quad \text{(no gap)} \tag{33}$$

where I have used equation (28) reduced by H/r to estimate the net torque (Ward 1986). This equation is appropriate for lower mass planets that do not tidally truncate the disk and form gaps. For planets massive enough to form gaps, the surface density in their vicinity is reduced dramatically, and equation (33) does not apply. Numerical calculations of the combined gap formation and tidal migration process (Lin & Papaloizou 1993) show that the inner and outer torques nearly cancel and the evolution time becomes the *local* viscous diffusion time, equation (8), applied at the orbit of the planet. Figure 11 illustrates a self-consistent calculation of gap formation and tidal migration in a protoplanetary disk.

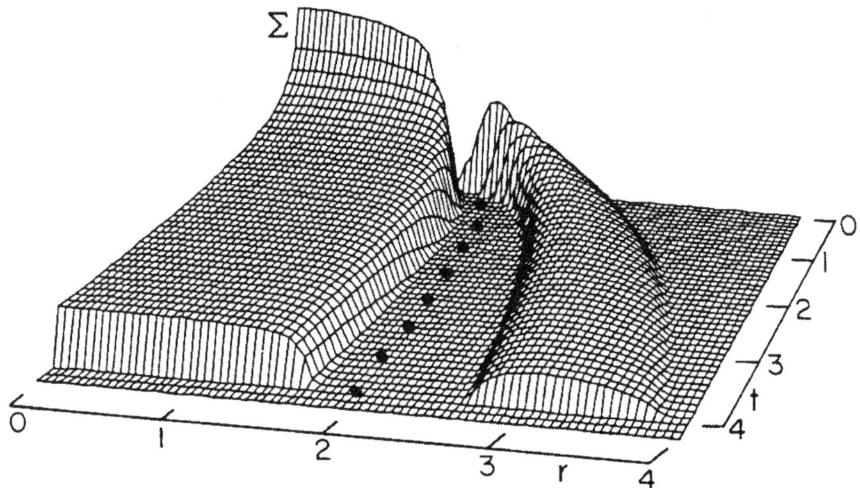

Figure 11. Tidal migration of a protoplanet in an evolving nebula from Lin & Papaloizou (1993). Surface density, Σ, is plotted versus distance from star, r, as a function of time, t. All variables are scaled and dimensionless. Notice the surface density decreases with time (due to accretion), and the outer disk radius increases. The position of the planet is marked by a dot. The mass of the planet is large enough that tides overcome viscous stresses, and a gap forms near the planet (the region where $\Sigma \longrightarrow 0$). In this example, the planet is near the outer disk radius and migrates outward as it absorbs angular momentum from the inner disk. Eventually the inner disk will be accreted by the star, and the planet will migrate inward. The migration timescale is the viscous timescale of the disk.

The tidal migration timescale is surprisingly short. Using values appropriate to the MMSN, the timescale (33), appropriate for planets not forming gaps, is $t_{\mathrm{mig}} \approx (M_*/m_{\mathrm{p}})(r/\mathrm{AU})$ years, which is about 3×10^5 years for the Earth. For planets massive enough to form gaps, the ratio of the viscous time at their orbital distance to the overall viscous time of the disk (having outer radius R_{D}) is $\sim (r_{\mathrm{p}}/R_{\mathrm{D}})^{3/2}$, from equation (8) assuming H/r

is roughly constant. Because Jupiter and Saturn were well within the outer radius of the primordial solar nebula ($r_p \ll R_D$), their migration times were much shorter than the $\sim 10^7$ year lifetime of the disk as a whole. Figure 12 shows a calculation of the orbital migration of Jupiter and Saturn in which a migration time of $\sim 10^5$ years is found.

5.4. THE SURVIVAL OF GIANT PLANETS

Tidal migration offers a neat way to explain the recently discovered Jupiter-mass planets having orbital distances of $\lesssim 0.1$ AU from their stars (Marcy & Butler 1998; also see the review by Marcy in this volume) if a mechanism can be found to *halt* the migration before the planets are swallowed by the star. The key is to find an outward torque that can counterbalance the inward tidal torque. Lin, Bodenheimer & Richardson (1996) have suggested that as the planet nears the central star, it will raise tidal bulges in the star that will transfer angular momentum from the spin of the star to the planetary orbit. This spin-orbit coupling can be effective at keeping the planet beyond the corotation radius (the distance in the disk where the Kepler orbital frequency equals the stellar rotation frequency). These authors also suggest that a star with a sufficiently strong magnetic field will have a magnetosphere that truncates the disk at a distance of several stellar radii. Once the planet migrates inside the inner disk edge, the inward tidal torques vanish because the surface density is negligible. Either mechanism can produce planets in orbits near the star.

Trilling, *et al.* (1998) have considered a mechanism involving mass loss from the planet. The size of the planetary Roche lobe (*c.f.* eq. [26]) is proportional to the distance from the star. As a giant planet migrates inward, it can fill its Roche lobe, leading to a transfer of mass from the planet to the star. Conservation of angular momentum of the system requires the planet to move outward. Stable mass transfer is achieved when the planet orbits at a distance where its radius equals its Roche radius. Planets in small orbits near stars can be produced if the disk is dissipated before the planet loses all its mass.

Our own Solar System with planets distributed from 0.4 - 30 AU does not shown signs of appreciable orbital migration, however. The question in our Solar System is to understand how the large orbital migrations apparently observed in other planetary systems were avoided. Because the migration timescale becomes the viscous timescale for planets massive enough to form gaps, a giant planet can avoid appreciable migration if it is born in a low viscosity nebula having evolutionary times comparable to or longer than the disk dispersal time. Before the planet has time to migrate far, the nebular gas is dispersed, and the migration is halted. Denoting the disk

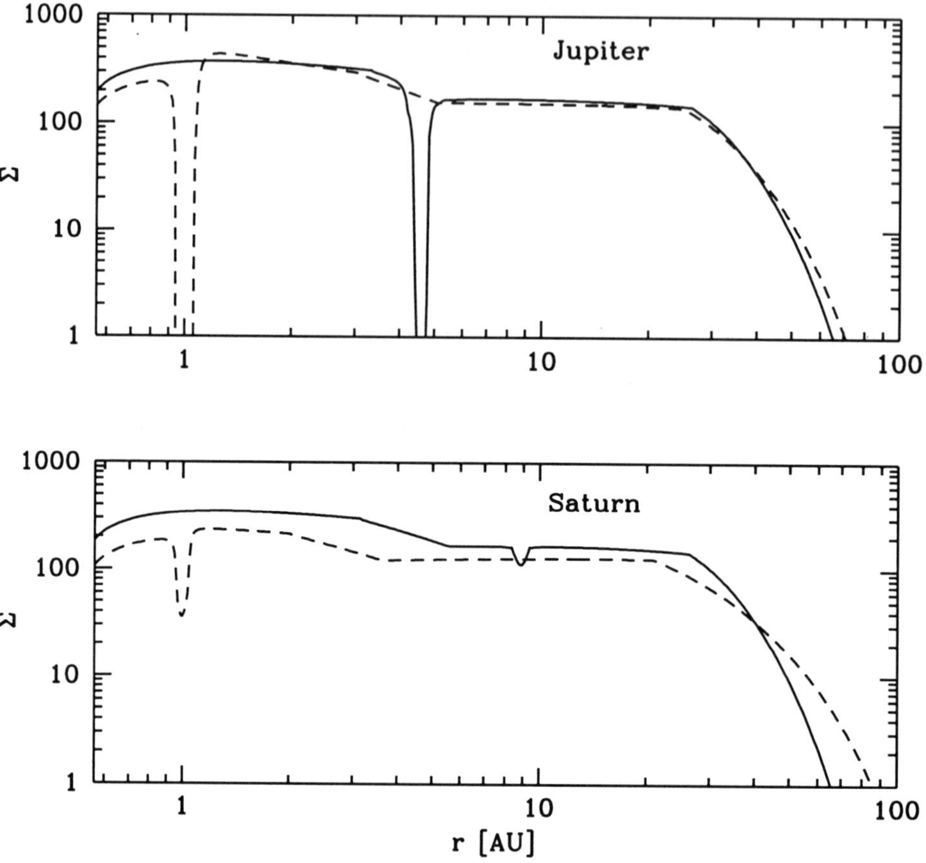

Figure 12. Tidal migration in the solar nebula models of Ruden & Lin (1986). Surface density in g/cm^2 is plotted versus distance from the Sun in AU. Jupiter is assumed to form instantaneously with its present mass and orbital radius at a time in the nebular evolution when it is just massive enough to form a gap (eq. [30]). Saturn is assumed to form coevally. **Top panel.** The solid curve is 10^3 years after Jupiter formation showing the gap (the region where Σ drops to zero) is formed rapidly. The dashed curve is 4×10^4 years later, showing Jupiter has migrated inward to 1 AU. **Bottom panel.** The solid curve is also 10^3 years after Saturn formation. Saturn does not form a gap, but the surface density is reduced there by 30%. The dashed curve is 1.5×10^5 years later. At this later time, the surface density in the gap is not zero but has been reduced by a factor of 5. The average inward migration speeds are about 40 cm/s for either planet.

dispersal time by $t_{\rm disp}$, we can estimate from equation (8) that this requires the viscous alpha parameter to be $\lesssim (r/H)^2/(\Omega t_{\rm disp})$. Using $r/H \approx 25$ and $t_{\rm disp} \sim 10^7$ years, I find $\alpha \lesssim 10^{-4}$ is sufficiently small that migration of Jupiter will not be severe. It is not unreasonable that giant planets form in a low viscosity environment. The lower panel in Figure 9 shows a model for the formation of Jupiter in which rapid gas accretion does not begin until $t \sim 8 \times 10^6$ years. In this model, the newly formed Jupiter would find itself in a nebular environment that had undergone significant evolution. The disk mass at that epoch could be quite small, with the corresponding low surface densities leading to small tidal torques and low migration rates. In fact, given that the MMSN contains only 10 – 20 Jupiter masses, formation of the giant planets at a late epoch when the gas mass may be appreciably less than in the MMSN may require the giant planets to accrete a large fraction of the remaining disk mass.

5.5. FUTURE RESEARCH QUESTIONS

- Is the core-instability model the only way to make giant planets?
  - Did the dispersal of the disk prevent Uranus and Neptune from going critical?
- What determines the final masses and orbital distances of the giant planets?
  - Does gap formation halt gas accretion onto giant planets?
  - What prevented significant orbital migration from occurring in our Solar System?
- How does the planetary mass spectrum merge into the brown dwarf mass spectrum?
  - Do brown dwarfs form in a fundamentally different manner than planets, *e.g.*, from collapse/fragmentation of the disk rather than from accumulation of dust?

Acknowledgments

I would like to thank Charlie Lada and Nick Kylafis for inviting me to the beautiful island of Crete for the second Star Formation summer school. This work has been supported in part by NSF AST–9157420 and NASA NAGW–5122.

References

Aarseth, S. J., Lin, D. N. C., Palmer, P. L. 1993, ApJ, 403, 351
Adachi, I., Hayashi, C., Nakagawa, K. 1976, Prog Theor Phys, 56, 1756
Adams, F. C., Lada, C. J., Shu, F. H. 1987, ApJ, 312, 788
Adams, F. C., Lin, D. N. C. 1993, in *Protostars & Planets III*, eds. E. H. Levy & J. I. Lunine, (Tucson: University of Arizona Press), 721
Artymowicz, P., Lubow, S. H. 1996, ApJ Lett, 467, L77
Balbus, S. A., Hawley, J. F. 1991, ApJ, 376, 214
Beckwith, S. V. W., Sargent, A. I. 1993, in *Protostars & Planets III*, eds. E. H. Levy & J. I. Lunine, (Tucson: University of Arizona Press), 521

Beckwith, S., Sargent, A. 1996, Nature, 383, 139

Beckwith, S. V. W., Sargent, A. I., Chini, R. S., Gusten, R. 1990, AJ, 99, 924

Binney, J., Tremaine, S. D. 1987, *Galactic Dynamics*, (Princeton: Princeton University Press)

Blum, J., Münch, M. 1993, Icarus, 106, 151

Bodenheimer, P., Pollack, J. B. 1986, Icarus, 67, 391

Bridges, F. G., Supulver, K. D., Lin, D. N. C., Knight, R., Zafra, M. 1996, Icarus, 123, 422

Brown, H. 1949, in *The Atmospheres of the Earth and Planets*, ed. G. P. Kuiper, (Chicago: Univ. of Chicago Press), 258

Butler, R. P., Marcy, G. W. 1996, ApJ Lett, 464, L153

Cameron, A. G. W. 1978, in *Protostars and Planets*, ed. T. Gehrels, (Tucson: Univ. of Arizona Press), 453

Chabrier, G., Saumon, D., Hubbarb, W. B., Lunine, J. I. 1992, ApJ, 391, 817

Chambers, J. G., Wetherill, G., Boss, A. 1996, Icarus, 119, 261

Cuzzi, J. N., Dobrovolskis, A. R., Champney, J. M. 1993, Icarus, 106, 102

Drazin, P. G., Reid, W. H. 1981, *Hydrodynamic Stability*, (Cambridge: Cambridge University Press), pp. 69

Dubrulle, B. 1993, Icarus, 106, 59

Goldreich, P., Tremaine, S. 1979, ApJ, 233, 857

Goldreich, P., Ward, W. R. 1973, ApJ, 183, 1051

Greenberg, R., Wacker, J. F., Hartmann, W. L., Chapman, C. R. 1978, Icarus, 35, 1

Greenzweig, Y., Lissauer, J. J. 1990, Icarus, 87, 40

Greenzweig, Y., Lissauer, J. J. 1992, Icarus, 100, 440

Hartmann, L., MacGregor, K. B. 1982, ApJ, 259, 180

Hayashi, C. 1981, Prog Theor Phys Supp, 70, 35

Hubbard, W. B., Marley, M. S. 1989, Icarus, 78, 102

Ida, S., Makino, J. 1993, Icarus, 106, 210

Ida, S., Nakazawa, K. 1989, Astron Ap, 224, 303

Koerner, D. W., Ressler, M. E., Werner, M. W., Backman, D. E. 1998, ApJ Lett, 503, L38

Lin, D. N. C., Bodenheimer, P., Richardson, D. C. 1996, Nature, 380, 606

Lin, D. N. C., Papaloizou, J. C. B. 1979, MNRAS, 186, 789

Lin, D. N. C., Papaloizou, J. C. B. 1980, MNRAS, 191, 37

Lin, D. N. C., Papaloizou, J. 1985, in *Protostars and Planets II*, eds. D. C. Black & M. S. Matthews, (Tucson: Univ. of Arizona Press), 981

Lin, D. N. C., Papaloizou, J. 1986, ApJ, 309, 846

Lin, D. N. C., Papaloizou, J. C. B. 1993, in *Protostars and Planets III*, eds. E. H. Levy & J. I. Lunine, (Tucson: Univ. of Arizona Press), 749

Lin, D. N. C., Pringle, J. E. 1987, MNRAS, 358, 515

Lissauer, J. J. 1987, Icarus, 69, 249

Lissauer, J. J. 1993, Ann. Rev. Astron. Ap., 31, 129

Lissauer, J. J., Stewart, G. R. 1993, in *Protostars and Planets III*, eds. E. H. Levy & J. I. Lunine, (Tucson: Univ. of Arizona Press), 1061

Lynden-Bell, D., Pringle, J. E. 1974, MNRAS, 168, 603

Marcy, G. W., Butler, R. P. 1996, ApJ Lett, 464, L147

Marcy, G. W., Butler, R. P. 1998, Ann. Rev. Astron. Ap., 36, 57

Markiewicz, W. J., Mizuno, H., Völk, H. J. 1991, Astron Ap, 242, 286

Mayor, M., Queloz, D. 1995, Nature, 378, 355

Miyake, K., Nakagawa, Y. 1993, Icarus, 106, 20

Mizuno, H. 1980, Prog Theor Phys, 96, 266

Mizuno, H. 1989, Icarus, 80, 189

Nakagawa, Y., Hayashi, C. 1981, Icarus, 45, 517

Nakagawa, Y., Nakazawa, K, Hayashi, C. 1981, Icarus, 45, 517

Ohtsuki, K., Ida, S. 1990, Icarus, 85, 499

Pepin, R. O. 1989, in *Origin and Evolution of Planetary and Satellite Atmospheres*, eds. S. K. Atreya, J. B. Pollock, M. S. Matthews, (Tucson: University of Arizona Press), 291

Pollack, J. B., Hubickyj, O., Bodenheimer, P., Lissauer, J. J., Podolak, M., Greenzweig, Y. 1996, Icarus, 124, 62

Pringle, J. E. 1981, Ann Rev Astron Ap, 19, 137

Prinn, R. G. 1982, Planet Sp Sci, 30, 741

Ruden, S. P. 1998, in preparation

Ruden, S. P., Lin, D. N. C. 1986, ApJ, 308, 883

Ruden, S. P., Pollack, J. B. 1991, ApJ, 375, 740

Safronov, V. S. 1972, *Evolution of the Protoplanetary Cloud and Formation of the Earth and Planets*, NASA TT-F-677

Shakura, N. J., & Sunyaev, R. A. 1973, Astron Ap, 24, 337

Shu, F. H., Adams, F. C., Lizano, S. 1987, Ann. Rev. Astron. Ap., 25, 23

Shu, F. H., Johnstone, D., Hollenbach, D. 1993, Icarus, 106, 92

Shu, F. H., Tremaine, S., Adams, F. C., Ruden, S. P. 1990, ApJ, 358, 495

Stepinski, T. F., Levy, E. H. 1990, ApJ, 350, 819

Stevenson, D. J. 1982, Planet Sp Sci, 30, 755

Strom, S. E., Strom, K. M., Edwards, S., Cabrit, S., Skrutskie, M. F. 1989, AJ, 97, 1451

Supulver, K. D., Bridges, F. D., Tiscareno, S., Lievore, J., Lin, D. N. C. 1997, Icarus, 129, 539

Takeuchi, T., Miyama, S. M., Lin, D. N. C. 1996, ApJ, 460, 832

Trilling, D., Benz, W., Guillot, T., Lunine, J. I., Hubbard, W. B., Burrows, A. 1998, ApJ, 500, 428

Ward, W. R. 1986, Icarus, 67, 164

Ward, W. R. 1997, Icarus, 126, 261

Ward, W. R., Hourigan, K. 1989, Icarus, 347, 490

Wardle, M., Königl, A. 1993, ApJ, 410, 218

Weidenschilling, S. J. 1977a, Astrophys Sp Sci, 51, 153

Weidenschilling, S. J. 1977b, MNRAS, 180, 57

Weidenschilling, S. J. 1984, Icarus, 60, 553

Weidenschilling, S. J., Cuzzi, J. N. 1993, in *Protostars and Planets III*, eds. E. H. Levy & J. I. Lunine, (Tucson: Univ. of Arizona Press), 1031

Weidenschilling, S. J., Spaute, D., Davis, D. R., Marzari, F., Ohtsuki, K. 1997, Icarus, 128, 429

Wetherill, G. W. 1988, in *Mercury*, eds. C. Chapman, F. Vilas, (Tucson: University of Arizona Press), 670

Wetherill, G. W. 1990, Ann Rev Earth Plan Sci, 18, 205

Wetherill, G. W., Stewart, G. R. 1989, Icarus, 77, 330

Wetherill, G. W., Stewart, G. R. 1993, Icarus, 106, 190

Whipple, F. L. 1972, in *From Plasma to Planets*, ed. A. Elvius, (New York: Wiley), 211

Wurm, G., Blum, J. 1998, Icarus, 132, 125

Zuckerman, B., Forveille, E., Kastner, J. H. 1995, Nature, 373, 494

Chris Matzner and Geoff Marcy relax after a long hike through the Sammaria Gorge and a swim in the sea.

EXTRASOLAR PLANETS:
TECHNIQUES, RESULTS, AND THE FUTURE

G.W. MARCY
San Francisco State University & UC Berkeley
Dept. of Physics and Astronomy, SFSU, San Francisco, CA,
USA 94132, gmarcy@stars.sfsu.edu

AND

R.P.BUTLER
Anglo-Australian Observatory
PO Box 296, Epping, NSW 2121 Australia, paul@aaoepp.aao.gov.au

1. Introduction

To date, we have not obtained direct images of planets around other stars. However, the small–amplitude Keplerian motion of stars, presumably caused by the gravitational forces exerted them by planets, now provides information on their masses and orbits. The first extrasolar planetary system was found around a neutron star, the 6.2–millisecond pulsar PSR 1257+12 (Wolszczan & Frail 1992, Wolszczan 1998). Its three terrestrial–mass companions indicate the ubiquity and diversity of planetary systems. The leading model for the formation of the pulsar planets involves the post–supernova recapture of material, perhaps from a pre–existing stellar companion, into an accretion disk. The accumulation of dust grains into the pulsar planets may have occurred by processes similar to the formation of rocky planetary cores in our Solar System. The existence of pulsar planets suggests that planet formation is robust and depends little on specific disk properties, as long as adequate heavy elements are present (Wolszczan 1998).

The discovery of a Doppler periodicity in 51 Peg by Mayor and Queloz (1995), followed soon thereafter by periodicities for 47 UMa and 70 Vir (Butler & Marcy 1996, Marcy & Butler 1996), seems to constitute the first detections of jupiter–mass companions to Solar-type stars. Any physical kinship of these companions to the "planets" in our Solar System must stem from the measured distributions of masses and orbital elements, along with

681

C.J. Lada and N.D. Kylafis (eds.), The Origin of Stars and Planetary Systems, 681–708.

models of formation. The models are undergoing rapid elaboration, as the diversity of Jupiter–mass companions reveals more complexity than was suggested by our orderly Solar-System architecture.

Doppler surveys of main sequence stars have revealed 17 companions to main sequence stars that are extrasolar planet candidates. Among these candidates, 15 have $M \sin i < 5$ $M_{\rm JUP}$ and constitute the best planet candidates. Here, $M \sin i$ represents the product of companion mass, M, and the (unknown) sine of the orbital inclination, i, where $i=90$ deg represents an orbit viewed edge–on. In addition, four main sequence stars are known to harbor Doppler companions that have $M \sin i = 15\text{--}75$ $M_{\rm JUP}$, which may represent the "brown dwarfs" (Mayor et al. 1998, Mayor et al. 1997).

The distinction between "planets" and "brown dwarfs" remains cloudy and rests on two formation scenarios. Planets form out of the agglomeration of condensible material in a disk into a rock–ice core (eg. Lissauer 1995). In contrast, brown dwarfs presumably form by a gravitational instability in gas (eg. Boss 1998, Burrows et al. 1998). Hybrid formation scenarios remain viable in which the relative importance of solid core growth and gas accretion within a disk lead to a continuum in internal structure. Subsequent collisions may lead to further growth and dynamical evolution. Some researchers prefer to differentiate planets from brown dwarfs based on mass alone, at 13 $M_{\rm JUP}$, below which Deuterium does not ignite (Burrows 1997). By either distinction of planets from brown dwarfs, this semantic segregation may explain the physics of substellar objects no more fully than "spiral" and "elliptical" summarize the physics of galaxies.

The most important statistical property of the substellar companions is their mass distribution. Companions having $M \sin i$ in the decade between 0.5–5 $M_{\rm JUP}$ outnumber those between 5–50 $M_{\rm JUP}$ by a factor of $\sim$3 (eg, Marcy & Butler 1998, Mayor et al. 1998). The poor detectability of the lowest–mass companions implies that the factor of 3 is a lower–limit to the cosmic ratio. This abundance of companions having Jovian masses suggests that distinct formation processes predominated, arguably similar to those associated with the giant planets in our Solar System (Lin et al. 1998).

The host stars of the extrasolar planet candidates have higher mean metallicity by a factor of $\sim$2 in abundance compared with field stars (Gonzalez 1998, Gonzalez et al. 1998). Equally interesting is that seven planets reside in orbits with a radius less than 0.12 AU, sometimes termed "51 Peg" planets (Mayor & Queloz 1995, Butler et al. 1998). These small orbits challenge us to explain their existence in a region where both the high temperatures and the small amount of protostellar disk material would inhibit formation *in situ*. The 51 Peg planets thus offer support for the prediction by Goldreich & Tremaine (1980) and by Lin (1986) that Jupiters may migrate inward from farther out (Lin et al. 1995, Ward 1997, Trilling et al.

1998).

Perhaps most intriguing about the planet candidates are the orbital eccentricities. The orbits of the 51 Peg planets may suffer some tidal circularization (Lin et al. 1998, Terquem et al.1998, Ford et al. 1998, Marcy et al. 1997), and indeed their orbits are all nearly circular (see Table 1). In contrast, the planets that orbit farther than 0.15 AU from their star all reside in non–circular orbits having $e > 0.1$, i.e. more eccentric than for Jupiter ($e=0.05$). Indeed, all but two have $e > 0.2$. This high occurrence of orbital eccentricity has lead to a variety of models in which Jupiter–like planets suffer gravitational interactions with a) other planets (Weidenschilling & Marzari 1996, Rasio & Ford 1996, Lin & Ida 1996, Levison et al. 1998), b) the disk (Artymowicz 1998), c) a companion star (Holman et al. 1996, Mazeh et al. 1997), and d) passing stars in the young cluster (de la Fuente Marcos and de la Fuente Marcos 1997, Laughlin and Adams 1999).

2. The Doppler Detection Technique

All of the orbiting planet candidates found to date were identified by precise Doppler measurements. Therefore we describe this technique in some detail here to illustrate its capabilities and limitations. The Sun "wobbles" around the Solar System center–of–mass at a speed of ~ 13 m s^{-1} , due to its mutual orbit with Jupiter and Saturn (12.5 and 2.7 m s^{-1} , respectively).

The semi–amplitude, K, of the radial velocity of a star induced by its orbiting companion is:

$$ K = \left(\frac{2\pi G}{P}\right)^{1/3} \frac{m_{\mathrm{p}} \sin i}{(M_* + m_{\mathrm{p}})^{2/3}} \frac{1}{\sqrt{1 - e^2}} $$

where P is the orbital period, m_{p} is the mass of the unseen companion, M_* is the stellar mass, and e is the eccentricity. For randomly oriented orbital planes, $< \sin i > = \pi/4$. The period, P, can be eliminated with Kepler's third law, $P^2 = \frac{4\pi^2 a^3}{G(M_* + m_{\mathrm{p}})}$, where a is the semimajor axis of the orbit of the companion relative to the primary. Figure 1 shows lines of constant amplitude, K, for the reflex velocity for a Solar–mass star, in a two–parameter plane of orbital radius, a, and $M \sin i$ for the companion mass. A Jupiter–mass at 1 AU causes a reflex amplitude of 28.0 m s^{-1} $\times \sin i$, which is easily detectable.

Orbital periods that are shorter than the duration of the velocity measurements for a given star can be revealed by periodogram techniques to identify the periodicity in the velocities. Velocity amplitudes, K, can be detected which lie at or below the velocity measurement errors. But for orbital periods that are comparable to (or longer than) the duration of observations, experience shows that a confident detection requires that the

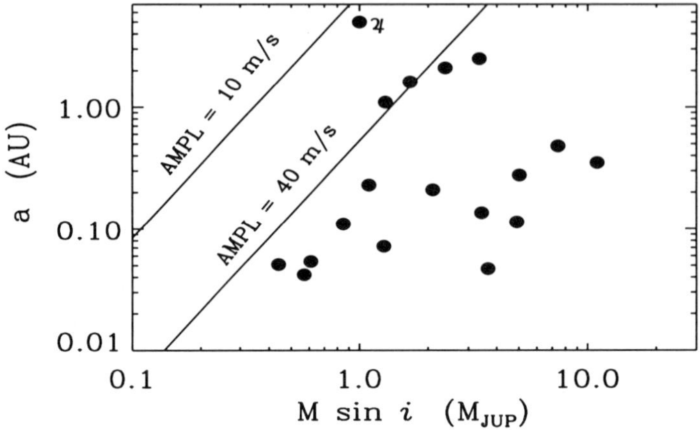

Figure 1. Detectability of companions by the Doppler technique for threshold velocities of 10 and 40 m s^{-1} . Known planet candidates are represented as dots.

amplitude be $\sim$4$\times$ the Doppler error. The stringent demand stems from the many ways to be fooled, such as by photospheric noise, multiple companions, or the occasional 3–σ data point which will appear among a multitude of target stars. Thus, for a precision of 3 m s^{-1} , amplitudes of 12 m s^{-1} are just detectable for the case of long orbital periods. Therefore, from Figure 1, with a Doppler precision of 3 m s^{-1} , companions having 1 M$_{JUP}$ at 5 AU are just detectable and as are saturn–masses at 1 AU.

The limit to the Dopper technique probably resides at a precision of 3 m s^{-1} , due to the intrinsic stability limit of stellar photospheres (Saar et al. 1998). Surface (photospheric) turbulence and pulsation could prevent detection of giant planets or produce false alarms. A detailed study of the velocity stability of F,G, and K dwarfs has been carried out by Saar and Donahue (1997) and Saar et al. (1997) who show that the photospheric flux spectra exhibit *intrinsic* velocity variability (not related to measurement errors) that depends on rotation period. For Solar-type dwarfs younger than 1 Gyr that rotate with $P_{rot} < 10d$, the intrinsic stellar velocities scatter by over 10 m s^{-1} . For rotation periods of 10 to 30 d, typical of main sequence stars in the general field, the RMS velocity variability declines from 15 m s^{-1} to 3 m s^{-1} , respectively. The likely causes of the variability are spots rotating across the stellar disk and short–term magnetically–controlled fluid flows on the stellar surface. Single G, K, and M type main sequence stars that are older than 3 Gyr have photospheric spectra that flutter by less than 5 m s^{-1} RMS, sufficiently quiescent to detect jupiter twins.

Precise Doppler measurements by four teams have provided detections

of companions plausibly below 10 M_{JUP} , namely by Cochran et al. (1997),
Mayor & Queloz (1995), Noyes et al. (1997), and by Marcy & Butler (1998).
Each team uses a somewhat different scheme for wavelength calibration, to
eliminate the systematic Doppler errors. Mayor and Queloz (1995) use a
stable, fiber–fed spectrometer, with a simultaneous Th-Ar lamp as wavelength calibration. An alternative approach is to pass the starlight through
iodine gas, to superimpose a comb of absorption lines that convey the wavelength scale directly on the stellar spectrum (Butler et al. 1996). For iodine
calibration, one must build a computer model of the composite (star and
iodine) spectrum which includes a floating PSF and Doppler shift, in a
least-squares fit to the observed spectrum. Resulting Doppler precision appears to be $\sim$10 m s^{-1} for all groups, but Butler et al. (1996) have achieved
3 m s^{-1} at Lick observatory on quiet stars. With this precision, Jupiter-
mass planets at 5 AU will be detectable within a few years. We note that
superior Doppler precision will not help, as the fundamental limit is set by
the stability of the stellar photospheres at $\sim$3 m s^{-1} .

3. Doppler Results: Properties of the Candidate Planets

The Doppler technique has revealed 17 companions to main sequence stars
which have $M \sin i < 15$ M_{JUP} , all listed in Table 1. This list is complete
as of 1999 Jan 1 (taken from Cumming et al. 1999).

A glance at Table 1 shows that among the 17 planet candidates, 4 have
semimajor axes greater than 1 AU. Two representatives of these long–
period planet candidates may be examined in Figure 2 which shows the
radial velocity versus time for 16 Cygni B and 47 Ursae Majoris, both
Solar-type stars. In both cases the value of $M \sin i$ near 2 M_{JUP} . Also,
the eccentricity is non–zero in both cases, 0.68 $\pm$0.03 for 16 Cygni B, and
e=0.10 $\pm$0.03 for 47 UMa. Indeed all long–period planetary companions
reside in eccentric orbits, compared to Jupiter and Saturn which both have
e=0.05.

Nonetheless, the most compelling analog to Jupiter detected to date is
the companion to 47 UMa (Butler and Marcy 1996). The velocities for 47
UMa exhibit one clear period of 3.0 yr, with a best-fit Keplerian eccentricity
of e = 0.10 $\pm$ 0.03, as shown in Figure 2.

The velocity residuals have an rms of 8 m s^{-1} (and only 4 m s^{-1} since
1994 November), consistent with the errors. These residuals give no evidence for additional companions, nor for non–Keplerian motion. The semimajor axis of 2.1 $\pm$ 0.1 AU and the companion mass of 2.4 $\pm$ 0.1 / $\sin i$
seems marginally consistent with expectations that giant planets form beyond the ice-condensation distance, at least several AU from the host star
(Lissauer 1995, Boss 1995). Orbital migration after formation may have

TABLE 1. Orbits and Masses of Extrasolar Planet Candidates

Star Name	Spec. Type	P (days)	a (AU)	K (m/s)	$M_p \sin i$ (M_J)	e	Reference
HD 187123	G3V	3.097	0.042	83	0.57	0.03	Butler et al. (1998)
τ Boo	F7V	3.31	0.047	468	3.66	0.00	Butler et al. (1997)
51 Peg	G5V	4.23	0.051	56	0.44	0.01	Mayor & Queloz (1995) Marcy et al. (1997)
Ups And[a]	F8V	4.62	0.054	71.9	0.61	0.15	Butler et al. (1997,99)
HD 217107	G7V	7.12	0.072	140	1.28	0.14	Fischer et al. (1999)
ρ^1 55 Cnc	G8V	14.7	0.11	75.9	0.85	0.04	Butler et al. (1997)
Gliese 86	K1V	15.8	0.114	379	4.9	0.04	Queloz et al. (1999)
HD 195019	G3V/IV	18.3	0.136	275	3.43	0.03	Fischer et al. (1999)
ρ CrB	G2V	39.6	0.23	67	1.1	0.11	Noyes et al. (1997)
Gliese 876	M4	61	0.21	217	2.1	0.27	Marcy et al. (1998) Delfosse et al. (1998)
HD 168443	G8IV	57.9	0.28	350	4.96	0.54	Marcy et al. (1999)
HD 114762	F7V	84.0	0.41	618	11.0	0.33	Latham et al. (1989) Marcy et al. (1999)
70 Vir	G2.5V	117	0.47	316	7.4	0.40	Marcy & Butler (1996)
HD 210277	G7V	437	1.20	42	1.28	0.45	Marcy et al. (1999)
16 Cyg B	G2.5	799	1.6	50.3	1.67	0.69	Cochran et al. (1997)
47 UMa	G0V	1092	2.1	47	2.38	0.11	Butler & Marcy (1996)
14 Her	K0V	>1620	>2.5	75	>3.3	0.36	Mayor et al. (1998)

[a]Probable 2nd companion, P=3.9 yr, e=0.58, $M \sin i$ =9.6 M_{JUP} (Butler et al. 1999)

brought 47 UMa b closer (Ward 1997, Trilling et al. 1998, Lin et al. 1999). Alternatively, 47 UMa b may have formed *in situ*.

For 16 Cyg B, the Doppler variations were detected independently at Lick and McDonald Observatories (Cochran et al. 1997), yielding similar orbital parameters. The velocity residuals, in both cases, are consistent with the measurement errors, with no indication of additional companions of jupiter–mass within 1 AU. The orbital parameters give P=803 $\pm$3 d, and eccentricity, e=0.687 $\pm$ 0.03 . The implied companion mass is M = 1.67 $\pm$ 0.1 M_{JUP} /sin i. This mass falls well within the range associated with "extrasolar giant planets", as defined structurally (Burrows et al. 1997).

The lithium abundance of 16 Cyg A is 5 times that of 16 Cyg B, despite their nearly equal masses and identical ages (King et al. 1996). The origin of this difference in Li–burning is not understood, but may result from the torque exerted on the star by the protoplanetary disk during the pre-main sequence phase (Strom 1994, Pinsonneault 1997, Martín 1997).

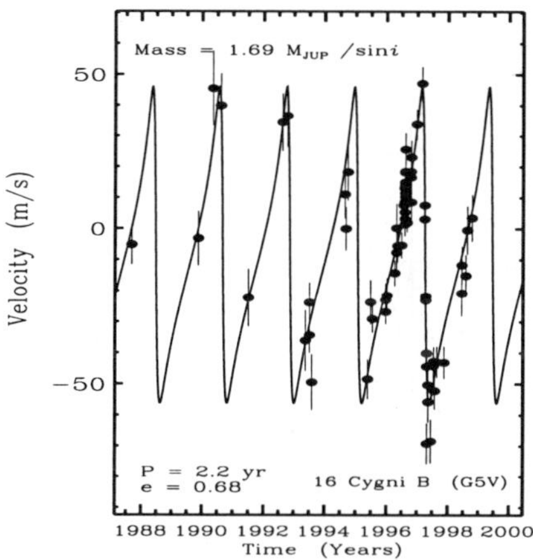

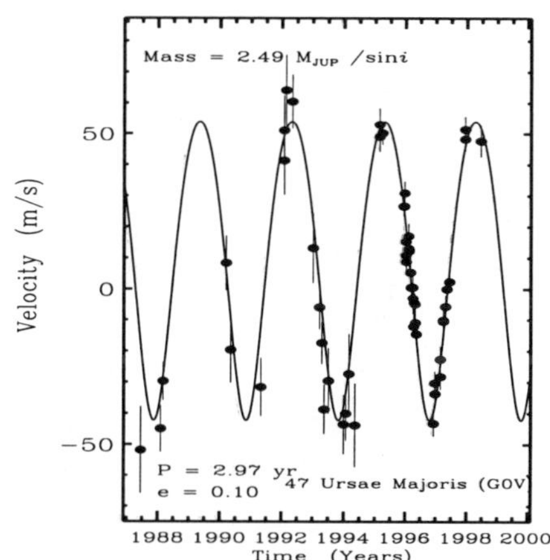

Figure 2. Measured velocity vs. time for two representative Jupiter–mass companions, 16 Cygni B (left) and 47 Ursae Majoris (right), in long orbits. The duration of the observations is 11 yr. The corresponding semimajor axes are 1.6 AU and 2.1 AU, and the eccentricities are 0.68 and 0.10, respectively. All 9 known Jupiter–mass companions that orbit beyond 0.2 AU reside in non–circular orbits.

For 16 Cyg B, the modest value of $M \sin i$ =1.7 M_{JUP} resides well below the lowest mass expected from "star formation" processes that produce brown dwarfs, though that mass limit remains poorly understood (Boss 1995). The orbital eccentricity of 16 Cyg B represents the first challenge to the predicted ubiquity of circular orbits for planets, with more violations soon to follow. Assuming that $\sim$2 M_{JUP} companions form in a dissipative disk, some dynamical process must have perturbed the original circular orbit.

3.1. GIANT PLANETS ORBITING WITHIN 0.1 AU

Eight jupiter–mass companions have orbital radii less than 0.15 AU (Table 1), often called "51 Peg" planets (Mayor & Queloz 1995, Butler et al. 1998, Marcy et al. 1997). Figure 3 shows the phased Doppler measurements for the two representative stars that apparently harbor short–period planetary companions.

These close jupiters may not represent the typical orbital positions of jupiter–mass planets, as detectability is enhanced by proximity to the star (Fig.1). Nonetheless, their existence raises possibility that inward orbital migration was important, as formation *in situ* seems unlikely (Lissauer

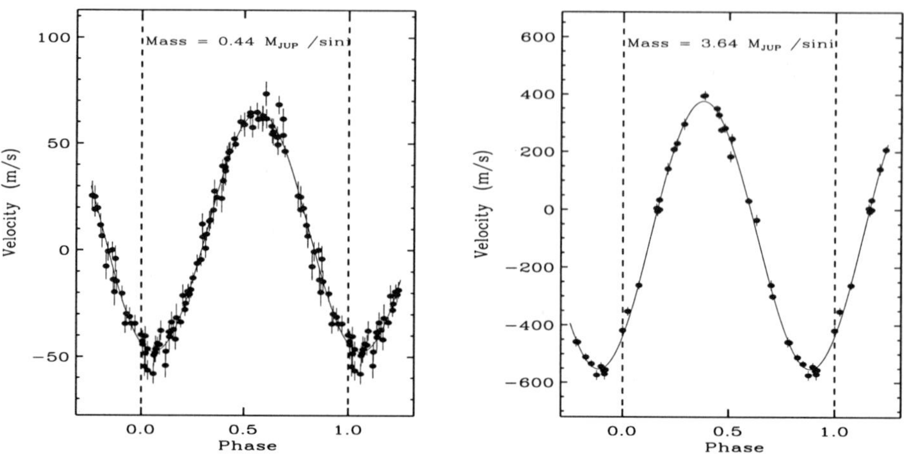

Figure 3.   Measured velocity vs time for two representative Jupiter–mass companions, 51 Pegasi and τ Bootes, in small orbits with $a=$ 0.05 AU. All 8 companions orbiting within 0.15 AU reside in nearly circular orbits, as evident from the sinusoidal velocity curves here. Tidal effects between the planet and star may circularize the orbit.

1995).

An upper limit to the mass of the companion to 51 Peg stems from its low X-ray flux (Pravdo et al. 1996). If the companion had a mass of more than 15 M_{JUP} , it would have tidally spun up the primary star, giving 51 Peg an active corona and a high X-ray flux, as seen for all tidally spun–up G Dwarfs (Marcy et al. 1997). It is plausible that the 51 Peg planets planets would have their orbits circularized theoretically on time scales of order 10^9 yr. (Rasio et al. 1996, Marcy et al. 1997, Terquem et al. 1998). Indeed all 51 Peg planets have low eccentricities, below 0.15 . Some of these planets may have their spins tidally locked to the orbit.

In particular, for τ Boo the sinusoidal Doppler variations yield a period, P = 3.3125 ± 0.0001 d, and eccentricity, e = 0.00 ± 0.01, and an amplitude of 470 m s^{-1} , implying a companion mass of 3.7 $M_{JUP}/\sin i$. The high companion mass and small orbital radius (a=0.045 AU), if correct, should force tidal synchronization of the *stellar rotation* with the orbit. Indeed, the spin period of τ Boo is 3–4 d from chromospheric diagnostics (Baliunas et al. 1997, Henry et al. 1997). Thus both the observed zero eccentricity and stellar spin period are consistent with (and demanded by) the existence of the companion around τ Boo.

Upsilon Andromeda has a Jupiter-mass companion with an orbital period of 4.6 d. If we subtract the orbital fit from the observed velocities, the residuals themselves appear periodic with a period of 3.9 yr, based on our velocities from 1987 to 1999. An orbital fit to these residuals yields P = 3.9 ±0.2 yr, e=0.58±0.1, K=195 m s^{-1} . The implied minimum mass of this putative companion is, $M \sin i = 9.8\pm1$ M$_{\rm JUP}$ (Butler et al. 1999). Periastron passage last occurred on JD 2451177 (29 Dec 1998). Hipparcos astrometry revealed no astrometric wobble to the star, implying an upper limit for the mass of this outer companion of 13 M$_{\rm JUP}$. Thus, Upsilon Andromedae appears to harbor two companions, one of Jupiter mass and the other of 10–13 M$_{\rm JUP}$. Further observations are necessary, in addition to further work on the chromosphere of the star (Saar et al. 1998, Noyes et al. 1984, Baliunas et al. 1997).

3.2. MASS DISTRIBUTION OF PLANETARY COMPANIONS

Three Doppler surveys are complete enough to reveal the mass distribution of substellar companions from 0 to 70 M$_{\rm JUP}$ within 5 AU. These three surveys include the modest precision (300 m s^{-1}) survey of 570 GK dwarfs by Mayor, Duquennoy, and Udry (Duquennoy and Mayor 1991, Halbwachs et al. 1998). The other two are high–precision Doppler surveys of Mayor et al. (1997) and Marcy and Butler (1998) which surveyed 140 and 107 GK stars respectively. The first survey of modest precision revealed 10 "brown–dwarf" candidates, characterized by $M \sin i = 15$ - 60 M$_{\rm JUP}$ (Mayor et al. 1997). However Hipparcos now enables the detection of wobbles for those cases in which the companion is massive enough and the semimajor axis is large enough that the wobble would be over a few milliarcsec. For every one of the 10 brown dwarf candidates for which the semimajor axis was large enough that a stellar companion could have been detected by Hipparcos, an astrometric wobble indicative of a pole–on orbital orientation was indeed seen. Thus, 7 of the 10 brown dwarf candidates have extreme values of $\sin i$, rendering the companion more massive than 0.075 M$_\odot$.

The $M \sin i$ values for the surviving substellar candidates are shown in Fig. 4, drawn from all three surveys mentioned above. This histogram shows 8 companions in the mass bin from 0 – 5 M$_{\rm JUP}$ (i.e., those planet candidates from Table 1 which came from only these three surveys). From 5 – 10 M$_{\rm JUP}$, there is only one companion (70 Vir b). The remaining mass range from 10 – 70 M$_{\rm JUP}$ contains only 4 companions. The dearth of companions from 15-70 M$_{\rm JUP}$ is surprising. Equally remarkable is the remarkable spike in the mass distribution for masses from 0 – 5 M$_{\rm JUP}$, in contrast to the "desert" of brown dwarfs of higher mass. Those 4 brown dwarf candidates may also be H-burning stars, but their periods are too short to permit astrometric detection of the wobble even by Hipparcos.

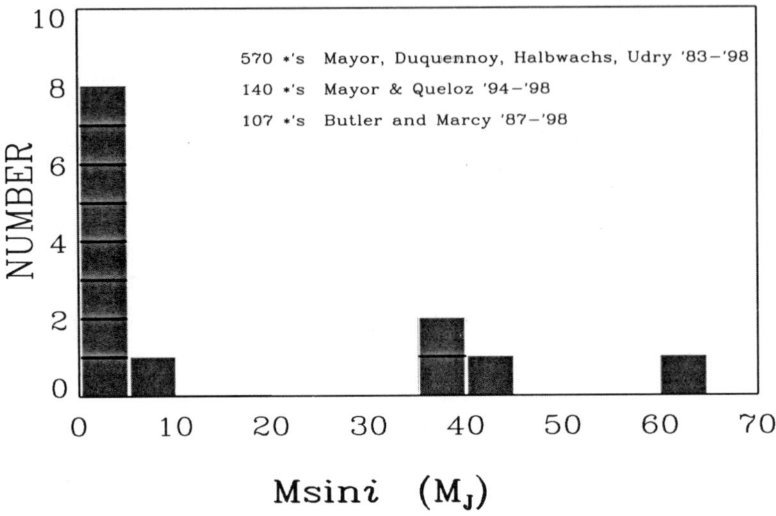

Figure 4.   Histogram of $M \sin i$ for all companions found by the two Doppler surveys at Geneva Observatory and Lick Observatory. The distribution of $M \sin i$ exhibits a discontinuous spike at the lowest masses, 0–5 M_{JUP} . This discontinuity suggests the presence of a new population of Jupiter–mass companions distinct from the more massive brown dwarfs that have larger masses. The companions from 0–5 M_{JUP} plausibly represent planets, not unlike the giant planets in our Solar System. Seven companions were removed from this histogram, as Hipparcos revealed the wobble directly, forcing $M \sin i > 70 \ M_{JUP}$, off scale.

Apparently, main sequence stars harbor more companions having masses below 5 M_{JUP} than from 5 - 70 M_{JUP} , within 3 AU. We note that all selection effects favor detection of the most massive companions. Therefore, the spike at low masses suffers incompleteness, and is likely to grow even taller as Doppler precision improves with time. This discontinuity in the mass distribution at 5 M_{JUP} suggests that two qualitatively different types of objects populate either side. A common nomenclature is found in the terms "planet" and "brown dwarf" for the two species. Clearly, the formation processes and internal structure for both types of substellar companions require further work. And it remains unknown to what extent the Jupiter–mass extrasolar companions share a physical heritage with the planets in the Solar System, if any.

A clue to any such relationship between extrasolar and Solar System planets may reside in the mass distribution. Figure 5 shows the histogram of Msini within the range 0<15 M_{JUP} for all known companions to main-sequence stars. The distribution of $M \sin i$ shows a rapid decline at roughly 4 M_{JUP} . There are no companions having $M \sin i = 7.5$–11 M_{JUP} and those

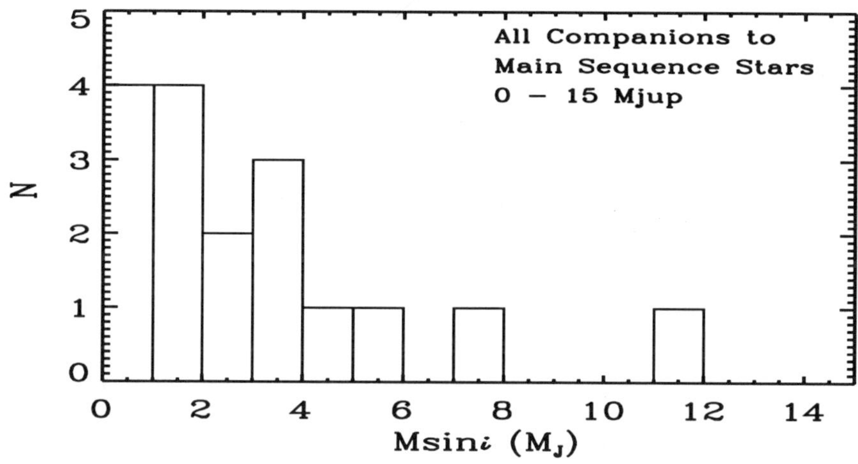

Figure 5.   Histogram of $M \sin i$ for all companions from all surveys, within the range $M \sin i$ = 0–15 M$_{JUP}$. All selection effects favor the detection of more massive companions. In contrast, the observed distribution of $M \sin i$ exhibits a rise from $\sim$6 M$_{JUP}$ toward smaller masses. Therefore, the actual mass distribution of companions must also rise. Apparently these companions represent the tip of the planet mass distribution.

massive companions would have been easily detected. This absence seems statistically significant relative to the 14 companions having $M \sin i$ =0.4–5 M$_{JUP}$ as all selection effects favor detection of the high–mass companions. Even accounting for the Poisson fluctuations within the mass bins and for the detectability of the companions, the distribution of $M \sin i$ must drop from 3 to 7 M$_{JUP}$. This drop implies that the distribution of companion masses, dN/dM, must decline beyond $\sim$5 M$_{JUP}$, for companions within 2.5 AU. Apparently, the most massive planets contain $\sim$5 M$_{JUP}$, with at most a few exceptions.

The largest $M \sin i$ among planet candidates (Fig 5, Table 1) is for HD114762 which has $M \sin i$ =11.02 M$_{JUP}$, well above the proposed decline in the planetary mass function at $\sim$5 M$_{JUP}$. The unknown $\sin i$ leaves uncertain its true mass and its affiliation with "planets." This companion may have a mass that is close to the H–burning threshold of 0.1 M$_\odot$ (Cochran et al. 1991, Hale, 1995).

3.3. SEMIMAJOR AXES OF JUPITER–MASS COMPANIONS

Prior to 1995, Jupiter–mass companions were expected to form and reside in circular orbits with a radius of $\sim$5 AU (Boss 1995, Lissauer 1995). Indeed,

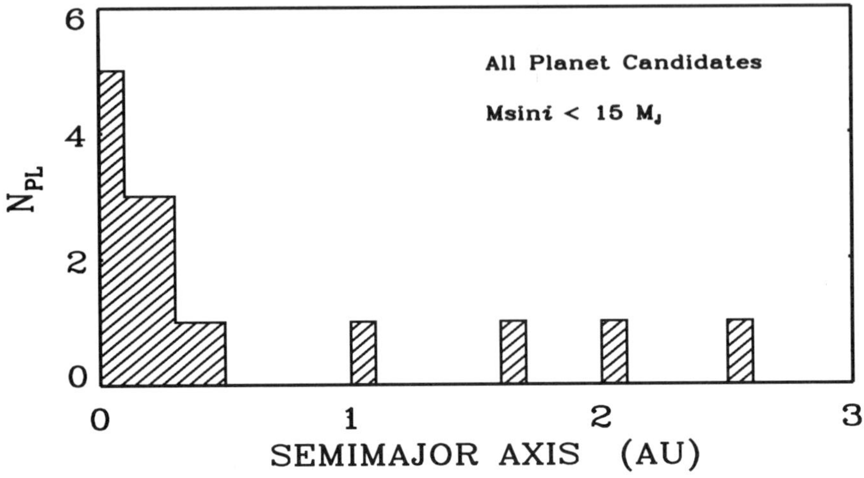

Figure 6.   Distribution of semimajor axes among all planet candidates. Detection of small orbits is favored, due to larger wobble of the star, causing a selection effect for them. However there is a pile-up of companions close to the host star: 11 of 17 have a=0–0.3 AU, but only 2 of 17 have a=0.3–0.6 AU, and none 0.6–0.9 AU. Selection effects cannot explain this occurrence in small orbits (Cumming et al. 1999).

planet–search strategies and funding decisions prior to 1995 were biased toward this expectation, revealing the treacherous symbiosis between theoretical expectation and the design of experiments. In sharp contrast, among the 17 extrasolar planet candidates detected to date, 11 reside in orbits with $a < 0.3$ AU. Figure 6 shows the histogram of semimajor axes of all extrasolar planet candidates. The surprisingly small orbits stand in contrast to the prediction that the first step in the formation of giant planets involves coagulation of ice grains, which exist only beyond $\sim$3 AU. Such grain growth provides the supposed requisite solid core, around which gas could accrete. In retrospect, orbital migration or planet scattering may explain these close orbits (e.g., Lin et al 1999, Rasio & Ford 1996).

One must consider the selection effect that favors detection of close companions (Figure 1). Jupiter–mass companions within 0.1 AU clearly occur around $\sim$3% of Solar–type stars. But do jupiters occur preferentially in close? The preliminary indication is yes, at least for the field–of–view within 3 AU of the host star. Figure 6 shows that 11 of 17 planet candidates reside within 0.3 AU, and the remaining 6 are spread from 0.3–2.5

AU, revealing a remarkable density gradient. Regarding the selection effect, giant planets that reside 0.5 - 1.5 AU from their star would impart a reflex velocity of $25 - 40$ m s^{-1} which is clearly detectable with current precision of 5 m s^{-1} (Figure 1). A precision better than 5 m s^{-1} has been achieved for 60 chromospherically quiet stars from Lick Observatory during the past 3 yr (Marcy & Butler 1998). None of the 60 stars exhibits a planetary-mass companion with a semimajor axis between 0.5 and 1.5 AU. Thus, the distribution of planetary semimajor axes contains an apparent maximum within 0.2 AU, in terms of dN/da. It remains unknown whether there is a population of Jupiter–mass companions beyond 3 AU.

3.4. ORBITAL ECCENTRICITY OF JUPITER–MASS COMPANIONS

All 5 extrasolar planets orbiting closer than 0.1 AU have small eccentricities (Table 1). These eccentricities are much lower than that found in stellar binary systems, for which the median is 0.4 , with very few lower than $e=0.1$ (Duquennoy and Mayor 1991). Stellar binary systems with periods under 15 d are found empirically to circularize within $\sim$5 Gyr.

Tidal circularization of orbiting planets may be due to dissipation in the planet rather than the host star. The tidal circularization time (due to dissipation in the planet) is given by Goldreich and Soter (1966):

$$\tau_{\text{circ}} = \frac{4Q}{63} \frac{m}{M} \left(\frac{1}{GM}\right)^{\frac{1}{2}} \frac{a^{6.5}}{R_p^5} \ .$$

where Q is the so–called damping factor, m is the mass of the companion, and a is the orbital radius. Rasio et al. (1996) show that for 51 Peg the tidal circularization time scale is $\tau_{\text{circ}} \approx 2 \times 10^9$ yr. The 8 extrasolar planets within 0.15 AU may have suffered some tidal circularization, as discussed further by Marcy et al. (1997), Terquem et al (1998), Ford et al. (1998) and Lin et al. (1999). The companions to 55 Cnc and Gliese 86 at $a=0.11$ AU seem too distant for circularization to be important and yet they have small eccentricities which remains a mystery. If the planetary companions were solid, the circularization time scale would be diminished (Marcy et al. 1997). Conceivably, planets could migrate inward, maintaining their circular orbits without need of tidal effect. Alternatively, planets could be gravitationally scattered inward, and their orbits circularized in a remnant inner gaseous disk, a proposal requiring further work.

Figure 7 shows orbital eccentricities vs. $M \sin i$ which exhibits no obvious trend. This suggests that orbital eccentricity is not correlated with planet mass, within the mass range 0.5–5 M_{JUP} . However, all 5 planets with $M \sin i < 1.1$ M_{JUP} reside in nearly circular orbits. This correlation may be a selection effect, as the lowest–mass planets are more easily detected close to

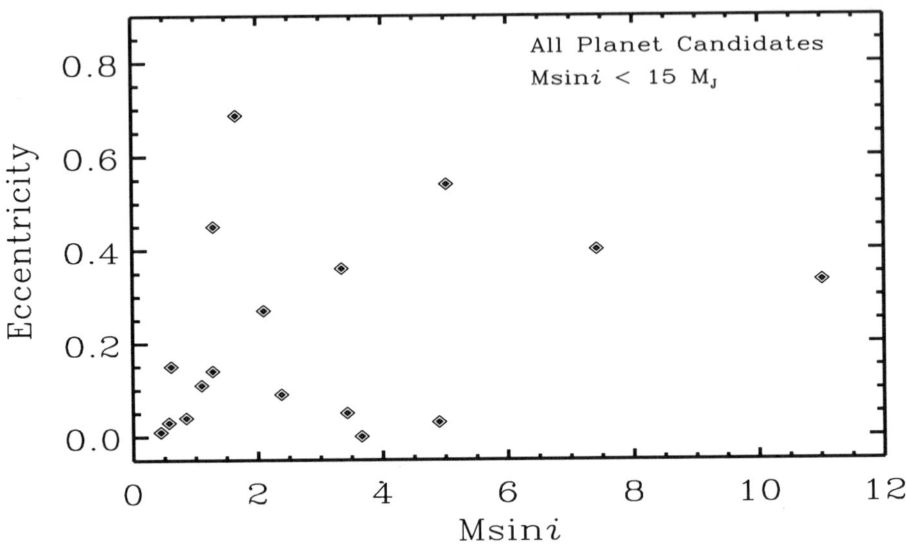

Figure 7.   Orbital eccentricity vs $M \sin i$ for all known companions to main sequence in the range $M \sin i = 0$–15 M$_{\text{JUP}}$. There is no evidence that orbital eccentricities are related to planet mass.

their host stars in order to induce a detectable Doppler reflex signal. These close planets are all subject to tidal circularization. Apparently, whatever mechanism that produces orbital eccentricities is not sensitive to the mass of the planetary companion, within the range 1–5 M$_{\text{JUP}}$.

In contrast to the small orbits that are nearly circular, all 9 planet candidates that have a$>$0.2 AU reside in eccentric orbits with $e > 0.1$, larger than the eccentricity for both Jupiter (e=0.048) and Saturn (e =0.055) . Figure 8 shows a plot of orbital eccentricities vs. semimajor axis. Apparently, Jupiter–mass companions orbiting from 0.2–2.5 AU (immune to tidal circularization) have large orbital eccentricities. Some mechanism commonly produces eccentric orbits in Jupiter–mass companions that reside from 0.2 – 2.5 AU in main–sequence stars. There is no selection effect operating here, as circular orbits would be easily detected as well. Thus, Jupiter–mass companions generally reside in eccentric orbits within the range a=0.2–2.5 AU, around stars of mass 0.3–1.2 M$_\odot$.

The explanation for eccentricities requires physical processes that are not explicitly included within the context of planet formation. Scattering of orbits by other planets, companion stars, or passing stars in the young star cluster offer mechanisms for producing eccentric orbits (Rasio and Ford 1995, Lin and Ida 1996, Weidenschilling and Marzari 1996, Laughlin and Adams 1999, Holman et al. 1997). General models with such scattering among planets are emerging (Levison et al. 1998). These mechanisms do not explicitly predict small orbits of <1 AU, because significant energy must be lost from the original orbits of ~ 5 AU.

One possibility is that planet–scattering continues to occur during the final era of the remnant gaseous protoplanetary disk. If the disk remains intact within the inner few AU where the original gas density was highest, the disk can serve as the reservoir into which the planet's orbital energy can be deposited, either by dynamical friction or by tidal interaction between planet and disk. In this scenario, scattered planets would reside in eccentric orbits subjecting them to dissipation during periastron passages. Clearly detailed models are required that include both planet scattering and the dissipative effects of a weak inner gaseous disk to determine the resulting planetary orbits.

Are the circular orbits of the giant planets in the Solar System unusual in context of general planetary systems? We have little information about giant planets that orbit beyond 3 AU. We expect such information in the coming years as Doppler programs extend their time baseline. Planets beyond 3 AU may well reside in predominantly circular orbits. A population of giant planets that never suffered significant scattering or migration could comprise these Jupiter analogs. But suddenly, it appears that entropy considerations may favor non–circular orbits for giant planets in general.

Figure 9 shows $M \sin i$ vs. semimajor axis for all 17 planet candidates. Apparently, the distribution of planet masses is not a strong function of semimajor axis from 0.05–2.5 AU for the range of detectable masses, 1–6 M_{JUP} . Some migration models predict that migration is initiated only above some planetary mass threshold (e.g., Trilling et al 1998), the mass at which tidal effects cause significant density perturbations and gaps in the disk. If so, this threshold does not lie from 1–5 M_{JUP} , which appears consistent with migration theory (Lin et al. 1999), but see Ward (1997). We conclude that if orbital migration within a gaseous disk brings the giant plants inward, neither that process nor the halting mechanism seem to depend on planet mass in the observed mass range. It is important to push the detection thresholds to smaller planetary masses, to search for this supposed mass threshold for the onset of migration.

3.5. METALLICITY

The spectra of main sequence stars are routinely analyzed to yield the abundance of the heavy elements relative to hydrogen in the atmosphere. The

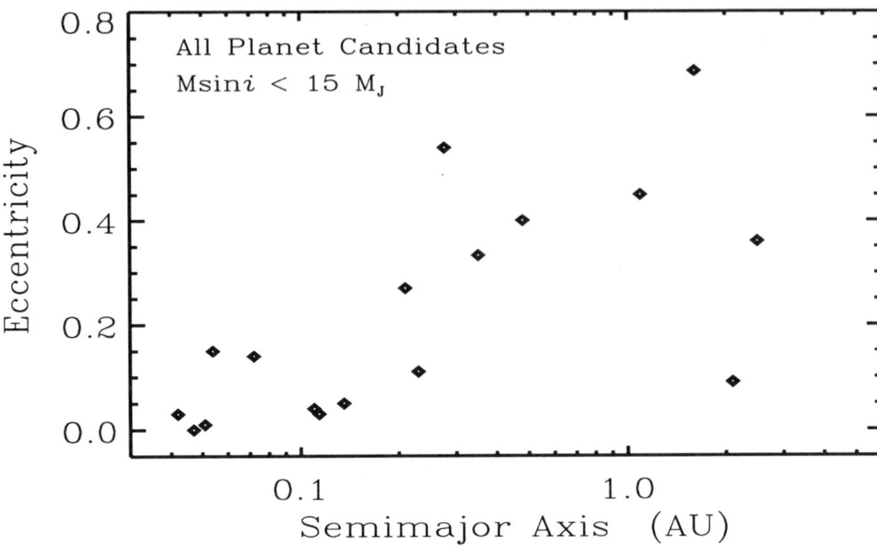

Figure 8. Orbital eccentricity vs semimajor axis for all known companions to main sequence in the range $M \sin i$ = 0–15 M$_{JUP}$. Appranently, planetary companions that are closer than ~0.2 AU to the star tend to exhibit nearly circular orbits. Remarkably, all 9 Jupiter–mass companions that orbit farther than 0.2 AU reside in non–circular orbits with e >0.1.

resulting metallicities of the planet–bearing stars are shown in Fig. 10 and come from detailed LTE spectroscopic synthesis of high resolution spectra (Gonzalez 1997, Gonzalez 1998, Gonzalez and Vanture 1998, Gonzalez et al. 1998). Figure 10 shows the histogram of metallicities of general G and K–type main sequence stars in the vicinity of the Sun. The planet–bearing stars are shown as tick marks at top, except for a few of the planet–bearing stars for which spectroscopic analysis has not yet been done. Compared to the average of main sequence stars, the planet–bearing stars appear to be metal rich. Figure 10 also includes a tick mark at top–left for the star HD114762 with its companion of $M \sin i$ = 11 M$_{JUP}$. That brown dwarf candidate indeed has low metallicity, [Fe/H] = -0.6 dex, in contrast to the metal rich status of the stars that bear the best planet candidates ($M \sin i$ < 5 M$_{JUP}$).

The most remarkable cases are 55 Cnc and 14 Her, which have metallicities of [Fe/H] = 0.45 ± 0.03 and [Fe/H] = 0.50 ± 0.05, respectively (Gonzalez 1998). Taylor (1996) finds [Fe/H] = 0.38 ± 0.07 for 14 Her. These two stars had been identified as remarkably metal rich prior to the

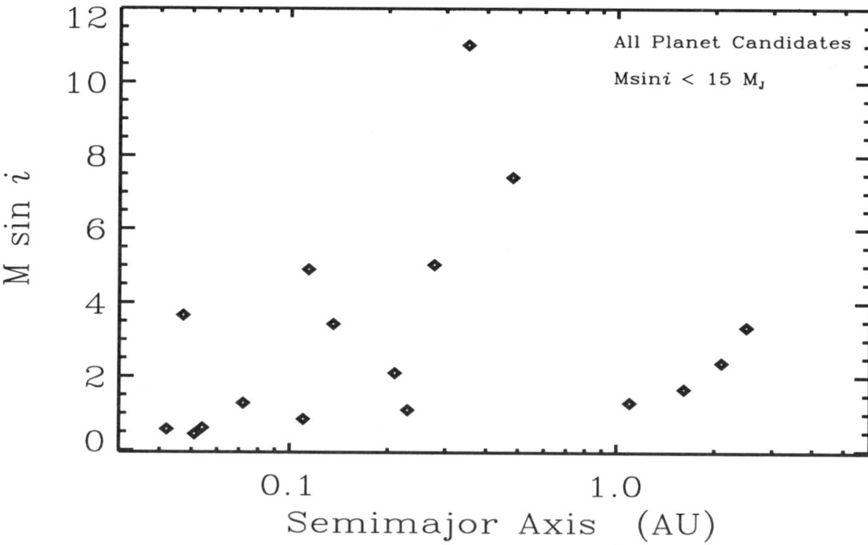

Figure 9. Measured $M \sin i$ vs semimajor axis for all known planet candidates having $M \sin i < 15$ M$_{\rm JUP}$. No strong correlation is apparent. This suggests that the planet formation process does not place the planets in orbits that depend on their mass. Indeed if a migration process plays a role in transporting planets, that process does not appear to favor companions according to mass within the range 1–5 M$_{\rm JUP}$, within which detection is robust from 0–2 AU.

planet detection (Taylor 1996), and they comprise two of the seven "Super Metal Rich" stars known in the Solar vicinity (Taylor 1996). The metallicity of HD187123 is also above Solar, [Fe/H] = 0.16. From Fig. 10, the Sun itself is metal rich compared to the field stars nearby.

The correlation between the presence of planets and metal abundance lends support to the standard models of giant–planet formation (Lissauer 1995, Boss 1995, Pollack et al. 1996). In all such models, the heavy elements, as well as water and CO, have condensed into dust. The condensible material coagulates into larger bodies, leading to solid cores which permits the eventual accumulation of gas in Jovian mass. Thus, the observed correlation between the abundance of heavy elements and the presence of planets supports the notion the observed Jupiter–mass companions represent objects similar to the giant planets as modelled.

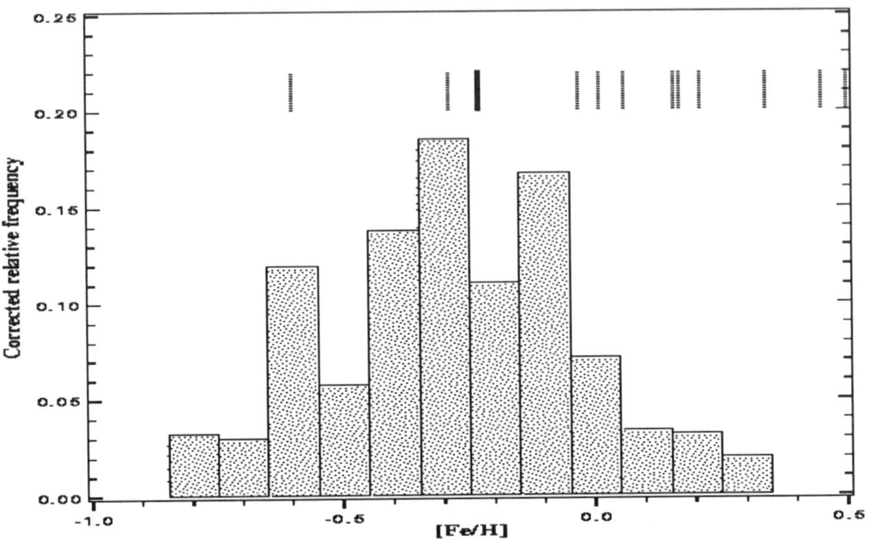

Figure 10.   Distribution of metallicities of G and K–type main sequence stars in the Solar neighborhood. The tick marks at the top show the metallicities of stars which have planetary companions, and for which a detailed spectroscopic analysis of the star has been done (Gonzalez et al. 1998). Apparently, planet–bearing stars are rich in heavy elements.

4. Formation of Giant Planets

The basic paradigm for the formation of the Solar System stems from its dynamical architecture, notably the correlated directions of the angular momentum vectors (orbital and rotational) and the nearly coplanar, circular orbits of the planets. There is a radial compositional gradient, from inner heavy element to outer light element constituents. The architecture suggests a model in which the Solar System formed in a Keplerian disk of gas and dust, as described in excellent reviews by Wetherill (1990), Lissauer (1995),Papaloizou and Lin (1995), Lin and Papaloizou (1996), Lin et al. (1999).

The standard Solar–System models often adopt as a benchmark initial condition, the "minimum–mass solar nebula", a 100 AU disk having mass of $\sim$0.02 $M_\odot$(cf Hayashi 1981, Shu et al. 1993, Lin and Papaloizou 1996). This total mass is set by the actual solid material in the nine planets, augmented by a cosmic complement of hydrogen and helium. Disks are indeed found around $\sim$50% of the $\sim$3 Myr-old T Tauri stars (eg, Strom 1994), with measured masses that are similar to the minimum-mass solar nebula (Beckwith and Sargent 1996). Figure 11 shows the measured distribution of disk masses, ranging from 0.005 - 0.2 $M_\odot$. These derived masses may contain systematic errors due to the poorly known gas-to-dust mass ratio (currently taken to be 100), the uncertain dust opacity, and the necessary extrapolation of disk density inward of 100 AU, about which little is known. Some suggest that the disk masses remain uncertain by a factor of $\sim$10, and surface mass densities are similarly uncertain within 20 AU where the giant planets may form. The model disks contain dust and molecular gas with a radial decline in temperature and density, such that ices (mostly water) condense outward of $\sim$4 AU, causing a factor of three rise in the density solids.

The formation of giant planets is thought to have occurred in six stages:

1. Infall of dust grains onto the disk, and their subsequent evaporation and condensation.

2. The formation and growth of solid particles from 1 mm to 1 km sizes, either by gravitational instability (Goldreich & Ward 1972) or by cohesive collisions of grains (Weidenschilling & Cuzzi 1993).

3. The run-away coagulation of planetesimals into prototerrestrial planets (Wetherill 1990).

4. Concurrent accretion of gas and planetesimals onto the terrestrial cores (Pollack et al. 1996).

5. Termination of accretion by gap formation in the disk (Lin et al. 1998, Artymowicz 1998).

6. Clearing of the disk material, by processes not well identified.

5. Future Planet–Detection Techniques

5.1. TRANSITS AND MICROLENSING

Planets may be detected as they transit in front of the disk of a star. The fractional reduction in the light from the star is simply the ratio of the areas of the planet and the star. For a benchmark Jupiter having radius 1 R_J, a Solar-type star will dim by 1% with a duration of order hours, depending on the orbital radius (Hale and Doyle 1994). Such photometry is possible from the ground, made efficient with automated wide–field telescopes that acquire CCD images of thousands of stars simultaneously (Borucki and

Summers 1984). The 1% dimming is easily distinguished from other effects such as starspots, flares, and fluctuations in photospheric granulation which would rarely exceed 1%. Actual transits would cause no change in color, would exhibit a flat–bottomed minimum (with limb-darkened ingress and egress), and would repeat like clock–work.

A transit requires a special geometry, such that

$$\tan i > \frac{a}{R_*}$$

where i is the orbital inclination, a is the planet's semimajor axis and R_* is the star's radius.

For randomly oriented orbital planes, the probability, P, that i will reside between 90 deg (edge–on) and i' is simply: $P(i' \text{to} 90°) = \cos i'$. For a population of Jupiters at 0.1 AU (51 Peg-like), 4.7% of them will transit and $\sim$2% of Solar-type stars have such close Jupiters (Marcy and Butler 1998). This implies that one in a thousand Solar-type stars should exhibit transits from close Jupiters. To enrich the stellar sample with edge-on planetary orbits, eclipsing binary stars may be selected. For CM Dra, companions larger than 2.5 Earth radii can be ruled out, for periods less than 60 days (Deeg et al. 1998).

A transit, followed by Doppler measurements of the star, would yield the planet mass directly since $\sin i \approx 1.0$. Transits also yield the radius of the planet and hence its density, which distinguishes gas giants from solid planets. The photometric transit approach should be pursued vigorously.

Transits by Earth-sized planets would dim the star 0.01%. The requisite photometric precision requires a space-borne platform, wide-field camera, and a detector capable of photometric precision of 3:100,000. Such a mission should reveal transits in $\sim$1% of Solar-type stars, if terrestrial planets at $\sim$1 AU are common (Borucki et al. 1996).

The French space mission COROT, to be launched in 2001, is planned to photometrically monitor about 25000 stars and permit the detection of a few dozen extrasolar planets (Schneider 1998).

5.2. ASTROMETRY

The angular wobble of a star in response to its companion is proportional to both planet mass and orbital radius, and inversely proportional to the distance to the star. Reviews of the astrometric approach to planet detection are provided by Gatewood (1987) and Colavita and Shao (1994). As a benchmark, a Jupiter analog orbiting 5 AU from a Solar-type star that is located 10 pc away would produce an astrometric amplitude of 0.5 milliarcsec (mas).

The astrometric technique offers two great prospects: 1. determination of an unambiguous mass and orbital inclination of a planet, 2. detection of sub–Jupiter–mass planets for future astrometric precision below 0.1 mas. An astrometric planet detection provides a secure mass which is not offered by direct detection.

Gatewood (1987) has demonstrated an annually-averaged astrometric precision of 1 mas, which he expects to improve by using the Keck II telescope. Pravdo and Shaklan (1996) have demonstrated a precision of 0.1 mas with direct, short CCD exposures from the Palomar 5-m and Keck telescopes. With the Palomar testbed interferometer, Colavita and Shao (1994) currently achieve precision of 60–70 μas/$\sqrt{hr}$, portending a bright future for next-generation interferometric astrometry. The two Keck telescopes and the European Very Large Telescope Interferometer should yield astrometric precision of 20 μas (Colavita and Shao 1994). A planned NASA space-born astrometric interferometer called the Space Interferometry Mission (SIM) has a goal of 4 μas for global astrometry (Unwin et al 1996; Boden et al. 1996), and perhaps better for planet searches. Due for launch in 2005, SIM astrometry should achieve many–sigma detections of planets having Neptune–like masses $(1/20\,M_{JUP})$ at 5 AU for stars within 10 pc. A mission lifetime of $\sim$12 years will be required. SIM Interferometric astrometry offers a clear path to statistically valuable ensembles of gas giants through the Neptune–mass regime, to constrain the planetary mass function.

To date, no definitive detection of a planetary companion has been accomplished by the astrometric method. Gatewood (1996) has noted strongly suggestive accelerations in Lal 21185 that may be due to two low–mass orbiting companions, but further data are required to confirm the orbits and masses (Gatewood, 1997, personal communication).

5.3. MICROLENSING

Gravitational microlensing of stars in the Galactic bulge may also reveal the presence of planets in orbit around the intervening lensing Angelobjects. Intensive follow-up photometry of microlensing events by a global telescope network can reveal the short-term perturbations on the standard microlensing light curve caused by an attendant planet (Peale 1997, Griest and Safizadeh 1998). Microlensing is most sensitive to planets at a projected distance from the lensing star of about an Einstein radius, which corresponds to 3-6 AU for a typical galactic bulge microlensing event. The duration of the planetary perturbation on the lightcurve is proportional to $\sqrt{M_P}$. Microlensing is unique in its ability to detect Earth–mass planets in orbits with semi-major axes of several AU around main sequence stars from the ground (hence inexpensively), and would yield statistics on the occurrence of such planets. Follow–up study of these planets would be difficult owing to their large distance ($\sim$5 kpc), and ambiguities in identifying the lensing object.

5.4. SPACE–BORNE NULLING INTERFEROMETRY

The acquisition of direct images of Earth–like planets will require a space–born interferometer that nulls, by destructive interference, the glaring light from the host star (Beichman 1998, Angel & Woolf 1997). Both the ESA in Europe and NASA in the United States are designing systems that woud have the following general properties. Four telescopes, operating at a temperature of 40K would be separated with a baseline of $\sim$100m. The telescope would operate at wavelengths of 3–30 μm where thermal emission from planets is maximum. The resulting angular resolution would be $\sim$0.5 milli-arcsec, depending on wavelength.

Ironically, the direct detection of planets is hindered by the zodiacal dust that both reflects and thermally emits light that competes with the signal from the planet (see, e.g., Beichman 1998, Becklin et al. 1998 and Fajardo-Acosta et al. 1999). These zodiacal dust disks are themselves being used to study the planetary systems in advance of the space–borne interferometry. For example, using ISO observations at 25 μm and 60 μm, Dominick et al. (1998) concluded that 55 Cnc (Table 1) has a disk roughly 60 AU in diameter. In addition, near IR coronographic observations by Trilling and Brown (1998) have resolved the 55 Cnc dust disk yielding its inclination which in turn sets the mass of the planetary companion (1.8 M_{JUP}). They estimate the mass of the disk to be 0.4 Earth-masses, 10$\times$ that of the dust in the Solar System. Plausibly this disk is related to the planet(s) around that star (Butler et al. 1997).

Most interestingly, a ring of dust was recently imaged at 850 μm around the nearby K2V star, ϵ Eridani(Jayawardhana et al 1998, Koerner et al. 1998, Greaves et al. 1998), showing that mid–IR and sub–mm imaging may routinely be able to detect dust disks. These images can provide the dust distributions and temperatures in the disks around stars of different ages and metallicities, hopefully conveying the relationship between disk properties and planets.

6. Summary

Precise Doppler measurements of $\sim$500 main sequence stars have revealed Jupiter–mass companions, yielding the first indications of their distribution of masses, orbital sizes and of eccentricities. Major theoretical questions remain about the origin of these properties. The mass distribution rises sharply at 5 M_{JUP} toward lower masses, indicating a robust formation process. Presumably these Jupiter–mass companions are high mass planets, evidently much more numerous than brown dwarf companions. To date, 17 Doppler companions are known that have $M \sin i < 11$ M_{JUP} . The 5 which orbit closer than 0.1 AU all reside in circular orbits, plausibly induced by tidal interactions with the host star. In contrast, all 9 planet candidates that orbit farther than 0.2 AU reside in eccentric orbits with $e > 0.1$, greater than that of Jupiter ($e=0.048$). Orbital eccentricity can be induced gravitationally by other orbiting planets, companion stars, passing stars in the dense star-forming cluster, or by the protoplanetary disk, but further theoretical work is needed. The planet–bearing stars are systematically metal–rich, as is the Sun, compared to the Solar neighborhood. This indicates that planet formation depends sensitively on the abundance of heavy elements. The occurrence of planets in the mass decade, 0.5–5 M_{JUP} , within 3 AU is $\sim5\%$ but that of "brown dwarf" companions in the next higher mass decade, 5 – 50 M_{JUP} , is at most 1% within 3 AU. Future planet searches will involve searches for planet transits by the dimming of starlight, giving planet radius. Ground– and space–based astrometry should detect Saturn– and Neptune–mass planets. Gravitational lensing can reveal Earth–mass planets, and future space–borne interferometry is planned to obtain direct images and spectra.

References

Angel R & Woolf N. 1997, ApJ, 475, 373

Artymowicz, P. 1998, On the Formation of Eccentric Superplanets. In Brown Dwarfs and Extrasolar Planets, ed. R. Rebolo, E. L. Martín, and M. R. Zapatero Osorio, pp. 152–161.

Baliunas S. L., Henry G. W., Donahue R. A., Fekel F. C., and Soon W. H. 1997. Properties of Sun-like Stars with Planets: ρ^1 Cancris, τ Bootis, and v Andromedae. Astrophys. J. Lett. 474:L119–L122.

Basri G, Marcy GW. 1997. in *Star Formation Near and Far (7th Astrophysics Conference)*, ed. by S.Holt & LG Mundy, AIP Conf. Proc., 393:228.

Becklin,E.E., Silverstone,M., Hare,J., Zuckerman, B., Spangler, C. Sargent,A., Goldreich,j P. 1998, "The Universe as seen by ISO", Paris: Abstracts.

Beckwith, S. V. W., and Sargent, A. 1996. Circumstellar disks and the search for neighbouring planetary systems. Nature 383:139–144.

Beichman, C.A. 1998. Terrestrial Planet Finder: The search for life-bearing planets around other stars. SPIE, 3350, 719

Bertelli,G., Bressan, A., Chiosi,C., Fagotto, F., Nasi, E. 1994, A&AS, 106, 275

Black, D.C. 1997, ApJL, 490, L171

Borucki WJ, Summers AL. 1984. *Icarus* 58:121

Boss, A. P. 1995, Science, 267, 360

Boss, A. P. 1997, Science, 276, 1836

Boss, A. P. 1998, ApJ 503, 923

Burrows, A., Sudarsky, D., Sharp, C., Marley, M., Hubbard, W. B., Lunine, J. I., Guillot, T., Saumon, D., Freedman, R. 1997. In Proceedings of the Workshop on Brown Dwarfs and Extrasolar Planets, held in Tenerife, Spain on March 17-20, 1997, ed. R.

Rebolo,E. Martín, M. R. Zapatero Osorio, pp 354-369

Burrows, A., Hubbard, B., Lunine, J., Marley, M., and Saumon,D. 1998, to appear in Protostars and Planets IV, eds. V. Manning, A. Boss. S. Russell.

Butler, R. P., Marcy, G. W., Williams, E., McCarthy, C., Dosanjh, P., & Vogt, S. S. 1996, PASP, 108, 500

Butler, R. P., & Marcy, G. W. 1996, ApJL, 464, L153

Butler, R. P., Marcy, G. W., Williams, E., Hauser, H., & Shirts, P. 1997, ApJ, 474, L115

Butler, R. P., Marcy, G. W., Vogt,S.S., Apps, K. 1998, PASP,110,1389

Butler, R. P., Marcy, G. W. & Fischer,D.A. 1999, submitted to ApJ Letters.

Carney, B.W., Latham, D.W., Laird, J.B., and Aguilar, L.A. 1994, AJ, 107, 2240

Cochran, W D., Hatzes, A.P., Hancock, T.J. 1991, ApJL,380, L35

Cochran, W. D., Hatzes, A. P., Butler, R.P., & Marcy, G. M. 1997, ApJ, 483, 457

Cumming, A., Marcy, G.W., and Butler, R.P. 1999, ApJ, in press.

Deeg HJ, Doyle LR, Kozhevnikov VP, Martín EL, Oetiker B, Palaiologou E, Schneider J, Afonso C, Dunham ET, Jenkins JM, Ninkov Z, Stone R, Zakharova PE. 1998, A&A, 338, 479

de la Fuente Marcos, C., de la Fuente Marcos, R. 1997, A&A 326, L21-24.

Delfosse, X., Forveille, T., Mayor, M., Perrier, C., Naif, D., Queloz, D. 1998, A&A, 338L 67

Dominck,C., Laureijs,R.J., Jourdain de Muizon,M. & Habing, H.J., 1998, A&A, 329, L53

Duquennoy, A. and Mayor, M, 1991, A&A, 248, 485

Fajardo-Acosta S.B., Thakur,N., Stencel,R.E., and Backman,D.E. 1999, ApJ, in press

Favata, F., Miccla, G., & Sciortino, S. 1997, A&A, 323, 809

Fischer, D.A., Marcy, G. W., Butler, R.P. Vogt, S.S., and Apps, K. 1999, PASP, 111, 50.

Ford, E. B., Rasio, F. A., & Sills, A. 1998, preprint, submitted to ApJ.

Fuhrmann, K. 1998, A&A, 338, 161

Gatewood GD. 1996. *Bull. Am. Astron. Soc.* 188, #40.11 ABSTRACT

Gonzalez G. 1997. *MNRAS* 285:403

Gonzalez G., and Vanture, A. D. 1998, A&A, submitted.

Gilmore, D. K., Rank, D. M. & Temi, P. 1994, Proc. SPIE Vol. 2198, p. 744-748, Instrumentation in Astronomy VIII, eds. David L. Crawford & Eric R. Craine.

Goldreich P, Ward WR 1973, *Ap.J* 183:1051-61

Goldreich P, Soter S. 1966, *Icarus* 5:375

Goldreich, P. & Tremaine, S. 1980, ApJ, 241, 425

Gonzalez, G. 1998, A&A, 334, 221

Gonzalez, G., Wallerstein, G., Saar,S. 1998, ApJL, in press

Greaves, J.S. et al. 1998, ApJL, in press.

Gray, D. F. 1992, in The Observation and Analysis of Stellar Spectra, 2nd Edition, Cambridge Univ. Press, Cambridge. p431

Halbwachs,J.L., Mayor,M., Udry,S. 1997, in "Brown dwarfs and Extrasolar Planets", Proceedings of a Workshop held in Puerto de la Cruz, Tenerife, Spain, 17-21 March 1997, ASP Conference Series #134, eds. Rafael Rebolo; Eduardo L. Martin; Zapatero Osorio, Maria Rosa, p. 308.

Hale A., Doyle L. R. 1994 The Photometric Method of Extrasolar Planet Detection Revisitied. Astrophys. Space Sci. 212:335-348.

Hale,A. 1995, PASP, 107,22

Hauser H., and Marcy, G. W. 1998. The Orbit of 16 Cygni AB. Pub. Ast. Soc. Pac., submitted.

Hayahsi, C. 1981. Structure of the solar nebula, growth and decay of magnetic fields, and effects of magnetic and turbulent viscosities on the nebula. *Prog. Theor. Phys. Suppl.* 70:35-53.

Heintz, W. 1994, AJ, 108,2338

Henry GW, Baliunas SL, Donahue RA, Soon WH, Saar SH. 1997. *Ap.J.* 474, 503

Holland,W.S., et al. 1998, Nature, 392, 788

Holman, M., Touma, J., and Tremaine, S. 1997. Nature 386, 264

Hatzes, A. P., Cochran, W. D., & Bakker, E. J. 1998, Nature, 391, 15

King JR, Deliyannis CP, Hiltgen DD, Stephens A, Cunha K, Boesgaard AM. 1996. *Astron. J.* 113:1871

Kirkpatrick, J. D. & McCarthy, D. W., Jr. 1994, AJ, 107, 333.

Kurucz, R. L., Furenlid, I., Brault, J., Testerman, L. 1984, National Solar Observatory Atlas No. 1 , (Tucson: NSO)

Jayawardhana,R., Fisher,S., Hartmann,L., Telesco,C., Pina,R. & Fazio,G. 1998, ApJL, 503, 79

Koerner,D.W., Ressler,M.E., WErner,M.W., & Backman,D.E. 1998, ApJL, 503, 83

Latham, D.W., Mazeh, T., Stefanik, R.P., Mayor, M., & Burki, G. 1989, Nature, 339, 38

Laughlin G., and Adams F. C. 1998. Possible Stellar Metallicity Enhancements from the Accretion of Planets. Astrophys. J., 491:L51–L54.

Laughlin, G. and Adams, F.C. 1999, "The Modification of Planetary Orbits in Dense Open Clusters", to appear in ApJ Letters

Levison, H. A., Lissauer J. J., & Duncan, M. J. 1998. AJ, 116, p.1998

Lin DNC. 1986. *The Nebular Origin of the Solar System* in The Solar System: Observations and Interpretation, ed. MG Kivelson. pp.68-69. Prentice-Hall, Englewood Cliffs, New Jersey,

Lin, D. N. C., Bodenheimer, P., & Richardson, D. C. 1996, Nature, 380, 606

Lin DNC & Papaloizou JCB. 1986. *Ap.J.* 309, 846-57

Lin DNC & Papaloizou JCB. 1996. *Annu. Rev. Astron. Astrophys.* 34:703-47.

Lin D. N. C. & Ida S. 1997, ApJ, 477, 781

Lin, D. N. C., Papaloizou, J.C.B., Bryden, G., Ida, S., and Terquem, C. 1999, to appear in Protostars and Planets IV, eds. V.Manning, A.Boss. S.Russell.

Lissauer JJ. 1995. Icarus, 114, 217

Marcy, G. W. & Butler, R. P. 1996, ApJL, 464, L147

Marcy, G. W., Butler, R. P., Williams, E., Bildsten, L., Graham, J. R., Ghez, A., & Jernigan, G. 1997, ApJ, 481, 926.

Marcy, G. W., and Butler, R. P. 1998. Detection of Extrasolar Giant Planets. Ann. Rev. Astron. Astrophys. 36, 57.

Marcy, G. W., Butler, R. P., Vogt, S. S., Fischer, D., Lissauer, J. J. 1998, ApJ, 505, L147

Marcy, G. W., Butler, R. P., Vogt, S. S., Fischer, D., Liu,M.C. 1999, ApJ. to appear Vol. 520, "Two New Candidate Planets in Eccentric Orbits"

Martín EL. 1998 *Lithium in Young Clusters and Pre−Main Sdquence Stars*, to appear in proceedings of the workshop."Cool Stars in Clusters and Associations", held in Palermo, Italy, September 1997, in the *Mem. Soc. Ast. Italy*, ed. G Micela, S Sciortino, R Pallavicini

Max, C. E., Olivier, S. S., Friedman, H. W. An, J., Avicola, K., Beeman, B. V., Bissinger, H. D., Brase, J. M., Erbert, G.V., Gavel, D.T., Kanz, K., Liu, M.C., Macintosh, B., Neeb, K. P., Patience, J., and Waltjen, K. E. 1997, Science 277, 1649.

Mayor, M., Queloz, D. 1995, Nature,378,355

Mayor M, Queloz D, Udry S, Halbwachs J.-L. 1997. in "Astronomical and Biochemical Origins and the Search for Life in the Universe", Proceedings of the 5th International Conference on Bioastronomy , held in Capri, Italy, IAU Colloquium No. 161, ed. C.Cosmovici, S.Bowyer, D.Werthimer . pp.313-30. Editrice Compositori:Italy

Mayor, M., Beuzit, J. L., Mariotti, J. M., Naef, D., Perrier, C., Queloz, D., & Sivan, J. P. 1998, ASP Conf. Ser. (in press), IAU Colloq. 170: Precise Stellar Radial Velocities, ed. J.B. Hearnshaw & C.D. Scarfe, (San Francisco: ASP)

Meyer C., Rabbia, Y., Froeschle, M., Helmer, G., Amieux, G. 1995, A&AS, 110,107

Mazeh, T., Latham,D.W., Stefanik, R. P. 1996, ApJ,466,415

Mazeh T, Krymolowsky Y, Rosenfeld G. 1997. ApJL 477, L103

Mclean, I.S., Macintosh, B.A., Liu, T., Casement, L.S., Figer, D. F., Lacayanga, F., Larson, S., Teplitz, H., Silverstone, M., Becklin, E. E. 1994, Proc. SPIE Vol. 2198,

p. 457-466, Instrumentation in Astronomy VIII, David L. Crawford; Eric R. Craine; Eds.

Ng,Y.K., Bertelli,G. 1998, A&A, 329, 943

Noyes, R. W., Hartmann, L., Baliunas, S. L., Duncan, D. K., & Vaughan, A. H. 1984, ApJ, 279, 763.

Noyes, R. W., Jha, S., Korzennik, S. G., Krockenberger, M., Nisenson, P., Brown, T. M., Kennelly, E. J., and Horner, S. D. 1997, ApJ, 483, L111

Queloz D., Mayor M., Weber L., Naeff D., Udry S., Santos N., Blecha, A., Burnet, M. and Confino, B. 1999, preprint (on discovery of Gliese86b).

Papaloizou J.C.B. & Lin, D.N.C 1995. *Annu. Rev. Astron. Astrophys.* 33:505.

Perryman, M. A. C., et al. 1997, A&A, 323, L49. The Hipparcos Catalog

Pinsonneault M. 1997. *Ap. J.* 480:303

Pollack,J., Hubickyj,O., Bodenheimer,P., Lissauer, J.J., Podalak,M., Greenzweig,Y., 1996, Icarus, 124, 62

Pravdo, S.H., Angelini, L., Drake, S.A., Stern, R. A., White, N.E. 1996, New Ast,1,171

Pravdo S. H., and Shaklan, S. B. 1996. Astrometric Detection of Extrasolar Planets: Results of a Feasibility Study with the Palomar 5 Meter Telescope. Astrophys. J. 465:264–277.

Rasio, F. A., & Ford,E. B. 1996, Science, 274, 954

Rasio, F. A., Tout,C.A., Lubow, S.H., & Livio, M. 1996, ApJ, 470, 1187

Saar, S. H. & Donahue R.A. 1997, ApJ, 485, 319.

Saar, S. H., Butler, R. P., & Marcy, G. W. 1998, ApJ, 403, L153.

Shirts, P. & Marcy, G. W. 1998, preprint to be submitted to PASP.

Shu FH, Najita J, Ruden SP, Lizano S. 1994. *Ap.J.* 429:797-807

Soderblom DR, Jones BF, Balachandran S, Stauffer JR, Duncan DK, Fedele SB, Hudon JD. 1993. *Astron. J.* 106:1059

Strom SE. 1994. in: *Cool Stars, Stellar Systems, and the Sun, Eighth Cambridge Workshop* p. 211. San Francisco: Astronomical Soc. Pacific

Szokoly, G. P., Subbarao, M. U., Connally, A.J., and Mobasher, B. 1998, ApJ,492,452

Taylor,B. 1996, ApJS, 102, 105.

Terquem, C., Papaloizou, J.C.B.,Nelson, R.P., Lin, D. N. C., 1998, ApJ, 502, 788

Trilling D., Benz W., Guillot T., Lunine J. I., Hubbard W. B., Burrows A., 1998, ApJ, 500, 428

Trilling D., Brown, R. preprint submitted to Nature

Unwin S, Boden A, Shao M. 1996, *Bull. Am. Ast.Soc.* 189, 121.02 ABSTRACT

Van de Kamp P. 1963. *Astrometric Study of Barnard's Star from Plates Taken with the 24-inch Sproul Refractor, Astron. J.*, 68:515-521.

Vogt, S. S. et al. 1994, Proc. Soc. Photo-Opt. Instr. Eng., 2198, 362

Walker, G.A.H., Walker, A.R., Irwin, A.W., Larson, A.M., Yang, S.L.S., Richardson, D.C. 1995. *Icarus* 116:359

Ward, W. R. 1997, ApJ, 482, L211

Weidenschilling S, Cuzzi. JJ. 1993. In *Protostars and Planets III*, ed. E Levy, J Lunine, M Matthews. pp.1031-60. Tucson: University of Arizona Press,

Weidenschilling, S. J. & Marzari, F. 1996, Nature, 384, 619

Wetherill GW. 1990. *Ann. Rev. Earth Planet Sci.* 18:205-56

Wolszczan A & Frail DA 1992. *Nature* 355:145

Wolszczan A 1998, to appear in *Reviews in Modern Physics*

"Geia sou...to Crete III!"

Phil Myers, Hans Zinnecker and Doug Johnstone have lunch by the sea (top). Anja Visser, Joao Alves, Gus Muench and Clare Martin at the afternoon coffee break (bottom).

3D MHD simulations, 199
47 UMa, 682
51 Peg, 682
70 Vir, 682
Abundances
  determination of, 117
  electron, 329-332
  grain, 329-332
  ion, 329-332
Accretion, 503
  in binary systems, 494
  in clusters, 503
  luminosity, 187, 358, 359
  rate, 187, 217
  shocks, 378
  spot, 350
  time, 380
  zone, 664
Accretion Disk, (See Disks, Accretion)
Adiabatic Index, 48
Alfven
  lengthscale, 334
  mach number, 46
  radius, 617
  velocity, 46,157
  waves, 199, 332, 333-334, 335
    damping, 307, 332
    torsional, 307, 319
Ambipolar Diffusion, 206, 210, 223, 305-339
  observations of drift speed, 332
Ambipolar Diffusion Time Scale, 61, 306, 316-318, 321, 325, 326
  effect on hydromagnetic waves, 324-334
  in star formation, 305-339
Angular Momentum, 306-308, 318-319, , 326, 327, 335, 336, 613, 616
  conservation, 318, 319, 325, 335
  scaling with mass, 327
  specific, 324, 327, 335
  transport, 419
Angular Momentum Problem, 308, 319, 324-327
Associations, (See OB Associations)
Astrometry, 701

B-ρ Relation, 307, 319, 320, 322-323, 326, 329, 330, 331
  plot, 320, 323, 330
Balbus & Hawley Instability, 221
Binary Stars, 146, 164, 455
  angular momentum/periods, 308, 326, 335
  binary-binary interactions, 508
  formation, 319, 335
  formation by capture, 482
  formation by collision, 506
  formation by fission, 481
  in clusters, 508, 455
  mass ratio, 494
Bipolar Outflows, 132, 177, 178, 183, 184, 203, 205, 216, 218, 223, 227, 631
  shell speed, 203
Boundary Layer, 617, 618, 619, 620, 623
Bow Shocks, 291-293, 632
Break-Up Velocities, 217, 366
Brown Dwarf, 400, 682, 691
  mass function, 168
C-Shocks, 214
Carbon-Chain Molecules, 108
Centrifugal Force, 306, 319, 324-326, 335
Chandrasekhar Limit, 193
Charge Transfer, 310, 336
Class 0 Source, 174, 240, 245
Class I Source, 173, 214, 466, 637
Class II Source, 178, 214, 466
Class III Source, 182
Close Binary, 492
Clouds, (See Molecular Clouds)
Clusters, 145, 161, 166, 451, 496
  bound, 451
  embedded, 451,
  dissolution, 507
  dynamics, 498
  formation, 461, 497
  gas accretion, 503
  morphology, 501
  relaxation, 499
  unbound, 452
Cold Magnetic Spots, 349
Cold Spots, 350